DIGITAL AND ANALOGUE INSTRUMENTATION

testing and measurement

Nihal Kularatna

The Institution of Electrical Engineers

Published by: The Institution of Electrical Engineers, London,
United Kingdom

The Institution of Electrical Engineers,
Michael Faraday House,
Six Hills Way, Stevenage,
Herts. SG1 2AY, United Kingdom

British Library Cataloguing in Publication Data

Kularatna, N.
Digital and analogue instrumentation –(IEE electrical
measurement series; no. 11)
1. Electronic instruments – Testing
I. Title II. Institution of Electrical Engineers
621.3′81548

ISBN 0 85296 999 6

Typeset in India by Newgen Imaging Systems
Printed in the UK by MPG Books Limited, Bodmin, Cornwall

This book is dedicated to my loving Amma and Thatha , who were not lucky to see any of my humble writings, and to my mother-in-law who left us recently.

With loving thanks to my wife, Priyani, and our two daughters, Dulsha and Malsha, who always tolerate my addiction to tech-writing and electronics.

Contents

Foreword

A few years ago, Nihal Kularatna's fist book, *Modern Electronic Test and Measuring Instruments*, gave me the opportunity to revive memories of my editorial days at the IEE. The success of that book has encouraged IEE Publishing to commission another one by Nihal – reminding us that the 'knowledge half-life' in electronics, communications and IT is barely half a decade.

In surveying this vast and changing field, Nihal demonstrates a sound grasp of technology assessment. He has, no doubt, drawn from his own rich and varied experiences in engineering and technology development in Sri Lanka – where he now heads the Arthur C Clarke Institute for Modern Technologies, set up by the government of Sri Lanka in 1983.

I wish this book and its author all success.

Sir Arthur C Clarke
Chancellor, University of Moratuwa, Sri Lanka
Chancellor, International Space University
Fellow of King's College, London
Patron, Arthur C Clarke Institute for Modern Technologies

19 November 2001
Colombo, Sri Lanka

Preface

In the new millennium, the *knowledge half-life* of electronic systems is less than four years. Consider Theo A.M. Classen's recent statement: "Usefulness of a product = Log (Technology)"[1]. This requires us to carefully ponder, then to become skilled in, the level of technology built in new electronic systems and products that we often take for granted. Simply stated, millions of transistors combined with millions of bytes of software equals the current level of technology developed over a period of just 50 years. All made possible only since the invention of the basic building block, the transistor.

A response from the readership of my first work on instruments (cir. 1996) encouraged me, and the IEE Publishing Director, Dr Robin Mellors-Bourne, to create this new volume, instead of merely reprinting the first edition of the book. I was astounded by the almost exponential increase in technological advances occurring in the five years following publication of the first book, making this a work of nearly 700 pages, compared with a mere 290 pages for the previous book.

Commencing with my initial position of an electronics engineer in the aviation industry in the mid-1970s, through a research career culminating in the directorship at the Arthur C. Clarke Institute in Sri Lanka, and now to a full-time academic position in New Zealand, I answered the call to dedicate time over this period of my life to share hard-won experience. In writing this work, I hope to encourage and inspire readers to develop greater depth of knowledge in today's complex instruments and instrumentation systems, and to more fully appreciate and understand the incredible assistance that these devices regularly provide in an increasingly complex world.

I wish to conclude this brief preface with a valuable statement of insight by Sri Sathyajith Sai Baba on teaching: "A good educator chooses the concept of inspiration compared to the possible explanation or complaining".

Nihal Kularatna
Dept of Electrical and Electronic Engineering
School of Engineering, University of Auckland
Private Bag 92019, Auckland, New Zealand

October 2002

[1] Cited by Brian Dipert: EDN magazine, 15 April 1999, p.52.

Acknowledgements

In the early 1990s, John Coupland, Library Manager of the IEE, encouraged me to work with the IEE on a book. Almost at the same time, the Late John R. Pierce, who added the important word "transistor" to the vocabulary, gave me the best advice in attempting to write technical books. Inspired by the ideas from them, as well as from many friends and colleagues, I decided to make an attempt to write a book on instruments for the IEE in 1992. Finally a print version came out in 1996. My readership . . . thanks to you all for the encouragement for me to write a new version with a wider content.

At the beginning of my career as an electronics engineer, I enjoyed the luxury of playing with the then "modern" instruments in the mid-1970s. All the transistors and the ICs combined in such an instrument may be much less than the transistor count inside a single modern VLSI chip today. Many instruments today are a combination of several complex VLSIs and application-specific ICs (ASIC) with millions of pieces of software stored in silicon memories. In appreciating these unprecedented developments, my first acknowledgement should go to the co-inventors of the integrated circuit, Jack S. Kilby and Robert Noyce. Very soon we may be able to enjoy the Instrument on a Chip concepts, while the System on a Chip (SoC) concepts are fully matured, reaping the benefits of submicron fabrication technologies.

The attempt in this book is to provide a broad approach to "understanding complete instrument systems and appreciating their internal designs" embracing many digital circuit blocks coupled with the mixed signal circuitry in instrumentation systems. In this work a large amount of published material from US industry and academia has been used, and the following organisations deserve acknowledgement.

Analog Devices, Inc., for almost 90 per cent of the material used for the Chapter 3 on Data Converters; Fluke Corporation of USA and their Wavetek Instruments Division for much of the material in Chapters 4 on Waveforms, Pulse Techniques and Multimeters, Chapter 8 on Signal Sources and Arbitrary Waveform Generators and Chapter 5 on calibration; Analog Devices, Inc., and IEEE for information sourced for Chapter 13 on Digital Signal Processors; Allegro Microsystems Inc., Analog Devices Inc, Motorola, Corporate Measurement (EG & G Sensors), Honeywell, NIST/IEEE working group on sensor integration and FiS Corporation, Japan for information on modern semiconductor sensors in Chapter 14; Tektronix Inc. and Agillent Technologies for information for Chapters 5 and 6 on oscilloscopes and Chapter 9 on spectrum analysers.

Living on a small island in the Indian Ocean while compiling the manuscript, and then making a move to beautiful New Zealand in the Pacific for the final stages of the work, I would not have completed this manuscript in a reasonable time if it were not for the great support extended by the following personnel for technical information.

James Bryant, Derek Bowers, Dan Sheingold, Walt Kester, Ethan Boudeaux, John Hayes, and Linda Roche of Analog Devices; Steve Chapman, Ed Chen and the staff of Fluke Corporation, USA, and Wong Leong Yoh and his colleagues of Fluke subsidiary in Singapore; Richard A. Roddis and Richard Wagner of Wavetek Instruments, San Diego, CA.; Karen Bosco and Christi Moore of Motorola; Taka Ito of FiS Japan; Kang Lee of NIST; Joan Lynch, Kathy Leonard and the editorial staff of the EDN magazine; Jon Titus and the editorial staff of T&M World magazine; Tom Lenihan, Amy Swanson and Bob O'Black of Tektroinx Inc., USA; Ward Elizabeth, Lisa Beyers, Nancy Trethewey and Deborah Prinster of Agillent Technologies, USA; Raymond Dewey of Allegro; Phil Brownlie of Rakon, New Zealand.; G.M. Hutchison of Honeywell, UK.; and Ms Candy Hall, Chris Honty and Carola Roeder of Butterworth Heinemann, USA.

I am very thankful for the tireless attempts by the members of staff of the Arthur C. Clarke Institute (ACCIMT) in Sri Lanka, Mrs Chandrika Weerasekera, and trainees Chamara, Kelum Bandara and Kapila Kumara in the preparation of the manuscript text and the graphics. Chandrika, you deserve a very special mention for the support for completing the manuscript during the last two years of mine at the ACCIMT as the CEO, working under pressure. Also I am very grateful to Indrani Hewage and Devika who helped me at my home environment, in addition to Nimalsiri and Kumara at ACCIMT who helped for many years in compiling collections of technical information.

I am very grateful to the Chairman, Mr Rohan Weerasinghe, former Chairman Professor K.K.Y.W. Perera and the Board Members and management of the ACCIMT for their encouragement for my research, CPD training and book writing. Particular gratitude is extended to the Board Members Dr Ray Wijewardene, Professor Dammika Tantrigoda, Mr M.D.R. Senananayke, Professor Uma Coomaraswamy, Mr H.S. Subasinghe, and Professor M.A. Careem for their greatest encouragement of my technical writing during my period as the CEO at the ACCIMT.

My special gratitude is extended to J.J. and D.J. Ambani, Erajh Gunaratne, and Niranjan de Silva and Mohan Weerasuriya and their staff (for my well maintained computing resources) of the Metropolitan Group of Companies in Sri Lanka who are always behind my work, giving great assistance and encouragement.

Also I am very grateful to my friends who encouraged me in these works, with special reference to Mr Keerthi Kumarasena (who also taught me the practical aspects of Buddhism for a peaceful and meaningful life), Padmasiri Soysa (the librarian who maintains and manages a vast array of industrial documentation in the ACCIMT library) and my friends such as Mohan Kumaraswamy (who helped with my first technical paper in 1977), and other friends such as Sunil, Kumar, Ranjith, Vasantha, Lakshman, Upali, Kithsiri, Lal/Chani and Shantha/Jayantha.

From the US side, my most sincere appreciation goes to Wayne Houser, formerly of Voice of America, for his support for the follow-up action on information collection and copyright clearances, and to Dr Fred Durant and Dr Joe Pelton of the Arthur C. Clarke Foundation for their support and encouragement for my work.

I am very thankful to the editorial and production staff at the IEE, with particular mention to Dr Robin Mellors-Bourne, Dr Roland Harwood, Diana Levy and Wendy Hiles. I was very impressed with the speed of the publication process.

After moving to New Zealand, to work at the University of Auckland, I was very impressed by the encouragement for my work by my former Head of the Department Professor John Boys (John, thanks for encouraging me to move away from my home land to which I served 23 years of 26 years of an engineering career and also leaving a CEO's job), Professor Allan Williamson (my present Head of Department) and my colleagues at the Department of Electrical and Electronic Engineering, University of Auckland. You all moved me to your spectacularly beautiful land to complete the final stages of the book in a friendly "kiwi atmosphere". It was a real difference for me in work environment too. At the last minute Mark Twiname helped me creating an urgent figure for Chapter 9 and Anja Hussman's simple suggestions saved me lot of time in index preparation. I am very thankful to both of them.

To Sir Arthur Clarke, Patron of the Clarke Institute ... my most sincere gratitude to you for your great encouragement for my little works.

Last, but not least, all this work was possibly because of the fact that all my most difficult family commitments are taken over by you ... Priyani, looking after the family needs and our two daughters Dulsha and Malsha, for allowing me lot of free time when I come home. Thank you.

Nihal Kularatna
10-15, Harrison Road
Ellerslie
Auckland
New Zealand

October 2002

Chapter 1

Introduction

1.1 The basis of measurement

Obtaining accurate, reliable and cost effective measurements depends on the instrument, the user and the mathematical treatment of the measured results. Proper selection, use of instruments and interpretation of measurement are the responsibilities of the user. A person's basic ability to make an intelligent selection and to use the instrument properly is greatly increased by an understanding of the basic theory of operation and the capabilities and limitations of the instrument families. Almost all families of test and measuring instruments give the user a set of values of a parameter that is of interest to the user. To interpret these parameter values more meaningfully some basic terms and techniques could be used.

The past two decades (1980–2000) could be cited as the maturity period for developments in digital instruments as well as the developments in interfacing, remote measurements and data management. During the 1990s, the ISO 9000 series of standards proliferated in the industry as a means of ensuring the quality and uniformity for procedures adopted in the production and service management. Adherence to quality standards such as the ISO 9000 series created a growing need for emphasis on measurement and calibration. With these developments many metrology laboratories are using personal computers to increase productivity and to collect and report information required by these standards. Raw or partly processed data available from the measuring instruments could be better interpreted when the user has a clear picture of definitions of the base units and the derivations together with a high level of confidence of the measurements.

This chapter gives a brief introduction to the basic terms, techniques and mathematical guidelines for a better interpretation of the measured quantities and provides an overview of more recent developments related to the interpretation of volt and ohm.

1.2 International unit system

The International System of Units (SI, after Système International d' Unités) is the modern form of the metric system agreed at an international conference in 1960. It has been adopted by the International Standards Organisation (ISO) and the International Electrotechnical Commission (IEC) and its use is recommended wherever the metric system is applied. It is now being adopted throughout most of the world and is likely to remain the primary world system of units of measurements for a very long time. The indications are that SI units will supersede the units of the existing metric system and all systems based on imperial units.

The adoption of SI presents less of a problem to the electronics engineer and the electrical engineer than to those concerned with other engineering disciplines as all the practical electrical units were long ago incorporated in the metre–kilogram–second (MKS) unit system and these remain unaffected in SI. The SI was developed from the metric system as a fully coherent set of units for science, technology and engineering. A coherent system has the property that corresponding equations between quantities and between numerical values have exactly the same form, because the relations between units do not involve numerical conversion factors. In constructing a coherent unit system, the starting point is the selection and definition of a minimum set of independent 'base' units. From these, 'derived' units are obtained by forming products or quotients in various combinations, again without numerical factors. Thus the base units of length (metre), time (second) and mass (kilogram) yield the SI units of velocity (metre/second), force (kilogram metre/second-squared) and so on. As a result there is, for any given physical quantity, only one SI unit with no alternatives and with no numerical conversion factors. A single SI unit (joule $\equiv$ kilogram metre-squared/second-squared) serves for energy of any kind, whether it is kinetic, potential, thermal, electrical, chemical, etc., thus unifying the usage in all branches of science and technology.

The SI has seven base units, and two supplementary units of angle. Certain important derived units have special names and can themselves be employed in combination to form alternative names for further derivations.

The metre is now defined in terms of the speed of light in vacuum; the kilogram, the only SI unit still defined by a physical artifact, is the mass of the international prototype of the kilogram; and the second is defined in terms of the transition between the two hyperfine levels of the ground state of caesium-133.

The SI unit of force, the newton, is derived from these three base mechanical units through Newton's second law; it is given by $\mathrm{N} = \mathrm{m} \cdot \mathrm{kg} \cdot \mathrm{s}^{-2}$. The joule, the SI unit of energy, follows from the definition of work: $\mathrm{J} = \mathrm{N} \cdot \mathrm{m} = \mathrm{m}^2 \cdot \mathrm{kg} \cdot \mathrm{s}^{-2}$. Finally, the SI unit of power, the watt, is defined as $\mathrm{W} = \mathrm{J\,s}^{-1} = \mathrm{m}^2 \cdot \mathrm{kg} \cdot \mathrm{s}^{-3}$.

The SI base electric unit, the ampere (A), is defined as 'that constant current which, if maintained in two straight parallel conductors of infinitive length, of negligible circular cross-section, and placed 1 metre apart in vacuum, would produce between these conductors a force equal to 2×10^{-7} newton per metre of length'. This definition implies that μ_0, the permeability of the vacuum, is exactly $4\pi \times 10^{-7}\,\mathrm{N/A}^2$.

The volt and ohm are then derived from the watt (and thus from the three base mechanical units) and the ampere, the base electric unit. Formally, $\mathrm{V} = \mathrm{W\,A^{-1}} = \mathrm{m^2 \cdot kg \cdot s^{-3} \cdot A^{-1}}$ and $\Omega = \mathrm{V\,A^{-1}} = \mathrm{W\,A^{-2}} = \mathrm{m^2 \cdot kg \cdot s^{-3} \cdot A^{-2}}$.

Each physical quantity has a quantity symbol (e.g. m for mass) that represents it in equations, and a unit symbol (e.g. kg for kilogram) to indicate its SI unit of measure. Note that the quantity symbol is printed in italic whereas the unit symbol is printed in upright roman type.

1.2.1 Base units

Definitions of the seven base units have been laid down in the following terms.

Length (l). The metre (m) is equal to the path travelled by light during a time interval of 1/299 792 458 of a second.

Mass (m). The kilogram (kg) is the mass of the international prototype (i.e., a block of platinum–iridium alloy preserved at the International Bureau of Weights and Measures at Sèvres, near Paris).

Time (t). The second (s) is the duration of 9 192 631 770 periods of the radiation corresponding to the electron transition between the two hyperfine levels of the ground state of the caesium-133 atom.

Electric current (i). The ampere (A) is the current which, maintained in two straight parallel conductors of infinite length, of negligible circular cross-section and 1 m apart in vacuum, produces a force equal to 2×10^{-7} newton per metre of length.

Temperature (T). The kelvin (K), unit of thermodynamic temperature, is the fraction 1/273.16 of the thermodynamic (absolute) temperature of the triple point of water.

Luminous intensity (I). The candela (cd) is the luminous intensity, in the perpendicular direction, of a surface of $1/600\,100\ \mathrm{m^2}$ of a black body at the temperature of freezing platinum under a pressure of 101 325 newtons per square metre.

Amount of substance (Q). The mole (mol) is the amount of substance of a system which contains as many elementary entities as there are atoms in 0.012 kg of carbon-12. The elementary entity must be specified and may be an atom, a molecule, an electron, etc., or a specified group of such entities.

1.2.2 Supplementary angular units

Plane angle ($\alpha, \beta, \ldots$). The radian (rad) is the plane angle between two radii of a circle which cut off on the circumference an arc of length equal to the radius.

Solid angle (Ω). The steradian (sr) is the solid angle which, having its vertex at the centre of a sphere, cuts off an area of the surface of the sphere equal to a square having sides equal to the radius.

1.2.3 Derived units

Table 1.1 lists nine of the more important SI derived units. Their respective units are defined as follows.

Table 1.1 Important SI derived units

Quantity	Unit name	Unit symbol
Force	newton	N
Energy	joule	J
Power	watt	W
Electric charge	coulomb	C
Electric potential difference and EMF	volt	V
Electric resistance	ohm	Ω
Electric capacitance	farad	F
Electric inductance	henry	H
Magnetic flux	weber	Wb

Newton. A newton is that force which gives to a mass of 1 kg an acceleration of $1\,\mathrm{m\,s^{-2}}$.
Joule. The work done when the point of application of a force of 1 N is displaced a distance of 1 m in the direction of the force is one joule.
Watt. The power which gives rise to the production of energy at the rate of $1\,\mathrm{J\,s^{-1}}$ is one watt.
Coulomb. The quantity of electricity transported in 1 s by a current of 1 A is one coulomb.
Volt. The difference of electric potential between two points of a conducting wire carrying a constant current of 1 A, when the power dissipated between these points is equal to 1 W, is one volt.
Ohm. The electric resistance between two points of a conductor when a constant difference of 1 V, applied between these two points, produces in this conductor a current of 1 A, is one ohm, this conductor not being the source of any electromotive force.
Farad. The capacitance of one farad occurs between the plates of a capacitor of which there appears a difference of potential of 1 V when it is charged by a quantity of electricity equal to 1 C.
Henry. One henry is the inductance of a closed circuit in which an electromotive force of 1 V is produced when the electric current in the circuit varies uniformly at a rate of $1\,\mathrm{A\,s^{-1}}$.
Weber. One weber is the magnetic flux which, linking a circuit of one turn, produces it in an electromotive force of 1 V as it is reduced to zero at a uniform rate in 1 s.

Some of the simpler derived units are expressed in terms of the seven basic and two supplementary units directly. Examples are listed in Table 1.2.

Units in common use, particularly those for which a statement in base units would be lengthy or complicated, have been given special shortened names (see Table 1.3). Those that are named from scientists and engineers are abbreviated to an initial capital letter: all others are in lower case letters.

Table 1.2 Directly derived units

Quantity	Unit name	Unit symbol
Area	square metre	m^2
Volume	cubic metre	m^3
Mass density	kilogram per cubic metre	$kg\,m^{-3}$
Linear velocity	metre per second	$m\,s^{-1}$
Linear acceleration	metre per second squared	$m\,s^{-2}$
Angular velocity	radian per second	$rad\,s^{-1}$
Angular acceleration	radian per second squared	$rad\,s^{-2}$
Force	kilogram metre per second squared	$kg\,m\,s^{-2}$
Magnetic field	ampere per metre	$A\,m^{-1}$
Concentration	mole per cubic metre	$mol\,m^{-3}$
Luminance	candela per square metre	$cd\,m^{-2}$

Table 1.3 Named derived units

Quantity	Unit name	Unit symbol	Derivation
Force	newton	N	$kg\,m\,s^{-2}$
Pressure	pascal	Pa	$N\,m^{-2}$
Power	watt	W	J s
Energy	joule	J	N m, Ws
Electric charge	coulomb	C	A s
Electric flux	coulomb	C	A s
Magnetic flux	weber	Wb	V s
Magnetic flux density	tesla	T	$Wb\,m^{-2}$
Electric potential	volt	V	$J\,C^{-1}$, $W\,A^{-1}$
Resistance	ohm	Ω	$V\,A^{-1}$
Conductance	siemens	S	$A\,V^{-1}$
Capacitance	farad	F	$A\,s\,V^{-1}$, $C\,V^{-1}$
Inductance	henry	H	$V\,s\,V^{-1}$, $Wb\,A^{-1}$
Luminous flux	lumen	lm	cd sr
Illuminance	lux	lx	$lm\,m^{-2}$
Frequency	hertz	Hz	$1\,s^{-1}$

1.2.4 Expressing magnitudes of SI units

To express magnitudes of a units, decimal multiples and submultiples are formed using the prefixes such as tera, giga, . . . , micro, . . . , pico, etc. This method of expressing magnitudes ensures complete adherence to a decimal system. For details, Reference 1 is suggested.

1.3 Measurement standards

To ensure that the units are consistently applied, a set of measurement standards is required. All instruments are calibrated at the time of manufacture against a measurement standard. Standards are defined in four categories. They are:

- international standards,
- primary standards,
- secondary standards,
- working standards.

International standards are defined by international agreement. These standards are maintained at the International Bureau of Weights and Measures at Sèvres near Paris. They are checked periodically by absolute measurements, in terms of the fundamental unit concerned. They represent certain units of measurement to the closest possible accuracy attainable by the science and technology of measurement.

Primary standards are maintained in the national standardising laboratories of different countries. These standards are not available for use outside the national laboratory, although they may be used to calibrate secondary standards sent to that laboratory. Primary standards are themselves calibrated at the various national laboratories by making absolute measurements in terms of the fundamental units.

Secondary standards are maintained in various laboratories in industry. Their prime function is to check and calibrate working standards. Responsibility for maintenance of the secondary standard is with the industrial laboratory concerned, although periodically these may be sent to national standardising laboratories for checking and calibration.

Working standards are the principal tools of a measurement laboratory. These standards are used to check and calibrate the instruments used in the laboratory or to make comparison measurements in industrial application. Working standards are periodically checked against secondary standards.

1.4 Electrical standards

The electrical standards exist for current, resistance, capacitance, voltage. Defining terms for ampere, ohm, farad and volt, etc., were given in section 1.2.3 and those are the official International System of Units (SI) definitions as adopted by the Comité International des Poid et Mesures. The main types of voltage standard are the Weston cadmium cell, the Josephson effect standard, and Zener diode standards.

Historically, representations of the volt have been based on electrochemical standard cells. However, at its thirteenth meeting, held in October 1972 [2], the Consultative Committee on Electricity (CCE), one of the principal international bodies concerned with such matters, suggested that the national standards laboratories should base their national representation of the volt on the Josephson effect in order to avoid the well-known problems associated with the EMFs of standard cells, for example, their variation with time or drift, severe dependence upon temperature, and occasional unpredictable abrupt changes. The Hall effect standard is for resistance. Details of these standards can be found in References 2–5.

1.4.1 The Josephson effect

The Josephson effect is characteristic of weakly coupled superconductors when cooled below their transition temperatures [4]. An example is two thin films of superconducting lead separated by an approximately 1 nm thick thermally grown oxide layer. When such a Josephson junction is irradiated with microwave radiation of frequency f, its current vs voltage curve exhibits vertical current steps at highly precise quantised Josephson voltage U_J.

1.4.1.1 Definition of the Josephson constant

When a Josephson junction is irradiated with microwave radiation of frequency f, its current vs voltage curve exhibits steps at highly precise quantised Josephson voltages U_J. The voltage of the nth step $U_J(n)$, where n is an integer, is related to the frequency of the radiation by

$$U_J(n) = nf/K_J, \tag{1.1}$$

where K_J is commonly termed the Josephson frequency-to-voltage quotient [6].

The Working Group on the Josephson Effect (WGJE) proposed that this quotient be referred to as the Josephson constant and, since no symbol had yet been adopted for it, that it be denoted by K_J. It follow from eq. (1.1) that the Josephson constant is equal to the frequency-to-voltage quotient of the $n = 1$ step.

The theory of the Josephson effect predicts, and the experimentally observed universality of eq. (1.1) is consistent with the prediction, that K_J is equal to the invariant quotient of fundamental constants $2e/h$, where e is the elementary charge and h is the Planck constant [6]. For the purpose of including data from measurements of fundamental constants in the derivation of their recommended value of K_J, the WGJE assumed that $2e/h = K_J$.

The Josephson effect provides an invariant 'quantum standard of voltage' that can be used to establish a highly reproducible and uniform representation of the volt based on an agreed upon value of K_J.

1.4.2 The quantum Hall effect (QHE)

The quantum Hall effect (QHE) is the characteristic of certain high mobility semiconductor devices of standard Hall-bar geometry when placed in a large applied magnetic field and cooled to a temperature of about 1 K. For a fixed current I through a QHE device there are regions in the curve of Hall voltage vs gate voltage, or of Hall voltage vs magnetic field depending upon the device, where the Hall voltage U_H remains constant as the gate voltage or magnetic field is varied. These regions of constant Hall voltage are termed Hall plateaus. Figure 1.1 indicates a QHE heterojunction device.

Under the proper experimental conditions, the Hall resistance of the ith plateau $R_H(i)$, defined as the quotient of the Hall voltage of the ith plateau to the current I,

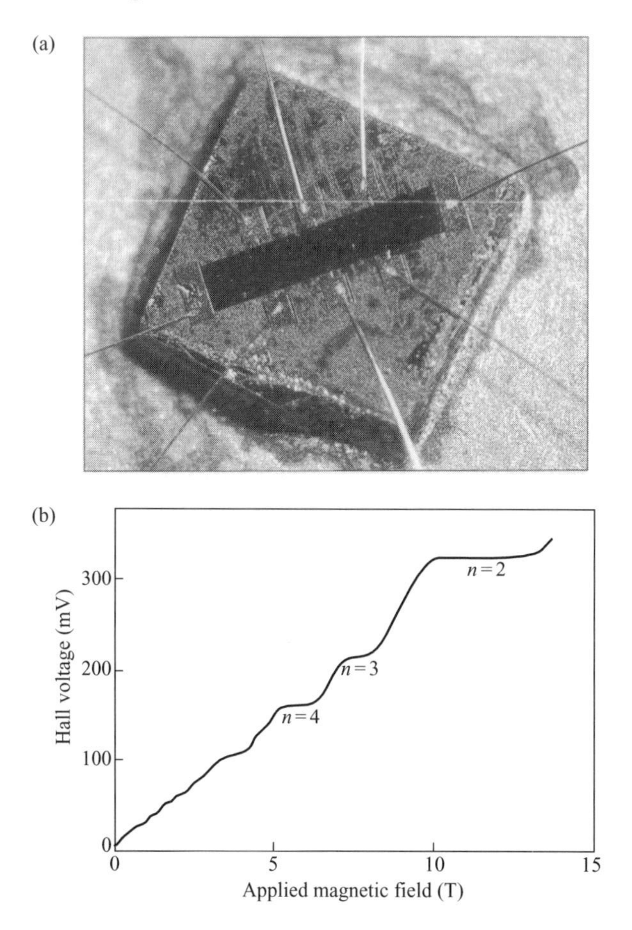

Figure 1.1 *QHE heterojunction device and characteristics: (a) GaAs–Al$_x$ Ga$_{1-x}$ As device in dark central portion; (b) plot of Hall voltage as a function of magnetic field (Source: IEEE Spectrum, July 1989, © IEEE 2002)*

is given by

$$R_{\mathrm{H}}(i) = U_{\mathrm{H}}(i)/I = R_{\mathrm{K}}/I, \tag{1.2}$$

where i is an integer.

1.4.2.1 The von Klitzing constant

Because $R_{\mathrm{H}}(i)$ is often referred to as the quantised Hall resistance regardless of plateau number, the Working Group on the Quantum Hall Effect (WGQHE) proposed that, to avoid confusion, the symbol R_{K} be used as the Hall voltage-to-current quotient or resistance of the $i = 1$ plateau and that it be termed the von Klitzing constant after the discoverer of the QHE. It thus follows from eq. (1.2) that $R_{\mathrm{K}} = R_{\mathrm{H}}(I)$.

The theory of the QHE predicts, and the experimentally observed universality of eq. (1.2) is consistent with the prediction, that R_K is equal to the invariant quotient of fundamental constants h/e^2 [7]. For the purpose of including data from measurements of fundamental constants in the derivation of their recommended value of R_K, the WGQHE assumed that $h/e^2 = R_K$.

1.4.3 New representations of volt and ohm

On 1 January 1990, by international agreement, the national standardising laboratories of most major industrial countries put into place new representations of the volt and ohm – practical units for the precise and highly reproducible measurement of voltages and resistances. Based on the Josephson and quantum Hall effects, these new representations changed the values of volt and ohm representations and, by derivation, those of the ampere and watt. Significant differences that existed among the values of some national representations of these units were to be eliminated [8].

The changes were large enough to require the adjustment of many thousands of electrical standards, measuring instruments, and electronic systems throughout the world in order to bring them into conforming with the new representations. For example, in the United States there were increases of 9.264, 1.69, 7.57, and 16.84 parts per million (p.p.m.) for the volt, ohm, ampere and watt representations, respectively.

The new standards answer important needs. They were to eliminate the differences that existed among the volt and ohm representations of different countries. The new quantum Hall effect standard eliminated the time variation, or drift, of any national ohm representations that were based (and most are) on wirewound resistors, which age. The new representations will also be highly consistent with the volt and ohm as defined in the International System of Units.

1.4.4 Josephson effect reference standard of voltage

The CCE reviewed the report from the WGJE and discussed at some length the draft recommendation E1 (1988), 'Representation of the volt by means of the Josephson effect,' prepared jointly by the WGJE and the Working Group on the Quantum Hall Effect. The CCE then agreed:

(i) to use the term 'Josephson constant' with symbol K_J to denote the Josephson frequency-to-voltage quotient;

(ii) to accept the WGJE's recommended value of K_J, namely $K_J = (483\,597.9 \pm 0.2)\ \mathrm{GHz\,V^{-1}}$, where the $0.2\,\mathrm{GHz\,V^{-1}}$ assigned one standard deviation uncertainty corresponds to a relative uncertainty of 0.4 p.p.m.; and

(iii) to use this recommended value to define a conventional value of K_J and to denote it by the symbol $K_{J\text{-}90}$ so that $K_{J\text{-}90} \stackrel{\text{def}}{=} 483\,597.9\,\mathrm{GHz\,V^{-1}}$, exactly.

The subscript 90 derives from the fact that this new conventional value of the Josephson constant is to come into effect starting 1 January 1990, a date reaffirmed by the CCE.

1.4.5 QHE reference standard of resistance

The CCE reviewed the report of the WGQHE and discussed the draft recommendation E2 (1988), 'Representation of the ohm by means of the quantum Hall effect', prepared jointly by the two Working Groups. Because of the similarities between the QHE and the Josephson effect, the review and discussion proceeded expeditiously. Also in analogy with the Josephson effect, the CCE agreed:

(i) to use the term 'von Klitzing constant' with symbol R_K to denote the Hall voltage-to-current quotient or resistance of the $i = 1$ plateau;
(ii) to accept the WGQHE's recommended value of R_K, namely, $R_K = (25\,812.807 \pm 0.005)\,\Omega$, where the $0.005\,\Omega$ assigned one standard deviation uncertainty corresponds to a relative uncertainty of 0.2 p.p.m.; and
(iii) to use this recommended value to define a conventional value of R_K and to denote it by the symbol $R_\text{K-90}$, so that $R_\text{K-90} \stackrel{\text{def}}{=} 25\,812.807\,\Omega$ exactly.

The same procedure was followed for draft recommendation E2 (1988) as for E1 (1988) regarding the Josephson effect.

Recently, economical versions of quantum Hall resistance (QHR) standards such as 'quant Q' have been developed. These systems are expected to assist the national laboratories, which are unable to afford expensive QHR primary standards. More details can be found in Reference 9.

1.5 Measurement errors

Measurement is the process of comparing an unknown quantity with an accepted standard quantity. All measurements are subject to errors, due to a variety of reasons such as inherent inaccuracies of the instrument, human error and using the instrument in a way for which it was not designed.

1.5.1 Error

Error is the difference between the result of the measurement and the true value of the quantity measured, after all corrections have been made. Error, which is not the same as uncertainty, has traditionally been viewed as being of two kinds: random error and systematic error. In general, although error always exists, its magnitude cannot be exactly known.

1.5.2 Statistics of errors

Statistical analysis is frequently used in measurements and four concepts commonly used in measurements are: averages, dispersion from the average, probability distribution of errors, and sampling.

1.5.2.1 Averages

The most frequently used averaging technique is the 'arithmetic mean'. If n readings are taken with an instrument, and the values obtained are $x_1, x_2, \ldots, x_n$, then the arithmetic mean is given by

$$\bar{x} = \frac{x_1 + x_2 + x_3 + \cdots + x_n}{n} \tag{1.3}$$

or

$$\bar{x} = \frac{\sum_{r=1}^{n} x_r}{n}. \tag{1.4}$$

Although the arithmetic mean is easy to calculate it is influenced unduly by extreme values, which could be false. An alternative averaging technique, called the geometric mean, is not overly affected by extreme values. It is often used to find the average of quantities that follow a geometric progression or an exponential law. The geometric mean is given by

$$x_g = \sqrt[n]{x_1 \times x_2 \times x_3 \times \cdots \times x_n}. \tag{1.5}$$

1.5.2.2 Dispersion from average

The 'average' represents the mean Figure of a series of numbers. It does not give any indication of the spread of these numbers. For example, suppose eight different voltmeter readings are taken of a fixed voltage, and the values read are 97, 98, 99, 100, 101, 102, 103 V. Hence, the mean value will be 100 V. If the readings are changed to 90, 90, 95, 100, 100, 105, 110, 110, the mean will still be 100 V, although now they are more widely separated from the mean. The three techniques most frequently used to measure dispersion from the mean are the range, mean deviation and the standard deviation.

The range is the difference between the largest and the smallest values. Therefore, for the first set of readings, the range is $(103-97) = 6$ V; for the second set of readings it is $(110 - 90) = 20$ V.

The mean deviation, M, is found by taking the mean of the difference between each individual number in the series and the arithmetic mean, and ignoring negative signs. Therefore for a series of n numbers $x_1, x_2, \ldots, x_n$, having an arithmetic mean $\bar{x}$, the mean deviation is given by

$$M = \frac{\sum_{r=1}^{n} |x_r - \bar{x}|}{n}. \tag{1.6}$$

For the first set of values, the mean deviation is

$$M = \frac{(3 + 2 + 1 + 0 + 0 + 1 + 2 + 3)}{8} = 1.50\,\text{V}.$$

For the second set of values, the mean deviation is

$$M = \frac{(10 + 10 + 5 + 0 + 0 + 5 + 10 + 10)}{8} = 6.25\,\text{V}.$$

Therefore, by comparing the mean deviation of the two sets of readings, one can deduce that the first set is more closely clustered around the mean, and therefore represents more consistent values.

Neither the mean deviation nor the range are suitable for use in statistical calculation. The standard deviation is the measure of dispersion that is most commonly used for this. The standard deviation of a series of n numbers $x_1, x_2, \ldots, x_n$ having a mean of $\bar{x}$, is given by

$$\sigma = \left(\frac{\sum_{r=1}^{n} (x_r - \bar{x})^2}{n} \right)^{1/2}; \tag{1.7}$$

because the deviation from the mean is squared before summing, the signs are taken into account, so that the calculation is mathematically correct. For the first set of values the standard deviation is 1.87 and for the second set of values standard deviation is 7.5.

1.5.2.3 Probability distribution of errors

If an event A, for example an error, occurs n times out of a total of m cases, then the probability of occurrence of the error is stated to be

$$p(\mathrm{A}) = \frac{n}{m}. \tag{1.8}$$

Probabilities vary between 0 and 1. If $p(\mathrm{A})$ is the probability of an event occurring then $1 - p(\mathrm{A})$, which is written as $p(\bar{\mathrm{A}})$, is the probability that the event will not occur.

There are several mathematical distributions that are used to define the spread in probabilities. The binomial, Poisson, normal, exponential and Wiebull distributions are commonly used. The normal distribution is most commonly used. The normal distribution curve is bell shaped, and is shown in Figure 1.2.

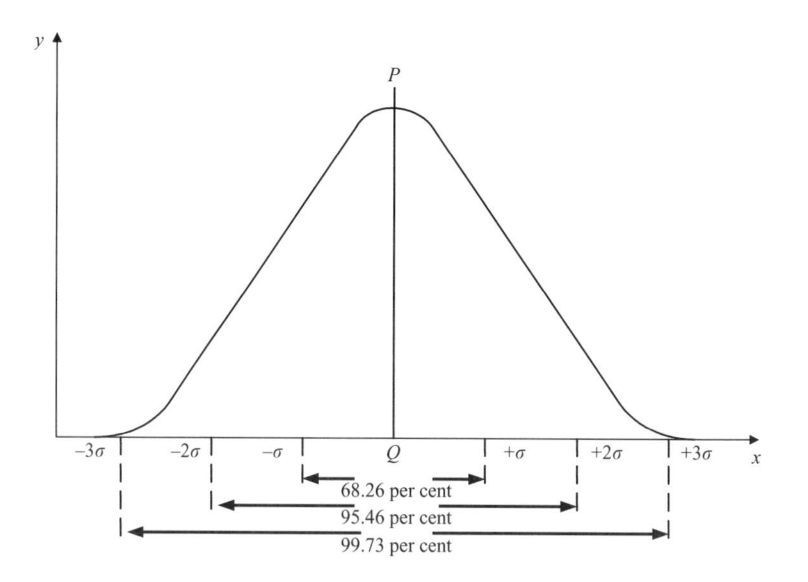

Figure 1.2 Normal curve

The x axis gives the event and the y axis gives the probability of the event occurring. If PQ represents the line of mean value $\bar{x}$ then the equation for the normal curve is given by

$$y = \frac{1}{(2\pi)^{1/2}} e^{-\omega^2/2}, \tag{1.9}$$

where

$$\bar{\omega} = \frac{x - \bar{x}}{\sigma}. \tag{1.10}$$

The total area under the normal curve is unity and the area between any two values of ω is the probability of an item from the distribution falling between these values. The normal curve extends from $\pm\infty$ but 68.26 per cent of its values fall between $\pm\sigma$, 95.46 per cent between $\pm 2\sigma$, 99.73 per cent between $\pm 3\sigma$ and 99.9999 per cent between $\pm 4\sigma$. For a detailed discussion on this, Reference 16 is suggested.

1.5.2.4 Sampling

Sampling techniques are often used in measurement systems. A small number of devices, from a larger population, are tested to give information on the population. For example, a sample of 20 resistors from a batch of 2000 may be tested, and if these are all satisfactory it may be assumed that the whole batch is acceptable. However, errors arise in sampling. These are usually evaluated on the assumption that sampling errors follow a normal distribution. Suppose a batch has n_b items, with a mean of $\overline{x_b}$. If a sample of n_s items is taken from this batch, and found to have a mean of $\overline{x_s}$ and a standard deviation of σ_s then

$$\overline{x_b} = \overline{x_s} \pm \frac{\gamma\sigma_s}{n_s^{1/2}}. \tag{1.11}$$

The value of γ is found from the normal curve depending on the level of confidence needed in specifying $\overline{x_b}$. For $\gamma = 1$ this level is 68.26 per cent, for $\gamma = 2$ it is 95.46 per cent and for $\gamma = 3$ it is 99.73 per cent.

1.5.3 Factors influencing measurement errors

Errors arise in measurement systems due to several causes, such as human errors or errors in using an instrument in an application for which it has not been designed. Several definitions are now introduced which define the factors that influence measurement errors.

1.5.3.1 Accuracy

Accuracy refers to how closely the measured value agrees with the true value of the parameter being measured. For electrical instruments the accuracy is usually defined as a percentage of full scale deflection.

1.5.3.2 Precision

Precision means how exactly or sharply an instrument can be read. It is also defined as how closely identically performed measurements agree with each other. As an example, suppose that a resistor, which has a true resistance of 26 863 Ω, is measured by two different meters. The first meter has a scale which is graduated in kΩ, so that the closest one can get to reading of resistance is 27 kΩ. The instrument is fairly accurate but it is very imprecise. The second instrument has a digital readout which gives values of resistance to the nearest ohm. On this instrument the same resistor measures 26 105 Ω. Clearly this instrument has high precision but low accuracy.

1.5.3.3 Resolution

The resolution of an instrument is the smallest change in the measured value to which the instrument will respond. For a moving pointer instrument the resolution depends on the deflection per unit input. For a digital instrument the resolution depends on the number of digits on the display.

1.5.3.4 Range and bandwidth

The range of an instrument refers to the minimum and maximum values of the input variable for which it has been designed. The range chosen should be such that the reading is large enough to give close to the required precision. For example, with a linear scale an instrument which has 1 per cent precision at full scale will have 4 per cent precision at quarter scale.

The bandwidth of an instrument is the difference between the minimum and maximum frequencies for which it has been designed. If the signal is outside the bandwidth of the instrument, it will not be able to follow changes in the quantity being measured. A wider bandwidth usually improves the response time of an instrument, but it also makes the system more prone to noise interference.

1.5.3.5 Sensitivity

Sensitivity is the degree of response of a measuring device to the change in input quantity. The sensitivity of an instrument is defined as the ratio of the output signal of response of the instrument to the input signal or measured variable.

1.5.3.6 Uncertainty

Uncertainty is an estimate of the possible error in a measurement. More precisely, it is an estimate of the range of values which contains the true value of a measured quantity. Uncertainty is usually reported in terms of the probability that the true value lies within a stated range of values.

Measurement uncertainty has traditionally been defined as a range of values, usually centred on the measured value, that contains the true value with stated probability. A measurement result and its uncertainty traditionally were reported as

$$\text{quantity} = \text{value} \pm U. \tag{1.12}$$

So the number usually reported and called 'uncertainty' was actually half the range defined here. The ISO Guide [10] redefines uncertainty to be the equivalent of a standard deviation, and thus avoids this problem.

1.5.3.7 Confidence interval and confidence level

1.5.3.7.1 Confidence interval

When uncertainty is defined as above, the confidence interval is the range of values that corresponds to the stated uncertainty.

1.5.3.7.2 Confidence level

Confidence level is the probability associated with a confidence interval. For example one could indicate that the true value can be expected to lie within $\pm x$ units of the measured value with 99 per cent confidence.

1.5.3.8 Repeatability

Repeatability is defined as the degree of agreement among independent measurements of a quantity under the same condition.

1.5.3.9 Reproducibility

Measurement reproducibility is the closeness of agreement between the results of measurements of the same measurand at different locations by different personnel using the same measurement method in similar environments.

1.5.4 Types of error

Measurement errors could be divided into four types: human, systematic, random, and applicational.

Human errors (gross errors) are generally the fault of the person using the instruments and are caused by such things as incorrect reading of instruments, incorrect recording of experimental data or incorrect use of instruments. Human errors can be minimised by adopting proper practices and by taking several readings, etc.

Systematic errors result from problems with instruments, environmental effects or observational errors. Instrument errors may be due to faults on the instruments such as worn bearings or irregular spring tension on analogue meters. Improper calibration also falls into instrumental errors. Environmental errors are due to the environmental conditions in which instruments may be used. Subjecting instruments to harsh environments such as high temperature, pressure or humidity or strong electrostatic or electromagnetic fields may have detrimental effects, thereby causing error. Observational errors are those errors introduced by the observer. Probably most common observational errors are the parallax error introduced in reading a meter scale and the error of estimation when obtaining a reading from a meter scale.

Random errors are unpredictable and occur even when all the known systematic errors have been accounted for. These errors are usually caused by noise and environmental factors. They tend to follow laws of chance. They can be minimised by taking many readings and using statistical techniques.

Applicational errors are caused by using the instrument for measurements for which it has not been designed. For example, the instrument may be used to measure a signal that is outside the bandwidth. Another common example is to use an instrument with an internal resistance that is comparable in value to that of the circuit being measured. Applicational errors can be avoided by being fully aware of the characteristics of the instrument being used.

1.5.5 Statistical process control

Statistical process control (SPC) is a tool for ensuring the quality of the output of an activity. In the metrology laboratory, SPC is most often used to control the quality of measurements, but has equally important applications in controlling many non-measurement activities of the laboratory.

SPC can be applied to any repeated operation that has a measurable output, including a measurement process. The simplest application of SPC in the metrology laboratory is in controlling the uncertainty obtained in operating measuring systems.

1.5.5.1 Controlling measurement uncertainties

Figure 1.3 illustrates the three parameters of interest to a metrologist.

- Error: the difference between the measured value and the true value.
- Uncertainty: the range of values that will contain the true value.
- Offset: the difference between a target (normal) value and the actual value.

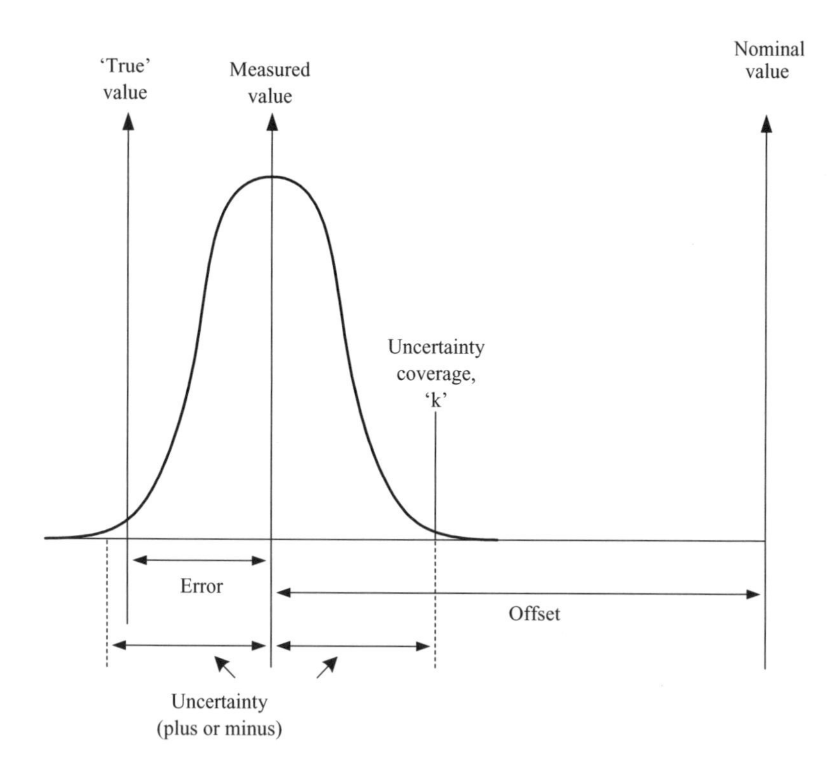

Figure 1.3 Error, uncertainty and offset

Offset is calculated by means of an experiment to determine the absolute value of a parameter, the simplest such experiment being a calibration against a superior standard.

Uncertainty is determined by an uncertainty analysis that takes into consideration the effects of systematic and random errors in all the processes that lead to the assignment of a value to a measurement result. Both standardisation and uncertainty analyses are relatively difficult and costly operations that should not be undertaken more often than necessary. It is most cost effective to perform these analyses once, and then to use process control techniques to provide evidence that the system used is the system analysed, that is, nothing has changed.

A process can change in two fundamentally different ways: the set point can change, or the variability of its process can change.

A change in set point corresponds to a change in the calibration of the system. A change in variability indicates that the uncertainty analysis is no longer valid, since the assumptions upon which it was based have changed. Either change could lead to measurement results outside the range of results expected, based on the uncertainty stated. For more details of SPC techniques in metrology and applications, chapter 23 of Reference 11 is suggested.

1.6 Summary

This chapter is prepared as a refresher module for the basics related to measurements and some developments in the past decade on volt and ohm representation, etc. Rigorous mathematical treatment of errors, statistical techniques and analysis of results, etc., are purposely eliminated. For these details many good references are available and some of these are listed under References 11–16. Reference 11 is an excellent reference for any one who plans to set up a calibration laboratory or a service.

1.7 References

1 KULARATNA, N.: 'Modern electronic test & measuring Instruments', *IEE*, 1996.
2 TAYLOR, B.N.: 'New measurement standard for 1990', *Physics Today*, August 1989, pp. 23–26.
3 TAYLOR, B.N.: 'New international representations of volt and ohm effective, January 1, 1990', *IEEE Transactions on Instrumentation & Measurement*, February 1990, **39** (1).
4 McCULLOUGH, R.: 'Coming soon from a calibration lab near you', *Electron. Test*, January 1990, pp. 36–40.
5 TAYLOR, B.N.: 'Preparing for the new volt and ohm', *IEEE Spectrum*, July 1989, pp. 20–23.
6 TAYLOR, B.N.: 'New internationally adopted reference standards of voltage and resistance', *Journal of Research of the NIST*, March–April 1989, **94**.
7 PRANGE, R.E. and GARVIN, S.M.: *The Quantum Hall Effect*, Springer-Verlag, NY, 1987.

8 BELECKI, N.B., DZIUBA, R.F., FIELD, B.F., and TAYLOR, B.N.: 'Guidelines for implementing the new representations of the volt and ohm'. *NIST technical note 1263*, US Department of Commerce.
9 INGLIS, D., *et al.*: 'A modular, portable, quantized Hall resistance standard', *Cal Lab Journal*, July–August 1999.
10 ISO: 'Guide to the expression of uncertainty in measurement' (Geneva, Switzerland (1993))
11 FLUKE CORP.: *Calibration: Philosophy in Practice*, 2nd edn, 1994.
12 JONES, L.D. and FOSTER CHIN, A.: *Electronic Instrument and Measurements*, Wiley & Sons, New York, 1983.
13 CERNI, R.H. and FOSTER, L.F.: *Instrumentation for Engineering Measurement*, John Wiley & Sons, New York, 1962.
14 LENK, J.D.: *Handbook of Electronic Test Equipment*, Prentice-Hall, New Jersey, 1971.
15 BENTLY, J.P.: *Principles of Measurement Systems*, Longman, London, 1983.
16 MAZDA, F.F.: *Electronic Engineer's Reference Book*, Butterworth Heinemann, London, 6th edn, 1994.

Chapter 2

Enabling technologies

2.1 Introduction

Following the invention of the transistor by the Bell Labs team – John Bardeen, William Schockley and Walter Brattain – electronic products and systems have gained much compactness, intelligence, user friendliness, inter-connectability and low power consumption. Half a century of developmental activity in the world of semiconductors has given the modern world several mature and growing technologies with enormous miniaturising power based on Si, gallium arsenide (GaAs) and newer materials such as SiGe. With compactness and power efficiency being key issues in the semiconductor world, other related technologies such as battery chemistries, magnetics, connectors, and packaging of components have also evolved in leaps and bounds. Modern flat panel display technologies have commenced the replacement of the cathode ray tube, which has served the industry for over a century.

Today we are heading towards systems on chip (SoC) concepts, with over 100 million transistor ultra large scale integration (ULSI) technologies entering into production at the beginning of the twenty-first century. While the SoC concepts and implementations providing dense ICs, research labs are always working hard to develop newer transistor structures. Some examples from US industry are the GaAs MOSFET [1] and indium phosphide (InP) high electron mobility transistor (HEMT) with a frequency of 350 GHz [2]. Meanwhile, Japanese researchers have achieved frequencies over 362 GHz for HEMT devices [3]. While CMOS based devices were at the heart of technological advances related to miniaturisation in the 1980s and 1990s, alternatives to CMOS logic devices are being investigated. Some of these approaches are single-electron transistors, quantum cellular automata (QCA) and chemically assembled electronic nanocomputers (CAEN) [4, 5]. One particularly interesting approach, molecular tunnelling devices, implements processes familiar to silicon device process engineers. Based on molecular switches, researchers at Hewlett Packard Labs and the University of California, Los Angeles, have developed logic gates and memory elements that fit between the wires used to address them.

This chapter provides an overview of important technologies enabling the development of test and measuring instruments and an update to chapter 2 of Reference 6.

2.2 Integrated circuits progressing on to SoC

Within the past five decades, the close interaction and dedication of solid state physicists, chemical engineers and creative circuit designers have given an enormous impetus to the micro miniaturisation power of silicon, GaAs and other semiconductor components. The early small scale integrated (SSI) circuits which had less than 10 to 100 transistors have now evolved into very large scale integrated (VLSI) circuits containing multimillion transistors on a single silicon integrated circuit. The developments have followed the path based on Moore's (or modified Moore's) curve [7] shown in Figure 2.1(a). With the advancements in semiconductor manufacturing techniques and process equipment, production today allows for high yields at volume production which, in turn, dramatically brings down the cost of the integrated circuits. Reference 8 is suggested as a guideline for microchip fabrication technology. Figure 2.1(b) depicts the cost path of integrated circuits over time.

It can easily be seen from the graph in Figure 2.1(a) that microprocessors and the memory components with higher component densities have been gradually progressing through the SSI, MSI, LSI and the ULSI eras. In the new millennium the SoC concepts with over 100 million transistors on a single piece of silicon are expected. Figure 2.7 indicates the trends of feature size reduction of dense electronic components, such as memories.

Based on presentations at the International Solid State Circuits Conference (ISSCC), it is interesting to see that at the start of the twenty-first century processor chips carry over 90 million transistors [3], compared with the early Intel microprocessor introduced in 1971, which had around 2300 transistors. All these VLSI/ULSI and SoC implementations follow three principles, namely higher performance, lower cost and better system reliability.

With the demand for high speed components operating in the region of several gigahertz, designers were faced with the limitations of silicon. In the early 1970s, however, researchers started working with much faster devices based on a more expensive, faster material, gallium arsenide (GaAs). Progress with gallium arsenide devices is also shown on Figure 2.1(a). GaAs chips can be characterised generally as a family that has higher speeds than CMOS and consumes less power than ECL despite being more expensive. Unlike silicon, GaAs is a direct compound semiconductor material with the inherent property of high electron mobility ($8500\,\mathrm{cm}^2\,\mathrm{Vs}^{-1}$), approximately six times that of silicon ($1350\,\mathrm{cm}^2\,\mathrm{V} - \mathrm{sec}$) and has a large bandgap of 1.424 eV compared with 1.1 eV for Si. These properties make it an ideal candidate for high frequency and high temperature applications. GaAs also has excellent optical properties that have been exploited in LEDs, lasers and optical detectors. However, it does have some properties that present a challenge for manufacturing. Application examples vary from high speed processors, optical communication subsystems allowing 2.5–10 $\mathrm{Gb\,s}^{-1}$ systems (compared with CMOS counter parts with 100–200 $\mathrm{Mb\,s}^{-1}$

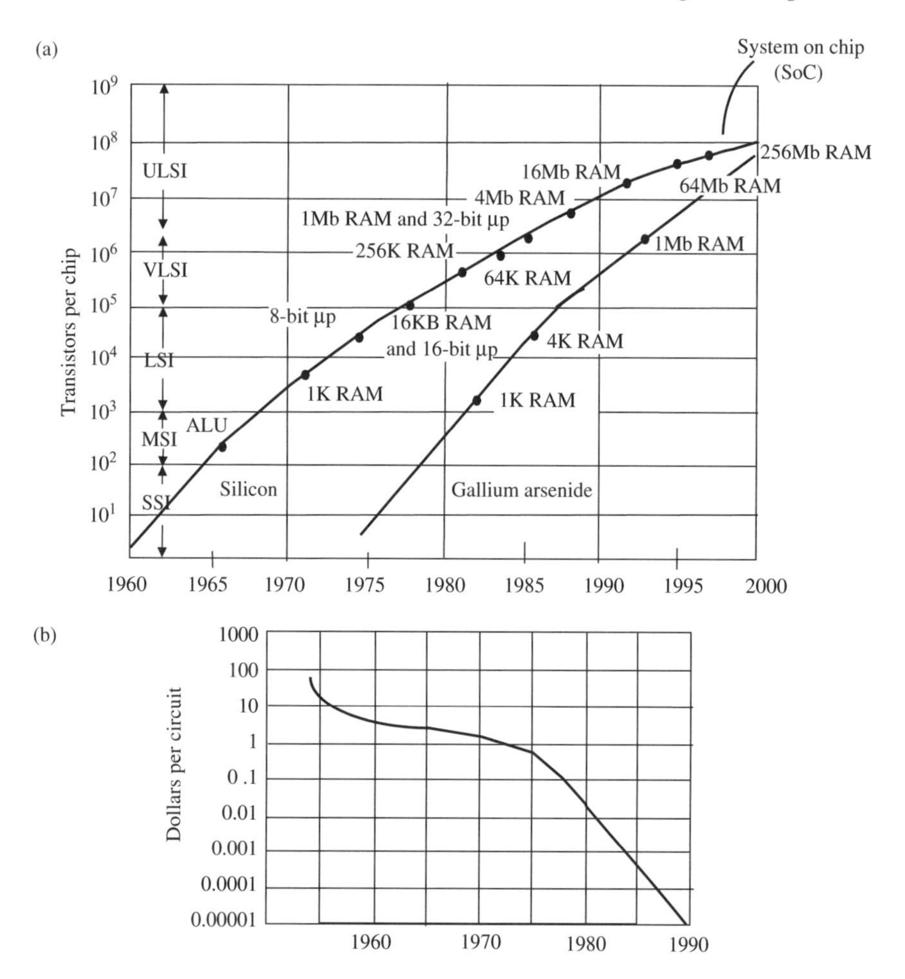

Figure 2.1 Progress of integrated circuits: (a) from SSI components to SoC; (b) cost path of ICs

speeds) and miscellaneous RF subsystems, etc. The proprietary high integration GaAs (H-GaAs) system developed by Vitesse Semiconductor, with its six inch wafer fabrication facility, seems to be opening up a new era for GaAs components [9].

Within the latter half of the past decade, SiGe components for high speed RF circuitry and analogue and mixed signal ICs have arrived, competing with the high speed components such as GaAs [10]. SiGe is a technology totally compatible with submicron CMOS, which can be used to build hetrojunction bipolar transistors (HBT) with very high transition frequency f_T. Other new materials such as SiC and GaN, with wide bandgap, are also gradually attempting to enter the market place [11–14]. Table 2.1 indicates the use of different semiconductor materials for different applications.

Table 2.1 Semiconductor materials and their common applications

Material	Applications	Remarks
Germanium (Ge)	Most basic transistors	Presently obsolete for basic components
Silicon (Si)	* Transistors * Power semiconductors * Integrated circuits	A mature and successful technology
Gallium arsenide (GaAs)	* High speed transistors and ICs * Opto components (infrared) * Power diodes for high speed applications	* Growing technology * More expensive than Si * Gigahertz order operations
Silicon germanium (SiGe)	High speed ICs	Newer material
Gallium aluminium arsenide (GaAlAs)	Red LEDs	Optical components
Gallium phosphide (GaP)	Green LEDs or red LEDs	Optical components
Silicon carbide (SiC)	Blue LEDs	Optical components
Gallium arsenide phosphide (GaAsP)	Broad colour range LEDs	Optical components
Gallium nitride (GaN) and Indium gallium nitride (InGaN)	Newer applications such as white LEDs, blue LEDs and lasers, etc.	Growing new area

Road map requirements for IC fabrication in the future are depicted in Table 2.2 and parameters of VLSI chip of the year 2005 are shown in Table 2.3 (using information in Reference 15).

2.3 Different categories of semiconductor components

Within about ten years from the invention of the transistor, the monolithic integrated circuit on germanium was demonstrated by Jack S. Kilby of Texas Instruments, and in 1961 Dr Robert Noyce of Fairchild Semiconductors demonstrated the digital monolithic IC on silicon. For historic developments related to electronics and communications, References 16 and 17 are suggested.

While the discrete components such as diodes and transistors, particularly the power semiconductor categories, still evolve with higher voltage and current ratings with enormous power handling capacities fuelled by the needs of power in the electronics world, modern integrated circuit families can be divided into three basic categories, namely (i) analogue, (ii) mixed signal, and (iii) digital. In an application

Table 2.2 Roadmap requirements for IC fabrication (Source: [15] © 2002, IEEE)

Year	1997–2001	2003–2006	2009–2012
Semiconductor technology needs			
Feature size, nm	250–150	130–100	70–50
Millions of transistors per square centimeter	4–10	18–39	84–180
Number of wiring layers	6–7	7–8	8–9
Assembly and packaging technology needs			
Pad pitch, nm	250–150	130–100	70–50
Power, W	1.2–61	2–96	2.8–109
Die size, mm^2	50–385	60–520	70–750
Performance, MHz	200–730	530–1100	840–1830
Voltage, V	1.2–2.5	0.9–1.5	0.5–0.9
Pin count	100–900	160–1475	260–2690

Source: National Technology Roadmap for Semiconductors, USA

Table 2.3 Characteristics of the VLSI chip of the year 2005 (reproduced by permission of IEEE)

Minimum feature size of process technology	0.1 μm
Total number of transistors	200 million
Number of logic transistors	40 million
Chip size	520 mm^2
Clock frequency	2.0–3.5 GHz
Number of I/O connections	4000
Number of wiring levels	7–8
Supply voltage	0.9–1.2 V
Supply current	~160 A
Power dissipation	160 W

point of view, the ICs can be regarded in three categories again such as (i) standard (catalogue) components, (ii) semicustom components, and (iii) custom parts. This division can be equally applied in analogue, mixed signal and digital areas.

Standard components are the devices that are fully functionally configured by the manufacturers; no changes to the internal design of the function blocks are possible or permitted to the users. Most common integrated circuits used by the system designers are identified by the semiconductor manufacturers and are produced in very large quantities to reduce the cost and improve the reliability. These standard components are manufactured by more than one company (usually with both function and the parametric performance being identical) and designers can easily obtain the part from several alternative suppliers.

In the case of semicustom components, manufacturers use various function blocks inside the integrated circuit so that those function blocks can be configured as decided by the end user, or some data can be stored within the blank device. Examples are programmable logic arrays, programmable read only memories (PROMs), and one-time programmable (OTP) microcontrollers, etc., and semicustom integrated circuits. In some device families the device can be reconfigured several times or an unlimited number of times, whereas some devices can be configured only once.

Semicustom and custom components are designed using special computer aided design (CAD) tools. The designers can configure their own circuits with a very large number of basic circuit blocks, such as simple logic gates, analogue function blocks (such as operational amplifiers or comparators, etc.) or even complex blocks, such as RAMs or processor cores. Among the advantages of using these components in an instrument or a system are the following: reduction in the 'real estate' needs of printed circuit board areas, improvement in reliability (owing to fewer interconnections and solder joints) and reduction of the overall cost of a product. With the VLSI and ULSI design capabilities, now available with the CAD layout packages and custom component manufacture services offered by manufacturers, many system designers are making use of these custom components in products. One disadvantage is that there may be heavy non-recurrent engineering (NRE) costs involved in the design process due to the needs of purchasing costly CAD layout tools and the initial fabrication costs of the sample chips. The application specific integrated circuits (ASICs) are, however, becoming popular with instrument manufacturers and most major instrument companies now own their integrated circuit fabrication facilities.

2.3.1 Analogue, mixed signal and digital components

These three basic categories cover almost all practical device families. Since human beings and nature behave in an analogue manner, while the advantage of high speed processing lies in the world of digital ICs, the above categorisation becomes a logical choice. In analogue components (such as the classic example of op amp and similar parts) the internal transistors behave as variable impedance devices within the lowest possible impedances to the cut-off region's high values. In digital IC families (such as logic gates, microprocessors, memories) the transistors take the two extremes of open and closed switch equivalents, allowing the lowest power dissipation, sub-micron feature sizes and integration advantages. Therefore these digital ICs lead to multimillion gate or transistor based VLSI/ULSI components.

Obviously, to work between nature and the digital systems, conversion processes – namely analogue-to-digital and digital-to-analogue conversion – are also necessary. This conversion need forces the designer community to develop mixed signal components. Most common mixed signal ICs are the A-to-D converters and D-to-A converters, and are called data converters.

2.3.1.1 Analogue components

Almost all practical systems have some parameters to handle in an analogue fashion, particularly at the input and output stages, etc. For this reason, the analogue world of components also grows gradually, though not as fast and dense as digital and mixed signal worlds.

In instrument systems, common types of analogue families used are:

- amplifiers,
- comparators,
- buffers and signal conditioners,
- function and waveform generators,
- voltage regulators and voltage references,
- analogue switches and multiplexers,
- data converters (analogue-to-digital converters and digital-to-analogue converters),
- filters,
- special functions, and
- radio frequency components.

The current trends in analogue components are towards:

(a) higher operational frequencies,
(b) lower operational voltages, and
(c) minimising secondary effects such as noise, distortion, non-linearities, etc.

In order to reach higher frequency capabilities when silicon components attain their fundamental limits, other materials such as GaAs and SiGe have entered the production facilities. For example, in 1989, GaAs based operational amplifiers with 10 GHz unity gain bandwidth were introduced [18].

Two emerging processes will provide a glimpse of things to come in analogue and mixed signal process technologies. One is a silicon–germanium heterojunction bipolar transistor (SiGe HBT) process. The other, a silicon-on-insulator (SOI) process, is an enabling wafer technology used at the heart of the most advanced complementary bipolar (CB) processes and potentially other advanced analogue and digital techniques. These two processes will join their CB, complementary bipolar CMOS (CbiCMOS) and ultra-fast biCMOS counterparts that now dominate analogue and mixed signal IC methodologies. While today's advanced analogue and mixed signal IC processes take advantage of technology originally developed for digital ICs, they have been adapted to their new tasks by sometimes minor and often major modifications. Examples include fine-line lithography, trench etching, and the SiGe HBT. In a new twist, some of the significant analogue IC improvements are now moving into the digital IC arena. Examples include RF processes on bonded wafers to handle digital signals in wireless applications, and in the future, in sub-0.25 μm CMOS.

A SiGe method conceived by IBM promises to be the most exciting new analogue/mixed-signal process. IBM has built practical npn HBTs, featuring unity-gain cut-off frequencies (f_T) of 113 GHz, that are compatible with a basic 0.25 μm silicon CMOS process. But more important, the beta Early-voltage product (βV_A) of these transistors is 48 000. A transistor's βV_A product determines the quality of its current sources (the heart of most analogue ICs) and should at least reach into

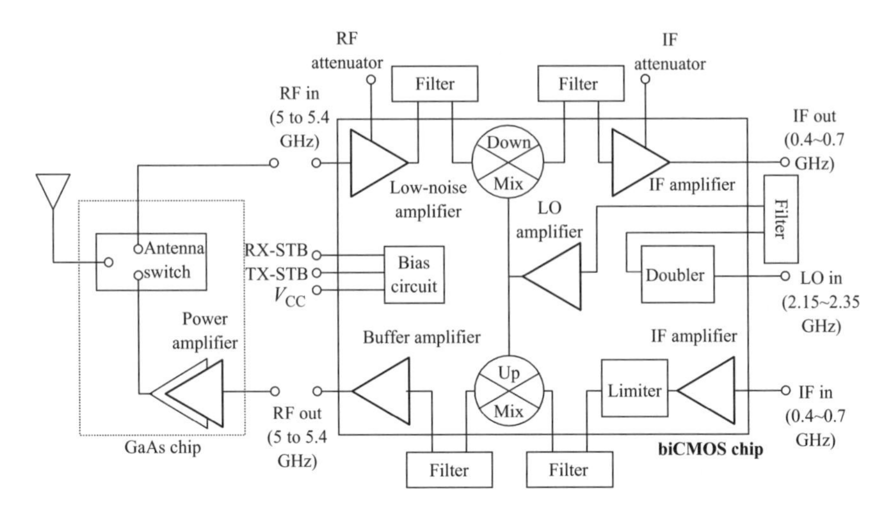

Figure 2.2 A 5 GHz band BiCMOS RF converter chip (source: Reference 20)

the thousands to be useful for precision analogue tasks. For details, Reference 10 is suggested.

During the 1990s, most manufacturers recognised the demand for very low power supplies, since the battery powered system design, green design concepts, etc., were high on the designer priority lists. So 3.3 V logic as well as analogue/mixed signal device families are very common now. Even as far back as 1992, there were amplifiers introduced with operating power supplies as low as 1 V [19].

Within the past few years, with the proliferation of cellular and wireless devices, RF system designs have shown several advances in analogue, mixed signal and RF systems. An example is the 5 GHz multifunctional up/down converter operating with a single bias voltage of 2.6–5.2 V [20]. The device uses a 22 GHz biCMOS process. When combined with a GaAs power amplifier/antenna switch chip, this device becomes a complete front-end system as depicted in Figure 2.2.

2.3.1.2 Variety of operational amplifiers

Bob Widlar, working at Fairchild, designed the first successful op amp back in 1965. The developments that continued from this experience of μA 709 provided several successful op amps such as μA 741 and LM 301, etc. In a modern approach, practical op amps available from the component manufacturers could be grouped into several categories, such as

(i) voltage feedback op amps (the most common form of op amp, similar to the 741 type);
(ii) wideband, high speed, high slew rate op amps (based on voltage feedback concepts);
(iii) current feedback op amps;
(iv) micro-power op amps;

(v) single supply op amps; and
(vi) chopper stabilised op amps.

The voltage feedback operational amplifier (VOA) is used extensively throughout the electronics industry, as it is an easy to use, versatile analogue building block. The architecture of the VOA has several attractive features, such as the differential long-tail pair, high impedance input stage, which is very good at rejecting common mode signals. Unfortunately, the architecture of the VOA provides inherent limitations in both the gain-bandwidth trade-off and the slew rate. Typically, the gain-bandwidth product (f_T) is a constant and the slew rate is limited. Figure 2.3(a) depicts early versions of VOA types.

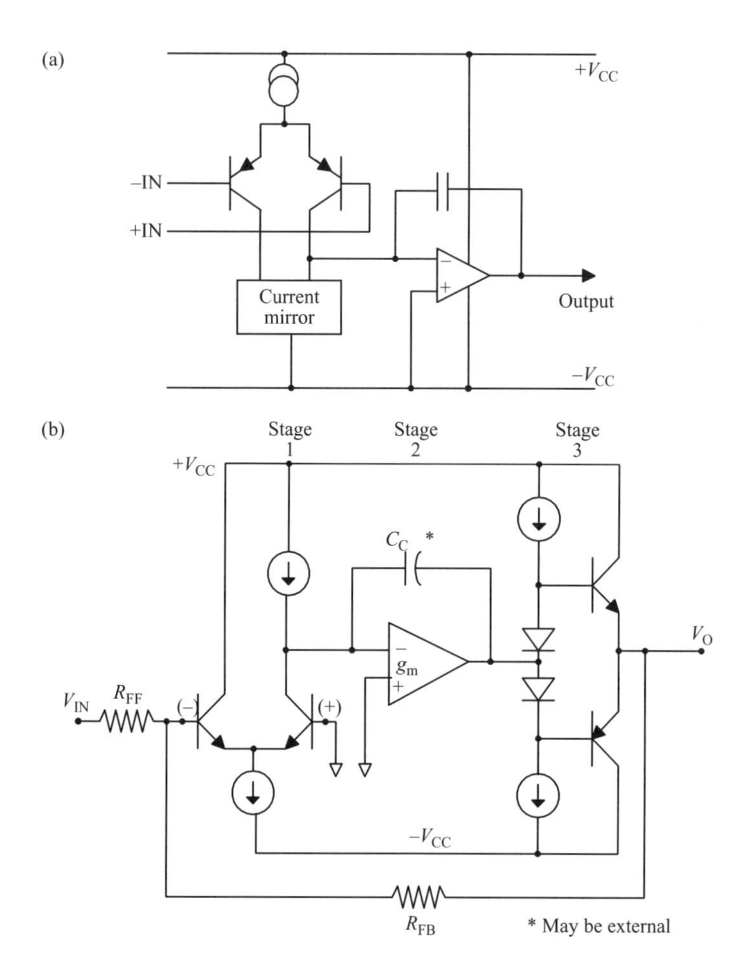

Figure 2.3 Voltage feedback op amp architectures: (a) some low speed versions; (b) simplified schematic of a high speed device such as AD 847 (courtesy: Analog Devices, Inc., USA)

Within the past decade, many op amps with high speed capability have entered the market to supply the demand from high speed A/D converters, video signal processing, and other industrial needs. Many of these op amps combine high speed and precision. For example, the devices introduced by Analog Devices in the late 1980s, such as the AD840, 841 and 842 devices, are designed to be stable at gains of 10, 1 and 2, with typical gain-bandwidths of 400, 40 and 80 MHz, respectively. A simplified schematic of a high speed VOA is shown in Figures 2.3(b) compared to a slow speed device. The example shown in Figure 2.3(b), AD847, is fabricated on Analog Devices' proprietary complementary bipolar (CB) process, which enables the construction of PNP and NPN transistors with similar values of f_T in the 600–800 MHz region.

The current feedback operational amplifier (CFOA) is a relatively new arrival to the analogue designer's tool kit. Elantec Inc. produced the first monolithic device in 1987. Current feedback operational amplifiers were introduced primarily to overcome the bandwidth variation, inversely proportional to closed-loop gain, exhibited by voltage feedback amplifiers. In practice, current feedback op amps have a relatively constant closed-loop bandwidth at low gains and behave like voltage feedback amplifiers at high gains, when a constant gain-bandwidth product eventually results. Another feature of the current feedback amplifier is the theoretical elimination of slew rate limiting. In practice, component limitations do result in a maximum slew rate, but this usually is very much higher (for a given bandwidth) than with voltage feedback amplifiers.

The current feedback concept is illustrated in Figure 2.4. The input stage now is a unity-gain buffer, forcing the inverting input to follow the non-inverting input. Thus, unlike a conventional op amp, the latter input is at an inherently low (ideally zero) impedance. Feedback is always treated as a current and, because of the low impedance inverting terminal output, R2 is always present, even at unity gain. Voltage

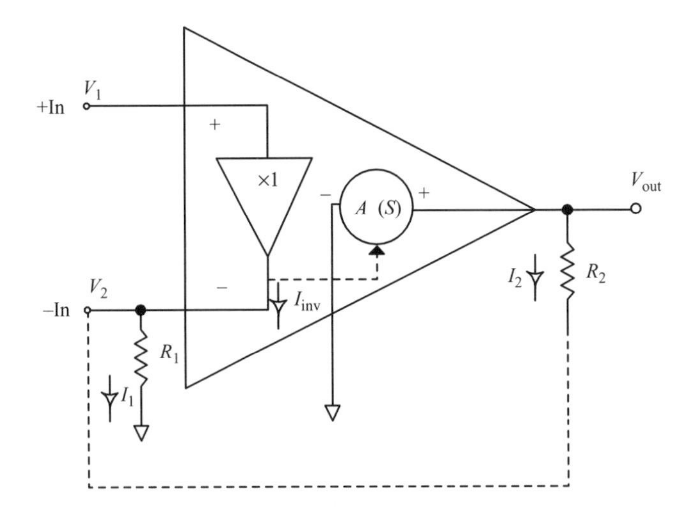

Figure 2.4 Current feedback op amp model

imbalances at the inputs cause current to flow into or out of the inverting input buffer. These currents are sensed internally and transformed into an output voltage.

The most significant advantages of the CFOAs are:

- the slew rate performance is high, where typical values could be in the range of 500–2500 V/μs (compared to $1 - 100$ V/μs for VOAs);
- the input referred noise of CFOA has a comparatively lower figure; and
- the bandwidth is less dependent on the closed-loop gain than with VOAs, where the gain-bandwidth is relatively constant.

For a more comprehensive discussion on modern op amps, Reference 21 is suggested. Design trends and issues in analogue CMOS devices are discussed in Reference 22.

2.3.2 Mixed signal components

Mixed signal ICs, which become the bridge between the real analogue world and the world of digital processing, is most interesting and difficult R & D inputs have been acknowledged. The most common mixed signal component families are the analogue-to-digital converters (ADC) and digital-to-analogue converters (DAC), which are generally grouped together as 'data converters'.

Until about 1988, engineers have had to stockpile their most innovative A/D converter (ADC) designs, because available manufacturing processes simply could not implement those designs onto monolithic chips economically. Prior to 1988, except for the introduction of successive approximation and integrating and flash ADCs, the electronics industry saw no major changes in monolithic ADCs. Since then, manufacturing processes have caught up with the technology and many techniques such a sub-ranging flash, self-calibration, delta/sigma, and many other special techniques have been implemented on monolithic chips. During the period 1985–1999, many architectures for ADCs have been implemented in monolithic form. Manufacturing process improvements achieved by mixed signal product manufacturers have led to this unprecedented development, which was fuelled by the demand from the product and system designers.

The most common ADC architectures in monolithic form are successive approximation, flash, integrating, pipeline, half flash (or sub-ranging), two step, interpolative and folding, and sigma-delta (Σ-Δ). While the Σ-Δ, successive approximation, and integrating types could give very high resolution at lower speeds, flash architecture is the fastest but with high power consumption. However, recent architecture breakthroughs have allowed designers to achieve a higher conversion rate at low power consumption with integral track-and-hold circuitry on a chip. The AD9054 from Analogue Devices is an example where the 200 MS s^{-1} conversion rate dissipation is 500 mW, which is an order less than comparable devices [23].

In order to summarise the trends of ADC development in the past, the 'figure of merit' (FOM) created by the ISSCC is a good indicator. The FOM is based on an ADC's power dissipation, its resolution, and its sampling rate. The FOM is derived by dividing the device's power dissipation (in watts) by the product of its resolution (in 2^n bits) and its sampling rate (in hertz). The result is multiplied by 10^{12}. This is

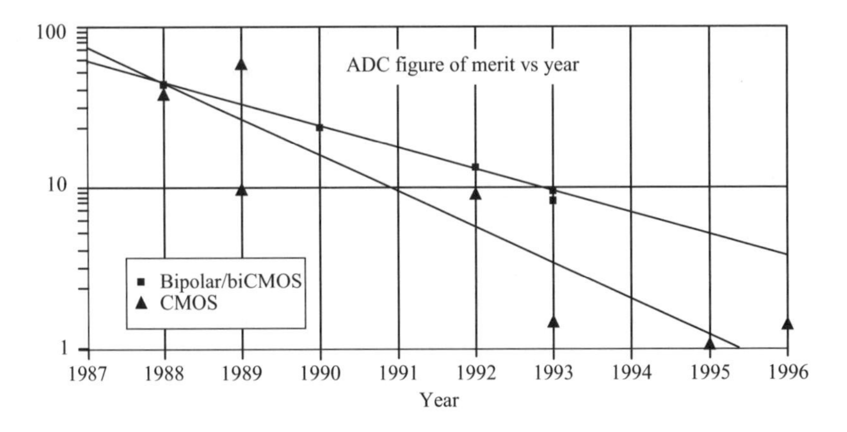

Figure 2.5 *General development trends of ADCs based on FOM (source: Electronic Design/ISSCC Conference Proceedings)*

expressed by the equation

$$FOM = \frac{PD}{R \cdot SR} 10^{12}, \tag{2.1}$$

where

PD = power dissipation (in watts)
R = resolution (in 2^n bits)
SR = sampling rate (in hertz).

Therefore, a 12-bit ADC sampling at 1 MHz and dissipating 10 mW has a figure of merit rounded off to 2.5. This figure of merit is expressed in the units of picojoules of energy per unit conversion [(pj)/conversion]. For details and a comparison of performance of some monolithic ICs, see Reference 24. Based on FOM, recent development trends [25, 26] of ADCs can be summarised as shown in Figure 2.5.

2.3.3 *Basic digital components*

At the disposal of system designers there are many off-the-shelf digital component families and the most common are the transistor transistor logic (TTL) family and its variations, complementary metal oxide semiconductor (CMOS) logic, emitter coupled logic (ECL) and bipolar CMOS (BiCMOS), etc. There are huge ratios between the highest and lowest values of several performance parameters: for speed (250 : 1), output drive (24 : 1) and power consumption (83 000 : 1).

The TTL family, which has been commercially available since the 1960s, is one of the most popular digital component families and it has its own variations, as shown in Table 2.1. The logic families that are viable product lines with projected long term availability are the ALS, the AS and the F families. The AS and F families represent the state-of-the-art in high speed, high performance TTL technology. The original

TTL family is seldom used in new designs but is frequently encountered in existing hardware. The outstanding characteristics of TTL compared with CMOS in the past have been its higher speed, output drive and transistor gain. These advantages are rapidly diminishing with the advances in the process technologies.

CMOS field effect transistors differ from bipolar transistors in both structure and operation. The primary advantages of CMOS are its low power dissipation and small physical geometry. Advances in design and fabrication have brought CMOS devices into the same speed and output drive arena as TTL. Again, enhancements have resulted in additional classifications: namely, MG (metal gate), HC (high speed CMOS) and National Semiconductor's 'FACT' (advanced CMOS).

ECL derives its name from common differential amplifier configuration in which one side of the differential amplifier consists of multiple input transistors with their emitters and tied together. The transistors operate in non-saturated mode and hence consume more power. There are few variations due to evolutionary advancements in ECL. Most common recent families include 100 k, 10 H and Motorola's ECL-in-PS.

Some representative logic families and their characteristics are shown in Table 2.4. For a detailed account of TTL and CMOS logic families the reader may refer to Reference 27.

Although many believe that the CMOS technology will be the choice of designers for a long time, a newer process named BiCMOS, where a bipolar power transistor output stage is coupled with CMOS logic stages, offers a combination of high speed, low power dissipation, high noise immunity and high fanout. This family may have a good chance of competing with the CMOS technologies. Philips and Motorola are already in this business and Philips/Signetics manufacture these devices using 13 GHz bipolar NPN devices, 1 μm (micron) NMOS and PMOS devices altogether with three layers of Al/Cu interconnect.

Philips Semiconductors has announced a silicon BiCMOS process technology that can be used to produce devices with f_{max} values of $\sim$70 GHz. These are roughly twice the typical values previously achieved for silicon production devices and are comparable with those of SiGe and GaAs ICs. Philips estimates that the new process can be used to fabricate devices at about 2/3 the cost of SiGe and GaAs products. This new Quality BiCMOS (QUBiC3) process can integrate high frequency RF bipolar circuits with high speed CMOS logic blocks onto a single chip. Philips introduced this process to produce a new range of high speed, high reliability RF front-end ICs that were scheduled for release in 1999 [28]. The company claims that these products will enable it to maintain a leading position in the aggressive mobile telecommunications market. Interestingly, within about 18 months Philips announced the QUBiC4 process, which allows f_{max} values of 90 GHz. The QUBiC4 process is expected to produce components for third generation mobile systems [29].

2.4 Processor based components and system components

The developments related to microprocessors, microcontrollers and digital signal processors have made the highest impact on electronic products and systems within

Table 2.4 Characteristics of logic families (source: Motorola)

Typical commercial parameter (0 °C to +70 °C)	TTL				CMOS		ECL		
	LS	ALS	FAST	MG	HC	FACT	10 H	100 K	ECLinPS
Speed									
OR propagation delay (t) (ns)	9	7	3	25	8	5	1	0.75	0.33
D flip-flop toggle rate (MHz)	33	45	125	4	45	160	330	400	1000
output edge rate (ns)	6	3	2	100	4	2	1	0.7	0.5
Power consumption									
quiescent (mW)	5	1.2	12.5	0.0006	0.003	0.003	25	50	25
operating (1 MHz)	5	1.2	12.5	0.04	0.6	0.8	25	50	25
Supply voltage (V)	+4.5 to +5.5	+4.5 to +5.5	+4.5 to +5.5	+3 to 18	+2 to 6	+2 to 6	−4.9 to −5.5	−4.2 to −4.8	−4.2 to −5.5 or −4.2 to −4.8
Output drive (mA)	8	8	20	1	4	24	50 Ω load	50 Ω load	50 Ω load
DC noise margin (%)									
high input	22	22	22	30	30	30	27	41	28/41
low input	10	10	10	30	30	30	31	31	31/31
Packaging									
DIP	yes	yes	yes	yes	yes	yes	yes	yes	no
SO	yes	yes	yes	yes	yes	yes	no	no	no
PLCC	no	yes	yes	no	no	yes	yes	no	yes
Functional device types	190	210	110	125	103	80	85	44	30
Relative 1–25 qty price/gate	0.9	1	1	0.9	0.9	1.5	2	10	28

a period of a quarter century. Although Intel Corporation's Marcian 'Ted' Hoff, an engineer there in the early 1970s, and Masatoshi Shima, an engineer for Busicom, a Japanese calculator manufacturer, are credited with developing the first microprocessor, they could not have envisioned the impact of microprocessors on products and systems we see today. By the twenty-fifth anniversary of microprocessors (μP) in 1996, the market for microprocessors in the USA was around $17 billion [30].

The most recent advances related to microprocessors and associated technologies can be summarised by Table 2.5, based on papers presented at International Solid State Circuits Conferences (ISSCC). This clearly indicates that, within the past several years, designers were making use of sub-micron geometries and the developments have followed 'Moore's Law' [7]. Some recent microprocessors were also using sub-micrometre, BiCMOS technology. For general-purpose computing, the complex-instruction set architectures have given way to reduced-instruction set approaches, which promise higher throughputs through simpler data path structures and streamlined instruction sets.

Even though most powerful microprocessor devices are aimed at personal computers and minicomputers, etc., and have high price tags (from a few to a few hundred US dollars), the prices of most microcontrollers have reduced to a few dollars owing to the volume of production. Microprocessors represent approximately 6 per cent of the total unit volume of microcontrollers and microprocessors sold, although they are highly priced compared with microcontrollers. The modular methods of producing microcontroller families are partly responsible for the lower cost and greater variety of microcontrollers. In this method, manufacturers begin with a CPU core and add modular on-chip peripherals such as RAMs, ROMs, and timers [31].

The most prominent of PC microprocessor parameters is the clock frequency (see Figure 2.6(a)), and pushing it above 1 GHz required not just a simple scaling of the devices to smaller sizes but material, process, and design changes as well.

In one processor clock cycle, the data ripples through a series of logic gates between registers, and covers a respectable distance across the face of the IC. (A logic depth of 12 gates was assumed in Figure 2.6(b)). In the past decade, shrinking line width down to the present 250 nm has steadily shrunk both transistor gate delay and on-chip interconnect delay between gates, since the logic gates are closer together. As a direct consequence of this scaling, microprocessors are speeding up. Beyond 250 nm, however, is a realm where the interconnect percentage of the delay, associated with sending the signals on narrower wires between the switching transistors, may reverse itself. Instead of the percentage decreasing with technology line width, the interconnect delay percentage will start to increase with line width reduction. As the interconnect shrinks to 100 nm, its increasing delay begins to dominate the overall gate delay. This delay depends on the time constant of the interconnect. For details, Reference 15 is suggested.

Below 250 nm, the narrow interconnect lines have huge losses and are reaching the limit of the current they can carry before electromigration starts to reduce yield and lifetime. Therefore the lines are designed to be as tall as 500–1000 nm at least in order to carry the required current. The tall, closely spaced lines have significant capacitance, primarily from side-to-side coupling between neighbouring lines.

Table 2.5 Progress of microprocessors (as presented at ISSCC conferences)

	1987	1989	1990	1991	1992		1995		1996	1997	1998	1999[5]	2000[6]
					CISC	RISC	CISC	RISC	RISC	RISC	RISC		
No. of transistors	150 k	283 k	580 k	1.2 M	1.7 M	1.02–3.1 M	3.5–6.7 M	5.2–9.3 M	3.8–9.8 M	7.5–15.2 M		25–116 M	25–28 M
Frequency (MHz)	16	32	61	100	200	40–250	93–167	66–300[1]	250–433[2]	300–600[3]	450–1100[4]	500–600	600–1000
Technology	CMOS	CMOS	CMOS	CMOS	CMOS	BiCMOS	CMOS/BiCMOS	CMOS	CMOS	CMOS	CMOS	CMOS	CMOS
Feature size (μm)	1.6	1.4	0.95	0.75	0.75	0.3–0.8	0.5–0.6	0.4–0.5	0.35–0.5	0.35	0.15	0.2–0.25	0.12–0.18

(1) Digital Equipment Corp's (DEC) Alpha processor.

(2) HP 8000 and DEC's Alpha processor.

(3) DEC Alpha.

(4) 64 bit Power PC processor.

(5) Extra low voltage power supply (<2 V). Super-scalar 15-stage pipelined processor (by AMD) was introduced.

(6) Power supply voltages in the range of 1.5 to 2.0 V with power consumption from 30 to 100 W.

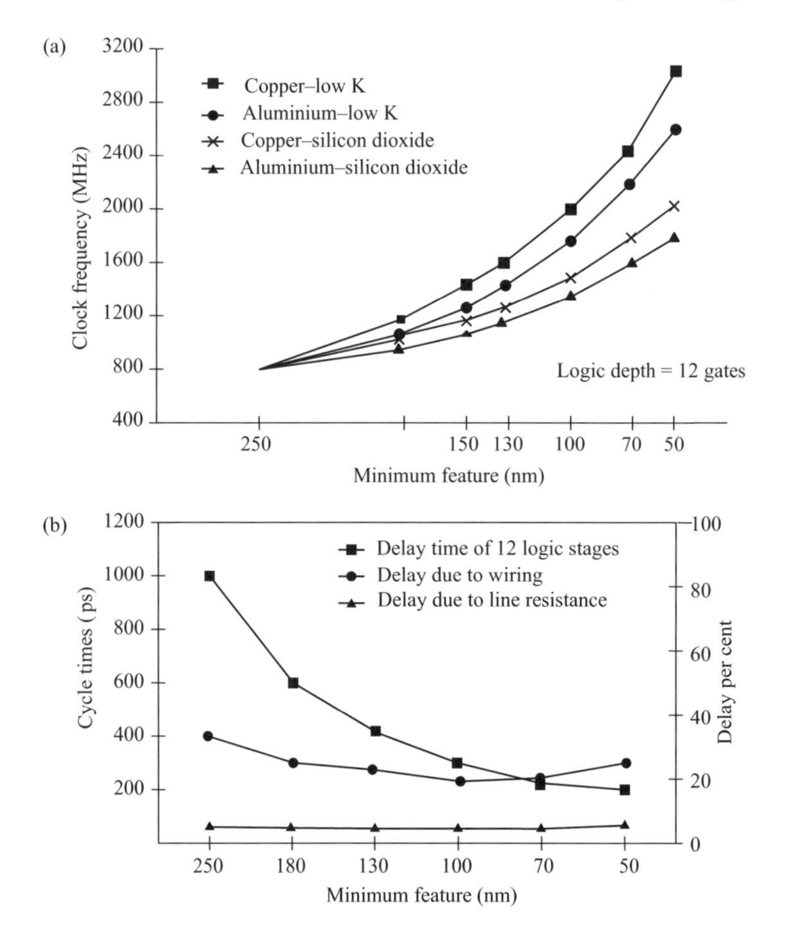

Figure 2.6 Speed related development parameters and feature size of microprocessors: (a) clock frequency vs feature size; (b) cycle time/delay percentage vs feature size (Source: [15] © 2002, IEEE)

Currently, some leading semiconductor manufacturers are in the process of replacing all or some of the aluminium of the interconnect lines with copper. Copper is an attractive replacement since it has a lower resistivity (about 2 $\mu\Omega$ cm compared with 4 $\mu\Omega$ cm for aluminium). More importantly, it has an order-of-magnitude higher electromigration current limit so that lines much smaller in cross-section can be formed. Line-to-line capacitance is reduced by shortening and narrowing the copper lines, but keeping the same spacing as for aluminium.

Once the transition to copper has been achieved, the next step will be the substitution of insulating films with low dielectric constant, K, for silicon dioxide as the interlayer dielectric for on chip interconnect. For a fixed geometry, the interconnect capacitance is proportional to K. Many companies such as IBM, Motorola, Texas

Instruments, and Advanced Micro Devices are exploring the use of low-K materials (K of 2 or below) to replace silicon dioxide (K of 4.5). The low-K materials are typically organic insulators; they have a lower process temperature, less mechanical strength, and lower thermal conductivity than silicon dioxide.

Between 1999 and 2003, microprocessors that are placed in a PC will undoubtedly go on improving in performance. By 2002, the clock frequency can be expected to be well above 1 GHz, even exceeding 2 GHz for top-of-the-range desktop systems. Such progress will be due to the reduction in IC feature size.

Microprocessors today employ a minimum lithographic line width of either 250 nm or 180 nm, and use additional process tweaks to push the effective gate length of the basic transistor to 150 nm or below. By 2003, the IC industry expects it will achieve 130 nm and be well on the way to 100 nm and minimum line widths with correspondingly shorter minimum gate lengths.

Every cycle of the digital IC's operation involves the charging or discharging of its internal circuit nodes. In the case of a CMOS microprocessor, the power thus dissipated is proportional to the product of the chip's effective capacitance, its clock frequency, and the square of its supply voltage. As the industry progresses from the 250 nm to the 100 nm chips, other challenges will be in packaging. Managing the power dissipation will be a huge challenge. Referring to Table 2.3, depicting the VLSI chip of year 2005, dissipating 160 W on a small area of silicon can be a real challenge.

Packaging is the other challenge. Most IC chip packaging today takes the form of either the quad flat pack (QFP), which has connections only on its periphery, or the pluggable pin-grid array (PGA), which has three to five rows of pin connections attached near its edges. Newer formats, in the course of being adopted by the industry, employ the full area under the chip package for connections, rather than just the peripheral region. These are the ball grid array (BGA) and chip-scale package (CSP).

To reduce the chip's power dissipation, it is desirable to lower the voltage as far as possible. In the past few years, the supply voltage of microprocessors has been reduced from 3.3 V, towards 2.5 V, and is expected to descend well below 2 V, and perhaps even 1 V in the near future. Moreover, to keep electrons from tunnelling through the thinner gate oxides of the CMOS transistors, the electric field must be reduced; accordingly, lower operating voltages are also desired to ensure long term reliability. It is expected that year 2003 devices will be operating at a supply voltage between 0.9 and 1.5 V (Table 2.3).

2.4.1 Digital signal processors

Early generations of 4- and 8-bit CISC processors have evolved into 16-, 32-, and 64-bit versions with CISC or RISC architectures. Digital signal processors can be considered as special cases of RISC architecture or sometimes parallel developments of CISC systems to tackle real-time signal processing needs. Over the past several decades, the field of digital signal processing has grown from a theoretical infancy to a powerful practical tool and matured into an economical yet successful technology.

At its early stages, audio and the many other familiar signals in the same frequency band appeared as a magnet for DSP development. The 1970s saw the implementation of special signal processing algorithms for filters and fast Fourier transforms by means of digital hardware developed for the purpose. Early sequential program DSPs are described by Jones and Watson [32]. The main differences with DSP chips include their ability to perform multiplications and multiply-and-accumulate (MAC) operations much faster than the CISC processor counterparts. Additionally, the DSP chips were able to execute instructions very quickly since, in a sense, the first DSPs were a precursor to the overall definition of a RISC chip, except with a more focused instruction set.

However, with today's ability to reduce the size of the CPU logic, integrate millions of gates, and piece together various functions on a single chip, it is getting harder to justify generic DSP chips. More common now are processors that offer merged architectures, combining the best of RISC and DSP on a single chip. In the late 1990s, the market for DSPs was generated mostly by wireless, multimedia, and similar applications. According to industry estimates, by year 2001, the market for DSPs was expected to grow to about $9.1 billion (Schneiderman) [33]. Currently, communications represents more than half of the applications for DSPs [33]. For a summary of architectural aspects and applications, Reference 21 is suggested.

2.4.2 Memories

Processor blocks provide the necessary base for embedding intelligence into electronic products and systems. Memories allow the human expertise (in the form of binary coded information) and data to be attached to embedded processor systems. Memory devices available in commercial production may be divided into two basic categories: namely, volatile components and non-volatile components. In volatile components, retention of information is guaranteed only when the power is on, and in the case of non-volatile memories the data retention is guaranteed even when the power is removed. For the purpose of data retention in non-volatile memories several primary techniques are used: for example, battery back-up, and charge retention in a capacitor or ferro-electric effect. All technologies related to memory devices are aimed at low cost per bit, high bit density, short access cycle times, random access read and write cycles of equal duration, low power, non-volatile low voltage operation over a wide temperature range, and a high degree of radiation hardness.

2.4.2.1 Random access memories

Random access memories (RAMs) can be divided into static RAMs (SRAMs) and dynamic RAMs (DRAMs). SRAMs are faster but of lower density and use fewer additional electronic components for memory management. At ISSCC conferences, by 1992, CMOS SRAMs with a 1 Mb capacity and access time of 7 ns using 0.3 μm technology operating at 3 V and consuming only 140 mW were reported [34]. Designs of 4 Mb SRAMs using bipolar technology have been reported with 6–9 ns access times. Furthermore, 16 Mb CMOS SRAMs with 15 and 12 ns access times have also been reported. These devices use 0.4 μm architectures with a 3.3 V power supply.

Other researchers have introduced basic memory cells which could be used in SRAMs with high speed and much lower (as low as 1 V) operating voltages. By 1999, at ISSCC, Toshiba Corporation has introduced 4 Mbit biCMOS SRAMS operating at 400 MHz operable on a 3.3 V power supply and fabricated on a 0.3 μm process [35]. Also, slightly slower but 4 Mbit SRAMS operating on 1.5 V rails were introduced during the same period.

Based on such research directions, commercially available SRAM families promise devices that keep pace with 350 MHz and faster processors. SRAM architectures have moved well beyond the simple 1-, 4-, and 8-bit-wide devices of 20 years ago. By 1998, designers were able to optimise their systems by selecting the ideal memory organisation from among 8-, 16/18-, 24-, 32/36-, and even 64/72-bit-wide data-bus options. In addition to the wide variety of data-path widths, SRAMs are available with word depths ranging from 32 to 512 kwords, depending on the chip's word width and total capacity.

The market is also shifting away from 5 V devices to chips that operate from 3.3 V supplies and offer 3.3 V low voltage TTL (LVTTL) interfaces rather than standard TTL or CMOS I/O (input/output) levels. And the voltages continue to drop as signal swings are reduced to improve access times. New generations of SRAMs offer 2.5 V input/output lines that meet the high speed stub-terminated logic (HSTL) interface specifications. Most companies feel that HSTL interfaces will provide sufficient bandwidth for memories that can operate at bus speeds of well beyond 300 MHz. These new generations of high performance SRAMs with double data rate (DDR) input/output capabilities or zero bus latencies (ZBT) have been making their way out of laboratories during the past few years [36].

The DRAM owes its popularity to the fact that it costs less than the SRAM, yet it has a higher bit density, an advantage that stems from the application of smaller and simpler memory cells. However, these devices require more electronics for housekeeping functions. Figure 2.7 shows the levels of integration of DRAMs over time. Most manufacturers are heading in the direction of lower supply voltage (around 3 V and lower) where easy battery back-up can be achieved. At the ISSCC conference in 1997, 4 Gb DRAMs were presented [37].

2.4.2.2 Non-volatile memories

Practical devices that store information on a non-volatile basis are (i) ROMs and PROMs, (ii) one time programmable (OTP) devices such as EPROMs, (iii) electrically erasable devices such as EEPROMs, (iv) flash EPROMs, (v) ferro-electric technology based devices, and (vi) battery backed-up devices. See also References 38 and 39 for further information.

2.4.2.2.1 PROMs and OTP devices

ROMs are frequently used in cost driven systems, where program stability and high volume demand a dense, inexpensive, permanent memory. But because ROMs are not reprogrammable, if you need to change the memory's contents, you must replace the ROM.

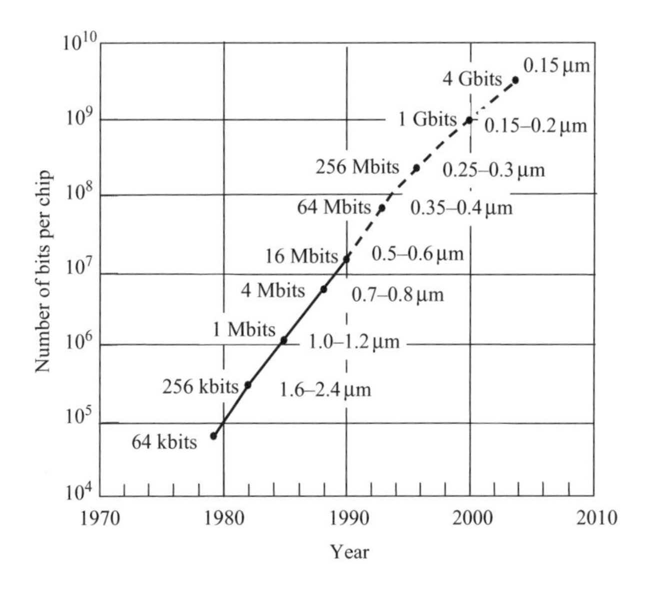

Figure 2.7 Integration level in DRAMs as a function of time (source: Reference 26)

To purchase a ROM, you provide a vendor with a truth table describing the circuit function you want to implement. The vendor converts your truth table to a mask pattern, which specifies the metallisation of the ROM. Thus, ROMs have either non-recurring engineering charges (NRE) or minimum volume requirements, and sometimes both. NREs are generally in the $2500 range, and minimum orders start at 1000 pieces or higher. On the other hand, densities as high as 4 Mbits are available and, as a group, ROMs have the lowest cost. Because of the mask step of ROMs, delivery of chips takes typically six to twelve weeks after the order is placed.

For shorter lead times, fuse link PROMs and OTP devices are used. Fuse link PROMs are bipolar or biCMOS devices. Bipolar PROMs are relatively fast and provide good radiation immunity, but they are also power hungry and, because of heat dissipation, limited in density. Testing is also a problem with PROMs. Because you do not want to blow a fuse to test it, you must take it on faith that the fuses work and limit your testing to the peripheral circuitry. PROMs have historically been used for storing macros and programs and in engine-control units; but, as high speed EPROMs evolve, many of these applications are being handled by fast EPROMs and their siblings, the OTP memories.

OTP memories are EPROMs (erasable programmable read-only memories), but instead of being packaged in expensive windowed ceramic packages, OTP devices come in plastic packages. Without a window to let UV light reach the die, OTP memories – as their name states – do not have the reprogrammability feature of EPROMs. Because you need not worry about removing OTP memories for reprogramming, you can even purchase them in surface-mount packages. OTP memories offer relatively inexpensive, permanent non-volatile memory and have short lead times.

2.4.2.2.2 Reprogrammable devices

Erasable PROMs (EPROMs), electrically erasable PROMs (EEPROMs) and flash memories are the reprogrammable families that are common in processor based systems.

The EPROM offers a floating gate sitting on an oxide film about 350 Å above the channel separating the source and drain of a transistor. The control gate of the device is above the floating gate. To program an EPROM cell with a 'one', the programmer applies about 20 V to the control gate and drain. The resulting channel current contains high energy electrons that flow through the insulating SiO_2 layer to the floating gate. When you drive the gate with a 5 V supply, the negative charge on the floating gate keeps the device off. Exposing the programmed EPROM to UV light with a wavelength centred around 253.7 nm neutralises the charge on the floating gate and erases the EPROM.

The programming of an EEPROM takes advantage of an effect known as Fowler–Nordheim tunnelling. This effect describes how a strong electric field can induce low energy electrons to tunnel through a thin oxide and into the floating gate. As with EPROMs, an EEPROM is programmed by applying a high voltage on the control gate. Capacitive coupling induces the voltage on the floating gate, which creates the electric field across the thin oxide. The tunnelling current of about 1 nA that results programs the cell with a smaller cumulative charge than that found on the EPROM and lets the EEPROM cell conduct current in either programmed state. A select gate isolates the EEPROM cell from the other cells in the array and sends the cell's data to a sense amplifier when the select gate is enabled. It is this select gate that takes up silicon area and limits the density of EEPROMs. The select gate permits byte erasure, which is, in effect, just a write operation. Burning an EEPROM involves grounding the program line for a row and then raising the column lines to the high programming voltage to match the data pattern that you want to load.

Flash memory technology, which was commercialised after 1988, uses a hybrid of the EPROM and EEPROM approaches. Table 2.6 compares EPROM, EEPROM and flash EPROMs.

With most flash memories, data are written into the flash cell by using hot-electron injection, as though writing an EPROM. Erasing flash memories utilises the Fowler–Nordheim tunnelling mechanism, where an electric field is created by grounding the control gate and applying a high voltage to either the transistor's drain or source. In general, flash memories do not have a separate select gate – either they do not have one at all, or the select gate is part of the control gate. Therefore, flash memories do not have byte erase capability.

Currently, the two main techniques vendors use to alter the data stored in a flash memory cell are channel-hot-electron (CHE) injection and Fowler–Nordheim (FN) tunnelling (Figures 2.8(a) and (b), respectively). In each case, an applied electrical field adds or removes electron charge from the transistor's floating gate, changing the threshold voltage. A subsequent read at a consistent reference voltage turns on some transistors, which may result in a one or zero at the memory output, and leaves off others. Both approaches use high voltages either from the outside world or from internal, on-chip charge pumps.

Table 2.6 Comparison of electrically erasable memory techniques

	EPROM	EEPROM	Flash EPROM
Relative size of cell	1	About 3	1.2–1.3
Programming	By external means	Internal	Internal
Programming			
Technique	Hot electron injection	Tunnel effect	Hot electron injection
Voltage	12.5 V	5 V	12 V
Resolution	Byte	Byte	Byte
Time taken	<100 μs	5 ms	<10 μs
Erasing	By external means	Internal	Internal
Technique	Ultraviolet light	Tunnel effect	Tunnel effect
Resolution	Whole chip	Byte	Whole chip or block
Time taken	15 min	5 ms	1 s

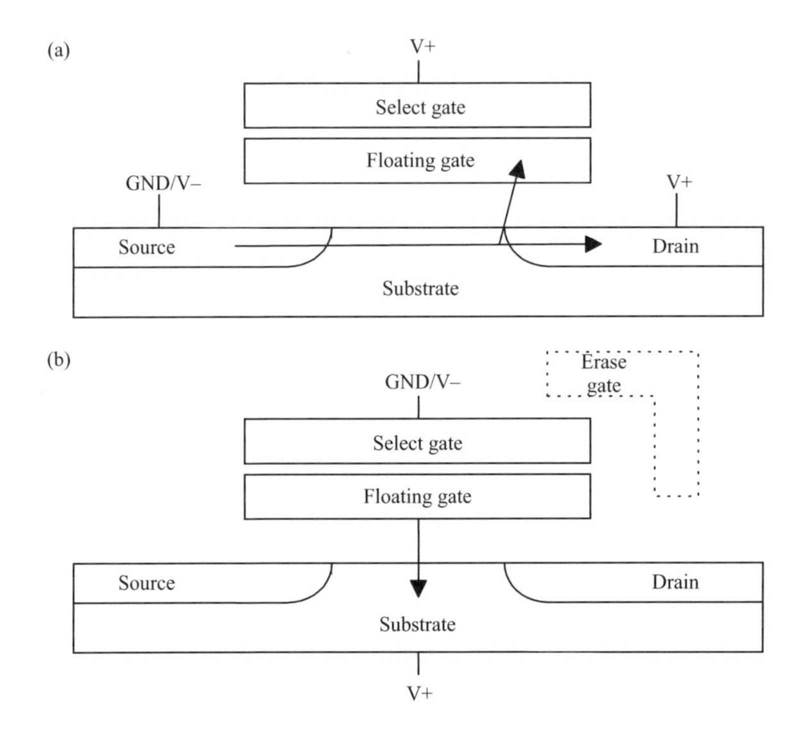

Figure 2.8 Flash memory techniques: (a) channel-hot-electron (CHE) injection; (b) Fowler–Nordheim (FN) tunnelling

In many portable devices such as personal digital assistants (PDA) and video cameras, etc., flash memories have become the primary storage medium since they allow upgradeability. For details of developmental aspects References 40–44 are suggested.

2.5 Semicustom and custom ICs

Until the late 1970s, most designers made use of standard component families (catalogue logic or fixed function devices) in designing electronic products and systems. However, these fixed function devices (SSIs to ULSIs) posed certain limits in achieving lower PCB real estate as well as product reliability levels owing to the use of many interconnections. In the early 1980s, designers realised the era of application specific integrated circuits (ASICs) where designers were able to configure their own logic functions using semicustom and custom devices. In its more generic usage, the ASIC also encompasses those families of components known as field programmable devices (FPD). The term FPD is the contemporary name for any IC that a user can customise in the field. That is, users having access to the appropriate electronic design automation (EDA) software utilities and associated physical device programming tools can customise an FPD to perform a task or function.

These devices could be divided into several primary categories, namely:

- simple programmable logic devices (SPLD),
- complex PLDs (CPLD),
- field programmable gate arrays (FPGA), and
- gate arrays and (full) custom ICs.

Around 1995, the term FPD generally embraced only digital logic such as simple PLDs (SPLDs), complex PLDs (CPLDs) and field programmable gate arrays (FPGAs). However, by the beginning of the twenty-first century it had grown to encompass field programmable analogue devices (FPADs) and field programmable mixed signal devices (FPMSDs) [45]. Between 1995 and 2000, programmable analogue devices also started to become commercialised [46–48]. Table 2.7 indicates some FPD terminology with architectural and commercial details.

Simple PLDs (sometimes termed simply as PLDs), are often used in place of 5–10 SSI/MSI devices and are the most efficient ASIC solution for densities up to a few hundred gates. (The gate count refers to the total number of two-input NAND gates that would be required to realise a logic function.) PLDs include a number of competing possibilities, all based on variations of AND-OR plane architectures. The primary limitations of the PLD architecture are the number of flip-flops, the number of input/output signals and the rigidity of the AND-OR plane logic and its interconnections.

2.5.1 Semicustom devices and programming technologies

Programming technology refers to the physical technique used to create user programmable switches for FPDs. The most common such technologies are fusible links, antifuses, EPROM and EEPROM cells and transistors, and SRAM cells.

Table 2.7 Field programmable devices (FPD)

Device family and subsets	Abbreviation	Description	Remarks
Simple programmable logic devices	SPLD	Commonly based on an AND array feeding into an OR array.	In addition to their core functionality, SPLDs are available with a variety of additional programmable options such as tristate and registered outputs.
SPLD variations			
Programmable array logic	PAL	AND array is programmable, the OR array is predefined.	• PALs operate faster. Most common of all the SPLDs. • PAL is a registered trade mark of Monolithic Memories, Inc.
Programmable logic arrays	PLA	User controls both AND and OR arrays.	More flexible than PALs.
Generic array logic	GAL	Sophisticated versions of electrically erasable PLDs. (EEPLDs)	In system programmability (ISP) is available.
Programmable read only memory	PROM	Predefined AND array driving a programmable OR array.	In reality a PROM's internal architecture is more akin to a decoder that is driving a programmable OR array.
Erasable PLD	EPLD	The devices where OTP type fusible links and antifuse technologies are replaced with EPROM transistors.	
Electrically erasable PLD	EEPLD	The devices where OTP type fusible links and antifuse technologies are replaced with EEPROM transistors.	ISP facility available.
Complex programmable logic devices	CPLD	A CPLD essentially comprises multiple SPLDs on a single chip.	Programmable switches can be based on fusible links, antifuses, EPROM transistors, EEPROM transistors or SRAM cells.
Field programmable gate arrays	FPGA	Comprise islands of programmable logic surrounded by programmable interconnect blocks.	Vendors field their own architectures.
Field programmable interconnect devices	FPID	FPID or Field Programmable Interconnect Chips (FPICs) are components that act as SRAM based switching matrices. Devices can be dynamically reconfigured.	FPIC is a trade mark of Aptix Corp., San Jose, California.
Field programmable analogue devices	FPAD	Electrically programmable analogue circuits which contain miscellaneous analogue modules such as log/antilog blocks, analogue multiplexers and sample-hold amplifiers etc.	Techniques such as electrically programmable analogue circuits/devices (EPAC/EPAD) and totally reconfigurable analogue circuits (TRAC) are used.
Field programmable mixed signal devices	FPMSD	A new breed of devices where both analogue and digital blocks are mixed in a configurable manner.	

The two main types of fusible link technologies employ either lateral or vertical fuses. A lateral fuse comprises a tungsten–titanium alloy wire in series with a BJT, which can pass sufficient current to melt the wire. This type of fuse commences as a short circuit and becomes an open circuit when you program it. By comparison, the diode at the base emitter junction of a BJT forms a vertical fuse. This type of link starts off as an open circuit, because the BJT acts like two back-to-back diodes, thereby preventing current from flowing. However, if you force a sequence of current pulses through the BJT's emitter, an avalanche effect occurs, and the emitter collapses and melts, creating a short circuit.

As an alternative to fusible links, some FPDs (predominantly, CPLDs and FPGAs) employ antifuse technology. Antifuse links comprise a via of amorphous (non-crystalline) silicon between two layers of metallisation. In its unprogrammed state, the amorphous silicon is an insulator with a resistance greater than 1 GΩ, but the user can program an antifuse link by applying signals of relatively high current (approximately 20 mA) to the device's inputs. The programming signal effectively grows a link by changing the insulating amorphous silicon into conducting polysilicon.

Both fusible link and antifuse technologies are known as one time programmable (OTP), because once you have programmed them there is no general possibility of reversal. In order to remove the limitation of one time programmability, the fusible links or antifuses can be replaced by EPROM or EEPROM transistors or SRAM cells. This provides the erasing or reprogrammable possibilities, including the in-system programmability (ISP). For more details on FPDs, References 45, 49–53 are suggested.

2.5.2 *SPLDs, CPLDs and FPGAs*

Categorising programmable logic devices as simple PLDs (SPLDs), complex PLDs (CPLDs), FPGAs, or some other label can be a difficult task. (Industry uses these terms in a rather non-standardised form.)

2.5.2.1 Simple PLDs

An SPLD is an integrated collection of macrocells (Figure 2.9). Each macrocell typically consists of a wide AND-OR structure (known as coarse grained logic) followed by an optional flip-flop that can be configured to one of several types (D, T, etc.). Most SPLDs also support programmable output polarity and internal signal feedback to the same or other macrocells, which can widen the product term implementation at the expense of speed; see also Reference 54. SPLDs also combine dedicated inputs, dedicated outputs, and programmable input or output pins to suit the needs of various designs. SPLDs are also known as programmable array logic (PAL) or generic array logic (GAL). They offer highly predictable timing as fast as 3.5 ns on some devices, and most design engineers understand development tools and programming languages such as ABEL well. Effective gate counts range as high as 500 or so, with as many as 28 pins.

Early PALs were bipolar, which offered speed at the expense of high power consumption, but most of today's devices are CMOS. The programmable elements can

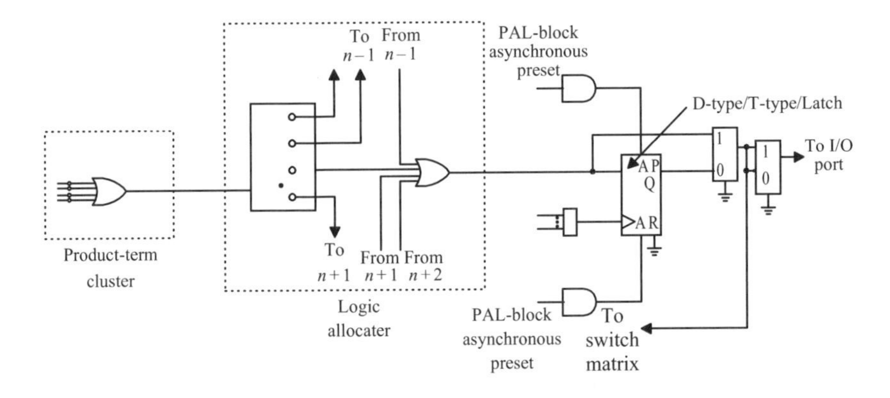

Figure 2.9 An SPLD macrocell showing wide fan-in with an optional register, signal feedback and configurable functions (Reprinted from EDN, May 22, 1997 © Cahners Business Information, 2002)

be PROM, EPROM, EEPROM, or flash memory. Lattice Semiconductor pioneered the concept of in-system programming (hence, the acronym switch from PAL to GAL) but, today, almost all vendors offer some form of this capability. Perhaps the most popular SPLD is the ubiquitous 22V10, which many vendors supply either in its generic form or with various superset enhancements.

2.5.2.2 Complex PLDs (CPLDs)

CPLDs, simply described, are collections of SPLD structures, combined on one die and interconnected by a central multiplexer or switch matrix that routes signals to macrocells from device inputs and other macrocell outputs. A typical SPLD structure, from Cypress Semiconductor, is shown in Figure 2.10. At higher gate counts, most CPLDs do not provide 100 per cent routing of all signals to all macrocells (to minimise interconnect complexity and the resultant impact on device performance and cost). Instead, CPLDs rely on statistical assumptions of device usage in implementing both the interconnect structure size for a given number of macrocells and inputs and the maximum number of signals that can enter each grouping of macrocells. CPLDs tend to provide less product compatibility among vendors than SPLDs, although, in some cases, superset features proliferate and diverge from a common base functionality.

With the exception of Altera's FLEX (Flexible Logic Element Matrix) architecture, SPLDs typically provide densities measured in thousands or a few tens of thousands of gates, and they also provide fewer registers at a given gate count compared with FPGAs. Most modern CPLD families allow the in-system programming capability to simplify field upgrades [55].

2.5.2.3 Field programmable gate arrays (FPGA)

A field programmable gate array (FPGA) is a collection of logic cells that communicate with each other and with I/O pins via horizontal and vertical routing channels (Figure 2.11(a)). Functions are built by connecting the logic cells. The logic contained

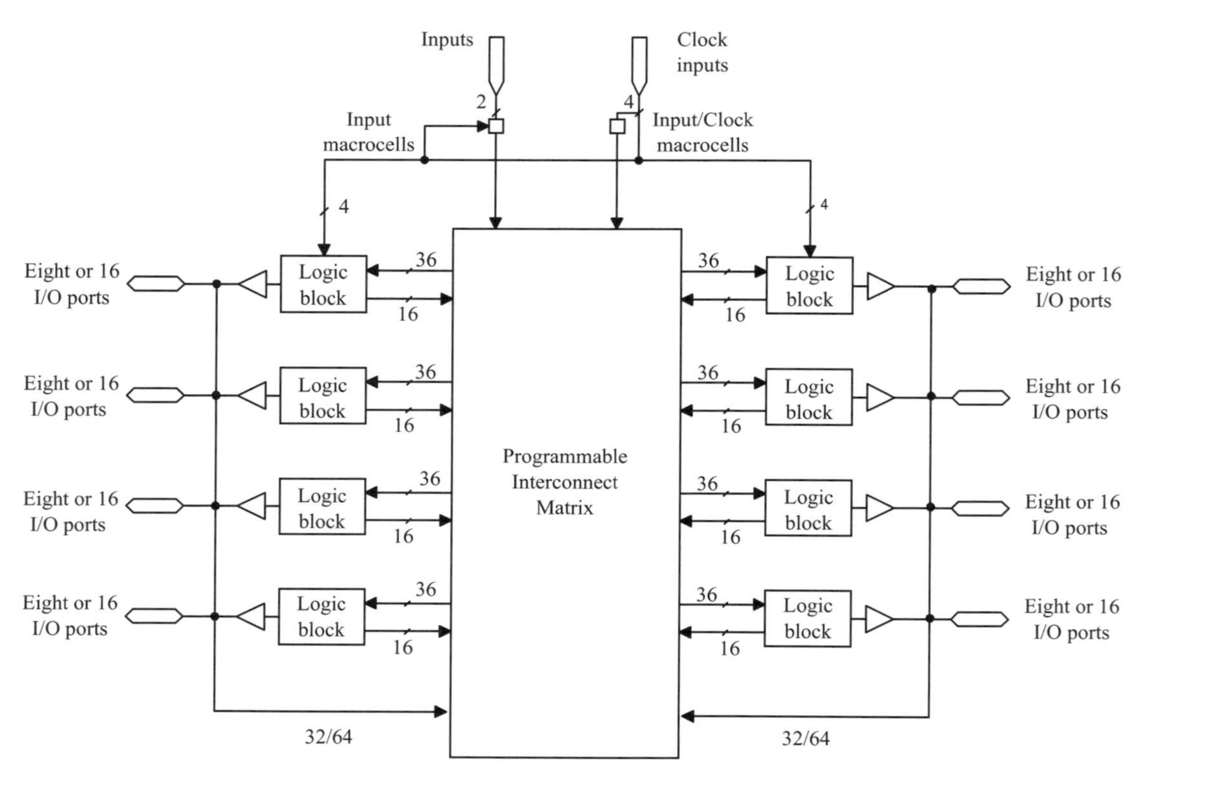

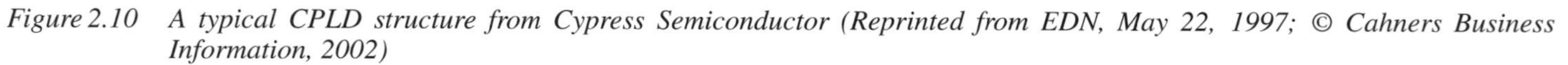
Figure 2.10 *A typical CPLD structure from Cypress Semiconductor (Reprinted from EDN, May 22, 1997; © Cahners Business Information, 2002)*

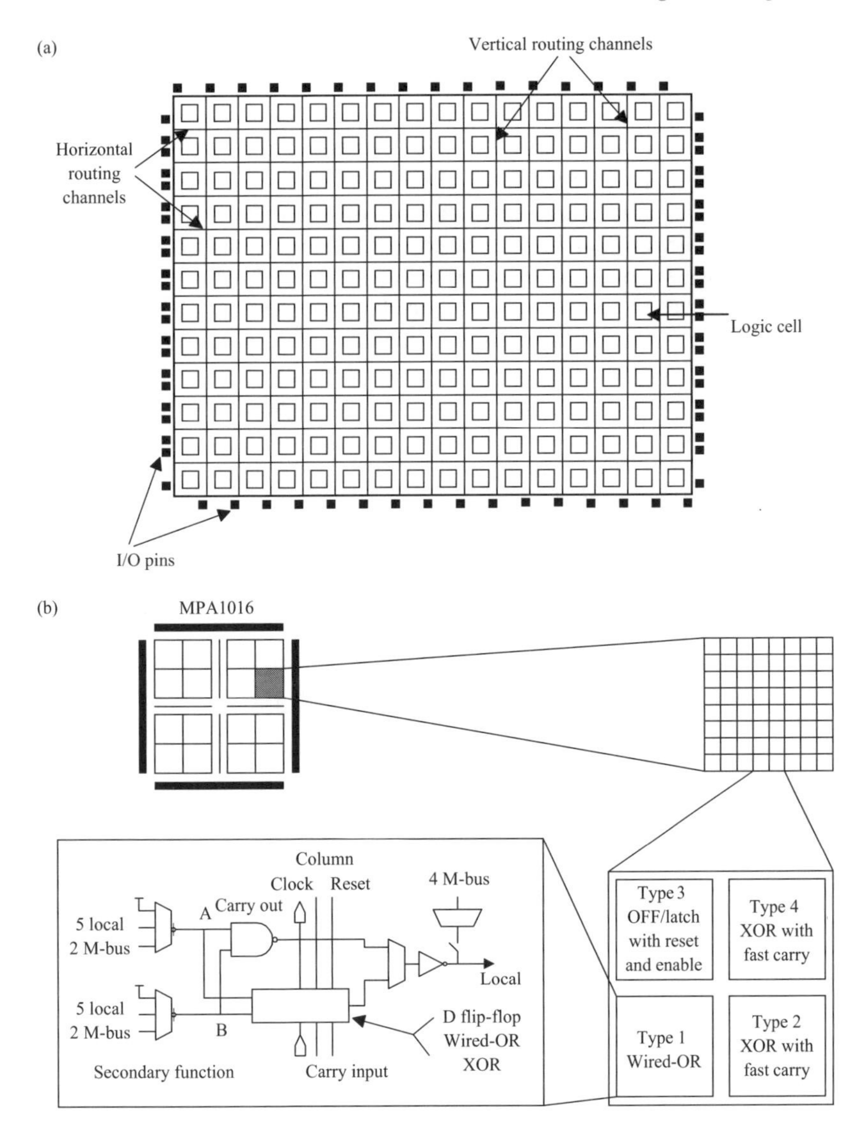

Figure 2.11 FPGA architecture: (a) basic concept; and (b) typical device, MPA 1000 from Motorola (reproduced with permission from Motorola, Inc., USA)

within FPGAs is typically more versatile or fine grained than that of simple PLDs or CPLDs. However, the flexibility of FPGAs comes at a price; FPGAs cannot provide fixed delays, so they are more complicated to use. Figure 2.11(b) indicates the collection of multiple cell types included in Motorola's MPA 1000, SRAM based FPGAs [56].

FPGAs have less predictable timing than CPLDs. However, FPGAs are generally more dense and contain more flip-flops and registers than CPLDs. For more details on FPGAs and comparison with other families, References 56–58 are suggested.

2.5.2.4 Gate arrays, standard cell and custom ICs

Gate arrays implement user logic by interconnecting transistors or simple gates into more complex functions during the first stages of the manufacturing process. Gate arrays offer densities up to 100 000 gates or more, with utilisation of 80–90 per cent for smaller devices and 40–60 per cent for the largest. Standard cell and custom ICs require unique masks for all layers used in manufacturing. This imposes extra costs and delays for development, but results in the lowest production costs for high volume applications. Standard cell ICs offer the advantages of high level building blocks and analogue functions.

CMOS is the most used technology for ASICs; companies like Altera, Fujitsu and Xilinx are developing proprietary processes, and some companies like Advanced Micro Devices act as alternative sources for these proprietary devices. Speed limitations of CMOS gate arrays reach around 200 MHz, and for this reason there are several vendors who offer ECL as well as GaAs solutions for higher speeds, sometimes at a higher power consumption.

Programmable logic chip costs have significantly decreased over the past few years; for densities of tens of thousands of gates or more, prices closely approximate those of gate array ASICs. At advanced process lithographies, the die size necessary to incorporate a given device pin count's bond pads may be larger than the amount of silicon needed to satisfy a target gate count. This characteristic, 'I/O bounding', means that, up to a certain gate count density and pin count, gate arrays and programmable logic have equivalent die sizes. ASIC vendors often charge NRE fees, which can reach tens or even hundreds of thousands of dollars. End users must buy a minimum quantity of parts in either units or dollars to satisfy the vendors' business criteria. This minimum quantity has increased over time, because advanced lithographies allow ASIC companies to squeeze more gates on each die and more die on each wafer. The cost differential per unit between a PLD or an FPGA and a gate array alternative depends on how many parts you contract to buy. For details, Reference 59 is suggested.

2.6 Display devices

Almost all test and measuring instruments use some type of a microprocessor or a microcontroller. This allows the instrument to use some type of a display device and to present measured results in a graphical form or textual base for user convenience. Electronic display devices can be classified into active and passive types. Figure 2.12 shows the most common electronic display devices; it also indicates the most basic forms of the technologies and many variations to these do exist. For example, in the ELD and LCD types, active matrix ELD and active matrix LCD (AMLCD) types are common.

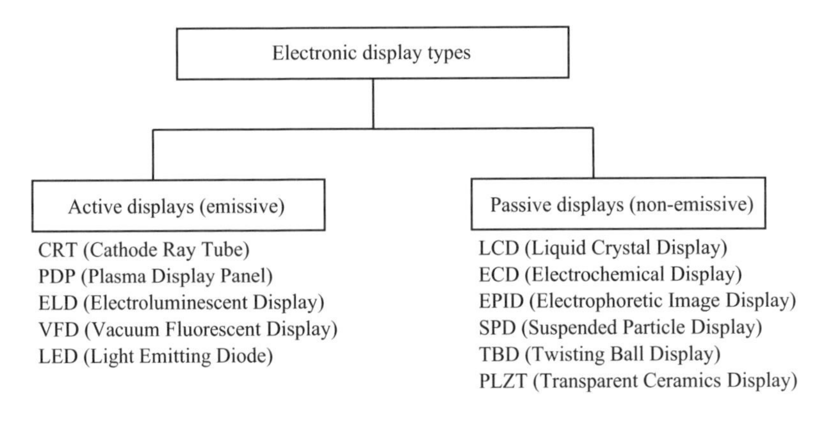

Figure 2.12 Display device technologies

Around the mid-1990s, CRT displays accounted for almost 90 per cent of the display device market, followed by LCD and LED displays accounting for the balance of 10 per cent. Today, full colour LCD videos and camcorders are common, and LCD television is entering the world markets. With the development of flat panel display technology, LCD displays account for over 90 per cent of the world wide flat panel display market. Most smaller instruments today use LCD displays, and gradually oscilloscope manufacturers are also replacing the CRT with LCD displays. Despite many improvements in CRT technology, CRTs are still larger, heavier and consume more power than LCD displays. More importantly, CRTs face a fundamental tradeoff between brightness and resolution.

In their simplest forms, LCDs offer users many advantages. They are inexpensive, thin, durable and lightweight, and they consume very little power. Their major disadvantages have been poor contrast, low brightness and a narrow viewing angle. Until the mid-1990s these disadvantages restricted the use of LCDs to low performance applications such as digital watches and calculators, simple alphanumeric displays and inexpensive laptop computers. Until the mid-1990s, high resolution displays capable of reproducing CGA, EGA and VGA standards were the exclusive province of the CRT. Although plasma and electroluminescent displays have been used in high resolution portable computers they do not offer the low power advantage of LCDs. Moreover, the plasma display exhibits particularly poor contrast, at least to the human eye, because of its characteristic orange-on-black colour. To answer the need for high contrast and high resolution, manufacturers of large area LCDs have turned to active matrix, thin film transistor technology and continue to improve double super twist technology.

The leading LCD technology is the thin film transistor (TFT) based active matrix LCD (AMLCD). Consisting of a liquid crystal sandwiched between two sheets of glass containing transparent electrodes, polarisers and back light, TFT LCDs position transistors at each pixel. By the mid-1980s, economic circumstances, coupled with sheer critical mass, created an environment in which the necessary LCD components came together. Dozens of companies, mostly in Japan, systematically addressed manufacturing yield, device structure/transistor architecture, power consumption (reduced from 25 W to 2 W for notebooks), viewing angle and speed. Fabricating a TFT LCD is similar to semiconductor processing, but with a fraction of the masking steps: approximately six and progressing to four or five. TFT LCD technology has progressed markedly in three critical areas: viewing angle, efficiencies and speed. To attain wider viewing angles, currently exceeding 100°, different techniques are used [60].

Other major new approaches to flat panel displays are (i) fludic self assembly (FSA), (ii) liquid crystal on silicon (LCOS), and (iii) organic light emitting diodes (OLED or OEL). For details, Reference 60 is suggested.

While flat panel displays continue to offer an increase in colour depth and resolution, both of which are equal to or better than CRT image quality, interfacing high resolution flat panel displays to PCs or processors is becoming a tough design issue [61, 62]. With LCD displays, electro-luminenscent lamps are used as a back lighting source, since it is the most practical way to illuminate a flat area evenly. For details, Reference 63 is suggested.

2.7 Software

The modern component families described in the previous sections, supported by powerful software, have almost revolutionised the design of modern instruments. As the microelectronic component families become cheaper and cheaper, allowing more memory power for storage of human designed algorithms at low cost, most instrument designers have been able to achieve the following.

(i) Elimination of expensive mechanically calibrated front panel controls with multiple contacts.
(ii) Introduction of auto-calibration facilities which may even be automatically activated at 'power-on'.
(iii) Addition of smart measuring aids (such as voltage and time cursors on modern scopes) and other similar features.
(iv) Addition of automatic fault diagnostic schemes to ease maintenance and performance checking of instruments.
(v) Allowing computerised control via standardised interfaces, such as IEEE-488.
(vi) Allowing storage of measurement data on a short and long term basis for non-volatile memories.
(vii) Transferring information via computer networks, etc.

In addition to assembler level software used inside instruments, all instruments today can communicate via various buses such as IEEE-488, USB, PCI and simple interfaces such as RS-232, etc. These facilities have gradually become standard features rather than options. In addition, instruments on a card concept have matured with the VME bus extension for instrumentation (VXI), allowing even the plug-and-play concept for multiple instruments in one card cage. The other important family is data analysis software.

2.8 Semiconductor sensors

With the development processes and design techniques applicable to silicon and other semiconductor materials, instrument designers were looking for smart sensors and transducers using silicon fabrication techniques including micromachining. Fairly inexpensive sensor devices are available for measuring temperature, pressure, force, acceleration, humidity, chemical composition of gases, magnetic field and many other physical parameters.

Improvements in silicon sensor fabrication technology and vertical integration let manufacturers call their devices 'smart sensors'. In these smart sensors one might identify five levels of sophistication: conversion, environmental compensation, communication, self-diagnostics, and logic actuation. With the advancement of silicon and microelectro-mechanical systems (MEMS) technologies, more 'smarts' are integrated into sensors. The emergence of the control networks and smart devices in the market-place may provide economical solutions for connecting transducers (hereafter specified as sensors or actuators) in distributed measurement and control applications; therefore, networking small transducers is seriously considered by transducer manufacturers and users. For anyone attempting to choose a sensor interface or networking standard, the range of choices is overwhelming. Some standards are open, and some are proprietary to a company's control products. The sensor market comprises widely disparate sensor types. Designers consume relatively large amounts of all types of sensor. However, the lack of a universal interface standard impedes the incorporation of 'smart' features, such as an onboard electronic data sheet, onboard A/D conversion, signal conditioning, device type identification, and communication hand-shaking circuitry, into the sensors. In response to industry's need for a communication interface for sensors, the IEEE with cooperation from the National Institute of Standards and Technology (NIST), decided to develop a hardware-independent communication standard for low cost smart sensors that includes smart transducer object models for control networks. The standard for the smart transducer interface for sensors and actuators, IEEE-P1451, is the result of this attempt by the IEEE Sensor Technology Committee TC-9.

The IEEE-P1451 standards effort, currently under development, will provide many benefits to the industry. P1451, 'Draft Standard for Smart Transducer Interface for Sensors and Actuators', consists of four parts, namely

(i) IEEE 1451.1 – Network Capable Application Processor (NCAP) information model,
(ii) IEEE 1451.2 – Transducer to Microprocessor Communications Protocols and Transducer Electronic Data Sheet (TDS) formats,
(iii) IEEE P1451.3 – Digital Communication and Transducer Electronic Data Sheet (TEDS) formats for distributed multidrop systems, and
(iv) IEEE 1451.4 – Mixed-mode Communication Protocols and Transducer Electronic Data Sheet (TEDS) formats.

For details, Chapter 7 of Reference 21 is suggested. The technology referred to as microelectro-mechanical system (MEMS) or microsystems technology is currently entering its maturity. Silicon micromachinery coupled with general semiconductor processing has given basic parts such as silicon accelerometers, which are common with systems for developing air bags. Looming on the horizon is a new class of microfabricated inertial sensors. These angular rate gyroscopes can measure the rate at which an object rotates. For details, Reference 64 is suggested.

Microelectro-mechanical systems (MEMS) technology is now where integrated circuit technology was about 30 years ago. It is clearly going to be pervasively important. Both commercial volume and patents on MEMS are growing at rates exceeding 20 per cent annually.

2.9 The future

Within the past decade, basic components such as transistors or the new versions of similar devices have been announced. This includes quantum transistors and similar devices allowing very high frequencies. For example, a device called the *double-electron layer tunnelling transistor* (DELTT) with over 700 GHz speed was introduced by Sandia Laboratories in the USA [65]. Based on recent achievements such as the *ballistic-nanotransistor*, faster silicon ICs are expected in the future [66, 67]. Microprocessors have reached clock speeds beyond 1 GHz while ADCs have reached high FOM values giving fast conversion possibilities with better resolution. With the dawn of organic electronics [68], organic full colour displays may eventually replace LCDs. Circuits on plastic is another possibility with organic electronics.

In addition to traditional common instrument families used in design and repair environments, the testing of complex ICs has special needs and several new standards for buses and test interfaces, etc., have also been finalised. Two of these are IEEE 1149.4 for mixed signal test bus and IEEE 1450, the standard test interface language (STIL), etc. [69].

With the demand for more user friendly interfaces, as seen in the recently formulated new law [70], *usefulness of a product = log(technology)*, future systems will be extremely complex from a hardware viewpoint. This demands the use of complex ICs, ASICs and DSP techniques for configuring the user friendly functions of instruments. Transferring measurement data over the Internet and by traditional communication systems will become standard features. However, owing to comparatively slower development of ADCs and other mixed signal components, some analogue instruments will also be continued for some years, particularly for cases such as high frequency spectrum observations, etc. One challenge for the test and measurement world is to cater for the growing demands of speeds and connection complexities of modern semiconductor chips.

2.10 References

1 Wafer Processing News: 'Bell Lab researchers demonstrate GaAs MOSFET', *Semiconductor International*, February 1997, p. 38.

2 SINGER, P.: '350 GHz transistor is world's fastest', *Semiconductor International*, November 1998, p. 42.

3 TEEGARDEN, D., LORENZ, G., NEUL, R.: 'World's fastest high electron mobility transistor developed', *MPT News*, February 21, 2000, **10** (23), pp. 1–2.
4 DE JULE, R.: 'A molecular alternative to CMOS', *Semiconductor International*, September 1999, p. 38.
5 GEPPERT, L.: 'Life after silicon: The solid state molecular IC', *IEEE Spectrum*, October 1999, pp. 24–25.
6 KULARATNA, N.: *Modern Electronic Test & Measuring Instruments*, IEE, London, 1996.
7 JOHNSON, H.: 'Keeping up with Moore', *EDN*, May 7, 1998, p. 24.
8 VAN ZANT, P.: *Microchip Fabrication – A Practical Guide to Semiconductor Processing*, McGraw Hill, New York, 1990.
9 BURROWS, I.: 'GaAs ICs: The future has arrived', *Semiconductor International*, April 1999, pp. 81–88.
10 GOODENOUGH, F.: 'SiGe, SOI to lead analog mixed-signal processors', *Electronic Design*, January 10, 1994, pp. 69–78.
11 DEJULE, R.: 'Wide bandgap materials draw interest', *Semiconductor International*, February, 1999, p. 34.
12 COOPER, J.A.: 'Advances in silicon carbide power switching devices', PCIM-1998 *Proceedings*, Santa Clara, USA, pp. 235–245.
13 DE JULE, R.: 'SiC supports blue leds and more', *Semiconductor International*, July 1999, p. 58.
14 DE JULE, R.: 'White LEDs: A better light bulb', *Semiconductor International*, May 1999, p. 42.
15 HERREL, D.: 'Power to the package', *Spectrum*, July 1999, pp. 46–53.
16 NIGHTINGALE, E.U.S.J.: 'Filings, filaments and FETs: Technologies for harnessing the waves', *Electronics & Communications Journal*, October 1997.
17 SCHWEBER, B.: 'The transistor at 50's: Not even considering retirement', *EDN*, December 18, 1997, pp. 83–88.
18 LARSON, L.E. *et al.*: 'A 10 GHz operational amplifier in GaAs MESFET technology'. Digest of technical paper, *ISSCC Proceedings*, USA, 1989, pp. 72–73.
19 FRANK, T.: 'A compact bipolar class AB output stage using 1 V power supply'. Digest of Technical Papers, *ISSCC Proceedings*, 1992, pp. 194–195.
20 AJLUNI, C.: 'CMOS technology prevails in analog designs at ISSCC', *Electronic Design*, February 9, 1998, pp. 64–68.
21 KULARATNA, N.: *Modern Component Families & Circuit Block Design*, Butterworth Heinemann, Boston, MA, USA, 2000.
22 FOTY, D.: 'Taking a deep look at analog CMOS', *Circuits and Devices Magazine, IEEE*, March 1999, pp. 23–28.
23 MCGOLDRICK, P.: 'Architectural breakthroughs moves conversion into mainstream', *Electronic Design*, January 20, 1997, pp. 67–72.
24 GOODENOUGH, F.: 'ADCs move to cut power dissipation', *Electronic Design*, January 9, 1995, pp. 69–74.
25 GOODENOUGH, F.: 'Analog technology of all varieties dominate ISSCC', *Electronic Design*, February 19, 1996, pp. 97–110.

26 MELLIAR SMITH, C.M.: 'Integrated circuits through the year 2010'. Proceedings of International Symposium on VLSI Technology, *System & Applications*, 1995, p. 84.
27 BUCHANAN, J.E.: *BiCMOS/CMOS Systems Design*, McGraw Hill, New York, 1991.
28 Semiconductor International: 'Phillips develops fastest silicon BiCMOS process', *Semiconductor International*, January 1999, pp. 22–23.
29 DANCE, B.: 'Silicon competes with SiGe in mobile phone market', *Semiconductor International*, April 2001, p. 50.
30 BURSKY, D.: 'The microprocessor: 25 years young and the best is still ahead', *Electronic Design*, December 2, 1996, pp. 36–38.
31 MOSLEY, J.D.: '8 & 16 bit microcontrollers', *EDN*, 28 September 1989, pp. 108–121.
32 JONES, N.B. and WATSON, J.D.M.: *Digital Signal Processing – Principles, Devices & Applications*, London, Peter Peregrinus/IEE, 1990.
33 SCHNEIDERMAN, R.: 'Faster, more highly integrated DSPs – Designed for designers', *Wireless Systems Designs*, November 1996, pp. 12–13.
34 SASAKI, K. *et al.*: 'A 7 ns, 140 MW, 1 Mb CMOS SRAM with current sense amplifier' Digest of Technical Papers, *ISSCC Proceedings*, 1992, pp. 208–209.
35 BURSKY, D.: 'Memories hit new highs & Clocks run jitter-free', *Electronic Design*, February 19, 1996, pp. 79–93.
36 BURSKY, D.: 'Next generation SRAMs deliver higher performance', *Electronic Design*, June 22, 1998, pp. 105–115.
37 BURSKY, D.: 'Innovation continues to thrive as circuits get denser and faster', *Electronic Design*, February 17, 1997.
38 MARKOWITZ, M.C.: 'Non-volatile memories', *EDN*, September 1, 1989, pp. 94–104.
39 SCHERER, T.: 'Flash EPROMs', *Elector Electronics*, October 1992, pp. 28–31.
40 NASS, R.: 'Portable-system users can take 40 Mbytes on the Road', *Electronic Design*, November 1, 1993, pp. 107–110.
41 BURSKY, D.: 'Flash-memory choices boost performance & flexibility', *Electronic Design*, May 30, 1995, pp. 63–75.
42 WRIGHT, M.: 'Tiny flash card battle floppy replacement', *EDN*, January 16, 1997, pp. 62–72.
43 DIPPERT, B.: 'Data storage in a flash', *EDN*, July 3, 1997, pp. 65–83.
44 BURSKY, D.: 'Serial flash memories rise to meet changing system needs', *Electronic Design*, March 22, 1999, pp. 77–82.
45 MAXFIELD, C.: 'Field-programmable devices', *EDN*, October 19, 1996, pp. 201–206.
46 FLETCHER, P.: 'Field programmable analog IC allows easy design of complex analog signal processors', *Electronic Design*, November 18, 1996, pp. 35–36.
47 BURSKY, D.: 'FPGAs and dense EPLDs challenge gate arrays', *Electronic Design*, July 10, 1995, pp. 69–80.
48 SCHWEBER, B.: 'Programmable analog ICs: Designer's delight or dilemma?' *EDN*, April 13, 2000, pp. 72–84.

49 TRIMBERGER, S.: (Guest Editor's Introduction): 'Field programmable gate arrays', *Design and Test of Computers (IEEE)*, Sept. 1992, pp. 3–5.
50 GALLANT, J.: 'ECL ICs play a role in high speed computers', *EDN*, 17 August 1989, pp. 73–85.
51 SMALL, C.H.: 'User programmable gate arrays', *EDN*, 27 April 1989, pp. 146–158.
52 QUINNEL, R.A.: 'Application tailored PLDs streamline designs bring speed and lower cost', *EDN*, 21 May 1992, pp. 81–96.
53 SMALL, C.H.: 'Programmable logic devices', *EDN*, 10 Nov. 1988.
54 DIPPERT, B.: 'Shattering the programmable-logic speed barrier', *EDN*, May 22, 1997, pp. 36–60.
55 JENKINS, J.H.: 'Use In-system programming to simplify field upgrades', *Electronic Design*, October 1, 1998, pp. 93–96.
56 BURSKY, D.: 'Gate arrays face onslaught of dense and flexible FPGAs', *Electronic Design*, June 26, 1995, pp. 85–96.
57 KAPUSTA, R.: 'Options do the programmable-logic landscape', *EDN*, July 6, 1995, pp. 107–116.
58 GALLANT, J.: 'Designing for speed with high performance PLDs', *EDN*, July 20, 1995, pp. 20–26.
59 DIPPERT, B.: 'Moving beyond programmable logic: if, when, how?', *EDN*, November 20, 1997, pp. 77–92.
60 DE JULE, R.: 'Directions in flat panel displays', *Semiconductor International*, August 1999, pp. 75–82.
61 NATH, M.: 'Interface issues deter faster adoption of flat panel displays', *EDN*, March 13, 1998.
62 MANJU, N.S.: 'Competing standards seek common ground for flat panel displays', *EDN*, June 10, 1999, pp. 103–110.
63 SCHWEBER, B.: 'Electroluminescent lamps: Sharp answer to flat lighting', *EDN*, February 18, 1999.
64 TEEGARDEN D., LORENZ, G. and NEUL, R.: 'How to model & simulate microgyroscope systems', *Spectrum*, July 1998, pp. 66–75.
65 GEPPERT, L.: 'Quantum transistors: Toward nanoelectronics', *IEEE Spectrum*, September 2000, pp. 46–51.
66 SINGER, P.: 'Engineering a better transistor', *Semiconductor International*, March 2000, pp. 66–69.
67 DESPOSITO, J.: 'Ballistic nanotransistor should yield smaller and faster silicon chips', *Electronic Design*, March 6, 2000, pp. 25–26.
68 FORREST, S., BURROWS, P. and THOMPSON, M.: 'The dawn of organic electronics', *IEEE Spectrum*, August 2000, pp. 29–34.
69 BRETZ, E.A.: 'Test and measurement', *IEEE Spectrum*, January 2000, pp. 75–79.
70 DIPPERT, B.: 'It's elementary', *EDN*, April 15, 1999, p. 52.

Chapter 3

Data converters[1]

3.1 Introduction

Modern design trends use the power and precision of the digital world of components to process analogue signals. However, the link between the digital/processing world and the analogue/real world is based on the analogue-to-digital and digital-to-analogue converter ICs, which generally are grouped together as the data converters.

Until about 1988, engineers had to stockpile their most innovative A-to-D converter (ADC) designs, because available manufacturing processes simply could not implement those designs onto monolithic chips economically[1]. Prior to 1988, except for the introduction of successive approximation, integrating and flash ADCs, the electronics industry saw no major changes in monolithic ADCs. Since then, manufacturing processes have caught up with the technology and many techniques such as sub-ranging flash, self-calibrating, delta/sigma, and many other special techniques have been implemented on monolithic chips.

High speed ADCs are used in a wide variety of real-time digital signal processing (DSP) applications, replacing systems that used analogue techniques alone[2]. The major reasons for using digital signal processing are that the cost of DSP processors has gone down, their speed and computational power have increased, and they are reprogrammable, allowing for system performance upgrades without hardware changes. DSP offers practical solutions that cannot be easily achieved in the analogue domain, for example, V.32 and V.34 modems.

This chapter provides an overview of design concepts and application guidelines for systems using modern analogue/digital and digital/analogue converters implemented on monolithic chips.

3.2 Sampled data systems

To specify intelligently the ADC portion of the system, one must first understand the fundamental concepts of sampling and quantisation and their effects on the signal.

[1] This chapter is an edited version of chapter 3 of the Butterworth/Newnes book 'Modern Component Families and Circuit Block Design', reproduced by permission of the publisher.

Let us consider the traditional problem of sampling and quantising a baseband signal whose bandwidth lies between d.c. and an upper frequency of interest, f_S. This is often referred to as Nyquist, or sub-Nyquist sampling. The topic of super-Nyquist sampling (sometimes called under-sampling) where the signal of interest falls outside of the Nyquist bandwidth (d.c. to $f_S/2$) is treated later. Figure 3.1 shows key elements of a baseband sampled data system.

3.2.1 Discrete time sampling of analogue signals

Figure 3.2 shows the concept of discrete time and amplitude sampling of an analogue signal. The continuous analogue data must be sampled at discrete intervals, t_S, which must be carefully chosen to ensure an accurate representation of the original analogue signal. It is clear that, the more samples taken (faster sampling rates), the more accurate the digital representation; and if fewer samples are taken (lower sampling rates), a point is reached where critical information about the signal is actually lost.

To discuss the problem of losing information in the sampling process, it is necessary to recall Shannon's information theorem and Nyquist's criteria. Shannon's information theorem states the following.

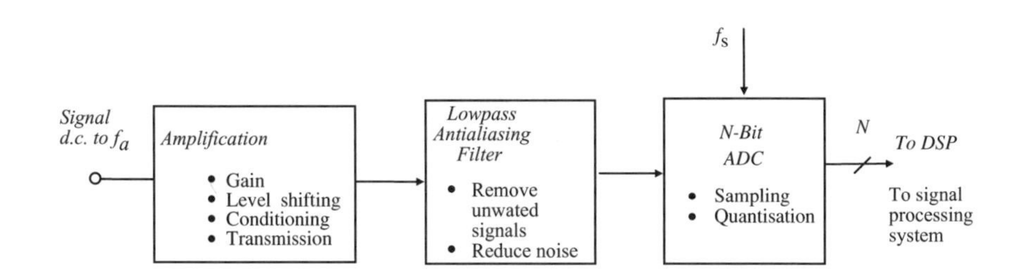

Figure 3.1 Key elements of a baseband sampled data system

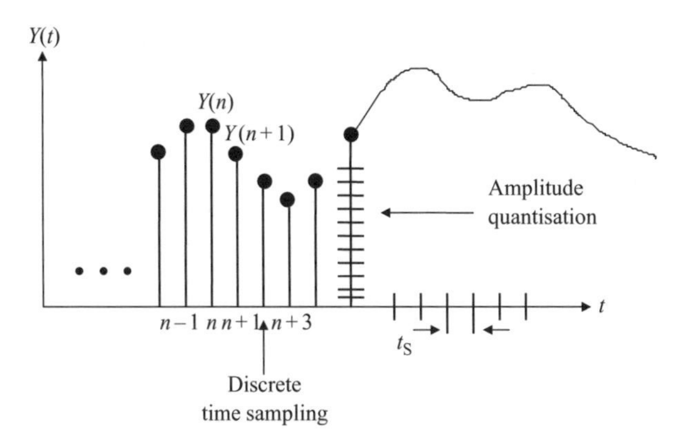

Figure 3.2 Sampling and quantising of an analogue signal

- An analogue signal with a bandwidth of f_a must be sampled at a rate of $f_s > 2f_a$ to avoid loss of information.
- The signal bandwidth may extend from d.c. to f_a (baseband sampling) or from f_1 to f_2, where $f_a = f_2 - f_1$ (under-sampling, or super-Nyquist sampling).

The Nyquist criteria are as follows.

- If $f_s < 2f_a$, than a phenomenon called aliasing will occur.
- Aliasing is used to advantage in under-sampling applications.

3.2.1.1 Implications of aliasing

To understand the implications of aliasing in both the time and frequency domains, first consider the case of a time domain representation of a sampled sinewave signal shown in Figure 3.3. In Figures 3.3(a) and 3.3(b), it is clear that an adequate number of samples have been taken to preserve the information about the sinewave. Figure 3.3(c) represents the ambiguous limiting condition where $f_s = 2f_a$. If the relationship between the sampling points and the sinewave is such that the sinewave is being

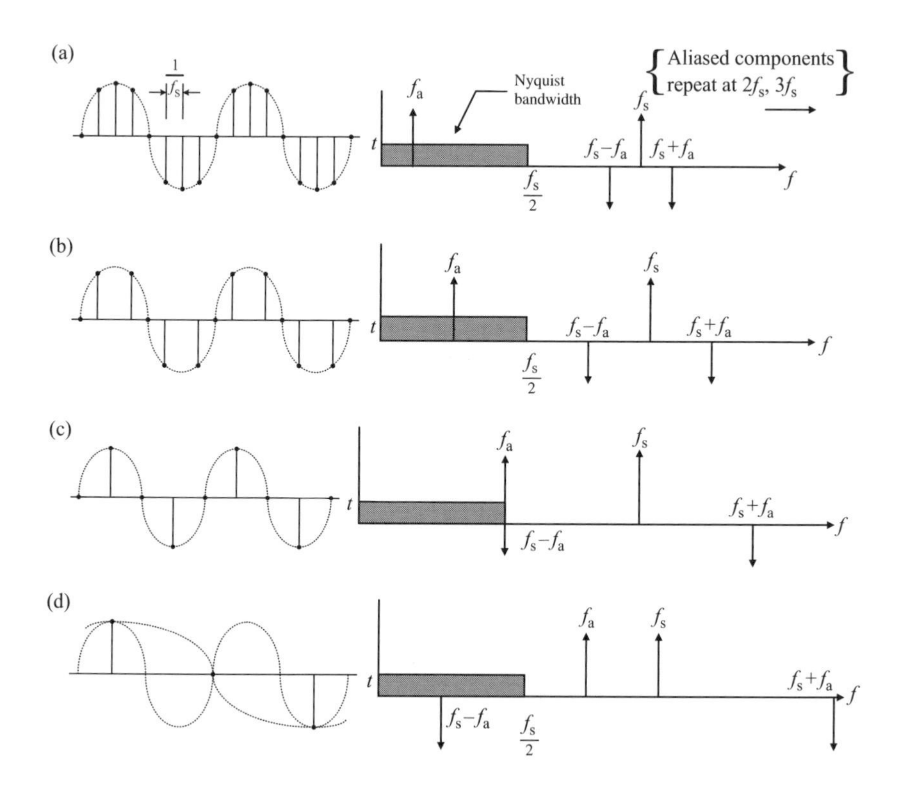

Figure 3.3 Time and frequency domain effects of aliasing: (a) $f_s = 8f_a$; (b) $f_s = 4f_a$; (c) $f_s = 2f_a$; and (d) $f_s = 1.3f_a$

sampled at precisely the zero crossings (rather than at the peaks, as shown in the illustration), then all information regarding the sinewave would be lost. Figure 3.3(d) represents the situation where $f_s < 1.3f_a$, and the information obtained from the samples indicates a sinewave having a frequency lower than $f_s/2$. This is a case where the out of band signal is aliased into the Nyquist bandwidth between d.c. and $f_s/2$. As the sampling rate is further decreased, and the analogue input frequency f_a approaches the sampling frequency f_s, the aliased signal approaches d.c. in the frequency spectrum.

Let us look at the corresponding frequency domain representation of each case. From each case of frequency domain representation, we make the important observation that, regardless of where the analogue signal being sampled happens to lie in the frequency spectrum, the effects of sampling will cause either the actual signal or an aliased component to fall within the Nyquist bandwidth between d.c. and $f_s/2$. Therefore, any signals that fall outside the bandwidth of interest, whether they be spurious tones or random noise, must be adequately filtered before sampling. If unfiltered, the sampling process will alias them back within the Nyquist bandwidth where they will corrupt the wanted signals.

3.2.1.2 High speed sampling

Now let us discuss the case of high speed sampling, analysing it in the frequency domain. First, consider the use of a single frequency sinewave of frequency f_a sampled at a frequency f_s by an ideal impulse sampler (see Figure 3.4(a)). Also assume that $f_s > 2f_a$ as shown. The frequency domain output of the sampler shows aliases or images of the original signal around every multiple of f_s, that is, at frequencies equal to

$$|\pm Kf_s \pm f_a|, \quad \text{where } K = 1, 2, 3, 4, \ldots. \tag{3.1}$$

The Nyquist bandwidth, by definition, is the frequency spectrum from d.c. to $f_s/2$. The frequency spectrum is divided into an infinite number of Nyquist zones, each having a width equal to $0.5f_s$, as shown.

Now consider a signal outside the first Nyquist zone, as shown in Figure 3.4(b). Notice that even though the signal is outside the first Nyquist zone, its image (or alias), $f_s - f_a$, falls inside. Returning to Figure 3.4(a), it is clear that, if an unwanted signal appears at any of the image frequencies of f_a, it will also occur at f_a, thereby producing a spurious frequency component in the Nyquist zone. This is similar to the analogue mixing process and implies that some filtering ahead of the sampler (or ADC) is required to remove frequency components that are outside the Nyquist bandwidth, but whose aliased components fall inside it. The filter performance will depend on how close the out-of-band signal is to $f_s/2$ and the amount of attenuation required.

3.2.1.3 Antialiasing filters

Baseband sampling implies that the signal to be sampled lies in the first Nyquist zone. It is important to note that, with no input filtering at the input of the ideal sampler, any frequency component (either signal or noise) that falls outside the Nyquist bandwidth

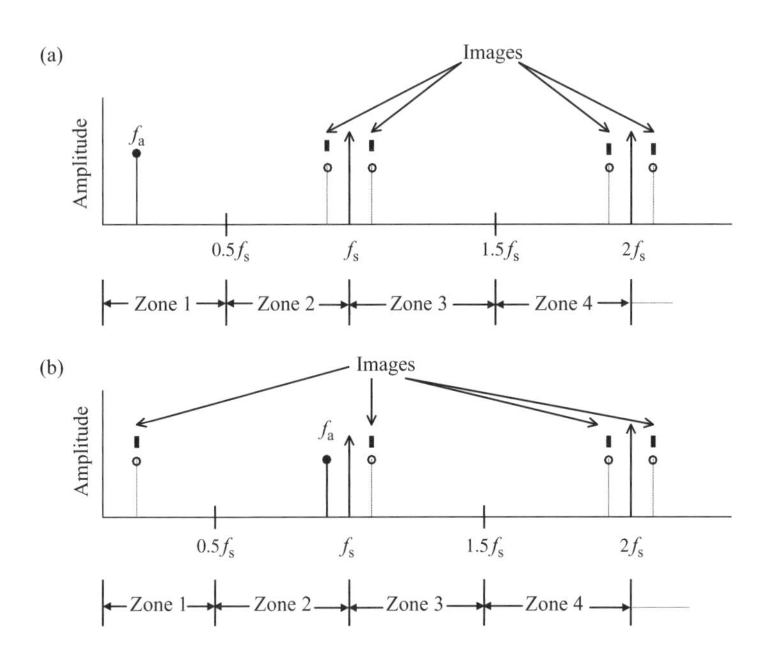

Figure 3.4 Analogue signal at frequency f_a sampled at f_s: (a) case of signal lying within $f_s/2$; (b) case of signal lying between $f_s/2$ and f_s

in any Nyquist zone will be aliased back into the first Nyquist zone. For this reason, an antialiasing filter is used in almost all sampling ADC applications to remove these unwanted signals.

Properly specifying the antialiasing filter is important. The first step is to know the characteristics of the signal being sampled. Assume that the highest frequency of interest is f_a. The antialiasing filter passes signals from d.c. to f_a while attenuating signals above f_a. Assume that the corner frequency of the filter is chosen to be equal to f_a. The effect of finite transition from minimum to maximum attenuation on system dynamic range (DR) is illustrated in Figure 3.5.

Assume that the input signal has full scale components well above the maximum frequency of interest, f_a. The diagram shows how full scale frequency components above $f_s - f_a$ are aliased back into the bandwidth d.c. to f_a. These aliased components are indistinguishable from actual signals and therefore limit the dynamic range to the value on the diagram, which is shown as DR.

The antialiasing filter transition band therefore is determined by the corner frequency f_a, the stop band frequency ($f_s - f_a$), and the stop band attenuation, DR. The required system dynamic range is chosen based on our requirement for signal fidelity.

Filters have to become more complex as the transition band becomes sharper, all other things being equal. For instance, a Butterworth filter gives 6 dB attenuation per octave for each filter pole. Achieving 60 dB attenuation in a transition region between 1 and 2 MHz (one octave) requires a minimum of ten poles. This is not a trivial filter,

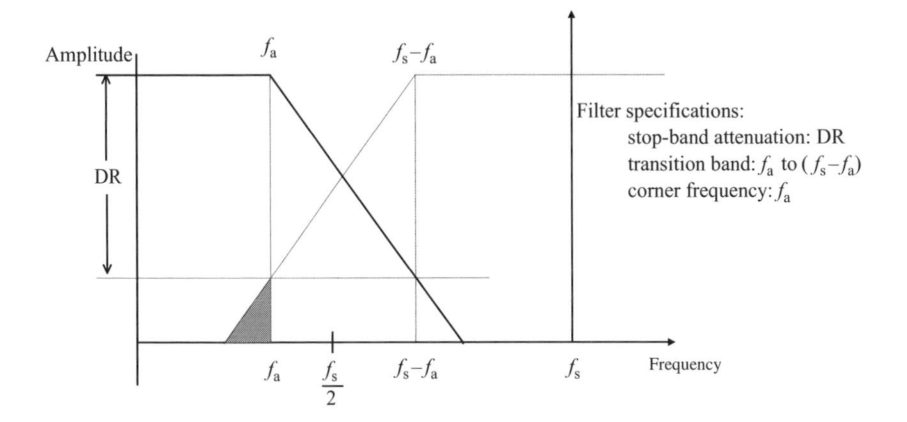

Figure 3.5 Effects of antialiasing filters on system dynamic range

and definitely a design challenge. Therefore, other filter types generally are more suited to high speed applications where the requirement is for a sharp transition band and in-band flatness coupled with linear phase response. Elliptic filters meet these criteria and are a popular choice.

From this discussion, we can see how the sharpness of the antialiasing transition band can be traded off against the ADC sampling frequency. Choosing a higher sampling rate (over-sampling) reduces the requirement on transition band sharpness (hence, the filter complexity) at the expense of using a faster ADC and processing data at a faster rate. This is illustrated in Figure 3.6, which shows the effects of increasing the sampling frequency while maintaining the same analogue corner frequency, f_a, and the same dynamic range, DR, requirement.

Based on this discussion one could start the design process by initially selecting a sampling rate of two to four times f_a. Filter specifications could be determined from the required dynamic range based on cost and performance. If such a filter is not realisable, high sampling rate, with a faster ADC, will be required.

The antialiasing filter requirements can be relaxed somewhat if it is certain that there never will be a full scale signal at the stopband frequency, $f_s - f_a$. In many applications, it is improbable that full scale signals will occur at this frequency. If the maximum signal at the frequency $f_s - f_a$ will never exceed X dB below full scale, the filter stop band attenuation requirement is reduced by that amount. The new requirement for stop band attenuation at $f_s - f_a$ based on this knowledge of the signal now is only (DR $- X$) dB. When making this type of assumption, be careful to treat any noise signals that may occur above the maximum signal frequency, f_a, as unwanted signals that also alias back into the signal bandwidth.

Properly specifying the antialiasing filter requires a knowledge of the signal's spectral characteristics as well as the system dynamic range requirements. Consider the signal (Figure 3.6(c)) that has a maximum full scale frequency content of $f_a =$ 35 kHz sampled at a rate of $f_s =$ 100 kSPS. Assume that the signal has the spectrum shown in Figure 3.6(c) and is attenuated by 30 dB at 65 kHz ($f_s - f_a$). Observe

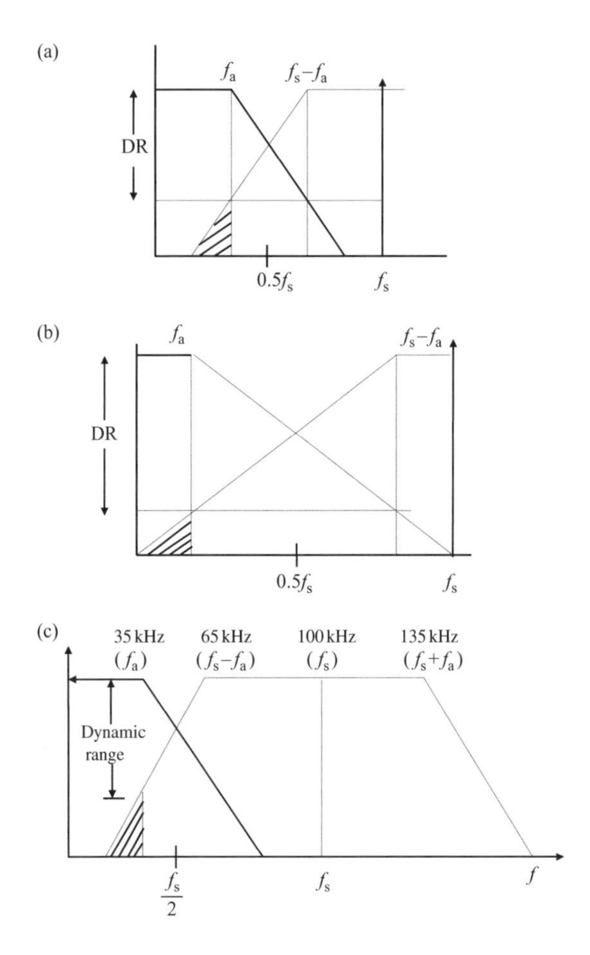

Figure 3.6 The relationship between sampling frequency and the antialiasing filter requirement: (a) low sampling rate with a sharper filter; (b) high sampling rate with relaxed filter specification; (c) a numerical example

that the system dynamic range is limited to 30 dB at 35 kHz because of the aliased components. If additional dynamic range is required, an antialiasing filter must be provided to provide more attenuation at 65 kHz. If a dynamic range of 74 dB (12 bits) at 35 kHz is desired, then the antialiasing filter attenuation must go from 0 dB at 35 kHz to 44 dB at 65 kHz. This is an attenuation of 44 dB in approximately one octave; therefore, a seven pole filter is required. (Each filter pole provides approximately 6 dB attenuation per octave.)

One must consider that broadband noise may be present with the signal, which also can alias within the bandwidth of interest. This is especially true with wideband op amps that provide low distortion levels.

Table 3.1 Bit sizes, quantisation noise, and signal-to-noise ratio (SNR) for 2.048 V full scale converters

Resolution (N bits)	1 LSB = q	% FS	r.m.s. quantisation noise, $q/\sqrt{12}$	Theoretical full scale SNR (dB)
6	32 mV	1.56	9.2 mV	37.9
8	8 mV	0.39	2.3 mV	50.0
10	2 mV	0.098	580 μV	62.0
12	500 μV	0.024	144 μV	74.0
14	125μV	0.0061	36 μV	86.0
16	31 μV	0.0015	13 μV	98.1

3.2.2 ADC resolution and dynamic range requirements

Having discussed the sampling rate and filtering, we next discuss the effects of dividing the signal amplitude into a finite number of discrete quantisation levels. Table 3.1 shows relative bit sizes for various resolution ADCs, for a full scale input range chosen as approximately 2 Vs, which is popular for higher speed ADCs. The bit size is determined by dividing the full scale range (2.048 V) by 2^N.

The selection process for determining the ADC resolution should begin by determining the ratio between the largest signal (full scale) and the smallest signal you wish the ADC to detect. Convert this ratio to dB, and divide by six. This is your minimum ADC resolution requirement for d.c. signals. You will actually need more resolution to account for extra signal headroom, because ADCs act as hard limiters at both ends of their range. Remember that this computation is for d.c. or low frequency signals and that the ADC performance will degrade as the input signal slew rate increases. The final ADC resolution actually will be dictated by dynamic performance at high frequencies. This may lead to the selection of an ADC with more resolution at d.c. than is required.

Table 3.1 also indicates the theoretical r.m.s. quantisation noise produced by a perfect N-bit ADC. In this calculation, the assumption is that quantisation error is uncorrelated with the ADC input. With this assumption the quantisation noise appears as random noise speed uniformly over the Nyquist bandwidth, d.c. to $f_s/2$, and it has an r.m.s. value equal to $q/\sqrt{12}$. Other cases may be different and some practical explanation is given in Analogue Devices [3].

3.2.3 Effective number of bits of a digitiser

Table 3.1 shows the theoretical full scale SNR calculated for the perfect N- bit ADC, based on the formula

$$\text{SNR} = 6.02N + 1.76\,(\text{dB}). \tag{3.2}$$

Various error sources in the ADCs cause the measured SNR to be less than the theoretical value shown in eq. (3.1). These errors are due to integral and differential non-linearities, missing codes, and internal ADC noise sources (some of which are discussed later).

In addition, the errors are a function of the input slew rate and therefore increase as the input frequency gets higher. In calculating the r.m.s. value of the noise, it is customary to include harmonics of the fundamental signal. This sometimes is referred to as the signal-to-noise-plus-distortion, S/(N + D) or SINAD, but usually simply SNR.

This leads to definition of another important ADC dynamic specification, the effective number of bits, or ENOB. The effective bits are calculated by first measuring the SNR of an ADC with a full scale sinewave input signal. The measured SNR (SNR_{actual} or SINAD) is substituted into the equation for SNR, and the equation is solved for N as shown next:

$$\mathrm{ENOB} = \frac{\mathrm{SINAD} - 1.76\,\mathrm{dB}}{6.02}. \tag{3.3}$$

This is shown for a typical ADC, AD676 from Analog Devices (a 16-bit ADC), in Figure 3.7.

For this device, the SNR value of 88 dB corresponds to approximately 14.3 effective bits (for 0 dB input), while it drops to 6.4 ENOB at 1 MHz. The methods for calculating ENOB, SNR and other parameters are described in References 4 and 5.

In testing ADCs, the SNR usually is calculated using DSP techniques while applying a pure sinewave signal to the input of ADC. A typical test system is shown in Figure 3.8(a). The fast Fourier transform (FFT) processes a finite number of time samples and converts them into a frequency spectrum, such as the one shown in Figure 3.8(b) for an AD676 type 16-bit 100 kSPS sampling ADC. The frequency spectrum then is used to calculate the SNR as well as harmonics of the fundamental input signal.

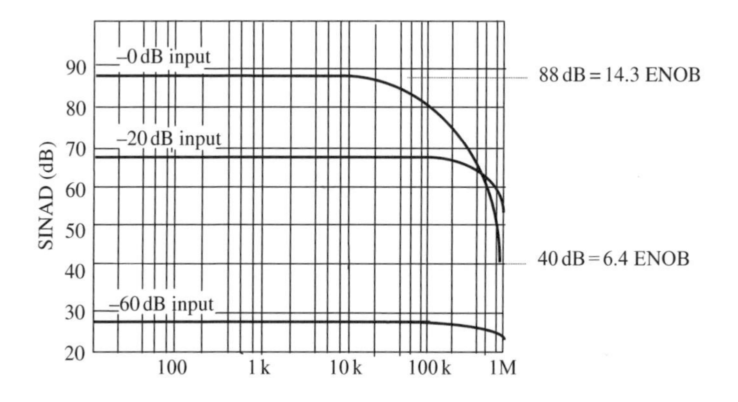

Figure 3.7 SINAD and ENOB for AD676 (reproduced by permission of Analog Devices, Inc.)

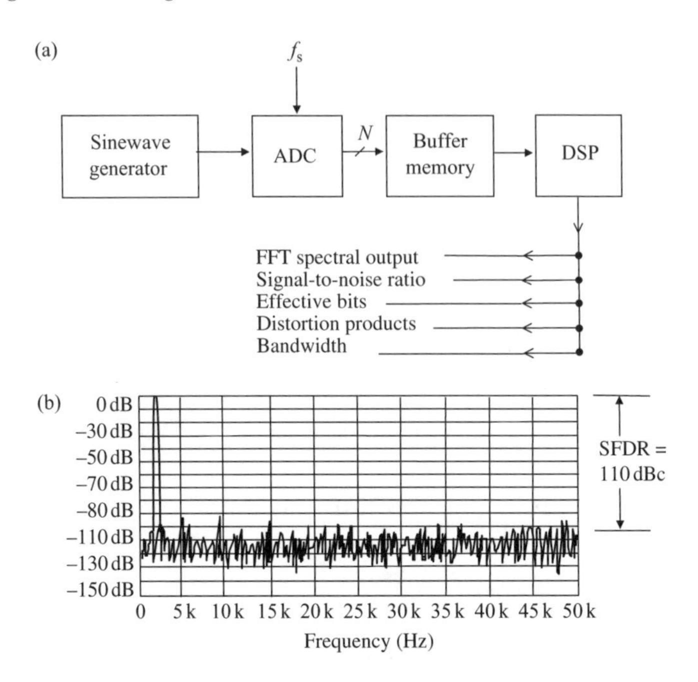

Figure 3.8 Testing of an ADC for its performance parameters: (a) test system; (b) typical FFT output for an AD676 ADC (reproduced by permission of Analog Devices, Inc.)

The r.m.s. value of the signal is first computed. The r.m.s. value of all other frequency components over the Nyquist bandwidth (this includes not only noise but also distortion products) is computed. The ratio of these two quantities, expressed in decibels (dB), is the SNR. Various error sources in the ADC cause the measured SNR to be less than the theoretical value, $6.02\,N + 1.76\,\text{dB}$.

3.2.3.1 Spurious components and harmonics

The peak spurious or peak harmonic component is the largest spectral component excluding the input signal and d.c. This value is expressed in decibels relative to the r.m.s. value of a full scale input signal as shown in Figure 3.8(a). The peak spurious specification is also occasionally referred to as spurious free dynamic range (SFDR). SFDR is usually measured over a wide range of input frequencies and at various amplitudes. It is important to note that the harmonic distortion or SFDR of an ADC is not limited by its theoretical SNR value. The SFDR of a 12-bit ADC may exceed 85 dB, while the theoretical SNR is only 74 dB. On the other hand, the SINAD of the ADC may be limited by poor harmonic distortion performance, since the harmonic components are included with the quantisation noise when computing the r.m.s. noise level. The SFDR of an ADC is defined as the ratio of the r.m.s. signal amplitude to the r.m.s. value peak spurious spectral content (measured over the entire first Nyquist zone, d.c. to $f_s/2$). The SFDR is generally plotted as a function of signal

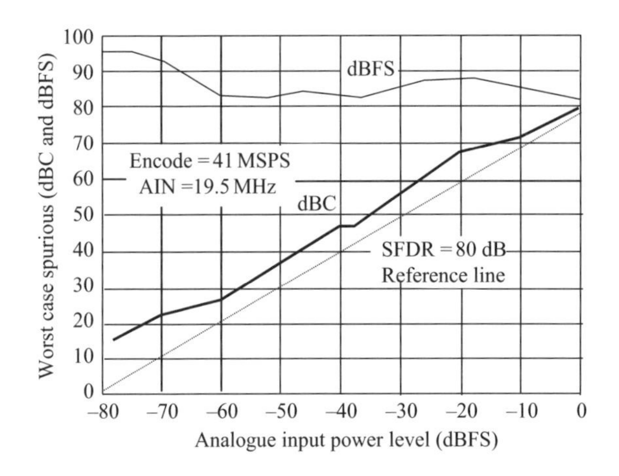

Figure 3.9 SFDR vs input power level for AD9042 (reproduced by permission of Analog Devices, Inc.)

amplitude and may be expressed relative to the signal amplitude (dBc) or the ADC full scale (dBFS).

For a signal near full scale, the peak spectral spur generally is determined by one of the first few harmonics of the fundamental input signal. However, as the signal falls several decibels below full scale, other spurs generally occur that are not direct harmonics of the input signal due to the differential non-linearity of the ADC transfer function. Therefore, the SFDR considers all sources of distortion, regardless of their origin.

The total harmonic distortion (THD) is the ratio of the r.m.s. sum of the harmonic components to the r.m.s. value of an input signal, expressed in a percentage or decibels. For input signals or harmonics above the Nyquist frequency, the aliased components are used in making the calculation. The THD is usually measured at several input signal frequencies and amplitudes.

Figure 3.9 shows the SFDR performance for a 12-bit, 41 MSPS wideband ADC designed for communications applications (AD 9042 from Analog Devices, Inc.).

Note that a minimum of 80 dBc SFDR is obtained over the entire first Nyquist zone (d.c. to 20 MHz). The plot also shows SFDR expressed as dBFS. SFDR is generally much greater than the ADC's theoretical N-bit SNR ($6.02N + 1.76$ dB). For example, the AD9042 is a 12-bit ADC with an SFDR of 80 dBc and a typical SNR of 65 dBc (the theoretical SNR is 74 dB). This is because of the fundamental distinction between noise and distortion measurements.

3.3 A-to-D converter errors

First, let us look at how bits are assigned to the corresponding analogue values in a typical analogue-to-digital converter (ADC). The method of assigning bits to the corresponding analogue value of the sampled point is often referred to as 'quantisation'

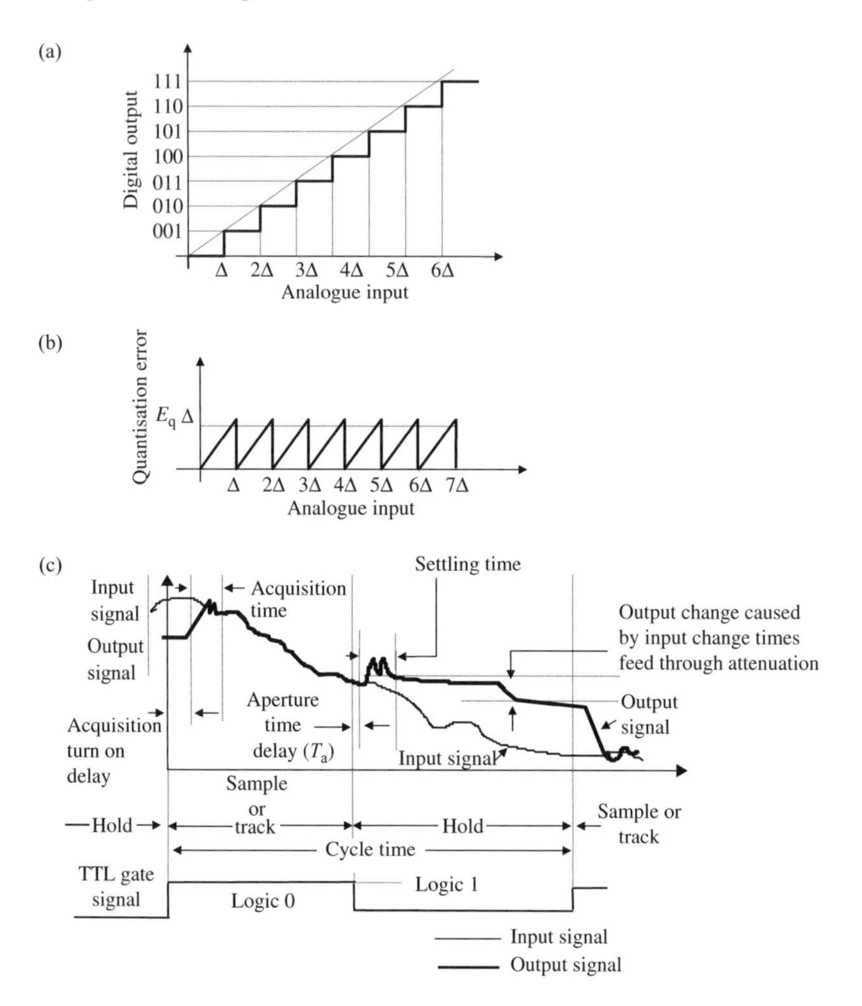

Figure 3.10 Quantisation process and quantising error: (a) basic input–output characteristics; (b) quantisation error; (c) basic timing of acquisition cycle of an ADC

(see Figure 3.10(a)). As the analogue voltage increases, it crosses transitions of 'decision levels', which causes the ADC to change state. In an ideal ADC, the transitions are at half unit levels, with Δ representing the distance between the decision levels. The Δ is often referred to as the 'bit size' or 'quantisation size'. The fact that Δ always has a finite size leads to uncertainty, since any analogue value within the finite range can be represented. This quantisation uncertainty is expressed as within the range of one least significant bit (LSB) as shown in Figure 3.10(b). As this plot shows, the output of an ADC may be thought of as the analogue signal plus some quantising noise. The more bits the ADC has, the less significant this noise becomes.

Certain parameters limit the rate at which an ADC can acquire a sample of the input waveform: the acquisition turn-on delay, acquisition time, sample or track time and hold time. Figure 3.10(c) shows a graphic representation of the acquisition cycle of a typical ADC. The turn-on time (the time that the device takes to get ready to acquire a sample) is the first event. The acquisition time is the next. This is the time the device takes to get to the point at which the output tracks the input sample, after the sample command or clock pulse. The aperture time delay is the time that elapses between the hold command and the point at which the sampling switch is completely open. The device then completes the hold cycle and the next acquisition is taken.

This process indicates that the real world of acquisition is not an ideal process at all and the value sampled and converted could have some sources of error. Most of these errors increase with the sampling rate.

The approximation or 'rounding' effect in A/D converters is called 'quantisation', and the difference between the original input and the digitised output, called the 'quantisation error', is denoted here by ε_q. For the characteristic of Figure 3.10(a), ε_q varies as shown in Figure 3.10(b), with the maximum occurring before each code transition. This error decreases as the resolution increases, and its effect can be viewed as additive noise ('quantisation noise') appearing at the output. Thus, even an 'ideal' m-bit ADC introduces non-zero noise in the converted signal simply owing to quantisation.

We can formulate the impact of quantisation noise on the performance as follows. For simplicity, consider a slightly different input/output characteristic, shown in Figure 3.11(a), where code transitions occur at odd (rather than even) multiples of $\Delta/2$. A time domain waveform therefore experiences both negative and positive quantisation errors, as illustrated in Figure 3.11(b). To calculate the power of the resulting noise, we assume that ε_q is (i) a random variable uniformly distributed between $-\Delta/2$ and $+\Delta/2$, and (ii) independent of the analogue input. While these assumptions are not strictly valid in the general case, they usually provide a reasonable approximation for resolutions above 4 bits. Razavi [6] provides more details and the derivations of eqs. (3.2) and (3.3).

Full specification of the performance of ADCs requires a large number of parameters, some of which are defined differently by different manufacturers. Some important parameters frequently used in component data sheets and the like are described here. Figure 3.12 could be used to illustrate parameters such as differential non-linearity (DNL), integral non-linearity (INL), offset error and gain error, all static parameters of the ADC process.

3.3.1 Parameters for ADCs

3.3.1.1 Differential non-linearity (DNL)

Differential non-linearity (DNL) is the maximum deviation in the difference between two consecutive code transition points on the input axis from the ideal value of 1 LSB. The DNL is a measure of the deviation code widths from the ideal value of 1 LSB.

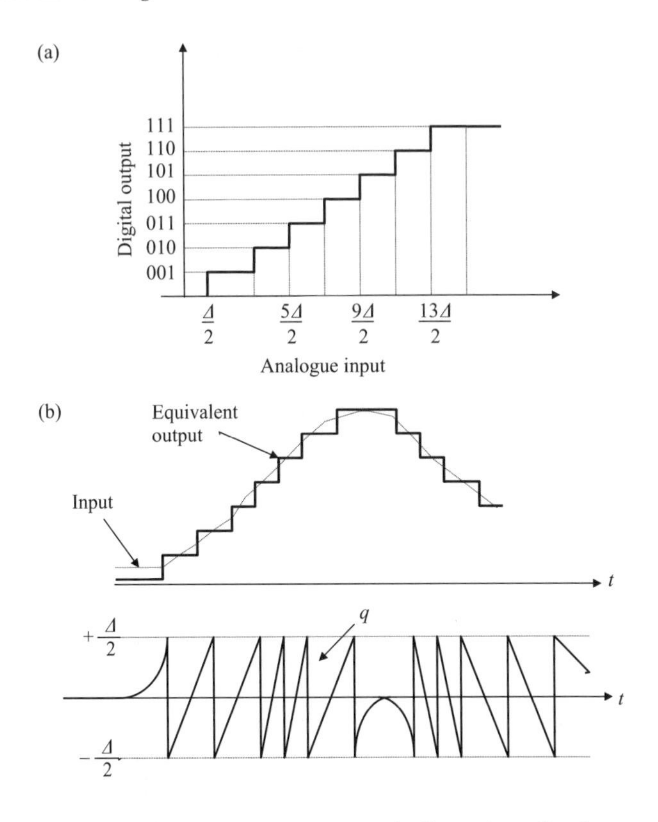

Figure 3.11 Modified ADC characteristics and effect of amplitude quantisation on a time domain waveform: (a) modified ADC characteristics; (b) effect of quantisation

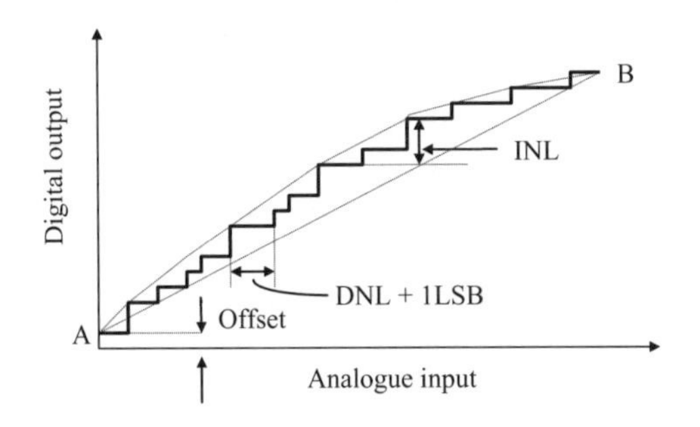

Figure 3.12 Static ADC metrics

3.3.1.2 Integral non-linearity (INL)

The INL is the maximum deviation of the input/output characteristic from a straight line passed through its end points (line AB in Figure 3.12). The overall difference plot is called the INL profile. The INL is the deviation of code centres from the converter's ideal transfer curve. The line used as the reference may be drawn through the end points or may be a best fit line calculated from the data.

The DNL and INL degrade as the input frequency approaches the Nyquist rate. The DNL shows up as an increase in quantisation noise, which tends to elevate the converter's overall noise floor. Theoretical quantisation noise for an ideal converter with the Nyquist bandwidth is

$$\text{r.m.s. quantisation noise} = \frac{q}{\sqrt{12}}, \tag{3.4}$$

where q is the size of the LSB.

At the same time, because the INL appears as a bend in the converter's transfer curve, it generates spurious frequencies (spurs) not in the original signal information. The testing of ADC linearity parameters is discussed in Reference 7.

3.3.1.3 Offset error and gain error

The offset is a vertical intercept of the straight line through the end points. The gain error is the deviation of the slope of line AB from its ideal value (usually unity).

3.3.1.4 Testing of ADCs

A known periodic input is converted by an ADC under test at sampling times that are asynchronous relative to the input signal. The relative number of occurrences of the distinct digital output codes is termed the code density. For an ideal ADC, the code density is independent of the conversion rate and input frequency. These data are viewed in the form of a normalised histogram showing the frequency of occurrence of each code from zero to full scale. The code density data are used to compute all bit transition levels. Linearity, gain and offset errors are readily calculated from knowledge of the transition levels. This provides a complete characterisation of the ADC in the amplitude domain.

The effect of some of these static errors in the frequency domain for high speed ADCs is discussed by Louzon [8]. ADC characterisation methods based on code density test and spectral analysis using FFT are provided in Reference 9.

3.4 Effects of sample-and-hold circuits

The sample-and-hold amplifier, or SHA, is a critical part of many data acquisition systems. It captures an analogue signal and holds it during some operation (most commonly during analogue-to-digital conversion). The circuitry involved is demanding, and unexpected properties of commonplace components such as capacitors and printed circuit boards may degrade SHA performance.

When the sample-and-hold amplifier is in the sample mode, the output follows the input with only a small voltage offset. In some SHAs, output during the sample mode does not follow the input accurately, and the output is accurate only during the hold period.

Today, high density IC processes allow the manufacture of ADCs containing an integral SHA. Wherever possible, ADCs with integral SHA (often known as sampling ADCs) should be used in preference to separate ADCs and SHAs. The advantage of such a sampling ADC, apart from the obvious ones of smaller size, lower cost, and fewer external components, is that the overall performance is specified. The designer need not spend time ensuring that there are no specification, interface, or timing issues involved in combining a discrete ADC and a discrete SHA.

3.4.1 Basic SHA operation

Regardless of the circuit details or type of SHA in question, all such devices have four major components. The input amplifier, energy storage device (capacitor), output buffer, and switching circuits are common to all SHAs as shown in the typical configuration of Figure 3.13(a).

The energy-storage device, the heart of the SHA, almost always is a capacitor. The input amplifier buffers the input by presenting a high impedance to the signal source and providing current gain to charge the hold capacitor. In the track mode, the voltage on the hold capacitor follows (or tracks) the input signal (with some delay and bandwidth limiting). Figure 3.10 depicts this process. In the hold mode, the switch is opened, and the capacitor retains the voltage present before it was disconnected from the input buffer. The output buffer offers a high impedance to the hold capacitor to keep the held voltage from discharging prematurely. The switching circuit and its driver form the mechanism by which the SHA is alternately switched between track and hold.

Four groups of specifications that describe basic SHA operation: track mode, track-to-hold transition, hold mode, hold-to-track transition. These specifications are summarised in Table 3.2, and some of the SHA error sources are shown in Figure 3.13(b). Because of both d.c. and a.c. performance implications for each of the four modes, properly specifying an SHA and understanding its operation in a system are complex matters.

3.4.1.1 Track mode specifications

Since an SHA in the sample (or track) mode is simply an amplifier, both the static and dynamic specifications in this mode are similar to those of any amplifier. The principle track mode specifications are offset, gain, non-linearity, bandwidth, slew rate, settling time, distortion, and noise; however, distortion and noise in the track mode often are of less interest than in the hold mode. Fundamental amplifier specifications are discussed in chapter 2 of Reference 10.

3.4.1.2 Track-to-hold mode specifications

When the SHA switches from track to hold, generally a small amount of charge is dumped on the hold capacitor because of non-ideal switches. This results in a hold mode d.c. offset voltage called pedestal error. If the SHA is driving an ADC, the pedestal error appears as a d.c. offset voltage that may be removed by performing a system calibration. If the pedestal error is a function of the input signal level, the

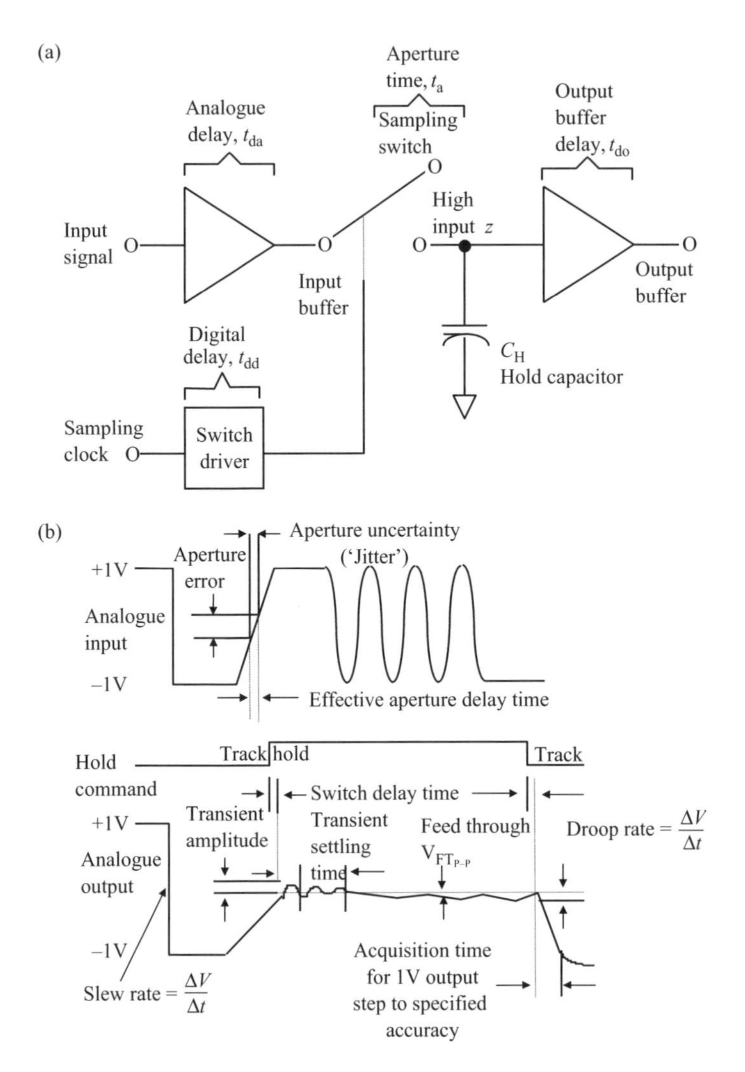

Figure 3.13 Sample and hold circuit and error sources: (a) basic sample and hold circuit; (b) some sources of error

resulting non-linearity contributes to hold mode distortion. Pedestal errors may be reduced by increasing the value of the hold capacitor with a corresponding increase in acquisition time and a reduction in bandwidth and slew rate.

Switching from track to hold produces a transient, and the time required for the SHA output to settle to within a specified error band is called the hold mode settling time. Occasionally, the peak amplitude of the switching transient also is specified; see Figure 3.14.

Table 3.2 Sample-and-hold specifications

	Sample mode	Sample-to-hold transition	Hold mode	Hold-to-sample transition
Static	• Offset • Gain error • Non-linearity	• Pedestal • Pedestal non-linearity	• Droop • Dielectric absorption	
Dynamic	• Setting time • Bandwidth • Slew rate • Distortion • Noise	• Aperture delay time • Aperture jitter • Switching transient • Settling time	• Feed through • Distortion • Noise	• Acquisition time • Switching transient

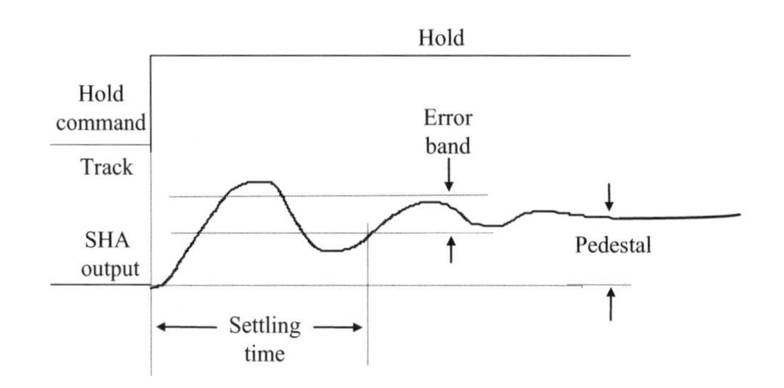

Figure 3.14 Hold mode settling time

3.4.1.3 Aperture and aperture time

Perhaps the most misunderstood and misused SHA specifications are those that include the word aperture. The most essential dynamic property of an SHA is its ability to disconnect quickly the hold capacitor from the input buffer amplifier (see Figure 3.13(a)).

The short (but non-zero) interval required for this action is called the aperture time (t_a). The actual value of the voltage held at the end of this interval is a function of both the input signal and the errors introduced by the switching operation itself. Figure 3.15 shows what happens when the hold command is applied with an input signal of arbitrary slope (for clarity, the sample to hold pedestal and switching transients are ignored). The value finally held is a delayed version of the input signal, averaged

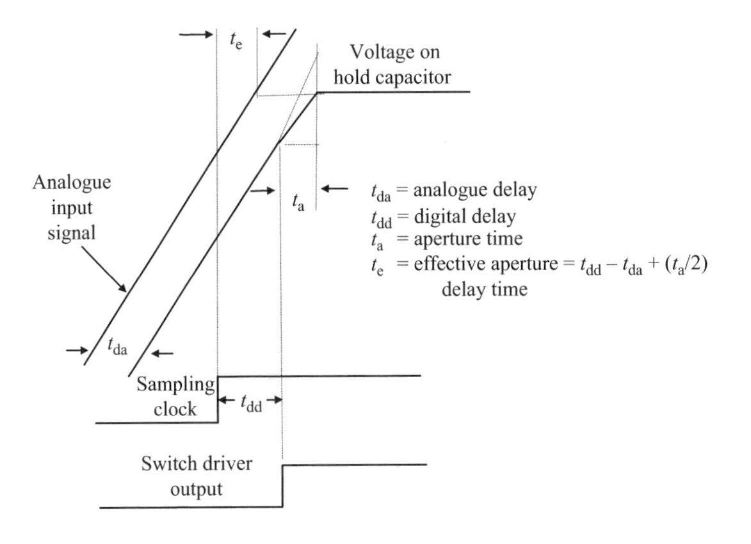

Figure 3.15 SHA waveforms

over the aperture time of the switch as shown in Figure 3.15. The first order model assumes that the final value of voltage on the hold capacitor is approximately equal to the average value of the signal applied to the switch over the interval during which the switch changes from a low to high impedance (t_a).

The model shows that the finite time required for the switch to open (t_a) is equivalent to introducing a small delay in the sampling clock driving the SHA. This delay is constant and may either be positive or negative. It is called the effective aperture delay time, or simply aperture delay (t_e), and is defined as the time difference between the analogue propagation delay of the front-end buffer (t_{da}) and the switch digital delay (t_{dd}) plus one-half of the aperture time ($t_a/2$). The effective aperture delay time is usually positive, but may be negative if the sum of one-half of the aperture time ($t_a/2$) and the switch digital delay (t_{dd}) is less than the propagation delay through the input buffer (t_{da}). The aperture delay specification thus establishes when the input signal actually is sampled with respect to the sampling clock edge.

Aperture delay time can be measured by applying a bipolar sinewave signal to the SHA and adjusting the synchronous sampling clock delay such that the output of the SHA is zero during the hold time. The relative delay between the input sampling clock edge and the actual zero-crossing of the input sinewave is the aperture delay time (see Figure 3.16).

Aperture delay produces no errors, but acts as a fixed delay in either the sampling clock input or the analogue input (depending on its sign). If there is sample-to-sample variation in aperture delay (aperture jitter) then a corresponding voltage error is produced, as shown in Figure 3.17. This sample-to-sample variation in the instant the switch opens is called aperture uncertainty, or aperture jitter and is usually measured in

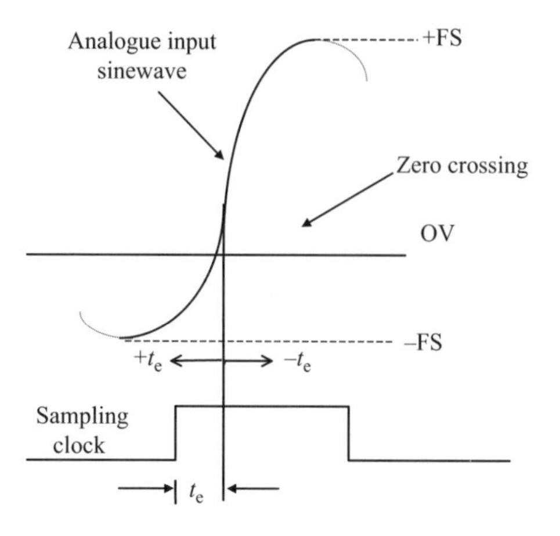

Figure 3.16 Measuring the effective aperture delay time

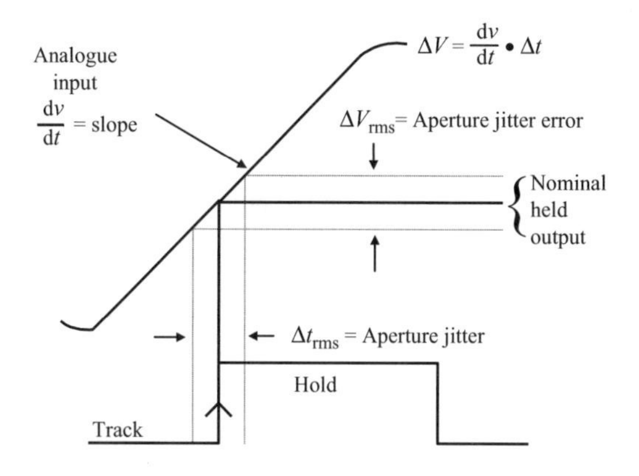

Figure 3.17 Effects of aperture jitter on SHA output

r.m.s. picoseconds. The amplitude of the associated output error is related to the rate-of-change of the analogue input. For any given value of aperture jitter, the aperture jitter error increases as the input dv/dt increases.

Measuring aperture jitter error in an SHA requires a jitter-free sampling clock and analogue input signal source, because jitter (or phase noise) on either signal cannot be distinguished from the SHA aperture jitter itself – the effects are the same. In fact, the largest source of timing jitter errors in a system most often is external to the SHA (or the ADC if it is a sampling one) and caused by noisy or unstable clocks, improper signal routing, and lack of attention to good grounding and decoupling techniques. SHA aperture jitter is generally less than 50 ps r.m.s., and less than 5 ps r.m.s. in high speed devices.

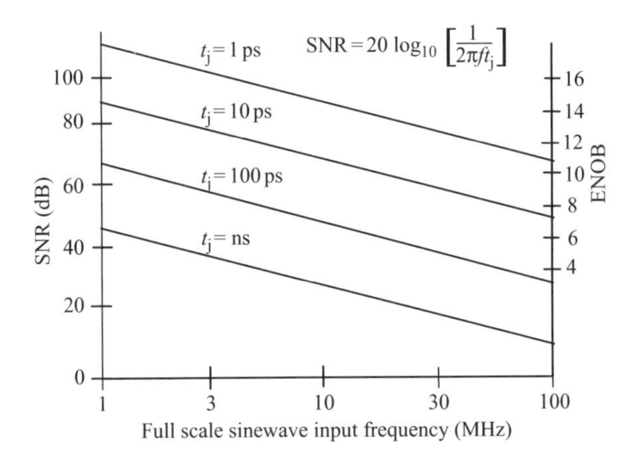

Figure 3.18 Effects of sampling clock jitter on the SNR (reproduced by permission of Analog Devices, Inc.)

Figure 3.18 shows the effects of total sampling clock jitter on the signal-to-noise ratio (SNR) of a sampled data system. The total r.m.s. jitter will be composed of a number of components, the actual SHA aperture jitter often being the least of them.

3.4.1.4 Hold mode droop

During the hold mode, there are errors due to imperfections in the hold capacitor, switch, and output amplifier. If a leakage current flows in or out of the hold capacitor, it will slowly charge or discharge and its voltage will change, an effect known as *droop* in the SHA output, expressed in V μs^{-1}. Droop can be caused by leakage across a dirty PCB if an external capacitor is used or by a leaky capacitor, but most commonly results from leakage current in semiconductor switches and the bias current of the output buffer amplifier. An acceptable value of droop is found when the output of an SHA does not change by more than 1/2 LSB during the conversion time of the ADC it is driving (see Figure 3.19).

Droop can be reduced by increasing the value of the hold capacitor, but this will increase acquisition time and reduce bandwidth in the track mode. Even quite small leakage currents can cause troublesome droop when SHAs use small hold capacitors. Leakage currents in PCBs may be minimised by the intelligent use of guard rings. Details of planning a guard ring are discussed in chapter 8 of Reference 3.

3.4.1.5 Dielectric absorption

Hold capacitors for SHAs must have low leakage, but another characteristic, which is equally important, is low dielectric absorption. If a capacitor is charged, discharged, and then left open circuit, it will recover some of its charge. The phenomenon, known as dielectric absorption, can seriously degrade the performance of an SHA, since it causes the remains of a previous sample to contaminate a new one, and may introduce

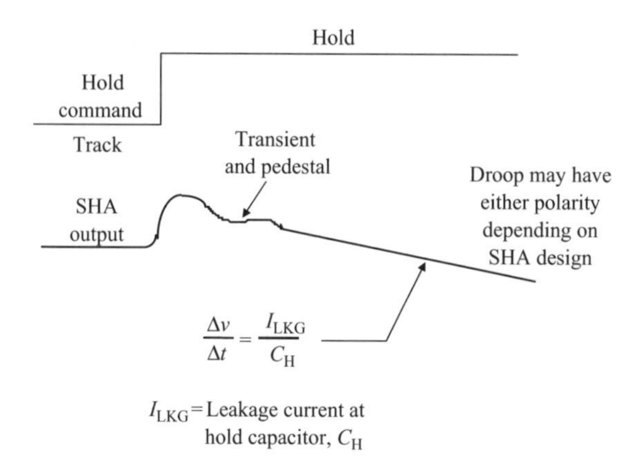

Figure 3.19 Hold mode droop

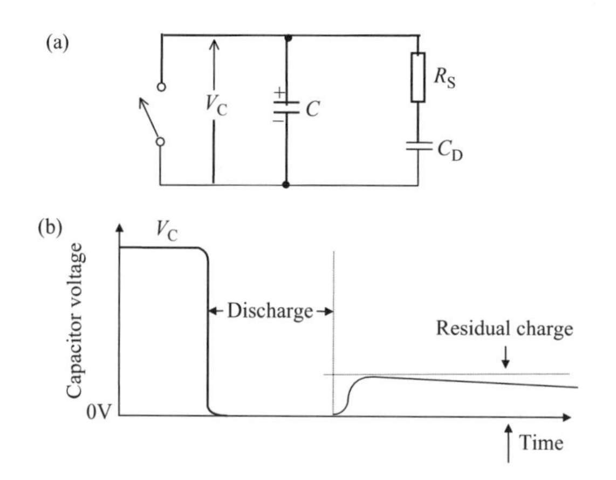

Figure 3.20 Dielectric absorption: (a) model; (b) waveform

random errors of tens or even hundreds of millivolts (see Figure 3.20). After discharge, C_D and R_S in the circuit could cause the residual charge.

Different capacitor materials have differing amounts of dielectric absorption – electrolytic capacitors are dreadful (and their leakage is high), and some high-K ceramic types are bad, whereas mica, polystyrene and polypropylene generally are good. Unfortunately, dielectric absorption varies from batch to batch, and even occasional batches of polystyrene and polypropylene capacitors may be affected. Measuring hold mode distortion is discussed in chapter 8 of Reference 3.

3.4.1.6 Hold-to-track transition specification

When the SHA switches from hold to track, it must reacquire the input signal (which may have made a full scale transition during the hold mode). Acquisition time is the interval of time required for the SHA to reacquire the signal to the desired accuracy when switching from hold to track. The interval starts at the 50 per cent point of the sampling clock edge and ends when the SHA output voltage falls within the specified error band (usually 0.1 per cent and 0.01 per cent times are given). Some SHAs also specify acquisition time with respect to the voltage on the hold capacitor, neglecting the delay and settling time of the output buffer. The hold capacitor acquisition time specification is applicable in high speed applications, where the maximum possible time must be allocated for the hold mode. The output buffer settling time must, of course, be significantly smaller than the hold time.

3.5 SHA architectures

There are numerous SHA architectures and we will examine a few of the most popular ones. For a more detailed discussion on SHA architectures, see Reference 6.

3.5.1 Open loop architecture

The simplest SHA architecture is shown in Figure 3.21. The input signal is buffered by an amplifier and applied to the switch. The input buffer may either be open or closed loop and may or may not provide gain. The switch can be CMOS, FET, or bipolar (using diodes or transistors), controlled by the switch driver circuit. The signal on the hold capacitor is buffered by an output amplifier. This architecture is sometimes referred to as *open loop* because the switch is not inside a feedback loop. Notice that the entire signal voltage is applied to the switch; therefore, it must have excellent common mode characteristics.

3.5.2 Open loop diode bridge SHA

Semiconductor diodes exhibit small on-resistance, large off-resistance, high speed switching, and thus potential for the switching function in sampling circuits. A simplified diagram of a typical diode switch is shown in Figure 3.22(a). Here, four

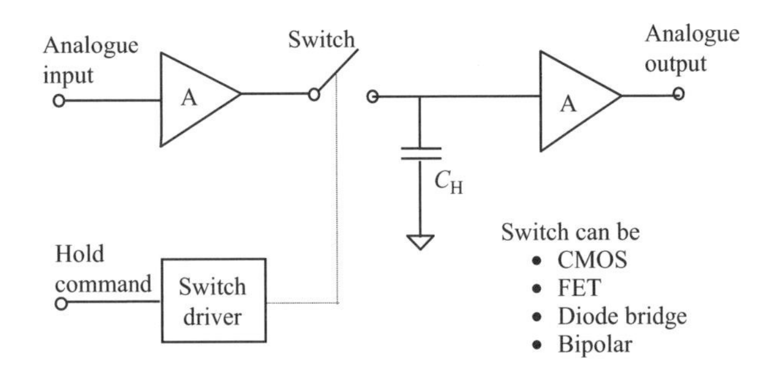

Figure 3.21 Open loop SHA architecture

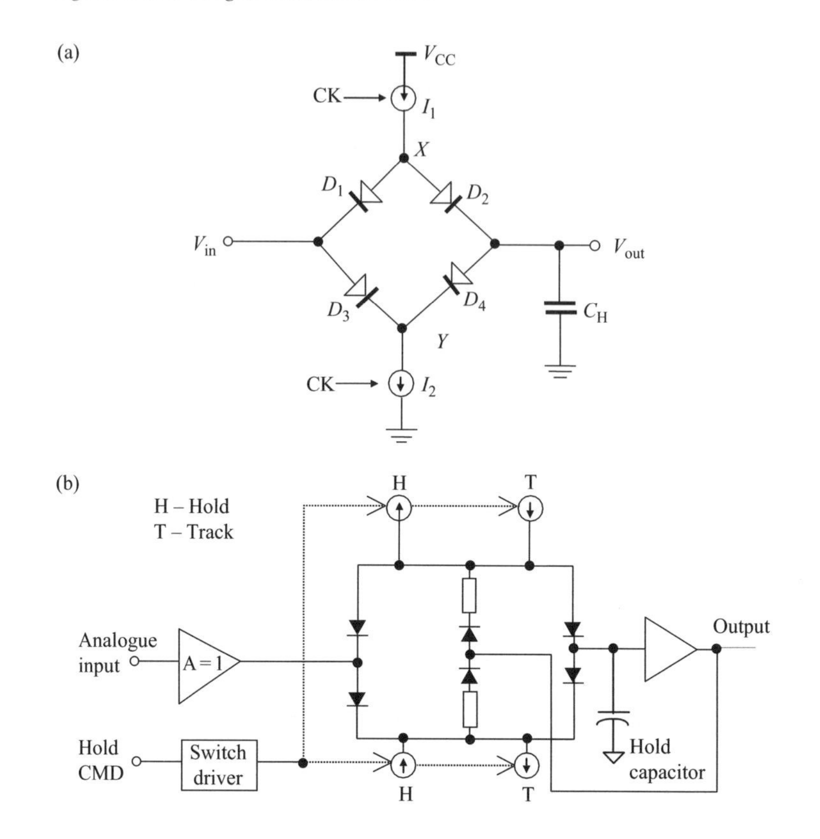

Figure 3.22 Diode bridge SHA: (a) basic diode bride; (b) implementation (reproduced by permission of Analog Devices, Inc.)

diodes form a bridge that provides a low impedance path from V_{in} to V_{out} when current sources I_1 and I_2 are on and (in the ideal case) isolates V_{out} from V_{in} when I_1 and I_2 are off. Nominally, $I_1 = I_2 = I$. Implementation is shown in Figure 3.22(b).

3.5.3 Closed loop architecture

The SHA circuit shown in Figure 3.23 represents a classical closed loop design and is used in many CMOS sampling ADCs. Because the switches always operate at virtual ground, there is no common mode signal across them. Switch S_2 (see also Figure 3.24) is required to maintain a constant input impedance and prevent the input signal from coupling to the output during the hold time. In the track mode, the transfer characteristic of the SHA is determined by the op amp, and the switches introduce no d.c. errors because they are within the feedback loop. The effects of charge injection can be minimised by using the differential switching techniques shown in Figure 3.24.

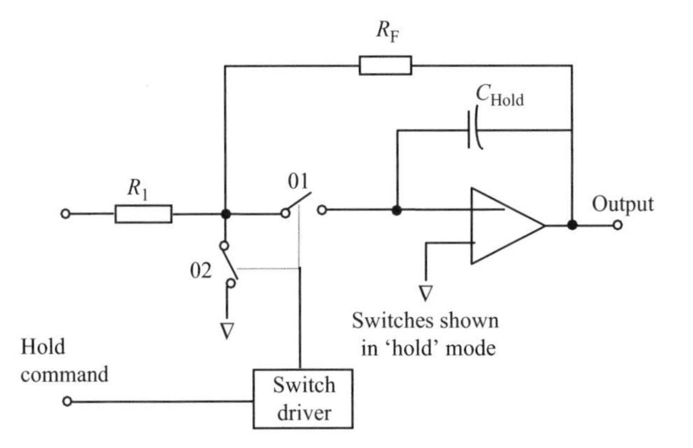

Figure 3.23 Closed loop SHA

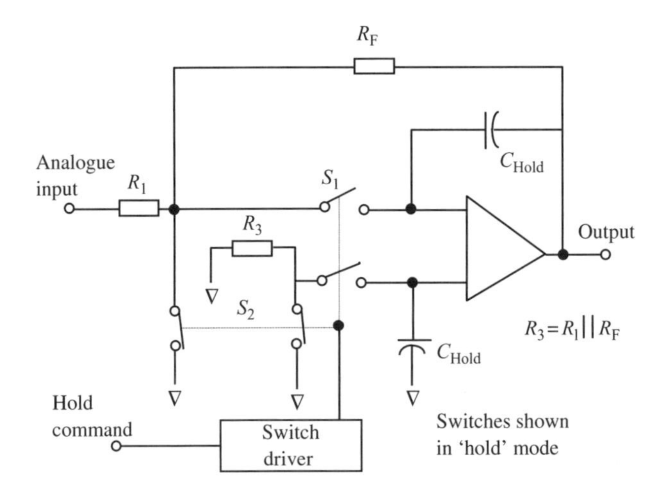

Figure 3.24 Differential switching for charge injection

3.6 ADC architectures

During the late 1980s and the 1990s, many architectures for A/D conversion were implemented in monolithic form. Manufacturing process improvements achieved by mixed signal product manufacture have led to this unprecedented development, which was fuelled by the demand from the product and system designers.

Most common ADC architectures in monolithic form are successive approximation, flash, integrating, pipeline, half-flash (or subranging), two step, interpolative and folding, sigma-delta (Σ-Δ). The following sections provide the basic operational and design details of these techniques.

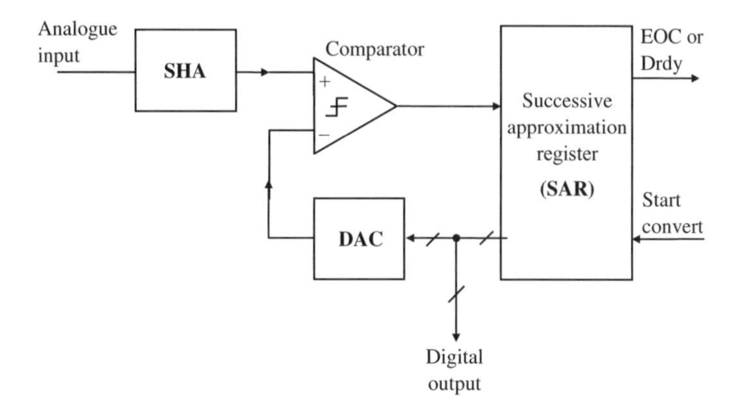

Figure 3.25 Block diagram of successive approximation ADCs

While Σ-Δ, successive approximation and integrating types could give very high resolution at lower speeds, flash architecture is the fastest but with high power consumption. However, recent architecture breakthroughs have allowed designers to achieve a higher conversion rate at low power consumption with integral track and hold circuitry on a chip [11]. The AD9054 from Analog Devices is an example.

3.6.1 Successive approximation ADCs

The successive approximation (SAR) ADC architecture has been used for decades and still is a popular and cost effective form of converter for sampling frequencies up to a few MSPS. A simplified block diagram of an SAR ADC is shown in Figure 3.25. On the 'Start convert' command, all the bits of the successive approximation register are reset to 0 except the most significant bit (MSB), which is set to 1. Bit 1 is tested in the following manner. If the ADC output is greater than the analogue input, the MSB is reset, otherwise it is left set. The next most significant bit then is tested by setting it to 1. If the digital/analogue converter (DAC) output is greater than the analogue input, this bit is reset; otherwise it is left set. The process is repeated with each bit in turn. When all the bits have been set, tested, and reset or not as appropriate, the contents of the SAR correspond to the digital value of the analogue input, and the conversion is complete.

An N-bit conversion takes N steps. On superficial examination, a 16-bit converter would have a conversion time twice as long as an 8-bit converter, but this is not the case. In an 8-bit converter, the DAC must settle to 8-bit accuracy before the bit decision is made, whereas in a 16-bit converter, it must settle to 16-bit accuracy, which takes a lot longer. In practice, 8-bit successive approximation ADCs can convert in a few hundred nanoseconds, while 16-bit converters generally take several microseconds.

The classic SAR ADC is only a quantiser – no sampling takes place – and for an accurate conversion, the input must remain constant for the entire conversion period. Most modern SAR ADCs are sampling types and have an internal sample and hold so that they can process a.c. signals. They are specified for both a.c. and d.c. applications. A SHA is required in an SAR ADC because the signal must remain constant during the entire N-bit conversion cycle.

The accuracy of a SAR ADC depends primarily on the accuracy (differential and integral linearity, gain, and offset) of the internal DAC. Until recently, this accuracy was achieved using laser-trimmed thin-film resistors. Modern SAR ADCs utilise CMOS switched capacitor charge redistribution DACs. This type of DAC depends on the accurate ratio matching and stability of on-chip capacitors rather than thin film resistors. For resolutions greater than 12 bits, on-chip auto-calibration techniques, using an additional calibration DAC and the accompanying logic, can accomplish the same thing as thin film, laser trimmed resistors, at much less cost. Therefore, the entire ADC can be made on a standard submicron CMOS process.

The successive approximation ADC has a very simple structure, low power, and reasonably fast conversion times (<1 MSPS). It is probably most widely used ADC architecture and will continue to be used for medium speed and medium resolution applications.

Current 12-bit SAR ADCs achieve sampling rates up to about 1 MSPS, and 16-bit ones up to about 300 kSPS. Examples of typical state of the art SAR ADCs are the AD7892 (12 bits at 600 kSPS), the AD976/977 (16 bits at 100 kSPS), and the AD7882 (16 bits at 300 kSPS).

3.6.2 Flash converter

Flash ADCs (sometimes called *parallel* ADCs) are the fastest type of ADC and use large numbers of comparators. An N-bit flash ADC consists of 2^N resistors and $2^N - 1$ comparators, arranged as in Figure 3.26. Each comparator has a reference voltage 1 least significant bit (LSB) higher than that of the one below it in the chain. For a given input voltage, all the comparators below a certain point will have their input voltage larger than their reference voltage and a 1 logic output, and all the comparators above that point will have a reference voltage larger than the input voltage and a 0 logic output. The $2^N - 1$ comparator output therefore behaves like a mercury thermometer, and the output code at this point is sometimes called a *thermometer code*. Since $2^N - 1$ data outputs are not really practical, these are processed by an encoder to an N-bit binary output.

The input signal is applied to all the comparators at once, so the thermometer output is delayed by only one comparator delay from the input and the encoder N-bit output by only a few gate delays on top of that, so the process is very fast. However, the architecture uses large numbers of resistors and comparators and is limited to low resolutions; if it is to be fast, each comparator must run at relatively high power levels. Hence, the problems of flash ADCs include limited resolution, high power dissipation because of the large number of high speed comparators (especially at sampling rates greater than 50 MSPS), and relatively large (and therefore expensive) chip sizes. In addition, the resistance of the reference resistor chain must be kept low to supply adequate bias current to the fast comparators, so the voltage reference has to source quite large currents (>10 mA).

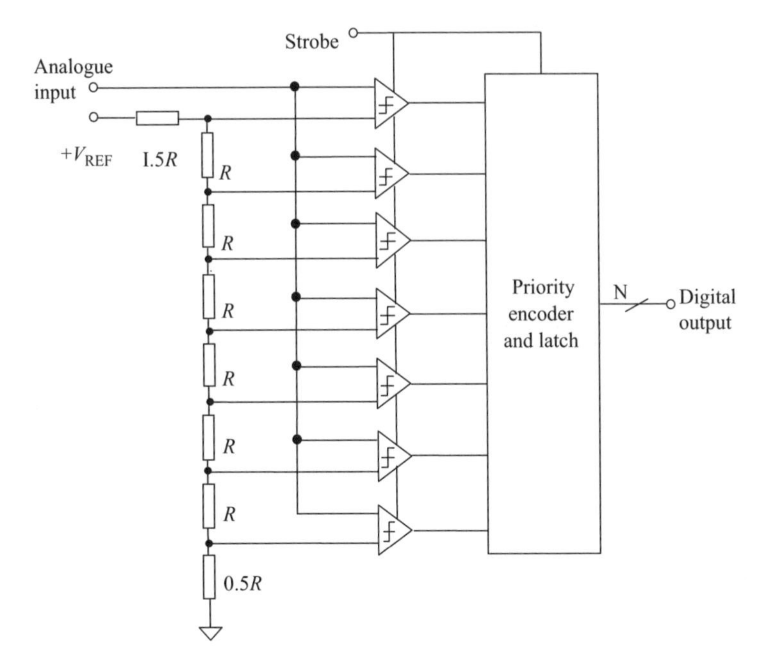

Figure 3.26 Flash or parallel ADC block diagram

In practice, flash converters are available up to 10 bits, but more commonly they have 8 bits of resolution. Their maximum sampling rate can be as high as 500 MSPS, and input full power bandwidths are in excess of 300 MHz.

However, as mentioned earlier, full power bandwidths are not necessarily full resolution bandwidths. Ideally, the comparators in a flash converter are well matched both for d.c. and a.c. characteristics. Because the strobe is applied to all the comparators simultaneously, the flash converter is inherently a sampling converter. In practice, delay variations between the comparators and other a.c. mismatches cause a degradation in ENOB at high input frequencies. This is because the inputs are slewing at a rate comparable to the comparator conversion time.

The input to a flash ADC is applied in parallel to a large number of comparators. Each has a voltage-variable junction capacitance, and this signal-dependent capacitance also result in all flash ADCs having reduced ENOB and higher distortion at high input frequencies. For more details see Reference 2.

3.6.3 *Integrating ADCs*

The integrating ADC is a very popular architecture in applications where a very slow conversion rate is acceptable. A classic example is the digital multimeter.

All the converters discussed so far can digitise analogue inputs at speeds of at least 10 kSPS. A typical integrating converter is slower than these high speed converters.

Useful for precisely measuring slowly varying signals, the integrating converter finds applications in low frequency and d.c. measurement.

Integrating converters are based on an indirect conversion method. Here the analogue input voltage is converted into a time period and later to a digital number using a counter. The integration eliminates the need for a sample/hold (S/H) circuit to 'capture' the input signal during the measurement period. The two common variations of the integrating converter are the dual slope type and the charge balance or multislope type. The dual slope technique is very popular among instrument manufacturers because of its simplicity, low price and better noise rejection. The multislope technique is an improvement on the dual slope method.

Figure 3.27(a) shows a typical integrating converter. It consists of an analogue integrator, a comparator, a counter, a clock and control logic. Figure 3.27(b) shows the circuit's charge (T_1) and discharge (T_2) waveforms. The conversion is started by closing the switch and thereby connecting the capacitor C to the unknown input voltage V_{in}, through the resistor R. This results in a linear ramp at the integrator output for a fixed period T_1, controlled by the counter. The control circuit then switches the integrator input to the known reference voltage, V_{ref}, and the capacitor discharges

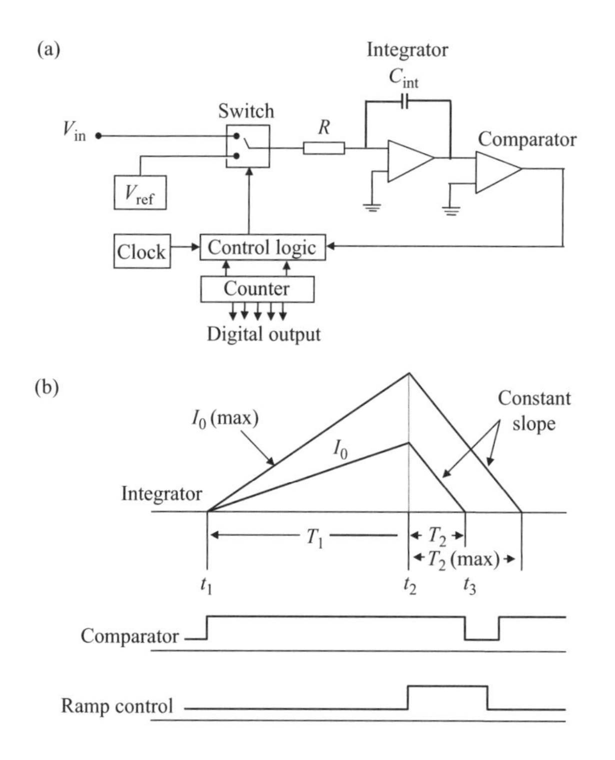

Figure 3.27 Diagram of an integrating type ADC: (a) block diagram; (b) timing diagram

until the comparator detects that the integrator has reached the original starting point. The counter measures the amount of time taken for the capacitor to discharge.

Because the values of the resistor, the integrating capacitor and the frequency of the clock remain the same for both the charge and discharge cycles, the ratio of the charge time to the discharge time is equal to the ratio of the reference voltage to the unknown input voltage. The absolute values of the resistor, capacitor and the clock frequency therefore do not affect the conversion accuracy. Furthermore, any noise on the input signal is integrated over the entire sampling period which imparts a high level of noise rejection to the converter. By making the signal integration period an integral multiple of the line frequency period, the user can obtain excellent line frequency noise rejection.

A charge balance integrating converter incorporates many of the elements as the dual slope converter, but uses a free running integrator in a feedback loop. The converter continuously attempts to null its input by subtracting precise charge packets when the accumulated charge exceeds a reference value. The frequency of the charge packets (the number of packets per second) the converter needs to balance the input is proportional to that input. Clock controlled synchronous logic delivers a serial output that a counter converts to a digital word in the circuit. Integrating converters in monolithic form typically are used in digital voltmeters due to their high resolution properties. Hybrid integrating converters with 22-bit resolutions were introduced to the market in the late 1980s. It is therefore possible to expect higher resolutions in the monolithic market as well. There could be very many variations of this technique as applied to digital multimeters; Reference 12 is suggested for details.

3.6.4 Pipeline architectures

The concept of a pipeline, often used in digital circuits, can be applied in the analogue domain to achieve higher speed where several operations must be performed serially. Figure 3.28 shows a general (analogue or digital) pipelined system. Here, each stage carries out an operation on a sample, provides the output for the following sampler, and, once that sampler has acquired the data, begins the same operation on the next sample. Thus, at any given time, all the stages are processing different samples concurrently; and hence the throughput rate depends only on the speed of each stage and the acquisition time of the next sampler.

To arrive at a simple example of analogue pipelining, consider a two step ADC where four operations (coarse A/D conversion, interstage D/A conversion, subtraction, and fine A/D conversion) must be performed serially. As such, the ADC cannot begin to process the next sample until all four operations are finished. Now, suppose an SHA is interposed between the subtractor and the fine stage, as shown in Figure 3.29, so that the residue is stored before fine conversion begins. Thus, the front-end SHA,

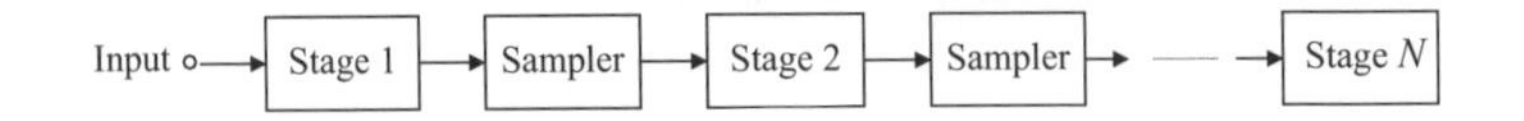

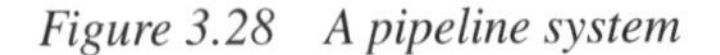
Figure 3.28 A pipeline system

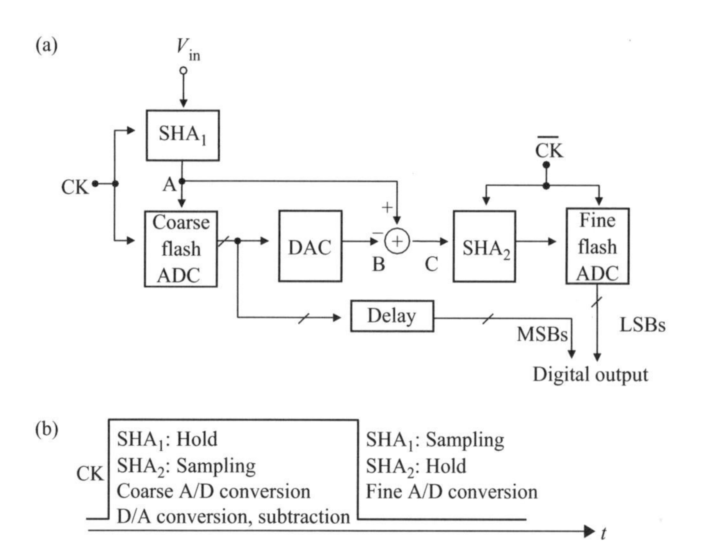

Figure 3.29 A two-step ADC pipeline: (a) block diagram; (b) clock waveform and related activities

the coarse ADC, the interstage DAC, and the subtractor can start processing the next sample while the fine ADC operates the previous one, allowing potentially faster conversion. More details on pipelined architectures can be found in Reference 8.

3.6.5 Half-flash ADCs

Although it is not practical to make them with high resolution, flash ADCs are often used as subsystems in 'subranging' ADCs (sometimes known as 'half flash ADCs'), which are capable of much higher resolutions (up to 16 bits).

A block diagram of an 8-bit subranging ADC based on two 4-bit flash converters is shown in Figure 3.30. Although 8-bit flash converters are readily available at high sampling rates, this sample will be used to illustrate the theory. The conversion process is done in two steps. The four most significant bits (MSBs) are digitised by the first flash (to better than 8-bits accuracy), and the 4-bit binary output is applied to 4-bit DAC (again, better than 8-bit accuracy). The DAC output is subtracted from the held analogue input, and the resulting residue signal is amplified and applied to the second 4-bit flash. The outputs of the two flash converters are combined into a single 8-bit binary output word. If the residue signal range does not exactly fill the range of the second flash converter, non-linearities and perhaps missing codes[2] will result.

[2] A converter must be able to correspond all possible digital outputs to analogue input. If it is unable to do so (owing to excessive DNL) it is said to have missing codes.

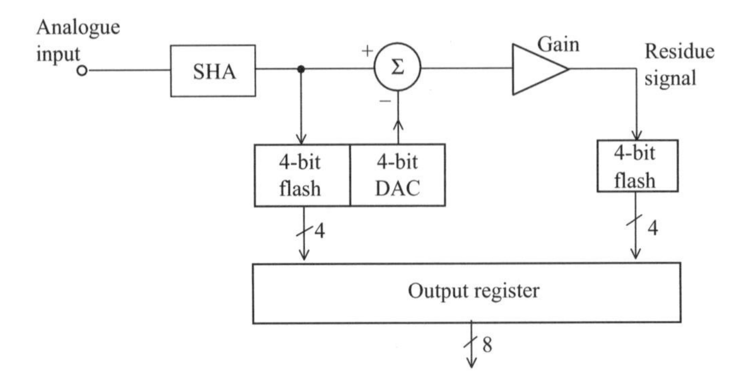

Figure 3.30 The 8-bit subranging ADC

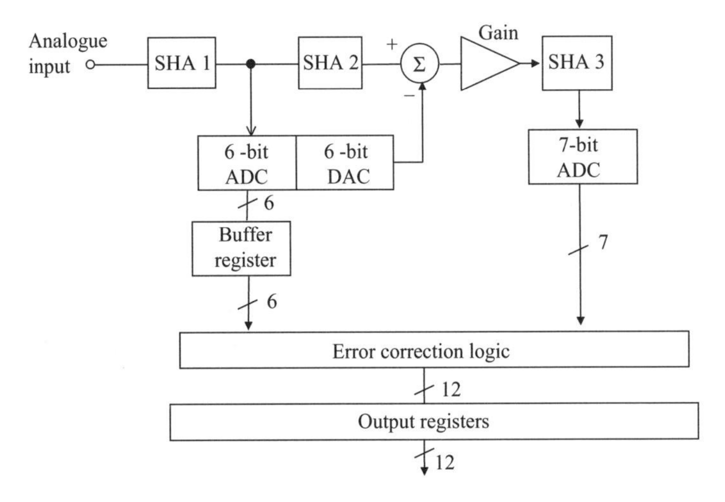

Figure 3.31 Pipelined subranging ADC with digital error correction (reproduced by permission of Analog Devices, Inc.)

Modern subranging ADCs use a technique called *digital* correction to eliminate problems associated with the architecture of Figure 3.30. A simplified block diagram of a 12-bit digitally corrected subranging ADC is shown in Figure 3.31. An example of such a practical ADC is the AD9042 from Analog Devices, a 12-bit, 41 MSPS device. Key specifications of AD9042 are given in Table 3.3.

Note that a 6-bit and 7-bit ADC have been used to achieve an overall 12-bit output. These are not flash ADCs, but utilise a magnitude-amplifier (MagAmp^{TM}) architecture. (See chapter 4 in Reference 2 for MagAmp^{TM} basics.) If there were no errors in the first stage conversion, the 6-bit 'residue' signal applied to the 7-bit ADC by the summing amplifier would never exceed one-half of the range of the 7-bit ADC. The extra range in the second ADC is used in conjunction with the error

Table 3.3 Key specifications of the AD9042 (reproduced by permission of Analog Devices, Inc.)

Parameter	Value
Input range	1 V peak-to-peak, $V_{cm} = +2.4$ V
Input impedance	250 Ω to V_{cm}
Effective input noise	0.33 LSBs r.m.s.
SFDR at 20 MHz input	80 dB minimum
SINAD at 20 MHz input	67 dB
Digital outputs	TTL compatible
Power supply	Single +5 V
Power dissipation	595 mW
Fabrication	High speed dielectrically isolated complementary bipolar process

correction logic (usually just a full adder) to correct the output data for most of the errors inherent in the traditional uncorrected subranging converter architecture. It is important to note that the 6-bit DAC must be better than 12-bit accurate, because the digital error correction does not correct for DAC errors. In practice, 'thermometer' or 'fully decoded' DACs using one current switch per level (63 switches in the case of a 6-bit DAC) are often used instead of a 'binary' DAC to ensure excellent differential and integral linearity and minimum switching transients [2].

The second SHA delays the held output of the first SHA while the first stage conversion occurs, thereby maximising throughput. The third SHA 'deglitches' the residue output signal, allowing a full conversion cycle for the 7-bit ADC to make its decision (the 6- and 7-bit ADCs in the AD9042 are bit-serial MagAmp ADCs, which require more settling time than a flash converter).

Additional shift registers in series with the digital outputs of the first stage ADC ensure that its output ultimately is time aligned with the last 7 bits from the second ADC when their outputs are combined in the error correction logic. A pipelined ADC therefore has a specified number of clock cycles of latency, or pipeline delay associated with the output data. The leading edge of the sampling clock (for example N) is used to clock the output register, but the data that appears as a result of that clock edge corresponds to sample $N - L$, where L is the number of clock cycles of latency – in the case of the AD9042, two clock cycles of latency.

The error correction scheme described previously is designed to correct for errors made in the first conversion. Internal ADC gain, offset, and linearity errors are corrected as long as the residue signal falls within the range of the second stage ADC. These errors will not affect the linearity of the overall ADC transfer characteristic. Errors made in the final conversion, however, translate directly as errors in the overall transfer function. Also, linearity errors or gain errors either in the DAC or the residue amplifier will not be corrected and will show up as non-linearities or non-monotonic behaviour in the overall ADC transfer function.

So far, we have considered only two stage subranging ADCs, as these are easiest to analyse. There is no reason to stop at two stages, however. Three-pass and four-pass subranging pipelined ADCs are quite common, and can be made in many different ways, usually with digital error correction. For details see Reference 2.

3.6.6 *Two-step architectures*

The exponential growth of power, die area, and input capacitance of flash converters as a function of resolution makes them impractical for resolutions above 8 bits in general. These resolutions calls for topologies that provide a more relaxed trade-off among the parameters. Two step architectures trade speed for power, area and input capacitance.

In two step ADC, first a coarse analogue estimate of the input is obtained to yield a small voltage range around the input level. Subsequently, the input level is determined with higher precision within this range. Figure 3.32(a) illustrates a two-step architecture consisting of a front-end SHA, a coarse flash ADC stage, a DAC, a subtractor, and a fine flash ADC stage. We describe its operation using the timing diagram shown in the figure 3.32(b).

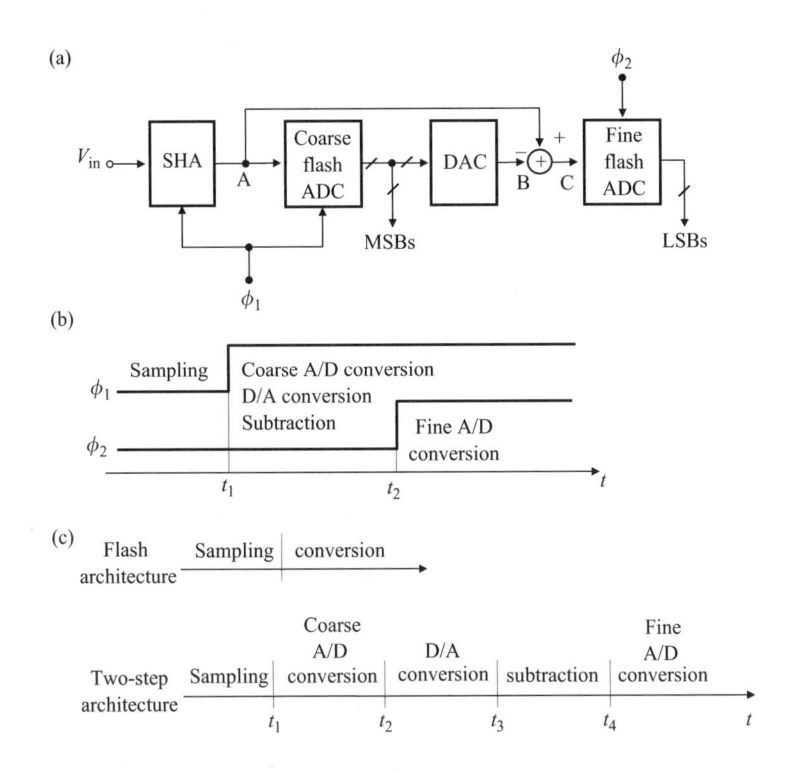

Figure 3.32 Two-step architecture: (a) block diagram; (b) timing; (c) comparison of timing in flash and two step architectures

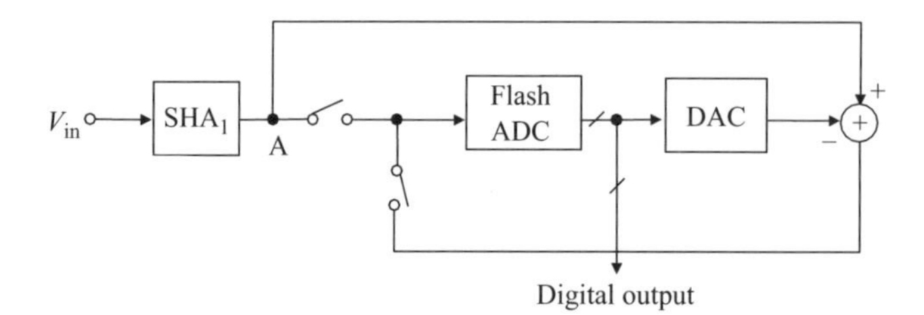

Figure 3.33 Recycling ADC architecture

For $t < t_1$, the SHA tracks the analogue input. At $t = t_1$, the SHA enters the hold mode and the first flash stage is strobed to perform the coarse conversion. The first stage then provides a digital estimate of the signal held by the SHA (V_A), and the DAC converts this estimate to an analogue signal (V_B), which is a coarse approximation of the SHA output. Next, the subtractor generates an output equal to the difference between V_A and V_B (V_C, called the 'residue'), which is subsequently digitised by the fine ADC. Comparison of timing in flash and two step architectures is shown in Figure 3.32(c). For more details see reference [6].

A two step ADC need not employ two separate flash stages to perform the coarse and fine conversions. One stage can be used for both, and such an architecture shown in Figure 3.33 is called 'recycling architecture'.

Here, during the coarse conversion, the flash stage senses the frond-end SHA output, V_A, and generates the coarse digital output. This output then is converted to analogue by the DAC and subtracted from V_A by the subtractor. During fine conversion, the subtractor output is digitised by the flash stage. Note that in this phase, the ADC full scale voltage must be equal to that of the subtractor output. Therefore, for proper fine conversion, either the ADC reference voltage must be reduced or the residue must be amplified.

While reducing area and power dissipation by roughly a factor of two, relative to two-stage ADCs, recycling converters suffer from other limitations. The converter must now employ either low offset comparators (if the subtractor has a gain of one), inevitably slowing down the coarse conversion, or a high gain subtractor, increasing the interstage delay. This is in contrast with two-stage ADCs, where the coarse stage comparators need not have a high resolution and hence can operate faster.

3.6.7 Interpolative and folding architectures

To maintain the one-step nature of the flash type architectures, without adding sample-and-hold circuits to the ADC, several other architectures are available. Among these techniques, interpolation and folding have proved quite beneficial. Earlier these techniques had been applied predominantly to bipolar circuits; recently CMOS devices have entered the market.

As a comprehensive discussion on these techniques is beyond the scope of this chapter only basic approach in the design is discussed here.

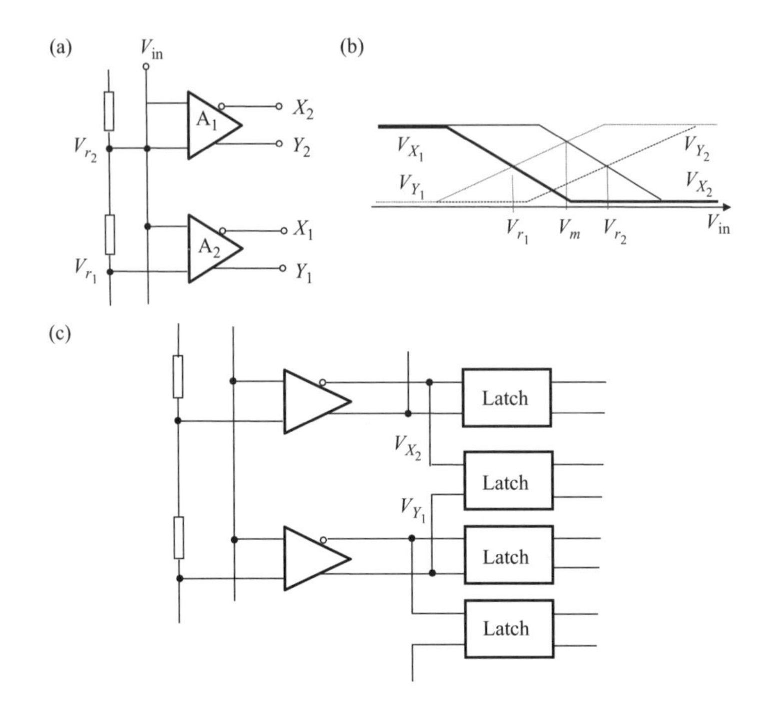

Figure 3.34 Interpolating architecture: (a) basic block; (b) interpolation between output of two amplifiers; (c) interpolation in a flash ADC

3.6.7.1 Interpolating architectures

To reduce the number of preamplifiers at the input of a flash ADC, the difference between the analogue input and each reference voltage can be quantised at the output of each preamplifier. This is possible because of the finite gain – and hence non-zero linear input range – of typical preamplifiers used as the front-end of comparators.

We illustrate this concept in Figure 3.34(a), where preamplifiers A_1 and A_2 compare the analogue input with V_{r1} and V_{r2}, respectively. In Figure 3.34(b), the input/output characteristics of A_1 and A_2 are shown. Assuming zero offset for both preamplifiers, we note that $V_{X_1} = V_{Y_1}$ if $V_{\text{in}} = V_{r_1}$, and $V_{X_2} = V_{Y_2}$ if $V_{\text{in}} = V_{r2}$. More importantly, $V_{X_2} = V_{Y_1}$ if $V_{\text{in}} = V_{\text{m}} = (V_{r_1} + V_{r_2})/2$; that is, the polarity of the difference between V_{X_2} and V_{Y_1} is the same as that of the difference between V_{in} and V_{m}.

The preceding observation indicates that the equivalent resolution of a flash stage can be increased by 'interpolating' between the outputs of preamplifiers. For example, Figure 3.34(b) shows how an additional latch detects the polarity of the difference between the single ended outputs of two adjacent preamplifiers. Note that in contrast with a simple flash stage, this approach halves the number of preamplifiers but maintains the same number of latches.

The interpolation technique of Figure 3.34(c) substantially reduces the input capacitance, power dissipation, and area of flash converters, while preserving the one-step nature of the architecture. This is possible because all the signals arrive at the input of the latches simultaneously and hence can be captured on one clock edge. Since this configuration doubles the effective resolution, we say it has an interpolation factor of two. For further details on this architecture see Reference 6.

3.6.7.2 Folding architectures

Folding architectures have evolved from flash and two-step topologies. Folding architectures perform analogue preprocessing to reduce hardware while maintaining the one step nature of flash architectures.

The basic principle in folding is to generate a residue voltage through analogue preprocessing and subsequently digitise that residue to obtain the least significant bits. The most significant bits can be resolved using a coarse flash stage that operates in parallel with the folding circuit and hence samples the signal at approximately the same time that the residue is sampled. Figure 3.35 depicts the generation of residue in two-step and folding architectures. In a two-step architecture, coarse A/D conversion, interstage D/A conversion, and subtraction must be completed before the proper residue becomes available. In contrast, folding architectures generate the residue 'on the fly' using simple wideband stages.

To illustrate the above principle, we first describe a simple, ideal approach to folding. Consider two amplifiers A_1 and A_2 with the input/output characteristics depicted in Figure 3.36(a). The active region of one amplifier is centered around $(V_{r_2} + V_{r_1})/2$ and that of the other around $(V_{r_3} + V_{r_2})/2$, and $V_{r_3} - V_{r_2} = V_{r_2} - V_{r_1}$. Each amplifier has a gain of 1 in the active region and 0 in the saturation region. If the outputs of the two amplifiers are summed, the 'folding' characteristic of Figure 3.36(b) results, yielding an output equal to $V_{\text{in}} - V_{r_1}$ for $V_{r_1} < V_{\text{in}} < V_{r_2}$ and $(-V_{\text{in}} + V_{r_2} + \Delta)$ for $V_{r_2} < V_{\text{in}} < V_{r_3}$, where Δ is the value of the summed characteristics at $V_{\text{in}} = V_{r_2}$. If V_{r_1}, V_{r_2}, and V_{r_3} are the reference voltages in an ADC, then these two regions can be viewed as the residue characteristics of the ADC for $V_{r_1} < V_{\text{in}} < V_{r_3}$. To understand

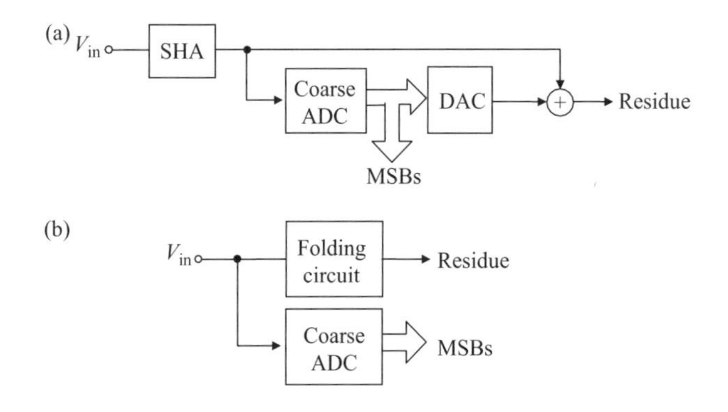

Figure 3.35 Generation of residue: (a) two-step architecture; (b) folding architecture

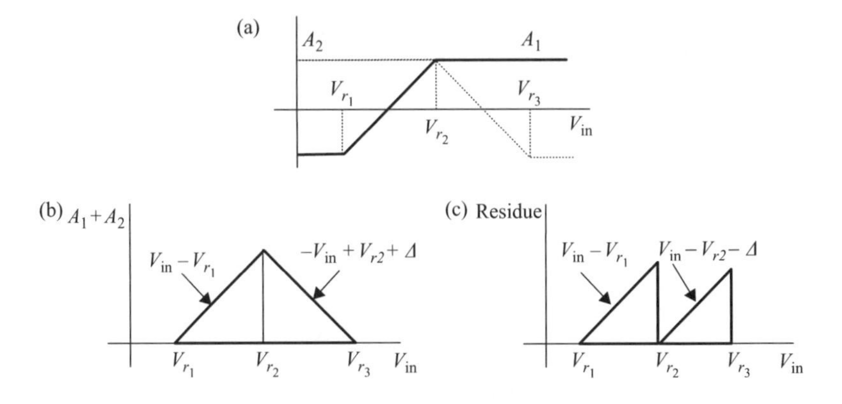

Figure 3.36 The concept of folding: (a) input/output characteristics of two amplifiers; (b) sum of characteristics in (a); (c) residue in two step ADC

why, we compare this characteristic with that of a two step architecture, as shown in Figure 3.36(c). The two characteristics are similar except for a negative sign and a vertical shift in the folding output for $V_{r_2} < V_{in} < V_{r_3}$. Therefore, if the system accounts for the sign reversal and level shift, the folding output can be used as the residue for fine digitisation.

An implementation of folding is shown in Figure 3.37(a). Here, four differential pairs process the difference between V_{in} and $V_{r_1}, \ldots, V_{r_4}$, and their output currents are summed at nodes X and Y. Note that the outputs of adjacent stages are added with opposite polarity; e.g., as V_{in} increases, Q_1 pulls node X low while Q_2 pulls node Y low. Current source I_5 shifts V_Y down by IR. To explain the operation of the circuit, we consider its input/output characteristic, plotted in Figure 3.37(b). For V_{in} well below V_{r_1}, $Q_1 - Q_4$ are off, $Q_5 - Q_8$ are on, I_2 and I_4 flow through R_{C_1}, and I_1, I_3 and I_5 flow through R_{C_2}. As V_{in} exceeds V_{r_1} by several V_T, Q_5 turns off, allowing V_X and V_Y to reach V_{min} and V_{max}, respectively. As V_{in} approaches V_{r_2}, Q_2 begins to turn on and the circuit behaves in a similar manner as before. Considering the differential output, $V_X - V_Y$, we note that the resulting characteristic exhibits folding points at $(V_{r_1} + V_{r_2})/2$, $(V_{r_2} + V_{r_3})/2$, and so forth. As V_{in} goes from below V_{r_1} to above V_{r_4}, the slope of $V_X - V_Y$ changes sign four times; hence we say the circuit has a folding factor of four.

The simplicity and speed of folding circuits have made them quite popular in A/D converters, particularly because they eliminate the need for sample-and-hold amplifiers, D/A converters, and subtractors. Nevertheless, several drawbacks limit their use at higher resolutions [6].

3.6.8 Sigma-delta converters

Sigma-delta analogue-to-digital converters (Σ-Δ ADCs) have been known for nearly thirty years, but only recently has the technology (high density digital VLSI) been developed to manufacture them as inexpensive monolithic integrated circuits. They

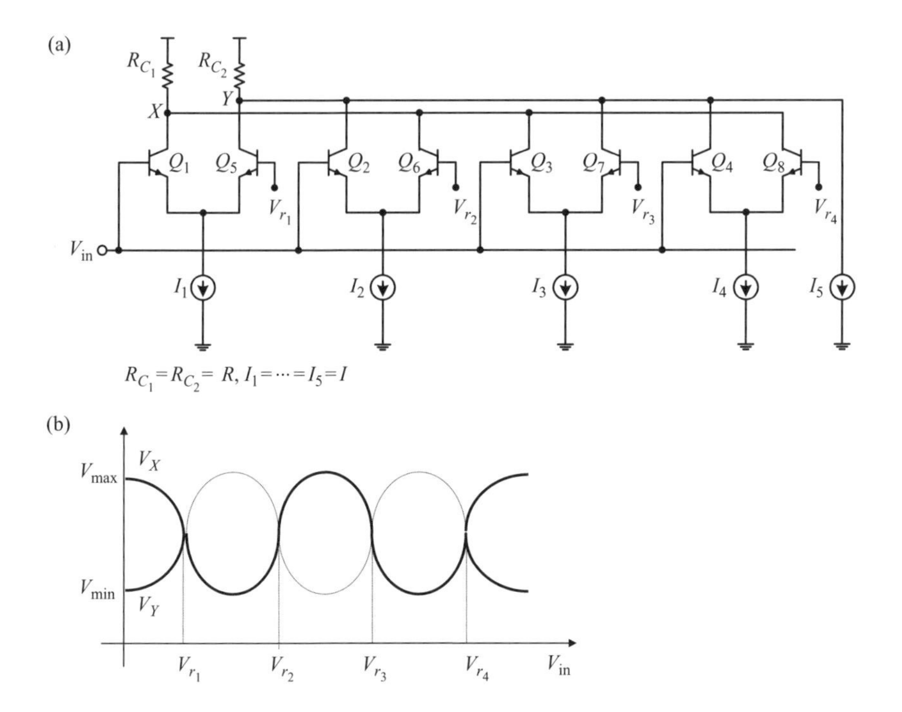

Figure 3.37 Folding circuit and its characteristics: (a) circuit; (b) characteristics

are now used in many applications where a low cost, low bandwidth, high resolution ADC is required.

There have been innumerable descriptions of architecture and theory of Σ-Δ ADCs [13]. A practical monolithic Σ-Δ ADC contains very simple analogue circuit blocks (a comparator, a switch, and one or more integrators and analogue summing circuits), and quite complex digital computational circuitry. This circuitry consists of a digital signal processor (DSP) which acts as a filter (generally, but not invariably, a low pass filter). It is not necessary to know how the filter works to appreciate what it does. To understand how a Σ-Δ ADC works one should be familiar with the concepts of over-sampling, noise shaping, digital filtering and decimation. This subject is described more in Reference 10.

3.6.9 Self-calibration techniques

Integral linearity of data converters usually depends on the matching and linearity of integrated resistors, capacitors, or current sources, and it is typically limited to approximately ten bits with no calibration. For higher resolutions, means must be sought that can reliably correct non-linearity errors. This often is accomplished by either improving the effective matching of individual devices or correcting the overall transfer characteristics. Since high resolution A/D converters typically employ a multistep architecture, they often impose two stringent requirements; small INL in their interstage DACs and precise gain (usually a power of two) in their interstage subtractors/amplifiers. These constraints in turn demand correction for device mismatches if resolutions above 10 bits are required.

ADC calibration techniques can be in two forms: use of analogue processing techniques for correction of non-idealities and digital calibration techniques. A description of these techniques is beyond the scope of this chapter. For details see References 6, 12 and 14.

3.6.10 Figure of merit for ADCs

The demand for lower power dissipating electronic systems has become a challenge to the IC designer, including designers of ADCs. As a result, a 'figure of merit' was devised by the ISSCC[3] Program Committee to compare available and future sampling type ADCs. The figure of merit (FOM) is based on an ADC's power dissipation, its resolution, and its sampling rate. The FOM is derived by dividing the device's power dissipation (in watts) by the product of its resolution (in 2^n bits) and its sampling rate (in hertz). The result is multiplied by 10^{12}. This is expressed by the equation:

$$FOM = \frac{PD}{(R \cdot SR)} 10^{12}, \tag{3.5}$$

where

$$\begin{aligned} PD &= \text{power dissipation (in watts);} \\ R &= \text{resolution (in } 2^N \text{ bits);} \\ SR &= \text{sampling rate (in hertz).} \end{aligned}$$

Therefore, a 12-bit ADC sampling at 1 MHz and dissipating 10 mW has a figure of merit rounded off to 2.5. This figure of merit is expressed in the units of picojoules of energy per unit conversion [pJ/conversion]. For details and a comparison of performance of some monolithic ICs, see Reference 15.

3.7 D/A converters

Digital-to-analogue conversion is an essential function in data processing systems. D/A converters provide an interface between the digital output of signal processors and the analogue world. Moreover, as discussed previously, multistep ADCs employ interstage DACs to reconstruct analogue estimates of the input signal. Each of these applications imposes certain speed, precision, and power dissipation requirements on the DAC, mandating a good understanding of various D/A conversion techniques and their trade-offs.

[3] ISSCC, International Solid State Circuits Conference.

3.7.1 General considerations

A digital-to-analogue converter produces an analogue output A that is proportional to the digital input D:

$$A = \alpha D, \tag{3.6}$$

where α is a proportionality factor. Since D is a dimensionless quantity, α sets both the dimension and the full scale range of A. For example, if α is a current quantity, I_{REF}, then the output can be expressed as

$$A = I_{REF} D. \tag{3.7}$$

In some cases, it is more practical to normalise D with respect to its full scale value, 2^m, where m is the resolution. For example, if α is a voltage quantity, V_{REF},

$$A = V_{REF} \frac{D}{2^m}. \tag{3.8}$$

From (3.7) and (3.8), we can see that, in a D/A converter, each code at the digital input generates a certain multiple or fraction of a reference at the analogue output. In practical monolithic DACs, conversion can be viewed as a reference multiplication or division function, where the reference may be one of the three electrical quantities: voltage, current or charge.

The accuracy of this function determines the linearity of the DAC, while the speed at which each multiple or fraction of the reference can be selected and established at the output gives the conversion rate of the DAC. Figure 3.38 shows the input/output characteristic of an ideal 3-bit D/A converter. The analogue levels generated at the output follow a straight line passing through the origin and the full scale point.

We should mention that, in some applications such as 'companding' (compressing and expanding) DACs, the desired relationship between D and A is non-linear [1], but in this chapter we discuss only 'linear' or 'uniform' DACs; that is, those that ideally behave according to (3.7) or (3.8).

The digital input to a DAC can assume any predefined format but eventually must be of a form easily convertible to analogue. Table 3.4 shows three formats often

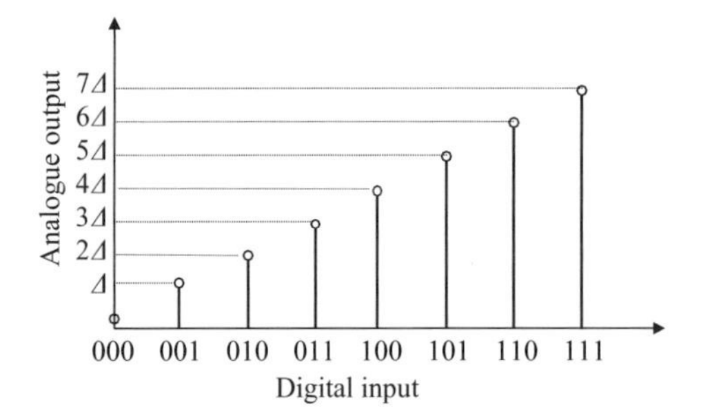

Figure 3.38 Input/output characteristic of an ideal 3-bit DAC

Table 3.4 Binary, thermometer and one-of-n codes

Decimal	0	1	2	3
Binary	00	01	10	11
Thermometer	0	0	0	0
	0	0	0	1
	0	0	1	1
	0	1	1	1
One-of-*n*	0	0	0	0
	0	0	0	1
	0	0	1	0
	0	1	0	0

used in DACs: binary, thermometer, and one-of-*n* codes. The latter two are shown in column form to make visualisation easier.

3.7.2 *Performance parameters and data sheet terminology*

In manufacturers' data books, many terms are used to characterise DACs. The following is a basic guideline only, and the reader should refer to manufacturers' data sheet guidelines for a more application oriented description. Figure 3.39 illustrates some of these metrics, which are listed in Table 3.5.

Among these parameters, DNL and INL are usually determined by the accuracy of reference multiplication or division, settling time and delay are functions of output loading and switching speed, and glitch impulse depends on the D/A converter architecture and design.

3.7.3 *Voltage division*

A given reference voltage V_{REF} can be divided into N equal segments using a ladder composed of N identical resistors $R_1 = R_2 = \cdots = R_N$ (N typically is a power of two) (Figure 3.40(a)). An m bit DAC requires a ladder with 2^m resistors, manifesting in the exponential growth of the number of resistors as a function of resolution.

An important aspect of resistor ladders is the differential and integral non-linearity they introduce when used in D/A converters. These errors result from mismatches in the resistors comprising the ladder.

The DACs most commonly used as examples of simple DAC structures are binary weighted DACs or ladder networks, but although simple in structure, these require quite complex analysis. The simplest structure of all is the Kelvin divider shown in Figure 3.40(b). An N-bit version of this DAC simply consists of 2^N equal resistors in series. The output is taken from the appropriate tap by closing one of the 2^N switches.

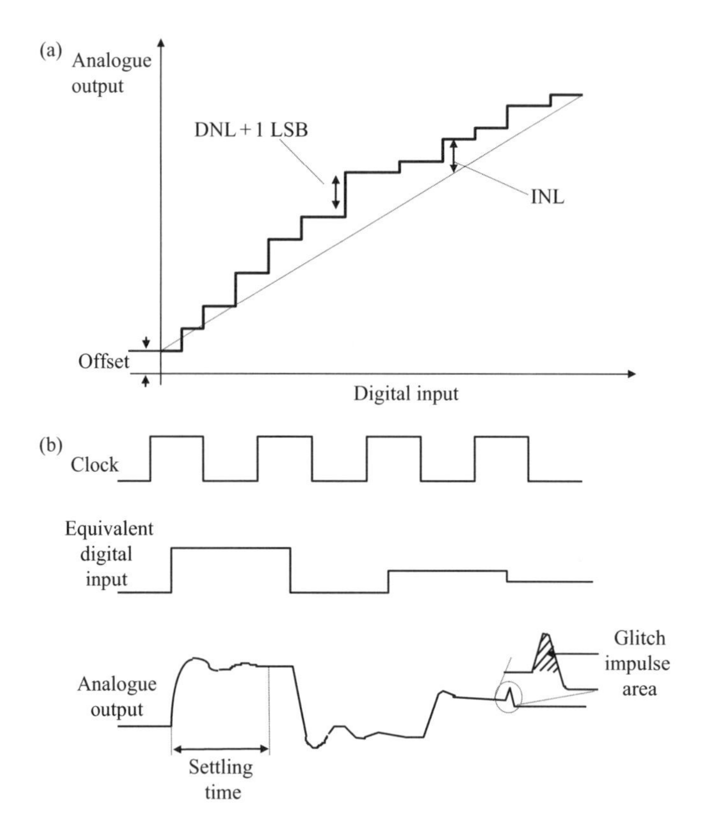

Figure 3.39 Parameters of DACs: (a) static parameters; (b) dynamic parameters

3.7.4 Current division

Instead of using voltage division, current division techniques can be used in DACs. Figure 3.41(a) shows how a reference current I_{REF} can be divided into N equal currents using N identical (bipolar or MOS) transistors. These currents can be combined to provide binary weighting as depicted in Figure 3.41(b) using a 3-bit case as the example. In this simple implementation, an m-bit DAC requires $2^m - 1$ transistors resulting in a large number of devices for $m > 7$.

While conceptually simple, the implementation of Figure 3.41(a) has two drawbacks: the stack of current division transistors on top of I_{REF} limits output voltage range, and I_{REF} must be N times each of the output currents. This requires a high current device for the I_{REF} source transistor. There are techniques for alleviating these problems [6]. DACs that employ current division suffer from three sources of nonlinearity: current source mismatch, finite output impedance of current sources, and voltage dependence of the load resistor that converts the output current to voltage.

Table 3.5 DAC performance parameters

Parameter	Description
Differential non-linearity (DNL)	Maximum deviation in the output step size from the ideal value of one least significant bit (LSB).
Integral non-linearity (INL)	Maximum deviation of the input/output characteristic from a straight line passed through its end points. The difference between the ideal and actual characteristics will be called the INL profile.
Offset	Vertical intercept of the straight line passing through the end points.
Gain error	Deviation of the slope of the line passing through the end points from its ideal value (usually unity).
Settling time	Time required for the output to experience full scale transition and settle within a specified error band around its final value.
Glitch impulse area	Maximum area under any extraneous glitch that appears at the output after the input code changes. This parameter is also called 'glitch energy' in the literature even though it does not have an energy dimension.
Latency	Total delay from the time the digital input changes to the time the analogue output has settled within a specified error band around its final value. Latency may include multiples of the clock period if the digital logic in the DAC is pipelined.
Signal-to- (noise + distortion) ratio (SNDR or SINAD)	Ratio of the signal power to the total noise and harmonic distortion at the output when the input is a (digital) sinusoid.

3.7.5 Charge division

A reference charge, Q_{REF}, can be divided into N equal packets using N identical capacitors configured as in Figure 3.42. In this circuit, before S_1 turns on, C_1 has a charge equal to Q_{REF}, while $C_2, \ldots, C_N$, have no charge. When S_1 turns on, Q_{REF}

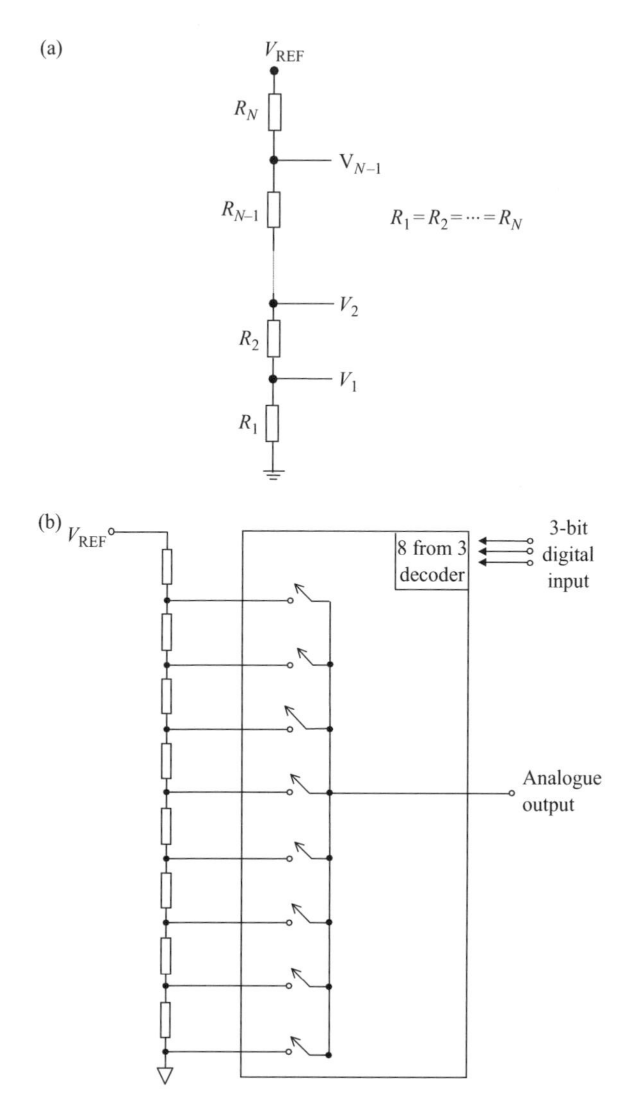

Figure 3.40 DAC using voltage division technique: (a) basic resistor ladder; (b) Kelvin divider (three bit DAC example)

is distributed equally among $C_1, \ldots, C_N$, yielding a charge of Q_{REF}/N on each. Further subdivision can be accomplished by disconnecting one of the capacitors from the array and redistributing its charge among some other capacitors.

While the circuit of Figure 3.42 can operate as a D/A converter if a separate array is employed for each bit of the digital input, the resulting complexity prohibits its use for resolutions above 6 bits. A modified version of this circuit is shown in Figure 3.43(a). Here, identical capacitors $C_1 = \cdots = C_N = C$ share the same top

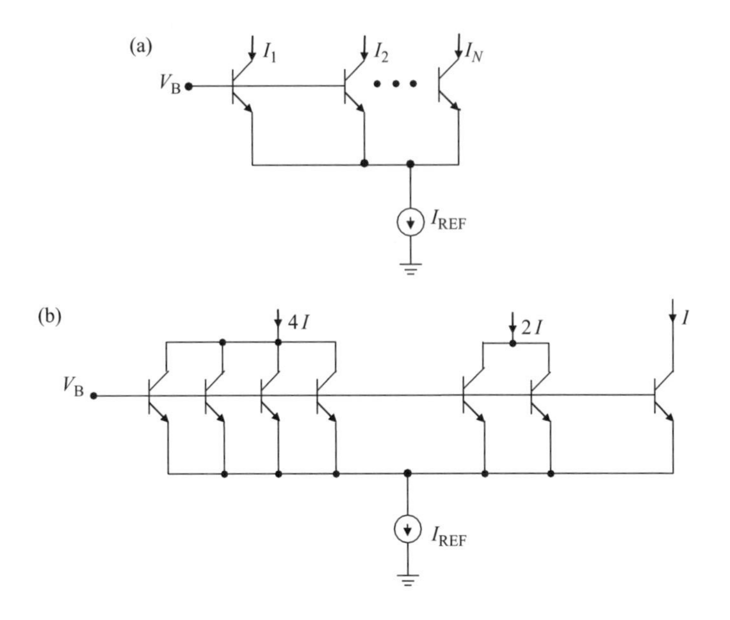

Figure 3.41 Current division: (a) uniform division; (b) binary division

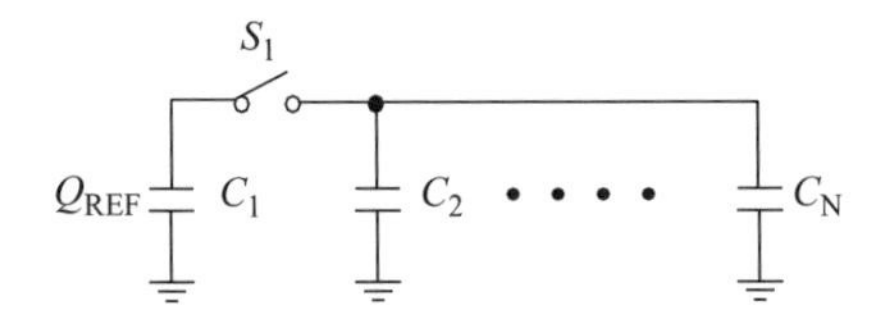

Figure 3.42 Simple charge division

plate, and their bottom plates can be switched from ground to a reference voltage, V_{REF}, according to the input thermometer code. In other words, each capacitor can inject a charge equal to CV_{REF} onto the output node, producing an output voltage proportional to the height of the thermometer code. The circuit operates as follows. First, S_P on the bottom plates of $C_1, \ldots, C_N$ is grounded, discharging the array to zero (Figure 3.43(b)). Next, S_P turns off, and a thermometer code with height j is applied at $D_1, \ldots, D_N$, connecting the bottom plate of $C_1, \ldots, C_j$ to V_{REF} and generating an output equal to jV_{REF}/N (Figure 3.43(c)). This circuit, in the strict sense, is a voltage divider rather than a charge divider. In fact, the expression relating its output voltage to V_{REF} and the value of the capacitors is quite similar to that of resistor ladders. Nonetheless, in considering non-linearity and loading effects it is helpful to remember that the circuit's operation is based on charge injection and redistribution.

The non-linearity of capacitor DACs arises from three sources: capacitor mismatch, capacitor non-linearity, and the non-linearity of the junction capacitance of any switches connected to the output code. For details and implementation of capacitor DACs, see Reference 6.

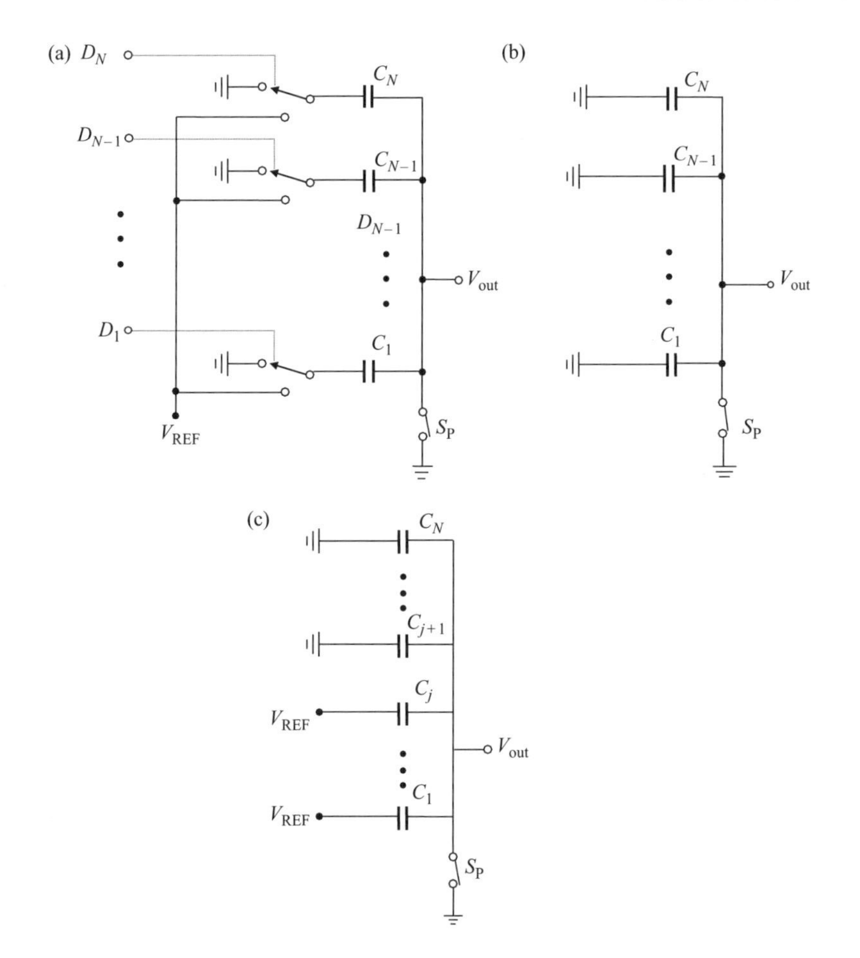

Figure 3.43 Modified charge division: (a) configuration; (b) circuit of (a) in discharge mode; (c) circuit of (a) in evaluate mode

3.7.6 DAC architectures

With the basic principles of D/A conversion explained, we can study this function from an architectural perspective. This section describes D/A converter architectures based on resistor ladders and current steering arrays, with an emphasis on stand alone applications. While capacitor DACs are frequently used in ADCs, they have not been popular as stand alone circuits.

3.7.6.1 Resistor–ladder DAC architectures

The simplicity of resistor–ladder DACs using MOS switches makes these architectures attractive for many applications. Simple ladder networks with simple voltage division as in section 3.7.3 have several drawbacks: they require a large number of resistors and switches (2^m, where m is the resolution) and exhibit a long delay at the output. Consequently alternative ladder topologies have been devised to improve the speed and resolution.

3.7.6.1.1 *Ladder architecture with switched sub-dividers*

In high resolution applications, the number of devices in a DAC can be prohibitively large. It is therefore plausible to decompose the converter into a coarse section and a fine section so that the number of devices becomes proportional to approximately $2^m/2$ rather than 2^m, where m is the overall resolution. Such an architecture is shown in Figure 3.44(a). In this circuit, a primary ladder divides the main reference voltage, generating 2^j equal voltage segments. One of these segments is selected by the j most significant bits of $(k + j) = m$. If $k = j$, the number of devices in this architecture is proportional to $2^m/2$. It is also possible to utilise more than two ladders to further reduce the number of devices at high resolutions.

Figure 3.44(b) depicts a simple implementation of this architecture using MOS switches that are driven by one-of-n codes in both stages [1]. Depending on the environment, these codes are generated from binary or thermometer code inputs. Details and drawbacks of this implementation are discussed in Reference 6.

3.7.6.1.2 *Intermeshed ladder architecture*

Some of the drawbacks of ladder DACs can be alleviated through the use of intermeshed ladder architectures [6]. In these architectures, a primary ladder divides the main reference voltage into equal segments, each of which is subdivided by a separate, fixed secondary ladder. Figure 3.45 illustrates such an arrangement [2], where all the switches are controlled by a one-of-n code.

The intermeshed ladder has several advantages over single ladder or switched ladder architectures. This configuration can have smaller equivalent resistance at each tap than a single ladder DAC having the same resolution, allowing faster recovery. Also, since the secondary ladders do not switch, their loading on the primary ladder is constant and uniform. Furthermore, the DNL resulting from finite on-resistance of switches does not exist here.

3.7.6.2 Current steering architecture

Most high speed D/A converters are based on a current steering architecture. Because these architectures can drive resistive loads directly, they require no high speed amplifiers at the output and hence potentially are faster than other types of DACs. While the high speed switching of bipolar transistors makes them the natural choice for current steering DACs, many designs have been recently reported in CMOS technology as well.

3.7.6.2.1 *R–$2R$ network based architectures*

To realise binary weighting in a current steering DAC, an R–$2R$ ladder can be incorporated to relax device scaling requirements. Figure 3.46(a) illustrates an architecture

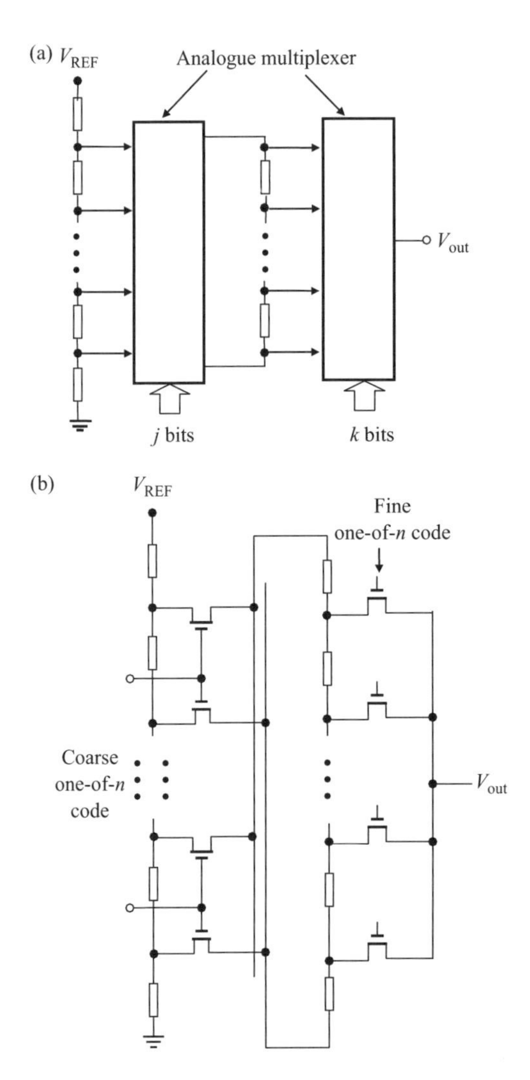

Figure 3.44 Resistor–ladder DAC with a switched sub-divider: (a) block diagram; (b) implementation

that employs an R–$2R$ ladder in the emitter network. A network with an R–$2R$ ladder in collector networks is shown in Figure 3.46(b). For details see References 6 and 16.

3.7.6.3 Other architectures

Other architectures for DACs include segmented current steering versions, multiplying DACs and Σ-Δ types. This chapter does not discuss these, so see References 3, 6 and 16 for further details.

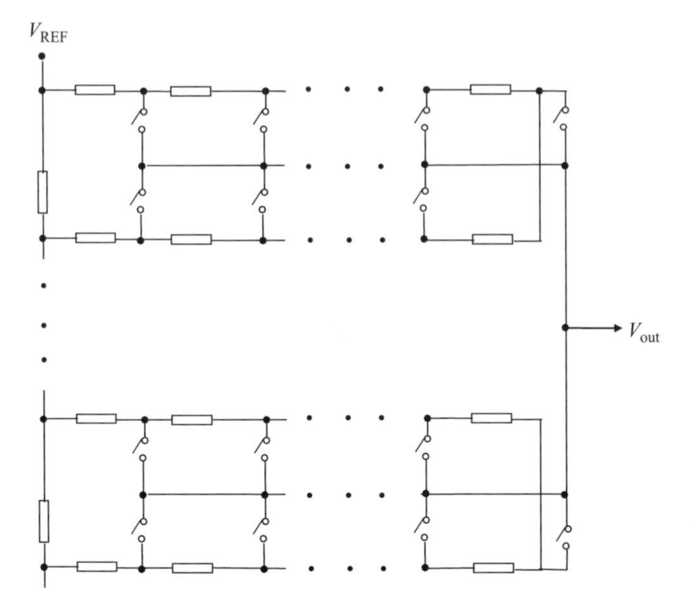

Figure 3.45 Intermeshed resistor–ladder DAC with one level multiplexing

3.8 Data acquisition system interfaces

3.8.1 Signal source and acquisition time

Continued demand for lower power, lower cost systems increases the likelihood that your next mixed signal design will operate from a single 3.3 or 5 V power supply. Doing away with traditional ±15 V analogue power supplies can help meet your power and cost goals, and will also eliminate some of your options.

Most low voltage ADC and DAS (data acquisition system) chips are designed for easy analogue and digital interfaces. The ICs' digital interfaces are generally compatible with popular microcontrollers, and the devices almost always can accept analogue input signals that ranges from ground to the positive supply voltage; the span is set by an internal or external bandgap voltage reference. Virtually all ADCs that operate from 5 V or less are CMOS devices that use arrays of switches and capacitors to perform their conversions. Although the architectural details vary from design to design, the input stage of this type of converter usually includes a switch and a capacitor that present a transient load to the input signal source. The simplified schematic of Figure 3.47 shows how these input stages affect the circuits that drive them.

R_{ON} is not a separate component; it is the on-resistance of the internal analogue switch. Sampling capacitor C_S connects to an internal bias voltage whose value depends on the ADC's architecture. In a sampling ADC, the switch closes once per conversion, during the acquisition (sampling) time. The on-resistance of the sampling switches ranges from about 5 to 10 kΩ in many low resolution successive approximation ADCs to 70 Ω in some multistep or half flash converters. The capacitors can be

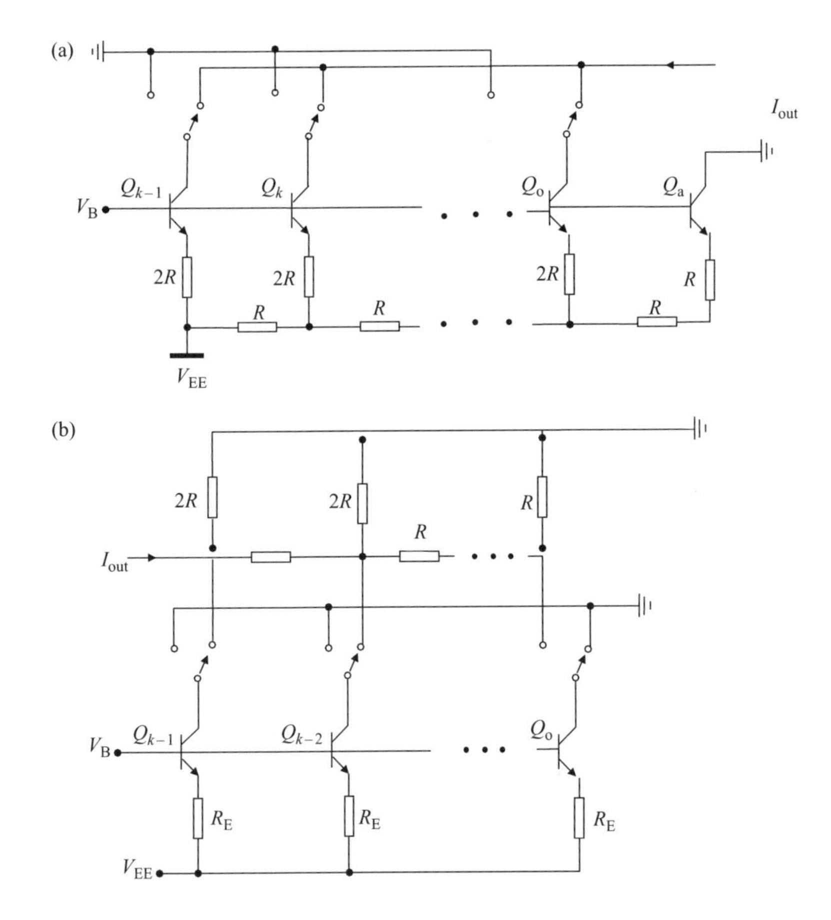

Figure 3.46 Current steering DAC with R–2R ladder: (a) R–2R ladder in the emitter; (b) R–2R ladder in the collector

as small as 10 pF in lower resolution successive approximation converters and 100 pF or more in higher resolution devices.

When the sampling switch closes, the capacitor begins to charge through the switch and source resistance. After a time interval that is usually controlled by counters or timers within the ADC, the switch opens, and the capacitor stops charging. The acquisition time described in Figure 3.10(c) is actually the time during which the switch is closed and the capacitor charges. As long as the source impedance is low enough, the capacitor has time to charge fully during the sampling period, and no conversion errors occur. Most input stages are designed conservatively and can work properly at their rated speeds with a reasonable source resistance (1 kΩ is common). Larger source impedance slows the charging of the sampling capacitor and causes significant errors unless you take steps to avoid them. Figure 3.48 illustrates this. Figure 3.48(a) indicates the case of insufficient acquisition time. Figure 3.48(b) shows

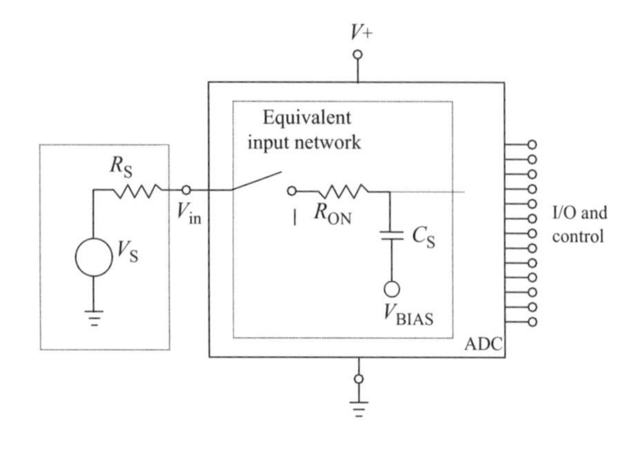

Figure 3.47 Simplified interface between a low voltage ADC and signal source

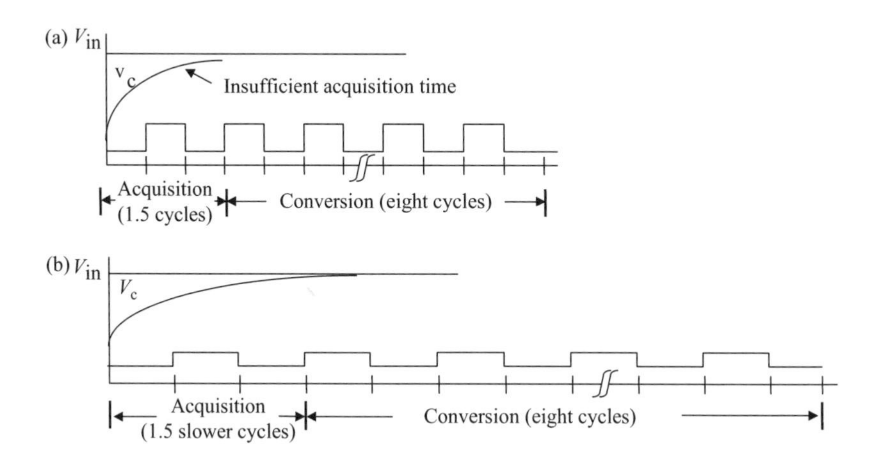

Figure 3.48 The effect of source resistance: (a) insufficient acquisition time; (b) slowing of clock to increase acquisition time

a case in which the problem could be solved by slowing the clock. For further details see References 17 and 18.

3.8.2 The amplifier–ADC interface

Operational amplifiers are nearly always present in data acquisition systems, performing basic signal conditioning ahead of the ADC. Their interactions with ADCs affect system performance. Although many amplifiers are good at driving a variety of static loads, the switched nature of the ADC input stage can introduce problems with some amplifiers, especially the low power, low speed devices most likely to be used in 3 and 5 V systems. Using the simple model in Figure 3.49(a), the load presented to the amplifier by the ADC input keeps switching abruptly between an

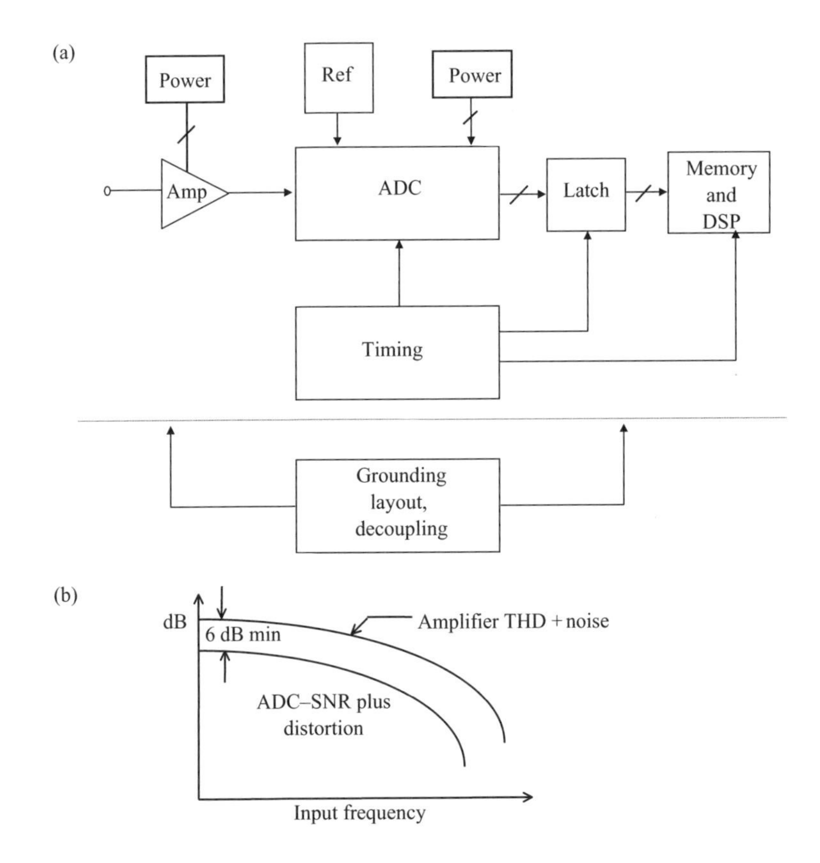

Figure 3.49 ADC–amplifier interface: (a) basic elements; (b) performance expected from ADC and amplifier (source: Analog Devices, Inc.)

open circuit and a series RC network connected to an internal voltage source. The op amp's response to the sudden load current and impedance change depends on several parameters. Among them are the device's gain-bandwidth product, slew rate, and output impedance.

Selecting the appropriate drive amplifier for an ADC involves many considerations. Because the ADC drive amplifier is in the signal path, its error sources (both d.c. and a.c.) must be considered in calculating the total error budget. Ideally, the a.c. and d.c. performance of the amplifier should be such that there is no degradation of the ADC performance. It is rarely possible to achieve this, however; and therefore the effects of each amplifier error source on system performance should be evaluated individually.

Evaluating and selecting op amps based on the d.c. requirements of the system is a relatively straightforward matter. For many applications, however, it is more desirable first to select an amplifier on the basis of a.c. performance (bandwidth, THD, noise, etc.). The a.c. characteristics of ADCs are specified in terms of SNR, ENOBs, and

Table 3.6 ADC drive amplifier considerations

Performance requirements	Parameter
a.c. performance	• Bandwidth, setting time • Harmonic distortion, total harmonic distortion • Noise, THD + noise
d.c. performance	• Gain, offset, drift • Gain non-linearity
General	• As a general principle, select first for a.c. performance, then evaluate d.c. performance • Always consult the data sheet for recommendations

distortion. The drive amplifier should have performance better than that of the ADC so that maximum dynamic performance is obtained (see Figure 3.49(b)). If the amplifier a.c. performance is adequate, the d.c. specifications should be examined in terms of system performance. Table 3.6 summarises the ADC drive amplifier considerations and further details can be found in Reference 3.

Other considerations in DAS interfacing are (i) input clamping and protection, (ii) drive amplifier noise configurations, (iii) ADC reference voltage considerations, and (iv) settling time considerations, etc. These are beyond the scope of this chapter and the reader is referred to References 2–4, 17 and 18.

Interfaces of data converters with DSPs are discussed in chapter 13.

3.9 References

1 SHEER, D.: 'Monolithic high-resolution ADCs', *EDN*, May 12, 1988; pp. 116–130.
2 ANALOG DEVICES, INC.: 'High Speed Design Techniques', 1996.
3 ANALOG DEVICES, INC.: 'Linear Design Seminar', 1995.
4 ANALOG DEVICES, INC.: *1992 Amplifier Applications Guide*, 1992, Chapter 7.
5 TEKTRONIX, INC.: 'Digital Oscilloscope Concepts', Engineering Note 37W-6136, 4/86.
6 RAZAVI, B.: *Data Conversion System Design*, IEEE Press, 1995.
7 SHILL, M.A.: 'Servo loop speeds tests of 16-bit ADCs', *Electronic Design*, February 6, 1995, pp. 93–109.
8 LOUZON, P.: 'Decipher high-sample rate ADC specs', *Electronic Design*, March 20, 1995, pp. 91–100.
9 DOERNBERG, J., LEE, H.-S. and HODGES, D.: 'Full-speed testing of A/D converters', *IEEE Journal of Solid-State Circuits*, **SC-19** (6), December 1984, pp. 820–827.

10 KULARATNA, N.: 'Modern component families and circuit block design', Butterworth-Newnes (2000).
11 MCGOLDRICK, P.: 'Architectural breakthrough moves conversion into main stream', *Electronic Design*, Jan. 20, 1997, pp. 67–72.
12 KULARATNA, N.: *Modern Electronic Test and Measuring Instruments*, IEE London, 1996, ISBN No. 0-85296-813-2.
13 CANDY, J.C. and TEMES, G.C.: *Over-sampling Delta–Sigma Data Converters – Theory, Design and Simulation*, IEEE Press, 1992, ISBN - 0-87942-285-8.
14 O'LEARY, S.: 'Self-calibrating ADCs offer accuracy, flexibility', *EDN*, June 22, 1995, pp. 77–85.
15 GOODENOUGH, F.: 'ADCs move to cut power dissipation', *Electronic Design*, January 9, 1995, pp. 69–74.
16 HOESCHELE, D.F.: *Analog-to-Digital and Digital-to-Analog Conversion Techniques*, John Wiley Interscience, 2nd edn, ISBN - 0-471-57147-4, 1994.
17 LACANETTE, K.: 'To build data acquisition systems that run from 5 or 3.3 V, know your ICs', *EDN*, September 29, 1994, pp. 89–98.
18 WATSON SWAGER, A.: 'Evolving ADCs demand more from drive amplifiers', *EDN*, September 29, 1994, pp. 53–62.

Chapter 4

Waveform parameters, multimeters and pulse techniques

4.1 Introduction

As the engineering community progresses with technological developments, sophisticated measuring instruments become a major requirement for the interpretation of performance, parameter comparison and improvements. However, in most engineering environments, basic instruments for the measurement of V, I, R, Z and temperature, etc., account for the major share of the instrument market.

In practical environments, such as the field service and maintenance workshops, etc., simple instruments, such as multimeters, are used to measure signals without paying any attention to the wave shape, signal frequency components, or to the instruments' capabilities or limitations. In digital system environments, pulse waveforms are used for device characterisation and performance checking. In such situations, pulse parameters and techniques play an important role, forcing the designer, service or the test engineer to understand pulse parameters and test techniques.

This chapter provides an overview of waveform measurements, multimeters and pulse techniques.

4.2 Waveform parameters and amplitude related measurements

Most waveforms may be divided into periodic and non-periodic signals. Furthermore, the non-periodic signals may be subdivided into finite and infinite energy signals. The chart in Figure 4.1 shows some practical signals grouped into these categories.

Each kind of waveform has its inherent features and qualities and several parameters have been defined to specify and evaluate the same. Table 4.1 indicates different parameters for different types of waveform, dividing them into primary and secondary categories in relation to the practical usage of these signals.

Measurement of waveform parameters is carried out using many different types of instrument, such as analogue multimeters, digital multimeters, oscilloscopes,

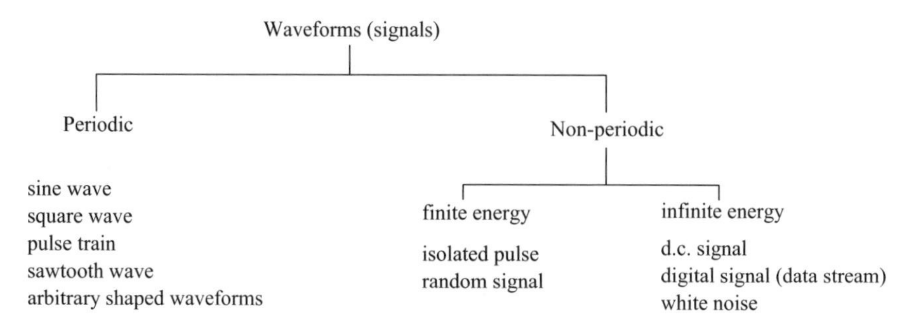

Figure 4.1 Signals and their categorisation

Table 4.1 Waveforms and their parameters

Signal	Primary parameters	Secondary parameters
Sinusoidal waveforms	Amplitude	Harmonic distortion
	Peak	
	Peak to peak	
	r.m.s.	
	Average	
	Frequency	
Square wave	Amplitude	Rise time
	Peak	Fall time
	Peak to peak	Pulse width
	r.m.s.	Duty cycle
	Average	Overshoot
		Undershoot
	Frequency	Pulse drop
	Mark space ratio	Slew rate
Pulse train	Amplitude	Rise time
		Fall time
	Baud rate (bits s^{-1})	Pulse width
		Duty cycle
		Overshoot
		Undershoot
		Pulse droop
Saw tooth wave	Periodic time, Max amplitude	Linearity
DC signal	Amplitude	Ripple
Digital data	Baud rate	Bit error rate

spectrum analysers, frequency counters, logic analysers, field strength meters and bit error rate meters. Most measurements are carried out in the time domain, although some measurements are carried out in the frequency domain using instruments such as spectrum analysers and selective level meters.

It is necessary to define and discuss several important parameters related to the use of the most common and inexpensive instruments, such as analogue and digital multimeters for amplitude measurements. These are:

- average value,
- root mean square (r.m.s.) value,
- rectified mean,
- crest factor, and
- form factor.

The **average value** of a time varying voltage waveform, V_{ave}, is defined as

$$V_{\text{ave}} = \frac{1}{T}\int_0^T v(t)\,\mathrm{d}t. \tag{4.1}$$

The **root mean square** value of a time varying voltage waveform, V_{rms}, is defined as

$$V_{\text{rms}} = \sqrt{\frac{1}{T}\left(\int_0^T v^2(t)\,\mathrm{d}t\right)}. \tag{4.2}$$

The **rectified mean**[1] or **mean absolute deviation** (or a.c. average) of a time varying voltage waveform, V_{mad}, is mathematically defined as

$$V_{\text{mad}} = \frac{1}{T}\int_0^T |v(t)|\,\mathrm{d}t \tag{4.3}$$

where $v(t)$ is the instantaneous voltage of the waveform, and T is the period of the waveform.

In practice, the r.m.s. value assigned to an a.c. current is the amount of d.c. current required to produce an equivalent amount of heat in the same load. For calibration purposes, this concept of heat generation is still used in the cases of thermal voltage calibrators (TVC) [1], while some multimeters use thermal conversion techniques for r.m.s. measurements.

The **crest factor** of a waveform is the ratio of its peak value to the r.m.s. value.

The **form factor** of a waveform is defined as the ratio of the r.m.s. value to the average value.

It is very useful to understand the practical meaning of these parameters in the context of analogue and digital multimeters in order to interpret the values read from

[1] The mean absolute deviation (m.a.d.) and mean absolute value (m.a.v.) are different terms used for rectified mean.

these instruments, particularly with non-sinusoidal waveforms. For example, in most analogue meter movements, which employ a permanent magnet moving coil (PMMC) deflection system, the meter responds to the average current flowing through the moving coil. When such a meter is used in measuring the r.m.s. value, the meter acts in a rectified mean sensing mode and the scale calibration uses the relation between the rectified mean and the r.m.s. value, usually with the assumption of a pure sine waveform at the input. Therefore, these defined parameters are very useful in interpreting the readings as the (analogue) meter dials or scales may be calibrated for only special cases of waveform, usually for an undistorted sinewave.

The r.m.s. value is a fundamental measurement of the magnitude of an a.c. signal. Unless the meter contains a special r.m.s. converter stage these parameters must be used carefully, bearing in mind the following guidelines.

(i) Identify the basic parameter to which the meter movement responds primarily.
(ii) Identify the factors used for calibration of the meter scale and the wave shapes to which the meter responds accurately.
(iii) For different waveforms measured use the knowledge about the wave shape to apply a correction factor.

For the purposes of calculating the correction factor in (iii) the instrument can be logically divided into two fundamental blocks (Figure 4.2) to which the information in Table 4.2 is utilised.

For example, if a PMMC analogue multimeter is designed for measuring an r.m.s. value, then when a pure sinewave is measured using the meter, the meter movement will respond to the average value of the waveform appearing at the point B and the scale will be calibrated accordingly. If the meter is designed with a half wave rectifier stage to convert the input a.c. signal to a d.c. value, the waveform appearing at point B will be a half sinewave. If the meter uses a full wave rectifier the response of the meter will be to the m.a.d. value of the waveform appearing at point B. Therefore, scale calibration will be different for the two cases. If a triangular waveform is observed on these two meters, the first meter will read a value V_1 against the second meter reading V_2, as calculated below.

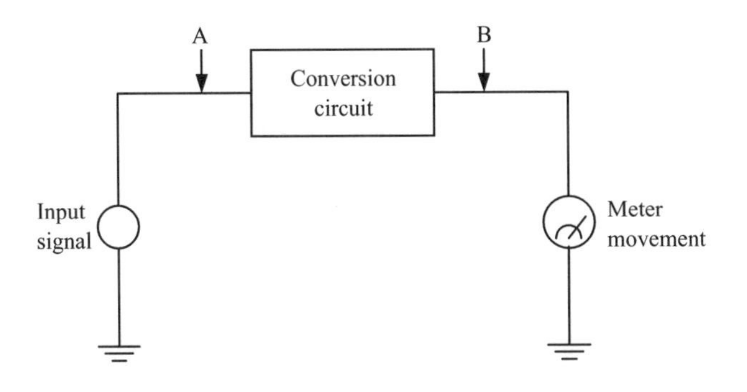

Figure 4.2 Basic block diagram of an analogue multimeter

Waveform	r.m.s.	m.a.d.	r.m.s./m.a.d.	Crest factor
Sinewave	$\frac{V_m}{\sqrt{2}}$	$\frac{2}{\pi} V_m$	$\frac{\pi}{2\sqrt{2}}$	$\sqrt{2}$
	$0.707\ V_m$	$0.637\ V_m$	1.111	1.414
Half wave rectified sinewave	$\frac{V_m}{2}$	$\frac{1}{\pi} V_m$	$\frac{2}{\pi}$	2
	$0.5\ V_m$	$0.318\ V_m$	1.571	2
Full wave rectified sine	$\frac{V_m}{\sqrt{2}}$	$\frac{2}{\pi} V_m$	$\frac{\pi}{2\sqrt{2}}$	$\sqrt{2}$
	$0.707\ V_m$	$0.637\ V_m$	1.111	1.414
Triangular wave	$\frac{V_m}{\sqrt{3}}$	$\frac{V_m}{2}$	$\frac{2}{\sqrt{3}}$	$\sqrt{3}$
	$0.577\ V_m$	$0.5\ V_m$	1.154	1.732
Amplitude symmetrical rectangular	V_m	V_m	1	1
Saw tooth pulse	$\sqrt{\frac{\eta}{3}} V_m$	$\frac{\eta}{2} V_m$	$\sqrt{\frac{4}{3\eta}}$	$\sqrt{\frac{3}{\eta}}$
Offset pulse	$V_m \sqrt{\eta(1 - A^2) + A^2}$	$V_m[\eta(1 - A) + A]$	$\frac{\sqrt{\eta(1 - A^2) + A^2}}{\sqrt{[\eta(1 - A) + A]}}$	$\frac{1}{\sqrt{\eta(1 - A^2) + A^2}}$
Exponential pulse	$\sqrt{\frac{\tau}{2T}(1 - e^{-2T/\tau})}$	$\frac{\tau}{T}(1 - e^{-T/\tau})$	$\sqrt{\frac{T}{2\tau}}$	$\sqrt{\frac{2T}{\tau}}$
	$\cong \sqrt{\frac{\tau}{2T}} V_m$	$\cong \frac{\tau}{T} V_m$		

For the general case,

$$V_{\mathrm{rms}} = k\ (V_{\mathrm{B}}), \tag{4.4}$$

where k is a constant used in calibrating the scale and V_{B} is the average value appearing at the point B.

When the input is a pure sinewave, and the conversion circuit is a half wave rectifier, $V_{\mathrm{B}} = 0.318\ V_{\mathrm{m}}$ and $V_{\mathrm{rms}} = 0.070\ V_{\mathrm{m}}$. As the meter is calibrated for a pure sinewave, $0.707\ V_{\mathrm{m}} = k \times 0.318$, hence, $k = 2.22$.

Similarly, in the full wave rectifier case,

$$V_{\mathrm{rms}} = k'(\mathrm{m.a.d.}).$$

When the input is a pure sinewave, $0.707 V_{\mathrm{m}} = k' \times 0.637 V_{\mathrm{m}}$. Hence, $k' = 1.11$.

Now, consider a triangular waveform measured with the two meters, using the parameters given in Table 3.2. For the first meter, the reading is

$$V_1 = 2.22 \times \frac{V_{\mathrm{m}}}{4} = 0.555 V_{\mathrm{m}}.$$

However, the actual r.m.s. value will be $= 0.577 V_{\mathrm{m}}$. Therefore, the percentage error is given by

$$\frac{0.577 V_{\mathrm{m}} - 0.555 V_{\mathrm{m}}}{0.577 V_{\mathrm{m}}} \approx 3.8 \text{ per cent.}$$

Similarly, the reading on the second meter $V_2 = 1.11 \times 0.5 V_{\mathrm{m}}$, and the error will be 3.8 per cent.

Similarly, if we apply a half sinewave to the two meters, it can be shown that the first meter will read

$$V_1 = 2.22 \times 0.318 V_{\mathrm{m}} = 0.707 V_{\mathrm{m}}.$$

The second meter reading will be

$$V_2 = 1.11 \times 0.318 V_{\mathrm{m}} = 0.353 V_{\mathrm{m}}.$$

The actual r.m.s. of the waveform will be $V'' = 0.5 V_{\mathrm{m}}$.

Therefore, errors in the first and second meters will be -41.4 per cent and 29.4 per cent, respectively. In calculating the errors, loading effects and other effects due to frequency limitations, also need to be considered. A treatment of these general effects and a mathematical analysis are to be found in References 2 and 3.

Figure 4.3 indicates the case of a precision rectifier and an averaging circuit used inside a meter.

The rectify-and-average scheme applies the a.c. input to a precision rectifier. The rectifier output feeds a simple gain scaled RC-averaging circuit to provide the output. In practice, you set the gain so that the d.c. output equals the r.m.s. value of a sinewave input. If the input remains a pure sinewave, accuracy can be good. However, non-sinusoidal inputs cause large errors. This type of voltmeter is accurate only for sinewave inputs with errors increasing as the input departs from sinusoidal.

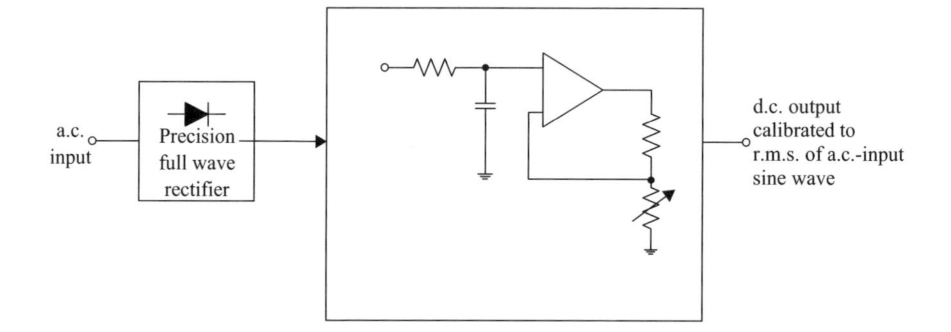

Figure 4.3 Rectify-and-average based measurement for sinusoidal inputs

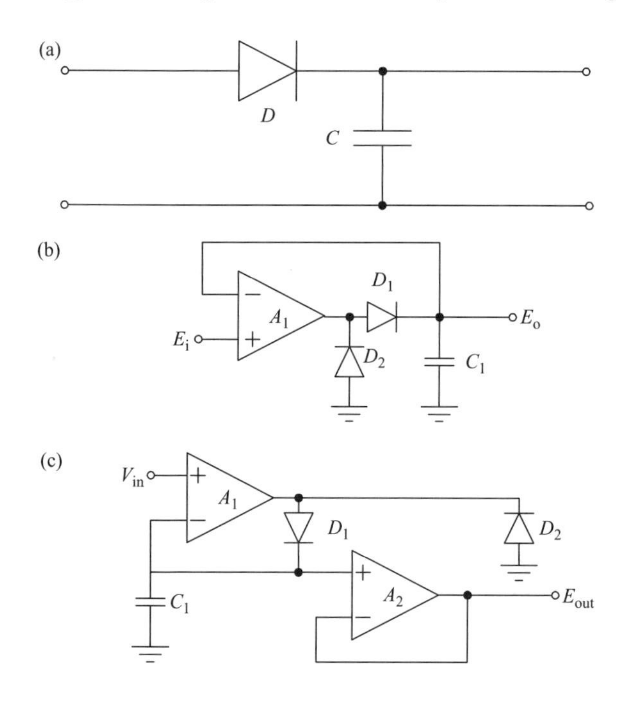

Figure 4.4 Peak detector circuits: (a) simple circuit; (b) single op amp based; (c) two op amp based

4.2.1 Peak value measurements

For measuring the peak value of a time varying waveform, peak detector circuits, given in Figure 4.4, are used in some instruments. The most basic circuit of a peak detector is shown in Figure 4.4(a).

First, let us consider the single-amplifier circuit shown in Figure 4.4(b). Since D_1 is inside the feedback loop, its forward voltage drop is divided by the open loop gain of the operational amplifier and can be expressed as an equivalent offset in series with the input of an ideal peak detector. A_1 serves two other useful purposes. The input signal source needs to supply only the input bias current of A_1. The output rise time is not determined by the time constant of the on resistance of D_1 times C_1 but is dependent only on the output current capability I_{max} of A_1. D_2 is necessary to prevent A_1 from overloading at negative saturation voltage when E_i is less than E_o. D_2, however, must withstand the short circuit current of A_1. A_1 should be an FET input amplifier to minimise the decay rate after detecting a peak since the input bias current of the inverting input will discharge C_1. Also, the input stage of the amplifier will not conduct when E_i is less than E_o. If the output is to be loaded, an output buffer amplifier is required to prevent the load from discharging C_1. A peak detector circuit using two operational amplifier stages is shown in Figure 4.4(c).

4.2.2 The r.m.s. measurements

From the preceding discussion it is clear that, when different waveforms are measured using different meters, the observed readings may not truly represent the r.m.s. values. To avoid this problem, some meters use special active circuit blocks, either using discrete components or monolithic ICs which convert the input signal to a d.c. voltage that is proportional to the true r.m.s. value.

In discussing these 'true' r.m.s. values it is very necessary that we clearly analyse the practical situations with a.c. and d.c. coupled signals. By definition, the r.m.s. takes the instantaneous values irrespective of the a.c. or d.c. components of the waveform separately. However, in the case of modern instruments, the trend is to use the monolithic r.m.s. to d.c. converter blocks, and some analogue and mixed signal IC manufacturers produce these catalogue components in monolithic form. In these component families some components calculate the r.m.s. value of the a.c. component of the waveform whereas some devices convert the input, in its d.c. coupled form, to the r.m.s. value. In the case of a.c. coupled conversion the value is '**true r.m.s. a.c. coupled**' and in the d.c. coupled case it is '**true r.m.s. d.c. coupled**'.

Practical a.c. voltmeter types could be divided into rectify-and-average, analogue computing, and thermal. The thermal approach is the only one that is inherently accurate for all input wave shapes. If the user intends measuring noise, this feature is relevant to determining the amplitude of r.m.s. noise. A fourth method for measuring the r.m.s. value of an input waveform uses sampling techniques, and involves taking a large number of samples of the input waveform and computing the r.m.s. value using digital techniques. Achievable accuracy for any given bandwidth varies with sampling rate and computational capability.

Users who select multimeters must be careful with the r.m.s. terminology. Many multimeters with the claim of 'r.m.s. capability' may accurately measure the r.m.s. component of an a.c. coupled waveform. On the other hand, an expensive bench multimeter may use processor based algorithms to take samples of a waveform and do a 'definition based' calculation of the r.m.s. value, providing the user with the 'true r.m.s.–d.c. coupled' value of a given waveform. Low cost hand held instruments claiming to have 'r.m.s. capability' may have only 'true r.m.s.–a.c. coupled' measurement accuracy.

4.2.3 The r.m.s. to d.c. converters

In monolithic IC converters different analogue computing techniques of conversion are used and these are divided into two basic categories, namely, the direct (explicit) computation method and the indirect (implicit) computation method. Both types are subsets of non-linear device families and use analogue multipliers and dividers [4].

4.2.3.1 Explicit method

The explicit method is shown in Figure 4.5(a). The input signal is first squared by a multiplier. The average value is taken by using an appropriate filter, and the square root is taken using an op amp with a second squarer in the feedback loop. This circuit has limited dynamic range because the stages following the squarer must try to deal with a signal that varies enormously in amplitude. This restricts the method to inputs with a maximum dynamic range of approximately 10 : 1 (20 dB). However, excellent

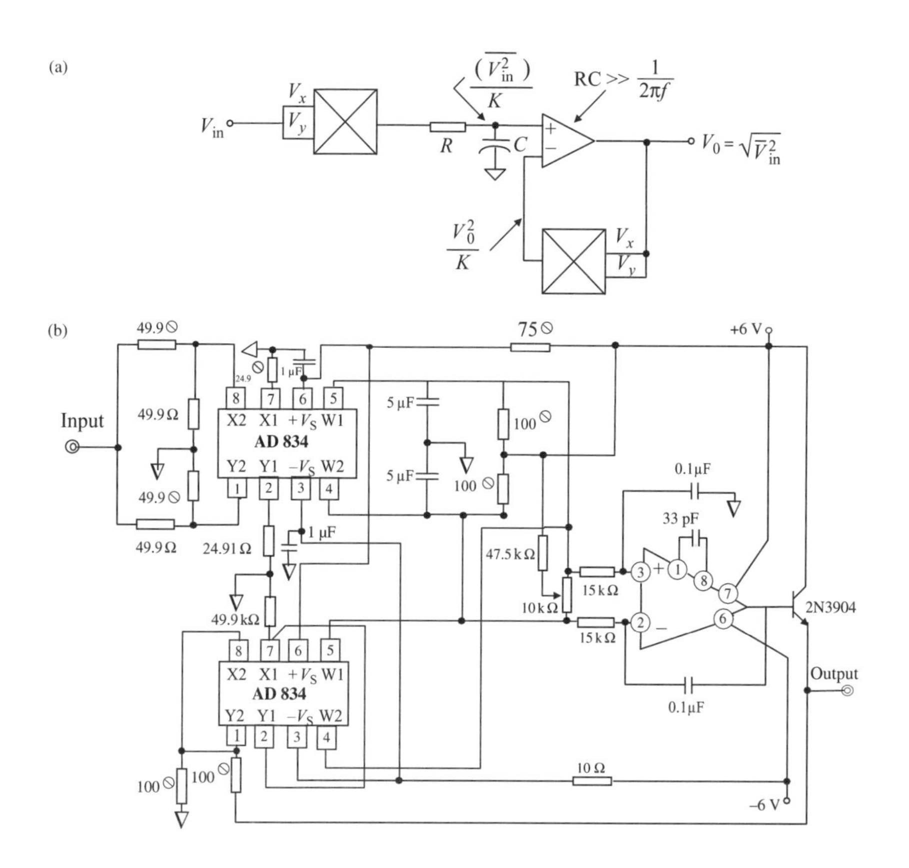

Figure 4.5 Explicit r.m.s. to d.c. converters: (a) basic technique; (b) wideband r.m.s. measurement using the AD-834 (reproduced by permission of Analog Devices, Inc.)

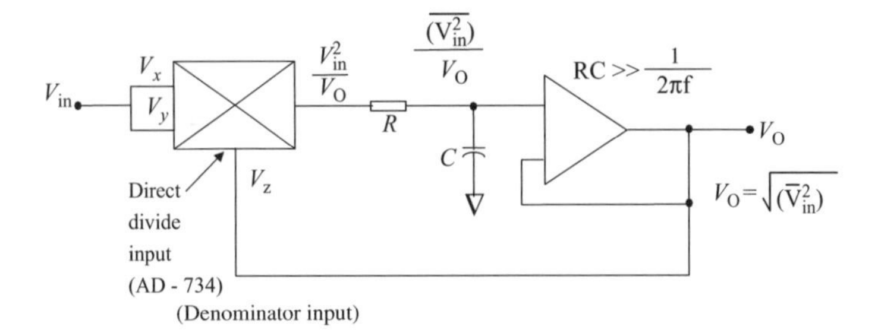

Figure 4.6 Implicit r.m.s. computation

bandwidth (greater than 100 MHz) can be achieved with high accuracy if a multiplier such as the AD-834 is used as a building block (see Figure 4.5(b)).

4.2.3.2 Implicit method

Figure 4.6 shows the circuit for computing the r.m.s. value of a signal using the implicit method. Here, the output is fed back to the direct divide input of a multiplier such as the AD-734. In this circuit, the output of the multiplier varies linearly (instead of as the square) with the r.m.s. value of the input. This considerably increases the dynamic range of the implicit circuit as compared to the explicit circuit. The disadvantage of this approach is that it generally has less bandwidth than the explicit computation.

Some advantages of the implicit conversion technique over the other methods are that fewer components are needed, a greater dynamic range is possible, and the cost is lower. A disadvantage of this method is that it generally has less bandwidth.

4.2.3.3 Monolithic r.m.s./d.c. converters

While it is possible to construct such an r.m.s. circuit from an AD-734, it is far simpler to design a dedicated r.m.s. circuit. The V_{in}^2/V_z circuit may be current driven and only one quadrant if the input first passes through an absolute value circuit. Figure 4.7(a) shows a block diagram of a typical monolithic r.m.s./d.c. converter such as the AD-536. It is subdivided into four major sections: absolute value circuit (active rectifier), squarer/divider, current mirror, and buffer amplifier. The input voltage, V_{in}, which can be a.c. or d.c., is converted to a unipolar current, I_{in}, by an absolute value circuit. I_{in} drives one input of the one-quadrant squarer/divider, which has the transfer function I_{in}^2/I_f. The output current, I_{in}^2/I_f, of the squarer/divider drives the current mirror through a low pass filter formed by R_1 and an externally connected capacitor, C_{AV}. If the $R_1 C_{AV}$ time constant is much greater than the longest period of the input signal, then I_{in}^2/I_f is effectively averaged. The current mirror returns a current, I_f, that equals the average value of I_{in}^2/I_f back to the squarer/divider to complete the implicit r.m.s. computation. Therefore,

$$I_f = \overline{\left(\frac{I_{in}^2}{I_f}\right)} = I_{in}(\text{r.m.s.}). \tag{4.5}$$

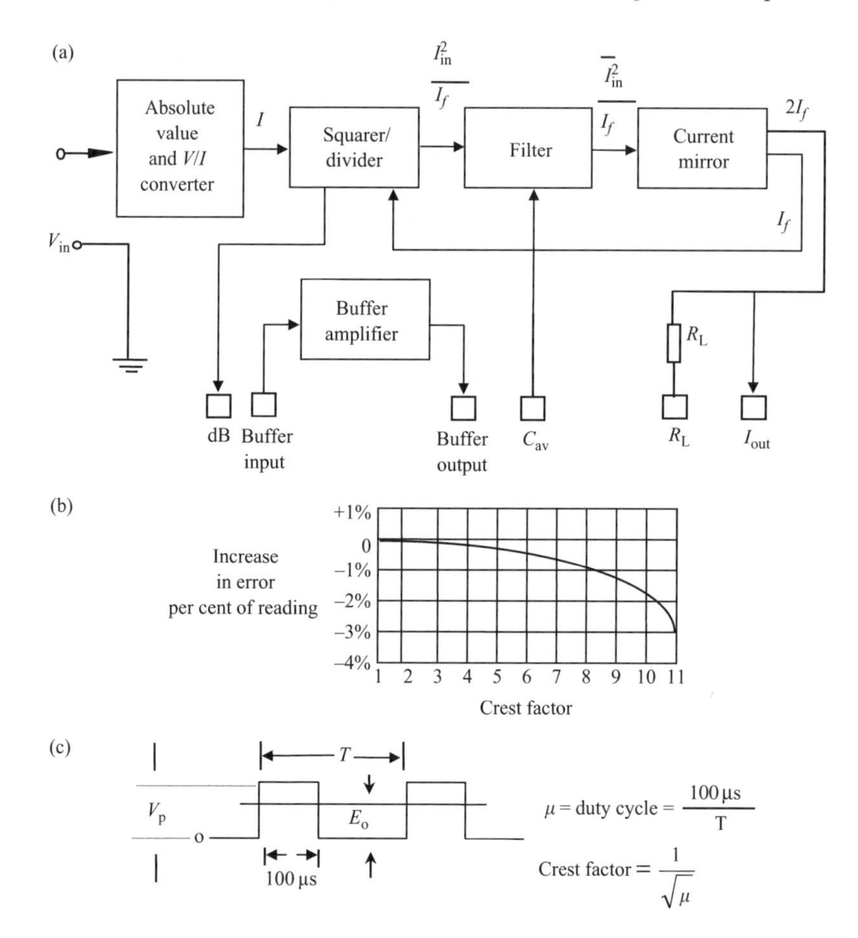

Figure 4.7 The AD-536 r.m.s./d.c. converter: (a) block diagram; (b) error against crest factor; (c) input waveform for (b) (reproduced by permission of Analog Devices, Inc.)

The current mirror also produces the output current, I_{out}, which equals $2I_f$. The circuit provides a decibel output also, which has a temperature coefficient of approximately 3300 p.p.m. K^{-1} and must be temperature compensated.

There are number of r.m.s./d.c. converters in monolithic form. A representative list from Analog Devices, Inc., is shown in Table 4.3. For practical application, design details and selection of these devices, References 4 to 6 are suggested.

When active components are used in multimeters, the user should be aware of the errors related to the waveform parameters in interpreting the actual results. In practical situations, these components use an averaging capacitor (such as C_{av} in Figure 4.7) and selecting the value of this capacitor is dependent on the waveform parameters such as waveform period, duty cycle and the d.c. offset. Also the percentage of the reading error is dependent on parameters such as crest factor. For example for AD536

Table 4.3 A representative set of r.m.s./d.c. converters from Analog Devices, Inc.

Part no.	Bandwidth	Full scale input voltage range	Remarks
AD-536	450 kHz	>100 mV input	±15 V rails
	2 MHz	>1 V	
AD-636	1 MHz	Up to 200 mV	Low power (±5 V) rails
	8 MHz		
AD-637	8 MHz	>1 V	Chip selector/power down function available (±3 to ±18 V rails)
	600 kHz	200 mV	
AD-736	350 kHz	100 mV	Low power precision converter ±5 to ±16 V power rails
	460 kHz	200 mV	
AD-737	170–350 kHz	100 mV	Low cost, low power true r.m.s. ±5 to ±16 V rails
	190–460 kHz	200 mV	

performance error vs crest factor is shown in Figure 4.7(b). This is one of the reasons why true r.m.s. multimeter specification sheets provide the crest factor behaviour.

4.2.3.4 Thermal conversion techniques

The thermally based a.c. voltmeter is inherently insensitive to input wave shape, making it suitable for measuring the amplitude of r.m.s. noise. Additionally, thermally based meters can achieve high accuracy bandwidths exceeding 100 MHz. Figure 4.8 shows the classic thermal scheme. This thermal converter comprises matched heater–temperature sensor pairs and an amplifier. The a.c. input drives a heater, warming it. The temperature sensor associated with this heater responds by biasing the amplifier. The amplifier closes its feedback loop by driving the output heater to warm its associated temperature sensor. When the loop closes, the heaters are the same temperature. As a result of this 'force balance' action, the d.c. output equals the input heater's r.m.s. heating value, which is the fundamental definition of r.m.s. Changes in wave shape have no effect because the scheme effectively down converts any wave shape into heat. This 'first principles' nature of operation makes thermally based a.c. voltmeters ideal for quantitative r.m.s. noise measurement.

4.2.3.5 Average vs true r.m.s. comparison

Average voltage measurements work well when the signal under observation is a pure sine wave, but errors mount as the waveform distorts. By using true r.m.s. measurements, however, one can measure the equivalent heating effect that a voltage produces, including the heating effects of harmonics. Table 4.4 shows the difference between measurements taken on averaging digital multimeters (DMMs) and

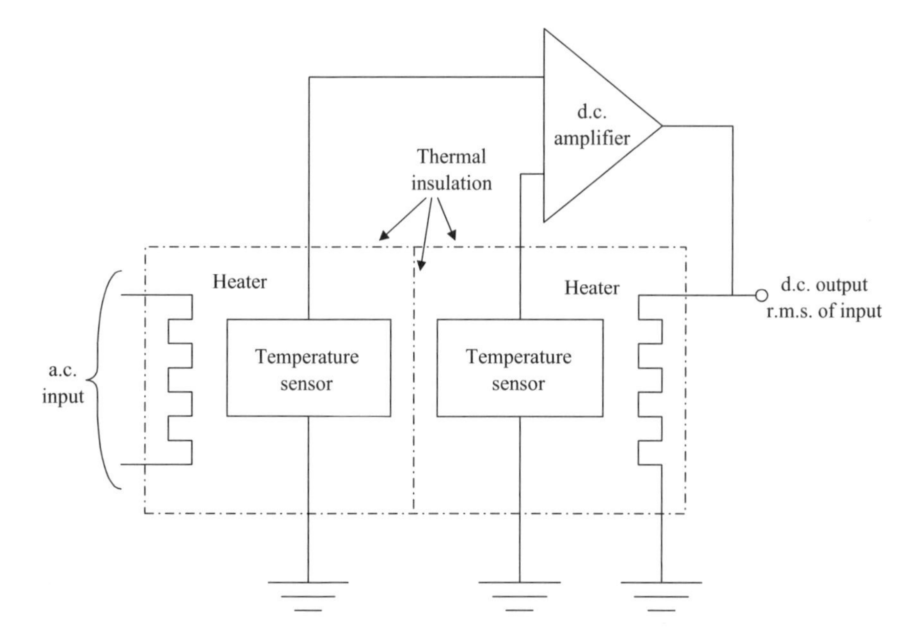

Figure 4.8 Thermal conversion technique (courtesy of EDN magazine)

Table 4.4 Average vs true r.m.s. comparison of typical waveforms

Waveform	Actual peak–peak	True r.m.s. reading	Average reading	Reading error (%)
Sine wave	2.000	0.707	0.707	0
Triangle wave	2.000	0.577	0.555	−3.8
Square wave	2.000	1.000	1.111	+11.1
Pulse (25% duty cycle)	2.000	0.433	0.416	−3.8
Pulse (12.5% duty cycle)	2.000	0.331	0.243	−26.5
Pulse (6.25% duty cycle)	2.000	0.242	0.130	−46.5

those taken on true r.m.s. DMMs. In each case, the measured signal's peak-to-peak value is 2 V.

For a 1 V peak sinewave, the average and r.m.s. values are both 0.707 V. But when the input signal is no longer a sinewave, differences between the r.m.s. values and the average reading values occur. Those errors are most prominent when you are measuring square waves and pulse waveforms, which are rich in harmonics. One limitation to making true r.m.s. measurements is crest factor, and the user should consider crest factor when making a.c. measurements.

A DMM's specifications should tell you the maximum crest factor that the meter can handle while maintaining its measurement accuracy. True r.m.s. meters can handle higher crest factors when a waveform's r.m.s. voltage is in the middle of the meter's range setting. Typically, a DMM may tolerate a crest factor of three near the top of its scale but it might handle a crest factor of five that is in the middle of the range. Therefore, if the user measures waveforms with high crest factors (greater than three), the DMM range should be adjusted so that the measured voltage is closer to the centre of the measurement range. For details, References 7 and 8 are suggested.

Some high performance digital multimeters, such as Philips models PM2525 and 2530, are capable of measuring the true r.m.s. value of a.c. or d.c. coupled signals, obtained by separately calculating the a.c. and d.c. components using the microprocessor subsystem. Other examples of true r.m.s. multimeters are the Tektronix TX3 and TX1. In the Fluke multimeter range, Models 187 and 189 display 'true r.m.s.–d.c. coupled' while Models 170 and 110 display 'true r.m.s.–a.c.'.

If a multimeter specification or design indicates that the internal r.m.s. conversion technique is taking an 'a.c. coupled' only approach (i.e. no d.c. component is measured in r.m.s. conversion), the user could do separate measurements for d.c. and a.c. components respectively and apply the formula below

$$\text{r.m.s. total} = \sqrt{(\text{a.c. r.m.s. component})^2 + (\text{d.c. component})^2}. \qquad (4.6)$$

An instrument's d.c. voltage function could measure the d.c. component and the a.c. voltage function could measure the r.m.s. a.c. component separately.

4.3 Digital multimeters

The multimeter may be one of the most common electronic test instruments in the world today. Used for anything from measuring currents to finding the value of a resistor, multimeters can be purchased in both digital and analogue form. Of these, digital multimeters are becoming more widely accepted as they have better accuracy levels and extra features.

The digital multimeter (DMM) industry has created something for every engineer or technician, ranging from simple $3\frac{1}{2}$ and $4\frac{1}{2}$ digit hand-held DMMs to very special system DMMs and bench types. The market trend is towards better accuracy, more features and low prices.

4.3.1 Basic block diagram

A simplified block diagram of a DMM is shown in Figure 4.9. In commercial versions, some of these blocks are built with LSI and VLSI components in order to reduce the size and improve the reliability while lowering the cost.

In a typical DMM, the input signal, the a.c. or d.c. voltage, current resistance or any other parameter such as temperature, is converted to a d.c. voltage within the range of the ADC. The ADC then converts the pre-scaled d.c. voltage into its equivalent

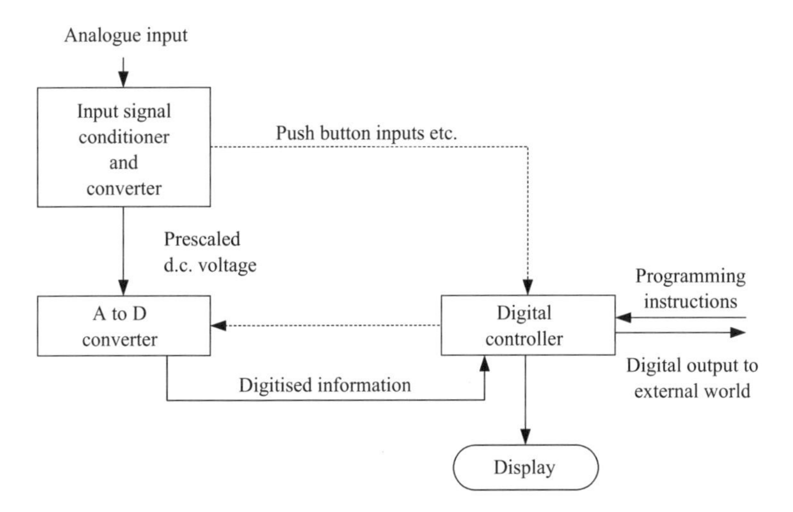

Figure 4.9 Simplified block diagram of a digital multimeter

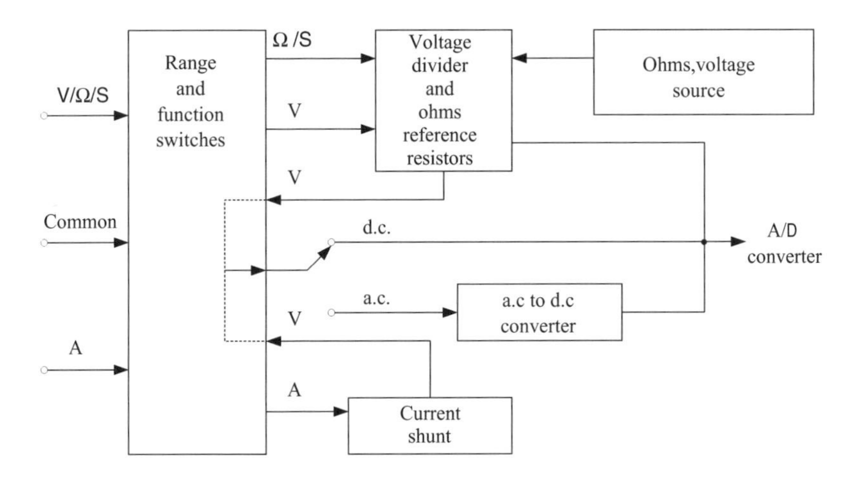

Figure 4.10 Function blocks of input signal conditioner

digital number, which will be displayed on the display unit. A digital control block, sometimes implemented using a microprocessor or a microcontroller, manages the flow of information within the instrument, coordinating all internal functions as well as transferring information to external devices such as printers or personal computers via industry standard interfaces. In the case of a simple hand-held multimeter, some or all of these blocks may be implemented in a single VLSI circuit. In such a case the A/D converter and display driver can be in the same IC. An example is the Intersil $3\frac{1}{2}$ digit single chip A/D converter [9].

The input signal conditioner and the converter could be further subdivided into function blocks as shown in Figure 4.10. Here the range and function switches, voltage divider and ohms reference resistors, current shunts, a.c. to d.c. converters and the power source for resistance measurements are blocked separately. This is a typical set up, but there can be many variations in practical instruments. Range and function switches scale down the excessive input voltages and route the input signals through corresponding circuit blocks ultimately to present a d.c. voltage to the ADC. Sometimes the voltage divider is implemented with an op amp, as shown in Figure 4.11(a). To convert a.c. or d.c. currents to a voltage, most DMMs also use simple current shunts as shown in Figure 4.11(b).

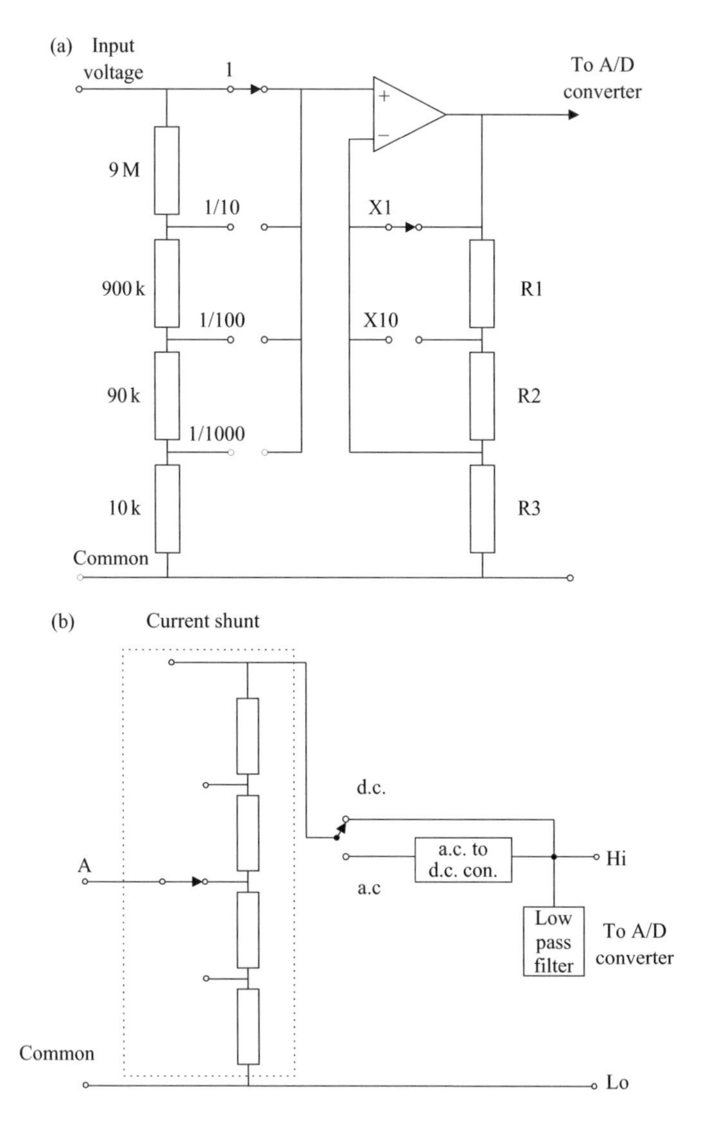

Figure 4.11 Range and function switches: (a) voltage divider with an op amp; (b) current to voltage conversion

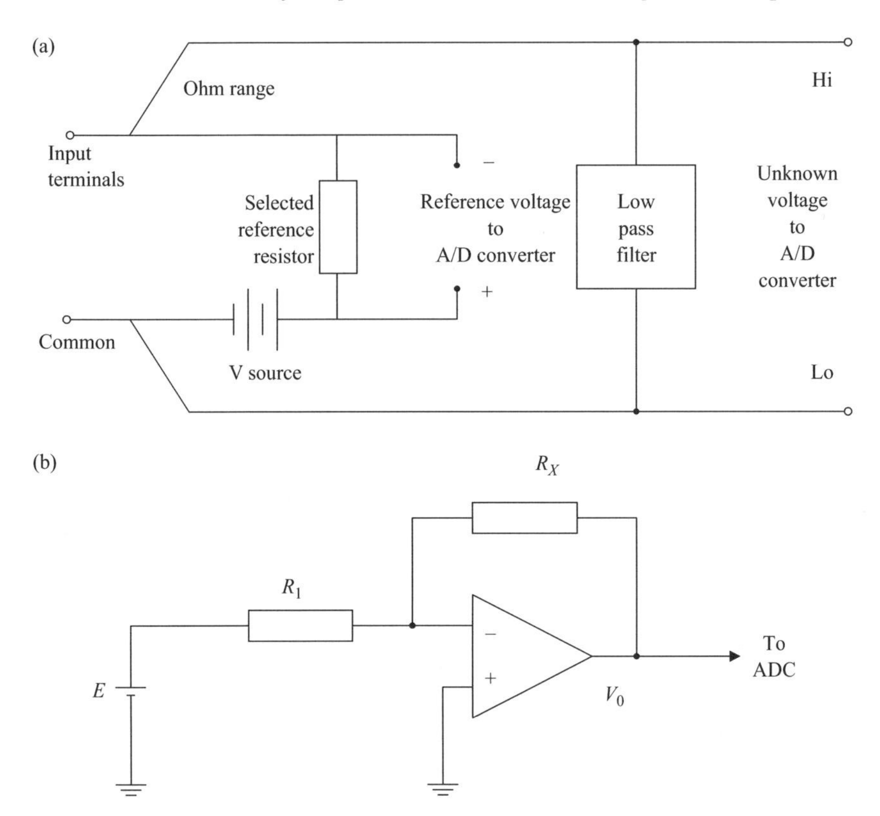

Figure 4.12 Resistance measurement techniques: (a) a typical resistance measurement technique; (b) use of an op amp circuit for resistance to voltage conversion

For a.c. voltage or current measurements, the input signal is passed through the a.c. to d.c. converter block. This block may be anything from a simple diode and capacitor combination to a true r.m.s. converter.

For measurement of resistance, some DMMs use the technique shown in Figure 4.12(a), where a reference voltage across a known resistor and the voltage across the unknown resistance (using a voltage source) are applied to the ADC stage of the multimeter. A low pass filter is used to filter out any residual a.c. components from reaching the ADC stage. In some multimeters a resistance to voltage converter is also used (see Figure 4.12(b)). This method is more straightforward as it provides an output voltage to the ADC that is proportional to the resistance measured, as the basic relationship of $V_0 = (R_x/R_1) \cdot E$ holds true.

4.3.2 Analogue-to-digital conversion process in practical DMMs

In most DMMs, the designers have used integration type ADCs or variations of the technique owing to its simplicity, low cost and performance parameters such as accuracy, elimination of propagation errors in the circuit and the ability to compensate for changes in clock frequency and integrator time constants. Most manufactures use the straightforward dual slope technique discussed in chapter 3, while some manufacturers, such as Fluke, Prema Precision and Hewlett-Packard, use some variants of this. Most of these manufacturers have developed their own VLSI components for the ADC process. This technique is preferred to other methods of analogue-to-digital conversion because the signals measured using DMMs are slowly varying and conversion times of a few hundred milliseconds to several seconds are tolerable. In most DMMs, the A-to-D measurement cycle is composed of several distinct time slots depending on the performance and features available in the instrument. For example, some DMMs have an auto-ranging facility and the ADC cycle is accordingly modified, whereas in some instruments, an average reading over a long period of time is observed using the output of the ADC block.

A basic dual slope ADC technique is described below using the block diagram in Figure 4.13(a). This type of technique is used in some DMMs manufactured by Fluke, Inc., such as the 8060A. Basic timing periods for switch operation during the complete measurement cycle are shown in Figure 4.13(b). Any given measurement

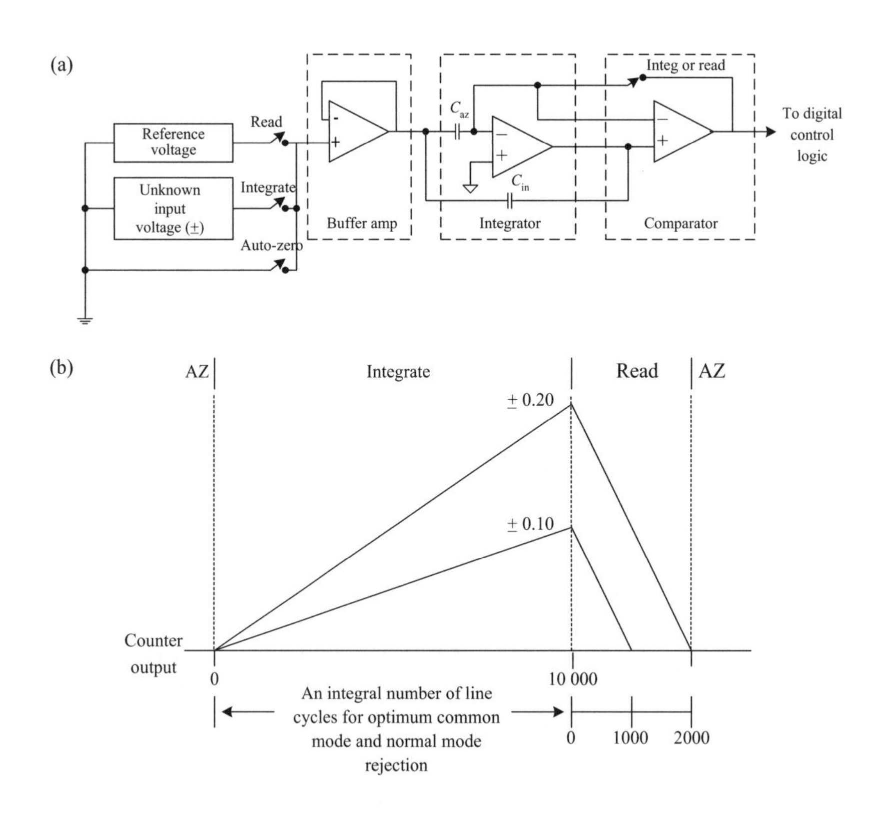

Figure 4.13 Basic A/D conversion process in a DMM: (a) block diagram; (b) measurement cycle

cycle can be divided into three consecutive time periods: auto-zero (AZ), integrate and read. Both AZ and integrate duration are fixed time periods. A counter determines the length of each time period by providing an overflow signal after a fixed amount of clock cycles. The read period is a variable time, which is proportional to the unknown input voltage. The value of the voltage is determined by the number of clock pulses that occur during the read period. During the AZ period, a ground reference is applied to the input of the ADC via a buffer amplifier. During this period, input offset error voltages accumulate in the amplifier loop and in turn are impressed across the AZ capacitor, where they are stored for the remainder of the measurement cycle. This stored level is used to provide offset voltage correction during the integrate and read periods. The integrate period begins at the end of the AZ period. At this time the AZ switches open and the integrate switch closes, applying the unknown input voltage to the input of ADC section. The voltage applied via the buffer amplifier determines the charge rate of the integrate capacitor, C_{in}. By the end of this fixed integrate period the capacitor C_{in} is charged to a level proportional to the unknown input voltage. During the read period the capacitor voltage is discharged at a fixed rate determined by a known reference voltage (of opposite polarity) until the charge is equal to the initial value at the end of the AZ period. Because the discharge rate is fixed during the read period, the time required for discharge is proportional to the unknown input voltage. The integrate period is usually taken as a multiple period of power line frequency to reduce power line noise.

In the example given in Figure 4.13(b), 10 000 counts are used for the integrating period. Therefore, reading two voltages with values of 0.20 V and 0.10 V gives counts of 2000 and 1000, respectively. At the end of the read period, a completely new cycle begins and the same process repeats.

To explain the practical use of the dual slope technique in a multimeter such as the Fluke model 8060A, let us refer to Figure 4.14. In the example of Figure 4.14(a), two major components make up the measurement system, namely (i) a four bit microcomputer chip, and (ii) a custom CMOS chip known as the measurement acquisition chip (MAC).

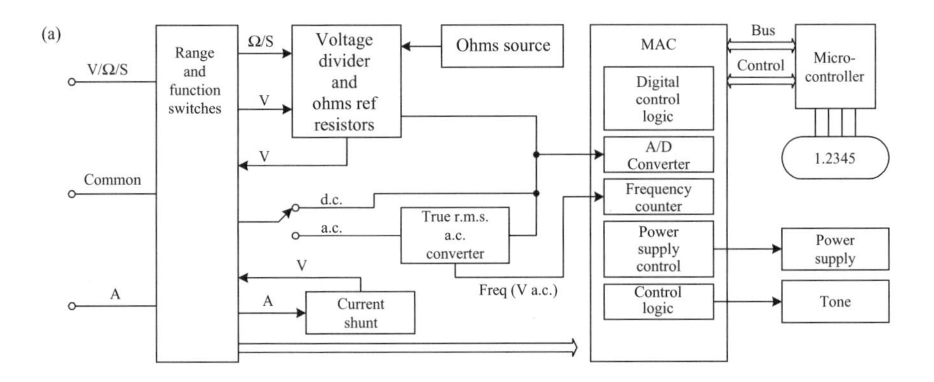

Figure 4.14 *Continued overleaf*

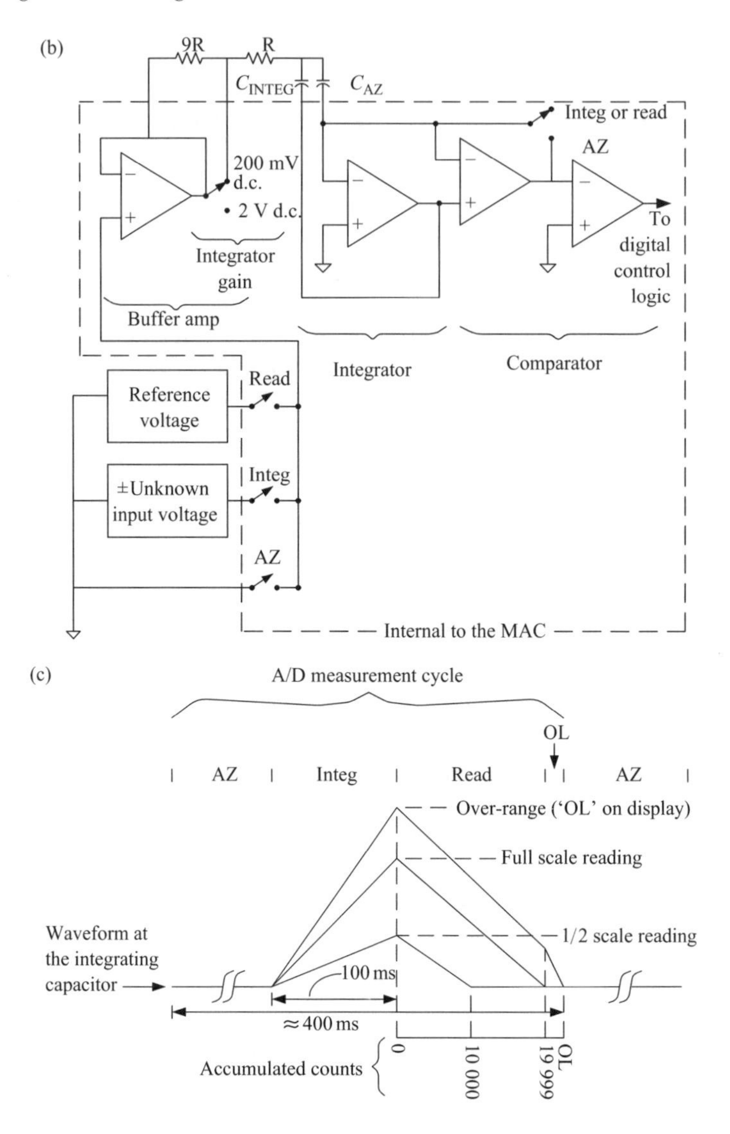

Figure 4.14 Fluke model 8060A DMM design (reproduced by permission of Fluke Corp., USA): (a) block diagram; (b) analogue portion of A/D converter; (c) A/D measurement cycle showing possible overload ('OL') condition

ADC resides inside the MAC and this is a dual slope ADC technique. A block diagram of the analogue portion of the A/D converter is shown in Figure 4.14(b). The internal buffer, integrator, and comparators work in conjunction with external resistors and capacitors to convert the d.c. analogue voltage to a digital number. The internal switches are FETs that are controlled by the microcomputer and the MAC digital control logic. The switchable integrator gain depends on the function and range selected.

The complete A/D measurement cycle is shown in Figure 4.14(c). It consists of three consecutive time periods: auto-zero (AZ), integrate (INTEG) and read. A fourth time period, overload (OL), is also used if an over range reading is taken. The total length of the measurement cycle is 400 ms. The length of the integrate period is fixed at 100 ms; this value is a multiple of the period of 50 Hz power, which helps to reduce the possible power line noise that might interfere with the measurement. The waveform at the INTEG capacitor is shown for three sample measurement readings: half scale, full scale, and over-range.

The measurement cycle begins with the auto-zero period. The AZ switches close, applying a ground reference as the input to the converter. Under ideal conditions the output of the comparator would also go to zero. However, input offset voltage errors accumulate in the buffer amplifier loop, and appear at the comparator output as an error voltage. To compensate for this error, the error is impressed across the AZ capacitor, where it is stored for the remainder of the measurement cycle. The stored level is used to provide offset voltage correction during the integrate and read periods.

The integrate period begins at the end of the auto-zero period. As the period begins, the AZ switches open and the INTEG switches close. This applies the unknown input voltage to the input of the converter. The voltage is buffered and then begins charging the INTEG capacitor. The waveform at the INTEG comparator is a ramp from near zero to some maximum value determined by the amplitude and polarity of the unknown input voltage.

As the read period begins, the INTEG switches open and the READ switches close. This applies the known reference voltage from a 'flying' capacitor whose polarity is chosen by the A/D converter to be the opposite of the polarity of the unknown input voltage. The INTEG capacitor begins discharging at a fixed rate while a counter begins counting. The counter stops counting when the INTEG capacitor voltage equals the initial auto-zero voltage. The count is proportional to the unknown input voltage, and is placed on the display by the microcomputer.

If during the read period the counter counts up to the maximum number of counts for a full scale reading (19 999 counts) and the INTEG capacitor charge has not yet reached the initial auto-zero voltage, the microcomputer knows that an over range reading has been taken. The microcomputer places 'OL' on the display and commands the A/D converter to go into the overload (OL) period, which rapidly slews the integrator voltage back to the initial auto-zero voltage.

The measurement cycle ends at the end of the read period for an on-scale reading, or at the end of the overload period for an over-range reading. A new measurement cycle then begins with the auto-zero period. The display update rate for measurement functions that use the A/D converter is approximately 0.4 s, or about $2\frac{1}{2}$ readings per second.

In some of these DMMs, custom ICs are used where the ADC process and the display driver requirements are incorporated onto a single chip (model 8026 by Fluke, Inc., uses such a custom IC). In more advanced versions, the custom IC may contain the ADC block, digital control logic, frequency counter blocks (when the DMM has frequency count capability) and other additional blocks for continuity testing.

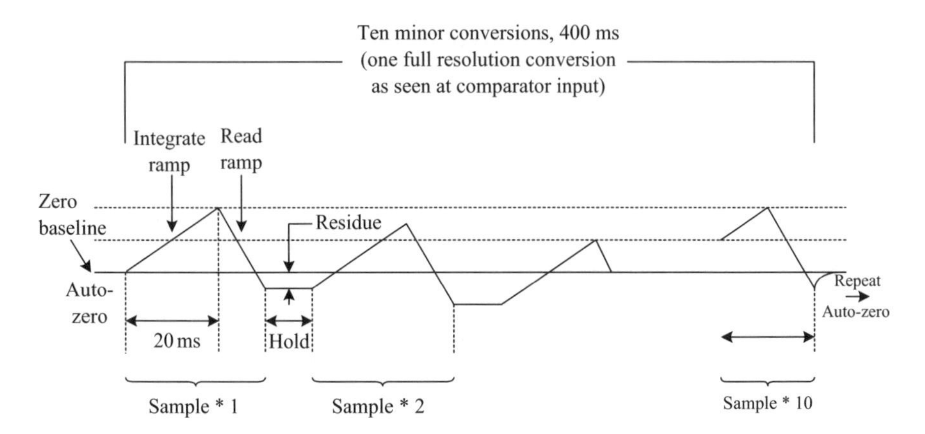

Figure 4.15 Measurement cycle of multiple slope technique (reproduced by permission of Fluke, Inc., USA)

There can be many variations to this basic dual slope technique. In a multimeter, such as the Fluke model 27 or model 79-III, for example, the basic technique is modified to achieve the complete conversion in several minor cycles. In Fluke 27, ten minor cycles are used, as shown in Figure 4.15. The method used is a charge coupled, multiple slope technique. Here a series of ten minor conversions occur at every 40 ms (typically), each at one tenth of the desired resolution without taking time for an auto-zero phase between conversions. These minor conversions occur at a rate of 25 per second and are used to provide a fast response bar graph display and fast auto-ranging. In this case, ten samples are used to sum and produce a full resolution digital display. A 100 ms auto-zero phase occurs following every ten sample sequence (typically).

As there is no auto-zero phase in between minor conversion cycles, a residual charge is retained by the integrator capacitor owing to overshoot past the true zero baseline. In the absence of an AZ phase, the residual charge would normally produce a significant error in the sample next taken. However, a digital algorithm is used to eliminate the error due to this residue as it propagates through all ten samples. In this type of multimeter, auto-range capability, touch–hold functions, etc., are achieved by varying the basic timing. Furthermore, a single custom IC is used for the ADC block, auto-range switching and other associated digital functions including a state machine for controlling the ADC process. Figure 4.16 shows a simplified block diagram of the overall multimeter, designed around such a custom IC.

4.3.2.1 Other multiple ramp techniques: Prema Precision Instruments

In some DMMs designed by Prema Precision Electronics, Inc. (USA), a patented technique called the multiple ramp technique is used where there is no switching of the input voltage (unknown). The basic technique and the integrator output voltage waveform are shown in Figure 4.17(a) and (b), respectively. As seen in Figure 4.17(a), the input voltage to the integrator is never switched off and therefore no information is

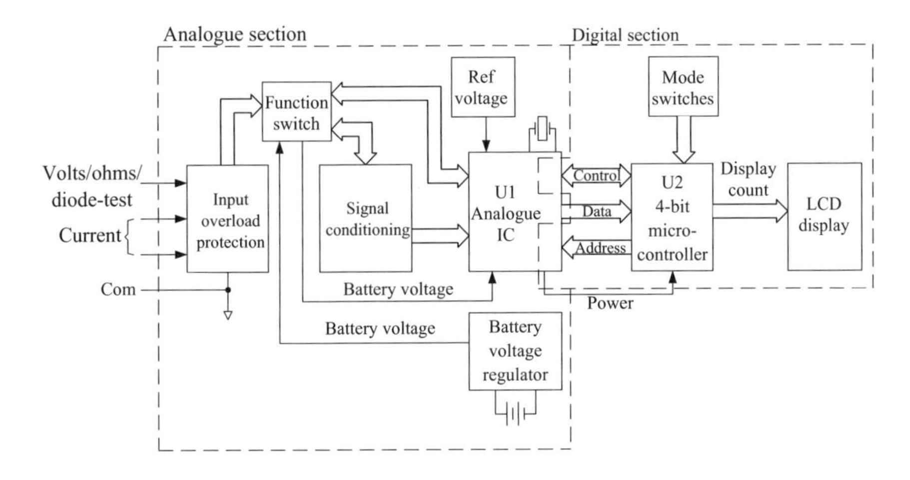

Figure 4.16 Block diagram of a DMM with a custom analogue IC (Fluke model 27) (reproduced by permission of Fluke, Inc., USA)

lost. This results in no errors being introduced during switching. The discharge of the capacitor, C, occurs during a predetermined measurement time, t, in periodic intervals of T_n using a current I_{ref} from a proportional voltage source with opposite polarity V_{ref}. During the discharge time of t_1 to t_n, the oscillator pulses are electronically added. The comparator monitors the polarity of the integrator output. When this signal coincides with the next pulse transition from the oscillator, the reference current is switched off.

Because the total output change of the capacitor is zero during this time, the following formula holds

$$\frac{1}{T}\int_0^T V_{\text{in}}\,\mathrm{d}t = -\frac{R_{\text{in}}}{R_0 T} V_{\text{ref}} \sum t_i. \tag{4.7}$$

With this method, capacitor dielectric losses or drifts do not affect the accuracy of the conversion. It also means that compensations that are generally prone to failure are not required. The results are also independent of the frequency of the clock oscillator. Prema Precision, Inc. (USA), manufactures their full custom ASICs (similar to ADC5601, an ADC with 25-bit resolution) for DMMs and similar applications based on this patented technique [10]. Hewlett-Packard also uses a patented technique termed multislope III for their DMMs.

4.3.3 Special functions

Most of the modern DMMs have several special measurement capabilities such as frequency, capacitance, true r.m.s. a.c. voltage, temperature, continuity test, diode and transistor tests in addition to basic measurements such as voltage, current, resistance or their derivatives such as conductance. In measuring these special parameters, the

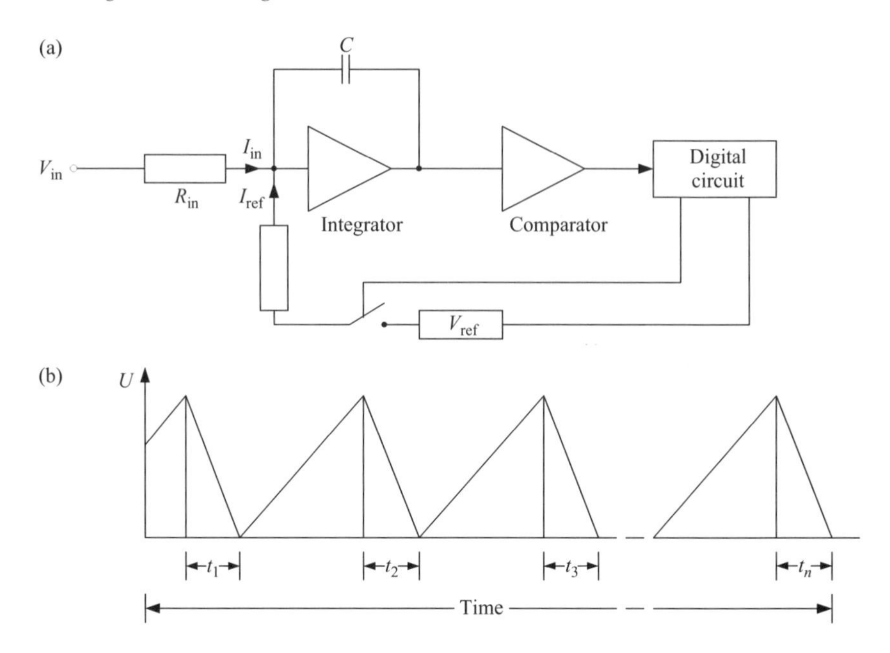

Figure 4.17 Multiple ramp technique by Prema Precision Instruments: (a) functional block diagram; (b) integrator output waveform (reproduced by permission of Prema Precision, Inc., USA)

manufacturers use modifications to the basic techniques or add special components to the basic system components.

For example, in DMMs such as the Fluke 8060A, true r.m.s. a.c. voltage or frequency measurement functions are added to the basic configuration by the use of a true r.m.s. a.c. converter block between the a.c. input and the custom IC (MAC). The overall block diagram and the circuit block for frequency measurements are shown in Figure 4.14(a) and (c), respectively. For frequency measurements, the a.c. signal is divided by the voltage divider and buffered by the r.m.s. converter. The signal is then applied to a comparator in the MAC for counting, where the counter gate is controlled by the microcomputer block. For very low frequency inputs the counter actually measures the period of the input signal. The microcomputer then inverts it to derive the corresponding frequency.

4.3.3.1 Measurement of capacitance

For capacitance measurements the basic dual slope technique is slightly modified in DMMs, such as Fluke models 83, 85, 87. Capacitance measurement is a very useful feature for troubleshooting. The measurement is carried out by measuring the charge required to change the voltage across the unknown capacitor from zero to reference voltage, V_{ref} (see Figure 4.18). This technique is referred to as a ballistic type of measurement.

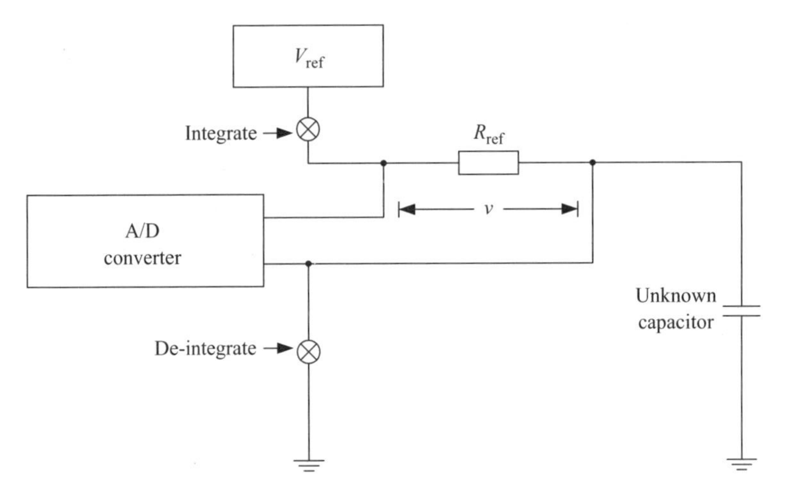

Figure 4.18 Capacitance measurement

The unknown capacitor is fully charged from zero up to V_{ref} during the integrate period. Hence, the charge Q is expressed as

$$Q = \int_0^T i \, dt \tag{4.8}$$

where

$$i = \frac{V}{R_{ref}}; \tag{4.9}$$

finally

$$Q = CV_{ref}, \tag{4.10}$$

so

$$C = \frac{Q}{V_{ref}}. \tag{4.11}$$

The A/D converter integrates the voltage (V) across the known resistor R_{ref} and latches the count. The capacitor is discharged during the de-integrate and hold period. The microcomputer calculates and displays the capacitance from the latched count, which is proportional to the unknown capacitance.

4.3.3.2 Noise measurements using voltmeters

There are many applications where a voltmeter can be used for noise measurements. One classic example is noise measurements of low noise, low dropout voltage regulators. The a.c. voltmeter not only must have adequate bandwidth, but also must faithfully respond to the r.m.s. value of the measured noise. Similarly, the voltmeter must have the crest factor capability to capture the noise signal's dynamic range.

Unfortunately, most a.c. voltmeters, including digital voltmeters with a.c. ranges and instruments with 'true r.m.s.' a.c. scales, cannot be accurate under these measurement conditions. Thus, selecting an appropriate instrument requires care. For details, Reference 11 is suggested.

4.3.3.3 Four wire technique for low resistance measurement

In the field and the design environments, users encounter low resistances in devices and circuits that include contacts found in switches, relays, and connectors. Likewise, users find low continuity resistances in PCB traces, cables, and vias. Typical low resistance measurements are within the range of 1 Ω to less than 1 mΩ. In some cases, users may want to measure resistances in micro-ohm ranges. Most multimeters, particularly the hand-held or the low cost bench-top versions, do not have the capability to measure low resistance values accurately.

To ensure accurate low level resistance testing, you must use the proper measurement technique. Because conductive paths usually have resistances of less than 1 Ω, you should use four wire measurement to eliminate the effects of your instrument's lead resistances.

The diagram in Figure 4.19 illustrates how two leads supply a current to the resistance under measurement, while the other two leads measure the voltage drop across the resistance. Although a small current may flow through the voltage sensing leads, it is usually negligible. Thus, the voltage measured by the meter is essentially the same as the voltage across the resistance. Eliminating thermally generated voltages

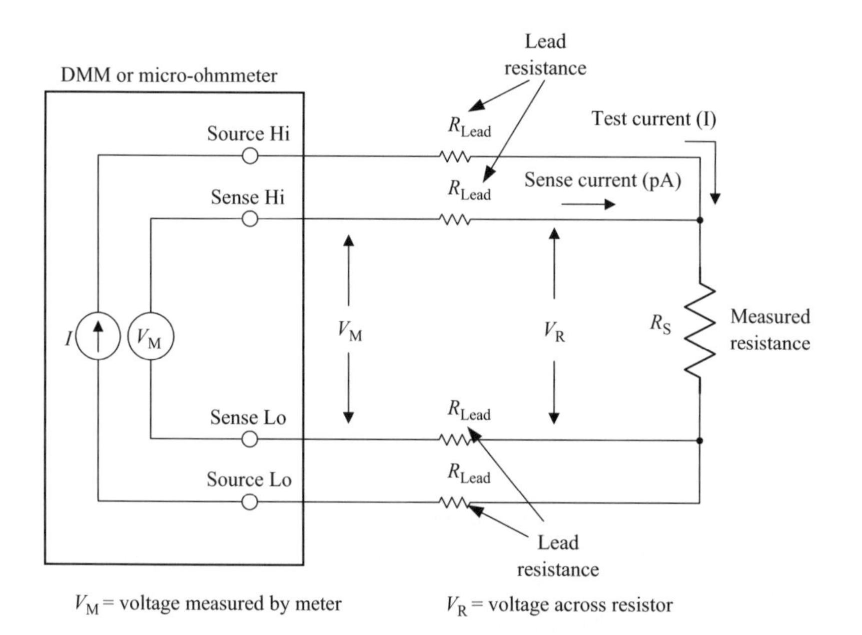

Figure 4.19 Four wire technique for low resistance measurement

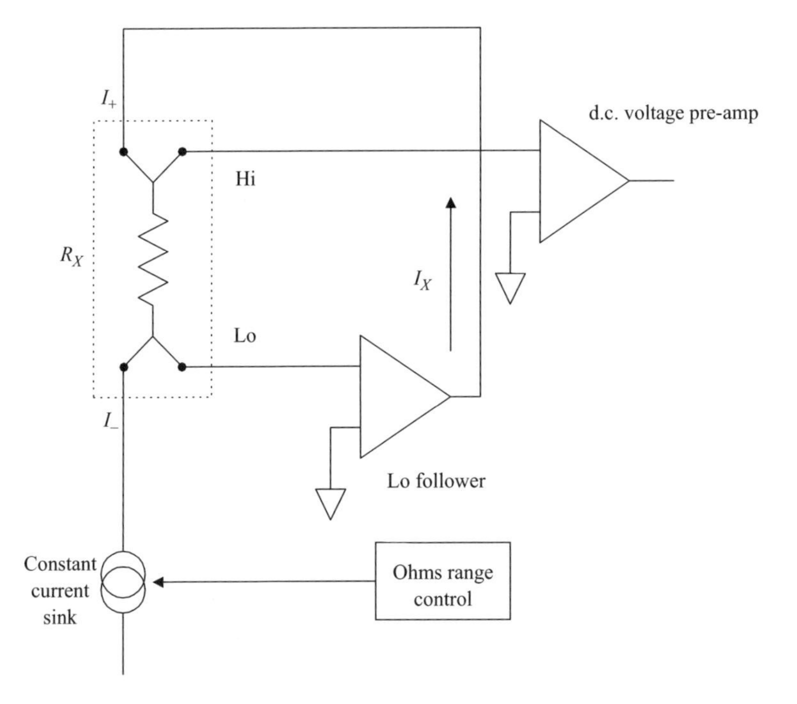

Figure 4.20 An ohms converter technique used in a high accuracy DMM (courtesy: Wavetek) [14]

is another essential part of making accurate measurements. There are two techniques you can use to compensate for thermal voltages; for details, References 12 and 13 are suggested.

Figure 4.20 shows the circuit configuration for resistance measurement used in high accuracy DMMS. The resistance option is primarily a range of selectable constant currents. A constant current generator forces a current I_x to flow through the test resistor. A true constant current source will generate a current independently of the voltage developed across its terminals, in this case designated I_+ and I_-. It therefore follows that if a known resistance is applied to the DMM and the display value noted, the insertion of an additional resistance in series with the I_+ lead should not significantly affect the DMM's reading.

4.4 High accuracy bench multimeters

Most hand-held DMMs are $3\frac{1}{2}$ to $4\frac{1}{2}$ digits and provide basic measurement functions such as a.c./d.c. voltage measurements, current and resistance measurements and a few additional functions such as frequency, temperature and capacitance. However, more accurate bench type or system multimeters are available at higher costs with better performance and powerful features. Most of these high performance system multimeters or bench types are offered in $4\frac{1}{2}$ to $9\frac{1}{2}$ digit versions with very high display update rates. In some high performance DMMs, readings could be obtained in 100 000 readings per second compared with the hand-held versions where only a few readings per second are only possible. Most of these have resolutions up to a few nanovolts and measurement accuracies up to 20 p.p.m. or better. Some of these bench models and system components also have the following special features:

(i) dual displays for two types of reading from a single input,
(ii) selectable count resolutions,
(iii) true r.m.s. a.c. or d.c. coupled measurements for a.c. parameters,
(iv) built-in self-tests and closed box calibration,
(v) menu driven operations,
(vi) built-in maths operations for measured parameters,
(vii) IEEE-488 and RS-232 computer interfaces,
(viii) standard programming languages such as SCPI, etc.,
(ix) storage for readings, and
(x) relative measurements using preset values.

In most of these high priced, high performance DMMs the instrument is designed around a sophisticated microcomputer subsystem and the ADC process may be achieved using custom components. The block diagram of such a bench DMM (Fluke model 45) is shown in Figure 4.21(a). Figure 4.21(b) shows how the customised component, the analogue measurement processor, is configured internally as well as in relation to the external components such as the true r.m.s. converter. This custom CMOS device performs the following functions under control of the main processor:

(i) input signal routing,
(ii) input signal conditioning,
(iii) range switching,
(iv) active filtering of d.c. type measurements (for fast reading this facility is disabled),
(v) ADC process, and
(vi) support for measurement functions.

Inside this analogue measurement processor IC it is easy to identify the basic blocks we have discussed in the previous paragraphs in relation to the basic measurement technique.

The DMM uses a modified dual slope minor cycle technique similar to the Fluke model 27 with a minor cycle period of 25 ms. For true r.m.s. a.c. measurements, an Analog Devices AD637 r.m.s. to d.c. converter chip is used to convert the input a.c. signals (a.c. coupled) to a d.c. value with an equivalent r.m.s. value. The instrument microprocessor subsystem does a calculation when the true r.m.s. value of the a.c. and d.c. components of a signal is needed.

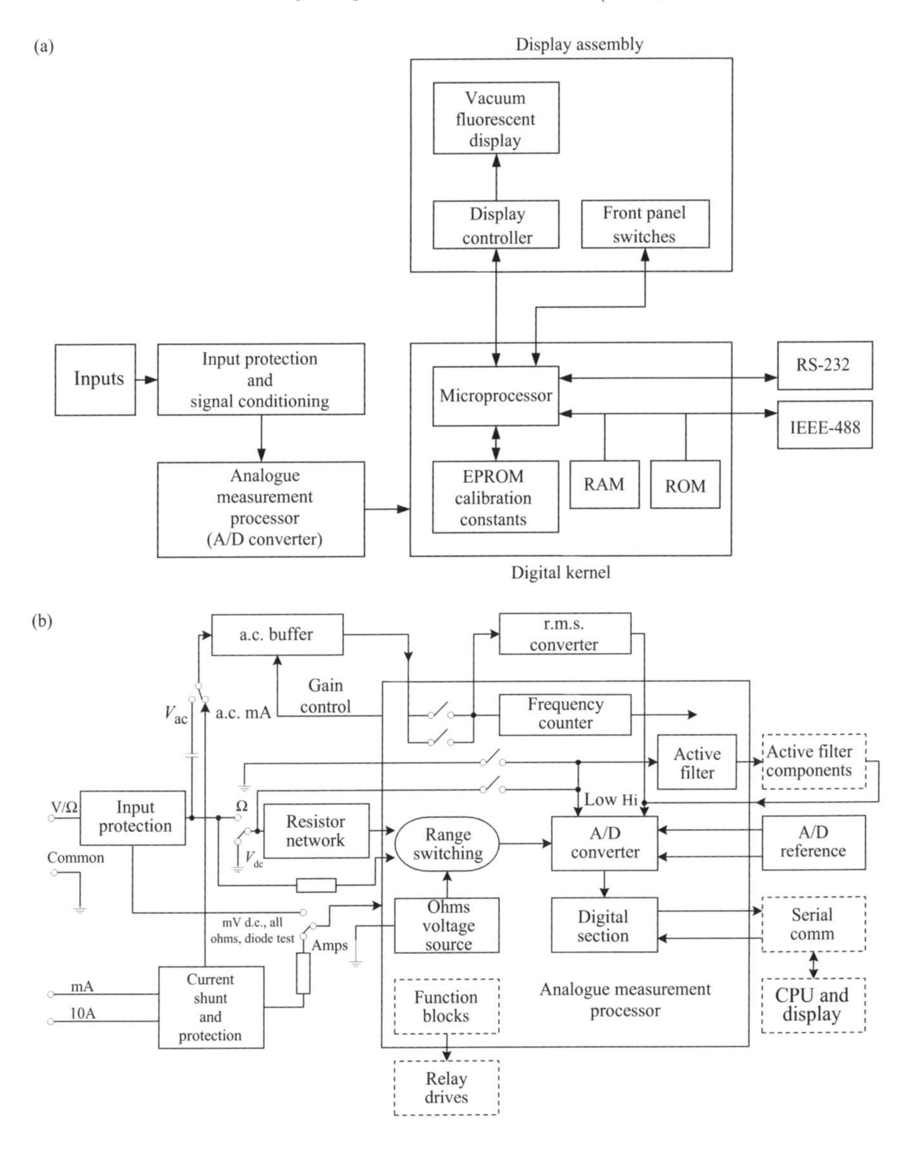

Figure 4.21 Design concepts in a bench multimeter: (a) block diagram; (b) analogue measurement processor (courtesy of Fluke Corporation, USA)

4.5 DMM specifications

A DMM manufacturer has two important goals when specifying a DMM. The manufacturer must ensure that the performance and functionality are clearly presented to the prospective purchaser so that the purchasing decision can be made, and ensure that the instrument's metrological performance is expressed to the best possible advantage in a competitive situation. The specification actually describes the performance of the instrument's internal circuits that are necessary to achieve the desired functionality. Each function will have an accuracy expressed in terms of percentage of the reading (±%R) with additional modifiers or adders such as ±digits or ±μV. This is known as a compound specification and is an indication that there are several components to the specification for any particular parameter. Sometimes this is seen as a way of hiding the true performance when, in reality, it is the only practical way to express performance over a wide parametric range with a single accuracy statement. Over the 20 years or so that DMMs have developed, the terminology has evolved to describe the performance parameters.

4.5.1 Performance parameters

It is convenient to think of the performance in terms of functional and characteristic parameters (see Table 4.5), where the former describes the basic capability or functionality (e.g. the DMM can measure voltage), and the latter is a descriptor or qualifier or its characteristics (e.g. with an accuracy of ±5 p.p.m.). One can easily see that it is essential to understand the significance of the characteristics if the instrument is to be verified correctly.

4.5.2 The d.c. voltage

Nearly all DMMs can measure d.c. voltage. This is because of the nature of the analogue-to-digital converter (ADC) used to convert from a voltage to (usually) timing information in the form of clock counts. The most common way of doing this is to use the dual slope technique or its variations. The basic dual slope method is used in low resolution DMMs, but longer scale length instruments require more complex arrangements to ensure better performance. DMMs are available with up to $8\frac{1}{2}$ digits resolution and these usually employ multi-slope, multi-cycle integrators to achieve good performance over the operating range. An ADC is usually a single range circuit,

Table 4.5 Performance parameters and characteristics of DMMs

Functional parameters	Characteristics
• Functions – V, A, Ω	• Stability with time and temperature
• Scale length/resolution	• Linearity
• Read rate	• Noise
• Amplitude range	• Frequency response or flatness
• Frequency range	• Input impedance
	• Compliance/burden
	• Common/series mode rejection
	• Crest factor

that is to say it can handle only a narrow voltage range (of say zero to ±10 V); however, the DMM may be specified from zero to ±1 kV. This necessitates additional circuits in the form of amplifiers and attenuators to scale the input voltage to levels that can be measured by the ADC. In addition, a high input impedance is desirable such that the loading effect of the DMM is negligible. Each amplifier and attenuator or gain defining component introduces additional errors that must be specified. The contributions that affect the specifications for d.c. voltage are given in Table 4.6, together with their typical expression in parentheses.

These contributions will be combined to give a compound specification expressed ±%R ±%FS ±μV. In order that the performance of the instrument can be verified by calibration, the above effects must be isolated. That is to say, it is not possible to measure *linearity*, for example, until the effects of *offset* and *gain* errors have been removed. Figure 4.22 shows these basic parameters.

Table 4.6 Contributions to DCV specifications

Specification	Typical expression
Reference stability	(Percentage of reading)
ADC linearity	(Percentage of scale)
Attenuator stability	(Percentage of reading)
Voltage offsets	(Absolute)
Input bias current	(Absolute)
Noise	(Absolute)
Resolution	(Absolute)

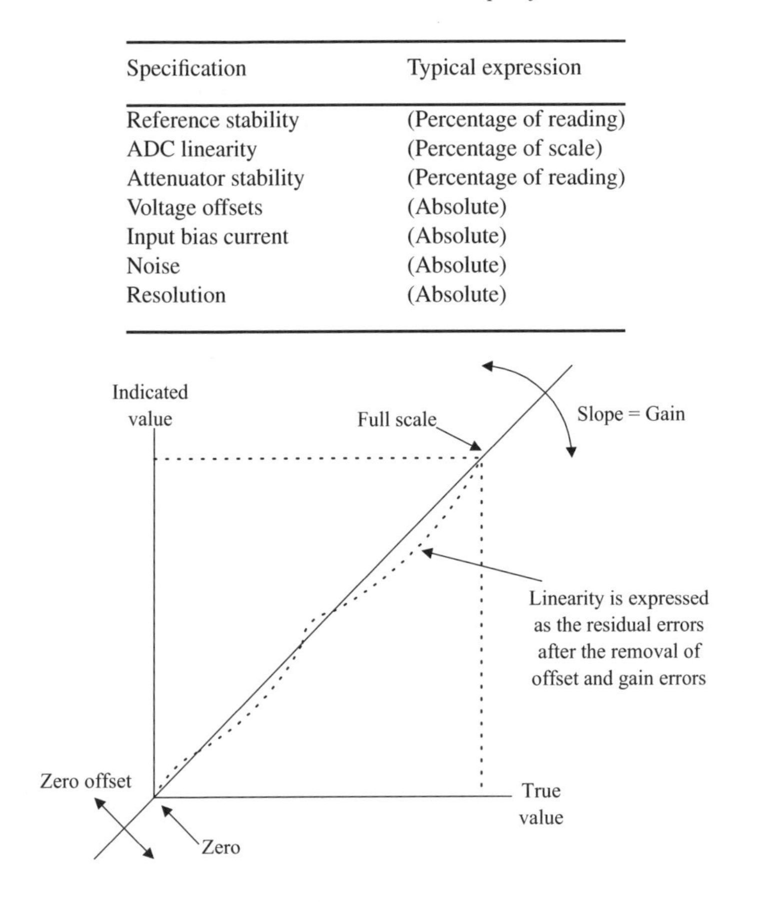

Figure 4.22 DMM offset, gain and linearity

The ADC is common to all ranges of all functions, therefore its characteristic errors will affect all functions. Fortunately, this means that the basic d.c. linearity need only be verified on the basic (usually 10 V) range. The manufacturer's literature should indicate the prime d.c. range. If this is not stated directly, it can be deduced from the d.c. voltage specification, i.e. the range with the best specification in terms of ± percentage R, ± percentage FS and ±μV will invariably be the prime range. Other ranges, e.g. 100 mV, 1 V, 100 V and 1 kV, will have a slightly worse performance because additional circuits are involved. At low levels on the 100 mV and 1 V ranges, the dominant factor will be noise and voltage offsets. For the higher voltage ranges, the effects of power dissipation in the attenuators will give a power law characteristic, the severity of which will depend on the resistor design and individual temperature coefficients. Knowledge of the design and interdependence of the DMM's functional blocks can greatly assist the development of effective test strategies; for details, Reference 14 is suggested.

4.6 Multimeter safety

In using multimeters, safety of the user and the instrument are equally important. The real issue for multimeter circuit protection is not just the maximum steady state voltage range, but a combination of both steady state and transient over-voltage withstand capability.

Transient protection is vital. When transients ride on high energy circuits, they tend to be more dangerous because these circuits can deliver large currents. If a transient causes an arc-over, the high current can sustain the arc, producing a plasma breakdown or explosion, which occurs when the surrounding air becomes ionised and conductive. The result is an arc blast, a disastrous event which causes more electrical injuries than the better known hazard of electric shock.

4.6.1 IEC 1010 standard

With respect to international standards, the most important standard applicable to transients and safety is the new IEC 1010-1 standard, which replaced IEC 348. This particular standard defines categories I–IV, often abbreviated as CAT I, CAT II, etc. See Figure 4.23.

The division of a power distribution system into categories is based on the fact that a dangerous high energy transient such as a lightning strike will be attenuated or dampened as it travels through the impedance (a.c. resistance) of the system. A higher CAT number refers to an electrical environment with higher power available and higher energy transients. Thus a multimeter designed to a CAT III standard is resistant to much higher energy transients than one designed to CAT II standards.

Within a category, a higher voltage rating denotes withstand rating: e.g., a CAT III–1000 V meter has superior protection compared with a CAT III–600 V rated meter. The real misunderstanding occurs if someone selects a CAT II–1000 V rated meter

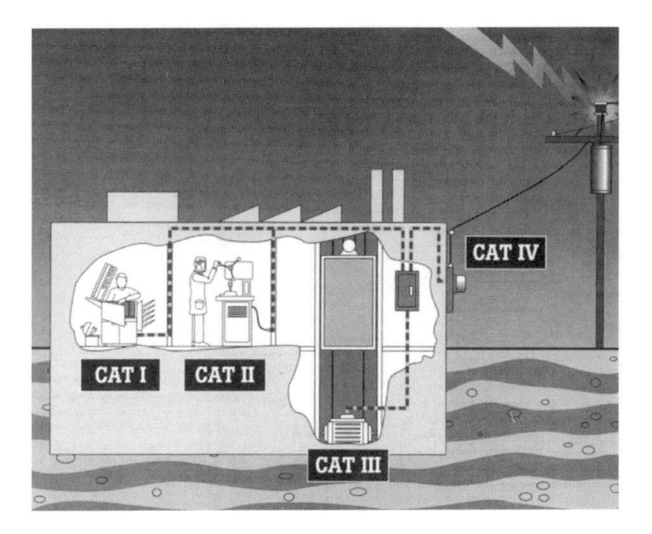

Figure 4.23 Locations and IEC 1010 categories (courtesy of Fluke, Inc., USA)

thinking that it is superior to a CAT III–600 V meter. Table 4.7 relates examples in the field and the over-voltage category.

4.6.2 Voltage withstand ratings and IEC 1010

IEC 1010 test procedures take into account three main criteria:

(i) steady state voltage;
(ii) peak impulse transient voltage; and
(iii) source impedance.

These three criteria together will tell you a multimeter's true voltage withstand value. Table 4.8 describes test values for over-voltage installation categories according to IEC 1010.

As described in Table 4.8, an instrument's true voltage withstand rating is dependent not only on the voltage ratings but also on the test source impedance. It is obvious from Table 4.8 that a CAT III–600 V meter is tested with 6000 V transient, whereas a CAT III–1000 V meter is tested with 8000 V transient.

However, a less obvious fact is the difference between the 6000 V transient for CAT III–600 V and the 6000 V transient for CAT II–1000 V. They are not the same and with a lower source impedance used for CAT III, it indicates that such a product offers superior transient protection. This is because of the fact that with a source impedance that is lower by a factor of six, the CAT III test source has 36 times the power applied to the meter in testing. In practical multimeters, to achieve this, safety features of transient absorbent devices such as metal oxide varistors, spark gaps or semiconductor types are used. For more details of safety, Reference 15 is recommended.

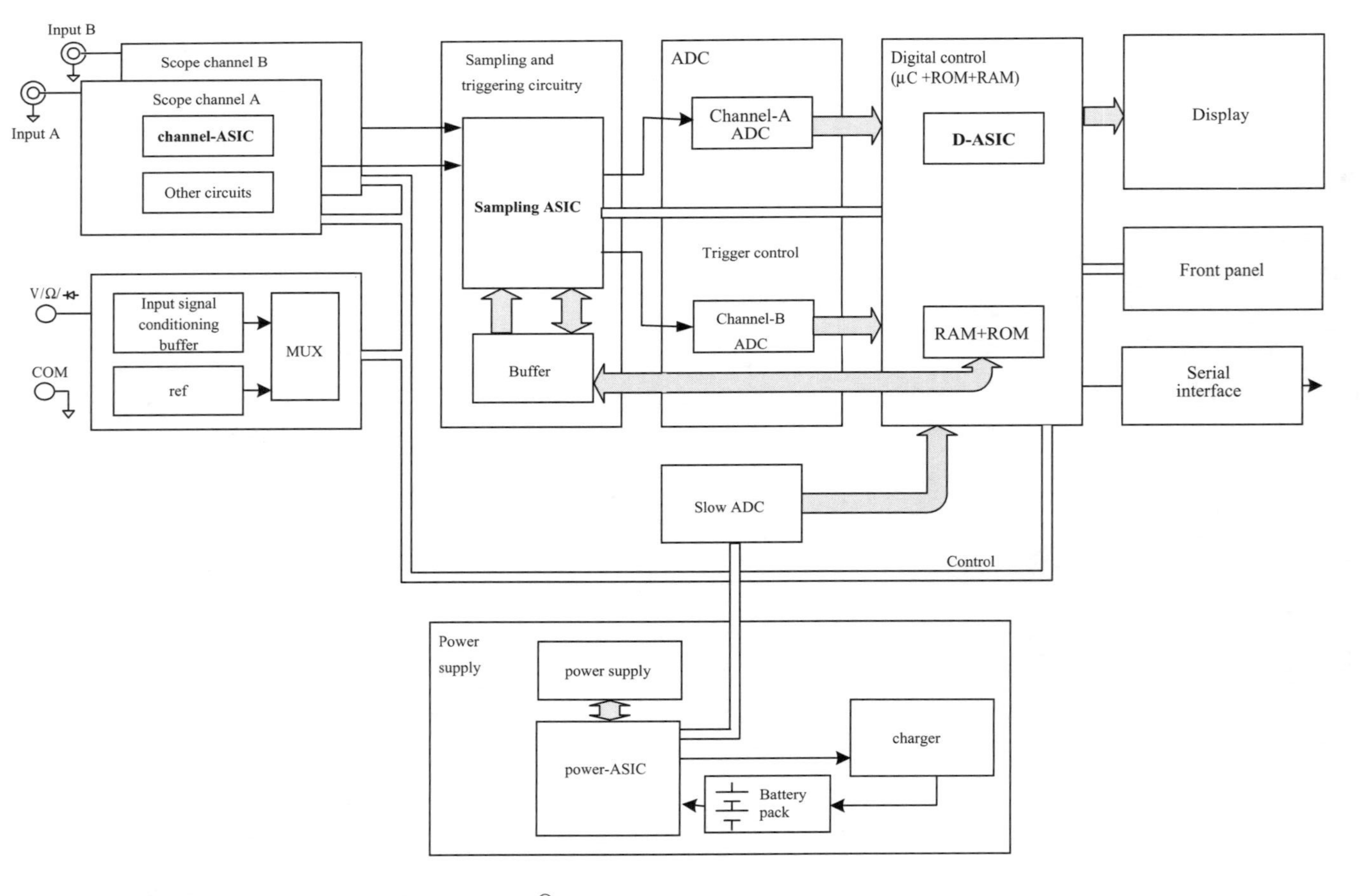

Figure 4.24 Simplified block diagram of Scope Meter® (courtesy: Fluke, Inc., USA)

Table 4.7 Over voltage installation categories where IEC 1010 applies to low voltage (<1000 V) test equipment (courtesy of Fluke, Inc., USA)

Over-voltage category	In brief	Examples
CAT IV	Three phase at utility connection, any outdoor conductors	• Refers to the 'origin of installation', i.e. where low voltage connection is made to utility power. • Electricity meters, primary over current protection equipment. • Outside and service entrance, service drop from pole to building, run between meter and panel. • Overhead line to detached building, underground line to well pump.
CAT III	Three phase distribution, including single phase commercial lighting	• Equipment in fixed installations, such as switch gear and polyphase motors. • Bus and feeder in industrial plants. • Feeders and short branch circuits, distribution panel devices. • Lighting systems in larger buildings. • Appliance outlets with short connections to service entrance.
CAT II	Single phase receptacle connected loads	• Appliance, portable tools, and other household and similar loads. • Outlet and long branch circuits. • Outlets at more than 10 m (30 feet) from CAT III source. • Outlets at more that 20 m (60 feet) from CAT IV source.
CAT I	Electronic	• Protected electronic equipment. • Equipment connected to (source) circuits in which measures are taken to limit transient over-voltages to an appropriately low level. • Any high voltage, low energy source derived from a high winding resistance transformer, such as the high voltage section of a copier.

4.7 Combination instruments – oscilloscope and DMM in a single package

Discussions so far have indicated that the accuracy of most waveform measurements are dependent on the waveform shape, the r.m.s. to d.c. conversion technique

Table 4.8 Transient test values for over-voltage installation categories (50/150/300 V values not included)

Over-voltage installation category	Working voltage (d.c. or a.c.-r.m.s. to ground)	Peak impulse transient (20 repetitions)	Test source ($\Omega = V/A$)
CAT I	600 V	2500 V	30 Ω source
CAT I	1000 V	4000 V	30 Ω source
CAT II	600 V	4000 V	12 Ω source
CAT II	1000 V	6000 V	12 Ω source
CAT III	600 V	6000 V	2 Ω source
CAT III	1000 V	8000 V	2 Ω source

employed, and the crest factor of the waveform, etc. In summary, the more complex the waveform, the more the reading's reliability and accuracy may be questioned. Some instrument manufacturers such as Fluke, Tektronix, etc., have introduced the combination instruments such as 'Scope Meter' and 'TekMeter'.

These instruments benefited from the ASIC technologies and the low cost LCD displays, etc., that have helped in creating a very useful, cost effective hand-held 'combination instrument'. For example, the Scope Meter® 190 series from Fluke provides the user with 200 MHz, 2.5 GS s^{-1} real time sampling and deep memory (of 27.5 K points per input channel) in a battery powered package. Scope Meter 123 is useful for industrial trouble shooting and installation applications. Some features of such an instrument are:

- a dual input 20 MHz digital storage scope,
- two 5000 count true-r.m.s. DMMs,
- a 600 V/CAT III safety level, and
- an optically isolated RS-232 interface.

4.7.1 Design concepts of Scope Meter®

This is a classic example of the use of ASICs in test instruments and the compact battery powered test solution is implemented in a portable case by the use of five ASICs and associated circuitry. A simplified block diagram is shown in Figure 4.24.

4.8 Pulse parameters and measurements

Use of pulse waveforms and measurement of their parameters requires an accurately defined set of pulse parameters taking into account various practical situations. An ideal pulse is shown in Figure 4.25(a), and a practical pulse is shown in Figure 4.25(b). Definitions of the pulse parameters and various terms are provided in IEC Standard

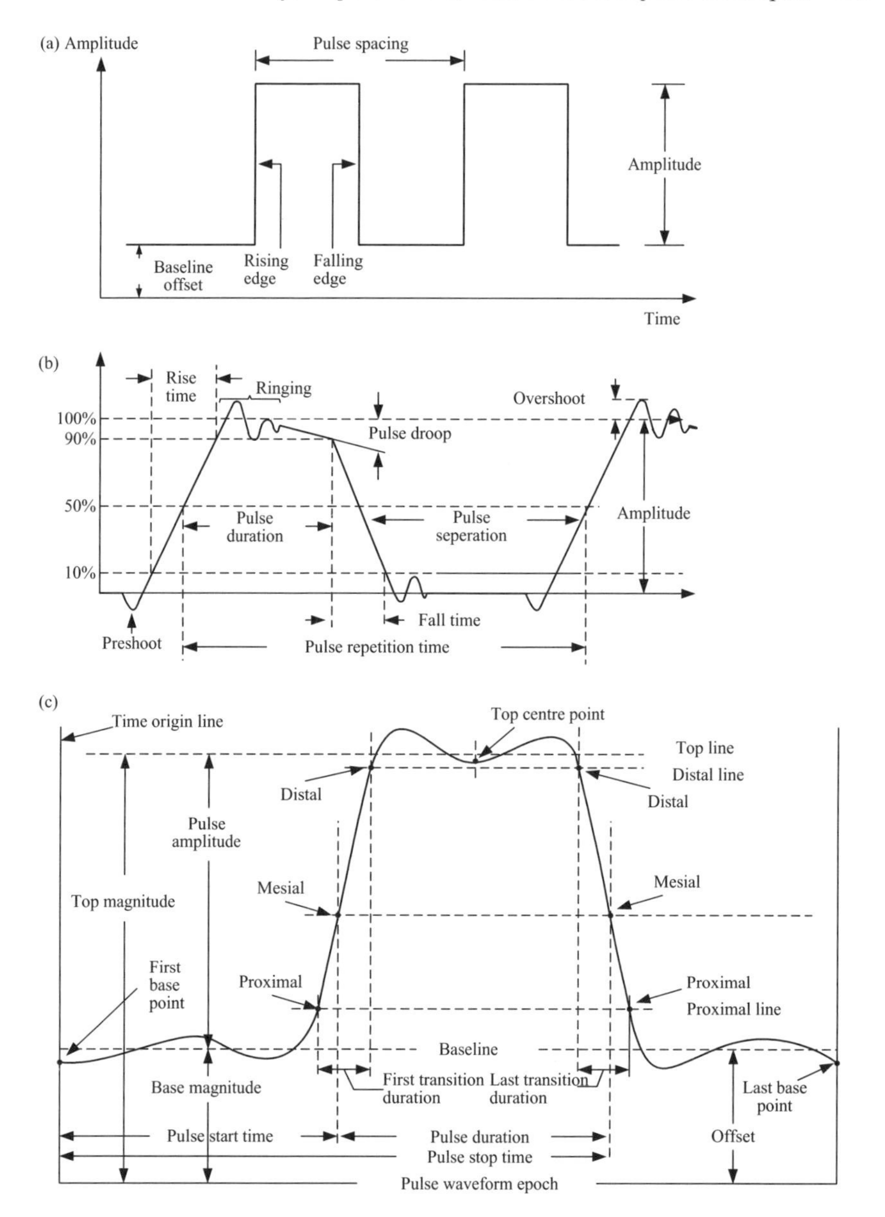

Figure 4.25 Pulse characteristics: (a) ideal pulse; (b) practical pulse; (c) single pulse waveform

469-1 (1987) and BS 5698 (1989); see Table 4.9. However, different practical terms are used in technical documentation, and in selecting pulse apparatus it is necessary to interpret these parameters carefully and accurately. In an ideal pulse, the rising or falling edges are vertical and the lower amplitude level (baseline) and the maximum

Table 4.9 Practical pulse parameters and their relation to IEC 469.1 terminology

Parameter		Practical definition	IEC 469 definition
Practical term	IEC 469.1 term		
1. Pulse amplitude	Same	The mean value of level difference between base line and top line	The algebraic difference between top magnitude and base magnitude
2. Pulse level	Base (top) magnitude	The level of base line (or top line) from zero voltage level	A magnitude of base (top) as obtained by a specified procedure or algorithm
3. Median	Mesian point	50% amplitude point on pulse edge	A magnitude referenced point at intersection of a waveform and mesial line
4. Pulse duration (pulse width)	Same	Time interval between leading edge and trailing edge medians	The duration between pulse start time and pulse stop time
5. Transition time	Transition duration	Time interval between 10% and 90% amplitudes on rising and falling edges of a pulse	The duration between proximal point and distal point on a transition waveform
6. Rise time	First transition duration	Transition time on leading edge	The transition duration of first transition waveform in a pulse waveform
7. Fall time	Last transition duration	Transition time on falling edge	The transition duration of last transition waveform in a pulse waveform

8. Preshoot	Same	Peak deviation from base line at rising edge (usually represented as a percentage)	A distortion which precedes a major transition
9. Overshoot	Same	Peak deviation from high level (100% amplitude) at leading edge (usually represented as a percentage)	A distortion which follows a major transition
10. Ringing	Same	This is defined as oscillation occurring at rising (or falling) edge	A distortion in the form of a superimposed damped oscillatory waveform which when present usually follows a major transition
11. Settling time		Time taken from 90% level on rising edge to settle down with amplitude accuracy band of pulse level	
12. Linearity		Peak deviation of an edge from a straight line through 10% and 90% amplitude points expressed as a percentage of pulse amplitude	
13. Jitter	Same	Short term instability of one event with regard to another	The instability of time characteristics of pulse waveform in a pulse train with regard to a reference time, interval or duration. Unless otherwise specified peak-to-peak jitter is assumed

amplitude level are horizontal. Thus, in a practical pulse, the following characteristics may be observed:

- the rising and falling edges are not vertical;
- at the base and peak levels, ringing, overshoots and preshoots, etc., can be observed;
- pulse base and peak levels are not uniform or horizontal;
- in repetitive pulse trains, the edges may shift with respect to one another about a mean value (jitter).

When practical instruments (oscilloscopes, etc.) are used as measuring instruments, the parameter errors, when added, may be quite high. It is important, therefore, that we evaluate the added errors due to the measurement process carefully in the process of interpreting the results.

Table 4.9 gives the practical definitions of the common pulse terms and their definitions according to IEC standard 469-1 (1987). See also Figures 4.25(b) and (c).

4.8.1 Use of oscilloscopes to measure rise and fall times

The most practical method of measuring the rise and fall times (first and last transition) is to use an oscilloscope. However, it is important to note that the measuring system (oscilloscope, connecting cables, etc.) adds a finite rise (or fall) time to the actual parameter. For example, if one can generate a pulse train with negligible rise (and fall) time compared with the oscilloscope rise time (discussed in chapter 5), the measured values will be almost equivalent to the oscilloscope rise time. If we consider a measuring system, as shown in Figure 4.26, with a pulse generator, cable and oscilloscope, the measured rise time t_{rm} will be given according to the following relationships

$$t_{rm}^2 = t_{rg}^2 + t_{rc}^2 + t_{ro}^2, \tag{4.12}$$

where t_{rg} is the rise time of the pulse generator, t_{ro} is the rise time of the oscilloscope, and t_{rc} the rise time of the cable.

Because vertical channel inputs can be simplified to an RC network (discussed in chapter 5) in most oscilloscopes, then

$$t_{ro} \cdot BW = 0.35, \tag{4.13}$$

where t_{ro} is the oscilloscope rise time (s) and BW is the vertical channel bandwidth (Hz). Also it is important to note that some oscilloscope designs may vary from this formula (where the constant is assumed to be 0.35). In such a case one has to use the

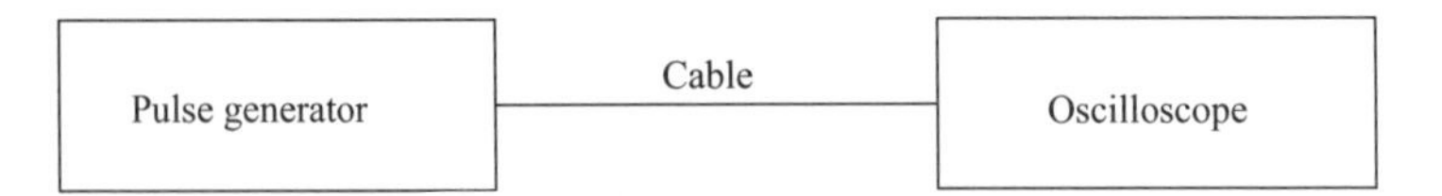

Figure 4.26 Pulse parameter measurement

constant that is specified by the scope manufacturer, or one should have the knowledge of the vertical channel characteristics. This is discussed in chapter 5.

By using the above two equations we can easily show that, when different oscilloscopes having different bandwidths (with negligible cable rise times) are used, the measured rise time for a 10 ns second pulse can vary from 10.01 ns (for a 1000 MHz oscilloscope) to 36.4 ns (for a 10 MHz oscilloscope). So, the percentage error may vary from 0.1 per cent to 260 per cent. Therefore, extreme care is necessary when selecting an oscilloscope for rise time measurements. Another criterion is that well screened cables which do not introduce unwarranted LC circuits, creating ringing, etc., should be used. Also, one should be extremely conscious about the measurement setup when observing high frequency waveforms beyond a few MHz.

4.8.2 *Observation of pulse jitter on oscilloscopes*

The jitter phenomenon of a pulse train is best observed with the aid of an oscilloscope. The scope has to be triggered at a suitable level and slope along the pulse train before observing the jitter on the relevant parameter. To this end, it is very useful to have dual time bases on the oscilloscope to aid in the measuring of the jitter parameters. (This is discussed in chapter 6.) Jitter, as applied to period, width and delay is shown in Figure 4.27.

In practical digital transmission systems, the measurement of jitter (considered as a high frequency error, >10 Hz) and wander (considered as low frequency instability, observable only after 100 ms or longer) requires special modulation domain analysers. (This is discussed in chapter 7.)

4.8.3 *Time domain reflectometry*

One useful application of pulse techniques is in time domain reflectometry (TDR), where a pulse or a step generator and an oscilloscope are used to observe the discontinuities along a transmission line. In this most elementary form of TDR a voltage step is propagated down the transmission line under investigation, and the incident and reflected voltage waves are monitored by the oscilloscope at a particular point on the line. This echo technique reveals, at a glance, the characteristic impedance of the line and it shows both the position and nature (resistive, inductive or capacitive) of each discontinuity along the line.

A time domain reflectometer set up is shown in Figure 4.28(a). The step generator produces a positive incident wave that is applied to the transmission system under test. The step travels down the transmission line at the velocity of propagation of the line, V_{p}. Thus,

$$V_{\mathrm{p}} = \frac{\omega}{\beta} \text{ unit length s}^{-1}, \tag{4.14}$$

where ω is the frequency in rad s^{-1} and β is the phase shift in rad unit length^{-1}.

The velocity of propagation approaches the speed of light, V_{c}, for transmission lines with air dielectric. For the general case, where ε_{r} is the relative permittivity

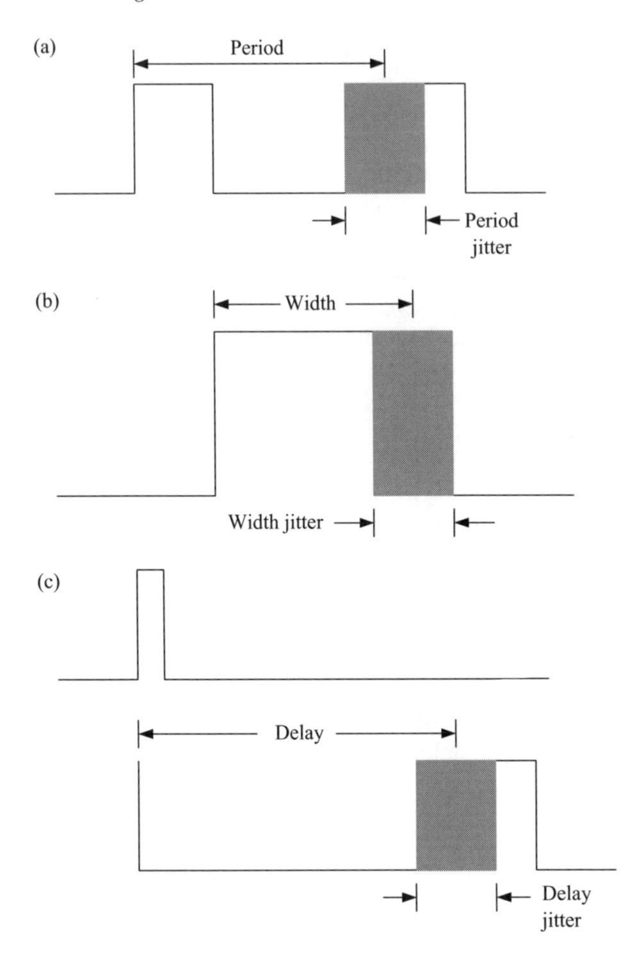

Figure 4.27 Different forms of jitter: (a) period jitter; (b) width jitter; (c) delay jitter

(sometimes called the dielectric constant),

$$V_{\mathrm{p}} = \frac{V_{\mathrm{c}}}{\sqrt{\varepsilon_{\mathrm{r}}}}. \tag{4.15}$$

When the transmission line is not terminated with the characteristic impedance Z_0, there will be a reflected wave, originating at the load and propagating back towards the source. Therefore, the quality of the transmission system is indicated by the ratio of this reflected wave to the incident wave originating at the source. This ratio, ρ, is called the voltage reflection coefficient, and is expressed as

$$\rho = \frac{E_{\mathrm{r}}}{E_{\mathrm{i}}} = \frac{(Z_{\mathrm{L}} - Z_0)}{(Z_{\mathrm{L}} + Z_0)}, \tag{4.16}$$

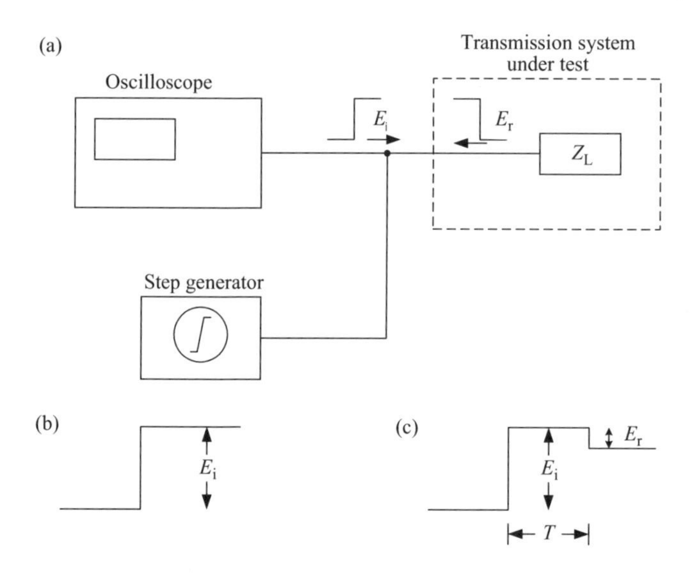

Figure 4.28 *Time domain reflectometry: (a) simple time domain reflectometer; (b) oscilloscope display where $E_r = 0$; (c) oscilloscope display where $E_r \approx 0$*

where E_r is the reflected voltage, E_i is the incident voltage, Z_L is the load impedance, and Z_0 is the characteristic impedance.

If the load impedance is equal to the characteristic impedance of the line, no wave is reflected and all that will be seen on the oscilloscope is the incident voltage step recorded as the wave passes the point on the line monitored by the oscilloscope (Figure 4.28(b)). If a mismatch exists at the load, part of the incident wave is reflected. The reflected voltage wave will appear on the oscilloscope display algebraically added to the incident wave as shown in Figure 4.28(c).

The reflected wave is readily identified because it is separated in time from the incident wave. This time is also valuable in determining the length of the transmission system from the monitoring point to the mismatch. If this distance is D, then

$$D = V_p \times \frac{T}{2}, \tag{4.17}$$

where T is the transit time from the monitoring point to the mismatch and back, as measured on the oscilloscope.

The velocity of propagation can be easily determined from an experiment on a known length of the same type of cable. The shape of the reflected wave is also valuable since it reveals both the nature and magnitude of the mismatch. Figure 4.29 shows a few typical oscilloscope displays and the load impedance responsible for each.

For a detailed discussion on time domain reflectometry principles, References 16–22 are suggested.

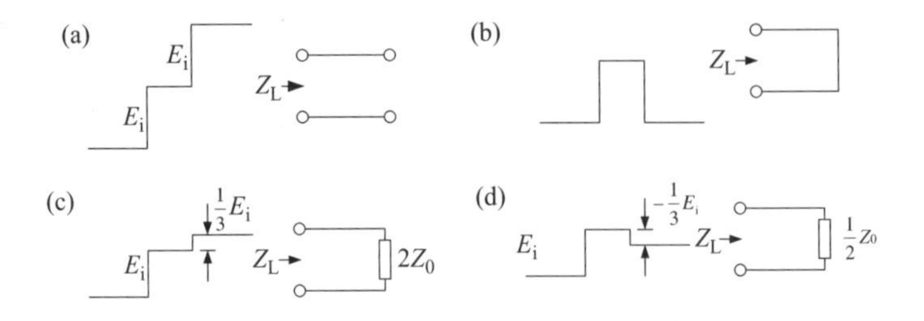

Figure 4.29 *TDR displays for different loads: (a) open circuit termination ($Z_L = \infty$); (b) short circuit termination ($Z_L = 0$); (c) line terminated in $Z_L = 2Z_0$; (d) line terminated in $Z_L = 1/2Z_0$*

Table 4.10 *Typical logic family voltage specifications [17]*

Logic type	TTL	Schottky	CMOS	ECL (10 K)	ECL (100 K)	GaAs
Rise time (ns)	4–10	1.5–2.5	10–100	1.5–4	0.5–2	0.2–0.4
V_{OHA} min (V)	2.7	2.7	5.9	−0.980	−1.035	−0.2
V_{IHA} min (V)	2.0	2.0	4.2	−1.105	−1.165	−0.9
V_{ILA} max (V)	0.8	0.8	1.2	−1.475	−1.475	−1.6
V_{OLA} max (V)	0.5	0.5	0.1	−1.630	−1.610	−1.9
Noise margin (high level)	0.7	0.7	1.7	0.125	0.115	0.7 (25 °C)

4.8.3.1 TDR applications and different types

Modern TDRs can be divided into two basic types, namely metallic time domain reflectometers (MTDR) and optical time domain reflectometers (OTDRs). MTDRs are used for testing metallic cables and OTDRs are used for optical fibre transmission systems.

4.8.3.1.1 MTDR applications

Metallic time domain reflectometers, or simply TDRs, can be used for many practical measurements. Some of these are

(a) testing of interconnects and signal paths in circuitry, and
(b) high speed digital design environments.

In RF and microwave systems, TDR usage is very common for identifying various discontinuities and these are well described in the application note in Reference 16.

With the processors and associated digital circuits reaching the clock speeds beyond several 100 MHz, most circuit interconnections on a printed circuit board (PCB) tend to act as transmission lines. Designers are then forced to apply distributed parameter concepts instead of simple lumped parameter concepts. To explain signal integrity constraints, let us refer to Table 4.10.

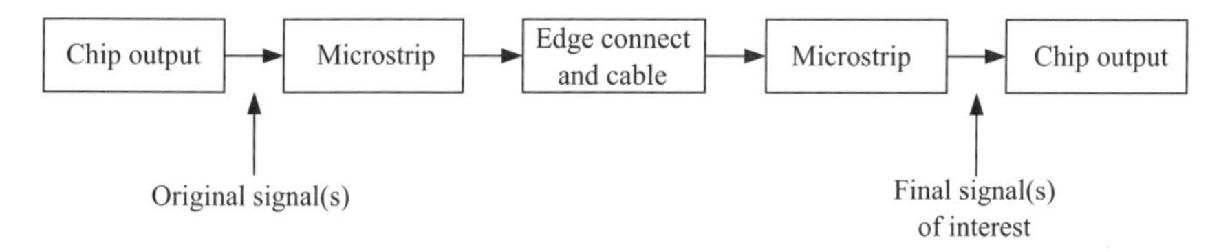

Figure 4.30 A typical digital transmission path

With reference to Figure 4.30, a worst case logic 1 output from a device (V_{OHA}) must pass through the transmission path and must maintain an adequate level for the next gate (V_{IHA}). This defines a noise margin for the high level, and similar requirements exist for the low level. As edge speeds increase depending on the logic family, transmission line factors like ringing, undershoot, reflections, and cross-talk can all become critical to maintaining these noise margins. Overall, the criteria is fairly forgiving for digital. Note that a 700 mV undershoot on a 1.7 V swing GaAs signal (45 per cent undershoot) is the margin. This 700 mV margin is at 25 °C and could approach almost zero at higher temperatures [17].

For the cases similar to the above, at very high frequencies of clock speeds, one has to be considerate of signal integrity issues, and TDR techniques become very useful. Details are discussed in Reference 17.

In testing interconnect paths in PCBs and other circuitry, TDR techniques greatly assist the designer. In these situations, an advanced technique called 'differential TDR' could be used. In this case, instead of driving a pulse waveform through the circuit path with respect to ground plane, a symmetric feed of a test signal is used as shown in Figure 4.31.

Single ended TDR techniques fail to provide complete information about the balanced signal propagation through the differential interconnect. When performing differential TDR measurement, you can make the following important assumptions.

- The lines in the device under test are symmetric. (This condition is true to an acceptable degree for a typical disk drive interconnect.)
- The TDR incident sources must be symmetric. (This condition is always true if designers use outputs of one TDR sampling head in the measurement.)
- The TDR incident steps arrive at the device under test at the same time. (Any skew between the TDR channels at the time that the TDR signals reach the device under test results in an erroneous differential measurement and model.)

This last point assumes that the measurement cables and probes interfacing the TDR instrument to the device under test must match reasonably well. You can correct some small variations in cable and probe length (on the order of a few hundred picoseconds, or 0.5 to 1 inches long) by using the internal capability of TDR instruments to adjust the relative position of the TDR step sources. Reference 18 discuss the application of the technique in flexible interconnect characterisation related to disk drives. For details, References 17–19 are suggested. Reference 20 provides a guideline for modelling interconnects and packages for TDR.

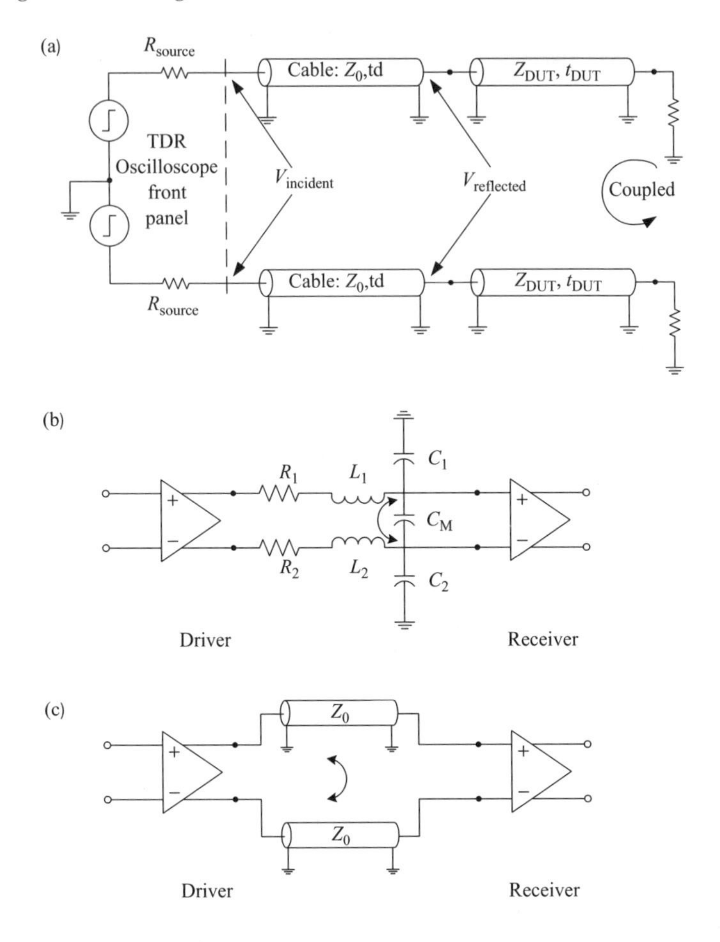

Figure 4.31 The differential TDR technique: (a) basic approach; (b) example of driver–receiver pair with lumped parameter approach; (c) driver–receiver pair with distributed parameter approach

4.8.3.1.2 OTDR applications

For several applications such as long distance telecom interconnection, metropolitan area networks, and basic telecom access to subscriber premises, fibre optic cables are extensively used today. Similar to MTDR applications, an OTDR sends light pulses down an optical fibre, measures the reflected and back scattered light as a function of time, and displays the measurement on a CRT or LCD display as a function of distance.

An OTDR works by transmitting pulses of light down a fibre and then observing the reflected pulses. This is a time based process. But by programming the OTDR with the index of refraction (IOR) of the fibre material, the display can be converted from what would otherwise be a time based distance. (The IOR is the ratio of the speed of light in a vacuum to the speed of light in the material. For glass fibre this value is 1.5.) In Figure 4.32(a) a pulse generator drives a laser diode that emits optical

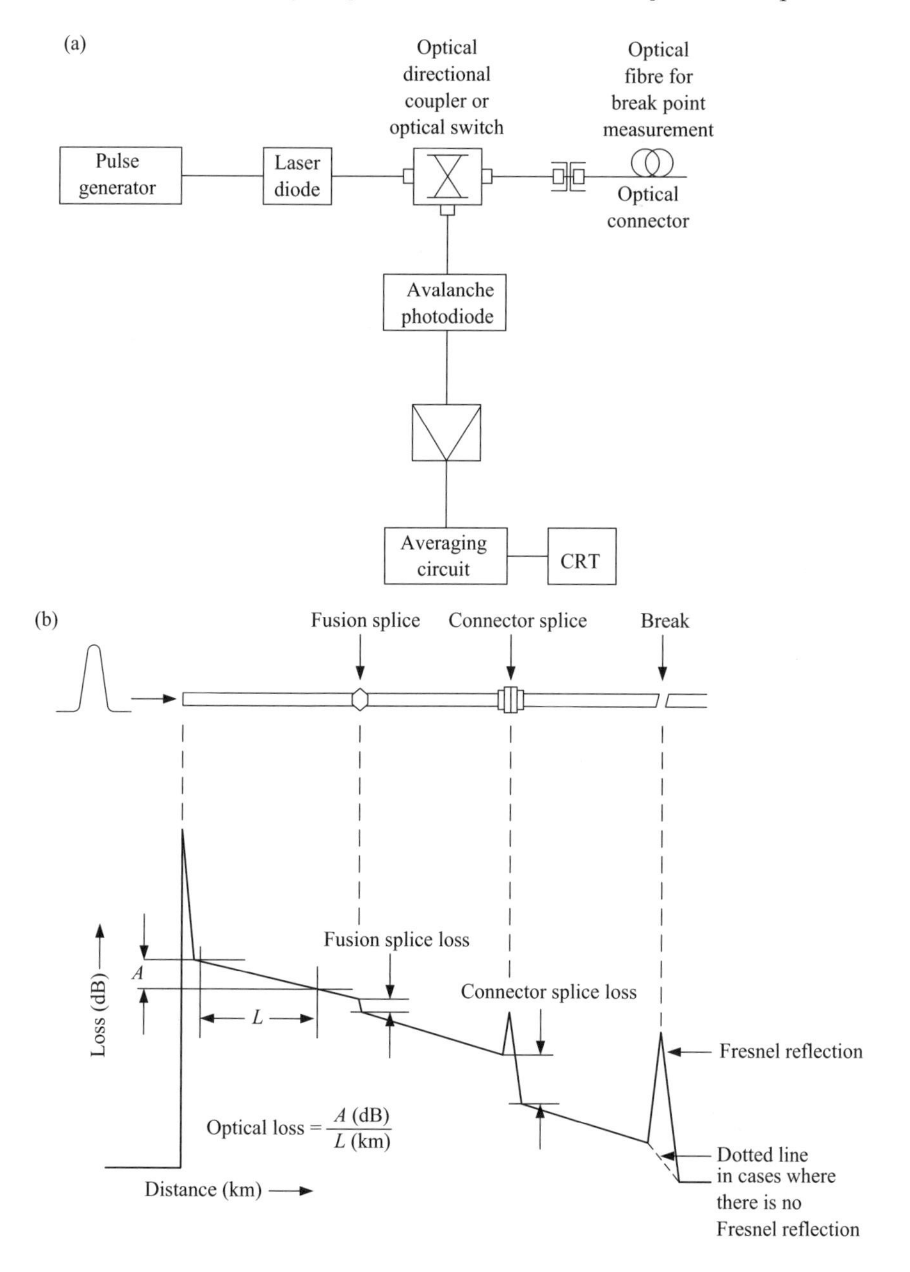

Figure 4.32 OTDR block diagram and a typical OTDR trace: (a) block diagram; (b) typical OTDR trace [21]

pulses of a specific pulse width and duty cycle. Wider pulse widths tend to increase dynamic range as they send a greater amount of light into the fibre. Narrower pulse widths, however, increase resolution. The OTDR user can program the pulse width to fit the application.

The light emitted by the OTDR then travels through a coupler or an optical switch that couples it to the fibre. An optical coupler can be much simpler and cheaper to use, but using an optical switch can improve performance and add a key function: optical masking of large reflections.

Optical masking is implemented by an electrically controlled switch which can be toggled to mask large reflections that might otherwise hide parts of the trace. For example, a relatively small fusion splice may be hidden in the display of a reflection from a large connector. Toggling the switch at the right time masks the reflection from the connector, allowing the much smaller reflection from the fusion splice to be seen. Without the masking process, this trace might be hidden by the amplifier's finite recovery time.

If the signal is not masked, as in normal OTDR operation, the reflected light returning from the fibre passes through the optical switch or coupler and is directed to an avalanche photodiode (APD), where it is detected. The signal, which is not electrical, passes through an amplifier. Finally, the averaging circuitry displays processing.

Figure 4.32(b) shows a typical OTDR trace of relative amplitude as a function of distance. Design considerations in OTDRs depend upon the two major types of phenomenon observed during fibre measurement. The first is known as Rayleigh backscatter. This can be seen on the trace in Figure 4.32(b) as the constant slope between connectors indicative of fibre loss. This is caused by light that has been 'backscattered' by imperfections in the fibre such as impurities, air bubbles, and moisture.

This backscattered light is usually very low level and is inversely proportional to the wavelength of the light raised to the fourth power. Thus, as the frequency of light increases from 850 to 1310 to 1550 nm, the backscatter level decreases dramatically.

The second phenomenon is known as Fresnel reflection. It occurs at all the connectors owing to the glass–air–glass dielectric interface. The level of this reflected light can be as high as 4 per cent of the incident light. The Fresnel reflected light level does not change nearly as much, as a function of frequency and mode, as does the Rayleigh backscatter level. For more details, References 21 and 22 are suggested.

4.8.4 Telecom testing and pulse marks

Telecom networks and their components transmit digitised voice and data over copper wires and fibre optic cables. To ensure that data transfers reliably, industry standards dictate the shapes of pulses that represent bits. Pulse masks as shown in Figure 4.33(a) described in the relevant standards set the limits for a given pulse shape to be accepted or rejected based on applicable tolerances for voltage and time axis parameters. Telecom equipment receivers must properly interpret pulses that fall within the mask limits, and transmitters must drive the signals along a medium so they arrive within tolerance.

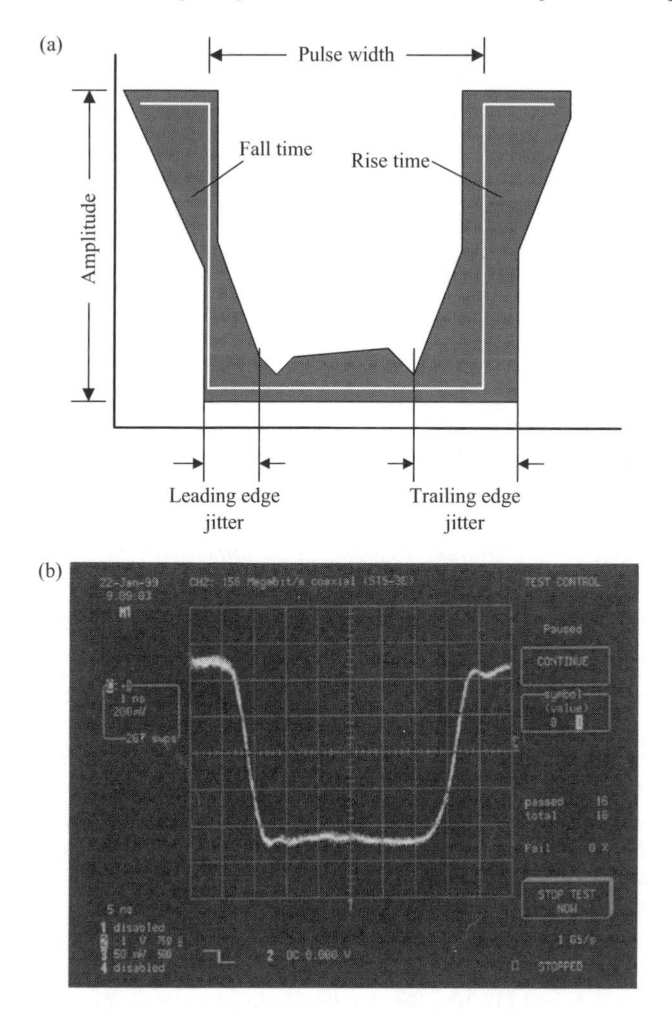

Figure 4.33 Pulse marks and oscilloscope measurement: (a) masks defining amplitude, rise/fall times and jitter; (b) a typical oscilloscope measurement (Courtesy: Test & Measurement World Magazine, USA)

Pulse masks as depicted in the example of Figure 4.33(a) set the limits for physical parameters such as rise time, fall time, pulse width, amplitude, overshoot, and undershoot. The entire pulse must fall within the mask for it to comply with a standard. Engineers use pulse masks as 'sanity checks' for transmitter and receiver designs. Later, the masks are used in tests to verify that a product complies with industry standards, and the masks may be used in pass–fail tests of production products.

Some digital scopes include mask options that allow the pulses to be measured. Figure 4.33(b) shows the mask for a 155.52 Mbps STS-3E binary-1 pulse together with an actual pulse captured by a digital storage scope. The mask is typical of that for other electrical signals such as T1, T3, and DS3 that travel over coaxial and twisted pair wires. Binary 1s may have two masks with inverse polarities. The polarity depends on the type of bit encoding and on the bit pattern used in the transmission system. In this example of Figure 4.33(b), the digital storage scope screen indicates that 16 pulses were captured with no failures, according to the information on the right hand side of the screen contents. Reference 23 is suggested for details. Chapter 12 provides details on digital transmission systems and associated multiplexing schemes, etc.

4.9 References

1 FLUKE CORPORATION: *Calibration: Philosophy in Practice*, 2nd edition 1994.
2 JONES, L. and FOSTER CHIN, A.: *Electronic Instrument and Measurements*, John Wiley & Sons, New York, 1983.
3 LENK, J.D.: *Handbook of Electronic Test Equipment*, Prentice Hall, Englewood Cliffs, NJ, 1971.
4 KULARATNA, N.: *Modern Component Families and Circuit Block Design*, Butterworth Heinemann (USA); 2000, Chapter 8.
5 KITCHIN, C. and COUNTS, L.L.: *RMS to DC Conversion Application Guide*, Analog Devices, Inc., USA, 1986, 2nd edition.
6 SHEINGOLD, D.H.: *Non-linear Circuits Handbook*, Analog Devices, Inc., USA, 1976.
7 KELLER, P.: 'Use true RMS when measuring AC waveforms', *Test & Measurement World*, October 1997, pp. 29–30.
8 CHARLES BUBLITZ: *True RMS Versus Average Responding DMMs*, Fluke Publication # 9599.
9 INTERSIL: ICL 7106/7107-3$\frac{1}{2}$ digit single chip A/D converter data sheet (Ref 11-77-008).
10 PREMA PRECISION ELECTRONICS INC. (USA): Data sheet ADC5601, '25 bit analog-to-digital convertor'.
11 WILLIAMS, J. and OWEN, T.: 'Understanding and selecting rms voltmeters', *EDN*, May 11, 2000, pp. 54–58.
12 TUPTA, M.A.: 'Measure low resistance accurately', *Test & Measurement World*, April 1994, pp. 35–39.
13 KIETHLY INSTRUMENT INC.: *Low Level Measurements*, 5th edition, 1998. Kiethly Books.
14 CRISP, P.: 'Getting the best out of long scale DMMs in metrology applications'. *Proceedings of National Conference of Standard Laboratories*, 1999.
15 FLUKE, INC.: 'ABCs of Multimeter Safety: Multimeter Safety and You', Application Note B 0317UEN Rev A, 12/96.
16 HEWLETT PACKARD: Application Note 62; 'TDR fundamentals', April 1988.
17 HEWLETT PACKARD: Application Note 62-3; 'Advanced TDR techniques', May 1990.
18 SMOLYANSKY, D.A. and COREY, S.D.: 'Use TDR for disk drive flexible interconnect characterization', *EDN*, 5th June 2000, pp. 133–142.

19 LVEKER, J.: 'Differential techniques move TDR into mainstream', *EDN*, 17 August 1989, pp. 171–178.
20 TEKTRONIX, INC.: 'TDR Tools in Modeling Interconnects & Packages', Application note 85w-8885-0(5/93).
21 GRANT, G.R.: 'OTDRs meet the challenge of fiber-optic test', *Electronics Test*, February 1990, pp. 40–42.
22 IRVING, R. and MELLAT, H.: 'Reflecting on fibre testing', *Telecom Asia*, August 1996, pp. 32–35.
23 MARTIN ROWE: 'Pulse marks define signal limits', *Test & Measurement World*, Sept. 1999, pp. 15–18.

Further reading

1 HART, B.: *Digital Signal Transmission Line Circuit Technology*, Chapman & Hall, London, 1988.
2 HEWLETT PACKARD: Application Note 62-1; 'Improving time domain network analysis measurements', April 1988.

Chapter 5
Fundamentals of oscilloscopes

5.1 Introduction

The oscilloscope, which provides an electronic version of the X–Y plotter, is perhaps the most popularly used laboratory instrument. Oscilloscope technology commenced with the development of the cathode ray tube (CRT). First applied in 1897 by Ferdinand Braun, the CRT predates most of what we consider to be active devices, including Fleming's diode valve, De Forest's triode and, by half a century, Bell Labs' transistor. By 1899, Jonathan Zenneck had added beam forming plates and applied a linear horizontal magnetic deflection field to produce the first oscillogram. During the first two decades of the twentieth century, CRTs gradually found their way into laboratory oscilloscopes. Early devices had various development problems, particularly owing to vacuum and hot cathode problems. In 1931, Dr V.K. Zworykin published details of the first permanently sealed, high vacuum, hot cathode CRT suitable for instrument applications. Featuring a triode electron gun, a second anode, and external magnetic deflection coils, the CRT operated at second anode voltages from 500 to 15 kV. Given a CRT that could be treated as a component instead of a process, instrument designers at General Radio introduced the first modern oscilloscope.

Between 1990 and 2000, designers were gradually turning towards liquid crystal displays (LCD), owing to the physical size of CRTs, and their fragility and manufacturing complexities, etc. In 1997, the Braun tube celebrated its hundredth anniversary, while the fabrication of the first active matrix liquid crystal display (AMLCD) was twenty-five years old. During the past decade, LCD technology has reached some major breakthroughs and has given hope to designers that within the first decade of the twenty-first century, CRT based instruments may gradually be replaced by other display technologies (chapter 2).

An oscilloscope display presents far more information than that available from other test and measuring instruments such as frequency counters or digital multimeters. With the advancement of the solid state technology now applied to the development of modern oscilloscopes, it is possible to divide the range of oscilloscopes into two major groups: namely, analogue oscilloscopes and digital storage oscilloscopes. Signals that can be handled by the modern instruments now reach 50 GHz for repetitive signals and beyond 1 GHz for non-repetitive signals. This chapter provides the essential basics for oscilloscope users. For further details, Reference 1 is recommended.

5.2 The cathode ray tube

Because oscilloscope technology evolved with the cathode ray tube (CRT), to understand the analogue oscilloscopes, it may be worth discussing the CRT first.

A large collection of sophisticated electronic circuit blocks are used to control the electron beam that illuminates the phosphor coated screen of a CRT. The basic components of a CRT are shown in Figure 5.1.

A simple CRT can be subdivided into the following sections:

- electron gun,
- deflection system (vertical and horizontal), and
- acceleration section.

The gun section is made up of the triode and the focus lens. The triode section of the CRT provides a source of electrons, a means of controlling the number of electrons (a grid) and a means of shaping the electrons into a beam. The triode consists of the cathode, the grid and the first anode. The cathode consists of a nickel cap coated with chemicals such as barium and strontium oxide. It is heated to assist in the emission of electrons. The grid, where a controlled positive pulse is applied for controlling

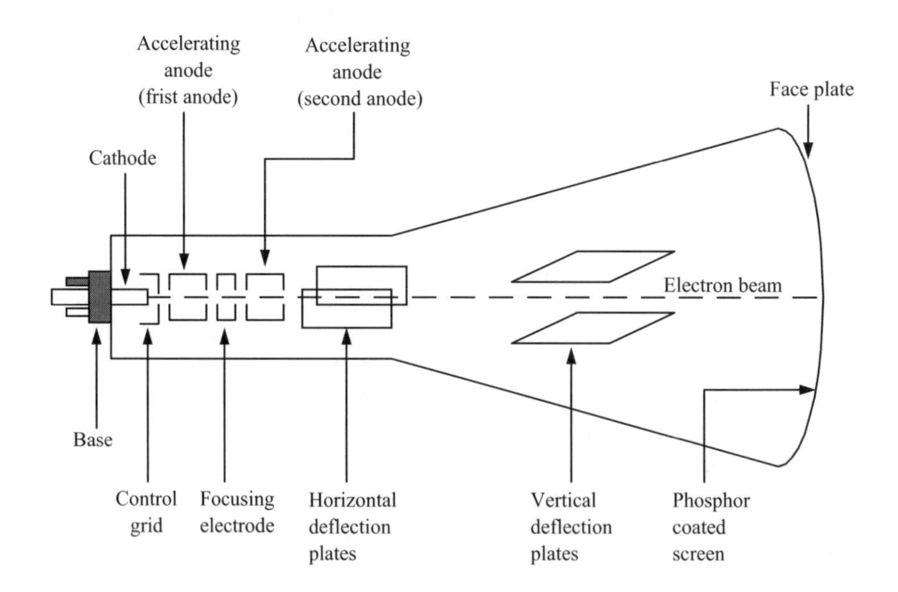

Figure 5.1 Cathode ray tube

the number of electrons, creates a beam of electrons passing through the first and second anodes and on through the focusing area. The first anode is located in front of the grid and is operated at several thousand volts above grid potential. In most situations the grid is operated at a more negative potential than the cathode. After the electrons pass through the first anode exit aperture they are converged to a sharp beam. The size and shape of the spot on the screen has two controls: that is, for focusing and for astigmatism. These controls adjust the voltages on elements comprising the focusing lens.

The purpose of the CRT deflection system is to deflect the electron beam vertically and horizontally with minimum deflection factor (maximum deflection sensitivity) and minimum distortion. The deflection factor (DF) of a CRT is the voltage required for one division of deflection (on the screen) and usually expressed in V cm^{-1}. Usually the first set of deflection plates deflects the beam vertically, and the second set of plates deflects it horizontally. In some CRTs, when the tube length must be short, magnetic deflection is used instead of electrostatic deflection.

The light output from the phosphor increases approximately according to the voltage through which the beam electrons have been accelerated. In the family of CRTs called 'mono accelerators', where electrons are accelerated between the cathode and first anode, as soon as the electrons have passed the second anode no other force is applied to change their axial velocity. To increase the light (brightness) output, various schemes are used where the beam electrons are kept at a relatively low voltage in the deflection region and then accelerated after deflection to a higher energy level. This concept is called the post deflection acceleration (PDA) where the prefix 'post' is in contradistinction to 'mono' (for mono accelerator). In these PDA schemes the electrons are further accelerated by passing through a greater voltage gradient created by an accelerating anode, which creates a higher potential difference ($\sim$10–25 kV) with reference to the cathode. The advantage of 'post' over 'mono' is the higher light output for viewing fast signals.

For a comprehensive description of CRT details, Reference 2 is suggested.

5.2.1 Phosphors: types and characteristics

The luminance or brightness of the phosphor coating on the inside of the CRT screen depends on the chemical composition of the phosphor and the accelerating potential. When the electron beam strikes the phosphor covered screen of a CRT, light and heat are emitted. The light is of primary interest, but the presence of heat and the possibility of burning the phosphor must also be considered. As some light can be lost as it is emitted back into the tube, some CRTs have their phosphors coated with a thin layer of aluminium to act as a mirror. These tubes are called 'aluminised tubes'.

The light output from the phosphor screen has several important characteristics that should be considered. Luminance is the production of light when a material such as a phosphor is excited with a source of energy. The luminance of a phosphor (or its total light output) is usually divided into two parts. Figure 5.2 shows the waveform exciting a phosphor and the total light output. The light produced while the source of energy is applied is known as fluorescence and the light output produced after

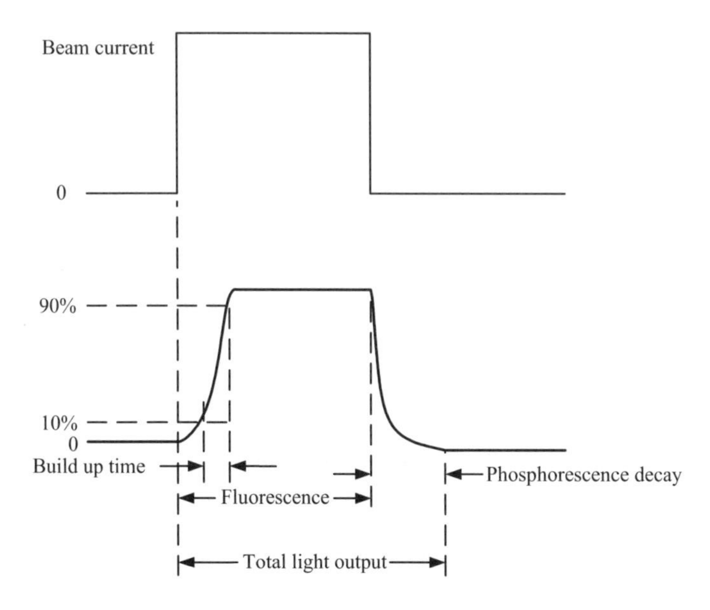

Figure 5.2 Waveform exciting a phosphor and the light output

the source of energy is removed is known as phosphorescence. When a phosphor is suddenly excited by an electron beam it requires some finite time for the light output to reach a constant level. The time required to reach 90 per cent of that constant level under specified excitation condition is called the build up time of the phosphor. The build up time decreases when the beam current density is increased. When the excitation is suddenly removed from a phosphor, a finite interval of time is required for the light output to drop to a low level. This time is known as decay time and is usually expressed as the time required for the light output to drop to (usually) 10 per cent of the original luminance level. This decay characteristic is sometimes termed persistence.

There are different types of phosphors used for different applications. For visual applications, a long decay phosphor allows the observer to see what has gone on even after the sweep has passed a given point. When a display is photographed long decay may fog the film or the display from a previous sweep may still remain. Phosphors are given various numbers based on their properties. P31 is most commonly used as this has a good compromise for visual and photographic needs. Table 5.1 lists some phosphors and their properties. Because of the difference in spectral sensitivity of the human eye and a photographic film, a phosphor having a high luminance may not necessarily have a high photographic writing speed. The writing speed figure expresses the maximum single shot velocity (in cm μs^{-1}), which may be recorded on film as a just visible trace. The factors, which influence writing speed, are related to three items: namely, the CRT, the instrument (oscilloscope) and the camera (a film).

The colour of the output peaks in different regions for different phosphors. For example, the commonly used phosphor P31 peaks more in the green region (~500 nm)

Table 5.1 Phosphor types and their properties

Phosphor		Fluorescence and phosphor-escence	Relative lumi-nance	Relative photographic writing speed	Decay	Relative burn resistance
WTDS	JEDEC					
GJ	P1	yellowish green	50%	20%	medium	medium
WW	P4	white	50%	40%	medium short	medium high
GM	P7	blue	35%	75%	long	medium
BE	P11	blue	15%	100%	medium short	medium
GH	P31	green	100%	50%	medium short	high
GR	P39	yellowish green	27%	—	long	medium
GY	P43	yellowish green	40%	—	medium	very high
GX	P44	yellowish green	68%	—	medium	high
WB	P45	white	32%	—	medium	very high

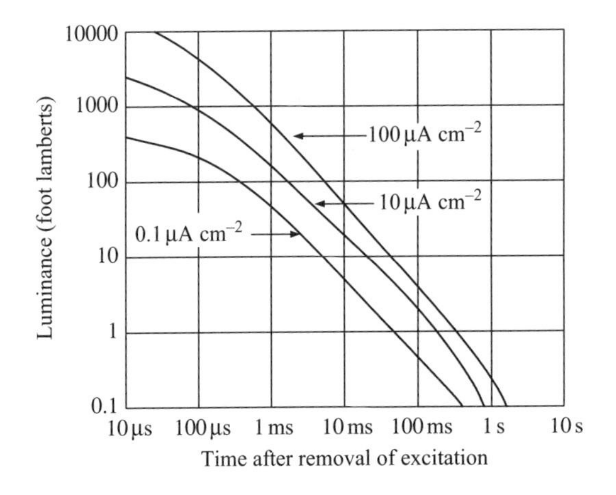

Figure 5.3 Decay chart for the phosphor P31 (courtesy Tektronix, Inc., USA)

and P11 peaks more in the blue region (~450 nm). The human eye responds in varying degrees to light wavelengths from about 400 nm (violet) to about 650 nm (deep red). Peak response of the human eye is about 555 nm. Figure 5.3 shows the decay chart for phosphor P31.

CRT manufacturers are now adopting the world phosphor type designation system (WDTS) as a replacement for the older JEDEC 'P' number system. Table 5.1 lists the comparable WTDS designations for the most common 'P' numbers.

5.2.2 *Liquid crystal displays*

Nowadays, instrument designers are turning away from CRTs in favour of liquid crystal display (LCD) devices. A few reasons for this are that CRTs are physically large, fragile glass jugs that require high voltages for operation. In addition, magnetic deflection coils and fly-back transformers can radiate electromagnetic fields that interfere with low level signals and possibly cause biological side effects. Bright lights can negatively affect the CRT's screen image.

LCDs also have their own problems. Off-axis viewing is difficult and, at low light levels, an LCD requires a backlight. In addition, LCDs are fragile and expensive (especially in large, high resolution formats) and turn black or become sluggish at low temperatures. Also, LCDs are strictly raster-format displays, while CRTs are available in raster or vector types. Most modern, low cost digital storage type scopes use LCD displays. More recent developments in flat panel displays (FPD) are described in Reference 3.

Compared with a CRT implementation, several possible variations of construction exist in a LCD display. In all of these a polariser, front glass, a liquid crystal layer with electrodes, rear glass and a polariser are used. Figure 5.4 shows the type called the twisted nematic construction. Other variations are super twist, double super twist and monochrome super twist, etc. The display technology of choice is an active matrix LCD (AMLCD), in which a backlighting unit with two or four cold cathode fluorescent lamps (CCFL) is used. As shown in Figure 5.4(b), in active matrix thin film transistor (TFT) displays each LCD pixel driven by its own transistor. The thin film transistors are usually FETs in which drain leads are connected to form column selecting terminals and the gate leads are connected to form row selecting terminals. Reference 4 provides details of LCD techniques.

Most recently, advances in colour filter materials, and the aperture ratio of active matrix designs, as well as improvements in edge lit backlight technology, have made it possible to reach a brightness on the order of 1500 cd m^{-2}. Placing a single CCFL at the long edge of the light pipe in a 10.4 inch diagonal, a thin film transistor (TFT) colour LCD produces a brightness level of 70 cd m^{-2} with a power consumption of less than 3 W.

References 5 and 6 provide details of LCD technologies and developments. Most modern, low cost digital storage type scopes use LCD displays. More recent developments in flat panel displays (FPD) are described in Reference 3.

5.3 The basic operation of an oscilloscope

The basic circuit blocks needed to display a waveform on the screen of a CRT are shown in Figure 5.5. The vertical deflection plates of the CRT are fed from the

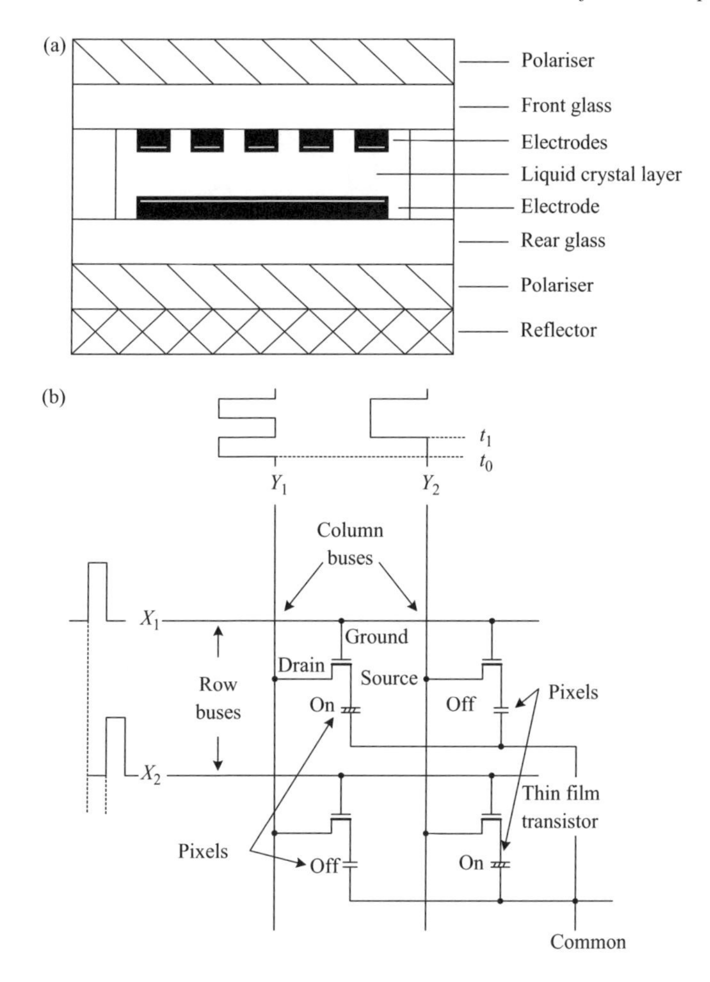

Figure 5.4 Liquid crystal display and pixel addressing: (a) an LCD in its twisted nematic form; (b) pixel addressing

vertical section. These plates receive the actual signal to be displayed as an amplified version or as an attenuated version. The horizontal system controls movements of the beam from left to right. The trigger section determines when the oscilloscope begins drawing by starting the horizontal sweep across the screen.

The electronic circuits of the oscilloscope move the electron beam across the phosphor coating on the inside of the CRT. The result is a glow that traces the path of the beam and that remains visible for a short time afterwards. A grid of lines etched on the inside of the faceplate, called the graticule, serves as the reference for measurement, as shown in Figure 5.6. To eliminate parallax errors the graticule is placed in the same plane as the screen on which the trace is drawn by the electron beam. The vertical and horizontal lines that create the major divisions and the tick

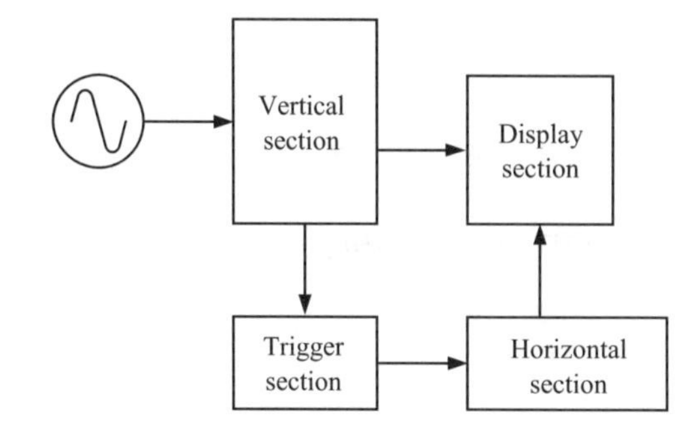

Figure 5.5 Basic circuit blocks needed to display a waveform on a CRT screen

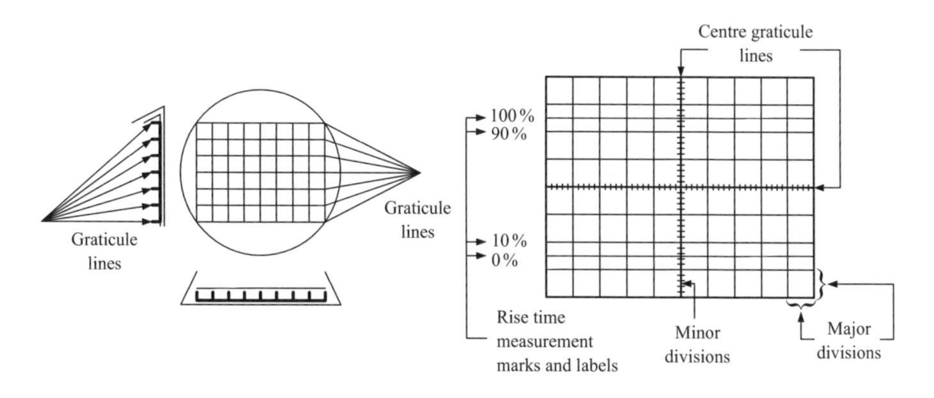

Figure 5.6 Graticule lines of a CRT

marks on each of the graticules are called minor divisions or subdivisions. Because oscilloscopes are often used for rise time measurements, most oscilloscopes have special markings such as 0, 10, 90 and 100 per cent to aid in making rise time measurements.

5.3.1 Display system

Figure 5.7 shows a simplified block diagram for an oscilloscope and the *Z* axis circuits. The *Z* axis circuits of a CRT determine the brightness of the electron beam and whether it is on or off.

The intensity control adjusts the brightness of the trace by varying the CRT grid voltage. It is necessary to use the oscilloscope in different ambient light conditions and with many kinds of signals. The electron beam of the scope is focused on the CRT faceplate by an electrical grid inside the CRT. Another display control on the front panel is the trace rotation control. This control allows electrical alignment of the horizontal deflection of the trace with the fixed CRT graticule to correct for the effects of the earth's magnetic field at different locations. Another common control

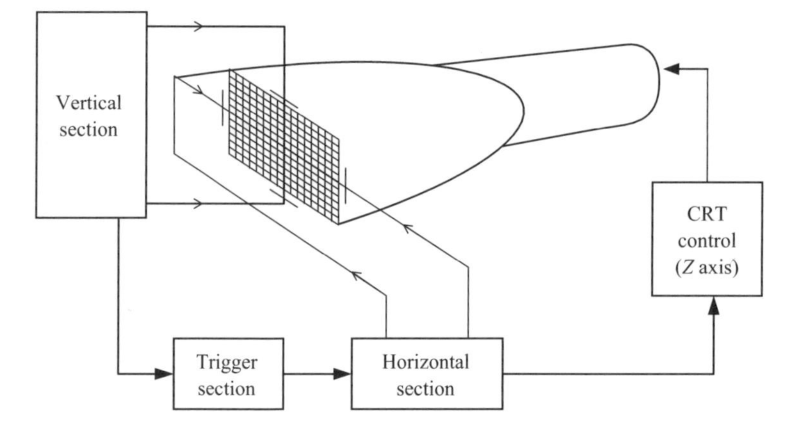

Figure 5.7 Simplified block diagram of the display system of an oscilloscope

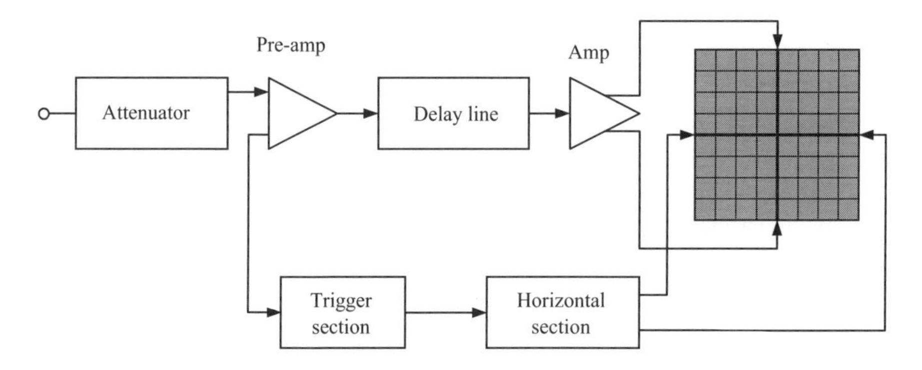

Figure 5.8 Block diagram of the vertical system of an oscilloscope

on the CRT control section of an oscilloscope is the beam finder. This control makes sure that the relative placement of the beam is easily located by limiting the voltage applied to the X and Y plates.

5.3.2 Vertical system

Figure 5.8 shows a block diagram of the vertical system of an oscilloscope. In its most basic form the vertical channel of an oscilloscope can be divided into attenuator, pre-amplifier, delay line and the amplifier sections. In the case of a multiple channel oscilloscope all the blocks except the amplifier section are duplicated and the final signal components are coupled to the amplifier section via appropriate switches.

The attenuator block suitably attenuates the signal using precision attenuators and allows the coupling of the input signal to be changed. When d.c. coupled, the signal is directly fed on to the attenuators and both the d.c. and a.c. components of the signal can be displayed. When a.c. coupled, the d.c. components of the signal are blocked by the use of a suitable high pass filter circuit. This allows the user to observe only the a.c. components of the signal with a magnified scale on the display.

The pre-amplification (pre-amp) section of the vertical channel allows for changing the position of the trace by suitably changing the d.c. component of the signal which is passed on to the next stages. It also allows for a change in the way different input signals are coupled (this is called the mode, which will be discussed later) to the final amplifier stage. In some cases, where one portion of the signal needs to be expanded in time, the relevant traces are isolated (trace separation) for conveniently displaying portions of the waveform.

Use of a delay line allows the display of the beginning of the signal. When a trigger starts drawing the trace on screen, owing to the signal delay introduced by this block the leading edge of the signal is always displayed. With suitable circuitry the voltage per division control can be continuously varied to fill the CRT graticule lines (for example, from the 0 per cent to 100 per cent marks) for practical measurements. From the pre-amp section signal components are picked up for the trigger system and three signal components are used as the internal triggers.

5.3.3 Bandwidth and rise time relationship for vertical pre-amplifiers

The pre-amplifier in the vertical section of an oscilloscope can be considered as a low pass filter which can be approximated by the circuit shown in Figure 5.9(a). The frequency response of this low pass filter is shown in Figure 5.9(b). The response of this circuit to a step input is shown in Figure 5.10. When a step voltage is applied to the input of the amplifier the output voltage rises exponentially. The rise time, t_r is defined as the duration for the output to rise from 10 per cent to 90 per cent of the final value.

If t_1 and t_2 are the times to rise up to 10 per cent and 90 per cent, respectively,

$$0.1A = A\left(1 - e^{-t_1/RC}\right) \tag{5.1}$$

$$0.9A = A\left(1 - e^{-t_2/RC}\right). \tag{5.2}$$

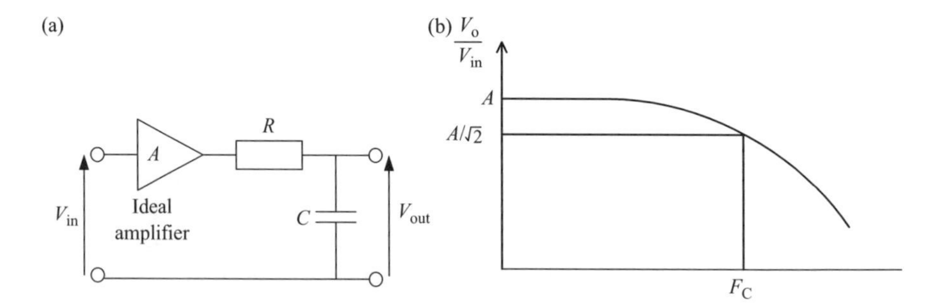

Figure 5.9 Low pass filter representation of vertical system: (a) equivalent circuit of pre-amp; (b) frequency response

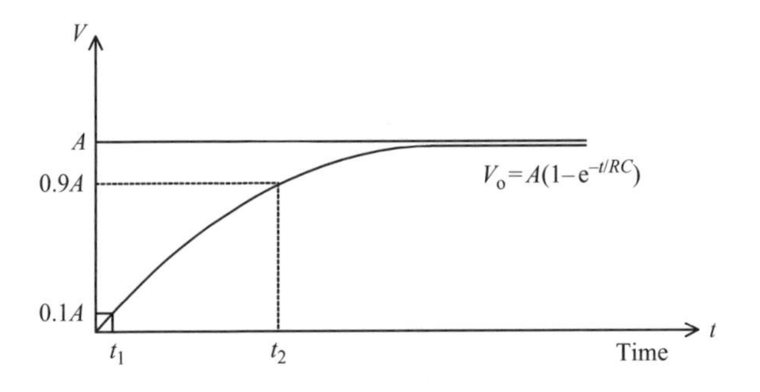

Figure 5.10 Response of low pass filter in Figure 5.9(a) to a step input

From eqs. (5.1) and (5.2) it can be derived that

$$t_2 - t_1 \cong 2.2RC = t_r, \tag{5.3}$$

where the rise time, t_r, is measured in seconds. Now, for a low pass filter with a 3 dB cut-off frequency,

$$f_c = \frac{1}{2\pi RC}, \tag{5.4}$$

where the cut-off frequency, f_c, is measured in hertz. Hence, from eqs. (5.3) and (5.4) it can be shown that

$$t_r \times f_c = 0.35. \tag{5.5}$$

This equation is sometimes called the *bandwidth–rise time relationship* and it is one of the primary relationships to be considered when selecting an oscilloscope. Most manufacturers specify a 3 dB bandwidth for the oscilloscope or more precisely a 3 dB bandwidth for the vertical channels.

For an input waveform with small rise times, the bandwidth–rise time formula can be used to estimate the rise time of the vertical channel itself. For example, if an oscilloscope with a 3 dB bandwidth of 75 MHz is used for rise time measurements, the oscilloscope rise time will be

$$T_r = \frac{0.35}{75 \times 10^6} = 4.7\,\text{ns}.$$

This gives an idea of the smallest rise times that can be measured with sufficient accuracy. The actual measurement situation may be worsened when probes and cables are used to couple the signal to channel inputs.

The bandwidth of an oscilloscope gives a clear idea of what maximum frequencies on practical waveforms can be measured. For example, bearing in mind the previous simplifications for an oscilloscope having a 3 dB bandwidth of 75 MHz, observation of a sinusoidal waveform of 75 MHz frequency will be attenuated by a factor of $\sqrt{2}$ by the vertical amplifier and the measured signal will hence be 0.707 times the input signal.

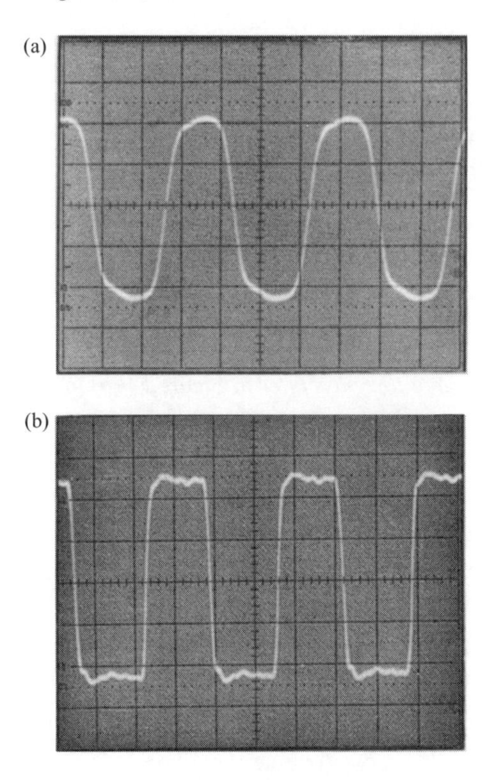

Figure 5.11 *A 15 MHz square wave shown on two different oscilloscopes: (a) on a 35 MHz oscilloscope; (b) on a 50 MHz oscilloscope*

The observed wave shape of a square waveform of 75 MHz fundamental frequency will be far from a square shape owing to the third, fifth and higher harmonics which will be heavily attenuated. The resultant wave shape on the CRT will therefore be quite close to a sine wave. For this reason, a user needs to select an oscilloscope with a bandwidth of at least five times the highest frequency of the signal to be observed. Failure to do this will lead to the display of seriously distorted wave shapes, particularly for square waveforms. Figure 5.11 indicates the effect of a 15 MHz square waveform observed on several scopes with different bandwidths. Clearly, the higher the bandwidth of the oscilloscope the more accurate will be the picture of the waveform.

It is important for the reader to realise that all oscilloscopes, analogue or digital, consist of vertical amplifier systems and the bandwidth–rise time concept applied in general.

5.3.4 *Vertical operating modes*

Almost all practical oscilloscopes available in the market have more than one vertical channel since, in practice, it is useful to compare the timing relationships between

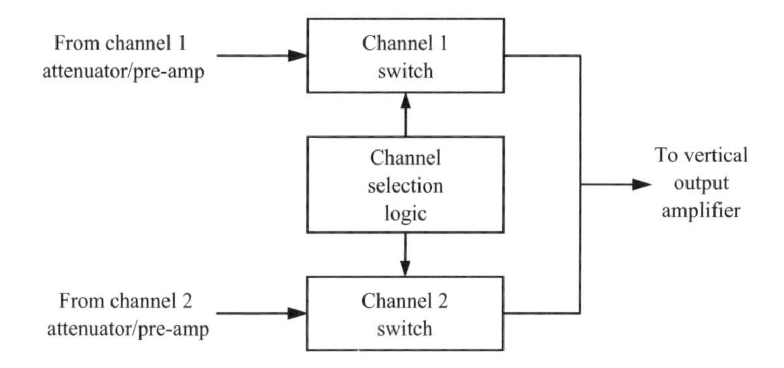

Figure 5.12 Channel switching

different waveforms. To this end, a two channel oscilloscope is employed. With the aid of front panel controls, the user can display any of the following:

- channel 1 only;
- channel 2 only;
- both channels simultaneously in alternate mode;
- both channels simultaneously in chop mode;
- two channels algebraically summed or subtracted.

This is achieved by switching the incoming signal from the attenuators selectively towards the vertical output amplifiers; see Figure 5.12.

When the user needs to display channel 1 or 2 only, the channel selector logic couples only one channel to the *Y* plates. When the user needs to display both channels alternately, the channel selector logic couples the signal from one channel until the beam draws the signal from one end to the other end of the CRT, and then couples the other channel signal similarly. If the displayed signals are fast enough due to the persistence of vision, one will see both channels on the screen simultaneously. However, if the signals are changing slowly it is not easy to see, and the chop mode must be used. In this case the channel selector logic couples the two channels alternately to the vertical plates, switching between them at a higher rate compared with the signal frequency, and the waveforms on the CRT appear continuous. This mode is typically used with slow signals where sweep speeds of 1 ms per division or slower are used. The switching rate (chopping rate) is in the region of several hundred kilohertz. An additional feature in practical oscilloscopes is the ability to algebraically add or subtract the individual signals and display as a single waveform.

5.3.5 Horizontal system

To display a graph on an oscilloscope requires both a horizontal and a vertical input. The horizontal system supplies the second dimension (usually time) by providing deflection voltages to move the beam horizontally. In particular, the horizontal system contains a sweep generator that produces a 'saw tooth' or a 'ramp' (Figure 5.13(b)).

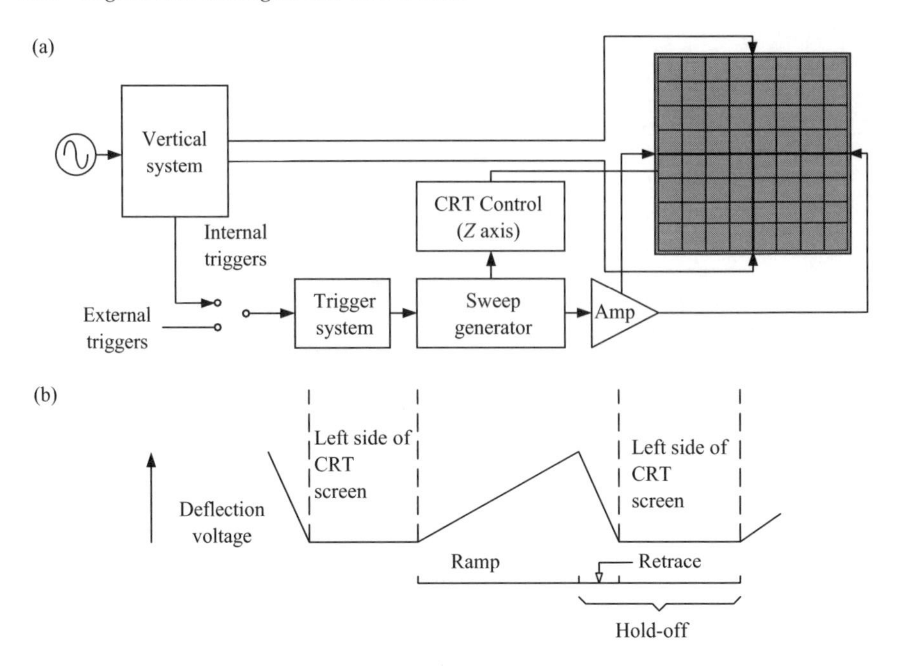

Figure 5.13 Horizontal system: (a) block diagram; (b) sweep waveform

The ramp is used to control the 'sweep rate' of the scope. The block diagram in Figure 5.13(a) shows the horizontal system of a scope.

The sweep generator is a circuit that produces a linear rate of rise in the ramp and this enables the time between events to be measured accurately. Because the sweep generator is calibrated in time it is termed the 'time base'. The time base allows the waveform of different frequencies to be observed because it ranges from the order of nanoseconds to seconds per horizontal division.

The saw tooth waveform is a voltage ramp produced by the sweep generator. The rising portion of the waveform is called the ramp, the falling edge is called the 'retrace' (or 'fly-back') and the time between ramps is called the hold-off. The sweep of the beam across the CRT is controlled by the ramp and the return of the beam to the electron beam to the left side of the screen takes place during the retrace. During the hold-off time the electron beam remains on the left side of screen before commencing the sweep. As for Figure 5.13(a) the sweep signal is applied to the CRT horizontal plates via horizontal amplifiers.

The horizontal system also is coupled to the *Z* axis circuit of the scope which determines whether the electron intensity is turned on or turned off. By applying suitable voltages to the horizontal plates, the horizontal position of the beam can be controlled.

5.3.6 *Trigger system*

The trigger system or trigger circuits inside the scope determine the exact time to start drawing a waveform on the CRT. 'Triggering' is the most important event in

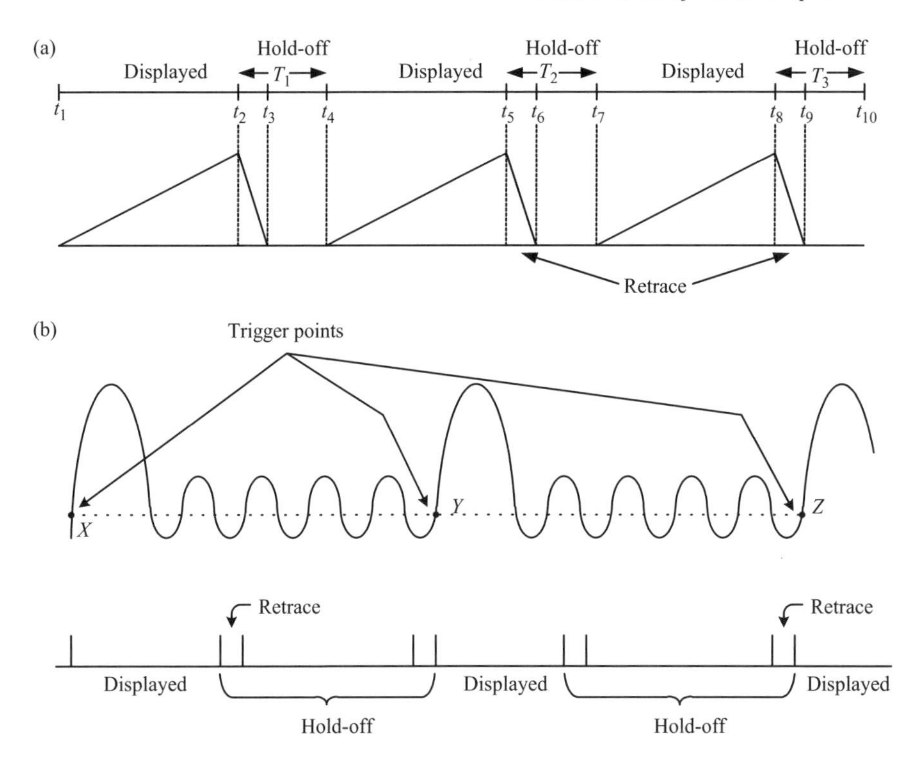

Figure 5.14 Triggering process: (a) sweep generator ramp; (b) trigger points on the waveform

an oscilloscope as observing time related information is one of the primary reasons for using an osilloscope. Any display on an oscilloscope screen is not static, even though it appears to be so. It is always being updated with the input signal and it is necessary that the relevant circuitry determines that the electron beam starts drawing the waveform repeatedly, commencing from the same starting point. Considering a repetitive waveform, as shown in Figure 5.14, it is important to start the sweep relative to the same point (recurring) on the waveform if it is to be displayed as stable. For this reason trigger circuits have *slope and level controls* that define the trigger points which commence the ramp.

The slope control determines whether the trigger points are selected on the rising or falling edge of a signal. The level control determines at which level the trigger occurs. The trigger source can be internal or external. In the case of internal triggering the trigger circuitry picks up a sample from one of the channel signals to be displayed and then routes that sample through an appropriate coupling circuit with a view to blocking or bypassing certain frequency components.

Looking at Figure 5.14, the ramp starts at points t_1, t_4, t_7, etc., repetitively. Therefore the graph on the CRT appears to display the signal from X to Y. As the waveform to be displayed is repetitive the screen will look stable while displaying a portion of

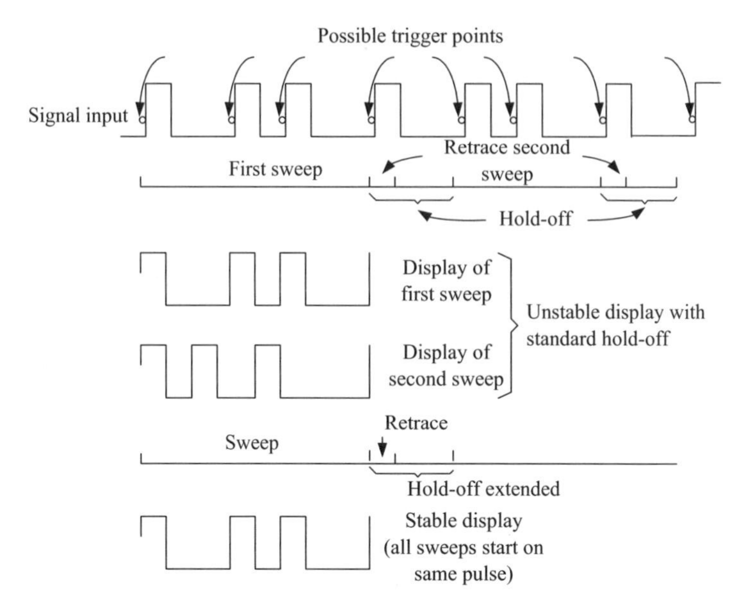

Figure 5.15 Triggering complications

the input. The trigger system recognises only one trigger during the sweep interval. Also, it does not recognise a trigger during the retrace and for a short time afterwards (the hold-off period). The hold-off period provides an additional time beyond retrace to ensure that the display is stable. Sometimes the normal (fixed) hold-off period may not be long enough to ensure a stable display. This possibility exists when the trigger signal is a complex waveform such as a digital pulse train; see Figure 5.15.

For a complex waveform having several identical trigger points (defined by slope and level only) the displayed sweep can be different, thus creating an unstable screen. However, if the scope is designed with a trigger system where hold-off time can be extended, then all sweeps could start on the same pulse, thereby creating a stable display. For this reason the more advanced oscilloscopes always have a variable hold-off possibility where the hold-off time is to be varied (typically) over 1 to 10 times.

In practical oscilloscopes the trigger pick up can be from different sources. For example, in the case of internal triggering either of the channels (in the case of a dual channel oscilloscope) could be used to trigger the sweeps. However, when both channels are displayed it may be necessary to trigger the oscilloscope from different channels alternately. For this purpose the trigger source switch has a 'vertical mode' position (or a composite signal) where the trigger source is picked from the vertical mode switch of the vertical section. For example, if channel 1 only is displayed via the vertical section, the trigger source automatically becomes channel 1.

When external triggering is selected, a totally external signal can be used to trigger the sweep. For example, in a digital system, the system reset signal can be used as an external trigger. Another useful trigger source is the 'line' trigger where the trigger signal pick up is from the power line; this is of immense help when observing power line related waveforms.

5.3.7 Trigger operating modes

A practical oscilloscope can operate in several different trigger modes such as normal, peak-to-peak auto, television field and single sweep. In the normal mode, trigger circuits can handle a wide range of triggering signals, usually from d.c. to above 50 MHz. When there is no trigger signal, the normal mode does not permit a trace to be drawn on the screen. This is used primarily for very low frequency signals less than about 100 Hz.

In the peak-to-peak auto-mode, a timer starts after a triggered sweep commences. If another trigger is not generated before the timer runs out, a trigger is generated and this keeps the trace drawn on the screen even when there is no signal applied to the channel input. The timer circuits are designed to run out after about 50 ms, allowing a trace to be generated automatically when the trigger signal frequency is less than about 20 Hz. In this mode the trigger level control automatically adjusts within the peak-to-peak amplitude of the signal. This eliminates the need for setting the trigger level control. This mode allows automatic triggering of the system when the input signal amplitude varies.

The single sweep mode operates exactly as its name implies. It triggers a sweep only once. After selecting the 'single sweep' trigger mode on the oscilloscope, the trigger system must be readied or armed to accept the very next trigger event that occurs. When the trigger event or signal occurs the sweep is started. When one sweep is completely across the CRT the trigger is halted. This mode is used typically for waveform photography and when 'baby-sitting' for glitches. (In practical oscilloscopes there is an indicator to confirm that the necessary single sweep occurrence has been completed to avoid waiting for the event with the scope.)

5.3.8 Trigger coupling

Similar to the possibility of selecting a.c. or d.c. coupling of input signals to the vertical system, the oscilloscope trigger circuit can pass or block certain frequency components of the trigger signals. For example, in most practical oscilloscopes we find the trigger coupling selection such as a.c., HF reject, LF reject and d.c. In the case of a.c. coupling, the d.c. component of the trigger signal is blocked and only the a.c. component is passed into the system. In the case of the HF reject and LF reject appropriate input filters are coupled to the trigger system and usually these filters have their roll-off frequencies around 30 kHz; see Figure 5.16. The HF reject removes the high frequency components passing on to the trigger system; the LF reject accomplishes just the opposite effect. These two frequency rejection features are useful for eliminating noise that may be riding on top of input signals which may prevent the trigger signal from starting the sweep at the same point every time.

Note that d.c. coupling passes both a.c. and d.c. trigger signal components to the trigger circuits. It is also important to note that pickups for the triggering system

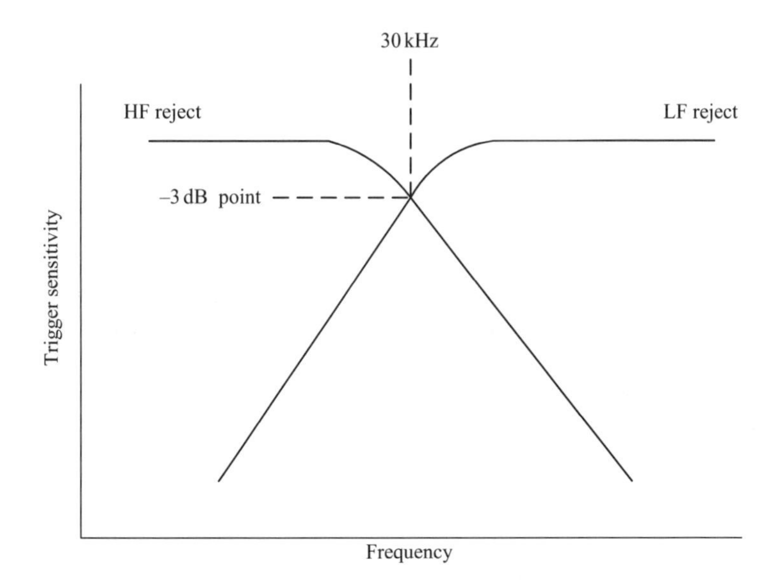

Figure 5.16 Trigger filters

generally occur after the vertical system input coupling circuit, and if a.c. coupling is selected on the vertical input coupling, or even if d.c. trigger coupling is selected, the d.c. component may not be passed on to the trigger circuits.

5.4 Advanced techniques

During the past two decades, CRT technology has advanced slowly, providing high speed CRTs. Over the past three decades, with the advances in semiconductor components providing inexpensive designs, designers were able to introduce many useful features to both analogue and digital oscilloscopes. Some of these were:

(i) multiple time bases,
(ii) character generators,
(iii) cursors, and
(iv) auto-setup and self-tests.

Use of custom and semicustom integrated circuits with microprocessor control has enabled the designers to introduce feature packed, high performance analogue scopes. In most of these scopes the use of mechanically complex parts has been reduced owing to the availability of modern microprocessor families at very competitive prices. Most of these microprocessor-based oscilloscopes incorporate calibration and auto-diagnostics to make them very user friendly.

In analogue scopes with bandwidths over 300 MHz, designers were using many advanced design techniques to achieve high CRT writing speed. One such technique used by Tektronix was the use of the micro-channel plate (MCP) for viewing high frequency real time waveforms between 400 MHz and 1 GHz. The technique is described in chapter 6 of Reference 1.

However, with the digital storage techniques with LCD displays advancing, these analogue scope techniques were becoming more expensive and less popular. With newer techniques such as digital phosphor oscilloscopes (DPO) by Tektronix, the prominence of high speed analogue scopes are gradually diminishing.

5.4.1 Multiple time bases

One important advance with modern oscilloscopes is the use of multiple time bases. Expensive oscilloscopes have at least two independent time bases, which can be controlled independently or in relation to each other. This helps the user to magnify selected parts of the waveform. The principle of dual time bases is explained using Figure 5.17. In these scopes there are two independent (and almost) similar sweep generators (A and B) where the trigger system can independently activate any one of the ramps from sweep generator A or sweep generator B, or from sweep generator B commenced after a time delay from the A sweep circuitry. Again, the practical advantage of such a multiple time base system is to observe portions of a waveform in detail.

The main time base A runs with a trigger signal received from the base trigger system of the scope. When a user needs to examine a waveform in detail on a selected area (Figure 5.18(a)), for example, t_1 to t_2, the Z axis control circuits can intensify the selected portion of the waveform on the CRT. By using the delay time control circuits the B sweep is started after a delay (via the delay gate signal). The B sweep generator usually runs at a faster rate than the A sweep generator, as determined by the time base B. This creates an expanded view of the area of special interest on the waveform. To view the magnified portion of the waveform clearly, trace separation signals are applied to the vertical sections. See Figure 5.18(b) for the display on the CRT.

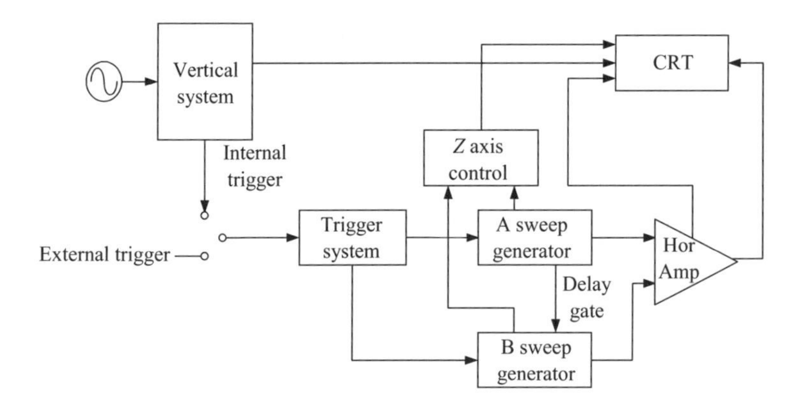

Figure 5.17 Technique of dual time bases

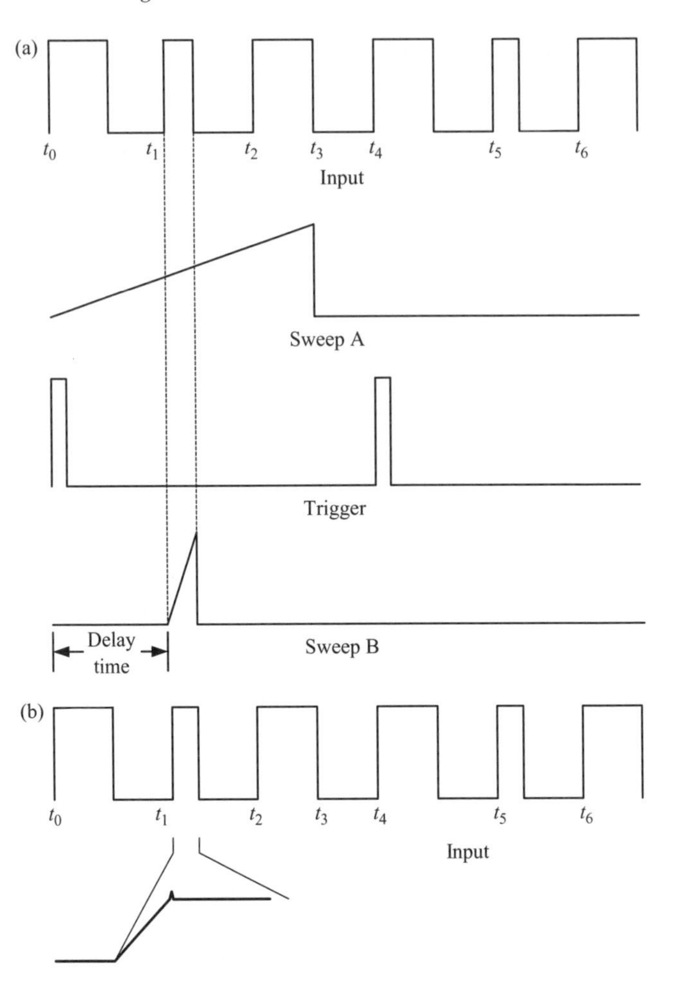

Figure 5.18 *Timing and display – dual time bases: (a) operation of two time bases; (b) expansion of selected portion of waveform using time base B showing the rising edge of section between t_1 and t_2*

In complex pulse trains, as encountered in digital systems, when a user needs to observe a certain portion of the waveform where jitter is present, time base B can be triggered independently using an independent trigger pulse. This can eliminate jitter and assist the user to observe an area of interest in a magnified form. See Figures 5.19(a) and (b), where time base B is triggered from input after the adjusted delay time. It is important to note that the Z axis circuits are appropriately controlled by the two sweep circuit blocks so that waveform display intensities are independently adjustable.

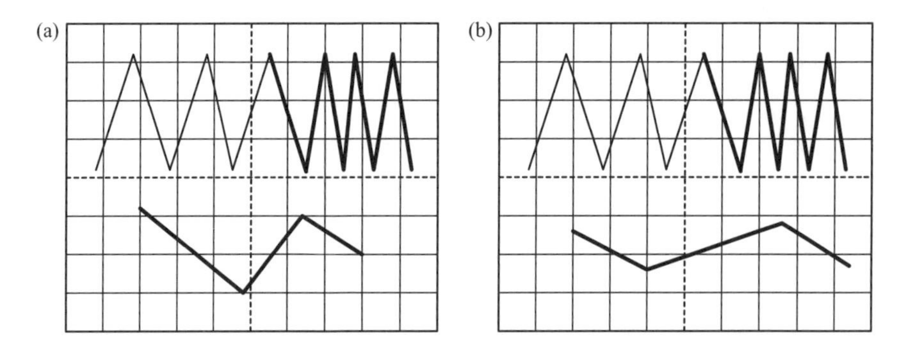

Figure 5.19 Elimination of jitter using dual time bases: (a) B sweep triggered from A; (b) B sweep triggered from input

5.4.2 Oscilloscopes with dual channels and dual time bases

A block diagram of a practical oscilloscope with two input vertical channels and two separate time bases is shown in Figure 5.20. Block 1 indicates the attenuators and associated circuitry of the two vertical channels. Note that both channels have identical circuitry.

Block 2 shows the vertical pre-amplifier and output amplifier circuits where the CH1 and CH2 pre-amplifier signals are switched towards the vertical output amplifier through the channel switch. From the front panel controls of the oscilloscope, the channel selection logic receives the vertical mode inputs (CH1, CH2, both CH1 and CH2, ADD, ALT, CHOP, etc.). The chop oscillator and associated blanking circuitry interacts with the channel switch (on block 2) and Z axis circuits on block 7.

The blocks in section 3 take care of generating the trigger signals, as determined by the front panel switches, which determine the trigger source selection. The selected trigger signal is applied to trigger level comparator circuits and the Schmitt trigger (or Schmitt trigger and TV trigger stage) for generation of the A trigger signal. Note that when the trigger source is selected as 'internal', out of the possible internal trigger sources (CH1, CH2, vertical mode), one signal is finally selected and passed on to the internal trigger amplifier circuit.

The circuit blocks shown in section 4 of the diagram take care of generating the ramp voltage necessary to sweep the X axis circuits of the oscilloscope. The Miller sweep block generates the ramp governed by the signals from the A sweep logic. The end of the sweep detect circuit, together with hold-off timing block, ensure that the ramp will start itself after the hold-off period.

When a user wishes to magnify sections of a waveform (on the time axis) the B sweep circuits are used and the delay time position comparator, shown in section 5, is activated. The comparator compares the A sweep ramp with an external voltage set by the B delay time position potentiometer and determines the time at which the B ramp can be started. Generation of the B ramp can be further delayed by another amount of time based on the B trigger level and the scope, using similar blocks as in section 3

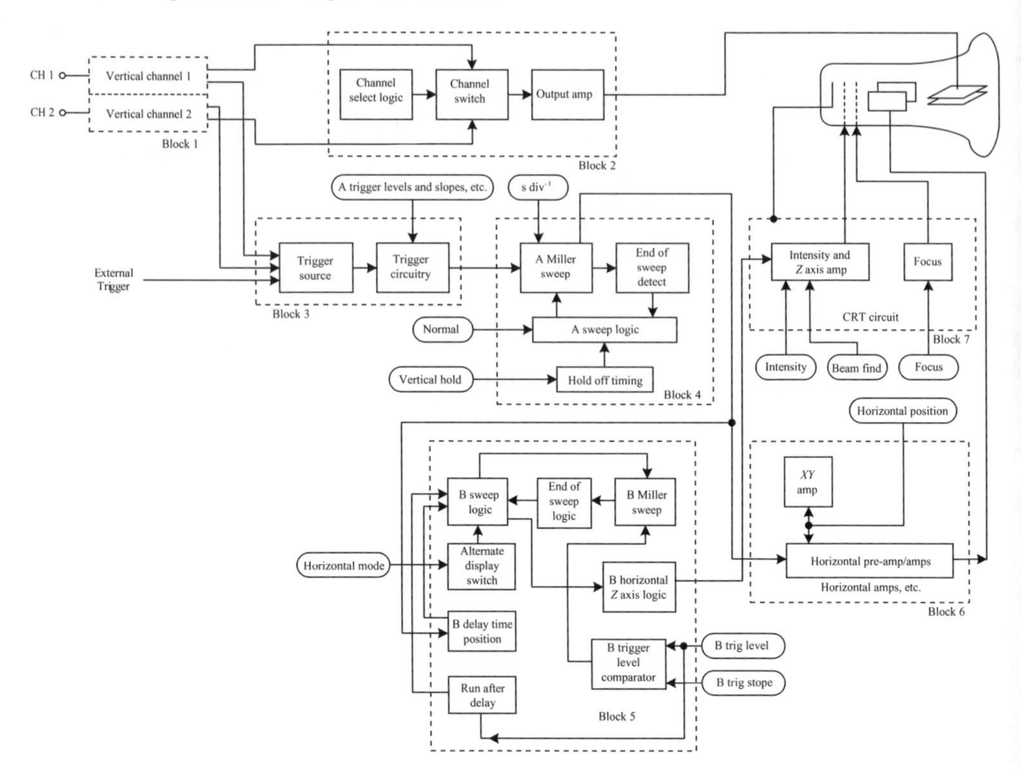

Figure 5.20 Block diagram of an oscilloscope with dual channels and dual time bases

(trigger level comparator and Schmitt triggers, etc). Use of A and B sweeps together can generate an expanded view of a portion of a signal as shown in Figure 5.21. Note that the *Z* axis logic block in section 5 interacts with the *Z* axis circuits and intensifies selected portions of the waveform.

5.4.3 Use of low cost microprocessor blocks

5.4.3.1 Counter timer and multimeter blocks

During the historic development of oscilloscopes, designers used the advantage of low cost microprocessor blocks to incorporate additional facilities such as digital multimeters, counters and timers, etc. An example is the Tektronix 2236, a 100 MHz scope with the counter–timer–multimeter (CTM). In these early designs, measurement values were displayed on vacuum fluorescent or LED displays.

Such an oscilloscope is capable of

- measuring frequency, period or width,
- totalising an input event, and
- providing a digital display of delay time/delta time.

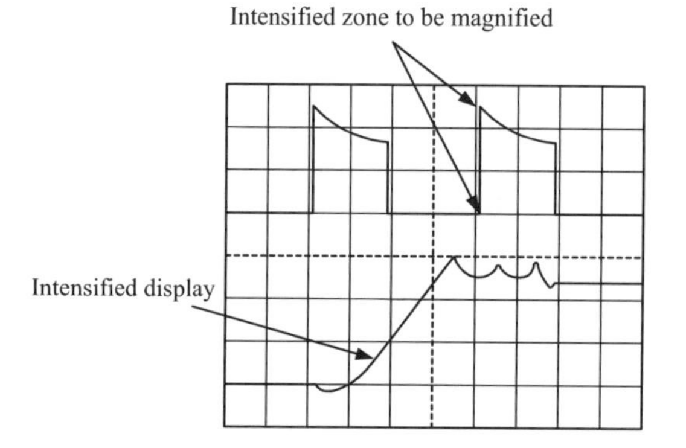

Figure 5.21 Intensifying portions of waveform and expanding intensified area

Figure 5.22 depicts the simplified block diagrams of a CTM, counter front-end of a CTM and a complex waveform where detailed measurements on parameters can be done.

In such designs, the gating signals of the A and B trigger systems are used for measuring frequency parameters of a complex waveform such as in Figure 5.22(c). For example, trigger A is used for measuring the basic frequency of a repetitive waveform. This measurement is called 'non-gated' measurement.

When the measurement is to be carried on a part of a complex waveform such as in Figure 5.22(c) (intensified zone), a B sweep is used to gate only a portion of a waveform to counter the front-end and such measurement are called 'gated' measurements. This is a very useful application of the dual time base. In such CTM type circuit blocks the microprocessor or microcontroller used is a simple 8-bit version and the oscilloscope inputs and oscilloscope trigger signals, etc., are coupled to counter front-ends in such a way that the CTM is independently designed. Synchronised signals (C_1/C_2), etc., are derived from scope inputs and are directed to CPU blocks, so that most voltage and time axis parameters can be conveniently displayed. A detailed explanation can be found in Reference 1 or the service manual of an oscilloscope such as the Tektronix 2236.

5.4.3.2 Character generators and cursors

With the use of microprocessor or microcontroller systems within an analogue oscilloscope, if the CRT writing speed is higher than the demand by waveforms to be displayed, then by the appropriate introduction of additional circuit blocks, the character generators and waveform cursors (horizontal and vertical pairs of straight lines which can be moved within the oscilloscope screen to ease the measurements) can be incorporated. These advanced techniques have eliminated the need for the use of separate display units, such as LCD, LED or vacuum fluorescent displays. However, in these additional tasks of coordinating the display of characters and waveform cursors, a fair amount of additional circuitry is necessary too. In modern oscilloscopes

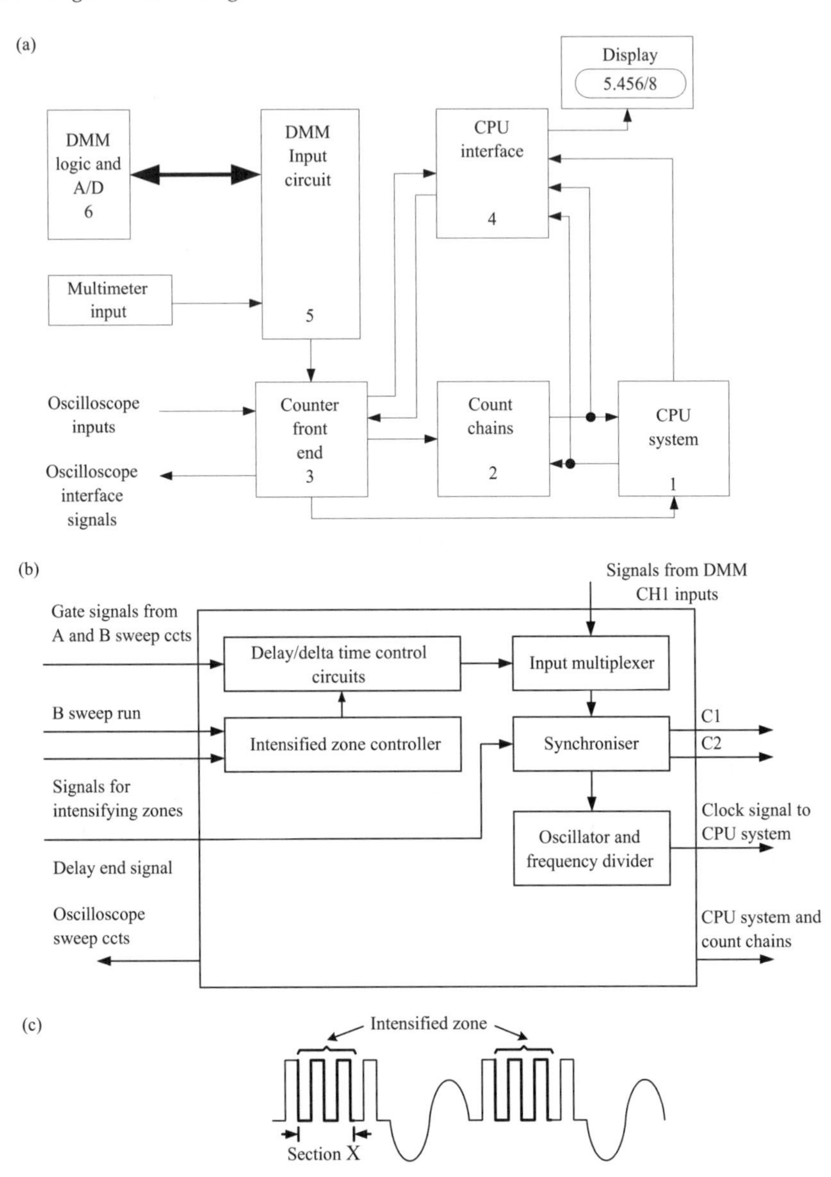

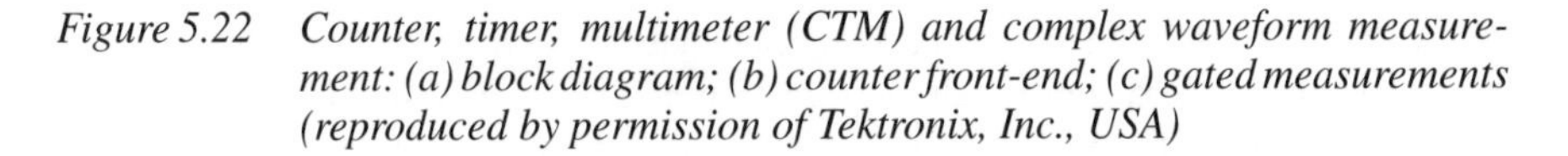

Figure 5.22 Counter, timer, multimeter (CTM) and complex waveform measurement: (a) block diagram; (b) counter front-end; (c) gated measurements (reproduced by permission of Tektronix, Inc., USA)

the manufacturers use custom VLSI circuits to save PCB real estate within the oscilloscope. By appropriate use of CPU blocks and associated peripherals in medium priced oscilloscopes these facilities are incorporated. Design concepts related to these techniques can be found in Reference 1. The high writing speed of an advanced CRT

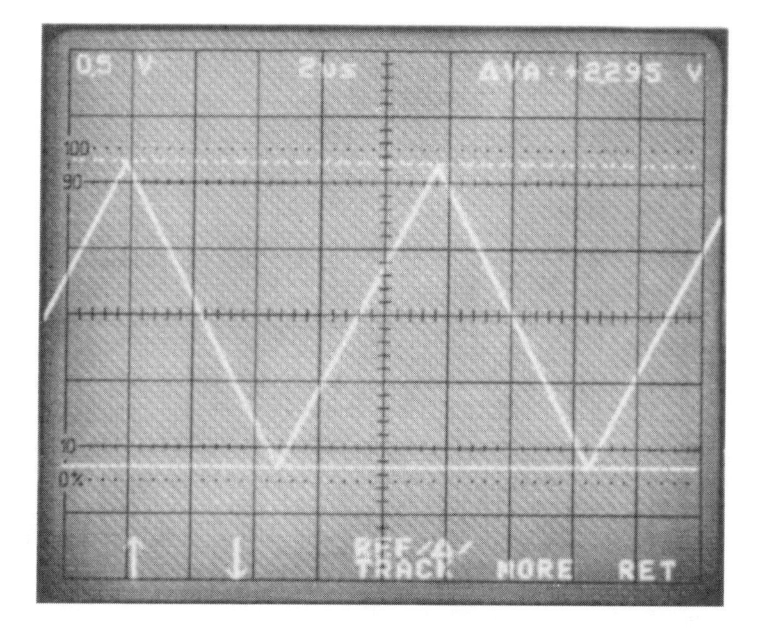

Figure 5.23 Cursor and character displays

design is used in such instruments, and Figure 5.23 indicates a typical display. In such scopes many custom VLSI chips are used to coordinate the scope display-coordination functions.

5.4.3.3 Simplification of front panel controls and addition of special features

In modern oscilloscopes, the processor and digital control circuitry directs the operation of most oscilloscope functions by following firmware control instructions stored in the memory. These instructions direct the microprocessor to monitor the front panel controls and send control signals that set up various circuit blocks such as vertical attenuators and time base. In such an instrument all front panel controls and potentiometer settings are scanned by the processor system every few milliseconds to check whether the user has changed the front panel controls. While executing the control program, the microprocessor retrieves previously stored calibration constant and front panel settings, and as necessary places program generated data for later use. The battery backed-up RAM provides these storage functions. When power is applied to the instrument a brief initialisation sequence is performed; the processor then begins scanning the front panel controls. The switch settings detected and the retrieved front panel data from the battery backed-up RAM cause the processor to set various control registers and control voltages within the instrument. These register settings and voltage levels control the vertical channel selection and deflection factors, the sweep rate, triggering parameters, readout activity and sequencing of the display. In these scopes most mechanically complex attenuator (volts per division) and time base (time per division) switches are replaced by much simpler switches associated with digitally driven magnetic latch relays. Self-celebration, auto-setup and '*power-on self-test*' (POST) capabilities are quite common attractive features of these microprocessor based scopes.

5.5 Digital storage oscilloscopes (DSOs)

With the rapid advancement of semiconductor technology, memories, data converters and processors have become extremely inexpensive compared with the price of CRTs, particularly high speed CRTs. Instrument designers have developed many advanced techniques for digital storage oscilloscopes, to the extent that most reputed manufactures such as Tektronix and Hewlett-Packard (now Agilent Technologies) have discontinued the production of analogue oscilloscopes. During the 1990s, the use of DSP techniques assisted replacing high speed CRT designs with technologies such as digital phosphor oscilloscopes (DPOs), etc. Among the primary reasons for using storage oscilloscopes are:

- observing single shot events,
- stopping low repetition rate flicker,
- comparing waveforms,
- unattended monitoring of transient events,
- record keeping, and
- observing changes during circuit adjustments.

Reference 1 and its associated references provide more details on early storage oscilloscopes.

Figure 5.24 depicts the block diagram of a simple DSO. It can be clearly seen that digital storage necessitates the digitising and reconstruction process using analogue-to-digital and digital-to-analogue converters in the digitising and reconstruction processes, respectively.

Based on the fundamental block diagram shown Figure 5.24, we will now discuss essential concepts related to sampling, display and bandwidth–rise time concepts, etc.

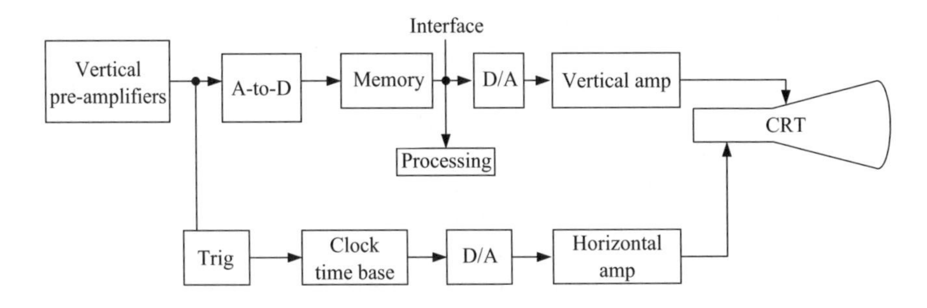

Figure 5.24 Digital storage oscilloscope block diagram

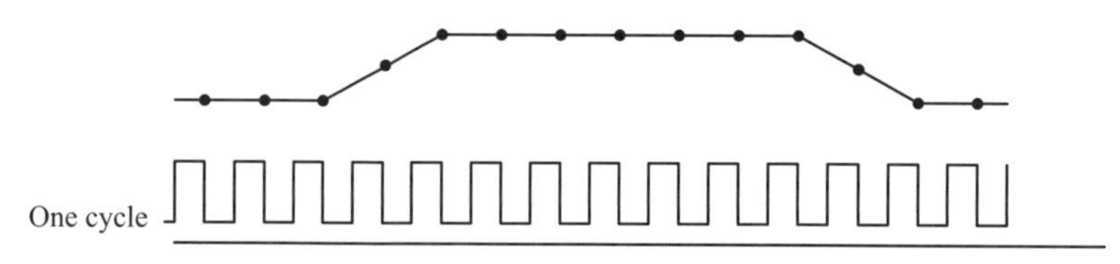

Figure 5.25 Real time sampling

5.5.1 Sampling techniques

In all digital storage scopes and waveform digitisers, the input waveforms are sampled before the A/D conversion process. There are two fundamental digitising techniques, namely (i) real sampling, and (ii) equivalent time sampling.

In real time sampling all samples for a signal are acquired in a single acquisition, as in Figure 5.25. In this process a waveform is sampled in a single pass and the sample rate must be high enough to acquire a sufficient number of data points to reconstruct the waveform. Real time sampling can be used to capture both repetitive and single shot events.

In equivalent time sampling, the final display of the waveform is built up by acquiring a little information from each signal repetition. There are two equivalent time sampling techniques: namely, sequential and random sampling. Figure 5.26(a) shows the technique of sequential sampling where it samples one point on the waveform every acquisition cycle. This is done sequentially and is repeated until enough points are acquired to fill the memory. If the memory is 1000 points long it would take 1000 passes to acquire the waveform.

In random sampling, signals are acquired at a random sequence in relation to where they are stored in the memory (Figure 5.26(b)). The points in time at which these samples are acquired are stored in reference to the trigger point. This type of equivalent time sampling has two advantages. First, since the points are reconstructed with reference to the trigger point, it retains the pre- and post-triggering capability, which sequential sampling cannot do. Second, because the sample acquisition is referenced to the trigger signal, the normal digital trigger jitter (discussed later) is not a factor.

Another variation of random sampling is multiple-point random sampling, where several points are obtained in one acquisition cycle. This reduces the acquisition time considerably. For example, if four points are acquired per cycle (as for the example in Figure 5.27), the time to acquire a complete waveform is reduced by a factor of four compared with a scope that acquires a single point on each cycle.

It is important to highlight here that real time sampling is necessary to capture transient events. Equivalent time sampling is used for repetitive signals and allows capturing of very high frequency components as long as the signal is repetitive.

5.5.2 DSO characteristics

Compared with analogue oscilloscopes, DSOs operate based on sampling concepts, as discussed in chapter 23. Primary parameters, which determine the capability of a

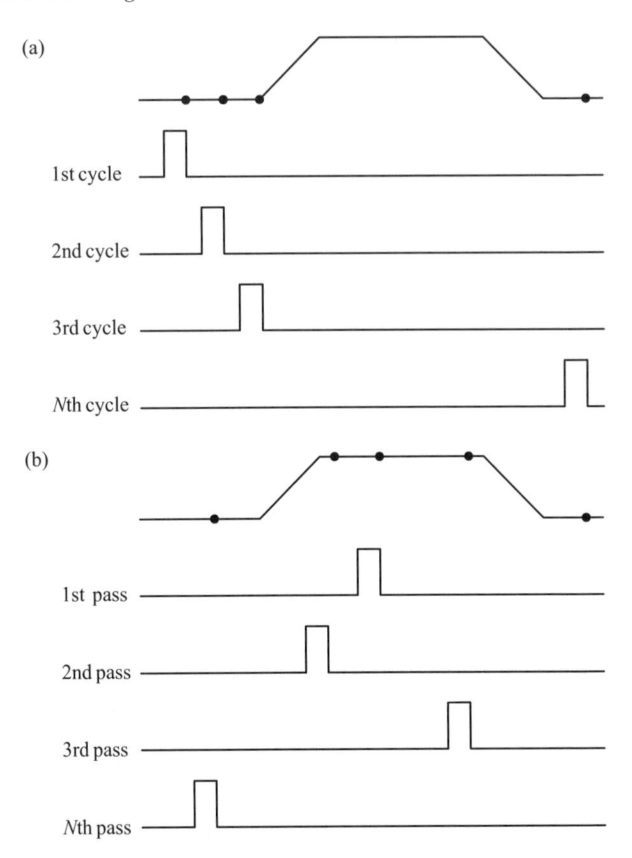

Figure 5.26 Equivalent time sampling: (a) sequential sampling; (b) random sampling

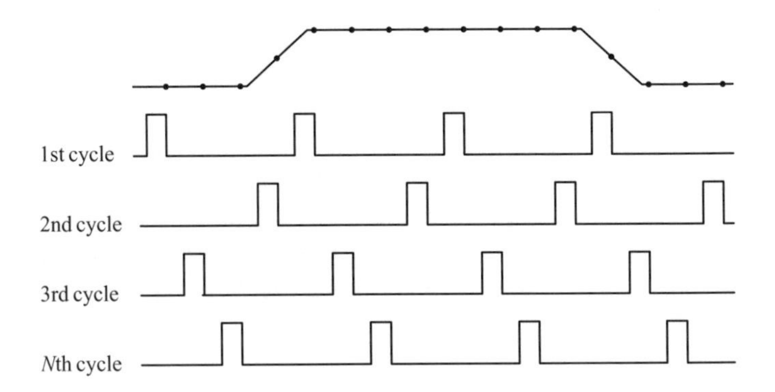

Figure 5.27 Multiple-point random sampling

DSO are:

- sampling rate and vertical resolution,
- display techniques,
- interpolation techniques,
- memory capacity, and
- signal processing capabilities.

Let us discuss some of the important parameters related to sampling process etc, which were not covered in chapter 3. For details, References 7 and 8 are suggested.

5.5.2.1 Sampling rate

The sampling rate (or digitising rate), one primary characteristic of the DSOs, is commonly expressed as a frequency such as 20 megasamples per second (20 Msamples s^{-1}). Another familiar expression would be as a 20 MHz sample rate. This could be expressed as the information rate, that is, the number of bits of data stored per second. (For example, 160 million bps for an 8-bit ADC and a sample rate of 20 Msamples s^{-1}.) In a practical oscilloscope the sample rate varies with the time base setting and the corresponding relationship is

$$\text{sample rate} = \frac{\text{record length of waveform}}{\text{time base setting} \times \text{sweep length}}. \tag{5.6}$$

When the sample rate is selected using this criterion it occupies the entire memory and fills the screen. As the time base is reduced (i.e. more time per division), the digitiser must reduce its sample rate to record enough signal samples to fill the display screen. However, reducing the sample rate also degrades the usable bandwidth (discussed later). Long memory digitisers maintain their usable bandwidth at slower time base settings compared with short memory digitisers. Because the cost of a DSO depends on the cost of memories and data converters used inside, this criterion is an important one in the selection of a DSO.

For example, if an oscilloscope has a 1024 waveform memory and sweep length of 10 divisions at 10 μs per division ($\mu s\, div^{-1}$) time base setting, then the sample rate (SR) is as follows,

$$\begin{aligned} \text{SR} = \frac{1024}{(10\,\mu\text{s}\,\text{div}^{-1}) \times 10} &= 10.24 \text{ samples}\,\mu\text{s}^{-1} \\ &= 10.24\,\text{MHz or } 10.24 \text{ Msamples}\,\text{s}^{-1}. \end{aligned}$$

For a digital storage oscilloscope with 50 ksample memory with 10 CRT divisions, the sample rate would be 1 gsample s^{-1} when the time base is set to 5 $\mu s\, div^{-1}$. For the same DSO, if the time base is set to 5 $ms\, div^{-1}$ the sample rate would be 1 Msample s^{-1}.

5.5.2.2 Display techniques

To view a waveform once it has been digitised, memorised and processed, there are several methods that may be used to redisplay the waveform. Basic techniques to

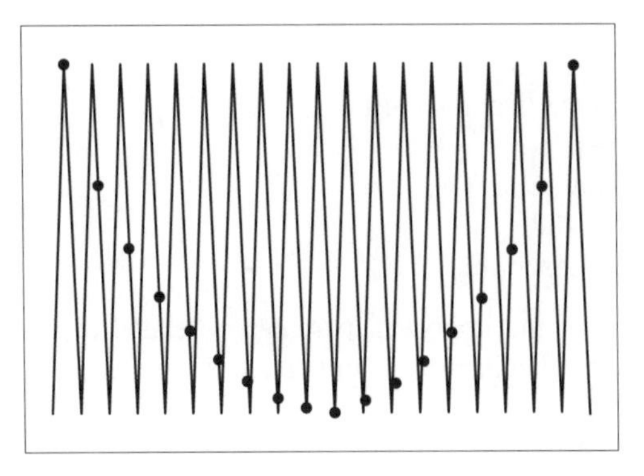

Figure 5.28 Effect of under-sampling

display the waveform include: (i) the use of dots, (ii) the use of lines joining dots (linear interpolation), and (iii) sine interpolation or modified sine interpolation.

All methods require a digital-to-analogue converter (DAC) to change the data back to a form that the human eye can understand. DACs, when used for reconstructing the digital data, do not require the performance characteristics of ADCs because the conversion rate could be much slower. The main purpose of the DAC is to take the quantised data and convert them to an analogue voltage.

In displaying the stored digitised data, aliasing may occur, unless the signal is sampled at more than twice the highest frequency component of the signal. The diagram in Figure 5.28 shows under sampling of a waveform where sampling is carried out only once per cycle. In this case the alias waveform is of a much lower frequency (approximately 1/19th of the actual frequency), as the information gathered via sampling is inadequate to represent the shape of the actual waveform.

More samples per period will eliminate aliasing. When the sample rate is calculated inside the oscilloscope using eq. (5.6), the bandwidth of the signal, which can be displayed, is estimated easily. Since we must always digitise twice as fast as the highest frequency in the signal, the simplest way to do so is to make sure that the user picks a time div^{-1} setting that determines a high enough sample rate. When this cannot be done, an antialiasing filter can be used to eliminate frequencies above the Nyquist limit.

5.5.2.2.1 Dot displays

Dot displays are, as their name implies, made up of points on the CRT. These displays are useful providing there are sufficient points to reconstruct the waveform. The number of points required is generally considered to be about 20 to 25 points per cycle (discussed later in section 5.5.2.4.1).

However, with dot displays fewer dots will be available to form the trace when the frequency of the input signal increases with respect to sample rate. This could result in perceptual aliasing errors, especially with periodic waveforms such as sine waves. This perceptual aliasing is a kind of optical illusion as our mind is tricked to

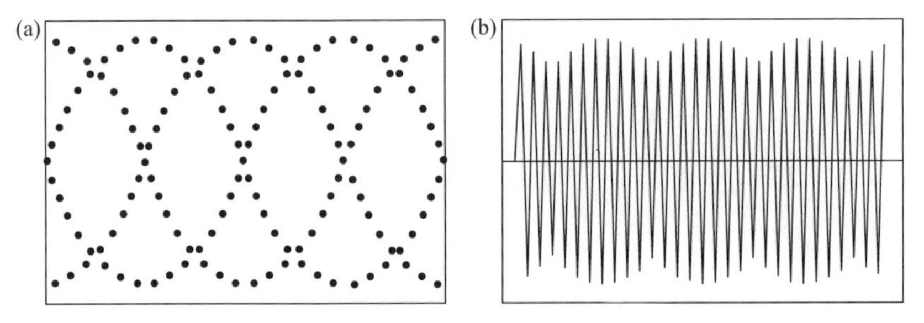

Figure 5.29 *Effect of aliasing and vector display: (a) perceptual aliasing; (b) vector display*

imagine a continuous waveform by connecting each dot with its nearest neighbours when viewing a dot display. The next closest dot in space, however, may not be the closest sample of the waveform. As shown in Figure 5.29(a), a dot display can be interpreted as showing a signal of lower frequency than the input signal. This effect is not a fault of the scope but of the human eye.

5.5.2.2.2 Vector displays

Perceptual aliasing, as in Figure 5.29(a), is easily corrected by adding vectors to the display, as shown in Figure 5.29(b). But this vector display can still show peak amplitude errors when the data points do not fall on the signal peaks because the vectors are only straight lines joining the points. When digital storage oscilloscopes use vector generators which draw lines between data points on the screen, perceptual aliasing is eliminated and only about ten vectors are needed to reconstruct a recognisable display.

5.5.2.3 Use of interpolators

Interpolators connect data points on the scope. Two different types of interpolator are available, both of which can solve the visual aliasing problem with the use of dots. These are linear interpolators and curve interpolators. Linear interpolators merely join the points with straight lines, as shown in Figure 5.29(b), which is a satisfactory procedure as long as enough points are available. Curve interpolators, which come in different forms, attempt to connect the points with a curve that takes into account the bandwidth limitations of the scope. Curve interpolators can make some very nice looking waveforms out of a very limited number of data points. It is important to note, however, that the available data points only constitute the real data and curve interpolators should be used with caution for waveforms with special shapes and high frequency components. Some manufacturers (e.g. Tektronix USA) who produce high speed analogue circuitry use interpolation (as well as averaging and envelope mode) during the acquisition process. Some other companies (e.g. LeCroy, Inc. (USA)) apply these techniques in the post-acquisition stage.

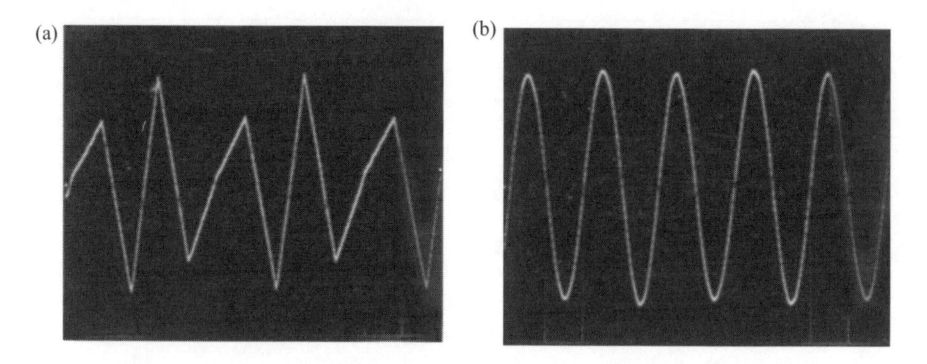

Figure 5.30 Interpolation: (a) line; (b) sine

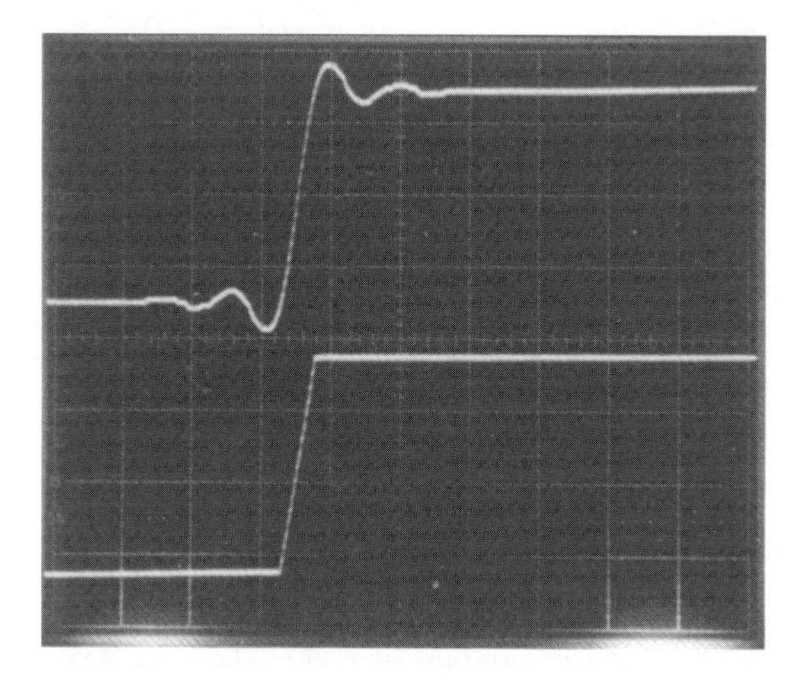

Figure 5.31 Effect of sine interpolation on pulse displays

5.5.2.3.1 Sine interpolation

A useful display reconstruction technique is 'sine interpolation', which is specifically designed for reproducing sinewaves. When using such a technique, 2.5 data words per cycle are sufficient to display a sinewave. Figure 5.30(a) and (b) show a 10 MHz waveform sampled at 25 MHz: (a) is displayed in a linear interpolated format and (b) is sine interpolated. However, as shown in Figure 5.31, an interpolator designed for a good sine wave response can add unnecessary pre-shoots and over-shoots to the displays of pulses.

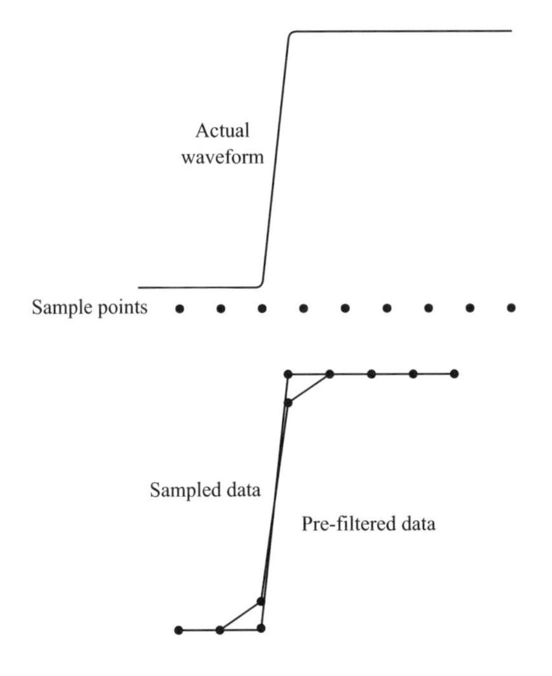

Figure 5.32 Use of a pre-filter

5.5.2.3.2 Modified sine interpolation

To eliminate the drawbacks of sine interpolation when displaying pulses, modified sine interpolation uses a digital pre-filter. The digital pre-filter coupled with the sine interpolator permits the reconstruction of the waveform to give a better representation. Figure 5.32 shows the principle. The pre-filter looks at the slope of two consecutive sets of three samples for discontinuity between slopes. Then, if the discontinuity is beyond a specified limit, the closest points to the discontinuity are adjusted by about 10 per cent of amplitude. This waveform is then processed by a sine interpolator.

5.5.2.3.3 Sine x/x interpolation

This interpolation technique used in Tektronix and LeCroy oscilloscopes uses rise time measurements where it fits curve segments between sample points to create a smooth waveform. However, since this technique only needs a minimum of four samples per cycle it is possible that noise can be misinterpreted as data. For details, Reference 9 is suggested.

5.5.2.4 Bandwidth and rise time

Having discussed analogue oscilloscope bandwidth–rise time implications, one needs to select the right DSO for a specific measurement. Once again to recap what we discussed earlier, it is necessary to remember that the vertical channels of an oscilloscope are designed for a broad bandpass, generally from some low frequency (d.c.) to a much higher frequency. This is the oscilloscope's bandwidth. The bandwidth specification

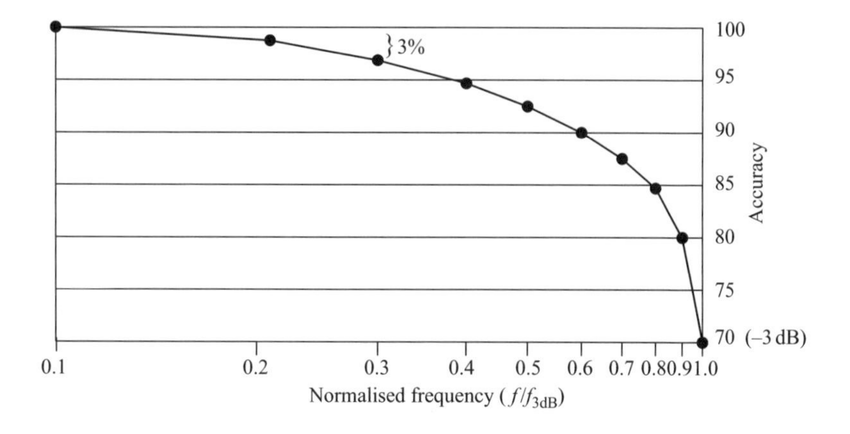

Figure 5.33 Bandwidth vs accuracy

gives you an idea of the instrument's ability to handle high frequency signals within a specified attenuation. Bandwidth is specified by listing the frequency at which a sinusoidal input signal has been attenuated to 0.707 of the middle frequencies. This is called the −3 dB point. At this point, vertical amplitude error will be approximately 30 per cent, as shown in Figure 5.33.

If 30 per cent attenuation is sufficient for your measurements (hardly the practical case), then you can acquire signals up to the full rated bandwidth of your scope. The maximum practical vertical amplitude error for most digital applications, however, is 3 per cent. This means, if we refer back to Figure 5.33, that you really only have about 30 per cent of the scope's rated bandwidth to work with.

A good rule of thumb for determining scope bandwidth, is to use the '5 times' rule. In other words, to calculate the bandwidth of the scope you need, multiply the highest bandwidth component of the signals you want to examine by 5. Using the '5 times' rule will typically give you better than 2 per cent accuracy in your measurements.

When one has to work with rise times, in previous discussions (eq. (5.5)) the assumption was a single pole filter. This equation turns out to be a very handy rule of thumb. Table 5.2 indicates the rise time related measurement requirements for common logic families together with typical signal rise times for different logic families and the necessary bandwidth requirements to make 3 per cent and 1.5 per cent accuracy measurements.

With the bandwidths of electronic circuits continually on the rise, however, oscilloscope designers are using different types of filter to expand the passband of their instruments. Table 5.3 lists some constants for several different filter types. As you can see, there are some large discrepancies from the 0.35 rule of thumb when special filters are used. For example, Tektronix model 7104 uses a second order Bessel filter; models TDS 540A and TDS 640A use fifth order Butterworth; TDS 684A and TDS 700A use brick wall filters.

Table 5.2 Measurement system requirements for different logic families [10]

Logic family	Typical signal rise times	Calculated signal bandwidth ($= 0.35/t_r$)	Measurement bandwidth for 3% roll-off error	Measurement bandwidth for 1.5% roll-off error
TTL	5 ns	70 MHz	231 MHz	350 MHz
CMOS	1.5 ns	230 MHz	767 MHz	1.15 GHz
ECL	500 ps	700 MHz	2.33 GHz	3.5 GHz
GaAs	200 ps	1.75 GHz	5.8 GHz	8.75 GHz

Table 5.3 Several different filters and their $t_r \times$ BW values (courtesy: Tektronix, Inc., USA)

Filter type	Rise time × BW
Gaussian	0.3990
Second order Butterworth	0.3419
Second order Bessel	0.3421
Single pole (common low pass filter)	**0.3497**
Sin(x)/x	0.3544
Third order Butterworth	0.3642
Third order elliptical	0.3704
Fifth order Butterworth	0.4076
Fifth order elliptical	0.4245
Brick wall	0.4458
Tenth order Butterworth	0.4880

It is very important to remember that the rise or fall time of a signal being displayed on a digitising oscilloscope is really a combination of the rise time of the circuit, the rise time of the probe, and the rise time of the oscilloscope. If the probe rise time is limited, the entire test system rise time will be limited. This is shown in the equation:

$$t^2_{\mathrm{r(system)}} = \sqrt{t^2_{\mathrm{r(oscilloscope)}} + t^2_{\mathrm{r(probe)}} + t^2_{\mathrm{r(source)}}}. \tag{5.7}$$

In digital storage scopes, while vertical channels dictate the above rules, for single shot events it is differently calculated.

5.5.2.4.1 *Useful storage bandwidth*

In the previous paragraphs, the importance of the sample rate was discussed in the case of dot displays as well as displays where interpolation techniques are used.

When it comes to digital storage oscilloscopes, most users prefer a single 'figure of merit' (like bandwidth or writing speed) that describes the maximum signal frequency these instruments can store. Useful storage bandwidth (USB) is a way to specify that maximum useful frequency. USB is defined for a digital storage oscilloscope for **single event capture** as

$$\text{USB} = \frac{1}{C} \times \text{maximum sample rate.} \tag{5.8}$$

The constant C depends on the method interpolation and the number of samples per cycle. Hence, for a dot display (where 25 samples per cycle is taken to eliminate perceptual aliasing), C is 25. When linear or vector interpolation is used at 10 samples taken per cycle, C is 10. For sine interpolation where 2.5 samples per cycle is taken, C is 2.5. It is important to note that the constant C can vary from technique to technique used by different manufacturers. However, for repetitive signals the DSO bandwidth is the analogue bandwidth of the scope. For example, the Tektronix model 11800 oscilloscope has a repetitive bandwidth of 50 GHz, with a sample rate of only 200 kHz.

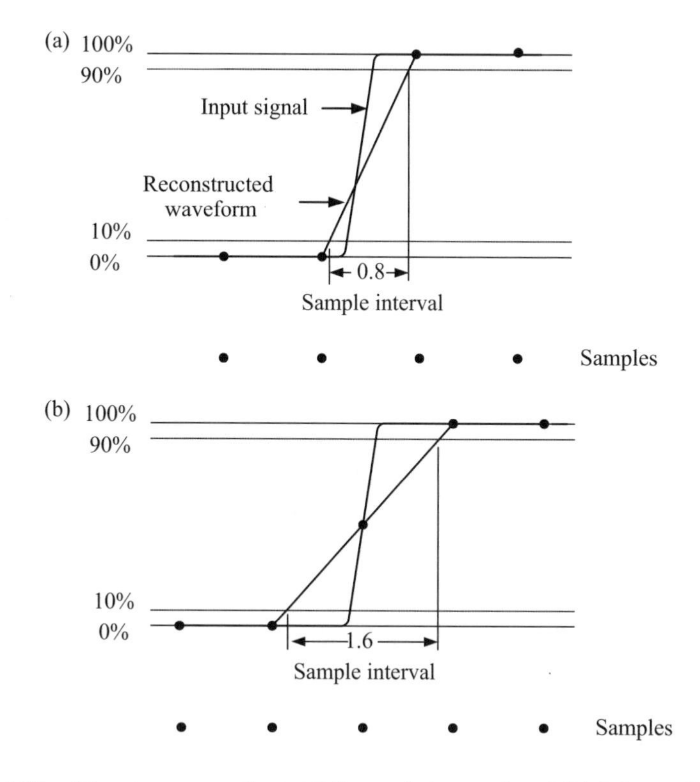

Figure 5.34 Rise time errors for (a) 0.8 sample interval and (b) 1.6 sample interval

5.5.2.4.2 *Useful rise time*

If a very fast signal is measured and displayed with a linear vector display on a digital storage scope, the simple geometry, as in Figure 5.34, shows that the displayed rise time can vary from (a) 0.8 to (b) 1.6 sample intervals. The displayed rise time depends entirely on where the sample falls on the input signal. This simple explanation shows that maximum positive rise time errors produced by a pulse interpolated display closely follow the form of an analogue scope when the analogue system has a rise time of 1.6 sample intervals. For this reason the useful rise time, U_{Tr} of a DSO may be defined as

$$U_{\mathrm{Tr}} = 1.6 \times \text{minimum sample interval}. \tag{5.9}$$

If, for example, a DSO has a sample rate of 10 Msample s^{-1} then

$$U_{\mathrm{Tr}} = 1.6 \times 0.1\,\mu\mathrm{s} = 160\,\mathrm{ns}.$$

It is important to note that DSO bandwidths and rise times change with the time base (time div^{-1}) setting. The useful bandwidth and useful rise time parameters of DSOs, however, give users an indication of the fastest signals that can be captured, for single shot and repetitive situations.

5.5.2.4.3 *Vertical accuracy and effective bits*

The effective bits, sometimes referred to as the effective number of bits (ENOB), combine all other digitiser performance factors into a single specification that describes the digitiser accuracy with respect to frequency. In essence, effective bits, or the ENOB, describes many specifications rolled into one. Most manufacturers measure effective bits in a similar fashion guided by an IEEE standard 1057–1994 to ensure consistency. This document [11] is a comprehensive publication for estimation of various errors in digitisers including the ENOB computation.

According to this method a pure sine wave is fed into the real digitiser. (Pure, in this case, means a sinewave whose harmonics are below the sensitivity of the digitiser.) Typical levelled sinewave generators produced by instrument manufacturers can be used here, provided their output is passed through an appropriate low pass filter. The signal is digitised and the error is calculated on the assumption that the signal has a high quality shape of unknown amplitude, offset and phase. For details, Reference 1 is suggested.

Figure 5.35 shows the dynamic performance of a practical DSO (old Tektronix model 2430). It should be noted that all this is based on the A-to-D converter techniques used in the instrument.

In this example we can clearly see that with the frequency increased the ENOB reduces sharply beyond 1 MHz. ENOB may be affected by the vertical amplitude of the signal.

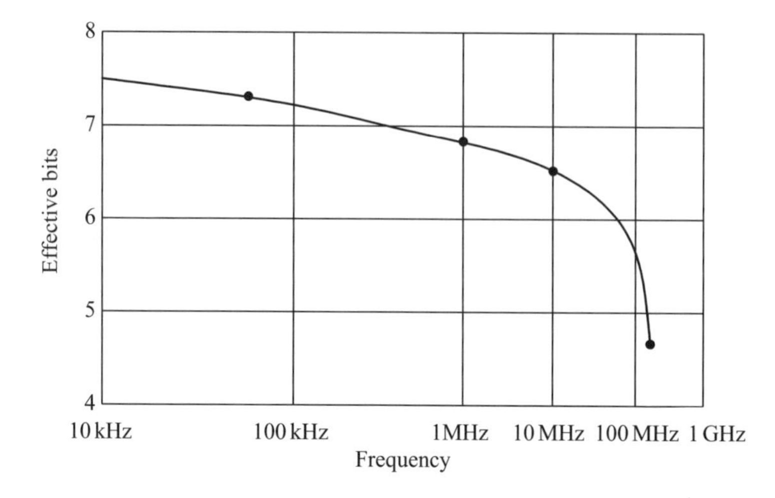

Figure 5.35 ENOB against input frequency for Tektronix 2430 (reproduced by permission of Tektronix, Inc., USA)

5.5.2.4.4 Other important DSO characteristics

There are many DSO characteristics and features that should be carefully observed when selecting a DSO. Principally, these may be listed as follows:

(i) analogue capability where DSO can act as an analogue scope,
(ii) complex triggering capabilities such as delay by events, logic triggering, etc.,
(iii) envelope display where the minimum and maximum values of a waveform are shown,
(iv) glitch/peak detection capability,
(v) interfacing to plotters, computers, etc.,
(vi) signal averaging to help filtering noise, etc., and
(vii) waveform mathematics and signal processing capabilities.

More details are provided in the next chapter with practical examples.

5.6 Probes and probing techniques

Probes are used to connect the measurement test point in a device under test (DUT) to the input of an oscilloscope. Achieving optimised performance in a measurement depends on the selection of the probe. The simplest connection, such as a piece of wire, would not allow the user to realise the full capability of the oscilloscope. A probe that is not appropriate for an application can mean a significant error in the measurement as well as undesirable circuit loadings. Many oscilloscopes claim bandwidths that exceed 1 GHz, but one cannot accurately measure signals with such high frequencies unless the probe can accurately pass them to the scope. Standard single-ended, 1 MΩ,

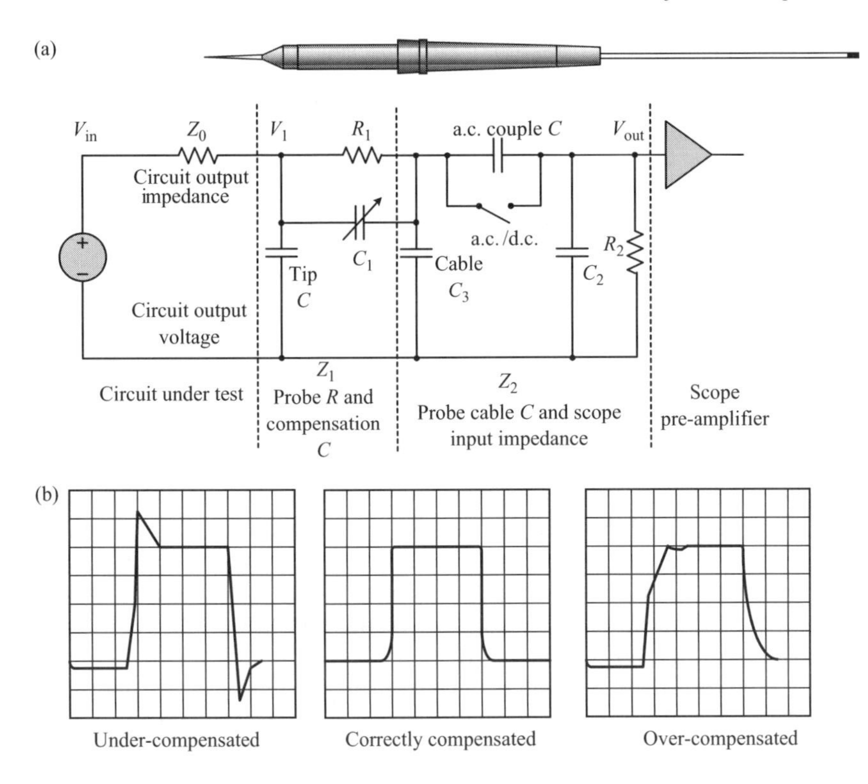

Figure 5.36 *Basic 10 : 1 type probe and compensation: (a) circuit; (b) effect of compensation*

passive probes work well in many applications, but their bandwidths cut off at around 500 MHz. Above that frequency, active probes are a better choice.

Using the right probe is critical to making accurate, safe measurements. Single-ended probes, whether passive or active, can't produce a meaningful measurement when you need to measure the difference between two non-zero voltages. Even worse, using a single-ended probe can be dangerous in some applications.

5.6.1 Performance considerations of common 10 : 1 probes

Passive, single-ended probes make up over 90 per cent of the probes in use today. These probes provide 10 : 1 signal division and provide a 10 MΩ input resistance when attached to a scope with a 1 MΩ input resistance. Figure 5.36 shows the equivalent circuit of a typical 10 : 1 passive probe and explains how correct probe compensation ensures a constant attenuation ratio regardless of signal frequency. The oscilloscope's input capacitance ($C2$) must lie within a probe's 'compensation range' or else you will not be able to adjust the probe to achieve the correctly compensated square corner.

The probe impedance and the scope input impedance form a voltage divider circuit where

$$V_{\text{out}} = \frac{V1 \cdot Z2}{Z1 + Z2}.$$

Correct probe compensation causes $R1C1$ to equal $R2 \cdot (C2 + C3)$. Under this condition,

$$V_{\text{out}} = \frac{V1 \cdot R2}{R1 + R2}.$$

This means that when a scope probe is correctly adjusted (compensated) the scope input signal is ideally a fixed fraction of the probe input voltage, for all frequencies. The deviation from the ideal will worsen with increasing source impedance. At high frequencies, second order effects ignored above will become more important.

The single adjustment compensation shown in Figure 5.36 is not perfect, either. The higher bandwidth probes contain a high resistance centre wire in their cable that attenuates ringing at high frequencies. The wire forms a distributed RC network with the capacitance of the cable shield and the dielectric. A trimmer capacitor tunes this network correctly, but you may need a potentiometer to adjust the probe's attenuation ratio. A 'tail-box' fitted between the probe's output BNC connector and the probe cable houses these and other components.

When a probe is used to monitor a high frequency continuous wave, the capacitive reactance of probes at the operating frequency must be taken into account. The total impedance with respect to the probe tip is designated R_{p} and is a function of frequency. R_{p} and the capacitive reactance X_{p} for a typical $10\times$ type probe are shown in Figure 5.37.

5.6.2 Low impedance passive probes

A 'low Z' passive probe offers a very low tip capacitance at the expense of relatively high resistive loading. A typical $10\times$ '50 ohm' probe has an input capacitance of about 1 pF and a resistive loading of 500 Ω. Figure 5.38 shows the circuit and equivalent model of this type of probe. This configuration forms a high frequency $10\times$ voltage divider and offers high bandwidth, typically up to few GHz, and rise times up to 100 ps. However, these have resistive loading effects; essentially they perform best if the impedance under test is 50 Ω or lower.

5.6.3 Probe bandwidth

All probes are ranked by bandwidth and they are like scopes or other amplifiers which are ranked by bandwidth. In these cases we can apply the following formula to obtain the system rise time:

$$\text{System rise time} = \sqrt{(T_{\text{r}}^{2}(\text{displayed}) - T_{\text{r}}^{2}(\text{source}))}.$$

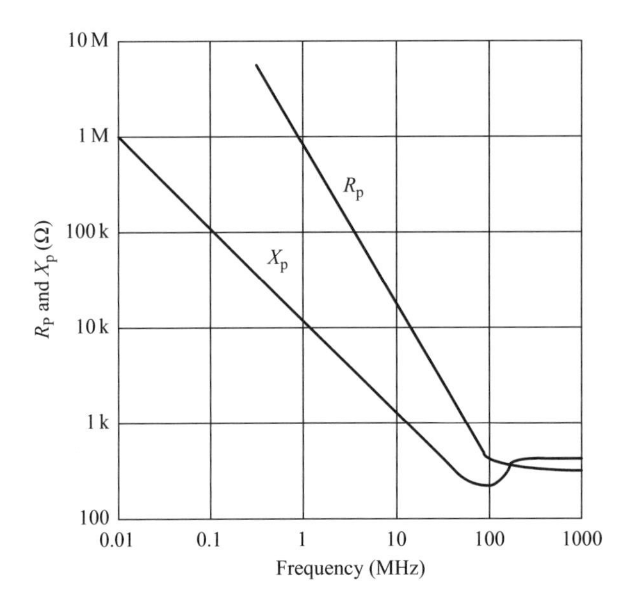

Figure 5.37 X_p *and* R_p *plotted against frequency for a typical* 10× *passive probe*

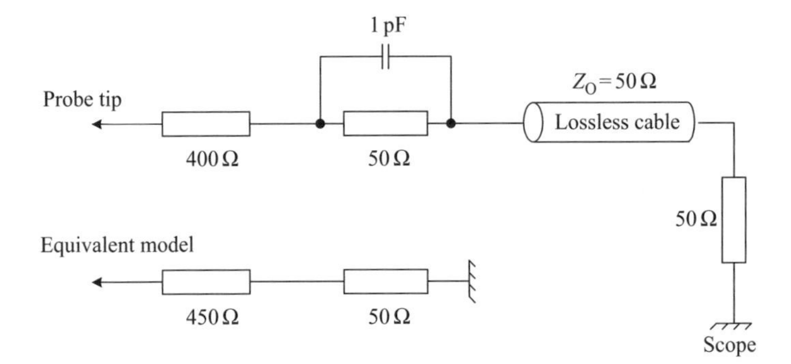

Figure 5.38 Circuit and equivalent model of 10×, 50 Ω probe

However, passive probes do not follow this rule and therefore should not be considered with this formula. General purpose oscilloscopes, which include standard accessory probes in the package, usually deliver the manufacturer's advertised scope bandwidth at the probe tip.

Another important consideration is the use of the probe ground or earth lead. A ground lead is a wire that provides a local return path to earth for the signal under observation. An inadequate ground lead (one that is too long or too high in inductance) can reduce the fidelity of the high frequency portion of the displayed signal. For a detailed practical account of these the reader should consult References 12 and 13.

5.6.4 Different types of probes

A wide variety of probes suitable for different oscilloscope measurements are available today. The basic types are as follows:

- voltage sensing,
- current sensing,
- temperature sensing,
- logic sensing, and
- light sensing (optical).

Most common are the voltage or current sensing types. These may be further subdivided into 'passive' and 'active' for voltage sensing and a.c. and d.c. for current sensing.

In addition, there are different varieties of temperature sensing and logic sensing probes available with the newer oscilloscopes. For example, word recogniser probes can be used to generate a trigger pulse in response to a particular logic state. Details of these are beyond the scope of this chapter and References 1 and 12–14 are suggested for details.

5.7 References

1 KULARATNA, N.: *Modern Electronic Test and Measuring Instruments*, IEE – London, 1996.
2 TEKTRONIX, INC.: *Cathode Ray Tubes: Getting Down to Basics*, 1985.
3 DE JUDE, R.: 'Directions in flat-panel displays', *Semiconductor International*, August 1999, pp. 75–81.
4 PRYCE, D.: 'Liquid crystal displays', *EDN*, October 12, 1989, pp. 102–113.
5 DANCE, B.: 'Europe's flat panel display industry is born', *Semiconductor International*, June 1995, pp. 151–156.
6 AJLUNI, C.: 'Flat-panel displays strive to cut power', *Electronic Design*, January 9, 1995, pp. 88–90.
7 HEWLETT PACKARD: 'Bandwidth and sampling rate in digitising oscilloscopes', Application Note 344, Hewlett Packard, April 1986.
8 TEKTRONIX, INC.: 'An introduction to digital storage', Tektronix Application Note 41W – 6051-4, 1992.
9 LECROY, INC.: *Reference Guide to Digital Waveform Instruments*, LeCroy, USA, 1993.
10 TEKTRONIX, INC.: 'How much scope do I need', Application Note 55W – 10063-1, 1994.
11 IEEE Standard 1057–1994: *IEEE standard for digitising waveform recorders*, December 30, 1994.

12 TEKTRONIX, INC.: 'ABCs of probes', Tektronix Application Note 60W – 6053-2, 1990.
13 WATTERS, E.S. *et al.*: 'Probing techniques become crucial above 500 MHz', *EDN*, 15th October 1987, pp. 165–174.
14 LECKLIDER, T.: 'Scope probes clinch signal integrity', *Test & Measurement World*, June 1998, pp. 22–30.

Chapter 6
Recent developments on DSO techniques

6.1 Introduction

Instruments based on cathode ray tube storage were the early analogue versions of the storage oscilloscopes. These instruments were justifiable, even with their limited performance as there were no alternatives based on semiconductor memories, A-to-D converters and processors. During the latter part of the 1970s and early 1980s, the unprecedented advances on the semiconductor memories and the data converter ICs provided sufficient design flexibility for the oscilloscope designers to develop processor based storage scopes.

As can be seen in Figure 6.1, the prices of digital storage oscilloscopes (DSOs) generally declined during the period from 1985 to 1994, while analogue scope prices remained steady [1]. For many applications, DSOs now provide all the necessary

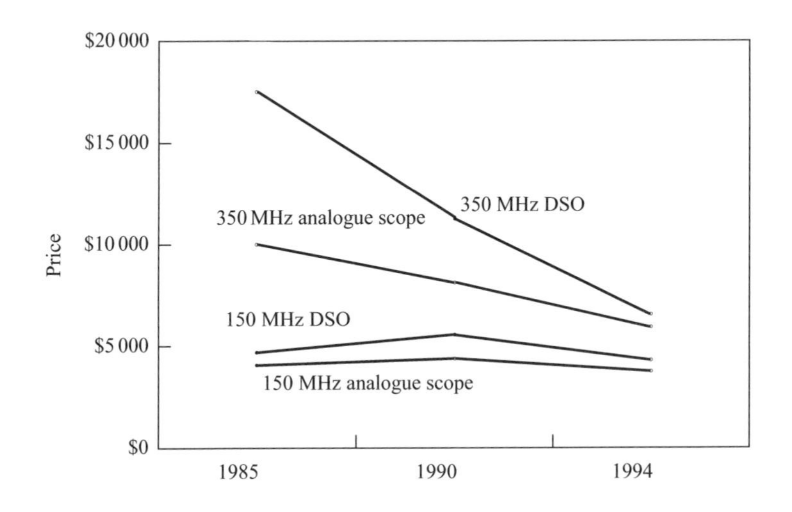

Figure 6.1 Representative price reduction of oscilloscopes (courtesy: Tektronix, Inc., USA)

functionality. Also, DSOs continue to offer the unique capabilities that have contributed to their rapidly increasing popularity. For these reasons and technological advances on semiconductors that justify the DSOs in place of long time popular fast analogue scopes, leading manufacturers such as Tektronix and Hewlett-Packard, etc., have now totally switched over to DSOs.

This chapter is a summary of recent advances on DSO techniques and applications, with a few examples of commercial versions.

6.2 Viewing complex waveforms

Video, data communications, and various modulated waveforms are complex because they contain both relatively low frequency signals and relatively high frequency signals (see Figure 6.2). Users analysing these waveforms have generally preferred an analogue scope because it can capture and display both high and low frequencies simultaneously, making it possible to analyse the entire waveform at one sweep-speed setting. Analogue scopes with sufficient bandwidth operate without compromising signal representation because they display an uncorrupted continuous waveform.

Early generations of DSOs failed at this task because they used to take the analogue signal and digitise it into several points spaced at regular intervals. During this process valuable data can be lost between the points. In such DSOs, too few sample points can cause aliasing if the sample rate is less than the Nyquist frequency. Very high sampling rates and long acquisition memories needed to solve these were expensive until recently.

6.3 Analogue oscilloscopes: advantages and limitations

At this point it may be appropriate for us to have a brief analysis of advantages and limitations on analogue oscilloscopes for field use, in order to appreciate the recent developments on DSOs [2, 3]. Analogue oscilloscopes used to remain the troubleshooting tool of choice because of a few useful characteristics. Let us review these, as well as a few of their limitations.

6.3.1 Highly interactive display

As the analogue scopes use a vector display technique, the display shows the changes in the signal in real time. This results from the use of the direct-view vector CRT. Adjustments are simplified because there is no delay between a change and the observation of the effect on the waveform. Another characteristic of the highly interactive display is that complex waveforms, such as TV, video and amplitude modulated RF, are displayed as expected.

6.3.2 Low price

Analogue oscilloscope technology is mature and price competition between suppliers has kept prices low. Compared with this, DSO prices have come down gradually only during the past decade or so, as shown in Figure 6.1.

(a)

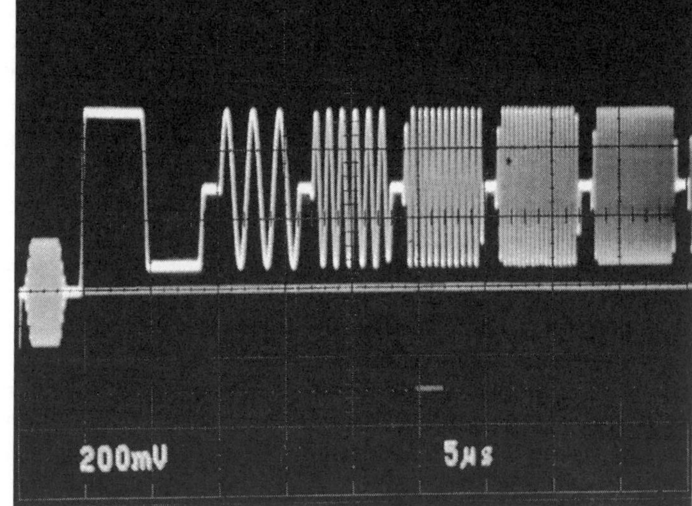

Analogue scope

(b)

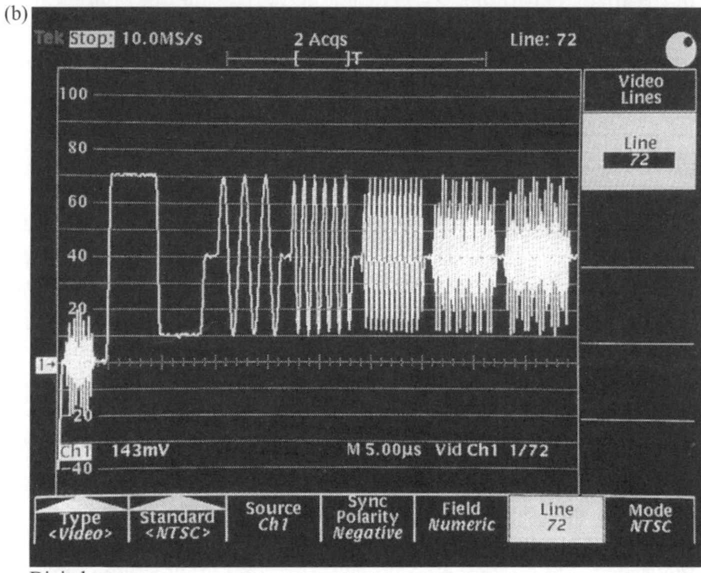

Digital scope

Figure 6.2 Quality of analogue oscilloscope display compared with early generations of DSOs: (a) Analogue scope; (b) Digital scope (courtesy of Tektronix, Inc., USA)

6.3.3 Limitations

Analogue oscilloscopes are not without limitations. The heart of the analogue oscilloscope is the vector CRT. This is the component that gives the analogue oscilloscope its highly interactive display. It is also the weak link in the analogue oscilloscope's ability to keep up with the shift to digital system troubleshooting and analysis. The light produced by the phosphor has a very short life time after the electron beam has passed. In the case of the most commonly used phosphor, P31, the light output decays to a level that is just discernible in normal room light in about 38 μs. This means that the electron beam must re-excite the phosphor at a rate that is fast enough to keep the light output at a level that is visible. If the refresh rate drops below 26 kHz there will be a dramatic drop in light output. This may not be a problem until something like a high resolution time interval measurement of a control or handshake line is encountered. In this case a fast sweep speed is required, and the signal repetition rate is less than the rate needed to keep the phosphor excited. The faster sweep speeds result in a lower light output from the phosphor because the electron beam does not spend enough time at any given location to cause the phosphor to produce its maximum light output. Thus, any decrease in refresh rate will have a greater effect on light output. Another factor that can reduce the light output of the CRT is the use of delayed sweep. As the magnification increases, the display repetition rate drops. Even in the case where the main sweep display is bright and crisp, the phosphor's light output may drop below the visible level as delayed sweep is applied to increase the timing resolution of the display. Tektronix developed the MCP technique to overcome some of these problems.

Another phenomenon that limits the usefulness of the CRT is flicker. When the refresh rate drops below a certain minimum, the display will appear to flash on and off (flicker). Although the flicker frequency is a function of the specific phosphor used in the CRT, the apparent rate seen by the eye is independent of the actual rate and is approximately 15–20 Hz. Flicker becomes more annoying as the display gets brighter. This often occurs when a fast signal must be viewed at a slow sweep speed, as is the case in many digital circuit applications.

Taking these two characteristics of the analogue CRT into account, a map of useful operating area can be developed. Figure 6.3 shows a plot of sweep speed as a function of input trigger repetition rate. Let us examine this map to see how the limits were established. Holding the input signal repetition rate constant at a point where a bright crisp display will be obtained and decreasing the sweep speed, we will reach a point where the display starts to flicker. Continuing to decrease the sweep speed will cause the flicker to increase to a point where the display is no longer useful. It will eventually decay to a moving bright dot. The point at which the flicker becomes so bad as to cause the display to be unusable is the left-hand limit of the operating area. Holding the sweep speed constant at a point where the light output and flicker problems are minimised and varying the signal repetition rate, we see that as the input signal repetition rate is decreased the display becomes dimmer. Once a point is reached where the display is no longer viewable in room light, the lower limit of the operating area is reached. This limit increases as the sweep speed is increased.

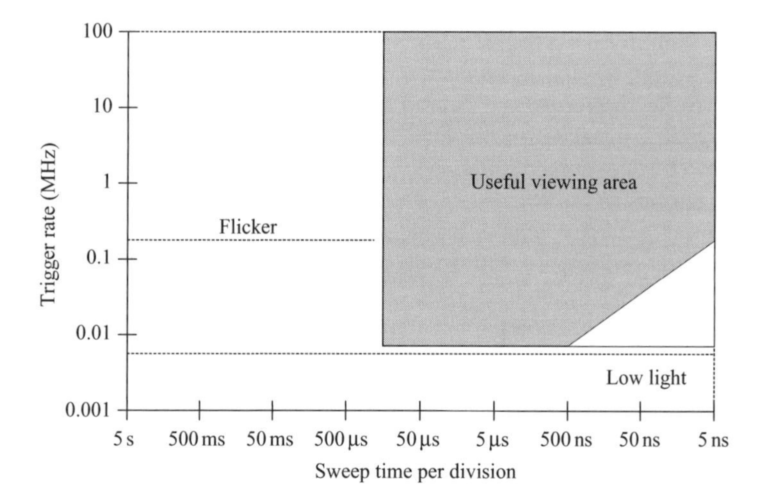

Figure 6.3 Analogue oscilloscope operating area (courtesy of Agilent Technologies, Inc.)

Another limitation of analogue oscilloscopes is their inability to look ahead of the trigger point. Laboratory quality analogue oscilloscopes have delay lines in their vertical systems to ensure that the trigger point is displayed, but this does not solve the problem when the trigger event is at the end of the event to be displayed. This is a very common case in digital system analysis.

Another limitation of analogue based oscilloscopes is that their measurements are based on analogue ramps with accuracy specified as a percentage of full scale. Often greater accuracy is required to determine the cause of problems in digital systems.

6.4 Modern DSO design techniques

In chapter 5, the basic concepts related to sampling, A-to-D conversion, display techniques and interpolation, etc., were discussed. In order to appreciate more modern techniques, it is necessary to have a little deeper analysis of user expectations together with the modern design approaches. DSO technology is the fastest moving area in test and measurement today. Users are reaping the benefits of advances in all categories of DSO technology: acquisition systems, the core processing systems, automated measurement capability, and displays.

6.4.1 The acquisition system

Typically, the first concern engineers have when they discuss digital oscilloscopes is the acquisition system. They want to know whether a scope that uses a sampling technique can accurately capture the signals they need to see. In a DSO, that accuracy takes two forms – amplitude (vertical) accuracy and timing (horizontal) accuracy. A major component of the former is the vertical resolution, which has pretty much standardised at 8 bits. However, 10- and 12-bit instruments are available, and averaging techniques can improve effective resolution. Overall vertical accuracies within 1 per cent are possible. However, as discussed in chapter 5, the effective number of bits (ENOB) reduces to much lower values at higher frequencies, owing to noise, non-linearities and other secondary reasons.

The timing accuracy is more problematic. Although such factors as clock precision and jitter are important, the specification most users concentrate on is the sampling rate. DSO manufacturers have made a strong effort to ease those concerns, with some major breakthroughs in sample rate performance. Sample rates as high as 1 Gsample s^{-1} were available at reasonable prices prior to 1995. Extremely high sample rates such as 5 Gsamples s^{-1}, etc., were also reported around 1993, which have allowed single shot performance beyond 500 MHz. For example, LeCroy Corporation has developed very special monolithic waveform digitisers. This proprietary technique, which makes use of a 'current sampling' technique [4], allows acquiring samples of the input voltage at 200 ps intervals [5].

6.4.1.1 Acquisition memory

Acquisition memory is also an area that has seen recent advances. In newer DSOs, memories of 2 or 4 kpoints per channel are the rule, replacing the once common ones of 1 kpoint per channel. More scopes are now available with 10 kpoint and longer memories as either standard or optional, and some instruments have megapoint-level storage capacities. Deep storage is important at today's faster sampling rates, because it eliminates the need to choose between a very short acquisition time or slowing down the sampling rate. While manufacturers such as Tektronix and Hewlett-Packard generally use kilobyte range memories, LeCroy promotes the advantage of deeper memories in the range of megabytes. It is important to understand that longer memories have obvious advantages, but add to the cost of the oscilloscope. In the DSO application fast memories are needed and those are more expensive than slower counterparts.

Record length determines either the duration or the resolution of the signal captured by the scope. A long record length might provide better representation of more complex waveforms. A long record length also allows the user to capture an event that occurs over a prolonged period and sift through the data for problems. Record lengths should be consistent at all sweep speed settings to ensure maximum resolution and ease of use.

From the preceding discussion on the acquisition memory, although the designs can store up to thousands or millions of waveform samples, the display screen has a lower number of columns (such as 500). This restricts the user to see about 500 points of data at a time on the display. If it is necessary to see more than 500 points at a time, compressed representation of data is mandatory.

6.5 Display update rate

Most digitising oscilloscope displays were not as responsive to signal changes as their analogue counterparts. Typically, the waveform was processed by software using a

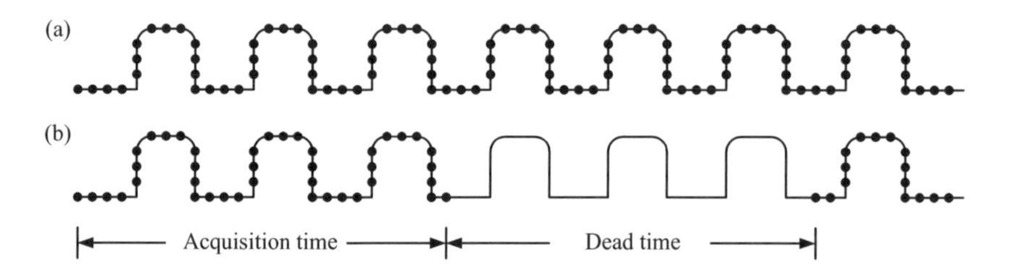

Figure 6.4 Effect of dead time on the display: (a) constant acquisition and no dead time – ideal case; (b) practical case with finite acquisition time

microprocessor that could cause dead times between acquisitions, limiting the number of waveforms displayed per second. Other display problems such as aliasing might leave the analogue oscilloscope user wondering whether the displayed waveform was telling the real story. This was especially important in troubleshooting applications, in which the user may not have known what the signal looked like.

Practical DSOs have a finite sample rate; moreover, they pause after each acquisition to process the acquired samples for display on the screen (Figure 6.4). The scope's sampling system determines the sample rate; faster sampling systems cost more. The time that the scope takes to process the acquired samples and prepare for the next acquisition is called *dead time* or *blind time*. The dead time depends on the required amount of data processing and the computational power of the microprocessor or ASIC that crunches the data.

Figure 6.4(b) shows the dead time as being relatively short – about the same duration as the acquisition time. The actual situation is usually much worse; the dead time can be orders of magnitude longer than the acquisition time. Another way to quantify this behaviour is as a duty cycle – the ratio of acquisition time to total time. The duty cycle, which varies greatly among scope models and also depends on the scope's sweep speed, can be surprisingly small – sometimes less than 0.0001 per cent when a scope with a long dead time operates at a fast sweep speed.

Another way to represent dead time is via the display update rate, expressed in samples per second or waveforms per second. Think of the display update rate as a measure of how efficiently a scope converts its maximum sample rate into displayed points. For example, the HP 54600B, a 100 MHz bandwidth oscilloscope with a very fast display update rate, has a maximum sample rate of 20 Msamples s^{-1} but a maximum display update rate of 1.5 Msamples s^{-1} [6]. So, when operating at its fastest sample rate, this scope can convert only 7.5 per cent of the possible samples into displayed points. Scopes with slower update rates take an even smaller percentage of the possible samples. Digital scopes do not always operate at their maximum sample rate, however, particularly on the slower sweep speed settings, so the efficiency is often higher. On most scopes, the dead time/display rate performance changes with time/division setting. Within the past few years (1991–1996) many oscilloscope vendors have developed special proprietary techniques for increasing the display update rate.

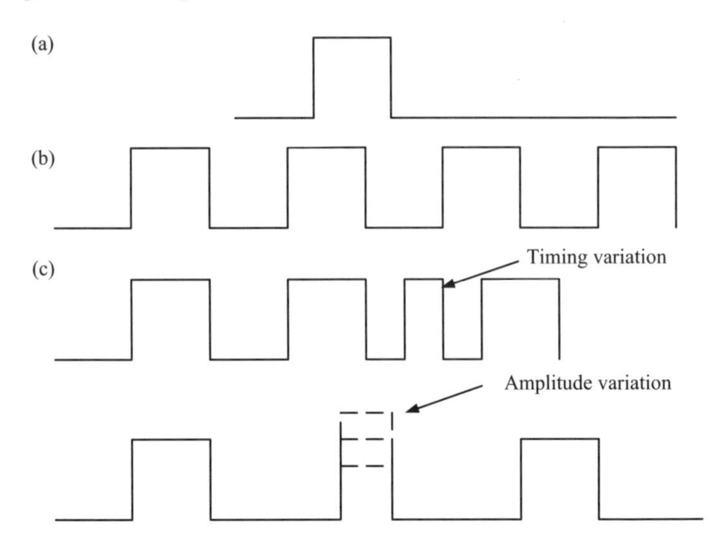

Figure 6.5 Different classes of signal: (a) single shot; (b) repetitive; (c) varying repetitive

6.5.1 Classes of signal

To appreciate the need for a fast display update rate one has to appreciate different kinds of waveform. For this purpose all waveforms in the measurement world can be divided into three basic categories, namely: (i) single shot waveforms, (ii) repetitive waveforms, and (iii) varying repetitive waveforms. Figure 6.5 indicates these different categories.

Some waveforms might not fit easily into one of these categories, but these waveform categories help you to understand how sample rate and dead time affect the quality of a measurement.

6.5.1.1 Single shot waveforms

Single shot waveforms occur only once, so you get only one chance to measure them. Alternatively, they occur so infrequently that it seems as though you get only one chance to measure them. Either way, the scope must have sufficient sample rate to capture the event and store it in its memory. Again, you could argue about how high the sample rate needs to be to capture the signal faithfully, but a sample rate above the Nyquist frequency is a minimum requirement.

If there is only one chance to capture a signal, the scope needs to know when the signal occurs, which leads to the issue of triggering. Triggering is a key feature in all scopes: if you cannot trigger on a single shot event, you will never see the event. If the event is something simple, such as a single pulse, edge triggering should be sufficient to capture the event.

If the event is more complex, you need a more sophisticated triggering capability. Suppose the single shot event is a single narrow pulse in a stream of wider pulses. This is a single shot event, but it is in a signal that hides it from an edge trigger. A trigger feature such as glitch trigger or time qualified pattern trigger could identify the one narrow pulse in the signal and trigger on it. Most high sample rate oscilloscopes for single shot capture include an advanced triggering system. Although triggering is an important issue for single shot signals, dead time is not. Because the event occurs only once, there is no need quickly to acquire a second event.

6.5.1.2 Repetitive waveforms

Repetitive waveforms are also easy to understand. The classic function generator waveforms – sine, square and triangular waves – all fall into the repetitive category. Most scopes display these waveforms well because the waveforms do not vary, and triggering on them is usually easy. Sample rate is not a big issue for these measurements as another cycle of the waveform will come along soon enough, so the scope can use repetitive sampling. Repetitive signals are not necessarily simple, though; complex signals, such as composite video, can be repetitive.

To view repetitive signals effectively, you need to trigger on them; otherwise, they are unstable and wander across the display. Simple signals require only edge triggering, but others, such as composite video, can require other trigger modes.

6.5.1.3 Varying repetitive waveforms

Varying repetitive waveforms are more difficult to understand than the previous two categories but represent an important and common class of waveforms that are often difficult to measure. Think of these waveforms as basically repetitive with some variation in the wave shape. For example, a sinewave that varies in amplitude as a function of time is basically repetitive but varies from cycle to cycle. This variation might be very slow, producing a scope display that shows a sinewave gradually changing in size. Alternatively, the variation could be very fast, occurring instantly on only one cycle of a waveform. The slowly varying case is easy to view with almost any scope, but the quickly varying signal presents a greater challenge. A scope with a short dead time and a high display rate is more likely to show the single short cycle, assuming that the scope triggers on the normal cycles of the sinewave. If the scope can trigger on the reduced amplitude cycle, the measurement problem degenerates to the single shot case.

Figure 6.6 shows some representative examples of varying repetitive waveforms. Often these signals are the unexpected ones that make the measurement difficult. These represent the hardware bugs that are unavoidable even though you never intentionally design them into a circuit.

Sample rate is a confusing issue for varying repetitive waveforms. The repetitive nature of the waveform means that you can use repetitive sampling to acquire the waveform. DSOs normally discard older sample points as new ones become available. This technique makes perfect sense in most situations, but when dealing with infrequent events, you might do better to keep all the sample points you can. Most digital scopes provide an infinite persistence mode that accumulates all samples on the display. (For example the HP 54600 series scopes' *Autostore mode* has the added

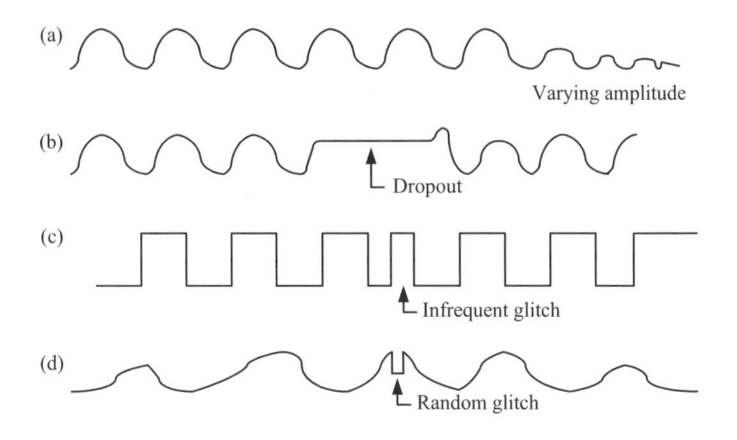

Figure 6.6 Representative examples of varying repetitive waveforms: (a) varying amplitude; (b) dropout; (c) infrequent glitch; (d) random glitch

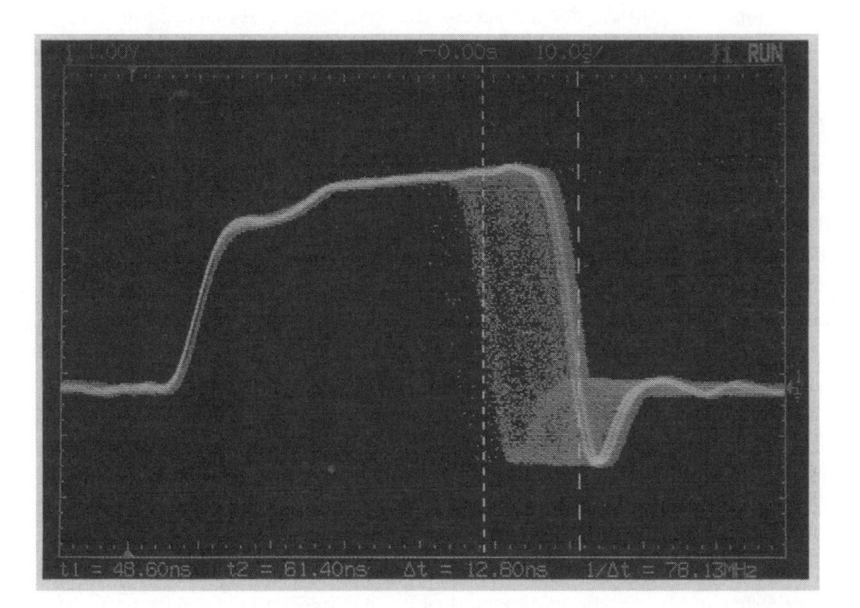

Figure 6.7 A jittery signal is shown with the range of jitter at half intensity and most recent signal at full intensity (courtesy: Agilent Technologies, Inc.)

benefit of showing the old samples in half-bright intensity and the most recent samples in full-bright intensity.) Regardless of the dead time, using the infinite persistence mode helps uncover infrequent variations in a waveform (see Figure 6.7).

More details in observing time varying signals can be found in References 7 and 8.

6.5.1.4 Triggering vs display rate

Triggering plays an important role in viewing infrequent events. If you can trigger on an event of interest, you essentially reduce the problem to the single shot case and eliminate the dead time issue. Triggering is often the ultimate solution to viewing a tough problem, provided that you can describe and trigger on the problem. In troubleshooting, the user often does not know in advance what the problem is. In such cases, a high update rate scope that does not miss much between acquisitions may be the best choice.

In summary,

- for single shot measurements, there is no substitute for a fast sample rate;
- for repetitive measurements, sample rate is less of an issue; most scopes display repetitive signals, scope dead time and display update rate become critical; and during its dead time, the scope cannot see variations in the signal;
- triggering can be a powerful way to measure infrequent events but requires some knowledge of the event on which you want the scope to trigger.

Some scopes specify their display's update rate or dead time, but most do not. In general, a scope manufacturer that designed a scope for a fast display update rate will specify the rate. If there is no specification, the user could determine the rate through experimentation.

6.6 DSO solutions for capturing infrequent events

Modern DSOs use many advance techniques to capture infrequent events. Some of these techniques are: (i) use of peak detect techniques, (ii) extremely fast sampling rates, and (iii) special trigger techniques.

6.6.1 Peak detect or glitch detect mode

Most digitising oscilloscopes reduce the effective sample rate at slow sweep speeds by simply throwing away sample points. This causes a problem because narrow pulses or glitches that are easily viewable on fast time base settings can disappear as the sweep speed is reduced. A special acquisition mode called *peak detect* (also known as *glitch detect*) maintains the maximum sample rate at all sweep speeds. In peak detect mode, any glitch that is above a minimum width is captured and displayed, regardless of the sweep speed. In this mode, modern oscilloscopes can capture glitches as narrow as one nanosecond. The Tektronix TDS series of oscilloscopes is a classic example of this. Figure 6.8 explains the technique graphically, showing its advantages.

In Figure 6.8 you could see that the glitch may or may not be captured depending on the sample rate, but gets acquired definitely in peak detect mode. Reference 8 provides some details related to this technique.

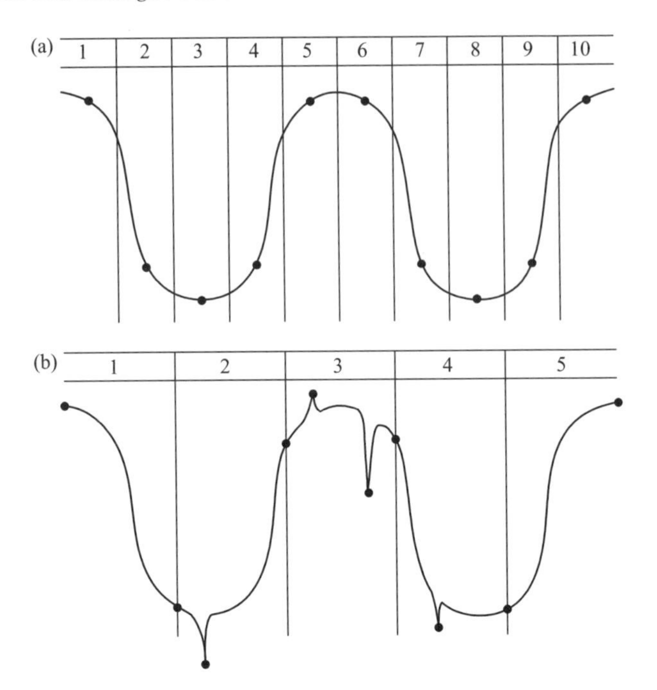

Figure 6.8 Sample mode and the peak detect mode: (a) sample mode; (b) peak detect mode

6.6.2 Fast sampling rates

The Nyquist criteria indicate to us that we have to sample a waveform at least at twice the rate of maximum frequency of waveform contents. However, building samplers running at very high frequencies is costly and requires more expensive technologies such as ECL or GaAs circuits. During the latter part of the 1990s, several manufacturers such as Tektronix, HP and LeCroy introduced oscilloscopes that allow sampling rates between 2 Gsamples s^{-1} to 10 Gsamples s^{-1}. This has allowed instruments with single shot bandwidths equivalent to over 500 MHz and running up to about 1 GHz. Fast sampling rate oscilloscopes do not have to depend on the repetitive nature of waveforms to capture high frequency components. Figure 6.9 indicates a fast sampling technique used in the LeCroy 936x series of oscilloscopes. Many DSOs that have an ADC per channel reassign unused channels' ADCs to active channels, thereby increasing the sampling rate. A four channel scope might digitise at 500 Msamples s^{-1} when you use all channels, at 1 Gsample s^{-1} when you use two channels, and at 2 Gsamples s^{-1} when you use a single channel. The converters take samples at staggered times and the scope interleaves the conversion results. For details, References 5 and 9 are suggested.

In LeCroy's proprietary multisample-and-hold circuitry, 500 sample-and-hold circuits are timed precisely to take samples of the input voltage at 200 ps intervals – that is, a 5 Gsamples s^{-1} rate. The 600 MHz inputs automatically send the input

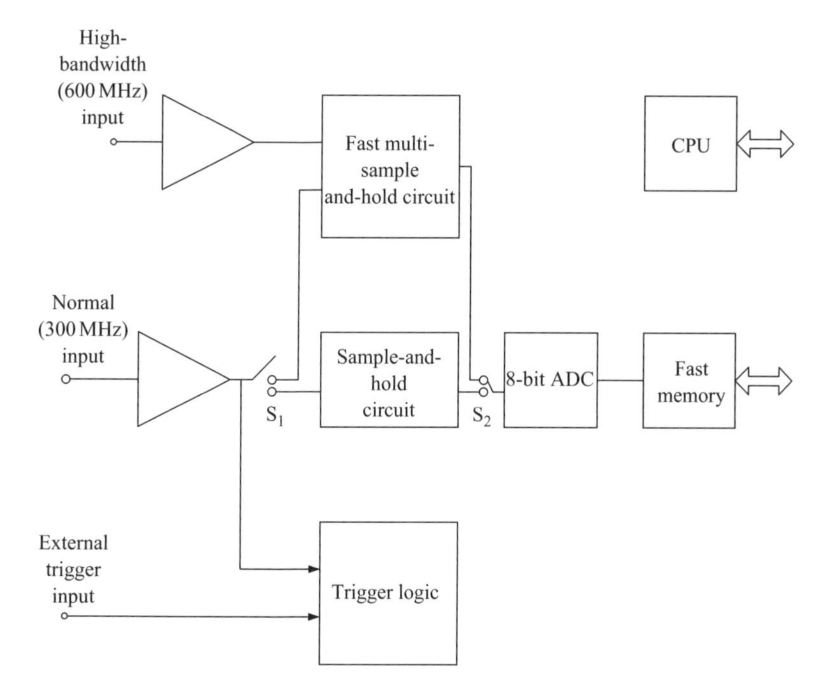

Figure 6.9 The 9360's fast sampling rate circuitry (courtesy of LeCroy Corporation)

through this circuitry (Figure 6.9). The 500 samples are then clocked sequentially through an 8-bit, 100 MHz ADC. It thus takes 10 ns to send each digitised sample to the memory. As a result, about 5000 ns, or 5 μs, are needed to digitise each 500 sample (100 ns long) acquisition.

Users who want to see a longer portion of the signal can switch to slower time bases. At the 9360's slowest sample rate, 100 Msamples s^{-1}, the oscilloscope will store 20 000 samples. At slower time bases, signals at the 300 MHz inputs are routed to a conventional sample-and-hold circuit.

Tektronix uses fast sampling rates between 500 Msamples s^{-1} to 5 Gsamples s^{-1} allowing most of their TDS series oscilloscopes to have single shot bandwidth specifications from about 50 MHz to 1 GHz.

6.6.2.1 Analogue bandwidth vs digital single shot bandwidth of a DSO

To create a waveform accurately, the DSO must gather a sufficient number of samples after the initial trigger. Based on the Nyquist criteria, a digital scope needs at least two samples per period (one full cycle of a regular waveform) to reproduce faithfully a sinewave; otherwise, the acquired waveform will be a distorted representation of the input signal. In practice, when the scope uses an interpolation such as $\sin(x)/x$ interpolation, the scope needs at least 2.5 samples per period.

This requirement usually limits the frequency a digital scope can acquire in real time. Because of this limitation in real time acquisition, most DSOs specify two

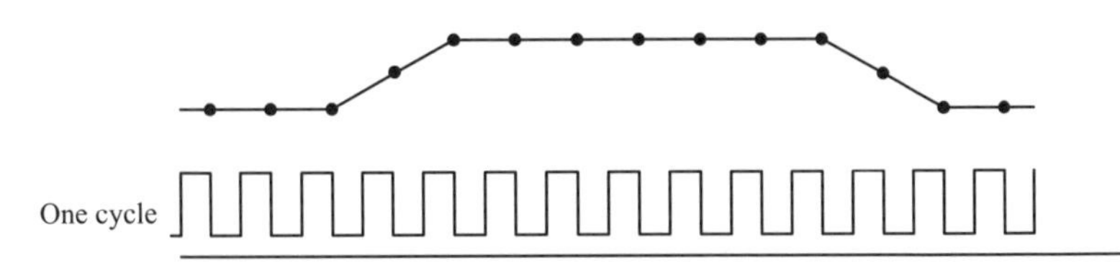

Figure 6.10 Digital real time sampling

bandwidths – analogue and digital real time. The analogue bandwidth, defined by the circuit composing the scope's input path (vertical channel amplifiers), represents the highest frequency signal a DSO can accept without adding distortion. The second bandwidth, called the *digital real time bandwidth* or the *single shot bandwidth*, defines the maximum frequency the DSO can acquire by sampling the entire input waveform in one pass, using a single trigger, and still gathering enough samples to reconstruct the waveform accurately. Figure 6.10 shows how this works. In a scope such as the TDS320, the signal is sampled 500 million times per second, and each sample becomes a record point.

For some DSOs, the calculated digital real time bandwidth exceeds the analogue bandwidth. (For example, the calculated digital real time bandwidth of the Tektronix TDS 320 is 500 million/2.5 = 200 MHz.) Even with the sampling rate suggesting a higher bandwidth, the actual input circuits of the vertical channels absolutely limit the scope's bandwidth.

So the analogue bandwidth defines the maximum frequency the input circuitry can handle without adding distortion; the digital real time bandwidth describes the highest frequency signal a DSO can accurately capture and display after just one trigger event. Although it is possible for the real time bandwidth to equal the analogue bandwidth, almost all low cost DSOs have a real time bandwidth that is less than half the published analogue or repetitive bandwidth.

For details, References 10 and 11 are suggested.

6.6.3 Special trigger techniques used in modern DSOs

In order to provide the same display advantages as high speed analogue oscilloscopes, modern DSOs provide many special trigger capabilities. These techniques address the problems created by single shot events, or varying repetitive signals or complex pulse trains occurring in digital systems. Extended triggering capabilities offered in most modern DSOs allow you to capture complex signals for device characterisation or troubleshooting. They also provide more selective trigger control than is available with conventional edge triggering. In addition, their sophisticated user interfaces allow rapid setup of trigger parameters to increase productivity.

For example, Tektronix DSOs allow you to trigger on pulses defined by amplitude (such as runt pulses) qualified by time (pulse, width, glitch, slew rate, setup-and-hold, and time-out), and delineated by logic state or pattern (logic triggering). Combinations of extended and conventional triggers also help display video and other difficult-to-capture signals. Advanced triggering capabilities such as these deliver a large degree of flexibility when setting up testing procedures and can greatly simplify measurement tasks.

Table 6.1 can be used to explain different possible trigger types available on modern DSOs. Conventional analogue scopes have the edge trigger, usually with

Table 6.1 Trigger types available on a typical DSO

Trigger type	Description	Typical range of settings
Edge (main and delayed)	Conventional level/slope driven trigger. Positive or negative slope on any channel or any auxiliary input.	Coupling selections: d.c., a.c., noise reject, HF reject, LF reject.
 Width	Trigger on width of positive or negative pulse either within or not within selectable time limits.	Time limit settable from 1 ns to 1 s.
 Glitch	Trigger on (or reject) glitches of positive, negative or either polarity.	Minimum glitch width from 1 ns to 200 ps.
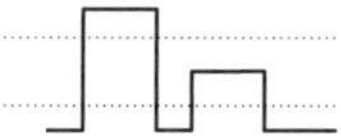 Runt	Trigger on a pulse that crosses one threshold but fails to cross a second threshold before crossing the first again.	Can be time qualified.
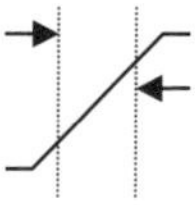 Slew rate	Trigger on pulse edge rates that are either faster or slower than a set rate.	Edges can be rising, falling or either.
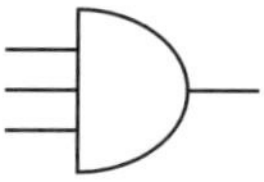 Pattern	Specifies a logical combination (AND, OR, NAND, NOR) of input channels (Hi, Lo, Don't Care). Trigger when pattern stays True or False for user specified time.	
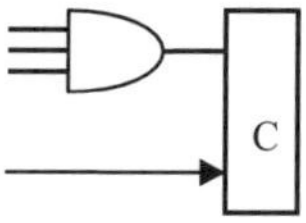 State	Trigger on violations of both setup time and hold time between clock and data which are on separate input channels.	Triggerable on positive or negative clock edge.
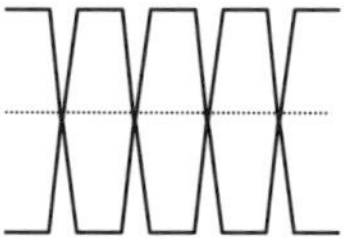 Setup and hold	Trigger on violations of both setup time and hold time between clock and data which are not separate input channels.	—

two time bases. With the modern signal processing techniques used inside the fast sampling DSOs, processor subsystems allow many special trigger types.

Another popular trigger mode is the dropout mode, where a trigger occurs when the signal disappears for a specified period of time. For details, Reference 12 is suggested.

6.6.4 InstaVu and exclusion triggering

Early DSOs were unable to compete with high bandwidth analogue oscilloscopes, because the instruments had to wait for long time to catch failures of transient types. This was because older DSOs had very low display durations (dead times were very long compared with acquisition times). *InstaVu* and *exclusion triggering* were two techniques perfected by Tektronix and LeCroy respectively to solve this problem.

6.6.4.1 InstaVu technique

In 1994, with the introduction of the proprietary InstaVu feature, Tektronix dramatically improved the speed at which DSOs capture transient failures. Instead of capturing 100 waveforms s^{-1}, Tektronix scopes in the InstaVu mode capture as many as 400 000 waveforms s^{-1}. (Only faster analogue scopes rival this number.) Because of the higher duty cycle, InstaVu scopes can capture phenomena in 1 s that conventional DSOs might wait an hour to acquire. InstaVu scopes can capture so many waveforms per second, because the scopes use a proprietary IC that rasterises the data as soon as the samples are digitised. All waveforms acquired between screen updates are, thus, stored in a pixel map and displayed at the next update.

Figure 6.11 indicates the InstaVu acquisition block diagram. When this mode is enabled, the data moved from the acquisition system are a complete, rasterised image of multiple triggered acquisitions of the captured signals. Transferring this pixel map requires more data to be moved between the two systems, but the raster is only moved at the refresh rate of the scope's display and contains information from tens of thousands of acquisitions. Alternatively, ten thousand 500 byte acquisitions would have to be moved to the display every 30 ms, requiring a data rate of 167 Mbytes s^{-1}. By comparison, a 500×200 raster with 1 bit per pixel moved to the display every 30 ms equals a data rate of only 417 kbytes s^{-1}.

Besides displaying many acquisitions as a single raster image, InstaVu acquisition achieves its rapid acquisition rates by allowing the system to rearm itself and acquire as soon as it has completed on acquisition rather than having the instrument firmware intervene on an acquisition-by-acquisition basis; the instrument firmware only shuts down the acquisition system occasionally – once every 30 ms to copy out the raster that shows the behaviour of the signal over the last 12 000 acquisitions.

To achieve this acquisition technique, Tektronix use a demultiplexing IC with 360 000 transistors in a 0.8 μm technology CMOS process. This 304 pin IC dissipates about 2.5 W. Ordinarily, the only function of the Demux IC would be to demultiplex data from the analogue-to-digital converter and store the data in high speed SRAM.

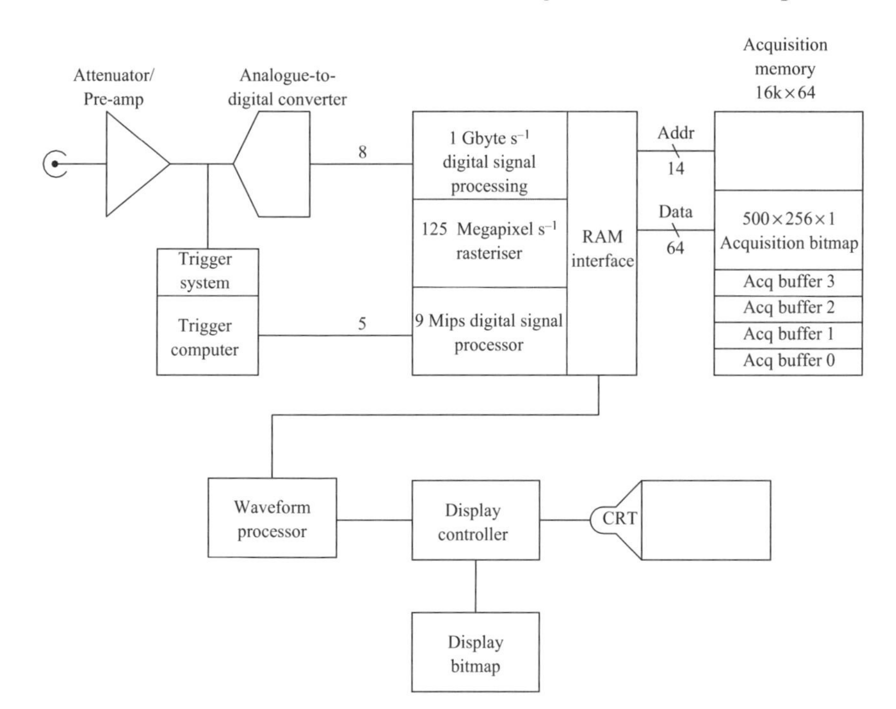

Figure 6.11 InstaVu acquisition block diagram (courtesy of Tektronix, Inc., USA)

One third of the Demux IC is devoted to this purpose (see Figure 6.11). The remainder is split evenly between a high-speed rasteriser and a digital signal processor (DSP). The DSP is included for local programmability, mathematical algorithms, and trigger position computations. The rasteriser is the primary enabler of the high live time DSO. For further details, Reference 13 is suggested.

6.6.4.2 Exclusion triggering

LeCroy offers a feature called 'exclusion triggering', which lets the DSO capture complete records only when a circuit misbehaves. The DSO then uses the records to compile histograms showing how often each type of failure has occurred, allowing the malfunctions to be fixed rapidly. LeCroy has added the exclusion triggering feature to the 9300 series of DSOs. Exclusion triggering keeps the scope from capturing normal waveforms; triggering occurs only on the waveforms whose time parameters are abnormal – and the scope can automatically determine what is abnormal. Moreover, the scope's store complete waveform records from which waveform parameters can be determined and statistical analyses can be performed. The scope displays the anomalous waveforms individually or in superimposed views and can display histograms of the parameters. Sometimes, the waveform statistics reveal several distinct failure modes that require separate corrective measures.

With InstaVu, if you want to analyse or keep copies of waveforms, you must return to the conventional real time acquisition mode and acquire additional records. In the exclusion trigger mode, a scope need not update its display after each acquisition. You can choose to have updates take place only after the acquisition of many anomalous waveforms. This display mode keeps delays associated with unnecessary screen updates from slowing waveform acquisition.

6.7 Signal processing functions

Most modern DSOs employ very complex digital signal processing techniques which allow very special types of waveform processing. Some common items that have become popular among the users and which are provided by practically all reputed manufacturers are: (i) averaging function, (ii) waveform mathematics, and (iii) fast Fourier transform (FFT). The following sections provide a brief introduction to these techniques.

6.7.1 Averaging

This useful function, which is available in many modern DSOs, allows the user to observe waveform data averaged over 2–10 000 cycles.

When viewed on a digitising oscilloscope, a signal often can be completely obscured by noise and will sometimes appear to contain no useful information. Because of its random nature, noise can be partly eliminated from the oscilloscope's display. More importantly, its random qualities also can be used as the means to increase digitiser resolution. Recent developments in fast waveform processing spurred the evolution of averaging techniques for efficiently reducing the noise seen on a display and improving the scope's measurement resolution.

The averaging process in a typical digital storage oscilloscope uses the differences between signals of interest and random additive noise sources. Because of the time locked positioning of samples relative to the trigger point in successive acquisitions, any given sample amplitude consists of two parts: a fixed signal component and a random noise component. The desired incoming signal contributes a fixed amplitude component to each given sample position in each triggered acquisition. Random noise, however, does not have a fixed time relationship to the trigger point. Because of this, noise may be viewed as numerous signals added together, with each noise signal being different in frequency from the desired signal. Noise signals move in time relative to the triggered signal, contributing positive and negative amplitudes equally to each sample in the acquisition record.

The amount of noise reduction increases with the number of acquisitions averaged, as the average of the noise amplitudes approaches zero. For more details on averaging, Reference 14 is recommended. The photographs in Figure 6.12 illustrate the principle. The averaging sequence of a square wave added to a slightly different frequency sinewave is shown in Figure 6.12(a), with the trigger on the square wave. Figure 6.12(b) and Figure 6.12(c) indicate the averaging effect for 16 and 256 acquisitions, respectively.

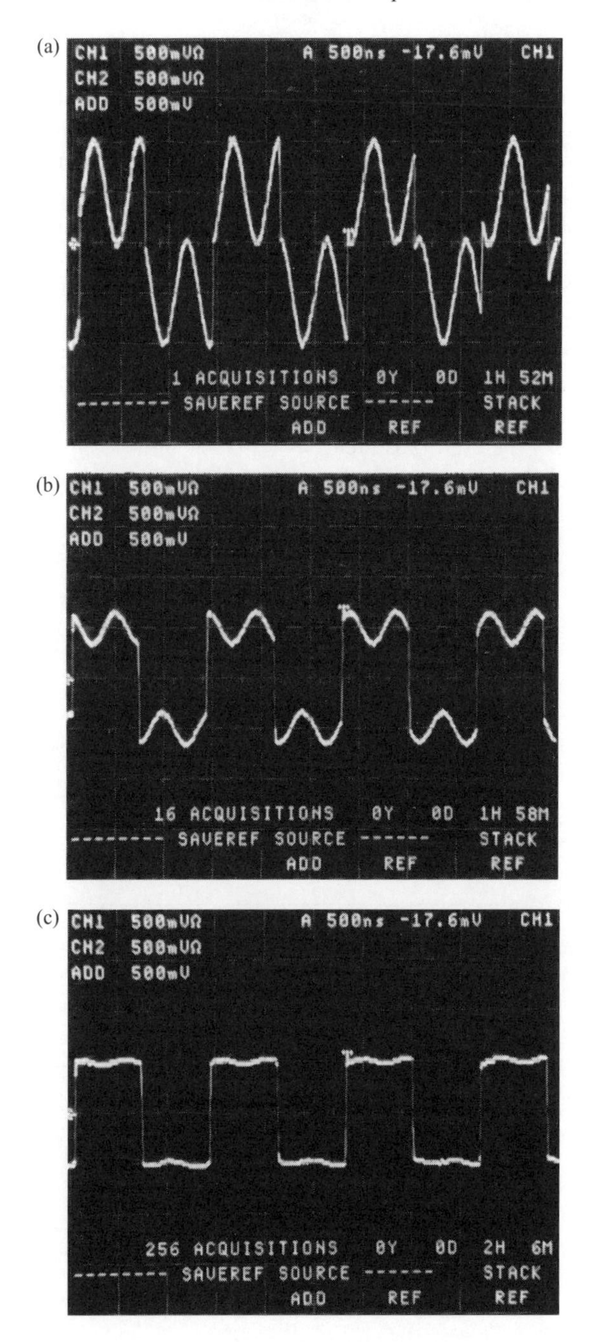

Figure 6.12 *Illustration of the averaging: (a) a sinewave superimposed on a square wave; (b)16 averages; (c) 256 averages (courtesy of Tektronix, Inc., USA)*

6.7.2 *Waveform maths*

All DSOs use microprocessors and DSPs, so the instruments can also perform various types of processing on the acquired waveform. For example, automatic calculation of pulse parameters and basic waveform mathematics (sum, difference, product, ratio) are standard features of some oscilloscopes. Some DSOs have options to perform more complex mathematics like integral, derivative, exponential, logarithm, extended averaging, digital filtering, extreme values, fast Fourier transform and more.

For example, the DSO can be used to compute power. This is shown in Figure 6.13, where the upper two waveforms are respectively voltage and current at the input of a switchmode power supply. These two waveforms are multiplied point by point to provide the power waveform at the bottom. Reference 15 indicates details of how modern DSO maths and other features can be used for power supply characteristics, illustrating some of the useful features.

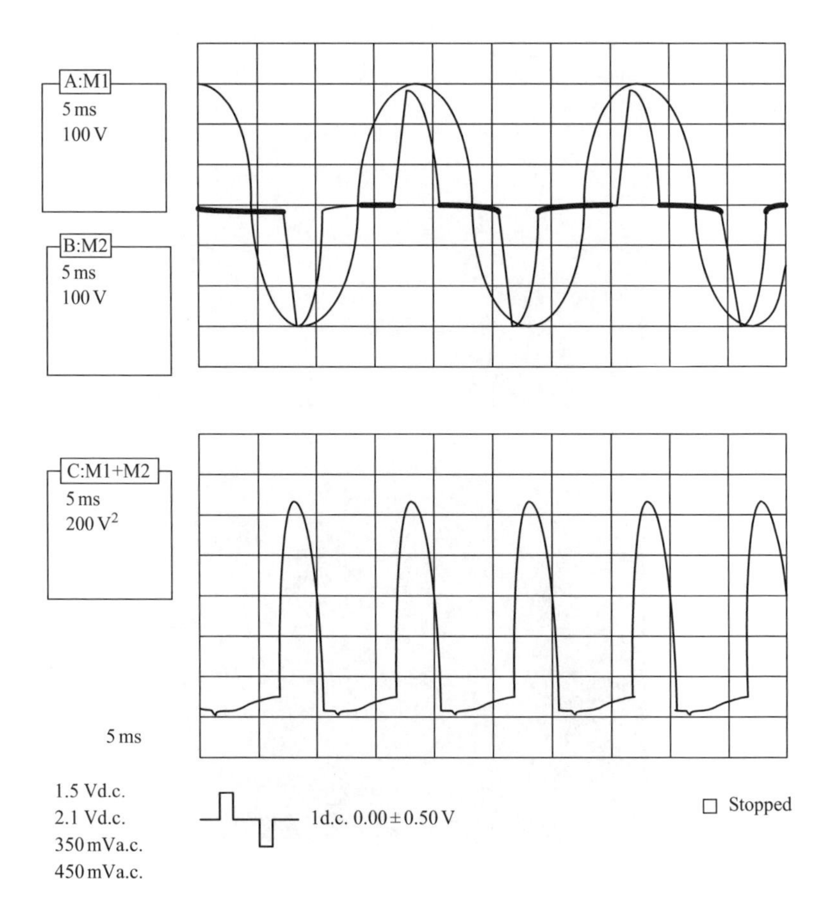

Figure 6.13 The use of DSO maths functions to indicate input power (Courtesy: Le Cray Corporation)

6.7.3 Fast Fourier transforms (FFT)

With the development of microprocessors, DSPs and the memory components, and with the prices of these components dropping, fast Fourier transform (FFT) calculations within the hardware systems of DSOs have become a reality. The ability to analyse the frequency domain with the FFT has opened up a whole new set of measurement opportunities in digital oscilloscopes. However, the user needs to understand the capabilities and constraints of both the measurement techniques and the instruments that perform them. An FFT's resolution, or how well it separates energy at closely spaced frequencies, depends on the number of samples used in the calculation. The basis of FFT calculations from the samples of an acquired waveform and display of the results are discussed in chapter 9. Chapter 3 provides the fundamentals applied to analogue-to-digital conversion process and vice versa. The reader is invited to follow these chapters carefully for a better understanding of the FFT techniques used for displaying the spectrum of a waveform captured by a DSO.

If you could perform a Fourier analysis on a signal that goes on forever, the results would represent perfectly the frequencies and the amplitudes of the frequencies present in the signal. DSOs, however, have limited space for data storage or samples. When sampling a waveform with an A/D converter, the system acquires discrete samples. When you execute an FFT algorithm on a limited number of waveform samples to obtain the frequency domain behaviour of the waveform, owing to multiplication in the time domain (the windowing operation), the result will be slightly different from the calculations performed on a continuous waveform or large number of samples of the waveform. Subject to these differences, a DSO can provide the users with a reasonable frequency domain representation of a waveform. In order to improve the results *windowing functions* are used in practical DSOs. (See chapter 9 for details.) A few operating hints for the FFT facility of a DSO may be summarised as

- keeping the effective sample rate greater than twice the signal bandwidth of interest,
- selection of a suitable window function (best amplitude accuracy is from a rectangular (flat top) window, while a Hanning window, etc., provides better frequency resolution),
- source coupling of the a.c. component of a waveform (the d.c. component of a waveform may cause errors in FFT calculations), and
- avoiding average acquisition mode, peak detect and envelope modes, etc.

DSOs may use anything from a few kilobytes of samples to megabytes for FFTs. For example, some oscilloscopes from LeCroy offers the option of 6 Msample FFTs. Some competing DSOs use only 10 000 or 50 000 samples in FFT calculations. When users of these scopes want high resolution FFTs, they must transfer waveform records to a computer and make off-line calculations. Figure 6.14 indicates an oscilloscope screen showing an FFT on a TDS210 scope.

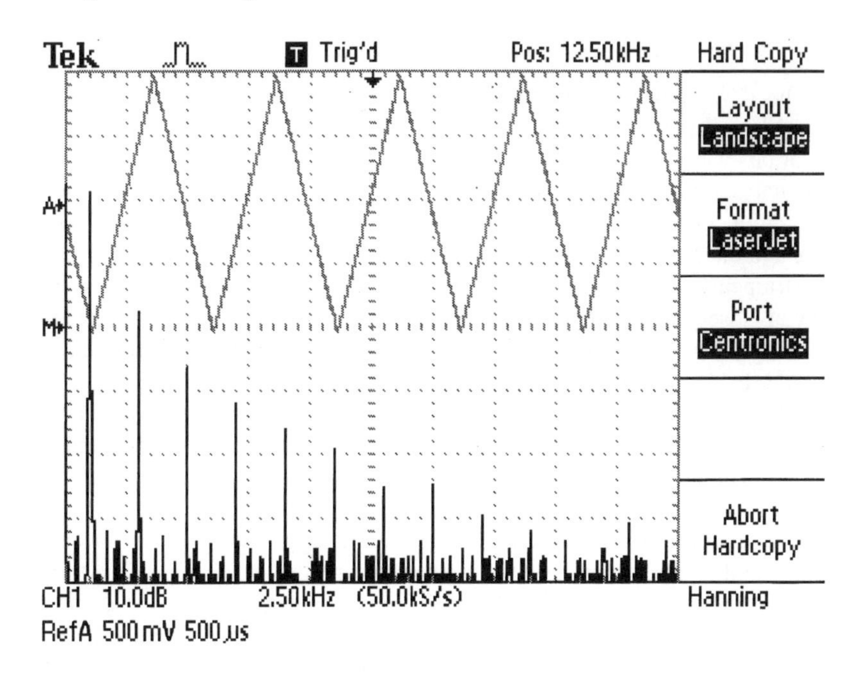

Figure 6.14 FFT display on a TDS210 scope

A detailed discussion on the use of FFT facilities on a practical DSO is beyond the scope of this chapter. References 16–18 provide some useful practical details, including the windowing functions.

6.7.4 Deep memories and waveform processing

Increasing the memory length of digital storage oscilloscopes brings many advantages, but not all of them are obvious. Among these are:

- no missed details on waveforms (effective sampling rate is very high),
- permanent glitch capture, without waveform distortion,
- better time and frequency resolution,
- reliable capture of events which are unpredictable in time, and
- no dead time between acquired events.

Deep memory certainly adds to a DSO's cost, but manufacturers are recognising that the feature's value outweighs the expense. LeCroy's 9300 series still offers the deepest memory in bench-top oscilloscopes. One model, the 9374L (500 MHz bandwidth), provides 2 Msample per channel memories that combine into an 8 Msample memory when only one channel is active. This feature nicely complements the interleaving of unused ADCs with ADCs of active channels. As you turn channels off, the memory depth and sample rate increase together. At a constant sweep speed, therefore, the records represent a fixed length of time. As discussed earlier longer memory provides better resolution FFT results too. For details, Reference 19 is suggested.

6.8 Practical DSO designs

In this section we could discuss the practical design of few modern DSOs which could compete with the facilities of wideband analogue scopes. Compared with older DSO techniques one could recognise that most advanced performance specifications and useful features are the result of revolutionary designs using modern VLSI components.

6.8.1 Low cost, 100 MHz digitising scopes – the HP 54600 series

The HP 5460X family 100 MHz oscilloscopes (see Figure 6.16(c)) represents a major improvement in digitising oscilloscope technology and product design. These oscilloscopes have two full range inputs (2 mV div^{-1} to 5 V div^{-1}) and the bandwidth of all channels is 100 MHz. A maximum sample rate of 20 Msamples s^{-1} provides a 2 MHz bandwidth for capturing single shot events. The 8-bit analogue-to-digital converter has a vertical resolution of 0.4 per cent. The designers' key objective in the development of the HP 54600 series digitising oscilloscopes was to design a low cost digitising oscilloscope that has the 'look and feel' of an analogue oscilloscope. In other words, the new oscilloscopes were to have familiar controls and traces on the screen that look almost as good as analogue oscilloscope traces, while maintaining the advantages of a digitising architecture [20].

In order to appreciate the design concepts used in the new architecture, it is worth comparing an older raster DSO architecture and the task flow with the HP54600.

6.8.1.1 Previous architecture

Figure 6.15 shows the simplified block diagram (a) and the task flow diagram (b) for a traditional DSO. Notice that the system microprocessor (a 68000 CPU in a case like the HP 54100) controls everything. It unloads points from the acquisition memories, reads the time interpolation logic, plots and erases points on the screen, services the front panel and the HP-IB connector, moves markers, manages memory, performs measurements, and performs a variety of system management tasks. Because the system microprocessor is in the centre of everything, everything slows down, especially when the system needs to be the fastest (during interactive adjustments). This serial operation gives rise to an irritatingly slow response. The keyboard feels sluggish, traces on the screen lag noticeably behind interactive adjustments, and even a small number of measurements slows the waveform throughput significantly.

6.8.2 HP 54600 architecture

In order to compare the differences between the traditional digitising oscilloscope architecture with a modern design such as the HP 54600 family, let us compare

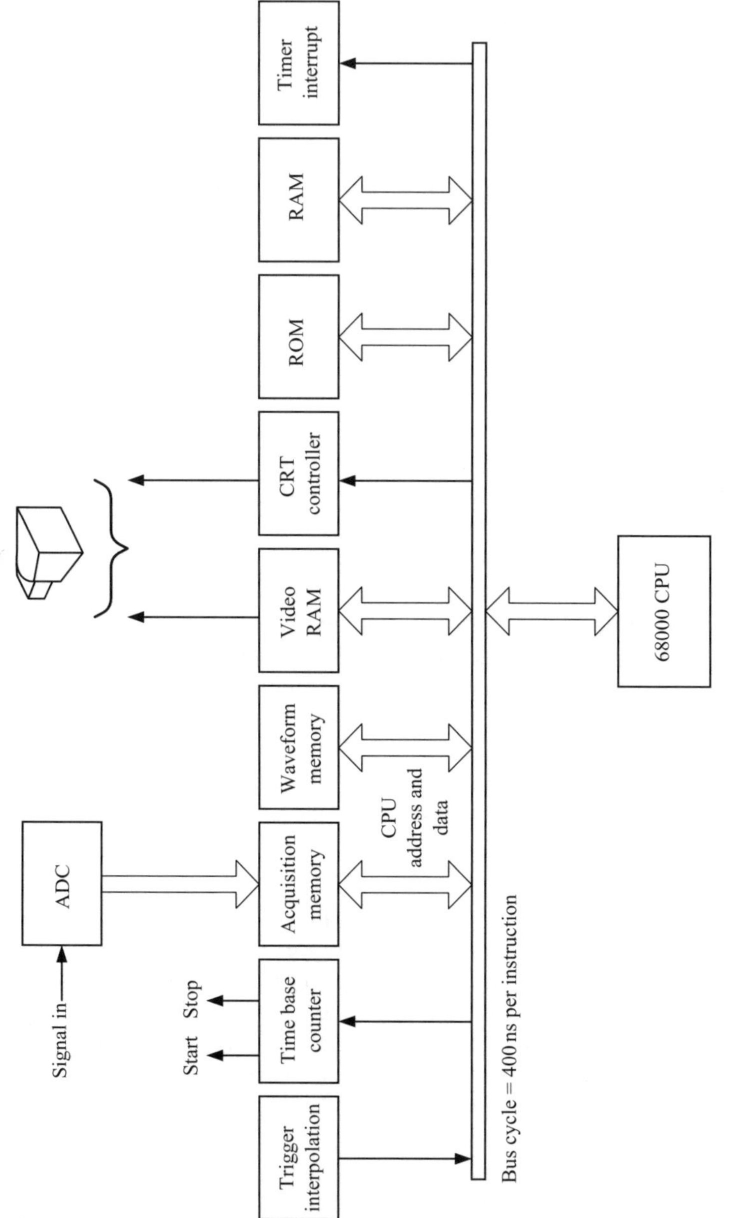

Figure 6.15 Continued overleaf

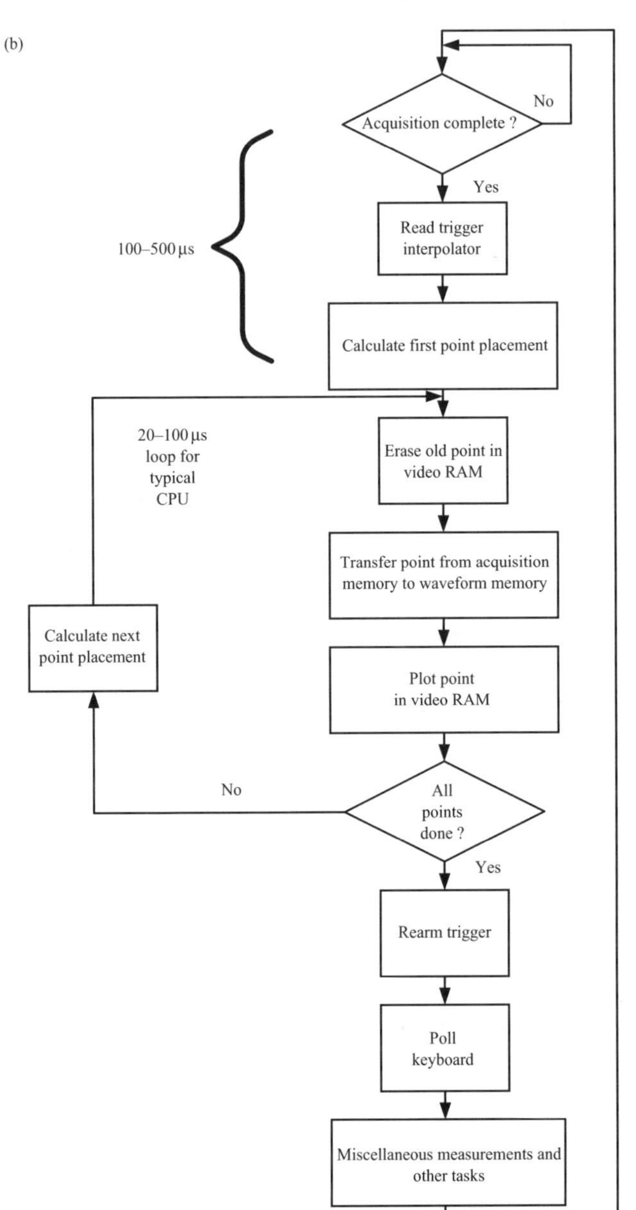

Figure 6.15 Simplified older architecture of a raster DSO: (a) block diagram; (b) simplified task flow diagram (Courtesy: Agilent Technologies, Inc.)

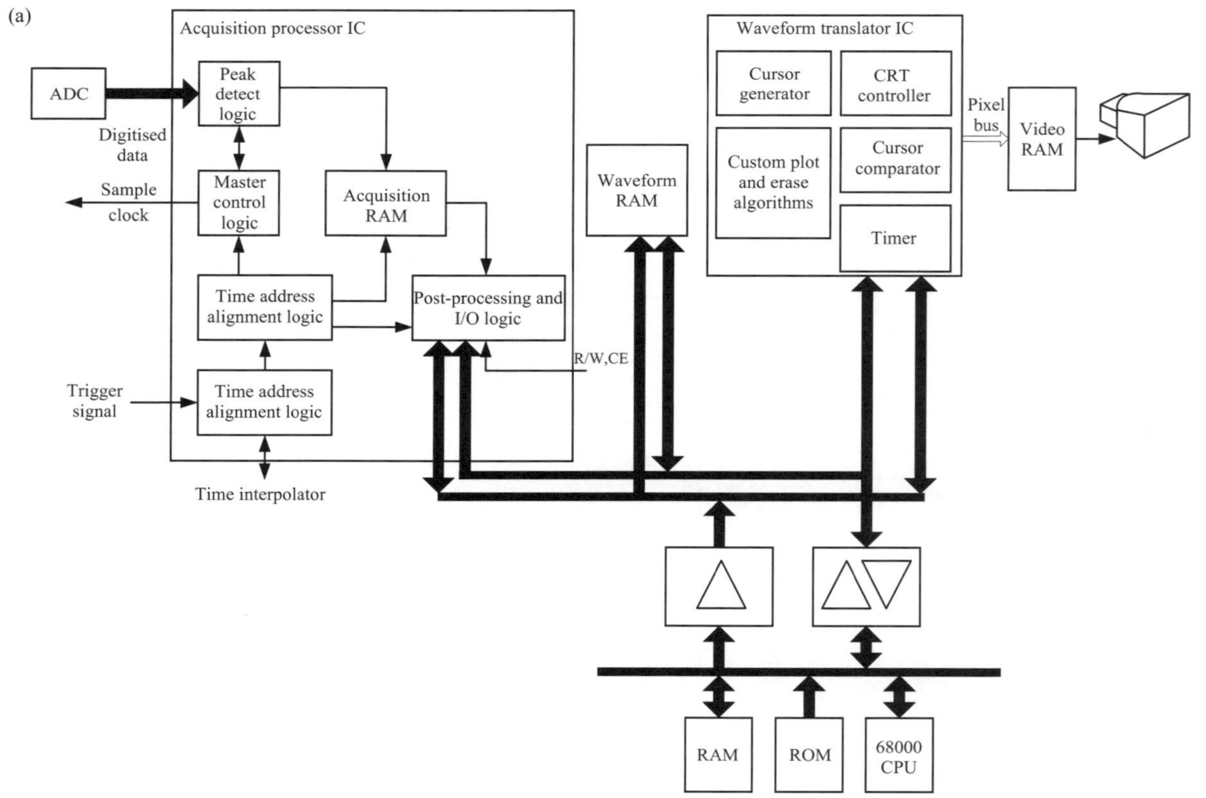

Figure 6.16 *Continued overleaf*

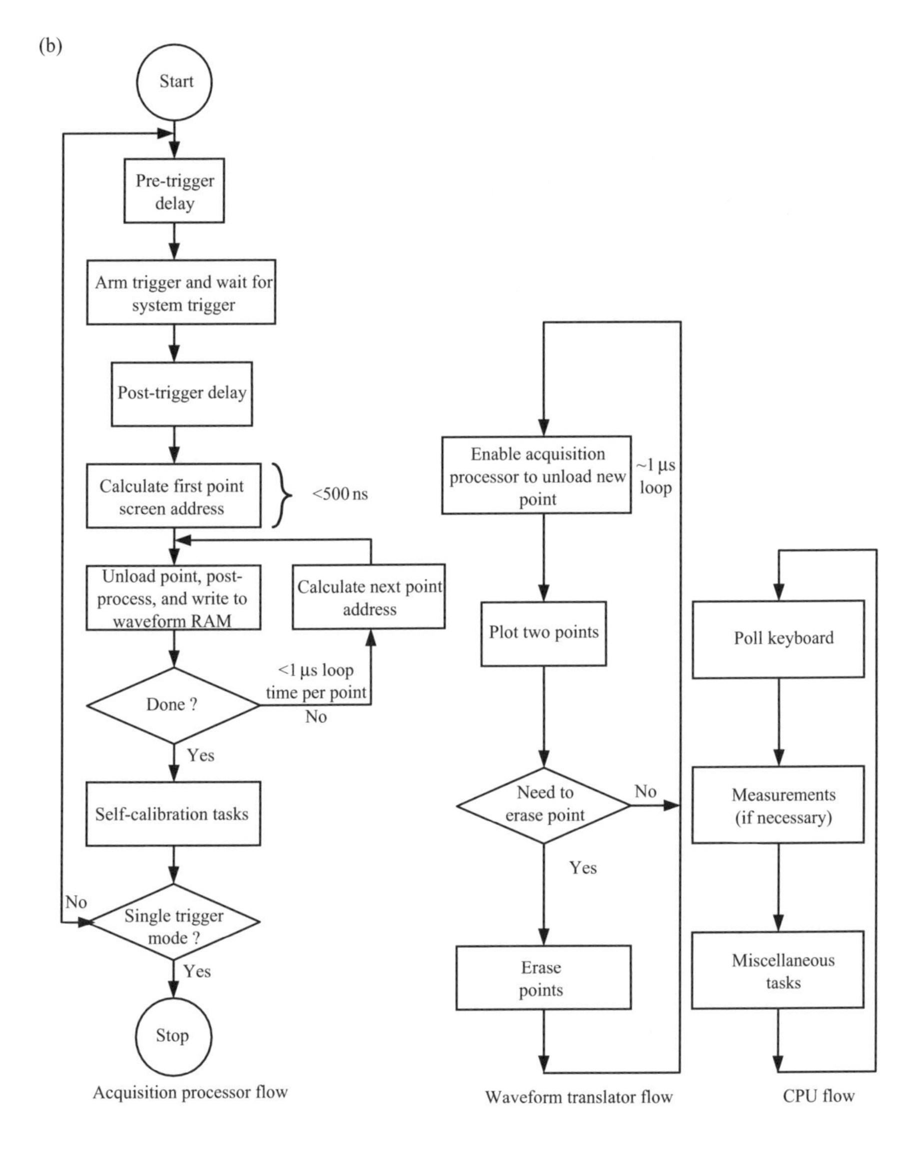

Figure 6.16 Continued overleaf

the two cases in Figures 6.15 and 6.16. In contrast to the central microprocessor topology shown in Figure 6.15, the new topology separates the acquisition and display functions – that is, the oscilloscope functions – from the host processor, and places these functions under the control of two custom integrated circuits. The acquisition processor IC manages all of the data collection and placement mathematics. The waveform translator IC is responsible for all of the waveform imaging functions. This leaves the main processor free to provide very fast interaction between the user and the instrument.

(c)

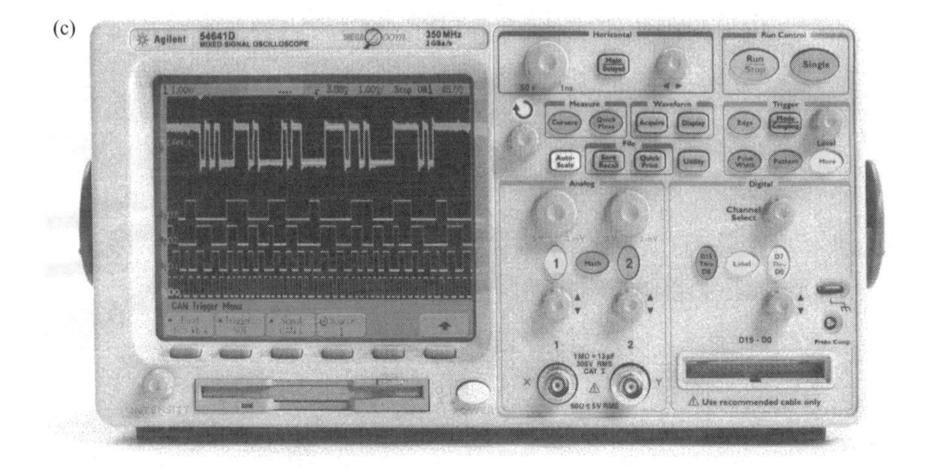

Figure 6.16 HP 54600 family design: (a) block diagram; (b) task flows; (c) HP 54641D photograph (Courtesy: Agilent Technologies, Inc)

6.8.2.1 Acquisition processor

The acquisition processor is a custom CMOS integrated circuit with roughly 200 000 transistors in an 84-pin PLCC package. This IC is essentially a specialised 20 MHz digital signal processor. It pipelines the samples from the analogue-to-digital converter into an internal acquisition RAM, and later writes these points as time ordered pairs into the external waveform RAM. It also contains the system trigger and time base management circuitry.

The external analogue-to-digital converters always run at 20 Msamples s^{-1}, regardless of the sweep speed setting of the oscilloscope. The digitised samples are brought onto the chip continuously over an 8-bit TTL bus. When the acquisition processor is set up to look at two channels, it alternately enables the analogue-to-digital converter output from each channel (still at a 20 Msamples s^{-1} rate), yielding a sampling rate of 10 Msamples s^{-1} per channel.

Internally, the vertical sampled data first passes through a dual-rank peak detection circuit. When the HP 54600 is set to a very slow sweep speed, there are many more samples available than can be stored into the final waveform record (2000 points). For example, because the analogue-to-digital converters are always digitising at a 20 Msamples s^{-1} rate, the acquisition processor may only be storing, say, one sample out of every hundred, yielding an effective sample rate of (20 Msamples s^{-1})/100 = 200 ksamples s^{-1}.

During this condition (known as over-sampling), the oscilloscope user may turn on the peak detect mode, enabling the acquisition processor's peak detection logic. Instead of simply ignoring the over-sampled points, the peak detection logic keeps track of the minimum and maximum analogue-to-digital converter values and passes these pairs downstream. This allows the user to watch for narrow pulses (glitches) even at the slowest sweep speeds. With one channel enabled, the minimum detectable glitch width is 50 ns. The samples to be stored are put into an internal 2 k × 8-bit acquisition RAM. When peak detection is enabled, the minimums are held in even addresses and the maximums are stored at odd addresses. Meanwhile, the trigger blocks track the trigger input signal, implementing the trigger functions such as trigger delay, hold off, pre-trigger and post-trigger delays, etc. Post-processing logic is responsible for transferring voltage/time signal pairs from acquisition RAM and to waveform RAM.

The host 68000 can program the acquisition processor by writing directly into its internal registers, using a traditional chip enable, read/write, register select protocol. Similarly, the 68000 can determine the acquisition status by reading a series of status registers.

6.8.2.2 Waveform translator

The waveform translator, implemented in a 1.5 μm CMOS process, is a gate array with a main clock speed of 40 MHz. The gate array has 6500 gates of custom logic packaged in an 84-pin PLCC package. The main function of the waveform translator is to take the time/voltage pairs from waveform memory and turn on the corresponding pixels on the display. The gate array also includes other features that reduce the number of parts in the microprocessor system and the number of microprocessor tasks on a system level. This results in a very responsive display system, while keeping the total system cost low.

The HP 54600 family uses a 7-inch-diagonal monochrome raster-scan display organised as 304 vertical pixels × 512 horizontal pixels. The area used for the oscilloscope graticule and waveform display measures 256 vertical pixels × 500 horizontal pixels, providing 8-bit vertical resolution and 9-bit horizontal resolution. The waveform translator acts as the display controller and provides the vertical sync, horizontal sync, and video signals to the display. The waveform translator has control over a 256 k × 4-bit video DRAM for image capture.

6.8.2.3 Achievements in the design of HP 54600 architecture

By two custom ICs controlled by a 68000 processor, this newer architecture has achieved the overall display ability shown in Figure 6.17. Compare this achievement with a typical operating area of an analogue scope shown in Figure 6.3. This design allows a new form of infinite persistence, called *autostore*. In autostore mode, half intensity is used to display all historical signal excursions while the current signal is shown at full intensity. Simple one-key operation allows the user to move into and out of autostore mode quickly.

6.8.3 TDS 210/220 oscilloscopes

The Tektronix TDS 2xx family (Figure 6.18(a)) of DSOs is based on the digital real time (DRT) design approach. It utilises Tektronix's proprietary fast-in-slow-out (FISO) ADC technology and in a low cost version of a DRT family based on a patented family of CMOS chips. This minimal chip set is a low power family for fast signal acquisition. The basic block diagram of such a system is shown in Figure 6.18(b). These scopes, such as TDS210 and TD220, have bandwidths of 60 and 100 MHz

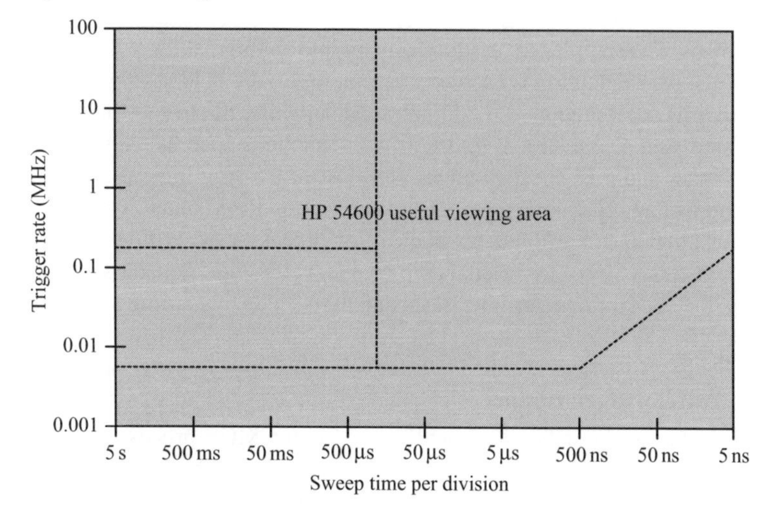

Figure 6.17 HP 54600 display capability (Courtesy: Agilent Technologies, Inc.)

respectively, with LCD display. The scopes are designed around ASICs for acquisition and display. The acquisition system carries signals from the two channels through attenuators to the acquisition ASIC, which contains (i) amplifiers and samplers for each input channel, (ii) A/D converter, and (iii) trigger logic.

The digitised waveform samples are transferred to the display ASIC on the display system, which consists of (i) display ASIC, (ii) DRAM memory, and (iii) system oscillator. Digitised acquisition samples are processed by the display ASIC and stored into DRAM. The display ASIC formats the waveform data and generates drive signals for the LCD display. Front panel scanning is also performed by the display ASIC. Figure 6.18(c) shows a block diagram of the TDS2xx family. The processor system consists of the microprocessor, instruction ROM and non-volatile memory. The processor system (i) interprets the front panel controls, programs acquisition and display parameters, (ii) computes waveform measurements, and (iii) manages extension module interface. The processor system shares DRAM with the display system.

TDS oscilloscopes can be coupled with several types of extension modules for (i) Centronix parallel interface (TDS2HM), (ii) GPIB and RS-232 interface (TDS2CM), and (iii) TDS2MM measurement module with RS-232/Centronix/GPIB interfaces and FFT facility. Hardware on these modules carry processor instructions to support these interface functions.

This scope allows sampling at a 1 Gsample s^{-1} rate allowing a single shot bandwidths of 60 MHz (TDS210) and 100 MHz (TDS220) respectively. The TDS2MM module transfers the 2048 points of the time domain waveform into a FFT waveform using several window types such as Hanning, flat top and rectangular. Figure 6.18(d) shows an FFT waveform obtained on a square waveform from an off line UPS. References 21–24 provide more details on the DRT-FISO based DSO and applications of the TDS 200 series.

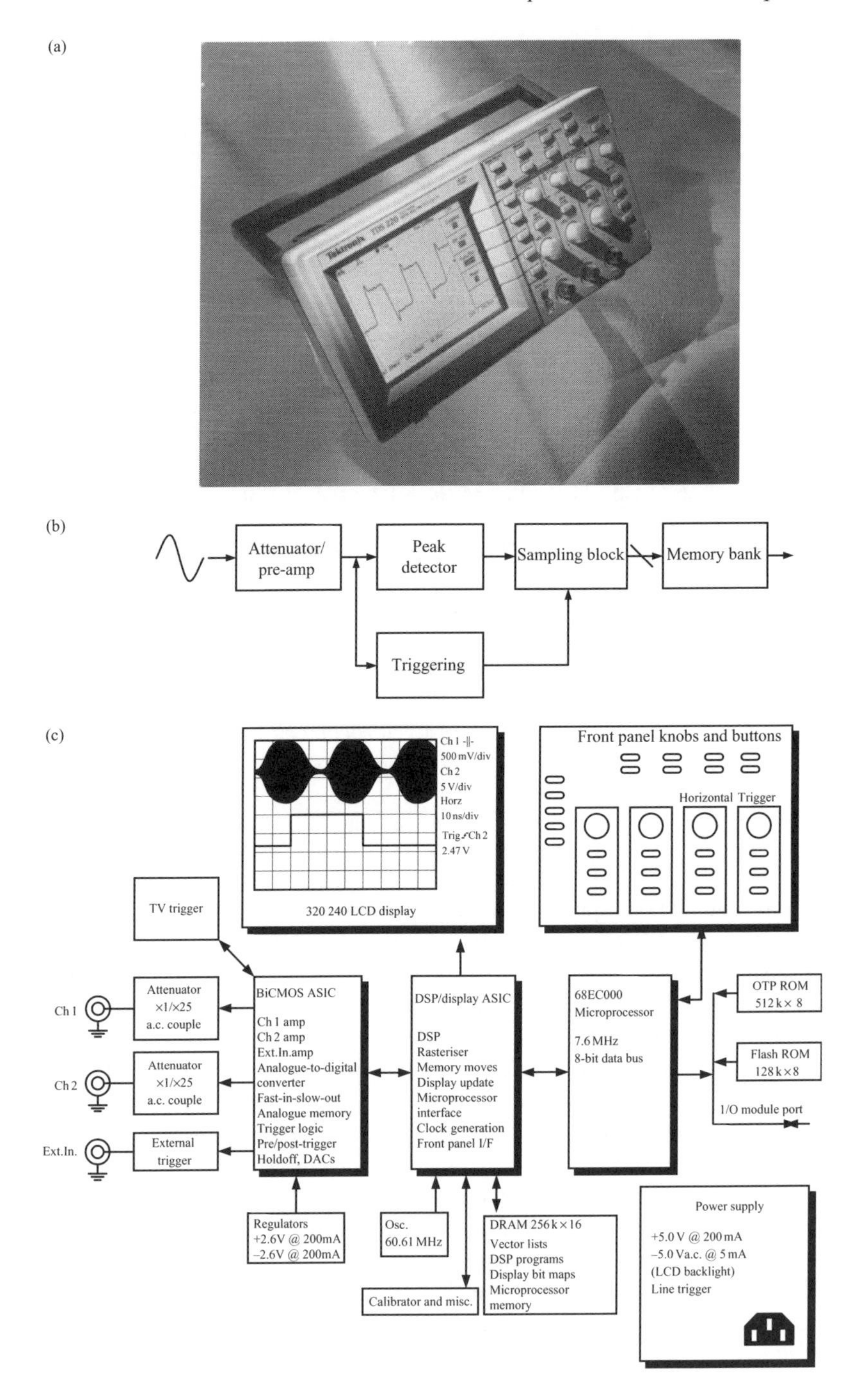

Figure 6.18 *Continued overleaf*

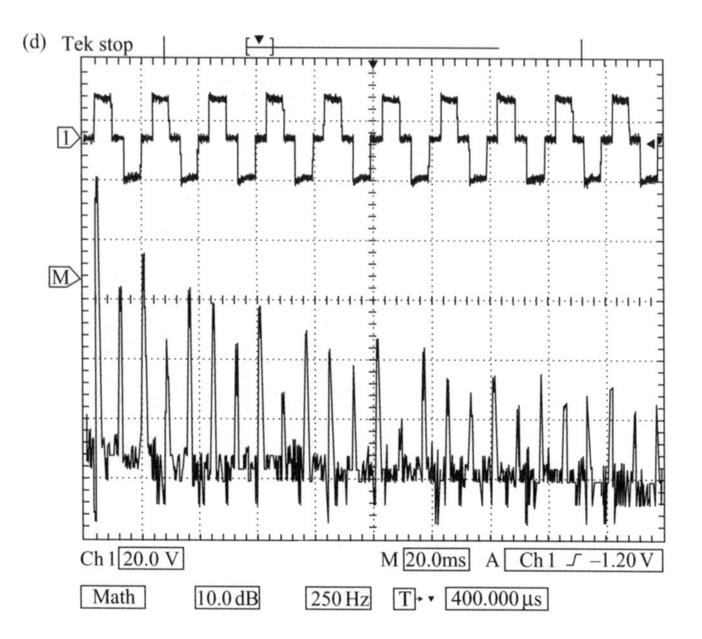

Figure 6.18 TDS 200 family oscilloscope: (a) TDS220/TDS2MM photograph; (b) block diagram of DRT-FISO based architecture; (c) block diagram of TDS2xx family; (d) FFT waveform of a 50 Hz off-line UPS waveform (lower trace) (courtesy of Tektronix, Inc.)

6.8.4 Digital phosphor oscilloscopes

In the conventional DSO approach, frequency-of-occurrence information is conveyed by simulating phosphor persistence with various post-processing techniques. But DSOs require acquisition over long periods of time, prohibiting immediate feedback. Analogue real time (ART) oscilloscopes (high speed analogue scopes) have a different set of problems. Although an ART can deliver qualitative insight into a signal's behaviour, it cannot provide detailed measurement information found in DSOs, nor permanent waveform storage. Furthermore, ARTs are limited by the writing speed of the CRT and can capture waveforms up to about 1 GHz only. However, even with all these limitations, ARTs, with persistence and intensity in their displays, are still the favourite among the engineers who measure modulated analogue signals. Some digital circuit designers and communications engineers also like ARTs because they provide information about a clock signal's jitter.

The digital phosphor oscilloscope (DPO), introduced in the latter part of 1998 by Tektronix, displays, stores and analyses complex signals in real time, using three dimensions of signal information – amplitude, time, and distribution of amplitude over time (Figure 6.19(a)) [6, 7]. The benefit of this third dimension is an interpretation of the signal dynamics, including instantaneous changes in the signal and the frequency of occurrence. Along with the three dimensions of waveform data, the DPO acquires waveforms up to a thousand times faster than a conventional DSO. That advantage protects against both aliasing and missing infrequent signal events.

Figure 6.19(b) shows a simplified block diagram of a DPO system. The heart of the DPO is the DPX^{TM} waveform imaging processor – a proprietary ASIC that rasterises the digitised waveform into a dynamic three-dimensional database called the *digital phosphor*. The DPX accumulates signal information in a 500×200 integer array. Each integer in the array represents a pixel in the DPO's display. If the signal traverses one point again and again, its array location will be updated repeatedly to highlight that fact. Over the time span of many samples, the array develops a detailed map of the signal intensity. The result is a waveform trace whose intensity varies in proportion to the signal's frequency of occurrence at each point – a type of '*grey scaling*' that is just like an analogue real time scope. But unlike an ART, the DPO

(a)

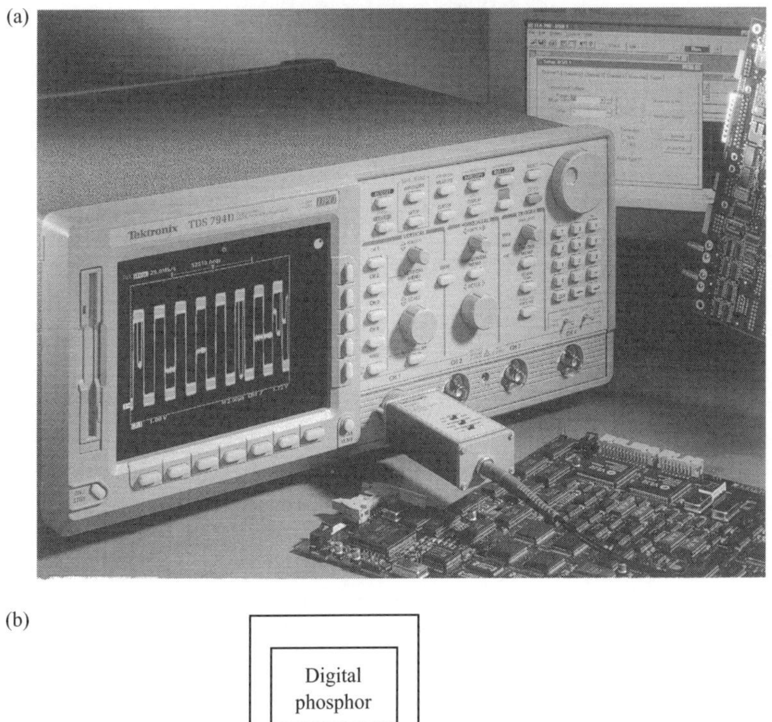

(b)

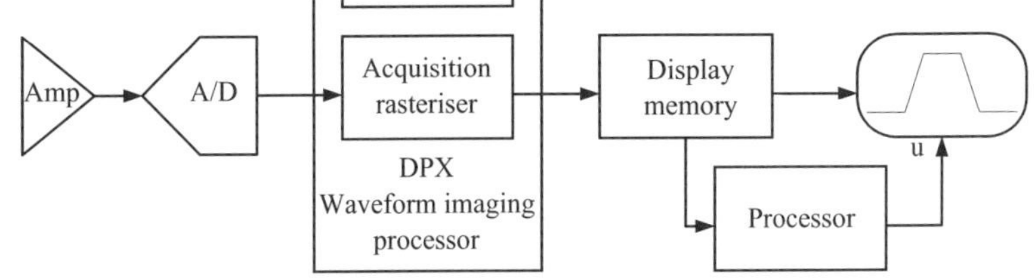

Figure 6.19 Continued overleaf

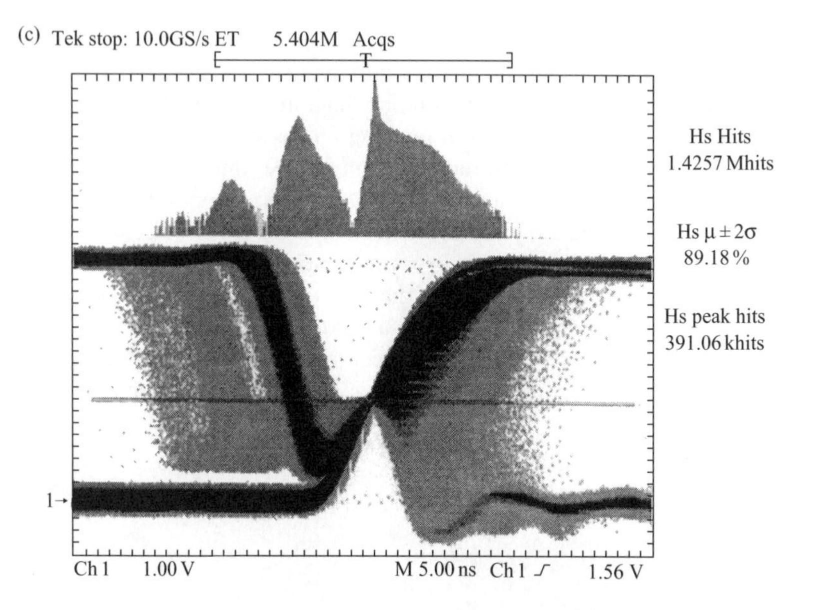

Figure 6.19 Tektronix digital phosphor oscilloscopes: (a) photograph of model TDS794D; (b) simplified block diagram of a DPO system; (c) DPO waveform image showing frequency of occurrence (Courtesy: Tektronix Inc.)

allows grey scale levels to be expressed in colour. Figure 6.19(c) uses a waveform from a metastable circuit to illustrate this effect. The intensity levels clearly express the frequency of occurrence at each point on the screen. The histogram above the main trace statistically represents the intensity information in the trace itself.

Tektronix has introduced several DPO models with up to 2 GHz bandwidth on four channels. Like the actual phosphor on an ART display, the DPO captures and remembers the frequency of events, resulting in a three-dimensional array that retains information for hundreds of millions of samples. In an ART, chemical phosphorescence creates a grey scale because of the decay in energy over time. Digital phosphor imitates this behaviour by digitally controlling the replacement of data in the three-dimensional array. As a result, DPOs can display signals as well as ARTs, while storing and analysing them like DSOs. For some useful applications of DPOs References 25–27 are suggested. Reference 28 is an excellent reference for DPO technique.

Like DPO, other techniques in DSOs that adjust colour or intensity go by different trade names such as 'analogue persistence' and 'TruTrace' (by Gould Instruments), etc. All of these gradient displays help the users to find signal problems and anomalies. For details of techniques such as *colour gradient displays* or *intensity gradient displays*, Reference 29 is suggested.

6.9 Colour displays, interfaces and other features

Scopes with colour displays offer obvious advantages, but until about 1993 those advantages commanded a high price. When you view several waveforms at once, colour makes them easy to tell apart. Depending on how a scope implements colour, a multihued display can supply information that is missing from most monochrome DSOs' raster displays, but is inherently present in an analogue scope's grey scale. A colour scope can also ease the task of telling when two waveforms do and do not coincide. In the past, the problem with colour display scopes was that most were expensive; some cost well over \$30 000; others cost close to \$60 000. Now most of these colour display type DSOs are within the \$10 000 price range. For these developments, colour LCD technologies also have helped a lot. For example, Gould's Datasys 700 series uses an active matrix LCD display.

Most scopes now come with optional or standard configurations to provide floppy disk drives, IEEE-488 or RS-232 interfaces, or even in some cases to carry a fax interface. Some manufacturers have started to offer a feature that expedites conformance testing against OSI-based data communications standards, particularly for the ISO model's physical layer [30]. This feature is called template matching, and one could use it to check pulse shapes, rise and fall times, and signal jitter specifications defined by communications standards. Some other recent advancements are the use of Java in Tektronix scopes and Hewlett-Packard's Windows based user interface (UI). For details of network connectivity and PC interfacing, Reference 31 is recommended.

6.10 References

1 MARTINEZ, C. and SHEMESH, E.: 'Analog or digital scope – which way to go', *Electronic Products*, July 1994.
2 MURPHY, J.B.: 'Comparing analog & digital oscilloscopes for troubleshooting', *Hewlett Packard Journal*, February 1992, pp. 57–59.
3 NOVELLINO, J.: 'DSOs take a technological leap', *Electronic Design*, April 14, 1994, pp. 95–97.
4 STRASSBERG, D.: 'Fast single-shot DSOs take varied design approaches', *Electronic Design*, July 8, 1993 pp. 47–53.
5 NOVELLINO, J.: 'Acquisition method gives scope 5-G samples/s speed', *Electronic Design*, July 8, 1993, pp. 86–88.
6 WITTE, R.A.: 'Low-cost, 100-MHz digitizing oscilloscopes', *Hewlett Packard Journal*, February 1992, pp. 6–10.
7 WITTE, R.A.: 'Use your DSO to measure elusine waveform variations', *EDN*, August 3, 1995, pp. 107–112.
8 KAHKOSKA, J.A.: 'Digital oscilloscope persistence', *Hewlett Packard Journal*, February 1992, pp. 45–47.
9 STRASSBERG, D.: 'Digital oscilloscopes: For best results, understand how they work', *EDN*, July 4, 1996, pp. 43–56.

10 TEKTRONIX, INC.: 'One bandwidth: Confidence, not confusion', Publication 5HW-8850-0, pp. 4/93.
11 LEVIT, L.B. and VINCELLI, M.L.: 'Characterising high speed digital circuits: A job for wideband scopes', *EDN*, June 10, 1993, pp. 153–162.
12 ROWE, M.: 'Triggers reel in waveform abnormalities', *Test & Measurement World*, August 2000, pp. 40–46.
13 TEKTRONIX, INC.: 'InstaVu^{TM} Acquisition – setting the benchmark for DSO performance', Document No: 55W-10341-0, pp. 12/94.
14 TEKTRONIX, INC.: 'Digital Oscilloscope Concepts', Engineering Note 37w-6136, 4/86.
15 LAUTERBACH, M.: 'DSO yeilds results difficult or impossible to achieve with an analogue oscilloscope', *PCIM*, September 1995, pp. 22–27.
16 TEKTRONIX, INC.: 'FFT Applications for TDS', Document No.: 55W-8815-0, pp. 2/93.
17 LYONS, R.: 'Windowing functions improve FFT results-Part I', *Test & Measurement World*, June 1998, pp. 37–44.
18 LYONS, R.: 'Windowing functions improve FFT results-Part II', *Test & Measurement World*, September 1998, pp. 53–60.
19 LECROY CORP: 'Benefits of long memories in digital oscilloscopes'; Reference guide to digital waveform instruments, 1993/94, Technical Tutorial, pp. 111–30 to 111–32.
20 HALCOMB, M.S. and TIMM, D.P.: 'A high throughput acquisition architecture for a 100 MHz digitizing oscilloscope', *Hewlett Packard Journal*, February 1992, pp. 11–20.
21 TEKTRONIX, INC.: 'A short course in Digital Real Time'.
22 TEKTRONIX, INC.: 'How can I observe non-repetitive information on a fast serial link?', Troubleshooting with the TDS 200 Series (Number 1).
23 TEKTRONIX, INC.: 'How do I eliminate circuit board noise', Troubleshooting with the TDS 200 Series (Number 2).
24 TEKTRONIX, INC.: 'How can I view video signals with the TDS 200?', Troubleshooting with the TDS 200 Series (Number 4).
25 TEKTRONIX, INC.: 'Testing Telecommunications Tributary Signals', Application Note 55W-12045-0; 5/98.
26 TEKTRONIX, INC.: 'Performing Jitter Measurements with the TDS 700D/500D Digital Phosphor Oscilloscopes', Application Note 55W-12048-1; 6/98.
27 TEKTRONIX, INC.: 'Baseband Video Testing with Digital Phosphor Oscilloscopes', Application Note 55W-12113-0; 4/98.
28 DEPOSITO, J. 'Digital Phosphor Oscilloscope Breaks New Measurement Ground', *Electronic Design*, July 6, 1998, pp. 61–64.
29 ROWE, M.: 'DSO displays: As good as analog', *Test & Measurement World*, February 2000, pp. 41–52.
30 LEIBSON, S.L.: 'DSOs and logic analysers inch their way up the ISO protocol stack', *EDN*, March 4, 1993, p. 43.
31 GROSSMAN, S.: 'Digital storage oscilloscopes are becoming PC friendly', *Electronic Design*, October 30, 2000, pp. 101–110.

Further reading

LECROY, W. and CAKE, B.: 'Advances is very high speed (5 Gsa/S) sampling technology', *Electronic Engineering*, February 1994, pp. 65–71.

MCCONNEL, E.: 'Choosing a data-acquisition method', *Electronic Design*, June 26, 1995, pp. 147–156.

OKADA, T.: 'Colour TFT LCD forms core of B5, Notebook Colour Digital Oscilloscope', *JEE*, February 1994, pp. 34–37.

ROWE, M.: 'T & MW evaluates handheld scopes', *Test & Measurement World*, April 1995, pp. 36–40.

TEKTRONIX, INC.: 'How much scope do I need', Application Note 55w-10063-1; 9/94.

TEKTRONIX, INC.: 'Digital Phosphor Oscilloscope Architecture Surpasses Analog, Digital Scope Strengths', Technical Brief 55W-12023-0; 4/98.

TEKTRONIX, INC.: 'How can I identify and capture fast glitches?', Troubleshooting with the TDS 200 Series (Number 3).

—— 'Scopes cleverly wed analog and digital', *IEEE Spectrum*, July 1998, pp. 77–79.

Chapter 7

Electronic counters

7.1 Introduction

Counting the occurrence of electrical events was a primary concern of electrical engineering, even in the era of vacuum tubes. The first- generation electronic counters were designed using vacuum tubes; these were bulky, heavy and power hungry. The second-generation counters introduced in the early 1960s were considerably smaller owing to transistorised circuitry even though the basic specifications of the instruments remained more or less the same as for the vacuum tube versions. The availability of digital integrated circuits at the end of 1960s led to the birth of a third generation with better performance and features. With the introduction of LSI and VLSI components, a fourth generation of powerful counters has appeared in the market within the past 15 years. Very recently, related families of instruments, such as modulation domain analysers, have also been introduced into the industry, thus bringing unique methods for viewing complex modulated signals in the modulation domain.

Basic concepts related to the electronic counters which can be used to measure frequency, time, phase, frequency ratio, time interval average, etc., will be discussed in this chapter with special reference to accuracy and modes of operation. Further, some techniques for frequency measurements in the microwave region will be discussed with a brief introduction to modulation domain analysis.

7.2 Basic counter circuitry

A counter can be divided into five basic blocks as shown in Figure 7.1. These are: (i) input circuit, (ii) main gate, (iii) decimal counting unit and display, (iv) time base, and (v) control circuit.

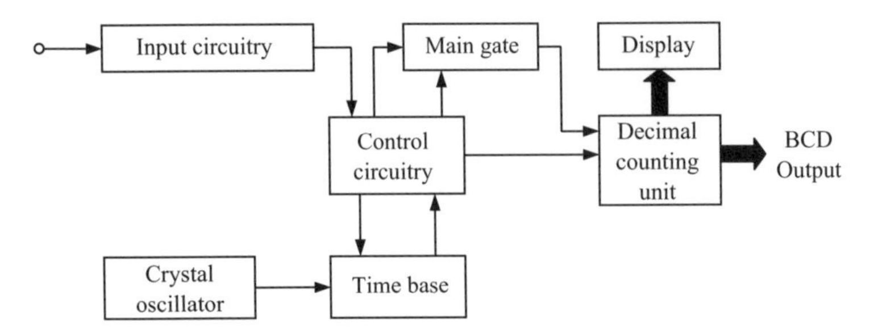

Figure 7.1 Block diagram of an electronic counter

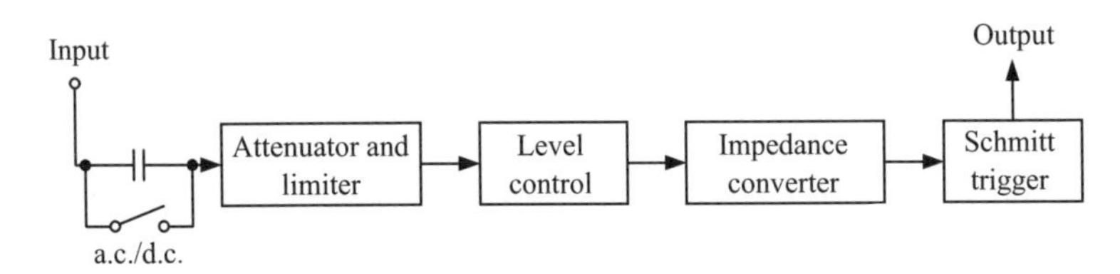

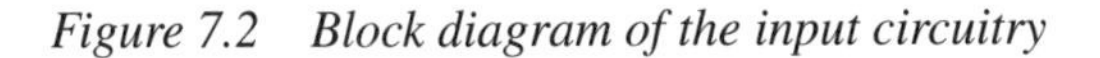

Figure 7.2 Block diagram of the input circuitry

7.2.1 *Input stage*

The input circuit is for signal conditioning of the input analogue signal and conversion into a digital form. It consists of the following main stages as shown in Figure 7.2:

- an a.c./d.c. coupling circuit,
- an input attenuator stage,
- a voltage limiter for input protection,
- an impedance convertor with level adjustment, and
- a Schmitt trigger circuit.

The a.c./d.c. block of the input stage takes care of coupling the input signal into the next stages with or without the d.c. component of the signal. The attenuator stage brings the signal level down to a safe level and the voltage limiter restricts the attenuated input signal to safe levels.

The impedance converter and the level adjust circuits match the impedances while adjusting the level of the signal to appropriate values so that the trigger point may be shifted from positive to negative values via zero. The final stage, the Schmitt trigger circuit, converts the input signal to a digital value. An important property of this block is that it determines the sensitivity of the counter and also provides noise immunity. Figure 7.3(a) and (b) show the effect of the trigger window on counter operation. We can clearly see here that if the input signal amplitude to the Schmitt trigger stage is not sufficiently large (to cross the trigger window) the output pulses will not be produced (Figure 7.3(b)). As discussed later, the trigger window adjustment is very important to the correct operation of the counter and it can be used to minimise the fault counts generated due to noise and other situations.

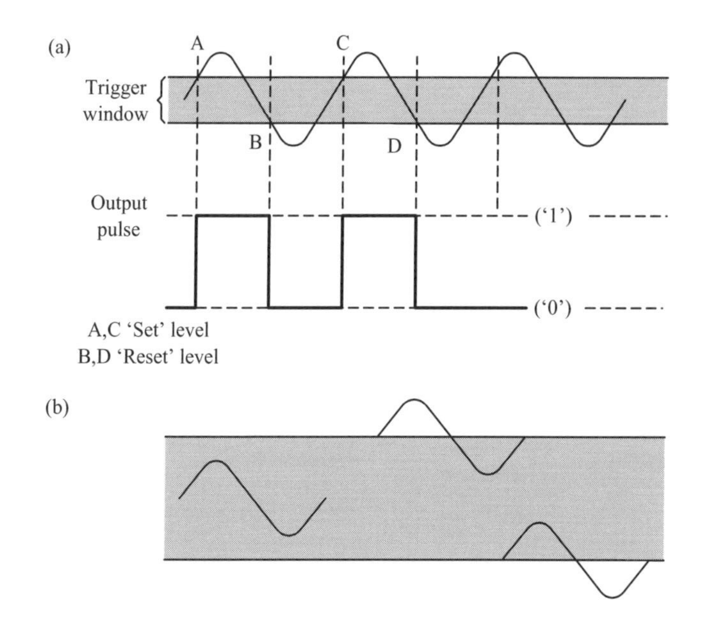

Figure 7.3 Trigger window: (a) correct operation; (b) incorrect setting of trigger levels

7.2.2 Main gate

The signal, conditioned by the input stage, is passed to the main gate, which is mostly a standard dual input logic gate. One input of the main gate receives the input signal to be counted and the other input receives the gate control signal. When the gate is turned on the incoming pulses pass through the gate to the next stage, the decimal counting unit. Both the Schmitt trigger and main gate circuits have a frequency limit beyond which the input signal can no longer be followed completely. Therefore, these stages of a counter which can handle high frequencies are designed using high speed logic gates. Usually silicon bipolar transistor ECL circuits or GaAs circuits are used for main gate circuitry of high speed counters.

7.2.3 Decimal counting unit and display

The heart of a digital counter is the decimal counting unit (DCU), which typically consists of a number of counting units in a cascade. Each decade consists of several basic units: a decade counter, a memory, a decoder/display driver and the display. Figure 7.4(a) shows a representative block diagram of a DCU and a seven segment display. In the case of a three digit counter, the input signal that is received at the first counting decade increments the counter up to nine counts and at the tenth count a 'carry' is presented to the next decade. The second decade hence counts in tens and the third stage counts in hundreds, etc. The decade counter stages present the counted pulses in the BCD form and the values counted in these stages are transferred to the memories at the reception of a transfer pulse from the control unit. The BCD

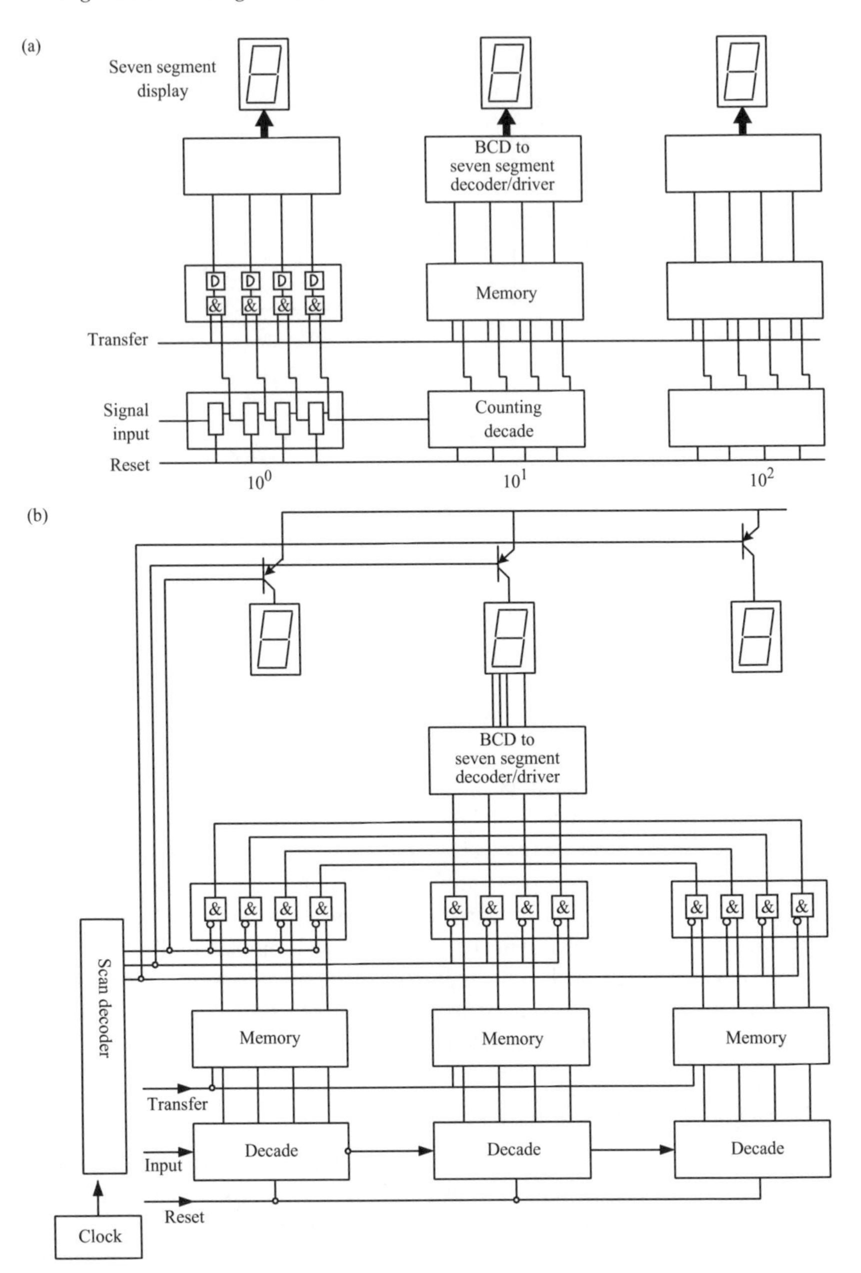

Figure 7.4 Block diagram of a counter: (a) DCU and display for a three digit counter; (b) dynamic display

to display driver transfers the counted values so that the final count is displayed on the display. The display could be LED, LCD or a vacuum fluorescent display in a modern counter.

In the above example (Figure 7.4(a)) each indicator is driven by a separate decoder/display driver and this method is called a 'static display'. Counters with large numbers of digits, usually beyond five, use a method called 'dynamic display' where one decoder/driver is shared by all the digits. When the speed of switching from one digit to the other is fast enough to eliminate flickering this method reduces the number of components to configure a multidigit counter. The technique is shown in Figure 7.4(b) for a representative three digit counter. This method of display has the advantage that only one decoder/driver is needed with some additional circuits to take care of scanning the display digits. As in the static display method, the counter decades first count the pulses, then a transfer pulse is given and the information in the counter decades is transferred to the memory. In this case all memory outputs are connected to a group of AND gates and, in turn, the outputs of the AND gates are connected (parallel) to the decoder/driver stage. The other inputs of the AND gates are connected to the scanning block. When the counts are registered the scan circuitry transfers each BCD output, together with the corresponding anode or cathode drives, sequentially, to the seven segment displays to transfer the final count to the display. For example, when the least significant digit (LSD) needs to be displayed, the scan block couples the corresponding BCD outputs to all LED displays but applies the anode or cathode voltage only to the LSD. When the next digit is to be displayed the corresponding memory output is transferred to the decoder/driver and the anode or cathode of corresponding digits will be driven. In the dynamic display method each digit of an N digit display will be illuminated for only $1/N$ of a period and this may decrease the brightness of the illuminated digits. However, in counters where the dynamic display method is used the displays are driven with higher current during the ON period to compensate for this drawback.

In most modern counters the display is much more complex than a simple collection of several seven segment displays and much more information may be displayed together with the count. Sometimes, modern counters use matrix display devices where all information to be displayed is created by pixels. In these cases the display driver circuits, which are governed by the control circuit, are designed using a microcontroller or a microprocessor. In low cost versions of modern counters a LSI or a VLSI is used with LCD displays.

7.2.4 *Time base circuitry*

In most counting operations, such as the measurement of frequency or time, a time reference of high accuracy and stability is needed. Frequencies are measured basically by counting a number of input pulses during a precisely defined time interval. For example, if the main gate is kept open for exactly 1 s and the number of pulses counted is 2330, the frequency is 2330 Hz. Time intervals are measured by means of trains of pulses of exactly known frequency, which are counted accurately during the time that the main gate is kept open by control signals related to the time interval

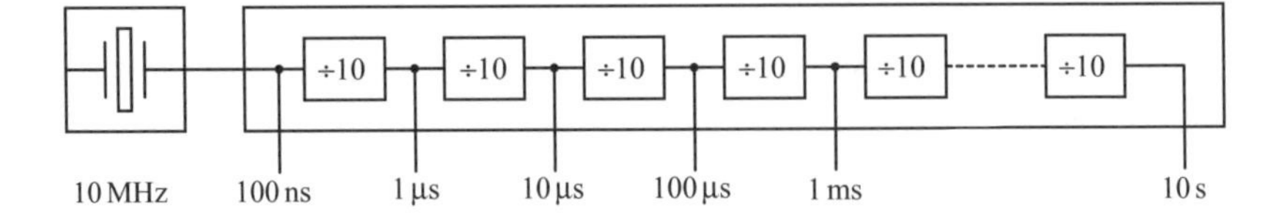

Figure 7.5 Crystal oscillator and time base divider

to be measured. For example, if the pulses to be counted have a pulse repetition frequency of 1 MHz and 6500 pulses are counted during the period between opening and closing of the main gate, the time interval measured is 6500 μs. It is clear that a time reference of high stability and accuracy is needed in all counters. Practically, a time base designed using a crystal oscillator is followed by a set of decade counters, as shown in Figure 7.5.

The performance of the time base circuitry is ultimately governed by the characteristics of the quartz oscillator, and the important characteristics of the oscillator that affect the overall performance are the temperature stability and the ageing. The accuracy of measurement and the performance of quartz oscillators are discussed later.

7.2.5 Control circuit

This block of the counter controls the overall operations within the instrument. Depending on the modes of operation various commands are given automatically, manually or by remote control. During a measurement cycle the operating conditions must be controlled in such a way that the measuring process can proceed without disturbance. The control circuit takes care of the following functions:

(i) control of the main gate,
(ii) generation of the reset pulses for counting decades and the time base dividers,
(iii) control of the display time,
(iv) generation of memory transfer pulses and timing pulses for dynamic display, and
(v) control of external devices such as printers or computer interface.

In a modern frequency counter which has multifunction capabilities the control circuit is designed around a microcontroller or a microprocessor. In low cost versions of frequency counters one VLSI circuit may contain the control circuit as well as other basic blocks such as the gate, time base dividers and the DCU and display drivers.

7.3 Modes of operation

Electronic counters can perform various functions such as totalising, measurement of frequency or time, frequency ratio, time averaging, measurement of pulse width or phase shift. The basic blocks described above could be configured in different ways to perform these functions and some important modes of operations are discussed here. For a more elaborate discussion on the modes of operation, Reference 1 is suggested.

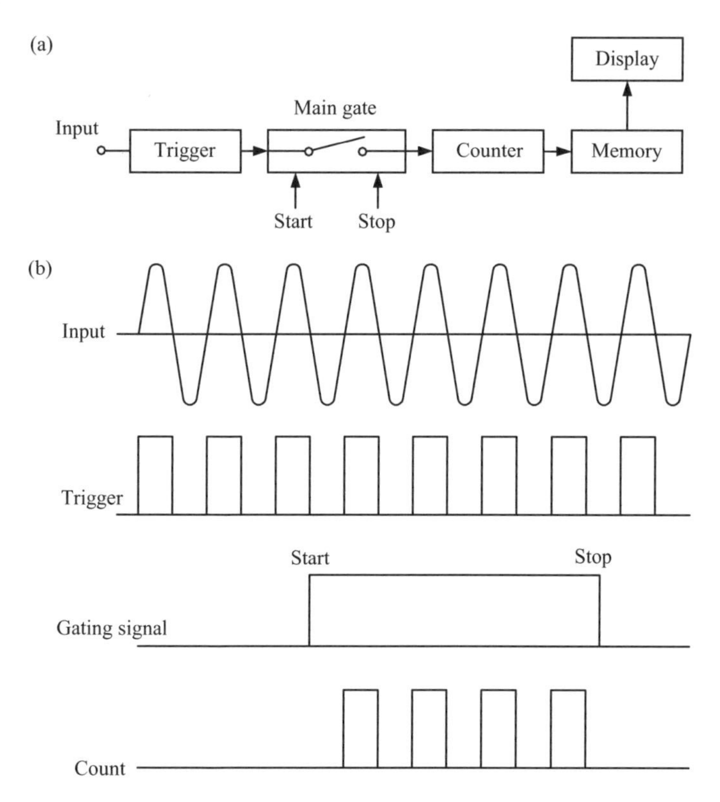

Figure 7.6 Totalise mode: (a) block diagram; (b) signal flow

7.3.1 Totalising

The 'totalise' or events mode of a counter is where the total number of input events are counted without reference to time. A block diagram of a counter in totalise mode is shown in Figure 7.6. In Figure 7.6(a) the input signal is converted to a pulse train and passed through the main gate, which works as a simple on/off switch to the decimal counting unit. The counter is generally started and stopped with a simple two position switch or with a start–stop pulse from a remote position in cases where the instrument is remotely controlled.

7.3.2 Frequency measurements

Because frequency is defined as the number of events per time interval, the frequency can be measured by keeping the main gate closed during a precisely known interval of time. For example, if the main gate is closed for 1 s the frequency is displayed in Hz and if it is closed for a period of 1 ms the frequency is read in kHz. The configuration for measuring the frequency is shown in Figure 7.7.

The crystal oscillator frequency is fed to the time base divider circuits to generate accurate time intervals of 1 s, 100 ms, 10 ms, etc., and then is used to operate the

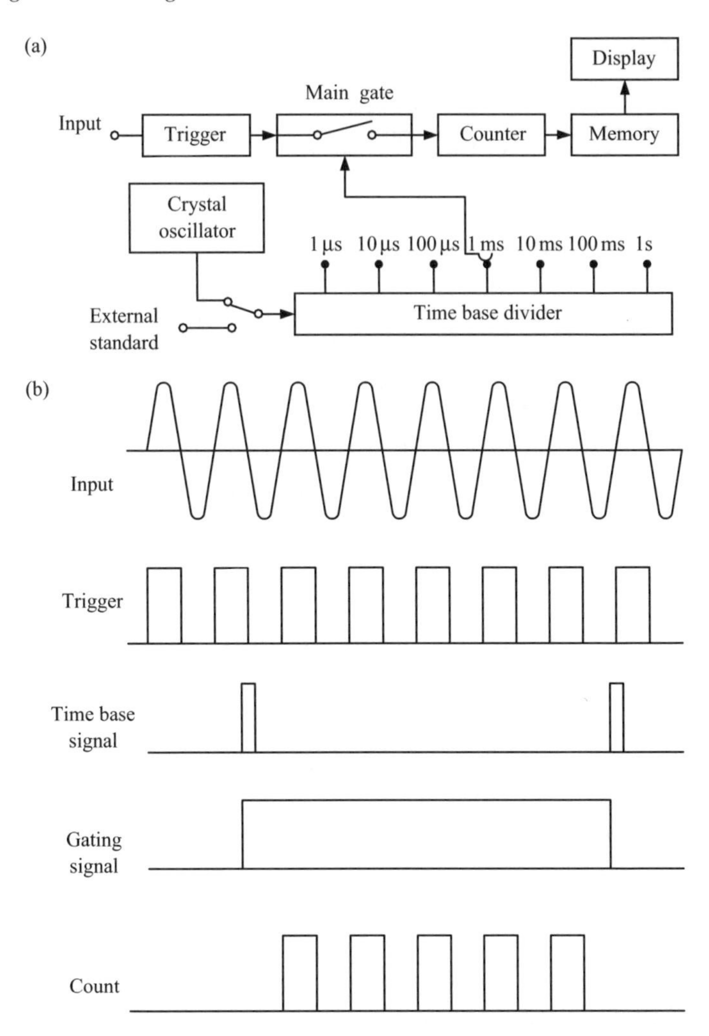

Figure 7.7 Frequency measurement: (a) block diagram; (b) signal flow

main gate. During this precise time interval the input signal is passed through the gate via the trigger circuits. Therefore, a count directly related to the frequency is passed on to the DCU and the display. In the above example we can clearly see that the accuracy of the measurement is dependent upon the accuracy of the crystal oscillator output. To obtain a more accurate readout, some of the frequency counters allow an external frequency standard to be coupled to the time base divider instead of the internal oscillator.

7.3.3 Frequency ratio measurements

This measuring mode provides a simple means of comparing two frequencies and can be considered as an extension of the frequency mode. As shown in Figure 7.8, two

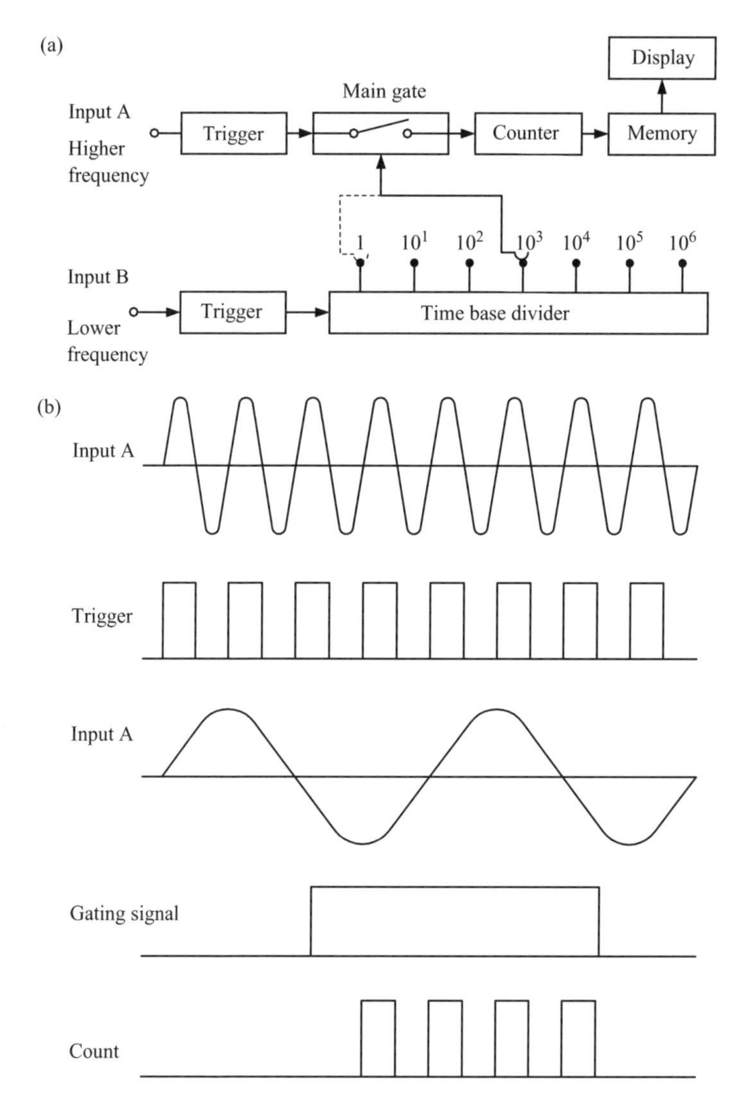

Figure 7.8 Frequency ratio measurement: (a) block diagram; (b) signal flow

independent input signal channels with independent trigger stages are coupled to the main gate and the time base divider. The internal crystal oscillator is not utilised for the measurement.

As shown in Figure 7.8(a), when the time base switch is in position 1 the signal from input B is not divided at all and the ratio between the two frequencies is shown directly. However, this method of direct comparison has the disadvantage of low accuracy when the two frequencies are close to each other. For instance, if input A has a frequency of 500 kHz against 100 kHz applied at the B input the ratio is 5 and, owing to the inherent uncertainty in digital measurement, the reading may be 4

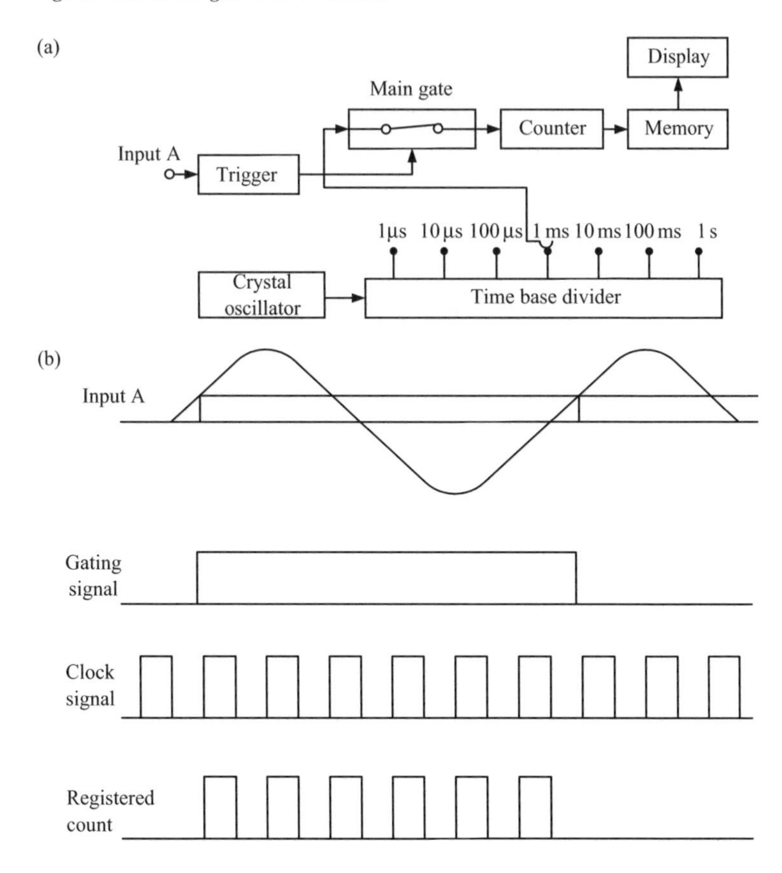

Figure 7.9 Period mode measurements: (a) block diagram; (b) signal flow

or 6, giving a possible error of 20 per cent. If we shift the time base divider to the 'divide by 10' position the main gate will be kept open 10 times longer and this will reduce the error to 2 per cent. When the control circuit has sufficient intelligence it can automatically shift the time base output to the appropriate position to increase the accuracy.

7.3.4 Period measurements

In practical frequency counters the period of a waveform can be easily measured by reconfiguring the basic circuit blocks. This mode is useful for measuring very low frequencies. For example, in frequency measurements of the order of 1 MHz requiring 1 p.p.m. resolution the gate time should be 1 s. However, if the frequency to be measured is 1 Hz then a gate time of 10^6 s will be required for the same 1 p.p.m. resolution. This type of impractical measuring condition can be easily avoided in the period mode where the signals applied to the main gate are reversed (Figure 7.9).

7.3.4.1 Single period measurements

In the period mode the signal is fed through the input stages and the trigger stages and is then passed on to the main gate control where it determines how long the gate will be open. The first trigger pulse opens the main gate and the next one closes the main gate. The time base signal is fed to the signal input of the main gate and, therefore, the gate output is related to the period of the input signal. If the measurement exercise needs a resolution of 1 p.p.m. for a frequency of 1 Hz, the time base output can be set to 1 MHz output, thus giving 1 μs timing pulses. In this case feeding a 1 Hz signal to the main gate keeps it open for 1 s, during which time 10^6 pulses are passed to the DCU and the display. The display will therefore indicate a reading of 1 000 000 μs with the desired resolution of 1 p.p.m. In period measurement mode the highest resolution is governed by the frequency limitation of the main gate and the first decade of the DCU. For example, a 1 MHz counter can have a maximum period mode resolution of 1 μs and a 100 MHz counter can have a resolution of 10 ns only.

7.3.4.2 Multiple period measurements

Even though (single) period mode measurements permit relatively fast measurements it can be shown that the trigger errors may affect the accuracy (discussed later). It is therefore desirable to average the reading over a number of periods in order to achieve a better accuracy and resolution. This is done in multiple period or period averaging mode. Figure 7.10 indicates a simplified block diagram for this mode.

Like the single period mode, the input signal is processed and trigger pulses are derived at the input stage. However, the trigger pulses are no longer fed directly to the main gate control circuit but are applied to the input of the time base divider chain. The main gate is opened by the first trigger from the divider and remains opened until the preset number of pulses (10^N) have passed through the time base divider and the closing pulse is delivered. The total count, which is (10^N) times greater than in the signal period mode, will be displayed. As our need is to measure the average value of a single period, the reading needs to be divided by (10^N). This can easily be achieved by shifting the decimal point on the counter display by N positions to the left by suitably configuring the control logic. In this mode it is very clear that the resolution is increased.

7.3.5 Time interval measurements

In all measurement modes discussed above the measurement of time is made between two identical points on the waveform. If it is necessary to measure the time between two different points on a waveform (or between two points on two different signals) a counter could be configured, as shown in Figure 7.11, using two separate input channels with independent trigger levels and polarity settings. For this kind of time interval measurement, the two input channels provide the start and stop pulses to the main gate, and the crystal oscillator via the time base divider feeds the counting pulses. The switch indicated with 'separate/common' is used to select the cases between same or different waveforms.

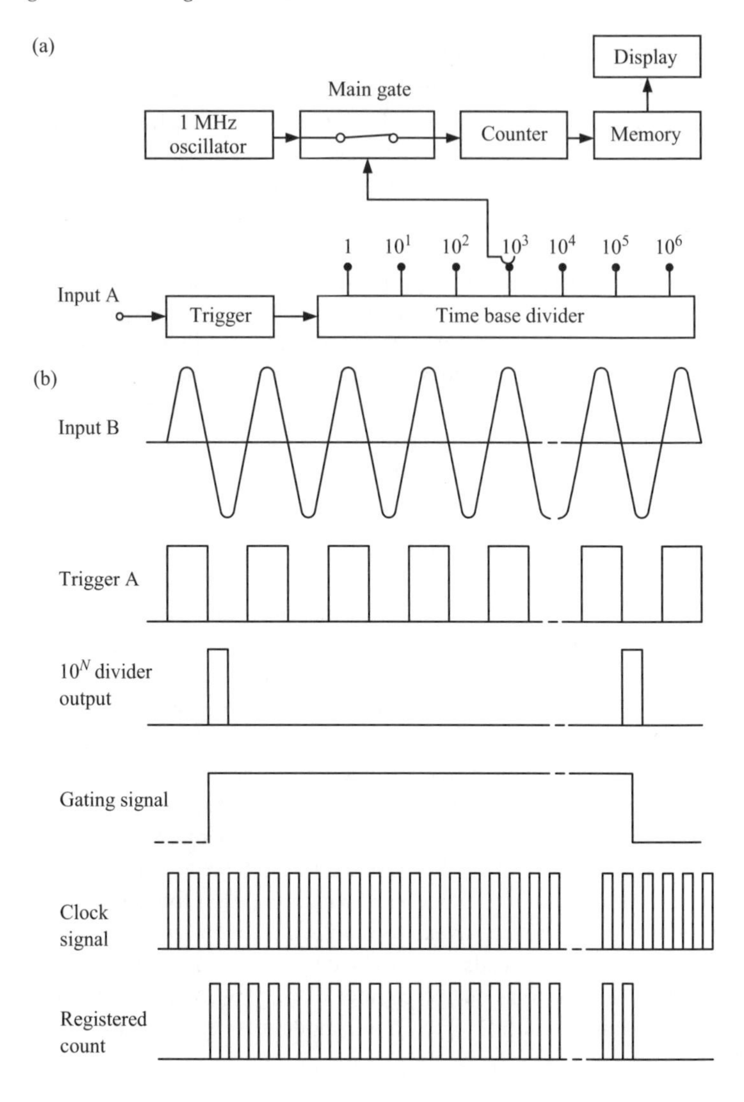

Figure 7.10 Multiple period mode: (a) block diagram; (b) signal flow

The time interval mode can be used to measure the phase difference between two sinewaves as shown in Figure 7.12. The measurement is made by determining the time interval between corresponding points on two identical waveforms. As a sine waveform has its steepest slope at the zero crossing it is better to measure the time interval between the points where the signals are at 0 V levels.

By a slight variation of the time interval mode, 'time interval averaging' is also possible, as shown in Figure 7.13. As in the case of Figure 7.11, the start/stop pulses are generated and the crystal oscillator frequency is applied to the main gate while the stop pulses generated are fed to the time base divider to the memory. In this case the counting pulses are applied to the counter between start and stop pulses but the

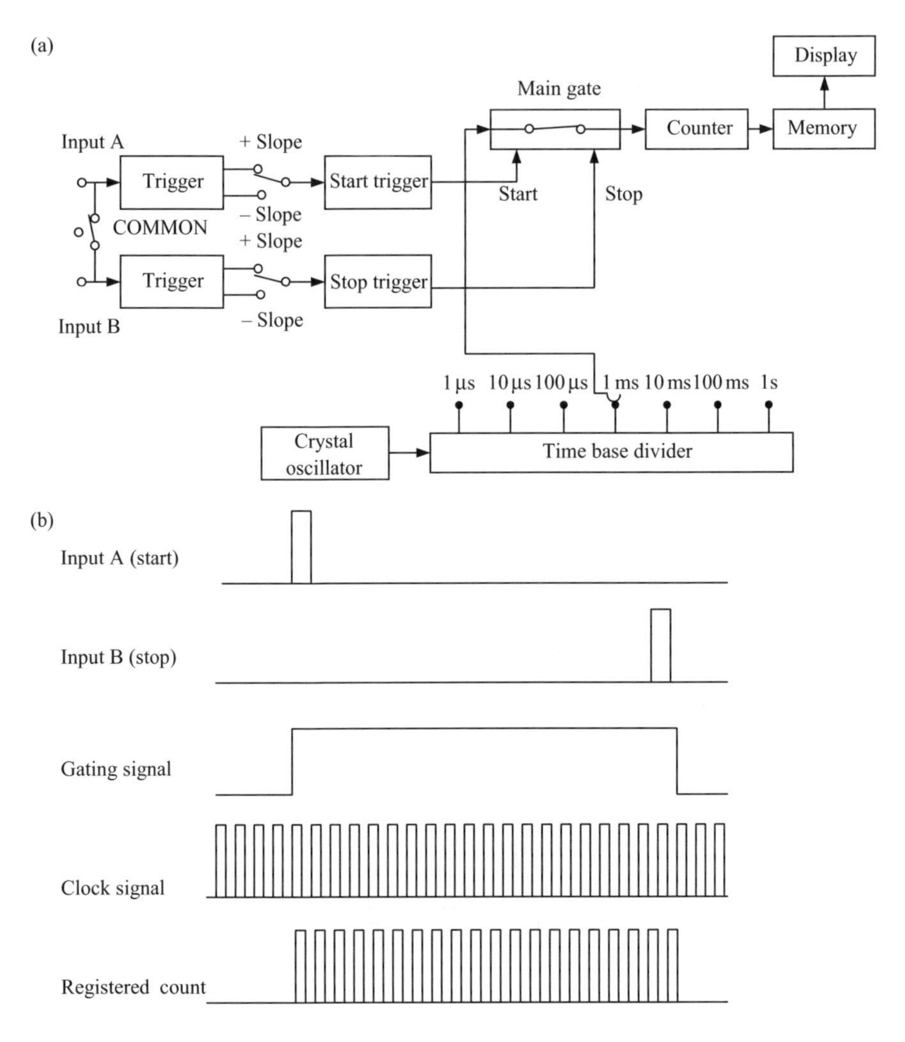

Figure 7.11 Time interval measurement: (a) block diagram; (b) signal flow

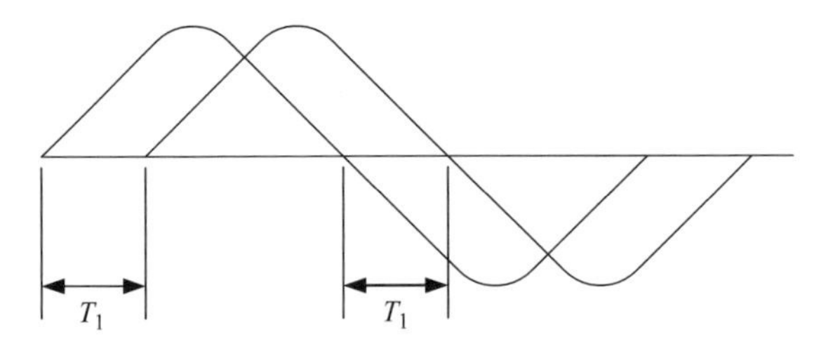

Figure 7.12 Phase difference measurement

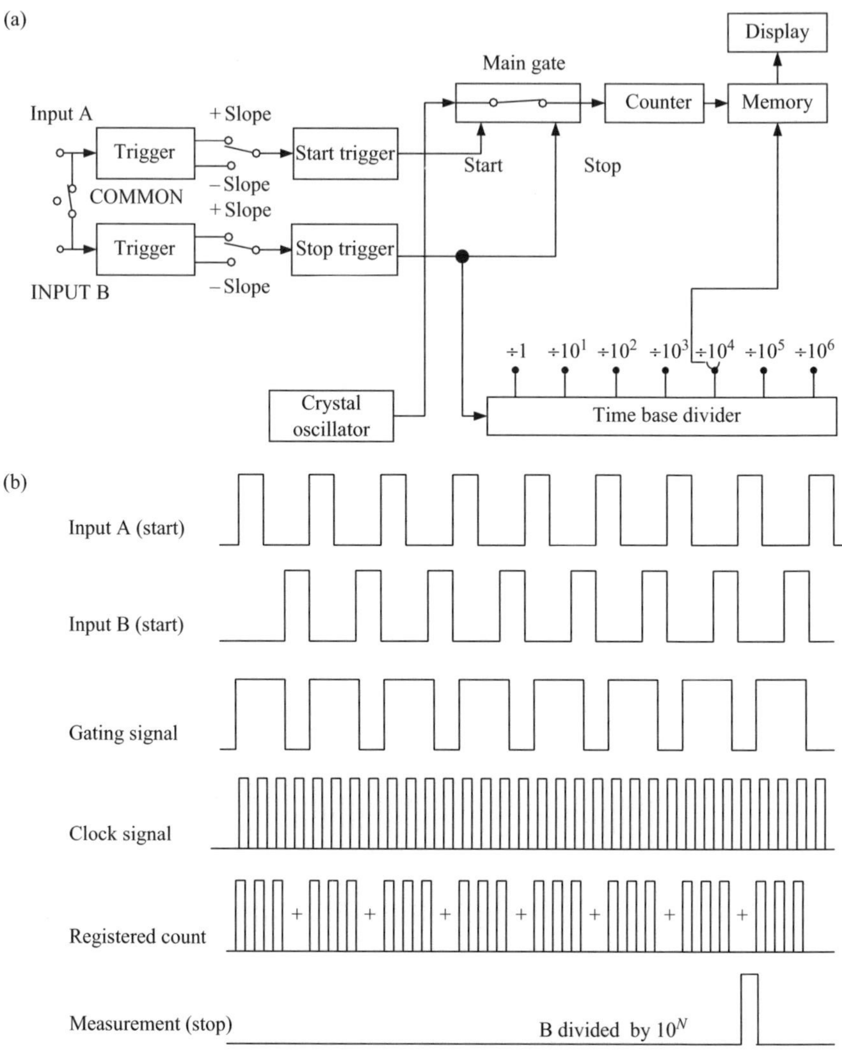

Figure 7.13 Time interval averaging: (a) block diagram; (b) signal flow

transfer and reset pulses are applied to the memory only at the end of last stop pulse, hence averaging the time interval over 10^N counts.

7.3.6 Other modes of operation

As for the above cases, the same basic blocks could be configured for various other situations such as pulse width measurements, scaling of an input frequency or even checking the internal crystal oscillator frequency. In some counters there is arming capability which allows the user to measure the time between two non-adjacent pulses. Figure 7.14 shows a simplified block diagram of a counter with independent arming

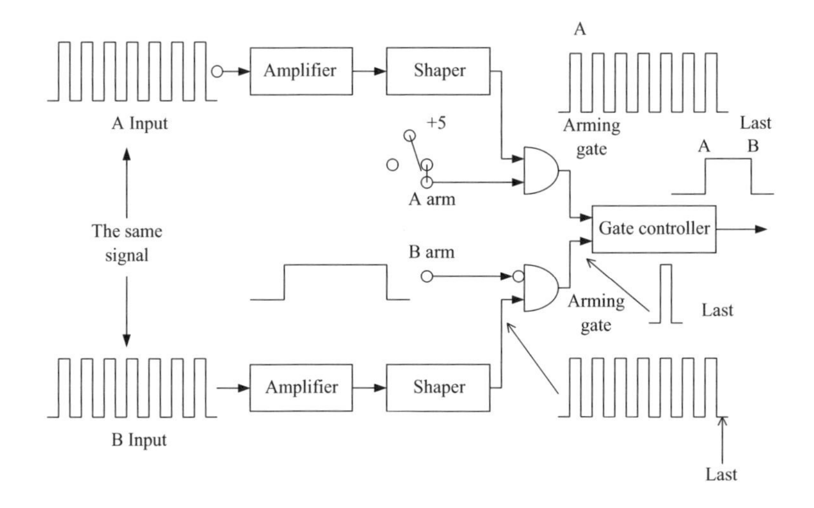

Figure 7.14 Block diagram of a counter with arming inputs

for two channels. In the example shown, the time is measured between the first and the last pulses of the input burst. The A arming input is not used because the first pulse of the A input signal will set the gate control signal high. However, a disarming signal is applied to the B arming input so that only the last negative transition of the input burst will set the gate control signal low. This results in a gate control pulse width equal to the width of the input burst.

7.4 Accuracy of counters

The errors to which digital counters are liable may be divided into two main categories, namely, (i) errors inherent in the digital system itself and (ii) errors specific to the mode of operation. Although modern counters often have large numbers of digits (such as 8 to 11) it does not give any guarantee of accuracy but gives the possibility of high resolution. Recognition and understanding of the sources of errors will help the user to find ways of reducing their effects.

7.4.1 Inherent errors

The basic errors for all digital instruments is the 'plus or minus' $\pm$ count error, because the two signals applied to the main gate are not usually synchronised. As may be clearly seen from Figure 7.15(a), if the gating is unsynchronised the number of pulses passed by two successive gating pulses may differ by one count. The relative error caused by the ± 1 count ambiguity is a function of the number of counts. This may be seen from the following equation:

$$\text{relative error} = \frac{1}{\text{number of counts}} \times 100\%.$$

This relation applies in the graph shown in Figure 7.15(b).

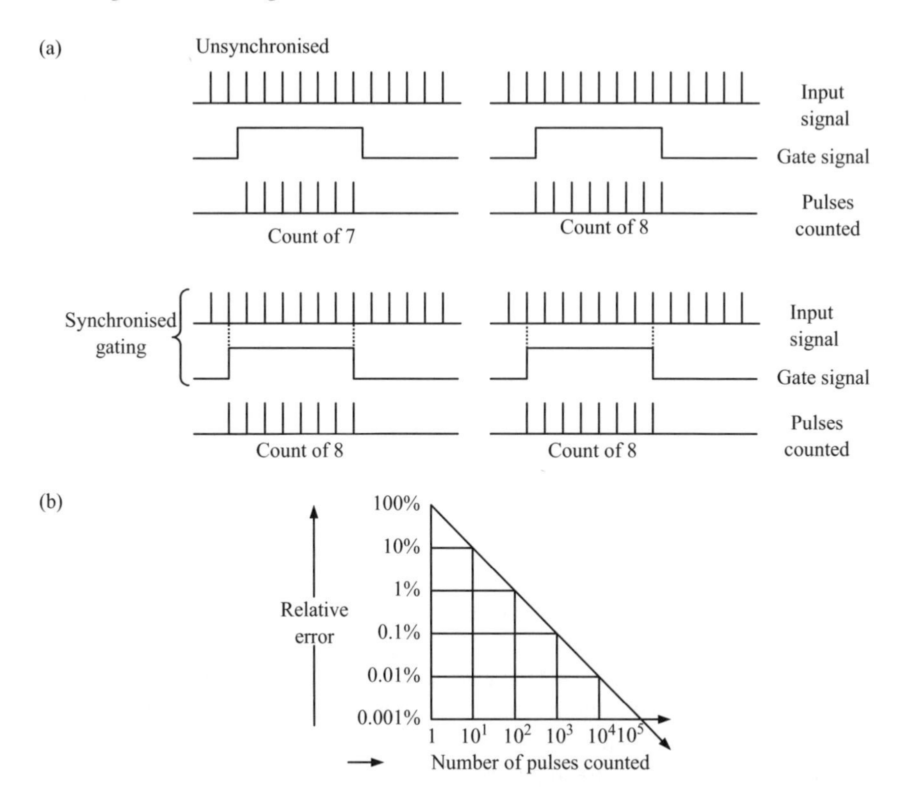

Figure 7.15 Plus or minus ± error: (a) ±1 count ambiguity; (b) percentage error due to ±1 count ambiguity against number of pulses counted

7.4.1.1 Time base error and crystal oscillator

The other major factor that governs the accuracy of a counter is the stability of the time base. Most counters use a crystal oscillator as the internal frequency source and a set of decade counters as the time base divider. A quartz crystal resonator utilises the piezoelectric properties of quartz. If a stress is applied to a crystal in a certain direction, electric charges appear in the perpendicular direction. Conversely, if an electric field is applied, it will cause mechanical deflection of the crystal. In a quartz crystal resonator, a thin slab of quartz is placed between two electrodes. An alternating voltage applied to these electrodes causes the quartz to vibrate. If the frequency of this voltage is very near the mechanical resonance of the quartz slab, the amplitude of the vibrations will become very large. The strain of these vibrations causes the quartz to produce a sinusoidal electric field which controls the effective impedance between the two electrodes. This impedance is strongly dependent on the excitation frequency and possesses an extremely high Q.

Electrically, quartz crystal can be represented by the equivalent circuit in Figure 7.16(a) where the series combination R_1, L_1 and C_1 represent the quartz and C_0 represents the shunt capacitance of the electrodes in parallel with the can

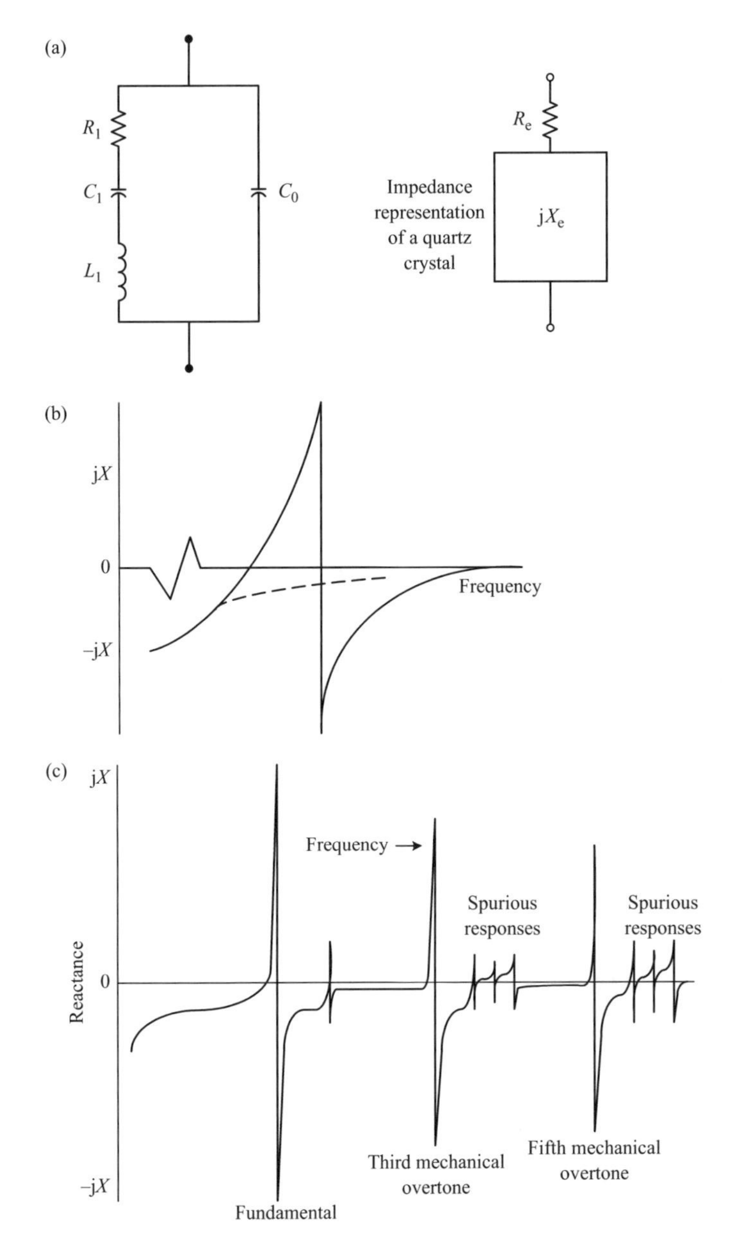

Figure 7.16 Continued overleaf

capacitance. A reactance plot of the equivalent circuit is given in Figure 7.16(b). The resonant frequency of the crystal is given by

$$F_s = \frac{1}{2\pi\sqrt{L_1 C_1}}, \tag{7.1}$$

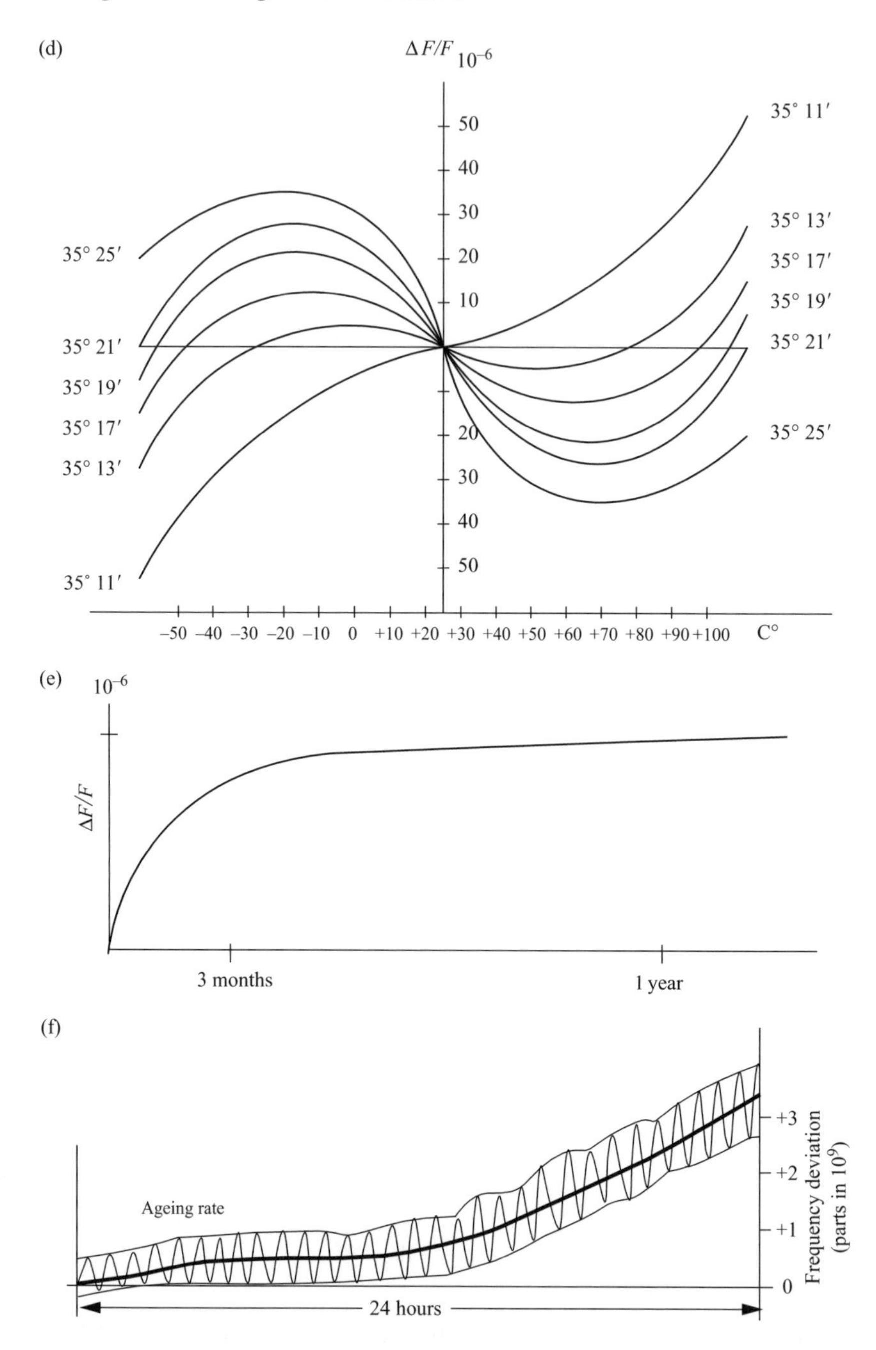

Figure 7.16 Properties of crystals: (a) equivalent circuit of a crystal; (b) reactance vs frequency; (c) overtone response; (d) relationship between temperature and frequency for an AT cut crystal; (e) ageing characteristics; (f) short term stability

where F_s is the series resonant frequency (in Hz), L_1 is the motional arm inductance (in H), and C_1 is the motional arm capacitance (in F).

There are many different ways that crystals can be cut. Some of the common types are AT, J, XY, SC, DT and BT. etc., from a typical manufacturer such as IQD Ltd [2]. In this example, J cut provides crystals below 10 kHz; XY between 3 and 85 kHz; DT cut from 100 to 800 kHz. In the frequency range of 1–200 MHz, AT cut crystals are normally chosen as they represent the best compromise between temperature related frequency stability, frequency accuracy and the pulling activity (adjustment of frequency by external capacitance). A crystal oscillator can be operating in the fundamental mode or its nth mechanical overtones such as third, fifth or seventh, indicated in Figure 7.16(c). SC cut crystals are normally used for ovenised oscillator applications. For details, References 2 and 3 are suggested. Use of the counter's external reference option provides the opportunity to use a reference which is locked to the Global Positioning System (GPS) atomic clocks. GPS long term accuracy is in parts in 10^{-11} to 10^{-12}.

7.4.1.2 Trimming

The main factors that affect the accuracy of the clock are trimming inaccuracy, drift due to temperature, long term and short term stability and power supply variations. In crystal oscillator circuits the final set frequency accuracy is dependent on the primary standard error and the trimming resolution. That is

$$\text{Trimming inaccuracy} = \text{primary standard error} + \text{trimming resolution.}$$

In practice, the crystal oscillator circuits are trimmed with a small capacitor. This approach is preferred as the mechanical production of quartz crystals with an exact frequency is not practicable. Even though crystal oscillator may have been set to an initial accuracy (usually within several p.p.m.) in the long run this accuracy may change owing to drifting.

7.4.1.3 Temperature effects

Temperature dependence might be reduced by choosing a crystal cut [4] with a low temperature coefficient or by operating the crystal at the temperature where the temperature coefficient is zero. See Figure 7.16 (d). Oven controlled oscillators (OCXO) use this technique.

A typical temperature coefficient of $5 \times 10^{-9\circ}C^{-1}$ could be achieved in a range of $\pm 1^\circ C$ around the turning point by suitably selecting the cut. When higher stabilities are required, special techniques are utilised in modern frequency counters. They are temperature compensated crystal oscillators (TCXO), oven controlled crystal oscillators (OCXO) and a few other variations such as digitally compensated crystal oscillators (DCXO). Table 7.1 indicates the characteristics and application of these variations. Figure 7.17 shows typical examples of these devices. Many package variations exist within these groups. Note that TCXO units typically have a voltage control option.

Table 7.1 Crystal oscillator variations, applications and characteristics (courtesy of Rakon Ltd, New Zealand and EDN Magazine, November 20, 1997 (reprinted with permission of Cahners Business imformation))

Type	Application	Temperature Stability (+/− ppm) 0–70°C	Ageing (+/− ppm/year)	Short Term Stability (ppb)
Basic quartz crystal	Miscellaneous applications	10 to 100	2 to 5	−Oscillator Dependent
Uncompensated crystal oscillators	Digital systems & microprocessor clocks	20 to 1000	1	> 1
Temperature compensated crystal oscillators (TCXO)	Telecomm, test instruments, satellite communication	0.1 to 5	1	0.2 to 1
Voltage controlled crystal oscillators (VCXO)	Telecomm, cellular, GPS, as a component in TCXOs and PLLs	20 to 100	1	0.2 to 1
Oven controlled crystal oscillators (OCXO)	Frequency counters, spectrum and network analysers, navigation and defence base stations	0.0001 to 5	0.1 ppb/day to 2 ppb/day	0.005
Digitally or Microprocessor compensated crystal oscillators (DCXO/MCXO)	Video, military, telecomm, high end base stations	0.05 to 1	0.1 ppb/day to 1 ppm/yr	0.5

Traditional TCXOs contain a network that compensates for temperature dependence of the quartz crystal by suitably configuring thermistors (either together with varactor diodes etc. directly or in oscillator loop), which, by changing the load presented to the crystal, creates a frequency change that is just the opposite of the frequency drift of the crystal. The residual drift is typically about ten times better than the frequency drift of the crystal. Recent developments have included the use of a dedicated function generator and VCO circuit IC to allow low cost mass production of TCXO units. In this case the function generator creates a voltage which is applied to VCO circuit exactly compensating the temperature effects. For really high stability requirements, proportionally controlled ovens are used to maintain the crystal at a

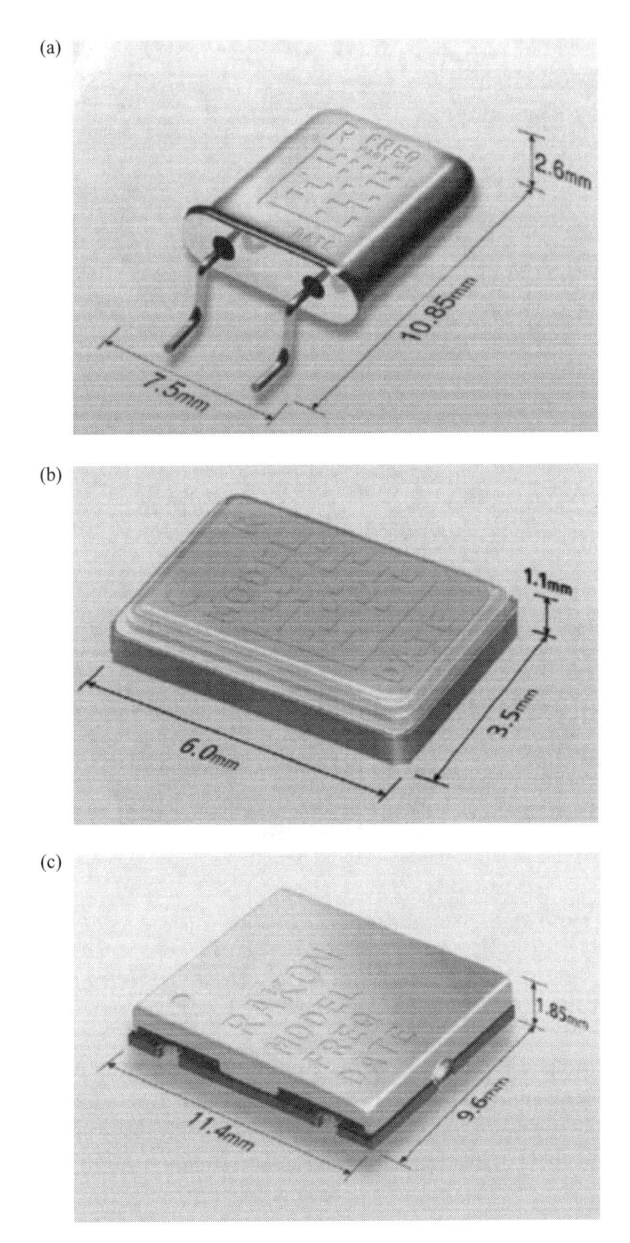

Figure 7.17 Continued overleaf

constant temperature at which there is a turning point (for example, in Figure 7.16(d), if the crystal is cut at 35°19′ and the oven is at 70°C). Usually this temperature is chosen about 20°C above the maximum ambient temperature. For a more detailed explanation, References 1 and 4–6 are suggested. Typically, TCXOs have drifts in

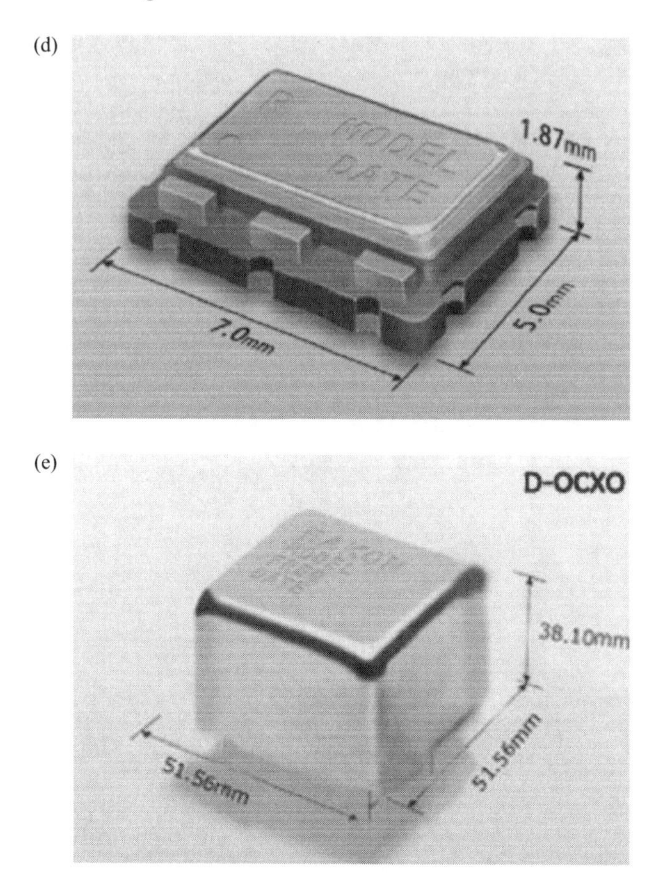

Figure 7.17 *Crystals and oscillator packages: (a) Bare crystal – leaded type; (b) Bare crystal – surface mount type; (c) Describe TCXO using thermistors; (d) TCXO using VCO IC; (e) Oven controlled oscillator (OCXO) (Courtesy: Rakon Limited, New Zealand)*

the range of 0.1–5 p.p.m.; a more costly solution is to use oven controlled versions which can give very much higher stabilities. Some oscillators include provisions for fine tuning the frequency by using a voltage controlled crystal oscillator (VCXO), where a varactor diode is used to pull the crystal frequency. For details, Reference 3 is suggested.

One practical way to produce an oscillator with the stability of ovenised units and the low current drain of thermistor compensated systems is to adopt digital compensation. The falling price of LSI integrated circuits and the development of high density assembly techniques have allowed designers to make use of DCXOs. For example, a digital solution, termed a mathematically temperature compensated crystal oscillator (MTCXO), is used in some of the frequency counters from Philips, such as the PM6662 and PM6669, to bring down the cost of the counters by eliminating the expensive oven controlled versions. In this unique technique, the temperature dependency curve for each individual crystal oscillator is factory measured and the frequency deviations across the temperature range, Δf, are stored in a non-volatile memory. During operation the Δf value for operating temperature is looked up in the memory and used to compensate the measuring result before it is displayed. This automatic temperature compensation also results in highly accurate measurements instantly, without the need for long warm-up times, with residual temperature stability of typically 2×10^{-7} over the temperature range from 0 to $50°C$.

7.4.1.4 Drift characteristics

Long term drift of crystal oscillators follows a typical curve as shown in Figure 7.16(e) and the short term stability is shown in Figure 7.16(f). Ageing rates depend strongly on the quality of the production techniques, but are higher during the first few months of operation. The ageing characteristic is exponential during the first part of the crystal life and gradually becomes linear. Short term stability is non-systematic and it causes the frequency to fluctuate about the long term mean value, as shown in Figure 7.16(f). It is therefore defined as the root mean square of the deviation over a certain averaging time. The better types of crystal oscillators have values of a few parts per 10^{10} per second of averaging time and a few parts per 10^9 in 24 h averaging times under constant environmental conditions and supply voltage variations. The other major factor influencing the accuracy of a crystal oscillator is the stability of the power supplies. It will be clear from the above discussion that all these factors should be taken into account when determining the accuracy of digital counters. Furthermore when the crystal oscillators are used over a considerable period of time the cumulative effect of the above mentioned factors may cause its frequency to drift beyond permissible limits and it will be necessary to recalibrate the oscillator from time to time.

7.4.2 Errors dependent on functional mode

In this section we discuss briefly the errors in the frequency and period mode measurements. For a more elaborate discussion (including errors on other modes), Reference 1 is suggested.

7.4.2.1 Frequency mode errors

In the frequency mode, the ± 1 count error and the time base accuracy can be combined as in the graph of Figure 7.18. A longer gate time offers higher accuracy in the frequency of the measured signal. However, the ultimate accuracy is governed by the time base accuracy.

Other errors in frequency measurements can be caused by amplitude modulation, frequency modulation, noise, etc. When the frequency of an amplitude modulated signal is to be measured, it is very necessary that the trigger window is set correctly so that there will be no pulses lost due to the envelope of the modulated signal crossing the trigger window. See Figure 7.19(a). If the trigger level is adjusted in such a way that the trigger window is placed within the $V_{\min}$ of the modulated signal the reading will be accurate.

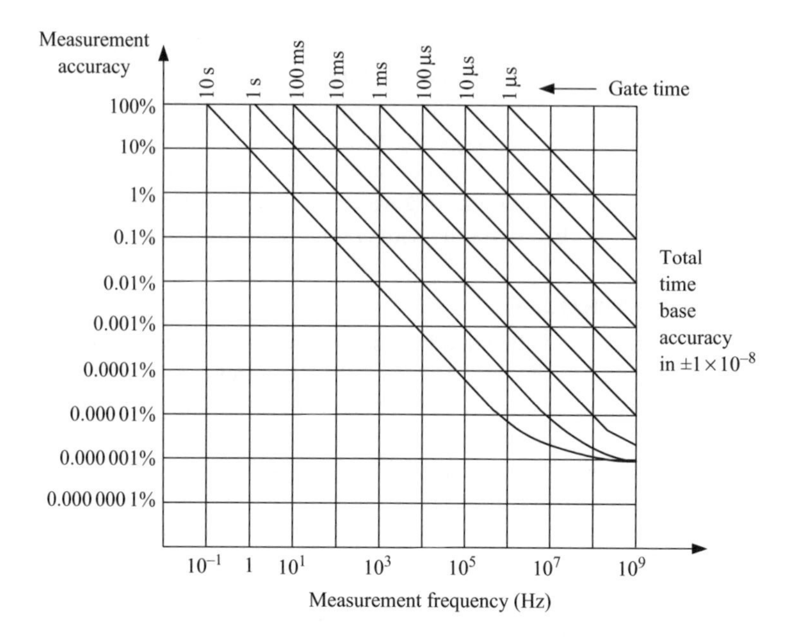

Figure 7.18 Frequency error as a function of gate time, ±1 count ambiguity and the clock accuracy

When frequency measurements are to be carried out on the signal bursts and the gate time t_g is long compared with the burst duration t_b, and not synchronised, enormous errors can occur; see Figure 7.19(b). Accurate measurements may be obtained when the gate time is synchronised and is just shorter than the burst duration. In the case of frequency modulated signals the frequency read by the counter will be an average of the frequency over the gate time chosen.

Another important factor that can greatly affect the accuracy of the counter measurements is noise. When the peak-to-peak value of the superimposed noise signal is large enough to cross the trigger window, false counts can be generated. Figure 7.20(a) shows the case of false counts generated due to noise crossing the trigger window. Figure 7.20(b) indicates how this is increased when the trigger level is towards the lower slope area (peak value of the signal) of the waveform.

In this case the absolute value of the noise is the one that affects the errors rather than the signal-to-noise ratio. To reduce the error the trigger window should be increased to a higher value than the peak-to-peak noise. Figure 7.20(c) illustrates this. In practical counters the measurements are commenced using maximum sensitivity and the attenuation can be gradually increased so that stable readings are obtained. Some counters use automatic gain control circuits to automatically adjust the input signal to values just larger than the trigger window. More versatile counters provide controls for varying the offset of the trigger window and observing the trigger level externally. Figure 7.21 shows a few cases of correct and incorrect trigger settings for situations where effects of pulse ringing, third harmonic distortion, noise spikes and d.c. offset level problems might be encountered.

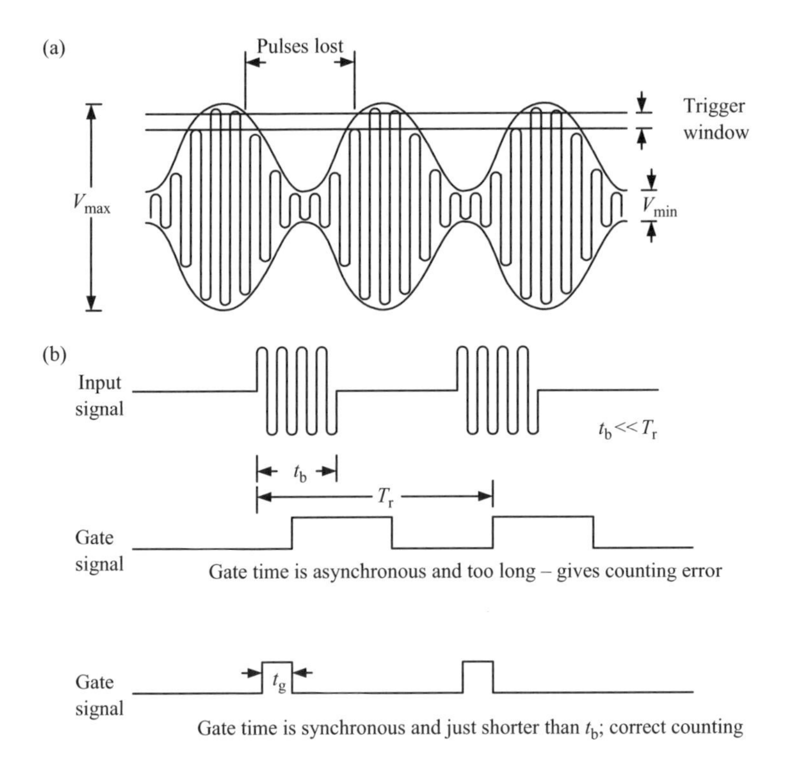

Figure 7.19 Frequency mode errors due to modulation: (a) due to amplitude modulation; (b) effect of burst frequency mode

7.4.2.2 Period mode errors

As in the frequency mode, in addition to the clock accuracy and the ±1 count ambiguity, trigger errors can severely affect the measurement in the period mode. As the main gate is opened and closed by the input signal, all noise, hum and interference spikes may operate the gate and these components superimposed on the input signal can affect the main opening time, as shown in Figure 7.22. It can be easily shown that the error is affected by the signal-to-noise ratio and the slope of the input waveform where the noise is superimposed.

For example, if the trigger level is shifted to the flat top of a sinewave, the error could be extremely high and it is excessive compared with the effect of the inherent error components. For the period measurements, if averaging is used the relative error could be very much reduced. For a more accurate measurement in period mode, the following criteria could be useful:

- set the trigger window in a clean part of the signal,
- use the multiperiod mode, and
- ensure high signal-to-noise ratio.

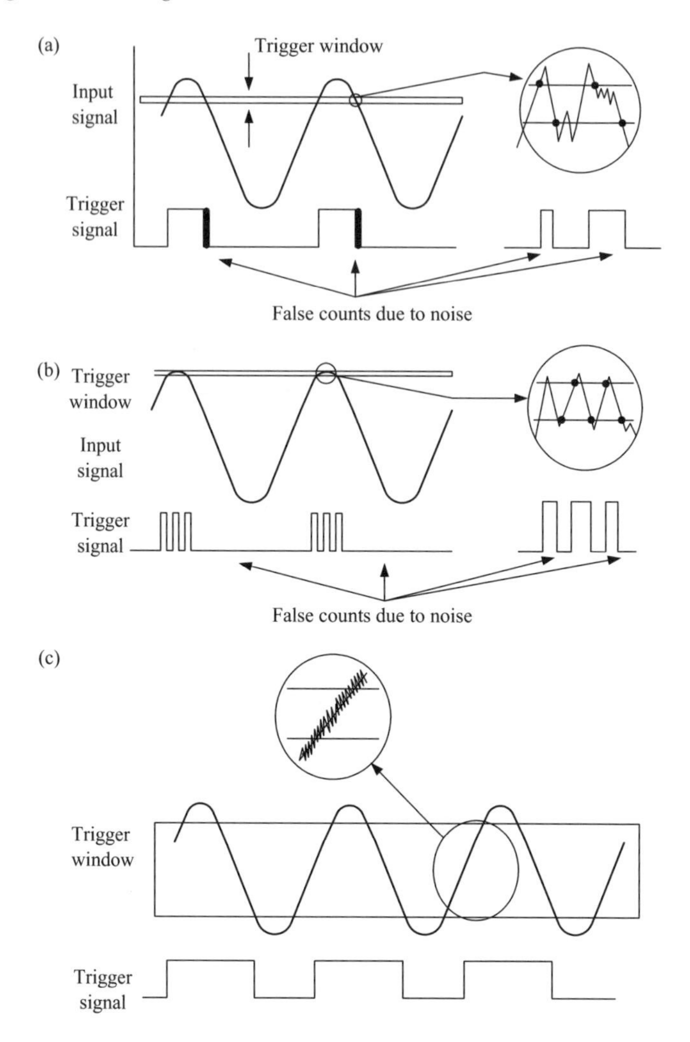

Figure 7.20 Influence of noise: (a) false counts due to noise crossing trigger window; (b) false counts increased as trigger level is changed; (c) false count eliminated by trigger adjustment

7.4.2.3 Practical measurements

The following summary may be useful to bear in mind during any counter measurements:

- a steady reading is usually correct,
- an unstable reading is definitely erroneous, and
- if in doubt, use of an oscilloscope to observe the trigger setting (in relation to input waveform) might be useful.

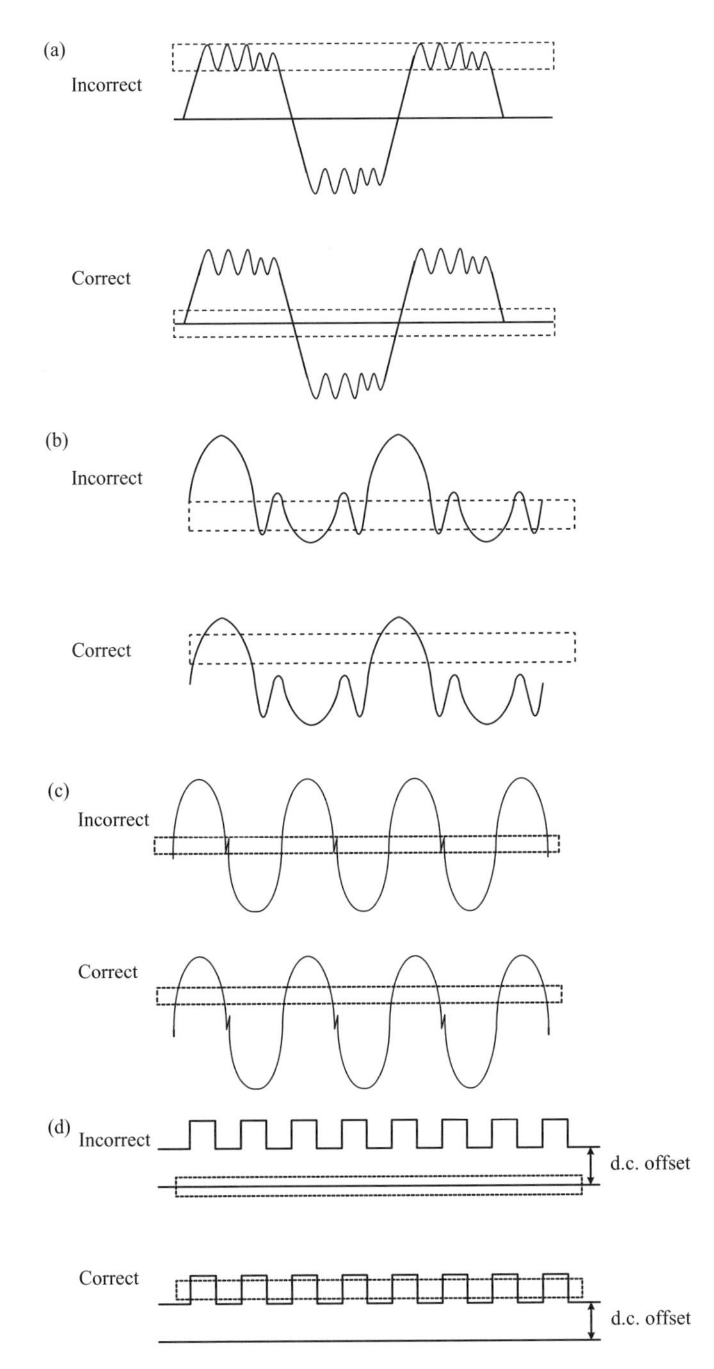

Figure 7.21 Correct and erroneous trigger settings: (a) effects due to pulse ringing; (b) effect of third harmonic distortion on a sinewave; (c) effect of signal with noise; (d) effect of d.c. offset

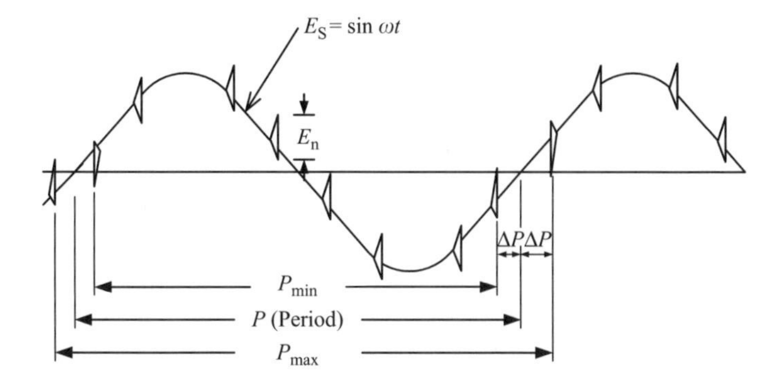

Figure 7.22 Influence of noise on period measurement

7.5 High frequency measurements and down conversion techniques

A counter, being a digital instrument, is limited in range by the speed of the logic circuitry employed in the design. Today, counters are capable of handling frequencies beyond 1 GHz with the modern logic families utilised in the design of the circuit blocks. Four basic techniques are utilised when the measurement range is required to extend up to microwave frequencies. These are: (i) pre-scaling, (ii) heterodyne conversion, (iii) use of transfer oscillators, and (iv) harmonic heterodyne conversion. These techniques easily extend the frequency measurement by digital means from 1 to 40 GHz. For details References 1 to 3 are suggested.

7.5.1 Pre-scaling

Pre-scaling involves simple division of the input frequency resulting in a lower frequency signal which can be counted by the digital circuitry. The frequency measured by the counter section is related to the input simply by the integer N. A display of the input frequency is accomplished by multiplying the content of the counter by N or by increasing the gate time by a factor of N, where N typically ranges from 2 to 16. A block diagram of a high frequency counter with pre-scaling is shown in Figure 7.23. The input signal is conditioned and shaped before applying to the pre-scaling block, which divides the frequency by N before applying it to the main gate. After this conversion process the same basic blocks of a counter are used with the exception that the crystal oscillator frequency is also divided by the same factor before applying to the timebase dividers. This is used to display the input frequency directly on the display. Pre-scaling has one drawback where the measurement becomes slow when high resolution is necessary.

7.5.2 Heterodyne frequency conversion

A heterodyne down conversion technique used in a microwave frequency counter is shown in Figure 7.24. From the time base circuits the oscillator frequency is applied

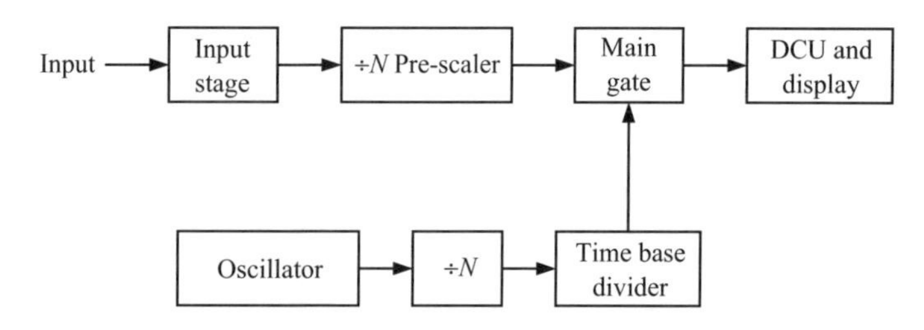

Figure 7.23 Pre-scaling

to a frequency multiplier and the output of the multiplier (F_x) is applied to a harmonic (comb) generator where the high frequency components are amplified and applied to an yttrium–iron–garnet (YIG) filter [1, 7]. By magnetic tuning of the YIG filter a harmonic component could be selected from the comb generator. This harmonic component is mixed with the incoming unknown frequency, F_{in}, and the output, which has frequency components $F_{in} \pm kF_x$ (k being the harmonic number) and the video amplifier amplifies the $F_{in} - kF_x$ component only and passes it on to the basic counter stages for counting. Control circuits under the command of a processor block take care of the YIG filter where it starts selectively tuning across the entire range of harmonic components commencing from the lowest, until the correct harmonic is mixed with the input unknown frequency F_{in}. This occurs until the video amplifier output frequency is within the measuring range of the basic counter circuit.

A key component in automating this heterodyne conversion process in modern counters is the YIG filter and the associated circuitry.

7.5.3 Transfer oscillator

In a way similar to the heterodyne convertor, the transfer oscillator mixes the incoming unknown frequency with an internally generated signal. The main difference between the two methods is that in the transfer oscillator the basic counter is used to measure the frequency of the internally generated signal compared with the measurement of mixer output in the heterodyne convertor. Figure 7.25 shows the block diagram of a frequency counter using a transfer oscillator with a phase locked loop.

With this technique, as the voltage controlled oscillator (VCO) is tuned across the range, one of its harmonics, generated by the harmonic generator, is mixed with the input frequency. This produces the difference frequency which is within the bandwidth of the video amplifier. The video amplifier output is applied to the phase detector and the phase detector output is used to adjust the VCO output to be phase locked. A basic counter block measures the frequency of the VCO. In practical counters this technique is suitably adapted with some additional blocks to detect the harmonic number so that a direct frequency reading is observed on the display.

7.5.4 Harmonic heterodyne converter

The harmonic heterodyne converter is a hybrid of the previous two techniques. In this case the synthesiser output frequency kF_s (where F_s is the frequency of a stable

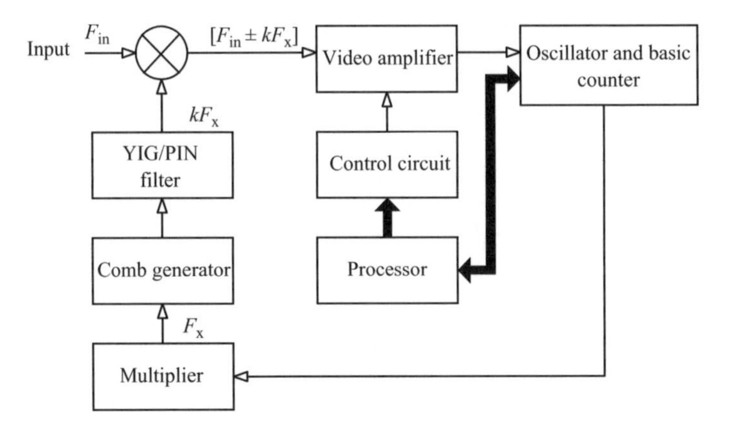

Figure 7.24 Heterodyne frequency conversion

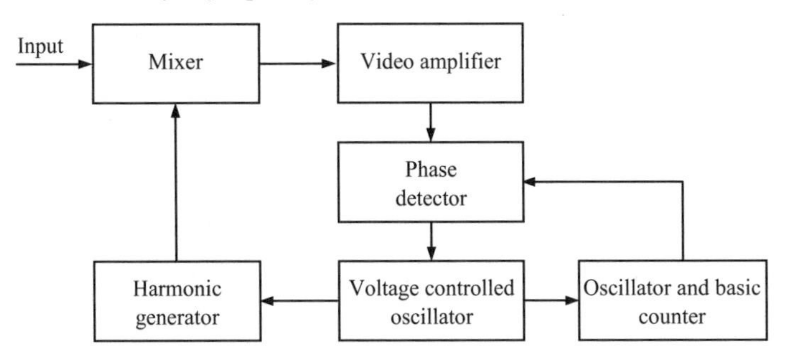

Figure 7.25 Frequency counter based on the principle of a transfer oscillator

reference source such as the reference oscillator of the basic counter) is generated using the synthesiser controlled by the processor block. See Figure 7.26. This frequency is mixed with the unknown incoming frequency, F_{in}, at the mixer. The mixer output via the video amplifier which amplifies one of the intermediate frequency components, $F_{if} = F_{in} - kF_s$, is measured by the basic counter. As soon as the harmonic number, k, is known, the input frequency can easily be calculated and displayed.

A practical method for calculating the harmonic number is to use the processor block to shift the synthesiser output frequency between two closely spaced frequencies and to read the counter value change, which can be used by the processor to calculate the harmonic number. The harmonic heterodyne convertor has the potential of lowering the cost of microwave counters because it can be designed with just one microwave component and a low cost microprocessor for control, decisions and calculations.

7.6 Modulation domain analysers

With the advances in frequency and time measurement capabilities of timers and counters towards the end of the 1980s, Hewlett-Packard developed the idea of the modulation domain, which allows users to record and display variations in time intervals, phase and frequency as a function of time. Time interval analysis, which can be carried out using products such as the HP53310A, E1725C and E1740, have

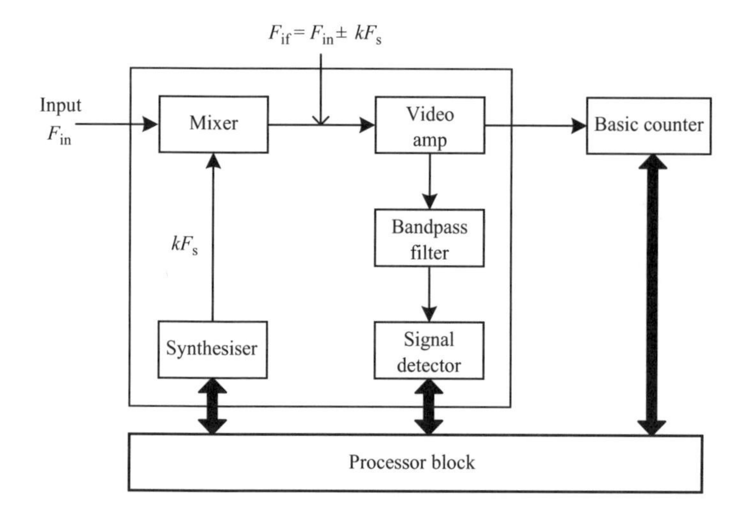

Figure 7.26 Harmonic heterodyne conversion

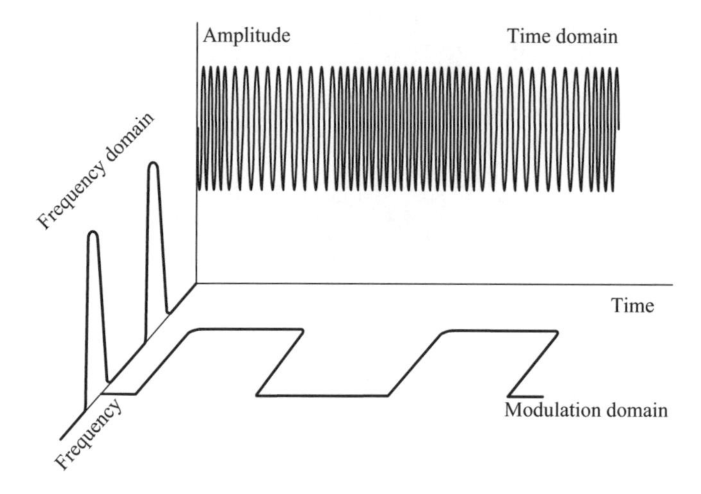

Figure 7.27 Time, frequency and modulation domains

allowed engineers to observe the modulation domain with respect to the frequency against time, phase against time, and the time interval against time. Figure 7.27 depicts the axes of time, frequency and amplitude indicating the practical relationships among time domain, frequency domain and the modulation domain. The spectrum analyser may not be able to provide a real time picture of rapidly changing signals because of limitations in sweep speed and resolution bandwidth, etc. (discussed in chapter 9). Modulation domain measurements have become a reality owing to developments related to the speed of frequency counters [4/8].

Modulation domain measurements are quite useful in magnetic discs, fibre optics, high speed modems, digital audio, video, sonar or satellite applications. These instruments can record and group the data and the time intervals (TI), etc., into various groups so that the variations of measured parameters over time can be easily analysed.

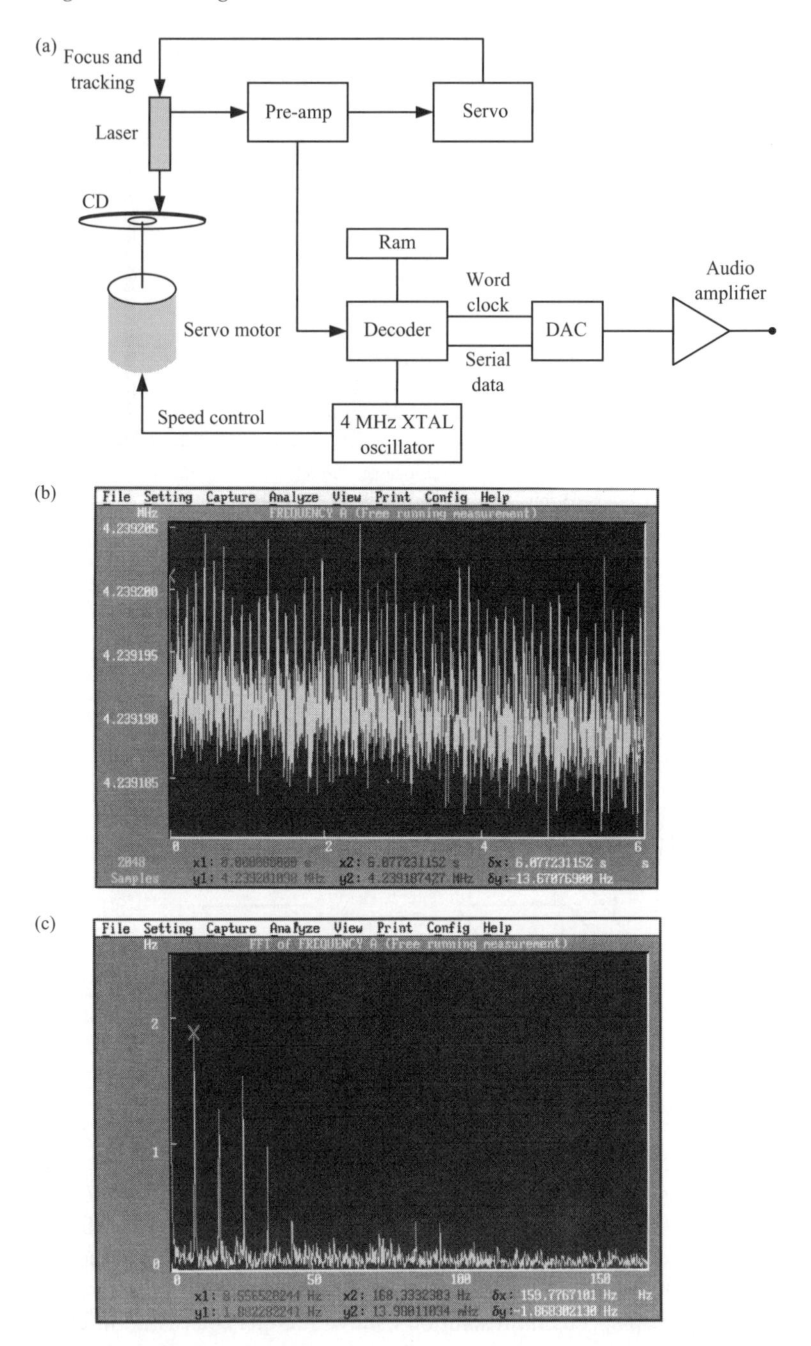

Figure 7.28 *CD player block diagram and measurements: (a) block diagram; (b) frequency vs time plot; (c) modulation components on oscillator (courtesy of Fluke Corporation, USA)*

These instruments can acquire and average data at very high rates and histograms or Allan variance parameters [9, 10], etc., can be calculated. Let us discuss a case of measurements on a CD player as in Figure 7.28.

Compact disc players are complex systems. They use electromechanical servo loops to rotate the disc and position the laser; they use digital filters and decoders to process the digital data; and they use an analogue audio chain downstream from a D/A converter for the right and left audio channels. See the block diagram in Figure 7.28(a). The operation of a CD player depends on a crystal oscillator that typically uses low cost AT crystal strip resonators to maintain system stability. All major clock functions and sampling rates are derived from this 4 MHz clock. When viewed with an oscilloscope, even a few samples of the output from the crystal oscillator show jitter on the signal. Thus the scope could easily indicate the short term stability of the signal, but it provides no further analysis capabilities. This is where a fast frequency counter excels.

The plot in Figure 7.28(b) shows frequency against time for the oscillator. In this application the frequency counter was set up for 2.5 ms measuring time per point, and Figure 7.28(b) shows the 2048 samples (about 5 s of data) that were collected using a Philips PM 6680 timer/counter and Time-View software. Although Figure 7.28(b) resembles a conventional oscilloscope plot, the information displayed is different. The plot displays calculated frequencies averaged over the measuring time between successive points (not voltages and the frequencies). Measuring times and sample intervals can be set independently. Notice that the oscillator appears to be frequency modulated at a periodic rate by an impulse waveform. The average height of the spikes is approximately 6 Hz off the main carrier of 4.239 MHz. We can analyse this behaviour in greater detail by using the fast Fourier transform function, which is also included in the Time-View software.

A Fourier transform of a plot of frequency vs time produces an entirely new plot called a frequency vs offset-frequency plot. See Figure 7.28(c). This transform is not difficult to understand; a Fourier transform operates just on waveforms, whether voltage vs time or frequency vs time. The tricky part is to assign meaning to the new axes. Taking the Fourier transform of a frequency vs time signal results in a direct demodulation of a signal. The vertical axis represents frequency deviation in Hz. The horizontal axis is the frequency offset from the carrier frequency. You do not need to convert the signal down to baseband in order to see the modulation products.

Notice that Figure 7.28(c) shows the carrier component with a 0 Hz offset and a 4.239 MHz 'deviation'. The carrier component is clipped to zoom in on the modulation components, which happen to be down in the 2 Hz range. The modulation is highly periodic over the 5 s time record shown, as evidenced by very little spreading in the sidebands. The fundamental frequency is 8.56 Hz. Notice that the modulation waveform is mainly an impulse: it is rich in both even and odd harmonics. If you were to check this frequency against other system elements in the CD player to find the source of the modulation, you would notice that the 8.56 Hz frequency exactly matches the rotation frequency of the compact disc (513 r.p.m.).

For a detailed account of these instruments and the measurement techniques, References 9–13 are suggested.

For modern trends in general frequency counters the reader may wish to consult the articles in References 14–16.

7.7 References

1 BOWENS, A.J.: *Digital Instrumentation*, McGraw Hill, New York, 1994.
2 IQD LTD: *Crystal Product Data Book*, 1993, pp. 193–205.
3 TRAVIS, B.: 'Looking good in wireless systems', *EDN*, November 20, 1997, pp. 39–48.
4 BENJAMIN, P.: *Design of Crystal and Other Harmonic Oscillators*, John Wiley & Sons, New York, 1983.
5 BREED, G.A.: *Oscillator Design Hand Book*, Cardiff Publishing Company, December 1991.
6 TEKTRONIX, INC.: *Digital Counter and Meter Concepts*, Tektronix, April 1980.
7 HELSZAJN, J.: *YIG Resonators and Filters*, John Wiley & Sons, New York, 1985.
8 PHILLIPS APPLICATION NOTE: 'Exploring the modulation domain', Application note number 9498 722 0211.
9 HOWE, D.A, ALLAN D.W., and BRANES, J.A.: 'Properties of signal sources and measurement methods'. *Proceedings of 35th annual frequency control symposium USAERADCOM*, Ft. Monmouth, NJ, May 1981.
10 ALLAN, D.W.: 'The measurement of frequency and frequency stability of precision oscillators', NBS technical note 669. *Proceedings of 6th PTTI planning meeting*, 1975.
11 HEWLETT PACKARD: 'HP5371A frequency and time interval analyser: jitter and wonder analysis in digital communications', Application note 358–5, June 1988.
12 HEWLETT PACKARD: 'Simplify frequency stability measurements with built in Allan variance analysis', Application note 358–12, January 1990.
13 STRASSBERG, D.: 'Time measurement instruments: a mature industry returns to innovation', *EDN*, 2 March 1989, pp. 47–60.
14 NOVELLINO, J.: 'Counter-timers', *Electronics Design International*, June 1989, pp. 81–86.
15 NOVELLINO, J.: 'Counter-timers advance into the digital age', *Electron. Des.*, 3 September 1992, pp. 81–82.
16 BARKER, D.: 'Counter-timers evaluate digital designs', *Electron. Des.*, 3 September 1992, pp. 8–90.

Further reading

HEWLETT PACKARD: 'Fundamentals of microwave frequency counters', Application note 200–1.

BRYANT, G.H.: *Principles of Microwave Measurements*, Peter Perigrinus, 1993.

HEWLETT PACKARD: 'HP5370B universal time interval counter: high speed timing acquisition and statistical jitter analysis', Application note 191–7, June 1987.

PHILLIPS APPLICATION NOTE: 'Exploring the modulation domain', Application note number PM 6680 Timer/counter.

Chapter 8

Conventional signal sources and arbitrary waveform generators

8.1 Introduction

Test systems often require you to create different types of waveforms. To meet this requirement the world of instruments uses different families of waveform, signal and data generators falling in to a family tree of signal sources as shown in Figure 8.1. Most traditional waveform or function generators and the signal generators provide you with either (i) sine, square, triangular or variations and combinations, or (ii) high frequency sine waveforms with miscellaneous types of modulation. The first type is used for testing miscellaneous types of analogue, mixed signal or digital circuit blocks, whereas the second category is used for testing communication systems in general. Reference 1 provides a detailed account of these analogue systems and design techniques.

In the world of digital instruments where the user is able to obtain an arbitrary waveform based on information stored in a memory, many new families have entered the market during the past two decades. This chapter provides an overview of the

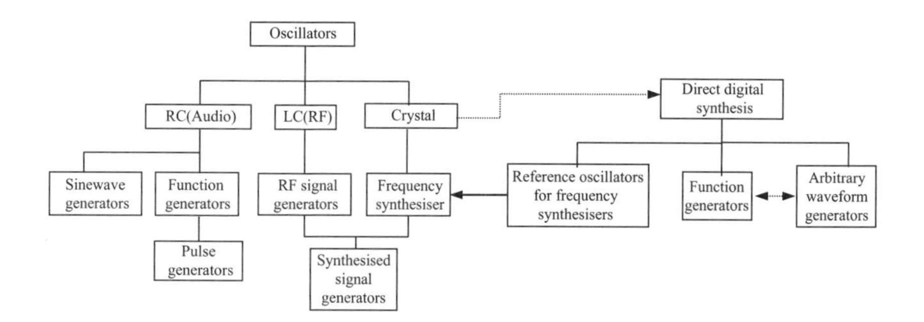

Figure 8.1 Family tree of signal sources

design techniques, advantages and limitations of conventional and arbitrary output instrument families and some applications.

8.2 Conventional signal sources

This traditional group includes function, pulse, audio and signal generators as well as swept frequency oscillators. The following is a summary of these conventional investments.

8.2.1 Function generators

Basic function generators supply sine, square and triangle waves with frequencies ranging from 1 mHz to over 50 MHz. A basic function generator using two operational amplifiers and other components is shown in Figure 8.2. The first stage, A_1, is an integrator that generates a triangular output and the second stage, A_2, is a voltage comparator that generates a square waveform. The triangular waveform generated passes through the sine shaper block and generates a sinusoidal output. Most of the inexpensive function generators are based on monolithic ICs and have the capability to generate sine, square and triangle waveforms.

As the price of function generators increases, so do such features as trigger output, increased frequency range, frequency stability, variable rise and fall times (for square waveforms), variable d.c. offset, increased frequency accuracy, increased output drive and lower distortion.

The need for increased performance in two of these parameters, output distortion and square wave rise time, has created two related classes of signal generators: the low distortion, low frequency function generators and the pulse generators. Signal sources optimised for low distortion often are used in low frequency applications where signal fidelity is important. High performance audio systems are the most

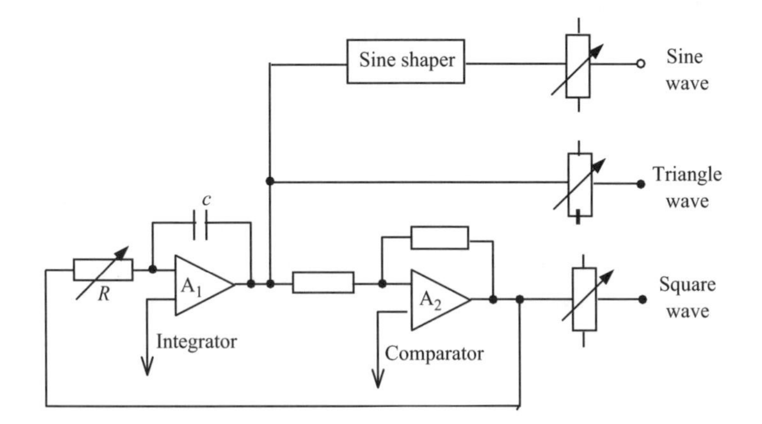

Figure 8.2 Basic principle of a function generator

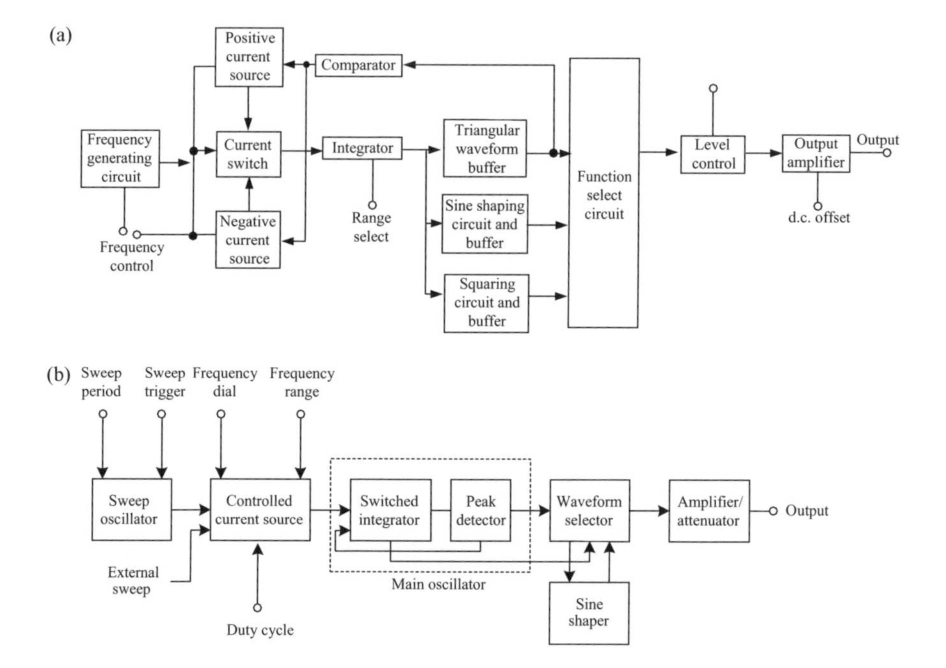

Figure 8.3 Function generator with constant current sources: (a) basic system; (b) with sweep capability

obvious example. The basic approach for achieving different waveforms by a resistor capacitor may be further refined by charging the capacitor using a constant current source and then discharging the capacitor using another constant current source. This technique improves the linearity of the slopes of the triangular wave generator. The frequency is determined by the selected range capacitor and the charging currents, which can be influenced by a control voltage input derived from the frequency dial setting and the sweep circuit inputs. A simplified block diagram of such a function generator is shown in Figure 8.3(a).

In the kind of function generator that includes the ability to sweep the frequency, the sweep oscillator generates a slowly varying ramp, which, in turn, is applied to the block that contains the controlled current sources. The controlled current sources are hence changed in such a way that the main oscillator frequency is changed between the start and stop values set by the sweep block. The principle of using switched current sources with integrating capacitor in a function generator might be further improved to include AM/FM/Ext. modulation facilities using an independent modulation generator. Furthermore an ADC block and a memory can be used to display the amplitude and frequency of the function generator output [1].

8.2.1.1 Low cost function generators

Most of the inexpensive (less than $50) function generators or generator kits in mail order catalogues are based on single chip function generators such as the Intersil 8038, which provides sine, square and triangle outputs. Another such waveform generator IC is the MAX 038 from Maxim Integrated products.

The MAX 038 is a precision high frequency function generator that produces accurate sine, square, triangle, sawtooth, and pulse waveforms with a minimum of external components. The internal 2.5 V reference (plus an external capacitor and potentiometer) allows varying of the signal frequency from 0.1 Hz to 20 MHz. An applied ±2.3 V control signal varies the duty cycle between 10 per cent and 90 per cent, enabling the generation of sawtooth waveforms and pulse-width modulation.

A second frequency control input used primarily as a VCO input (in phase locked loop applications) provides ±70 per cent of fine control. This capability also enables the generation of frequency sweeps and duty cycle controls to have minimal interaction with each other. Figure 8.4 depicts the MAX 038's block diagram and operating circuit [2].

8.2.1.2 Use of microprocessors and custom ICs for function generation

Most modern function generators are designed using custom ICs for waveform generation and a microprocessor for overall control of the instrument. For example, the HP8116A, a function generator with 50 MHz maximum frequency, uses three custom ICs for the basic waveform generation functions and the overall control is based on an MC6802 microprocessor. The simplified block diagram is shown in Figure 8.5.

The microprocessor controls the operation of the instrument by reading inputs from the front panel keyboard or an interface such as an IEEE-488 bus and sending appropriate data to DACs which control the generator hardware. It also updates the front panel LEDs and display in response to the keyboard and IEEE-488 inputs. The generator hardware contains three custom ICs. They are voltage controlled oscillator (VCO) IC, timing IC and a shaper IC. The VCO (or the slope) IC is a triangular waveform generator up to 50 MHz designed around the basic principle similar to previous cases. The output could be continuous, gated or triggered. The timing IC is used as a triggerable pulsewidth generator and a trigger source for an internal burst mode. For pulsewidth modulation, the external signal applied at the control input controls the pulsewidth. A shaper IC is a linear pre-amplifier and triangle-to-sine converter. Another practical example is the HP model 8904A multifunction synthesiser based on a custom IC using DSP techniques [3].

8.3 Pulse and data generators

Although function generators are capable of producing square waves, many digital and some analogue systems require square waves with varying duty cycles and fast rise times. Pulse generators provide these waveforms. A basic pulse generator provides square wave with a variable period or repetition rate and variable duty cycle. The fastest units are further characterised by a jitter specification that indicates uncertainty or variation in the position or period of a pulse. Pulse generators with rise times of 1 and 2 ns are fairly common. Pulse generators should also be able to produce output logic levels compatible with standard logic families. One useful feature is a programmable

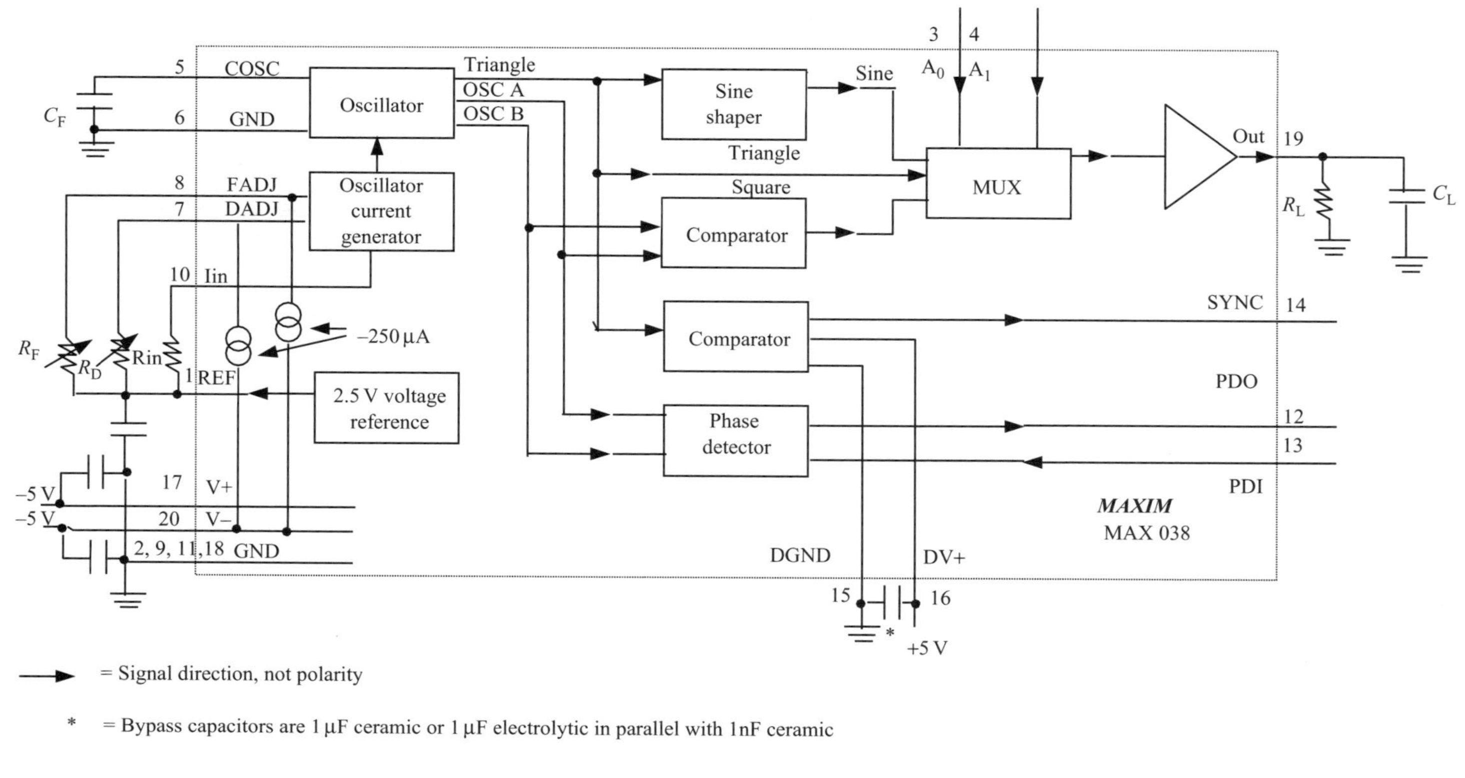

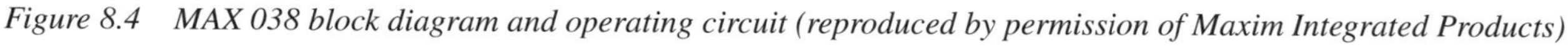
Figure 8.4 MAX 038 block diagram and operating circuit (reproduced by permission of Maxim Integrated Products)

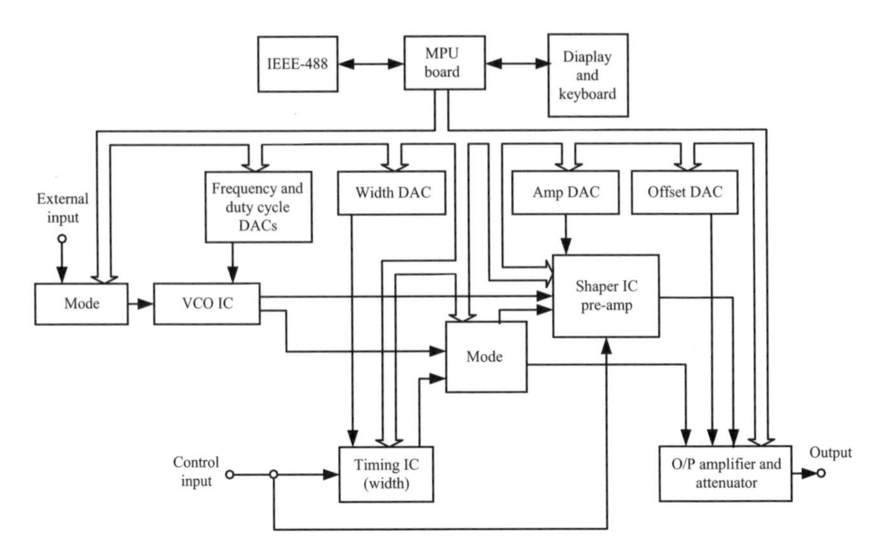

Figure 8.5 Simplified block diagram of a microprocessor controlled function generator using custom ICs (reproduced by permission of Hewlett-Packard)

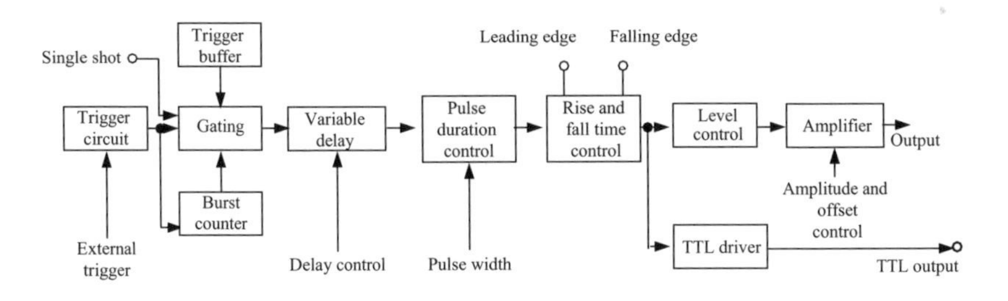

Figure 8.6 Block diagram of a pulse generator

output level that allows the users to evaluate speeds of circuits as a function of the input signals or logic level of the input signals.

The cost of pulse generators increases with the rise time: the faster the rise time, the more expensive the instrument. The key evaluation parameter is that the pulse generator must be faster than the rise time of the device under test (DUT). A basic block diagram of a pulse generator shown in Figure 8.6. As shown, a pulse generator usually has the single shot capability, baseline offset, are triggered operation as well as TTL compatible outputs in general.

Traditional pulse generator applications focus on the analogue response of a circuit. By running a pulse through a circuit, a designer can measure characteristics like group delay. But pulse generators are also useful for evaluating digital circuits, especially at high speed, where maintaining signal integrity is difficult. In such applications, users may have to stimulate the circuit with marginal clock signals to test the design's tolerance to variations in the clock.

In many applications, engineers need both a pulse generator and its cousin, a data generator, to evaluate fully their fast, complex digital circuits. Pulse generators generally provide only a periodic pulse waveform. They are unable to supply a string of one and zeros that form a data stream needed to exercise sequential logic or initialise a circuit under test. Such tasks generally fall to a data generator. Although some people describe a data generator as a pulse generator with memory (to store the patterns needed), data generators typically do not provide the same level of control over pulse parameters as do pulse generators.

Data generators come in two types, serial and parallel. The terms word generator and pattern generator are often used to denote parallel data generators, but there is no standard definition. The serial version generally has one or two lines supplying serial outputs. A parallel data generator may have 16, 32, or more, lines clocking out wide words.

Pulse instruments useful in characterising semiconductor components naturally demand higher frequencies or repetition rates and, more importantly, faster rise times. Rise times of 1 ns, near the top of the performance range, may not provide the precision and resolution needed for some characterisation jobs. For instance, a system with a clock rate of 50 MHz (a 20 ns period) may have transition times of about 1 ns. But for characterisation purposes, engineers want performance at least 10 times that of the system under test. That translates to an instrument rise time of 100 ps. As a result, a 3 GHz pulse generator with 100 ps rise time may be necessary to evaluate 50–100 MHz systems.

Pulse generators are fundamentally analogue boxes and they are still based on RC circuits that generate the timing parameters. A pulse generator provides lot of range and a lot of control over timing parameters but has the disadvantage of not providing the stability of a digital synthesiser. In some pulse generators built-in calibrators are used to monitor the timing parameters and compensate for the inaccuracies of the analogue circuitry; see also Reference 4.

8.4 Signal generators

Although fast signals are useful in characterising high speed digital systems, the RF and microwave world requires high frequency sinewave instruments that produce sinewaves from a few kHz to several GHz, with modulation.

Generation of sinewaves for RF and microwave applications can be divided into two primary cases. These are (i) synthesis in the frequency domain, and (ii) synthesis in the time domain.

8.4.1 Frequency domain synthesis

The techniques listed below fall into the category of frequency domain synthesis.

- Tuneable free running LC or RC oscillators.
- Switchable crystal oscillators.

- Frequency domain synthesis (addition and subtraction) based on a stable set of references.
- PLL using VCXO, LC or RC oscillator and variable divider.
- Harmonic selection.
- Any combination of the above techniques.

For a detailed account of these techniques References 5–8 are suggested.

Older, cheaper designs still use LC oscillators. These lack keypads and digital readouts, but often have a better spectral purity than the newer synthesised units; moreover, they are easier to use. Some of these generators have almost no measurable spurious outputs.

Current terminology divides the new units into two classes: RF synthesisers and synthesised signal generators. Both types use digital synthesisers to generate the RF waveform but differ in features. A synthesiser produces an RF output but lacks internal modulation capabilities. Synthesised signal generators, on the other hand, are RF synthesisers with added modulation capabilities including amplitude, frequency and phase.

Frequency synthesis in the frequency domain may be divided into two basic types: namely, direct synthesis and indirect synthesis. The direct synthesis method uses a reference crystal oscillator, the frequency of which is divided and multiplied, passed through filters and synthesised using mixers for each frequency components; see Figure 8.7. The advantages of direct synthesis techniques are high signal purity and fast frequency switching while the disadvantages include the difficulty of suppressing the generation of spurious signals caused by the mixer and higher cost. Details of these methods are to be found in References 5 and 7.

Modern synthesised signal generators use indirect digital frequency synthesis where a voltage-controlled oscillator (VCO) is phase locked to a reference oscillator;

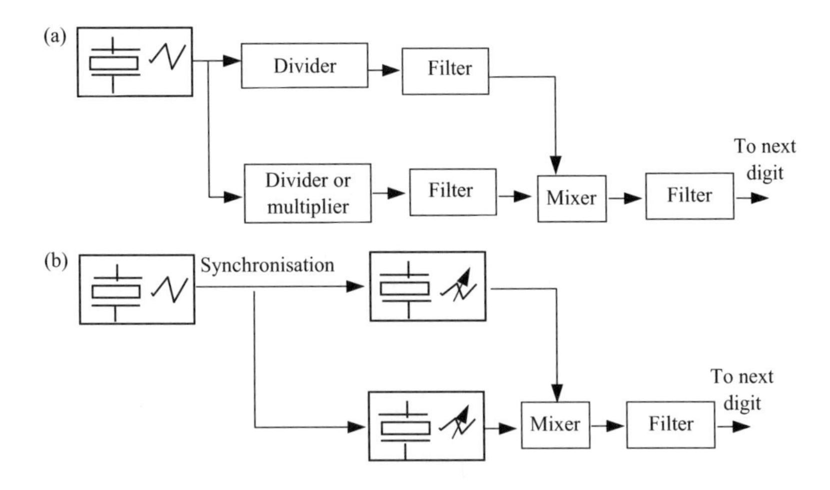

Figure 8.7 Direct frequency synthesis: (a) using multipliers and dividers; (b) using mutually synchronous oscillators

see Figure 8.8. The type of synthesiser design used and the accuracy of the reference oscillator determine the degree of frequency resolution, the amount of phase noise and the cost. The reference oscillator is usually a stable crystal controlled oscillator in a temperature controlled oven.

8.4.2 Time domain synthesis: direct digital synthesis

With the advancement of high speed digital circuits, it is now possible to consider the synthesis of waveforms directly in the time domain (i.e. as waveforms) rather than in the frequency domain.

To understand the principle of DDS systems let us refer to the simple circuit in Figure 8.9. Each location in the PROM corresponds to a discrete sample of the

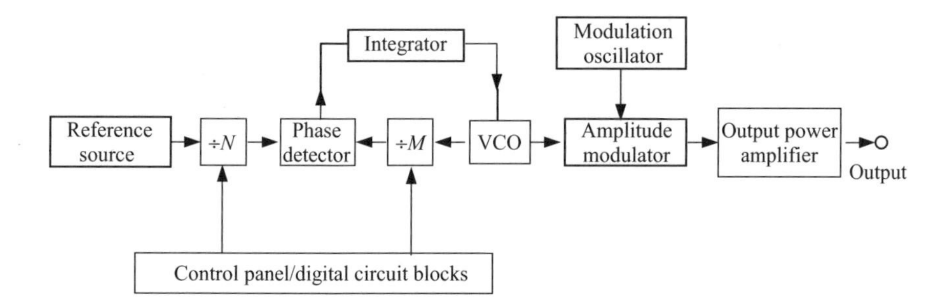

Figure 8.8 Indirect digital frequency synthesis used in signal generators with amplitude modulation capability

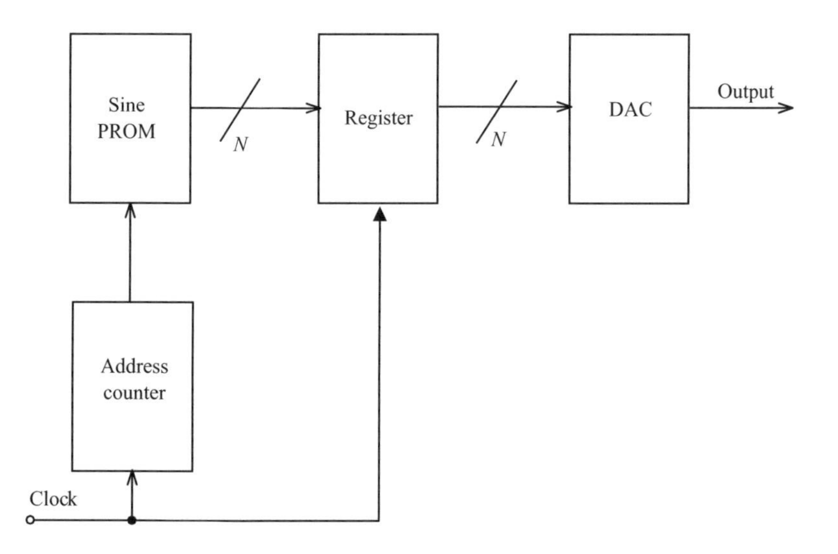

Figure 8.9 A simple DDS system

sinewave. The PROM must contain an integral number of cycles in order to prevent a discontinuity when the PROM rolls over. This approach is limited, however, because the sinewave frequency can only be changed by varying the clock rate or by reprogramming the PROM.

A much more flexible scheme is shown in Figure 8.10 and is the basis of modern DDS techniques. The circuit driving the DAC is often referred to as a numerically

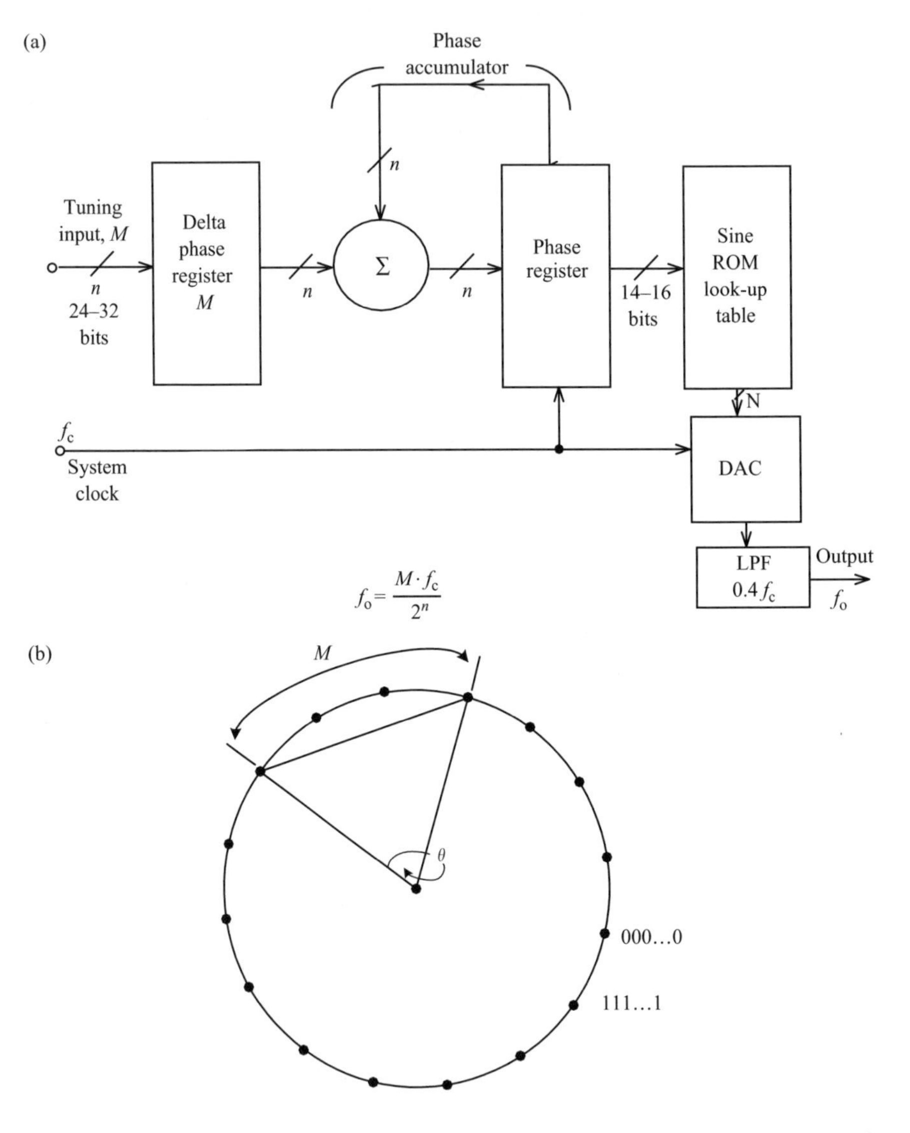

Figure 8.10 A more flexible DDS system and phase circle: (a) block diagram; (b) phase circle

controlled oscillator (NCO) and serves the same function as the PROM in the simple DDS system described above.

In order to understand the system, first consider a sinewave oscillation as a vector rotating around a phase circle as shown in Figure 8.10(b). Each point on the phase circle corresponds to a particular point on the output waveform. As the vector travels around the phase circle, the corresponding output waveform is generated. One revolution on the phase circle corresponds to one cycle of the sinewave. A phase accumulator is used to perform the linear motion around the phase circle. The number of discrete points on the phase circle is determined by the resolution of the phase accumulator. For an n-bit accumulator, there are 2^n points on the phase circle. The digital word in the delta phase register (M) represents the '*jump size*' between updates. It commands the phase accumulator to increase by M points on the phase circle at each time the system is clocked. If M is the number stored in the delta phase register, f_c is the clock frequency, and n is the frequency of rotation, then the frequency of rotation around the phase circle (the output frequency) is given by

$$f_o = M \cdot f_c/2^n. \tag{8.1}$$

This is known as the '*tuning equation*'. The frequency resolution of the system is $f_c/2^n$ which represents the smallest incremental frequency capable of being produced.

The delta phase register and the accumulator are typically 24–32 bits wide. Table 8.1 indicates the number of points vs M. The output of the phase accumulator drives the address input of a sinewave ROM look-up table in which is stored amplitude information for exactly one cycle of a sinewave. The phase data is usually truncated in order to minimise the size of the ROM and the resolution of the DAC. The phase resolution (corresponding to the number of locations in the ROM) directly affects the spectral purity of the output. For example, if the phase information is truncated to 15 bits, the theoretically largest phase spur is about 90 dB below full scale when the DAC spurs are neglected.

Table 8.1 Number of bits in the delta phase register vs number of points in the phase circle

Number of bits in the delta phase register (n)	Number of points
8	256
12	4 096
16	65 536
20	1 048 576
24	16 777 216
28	268 435 456
32	4 294 967 296

State-of-the-art chips realised in gallium arsenide now provide an output frequency of over 400 MHz. Silicon technology delivers output frequencies in excess of 300 MHz. For these component families, when operating at the top end of their frequency range, the worst case levels of spurious outputs are in the region of 40 or 45 dB below the wanted output. For details, References 9–13 are suggested.

8.4.3 Key signal generator specifications

The key specifications for signal generators fall into six major categories: (i) frequency, (ii) output, (iii) modulation, (iv) spectral purity, (v) control, and (vi) switching speed. Out of these, spectral purity is one of the most important specifications for the signal generators used for communications. The following sections provide a summary, while Reference 14 is suggested for further details.

8.4.3.1 Frequency

Frequency specifications start with the range covered by the signal generator. Some generators use doublers or triplers to extend their frequency range, often with some degradation (subharmonics, spectral purity) in specifications. In addition to the frequency range, manufacturers specify the resolution of the dials or digital readouts of their signal generators.

The frequency accuracy of a signal generator depends on the accuracy and ageing rate of the reference oscillator. The reference oscillator is usually a crystal controlled oscillator in an oven, but it may also be a direct digital synthesis oscillator. Some generators also allow use of an external reference oscillator, which is not only useful for providing a more accurate reference, but also allows the phase of two or more signal generators to be synchronised by using a single reference source. This feature is useful when checking mixers.

8.4.3.2 Output level

A signal generator's output level is calibrated over a typical range, such as +13 dBm to −145 dBm into a characteristic impedance of 50 or 75 Ω. High level passive mixers may require a local oscillator drive levels as high as +16 dBm, while receiver sensitivity tests may require power levels at least as low as −120 dBm. The signal generator's output level is usually measured at the factory with correction factors programmed into an on-board EPROM to maintain accuracy in the field.

8.4.3.3 Modulation

Signal generators usually feature built-in amplitude, frequency and phase modulation. Pulse modulation usually comes as an option. A built-in modulator requires a built-in signal source. Hence, signal generators themselves contain additional signal sources, usually a low frequency oscillator with its own amplitude, frequency and distortion specifications. These low frequency oscillators typically have a range from 10 Hz to 100 kHz.

Each modulation mode itself has specifications. Amplitude modulation, for example, is characterised by such specifications as modulation depth, bandwidth, distortion, depth accuracy and incidental FM. The modulation depth is usually specified as a percentage. For FM, the criteria are peak deviation, bandwidth, distortion and incidental AM.

D.C. coupling adds to the usefulness of both amplitude and frequency modulators. For AM, d.c. coupling allows the carrier to be swept by an external or internal source. This method permits swept frequency characterisation of narrow band filters. Phase modulation is similar to frequency modulation; the key specifications are peak deviation, bandwidth and distortion.

8.4.3.4 Spectral purity and Electromagnetic Interference (EMI)

Although versatile and packed with features, synthesised signal sources have a major drawback, namely, lack of spectral purity. For example, synthesised signal generators generally suffer from the phase noise generated in their synthesisers. In addition, the numerous dividers and associated digital circuitry can generate spurious signals that diminish the purity of the spectral output.

Figure 8.11 shows the spectral output of a synthesised signal generator which has phase noise, harmonics and spurious emissions that affect the spectral purity. Note that the output of a synthesised signal generator is not the ideal case of a single peak at the output frequency, but rather one peak and a bunch of little peaks. Phase noise (specified as dBc MHz^{-1}) shows up when the signal begins to spread at the base of the carrier output.

8.4.3.5 Control

Synthesised signal generator manufacturers have adopted the keypad and digital readout as the standard way of controlling synthesised signal generators. The process starts by using one set of keys to program the numerical value of a parameter (say, 2.885), then another set of keys to select the parameter (MHz), for example. For automated

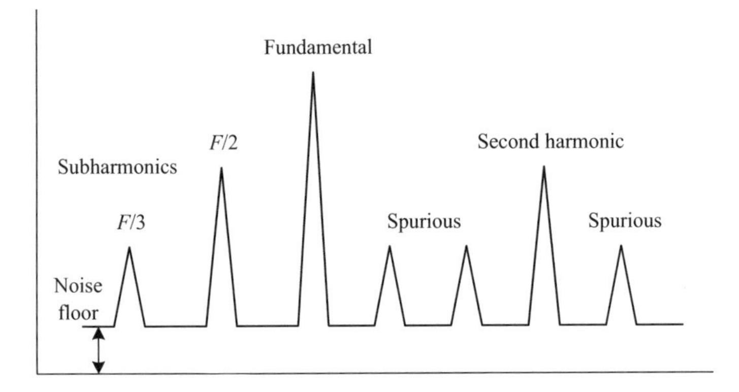

Figure 8.11 Generalised spectrum of a synthesised signal source

control, some synthesised signal generators allow storage of front panel set-ups, command sequences or both. Many now offer a feature called digital sweep that allows a user to select upper and lower frequency limits, a step size and the number of steps. When operating under IEEE-488 control consider whether the interface is standard or an option and whether the signal generator can operate as a talker or a listener.

8.4.3.6 Switching speed

Switching speed can best be described as the time from the initiation of a command until the output of the synthesised signal generator reaches the required parameters. Although switching speed is treated separately it is really a subset of control specifications. One of the problems with switching speed specifications is that different manufacturers use different methods to define the beginning and end of this interval.

Ideally the time should include the data transfer time on the control interface (such as the IEEE-488 bus), the signal generator's internal switching times and the time the output takes to settle to its desired parameters. The problem with non-standard specifications for settling times is that a signal generator may settle to within 100 Hz of the final frequency in 100 ms but take another 300 ms to settle within 50 Hz.

8.5 Microwave signal sources

Modern microwave signal sources have a broad range of capabilities. Two to three GHz sources will cover most telecom product and wireless LAN testing needs. Satellite communications and microwave links, etc., commonly operate over narrow bandwidths at frequencies up to about 30 GHz. The frequency range is often the main factor in the base price of an instrument – the greater the range, the higher the price. Generally the price range varies from US$3000 to nearly $50 000. Generally the lower frequency end of the instrument must not be over-specified to cut the costs down. Table 8.2 provides a representative collection of microwave sources.

Power output of commercial microwave sources varies from about -144 dBm to over $+17$ dBm. On many occasions it is possible to buy accessories to boost the output by 6–10 dB rather than selecting a more powerful signal source. The raw unlevelled output of a basic source may be adequate if the user needs a constant, but not an absolute, value signal. Many options in the most complex microwave sources often reduce the output power level by as much as 10–15 dB because the modulation process attenuates the carrier that it modulates.

Most commercial products have sweep capabilities, both in frequency and power level. Most sources still have analogue sweep capability and when absolute sweep resolution is needed (no steps) it is the only choice.

When precise control over frequency or level, step size, and dwell time are required, one should specify a digital sweep capability. With digital sweep, one can define such a small step size that the sweep seems continuous, but the output frequency requires between 10 ms and 25 ms to 'lock' after each step. So we cannot have a smooth sweep and a fast sweep at the same time. If you must have both the precision of digital sweep and the sweep and smoothness of analogue sweep, you must select an instrument that has both analogue and digital sweep capabilities.

Most analogue instruments have always had simple modulation capabilities, such as AM, FM, PM (pulse modulation) and ϕM (phase modulation). In modern instruments an increasing variety of complex modulation techniques forces suppliers of signal sources to include new digital modulation capabilities.

A digital modulation capability generally comes with only a modest increase in price. But signal sources that rely on digital modulation techniques offer an advantage over their analogue counterparts: manufacturers can upgrade their digital instruments' modulation capabilities through software often in the field. Also, many sources have an external modulation input that can accommodate complex signals; even sources without a digital modulation capability may effectively produce an output with digital modulation using the external modulation input.

Spectral purity and noise specifications are often confusing because manufacturers state them in so many different ways. Fortunately, most instruments perform well enough that you will not have to delve into evaluating or comparing these specifications in most applications. But a few specifications are standard, so it is worthwhile for the user to be familiar with them; see also Reference 15.

Low phase noise is critical in applications such as high frequency (10–20 GHz) digital clocks, and Doppler radar for the detection of extremely slow moving targets for electronic countermeasures (ECM). Most microwave sources have a specification for single sideband phase noise at a 20 kHz offset from centre frequency. For most sources, values range between −100 dBm and −120 dBm. Some lower priced instruments have readings of about −80 dBm, while the more expensive ones have readings of −130 dBm.

Harmonics, subharmonics, non-harmonics, and residual FM and AM are generally inconsequential. If they are important in a given application, though, the user should be aware that most sources seem to have harmonic readings of −30 dBc for output frequencies around 2 GHz. Users can also use external filters to attenuate harmonics specifications.

Most sources have standard communications interfaces, with most having both IEEE-488 and RS-232C ports. Reference 16 provides some details on microwave sources.

8.6 Arbitrary waveform generators (AWG)

An arbitrary function generator (AFG), arbitrary waveform generator (AWG), or sometimes called an ARB, provides the user with the ability to generate custom waveforms. Direct digital synthesis (DDS) is the basis for AFGs. Compared with a modern signal generator based on the DDS technique, the look-up table in an AFG contains the information necessary to create an arbitrary wave shape. As such, it is ideally suited to automated testing and circuit characterisation during the design or test phases. For example, AFGs can simulate such waveforms as encoded radar signals, data signals from magnetic disks, mechanical vibration transients, video signals and

Table 8.2 A representative set of microwave signal sources (Courtesy of Test & Measurement World magazine)

Company	Model	Frequency range	Power output	Modulation	Resolution
April Instrument	8001	2–8 GHz	> +10 dBm	ext. FM	1 MHz
	8002	2–10 GHz	> + 10 dBm	ext. FM	1 MHz
	8003	2–20 GHz	> + 10 dBm	ext. FM	1 MHz
Anritsu Wiltron	69147A	10 MHz–20 GHz	−120 to + 17 dBm	AM/FM/ext. PM	0.1 Hz/0.01 dB
	69247A	10 MHz–20 GHz	−120 to + 17 dBm	AM/FM/PM/ΦM	0.1 Hz/0.01 dB
	69347A	10 MHz–20 GHz	−120 to + 17 dBm	AM/FM/PM/ΦM	0.1 Hz/0.01 dB
EIP Microwave	1140A	10 MHz–20 GHz	−90 to + 13 dBm	AM/FM/ext. PM and complex ones	0.1 Hz/0.1 dB
Giga tronics	GT9000/2-26	2–26.5 GHz	−120 to + 11 dBm	AM/FM/PM	0.1 Hz/0.01 dB
	GT9000S/2-26	2–26.5 GHz	−120 to + 11 dBm	AM/FM/PM	0.1 Hz/0.01 dB
	50220A	2–20 GHz	−100 to + 10 dBm	AM/FM/PM	0.1 Hz/0.1 dB
Hewlett-Packard	ESG-D4000A	250 kHz–4 GHz	−136 to + 7 dBm	AM/FM/PM/ΦM DQPSK/QPSK/I-Q NADC/PDC/PHS GSM/DECT/TETRA	0.01 Hz/0.02 dB
	8665B	100 kHz–6 GHz	−139.9 to + 13 dBm	AM/FM/PM	0.1 Hz/0.01 dB
	83650B	10 MHz–50 GHz	−110 to + 10 dBm	AM/FM/PM, ScanAM	0.1 Hz/0.02 dB
Marconi Instruments	2032	10 kHz–5.4 GHz	−144 to + 13 dBm	AM/FM/PM/ΦM	0.1 Hz/0.01 dB
	2042	10 kHz–5.4 GHz	−144 to + 13 dBm	AM/FM/PM/ΦM	0.1 Hz/0.01 dB
	2052	10 kHz–5.4 GHz	−144 to + 13 dBm	AM/FM/PM/ΦM DQPSK/QPSK/I-Q NADC/PDC/PHS GSM/DECT/TETRA GMSK	0.1 Hz/0.01 dB

Programmed Test Sources	PTS 3200DJT10	1 MHz–3.2 GHz	+3 to + 13 dBm	Digital ΦM option	1 Hz/1 dB
Tektronix	R&S SMT06	5 kHz–6 GHz	−144 to + 13 dBm	AM/FM/PM/ΦM/Stereo Multiplex/VOR/ILS	0.1 Hz/0.01 dB
	R&S SME06	5 kHz–6 GHz	−144 to + 13 dBm	AM/FM/PM/ΦM/Stereo Multiplex/VOR/ILS FSK/GFSK/GMSK/QPSK and Paging	0.1 Hz/0.01 dB
	R&S SMP04	10 MHz–40 GHz	−130 to + 10 dBm	AM/FM/PM/ΦM/Stereo Multiplex/VOR/ILS FSK/ASK and chirp, fading and Doppler simulation	0.1 Hz/0.01 dB

Table 8.3 Comparison of AWGs and conventional waveform generators

Conventional function generators	Arbitrary function generators
Standard waveforms only	Any wave shape can be obtained
Design is predominantly analogue	Design is digital
Wave shape cannot be stored	Wave shape can be stored
Specifications are analogue based	Specifications are digital based
Wave shape adjustment under computer control is generally not possible	Wave shape adjustment under computer control is possible with variety of techniques

a variety of impulses. They can also play back a waveform captured by a DSO, a feature unique to AFGs and impossible with its analogue counterparts. Furthermore, it allows the users to vary the size and shape of a small section of the waveform digitally.

This interesting family of instruments which proliferated during the past decade has entirely changed the way designers have been thinking of creating the waveforms. Table 8.3 briefly compares the characteristics of conventional function generators against AFGs.

AFGs allow the user to emulate virtually any system output and the wave shape is created by a system based on processors and memories that therefore provides the programmability, repeatability and reliability synonymous with today's digital technology. Situations for using an AFG may be as listed below.

- Need for non-standard real world stimulus.
- Need for an ATE programmable stimulus.
- When the repeatability of memory storage is important.

A traditional function generator is based on RC circuits and analogue techniques. The whole concept of waveform creation inside an AWG is therefore totally different from the waveform generation concepts used in traditional signal sources.

The first commercially available arbitrary waveform generator, the model 175 from Wavetek, was introduced in 1975. By today's standards it was a modest performer with a maximum clock rate of around 3 MHz and a mere 256 bytes of waveform memory.

8.6.1 Fundamentals

Having discussed the fundamentals related to DSOs, AWGs can be considered as the inverse implementation of the DSO. In a DSO we receive the signal, and pass it through an ADC which converts the signal into a set of digital values and stores in a semiconductor memory. In the display process we do the reverse to have the waveform shape displayed on the CRT or the LCD screen. In its most basic form

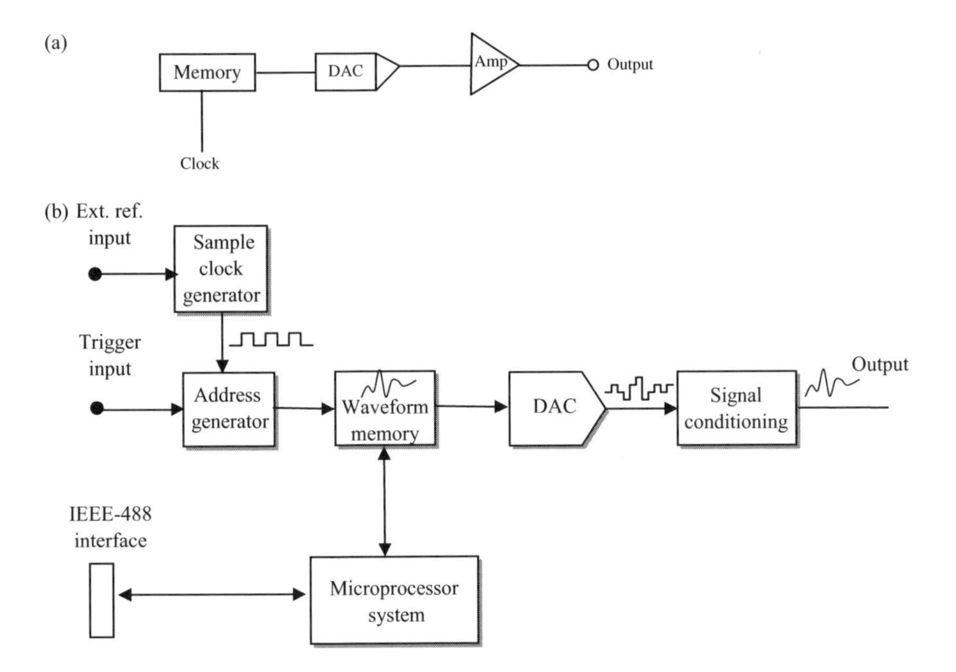

Figure 8.12 Basic concept of arbitrary waveform generators: (a) basis of implementation; (b) practical implementation

AWG can be conceptually considered as shown in Figure 8.12(a). The same concept can be elaborated into a more practical block diagram as shown in Figure 8.12(b).

As shown in Figure 8.12(a), in an AWG the user stores the waveform in the memory, clocks it and transfers it to a DAC and then passes it through a filter to have the desired output waveform. This concept indicates that the signal and its shape are governed by several fundamental items such as

- sampling rate,
- size of memory representing the number of points (time samples),
- vertical resolution and DAC characteristics, and
- signal conditioning blocks and filters, etc.

8.6.1.1 Sample rate and number of points per circle

With respect to AWGs, the Nyquist theorem also applies to the sample rate. Figure 8.13 shows the highest signal repetition rate possible from an AWG with a 1 kHz sampling rate (1 ms sample^{-1}). Because you need at least two points to create a complete cycle of the signal you must divide the sampling rate by two to get the highest possible signal repetition rate. In this case 500 Hz is the best achievable. Figure 8.13(a) shows a signal that is a square wave, but the bandwidth of the AWG may distort the signal, particularly at the corners, so it may look more sinusoidal than square.

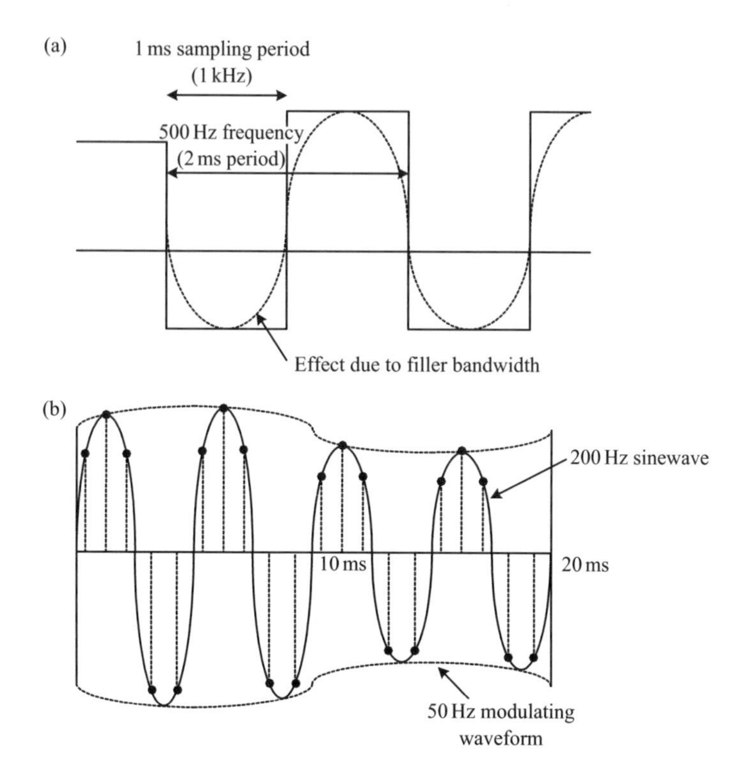

Figure 8.13 Sample rate and signal repetition rate: (a) clock rate and Nyquist criteria; (b) an amplitude modulated signal and sample rate

Let us now assume that we need to create a more complex signal such as an amplitude modulated signal with a 50 Hz sinewave as the modulation signal (Figure 8.13(b)). With a 1 kHz sample rate, you need 20 samples to get one cycle of the modulation signal. Assume that 5 points per cycle is sufficient for the carrier. So, with a 1 kHz sample rate, the maximum carrier frequency is 1 kHz/5, or 200 Hz. However, the repetition rate for the modulation signal is 1 kHz/20, or 50 Hz. This indicates that, to calculate the repetition rate for any signal, the following formula can be used

$$\text{signal repetition rate} = \frac{\text{sample rate}}{\text{number of points per cycle}}. \quad (8.2)$$

8.6.1.2 Memory

As equally important as the sample rate is the size of the AWG's memory. It is the memory, or lookup table, that contains the actual output signal amplitude. The most obvious way to set up an AWG's lookup table is point-by-point, from one of the output sample signal to the other. When a traditional AWG is used, user could program the instrument with an amplitude for each point in the waveform, then set the sample rate. In an AWG based on direct digital synthesis (DDS), the sample rate is fixed, so the number of samples used to build a waveform changes.

Memory depth places limitations on the complexity and time duration of a waveform. For example, consider an AWG being used to simulate speech in telephony. Since speech is limited in telephony to about 4 kHz in bandwidth, a sample clock of 8 kHz is adequate to play out speech with good fidelity. But if the memory depth of the AWG is 8000 points, for example, then about one second of speech can be stored:

$$\text{speech duration} = (125\,\mu\text{s point}^{-1}) \times (8000\text{ points}) = 1\text{ s}.$$

There are two types of memory architecture used in AWGs: direct and multiplexed. In a direct memory system (the simpler of the two) there is a one-to-one correspondence between an address presented to the waveform memory and a waveform point in the memory. Direct memory architecture has limited speed. For example, if a clock rate of 100 MHz is desired, then the access time of the waveform must be less then 10 ns. The lower the access time of a memory chip, the more expensive it is.

For reasons of cost minimisation and increased speed, multiplexed memory architectures could be used as shown in Figure 8.14. This example indicates a case for 100 MHz.

The four RAMs driven by one address generator allow the address generator to be operated at one-quarter of the sample clock frequency. The access time requirement of each RAM is therefore increased to four times the sample clock period, or 40 ns. Increasing the access time means that the RAMs can be denser, lower cost, and lower power. For the system shown in Figure 8.14, 128 kpoints of waveform memory are available for about the same power and cost as 16 kpoints of memory in a direct memory system.

There are some disadvantages to using multiplexed memory. One disadvantage is that waveform size must be in increments of the multiplex ratio. In other words, with a multiplex ratio of 4 : 1, waveform size will always be in multiples of four. In practical applications this limitation is usually not a problem. Another disadvantage to the multiplexed memory architecture is that waveform access is no longer random. The waveform must be played out from adjacent memory locations, which implies that the address increment must be equal to one. This presents a problem with direct digital synthesis (DDS) techniques (discussed later).

8.6.1.3 Vertical resolution, DAC output and quantisation

In an AWG, vertical resolution determines how accurately you could design the waveform with respect to its amplitude. For example an n-bit system allows you to distinguish amplitude changes for $V_{max}/2^n$, where V_{max} is the maximum amplitude of the waveform. In practical instruments vertical resolution varies from 8 to 16 bits.

The digital-to-analogue converter (DAC) in an AWG converts the digital representation of the waveform to voltage. Because the digital input to the DAC is a series of discrete numbers, the output of the DAC is a series of discrete voltages. In other words, the DAC output is quantised. Because the digital value has a specific

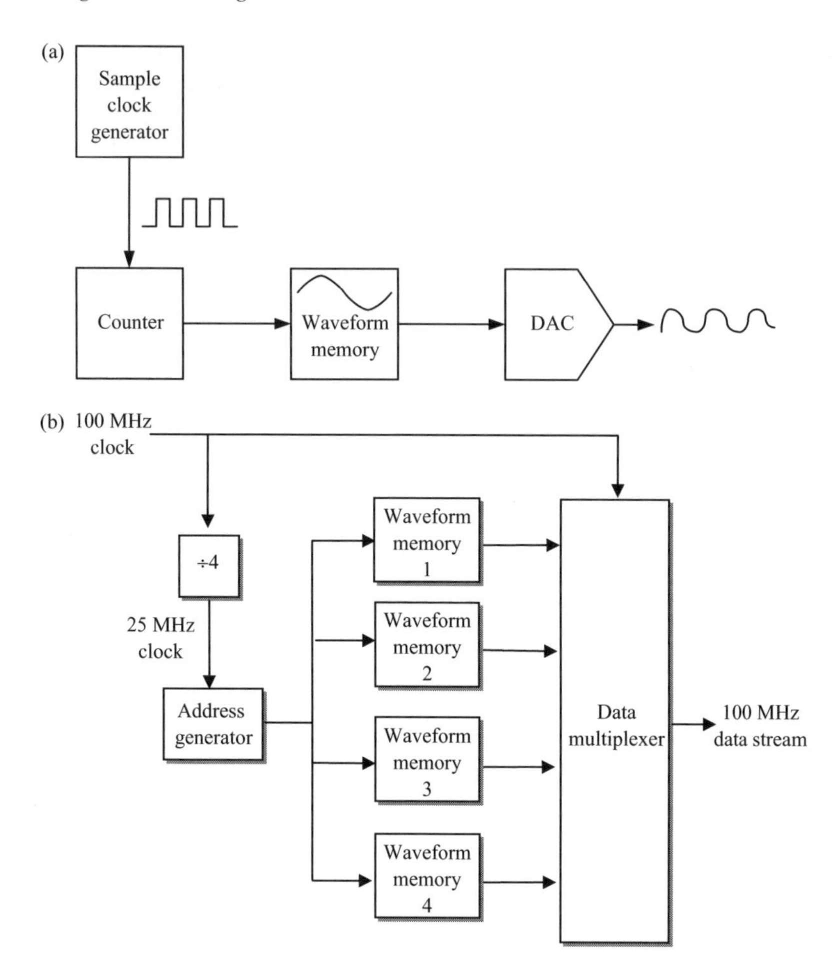

Figure 8.14 Memory architectures in arbitrary waveform generators: (a) simple memory; (b) multiplexed memory (reproduced by permission of Fluke Inc., Wavetek Instruments, USA)

resolution, the voltage output has the same resolution (or quantisation). For example, an 8-bit DAC has 2^8 (or 256) possible input codes and therefore has the same number of possible output voltages. The output of the DAC is also quantised in time. The DAC output could change from one voltage to the next only at discrete times defined by the sample clock. As an AWG is quantised in time and voltage, it approximates a continuous signal. DAC characteristics such as quantisation error, linearity and other parameters are discussed in chapter 3. DAC characteristics determine the overall quality and characteristics of the AWG output. For details, Reference 13 is suggested.

As the resolution of the DAC and the speed of the sample clock increase, the quantisation error decreases. The user of an AWG must know if the approximation is

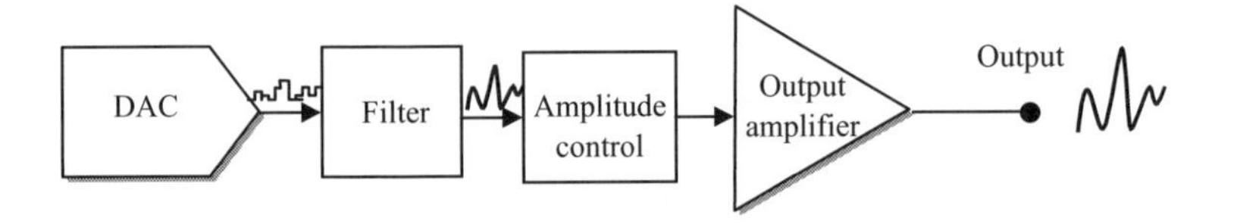

Figure 8.15 Signal conditioning in AWGs

close enough to the desired signal so that test results are not adversely affected in a given application.

8.6.1.4 Signal conditioning

Once the digital representation of a waveform has between converted to an analogue signal by the DAC, additional processing or conditioning is required before the signal can be used. A typical signal conditioning system is shown in Figure 8.15.

Various types of filtering are often desirable to remove high frequency spurious signals that result from the quantisation at the DAC output. Filter types and bandwidths are often selectable to allow the user to tailor the filtering performance to the application.

Amplitude control is necessary to provide maximum vertical resolution for all amplitudes of arbitrary waveforms. The amplitude of an arbitrary waveform could be controlled by defining the waveform to use more or less of the full scale range of the DAC, but this would cause low amplitude signals to be defined with very poor vertical resolution. Much better performance can be achieved by always defining arbitrary waveforms to use the DAC's full scale range and then setting the desired output amplitude using the amplitude control.

8.6.2 AWGs as sampled data systems

The sampled data system described in chapter 3 applies to all AWGs. The sampling process has a significant effect on the original continuous signal. If you think of the original continuous signal as information, then the sampling process modifies the original information. In summary, AWG performance is governed by fundamental items such as (i) sampling rate, (ii) size of memory, (iii) vertical resolution (number of bits), (iv) no of points (time sample) used to create the waveform, and (v) output filter characteristics.

8.6.2.1 The spectrum of a sampled signal

The sampling process modifies the frequency spectrum of the original continuous signal in predictable ways. Sampling causes the continuous signal spectrum to be translated to the clock frequency and its entire harmonics. This translation occurs within an amplitude envelope defined by the $(\sin x)/x$ function (Figure 8.16). These translated spectral lines are called sampling spurs or sampling aliases.

Each spectral line in the original continuous signal will have a corresponding pair of sampling spurs mirrored in the clock and its entire harmonics. For example, if a

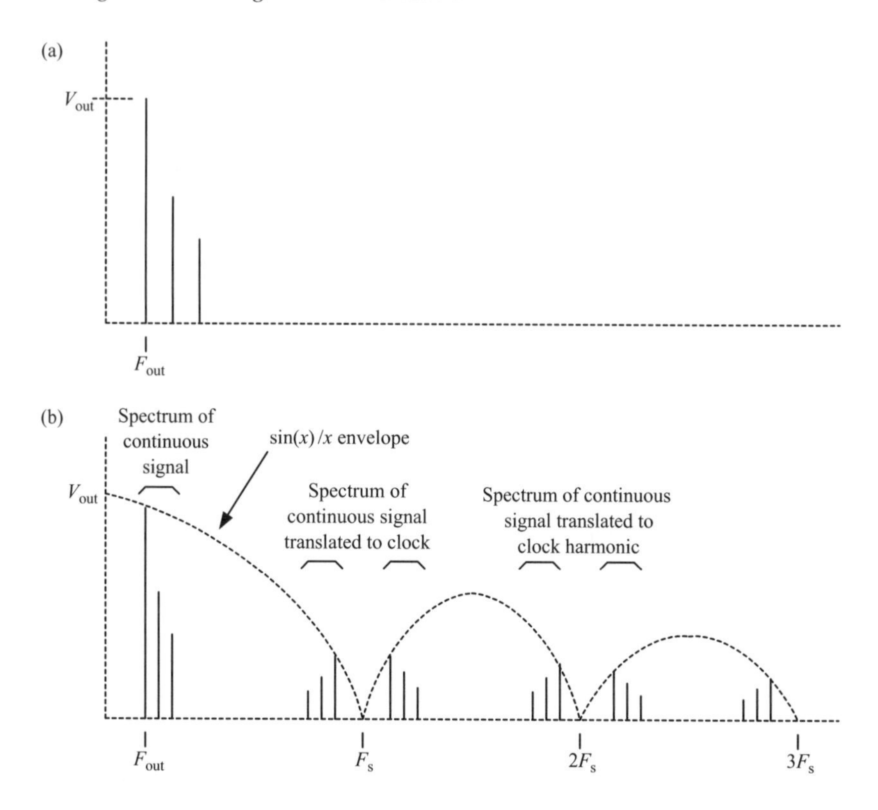

Figure 8.16 Spectrum of continuous and sampled signals: (a) continuous signal with harmonic; (b) sampled signal

spectral line in the continuous signal occurs at 1 kHz and the sample clock is 1 MHz, then a sampling spur will occur at 0.999, 1.001 and 1.999 MHz, and so on. The amplitude of each of the sampling spurs follows the $(\sin x)/x$ envelope determined by the amplitude of the original continuous (unsampled) signal. Notice that the amplitude of the sampled signal fundamental is reduced from the original unsampled signal. As the fundamental moves towards the clock (fewer and fewer samples), its amplitude decreases as it follows the $(\sin x)/x$ envelope.

By comparing the spectrums of the continuous signal and its sampled version in Figure 8.16, it is clear that the sampling process adds unwanted spectral lines to the original signal. The sampled signal could easily be restored, however, by applying a low pass filter that would pass the original signal but filter out all of the sampling spurs. Filter types and their effect are discussed later.

8.6.2.2 Nyquist frequency and inband spurs

As discussed in chapter 3, in a sampled signal as the signal increases in frequency, its first alias decreases in frequency.

When a continuous signal with harmonics is sampled, the sampling process can mirror the harmonics to create 'in-band' spurious signals (signals that are close to the fundamental signal and not easily filtered). Consider a sampling system with a sample clock of 10 MHz. We are going to sample a sinewave with a fundamental frequency of 3 MHz and harmonics at 6 and 9 MHz. Remember, the sampling process will mirror (alias) each of the spectral lines in the clock and 'reflect' them back down in frequency. The frequency of each of the clock alias signals can be predicted by subtracting the frequency of the original signal from the clock frequency. This spectrum is illustrated in Figure 8.17.

The spectrum of the sampled signal now has additional spectral lines at 1 MHz, 4 MHz and 7 MHz. Constructing a filter to eliminate the spurious signals and recover the original spectrum would be extremely difficult and impractical.

In-band spurious signals occur whenever the sampled signal approaches or exceeds the Nyquist frequency. The signal shown in Figure 8.17(a) is relatively simple, with only three spectral lines. Sampling of more complex signals will generate even more unwanted spurious signals. For more details, Reference 17 is suggested.

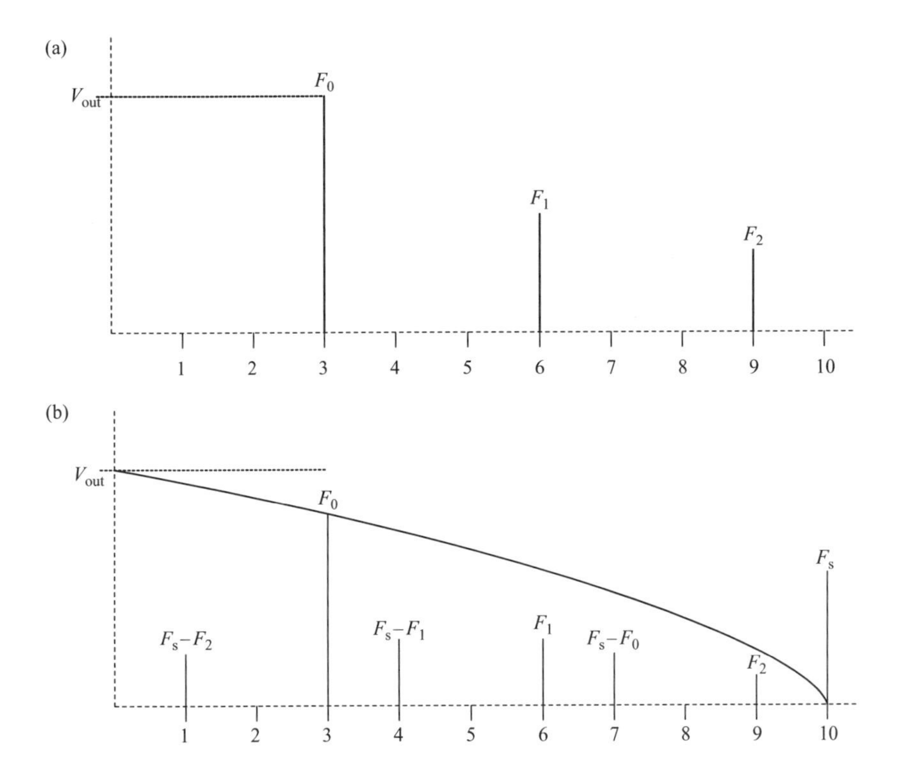

Figure 8.17 In-band spurs and sampling: (a) spectrum of a continuous signal; (b) in-band spurs due to sampling

8.7 Different kind of ARB

Commercial versions of arbitrary waveform generators were introduced in the mid-1970s [17]. With over two decades of technology advancement many different versions have entered the market with various acronyms such on ARB, AWG and AFG. Not all of these instruments that use digital technology to create baseband analogue waveforms are alike, even though all types use clocks, memories, DACs and filters, etc. Two fundamental types of AWG are available today, namely:

(i) traditional AWGs, and
(ii) AWGs based on direct digital synthesis (DDS).

Both types use a clock and a look-up table (the instrument's memory). A traditional AWG uses a variable sample clock and segments of a look-up table. The look-up table contains information that represents the output signal's amplitude, and each location in the table represents one point in a waveform. A DDS based AWG uses a fixed clock, a phase increment register, a phase accumulator and a latch. The traditional AWG requires several low pass filters, while the DDS based AWG needs only one. Compare the two Figures 8.18(a) and 8.18(b).

8.7.1 Traditional AWGs

A traditional AWG's clock increments a counter that activates an address in the table (see Figure 8.18(a)). The counter moves sequentially through the table's addresses until it reaches the end of the segment. The data at each address drive the digital-to-analogue converter (DAC). The DAC's output voltage passes through one of several

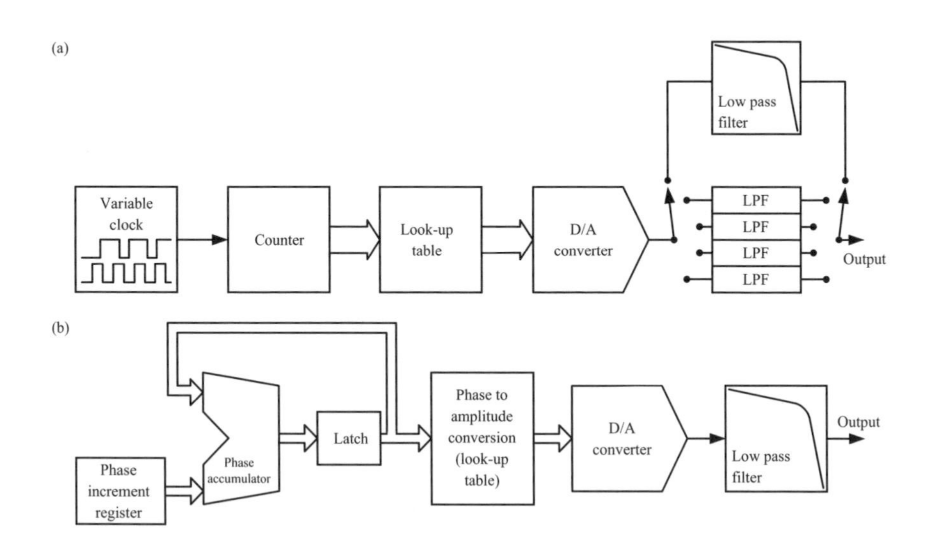

Figure 8.18 Comparison of AWGs: (a) traditional AWG; (b) DDS based AWG

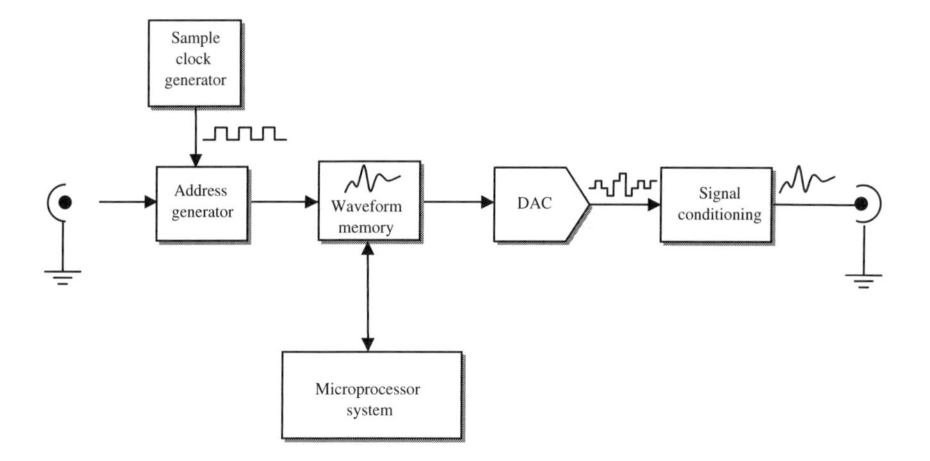

Figure 8.19 Block diagram of a traditional AWG

low pass filters that smoothes the sharp edges created by the DAC's discrete voltage levels.

To set the frequency of waveform, the AWG will adjust its clock rate according to the frequency of the output signal you specify. For example, if you have a 10 kHz signal made up of 250 points, the AWG's sampling rate will be 2.5 MHz. Because the traditional AWG's sampling rate can vary, so must the cut off frequency of its low pass filter. Therefore traditional AWGs have several low pass filters. The filter that is active depends on the sampling rate and the size of the table. If a filter's cut-off frequency does not vary with the sampling rate, then it will fail to filter (if the signal's frequency is too low), or it will attenuate the wanted signal (if the signal's frequency is too high).

An arbitrary waveform generator can best be described as a digital waveform synthesiser. The waveform definition is stored digitally in a random access memory (RAM). The digital representation is converted to an analogue representation by playing out the digital information sequentially into a digital-to-analogue converter (DAC). The speed at which the waveform is played out is controlled by the clock rate (sample clock generator) applied to the address generator (Figure 8.19). Changing the digital representation of the waveform stored in RAM can easily modify the analogue signal definition. Since the contents of RAM can be modified to any set of values, the user has total control over the waveform definition.

8.7.1.1 Address generator

The address generator is the waveform 'playback' mechanism in an AWG. If you think of the contents of the waveform memory as a tape recording, then the address generator is like a tape recorder that can play the recording forwards or backwards or repeat sections of it. The address generator accomplishes this by outputting a sequence

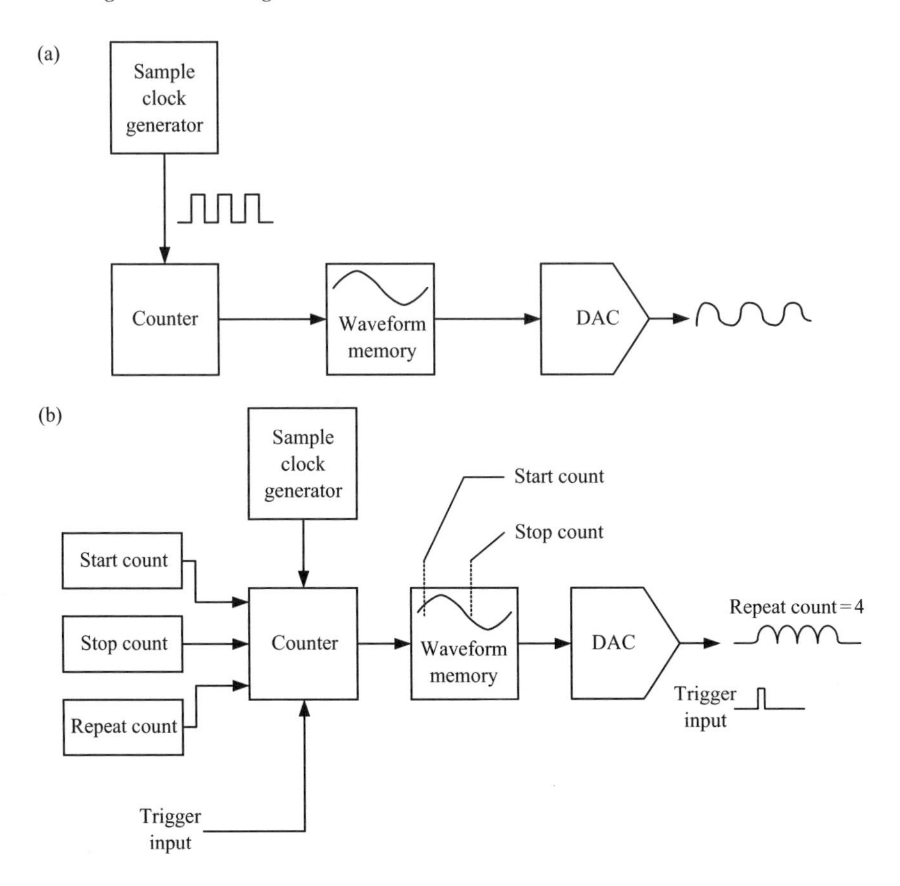

Figure 8.20 *Address generator concepts: (a) simple counter based; (b) with start, stop and repeat count capability (reproduced with permission of Fluke Inc., Wavetek Instruments, USA)*

of addresses to the waveform memory, supplying a new address on each clock cycle (the sample clock).

The simplest type of address generator is a counter (Figure 8.20(a)). Each time a counter is clocked it will increment by one. When the maximum count is reached, the counter will start over at zero. If the output of the counter is provided as the address to the waveform memory, the waveform will be played over continuously. The rate at which the waveform is played out is directly controlled by the sample clock frequency.

The counter based address generator is simple but very limited in functionality because the entire contents of waveform memory must be played out continuously.

A more sophisticated and flexible approach is shown in Figure 8.20(b). It allows the counter to start and stop at any address, as well as repeat the address sequence any number of times in response to a trigger signal. The architecture shown in Figure 8.20(b) allows several different waveforms to be stored in the waveform memory and accessed by changing the start and stop values to match the beginning and ending addresses of the waveform in memory.

The output frequency (F_{out}) is a function of the waveform size and the sample clock frequency (F_{clk}) and is computed as

$$F_{out} = F_{clk} \div (\text{stop count} - \text{start count}). \tag{8.3}$$

This equation assumes that the counter always increments by one when clocked, thereby ensuring that each point in the waveform memory is accessed. But what if the counter could increment by any number? For example, if the counter incremented by two each time it was clocked, then every other point in the waveform memory would be accessed. Only half the points would appear at the DAC output, but the waveform would be played out twice as fast. Under these conditions the formula for computing output frequency is modified to include the counter increment:

$$F_{out} = F_{clk} \div (\text{stop count} - \text{start count}) \times (\text{increment}). \tag{8.4}$$

Now the output frequency can be changed by changing the increment. But, if the increment is always an integer, then the precision with which frequency can be controlled is limited. If the increment could be a fraction, however, then greater precision in frequency control can be achieved.

These fractional increment counters are the phase accumulators described under DDS based AWGs and can be used to advantage as address generators in AWGs.

8.7.1.2 Detailed block diagram description of a traditional AWG

Figure 8.21 is a typical block diagram of an traditional AWG. This diagram may be subdivided into four basic blocks: signal path (block 1), timing and triggering (block 2), instrument control (block 3) and communication option (block 4).

The signal path consists of the waveform address counter, waveform address selector, waveform memory, waveform latch, waveform synthesiser, amplitude and offset controls, output amplifier, output attenuator and Z axis drive. This block takes care of storing the digital waveform data that originate from the microprocessor. The information to be accessed is determined by the waveform address counter which is updated at a rate determined by the timing and triggering block. The digital waveform information is sent through the waveform latch to the waveform synthesiser which converts the data back to an analogue waveform. The waveform is then run through the amplitude control blocks, which are also under microprocessor control. In addition the waveform could be shifted by a d.c. offset under microprocessor control. The output attenuator block, which uses several decade steps, is also under microprocessor control and maintains a fixed output impedance.

The timing and triggering section (block 2) uses a frequency synthesiser for generating the waveform clock used by the waveform address counter for generating the final waveform. In addition this block contains circuitry for controlling the waveform trigger functions and creating bursts, etc.

The instrument control section (block 3) takes care of generating the instrument parameters and waveform information entered by the user through the front panel or

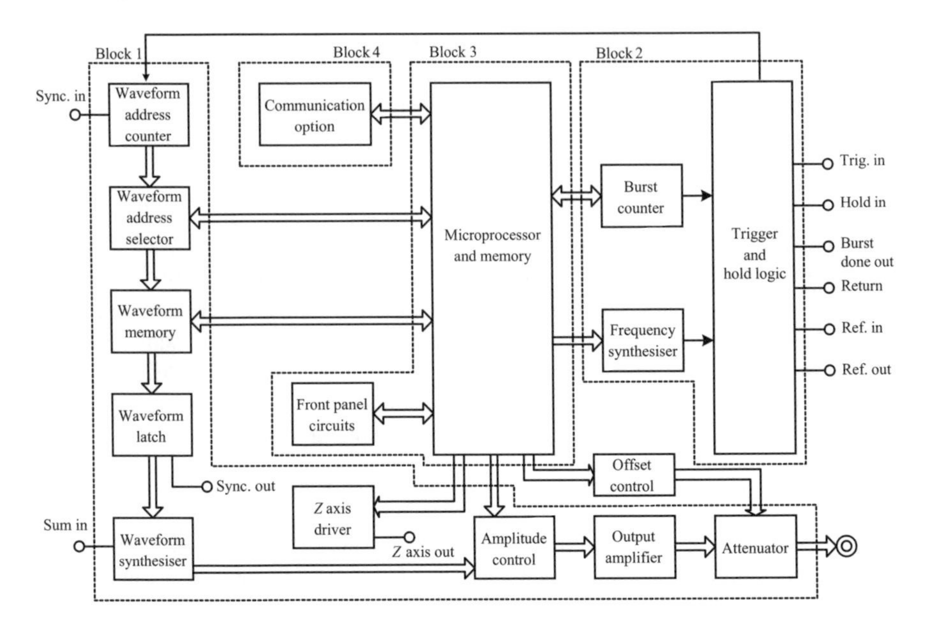

Figure 8.21 Block diagram of a traditional AWG (reproduced with permission from Wavetek Inc. USA)

through the communication port such as an IEEE-488 interface. This particular block is the central control block which takes care of the overall control of the instrument. Almost all AWGs have the computer interface to control the instrument and this is generally an IEEE-488 or an RS-232 interface.

8.7.2 *AWGs based on DDS techniques*

As discussed earlier, DDS techniques have been used for years to generate sinewaves with very high frequency resolution, agility and stability at a relatively low cost.

DDS systems with 10–14 digits of frequency resolution and capable of changing from one frequency to another in nanoseconds are possible. The basic building blocks of a DDS system are very similar to those of a traditional waveform generator. In a DDS system, the sinewave signal is first generated digitally and then converted to analogue through a DAC. A block diagram of a DDS system is shown in Figure 8.22.

One of the primary advantages to a DDS system is that it uses a fixed clock frequency. The clock typically is derived directly from a crystal oscillator, which means that frequency accuracy, stability, and phase noise will be excellent. Operating from a fixed frequency also simplifies filtering requirements. If the sample clock is a fixed frequency, then the sampling spurs also are in a fixed and predictable location allowing for a fixed frequency low pass filter to be used. This filter can be constructed with very steep attenuation and stop band characteristics for high performance.

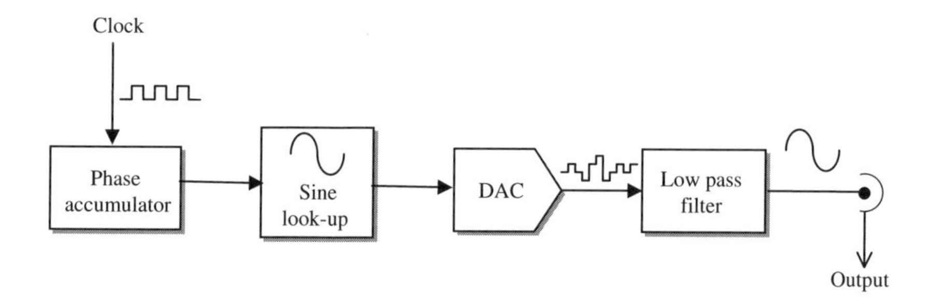

Figure 8.22 A DDS system: block diagram for sine output

If the sine look-up table in the block diagram of Figure 8.22 is replaced by a RAM, then any waveshape could be stored, rather than sinewaves only. The block diagram would then look much like that of an AWG with the address generator implemented as a phase accumulator. Such a DDS based AWG has many advantages over a counter based AWG, as well as some subtle restrictions.

In a typical DDS system, a very steep low pass filter is used to pass the fundamental and reject all of the sampling aliases. If the low pass filter frequency cut-off is kept below the Nyquist frequency, then the DDS system can output an undistorted sinewave up to the filter cut-off. The same is basically true for an arbitrary waveform. As long as all of the spectral lines that define the arbitrary waveform are below the cut-off frequency of the low pass filter, the arbitrary waveform can be output undistorted. Once frequency components of the arbitrary waveform are pushed beyond the cut-off of the filter, waveform information is lost and distortion will occur.

As shown in Figure 8.23, a ramp waveform that is rich in harmonics can be generated with a phase accumulator, but at a much lower fundamental output frequency than a sinewave. The ramp waveform will require the first 12–15 harmonics to be reproduced without significant distortion. Therefore all of the required harmonics must be below the cut-off of the low pass filter. With a sample clock of 10 MHz and a low pass cut-off frequency of 4 MHz, the ramp frequency is limited to about 300 kHz before significant distortion is noticed.

With reference to Figure 8.22 the heart of a DDS system is the phase accumulator, which is nothing more than a long register that keeps track of the relative phase of the signal. Controlling the value in the phase increment register sets the output frequency. Each phase increment is added to the value in the phase accumulator at a rate set by the instrument's crystal clock oscillator. The phase increment register determines how many clock cycles are needed to step through the signal's entire phase, which in turn determines the output frequency. The larger the increment value, the higher the output frequency, seen from eq. (8.1).

The number of bits in the phase accumulator determines the instrument's frequency resolution. For example, a 24-bit phase accumulator provides 10 Hz resolution at 20 MHz (seven digits). The most significant bits of the accumulator form the address that is strobed out to the waveform memory. A 32 kbyte waveform memory depth

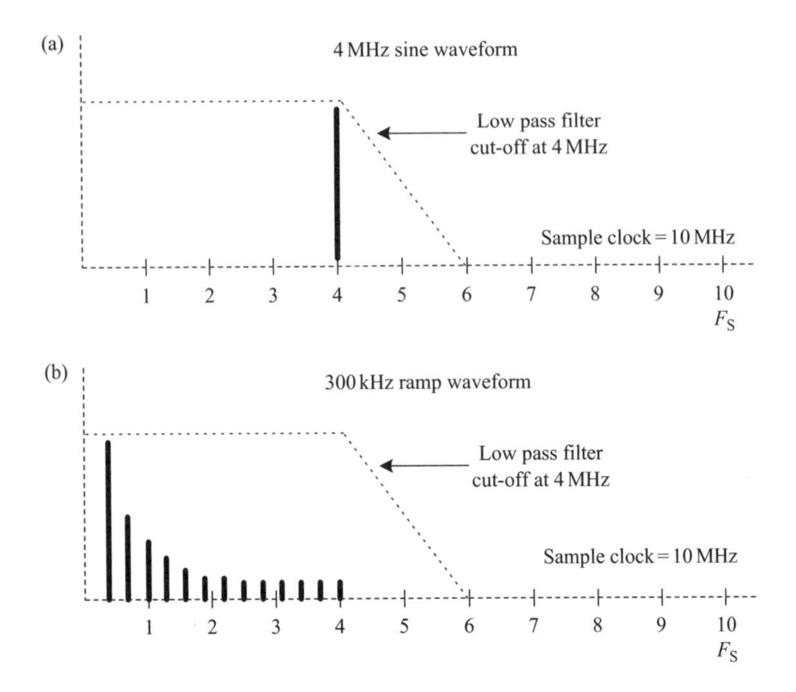

Figure 8.23 Maximum frequency of DDS: (a) sinewave; (b) ramp waveform

corresponds to a 15-bit-wide address bus. The contents of the waveform memory are clocked out to the digital-to-analogue converter to generate the waveform. An analogue low pass filter may be switched into suppress sample rate noise present at the output. The main benefits of DDS are increased frequency resolution and stability (owing to the crystal oscillator), and the ability to generate very fast, phase continuous frequency hops. Also, DDS facilitates the creation of FM and swept waveforms. For a more detailed explanation, Reference 18 is suggested.

One primary problem with the traditional AWG is the need for different kinds of filter at the output stage. An arbitrary waveform generator that uses DDS gets around this situation by using a fixed clock rate. With a fixed clock rate, a DDS based AWG can use a single low pass filter, which simplifies the instrument's design and allows the DAC to be optimised for that sampling rate.

DDS based AWGs change output signal frequency by varying the phase increment value (degrees per clock cycle). In the Figure, the phase accumulator adds the value from the increment register to its total each time the latch receives a clock pulse. After it is latched, the accumulator's output corresponds to an address in the look-up table. The address contains the amplitude of the waveform at that specific point. At the next clock pulse, the accumulator increments again by the value in the register, landing on another point in the waveform where the look-up table contains the amplitude at that point.

Eventually, the accumulator hits enough points to construct the entire waveform. At low frequencies, the AWG may remain at the same memory address for multiple clock cycles. To create high frequency signals, the instrument may skip addresses. Either way, a DDS based AWG uses the entire capacity of its look-up table.

Because DDS based AWGs use the entire table, they are unable to perform sequencing as traditional AWGs do.

8.8 Waveform editing features in AWGs

The biggest advantage of AWGs is the ability to create arbitrary shapes of waveforms by different methods. To simulate 'real world' conditions, test engineers need to synthesise waveforms that their equipment will encounter in the field. Real world waveforms are random and they can also change arbitrarily. In modern AWGs many waveform editing methods could be used as seen from Table 8.4.

8.8.1 Sequencing, looping and linking of waveforms

You can break up a traditional AWG's look-up table into segments. You can program the AWG to repeat the waveform some number of times or to connect segments together to form a longer waveform. The number of segments you can store and the number of times you can use them depend on the capabilities of your AWG. Some traditional AWGs include a sequencing function; on other AWGs this function is optional. DDS based AWGs do not have a sequencing function because they must use the entire look-up table. Sequencing segments by linking and looping can produce a repetitive waveform and greatly increase your amount of effective memory. To repeat a segment, you need only program that set of points once. Reference 19 provides a practical guide to equation editing.

8.9 Performance factors of AWGs

One of the most important aspects of an AWG's performance is the ability accurately to simulate or duplicate a desired signal. This accuracy can be quantified by comparing the desired ideal signal to the actual AWG output and analysing the differences. We

Table 8.4 AWG waveform editing methods

Technique	Function
Graphical edit	Cut, paste and insert, etc.
Direct waveform edit	Transfer of signals from DSO to AWG
Timing and table edit	Edit by manipulating points
Equation edit	Edit by input of a mathematical equation
Sequence edit	Edit large waveforms by small sequences
FFT edit	Edit by changing frequency domain components

have already discussed how the sampling process affects the spectrum of a signal. Now let us look at some other error sources. To simplify this analysis we will use a sinewave as the ideal waveform and then analyse how error sources cause noise and distortion.

8.9.1 *Quantisation error*

As we have seen, the output of an AWG is quantised in both time and voltage. Each of these quantising effects results in errors that distort the ideal sinewave.

8.9.1.1 Time quantisation

As discussed in subsection 8.6.2.2, the sampling process of a signal could generate unwanted spurs into the base band of the signal. Also the Fourier mathematics (discussed in chapter 9) indicates that the alias frequency components are represented by $(\sin x)/x$ values. Practically, time quantisation in an AWG is a result of the sample clock, which only allows the signal to change voltage at specific time intervals. The effect is to create a sampled signal. As time quantisation gets bigger (i.e. there are fewer samples in the waveform) the sampling aliases increase. In general the sampling aliases increase by 6 dB each time quantisation error doubles.

To explain this situation, let us consider a 1 kHz, 1000 point sinewave as an example. The first sampling alias will be at a frequency 999 times the fundamental (999 kHz) at a level of 59.99 dBC predicted by the $(\sin x)/x$ function. The calculations are shown below:

$$V_{\text{Fund}} = \frac{\sin(\pi F_{\text{Out}}/F_{\text{Clk}})}{(\pi F_{\text{Out}}/F_{\text{Clk}})} = \frac{\sin(\pi 1\,\text{kHz}/1000\,\text{kHz})}{(\pi 1\,\text{kHz}/1000\,\text{kHz})} = 0.999\,998,$$

$$V_{\text{Alias}} = \frac{\sin(\pi F_{\text{Alias}}/F_{\text{Clk}})}{(\pi F_{\text{Alias}}/F_{\text{Clk}})} = \frac{\sin(\pi\, 999\,\text{kHz}/1000\,\text{kHz})}{(\pi\, 999\,\text{kHz}/1000\,\text{kHz})} = 0.001\,000\,9.$$

The level of the sampling alias expressed in dBc is

$$\text{dBc} = 20\log\left(\frac{V_{\text{Alias}}}{V_{\text{Fund}}}\right) = 20\log\left(\frac{0.001\,000\,9}{0.999\,998}\right) = -59.999.$$

If the number of points in this sinewave is reduced to 500 points (i.e. the time quantisation is increased by a factor of 2) then the first alias will increase to −53.96 dBc, which is almost 6 dB greater than the 1000 point version.

As shown by the previous example, time quantisation distortion decreases as the number of points in a waveform increases. Table 8.5 shows the relative amplitude of the first sampling alias as a function of waveform size. These values are indicative of unwanted spurs generated in the sampling process of an AWG and demand suitable filtering to keep the output waveform reasonably clean of harmonics related to sampling process.

8.9.1.2 Voltage quantisation

Voltage quantisation results from the discrete nature of the DAC itself. An 8-bit DAC for example has only 256 possible output voltage values. As discussed in Chapter 3,

Table 8.5 Sampling alias level vs number of waveform points (source: [20])

n	Number of waveform points	Sampling spur (dBc)
8	256	−48.1
9	512	−54.2
10	1024	−60.2
11	2048	−66.2
12	4096	−72.2
13	8192	−78.3
14	16 384	−84.3
15	32 768	−90.3
16	65 526	−96.3
17	131 072	−102.4
18	262 144	−108.4
19	524 288	−114.5
20	1 048 576	−120.5

for an ideal DAC there may be up to 1/2 LSB error at any point when comparing an ideal waveform to its discrete approximation.

The amount of distortion due to voltage quantisation is dependent upon the number of bits in the DAC. For a sinewave, spurious signals caused by voltage quantisation will improve by about 8.5 dB per bit of DAC resolution. Table 8.6 shows the distortion levels that are predicted for DACs of different resolution. These values are for an ideal DAC with perfect accuracy. Other error sources will further degrade the predicted performance shown in Table 8.6.

8.9.2 DAC related error

In an arbitrary waveform generator, DAC performance will generally be the limiting factor above all others in determining the overall performance. The quantisation errors discussed above assume that the performance of the DAC is perfect. But, of course, the DAC's performance will not be perfect and in most situations the DAC performance will degrade beyond that predicted by quantisation errors alone. For example, in Table 8.6 the spurious contribution of a 12-bit DAC due to voltage quantisation is −102 dBc. In reality, this distortion level will never be achieved using a 12-bit DAC, because of other more significant error sources in the DAC.

8.9.2.1 DAC accuracy

The first and most obvious source of errors that will cause unwanted spurious signal is the accuracy of the DAC itself. DAC accuracy is usually quantified as '*differential non-linearity*' and '*integral non-linearity*'. Differential non-linearity is the

Table 8.6 Spurious level vs number of DAC bits (source: [20])

DAC bits	Distortion level (dBc)
8	−68.0
9	−76.5
10	−85.0
11	−93.5
12	−102.0
13	−110.5
14	−119.0
15	−127.5
16	−136.0
17	−144.5
18	−153.0
19	−181.5
20	−170.0

deviation from an ideal 1 LSB change in the DAC output from one adjacent output state to the next. Integral non-linearity is the maximum deviation of the DAC output from a perfect straight line drawn between the two end points. As an example, Figure 8.24(a) illustrates differential and integral non-linearity for a 4-bit DAC (16 output states).

Notice the error plot in Figure 8.24(b). It illustrates the difference between the ideal DAC and a real DAC. If the real DAC output is considered as being the sum of the ideal DAC and the error plot, then it is easy to understand how the errors produce unwanted spurious signals. For any waveform, the spectral lines form the signal created by the ideal DAC and the spectral lines from the error signal. This indicates why the location of unwanted spurs due to DAC accuracy errors is not easy to predict. Spurs will depend upon the error characteristics of each individual DAC, as well as the specific DAC codes that define a waveform.

Although not easy to predict exactly, we can put some upper bounds on the distortion caused by DAC accuracy errors. Linearity specifications are typically given in LSBs. If we consider the worst case error signal that will fit within the linearity specification we have an upper bound. For example, consider a 12-bit DAC with an LSB integral linearity specification. We can compute the effect of a worst case error signal with a 2-bit amplitude by comparing it to a full scale signal of 4096 bits:

$$\text{maximum error signal amplitude} = 20 \times \log(2/4096) = -66\,\text{dBc}.$$

So in this case, we can say that distortion due to 2 LSB integral non-linearity will be *less than* −66 dBc.

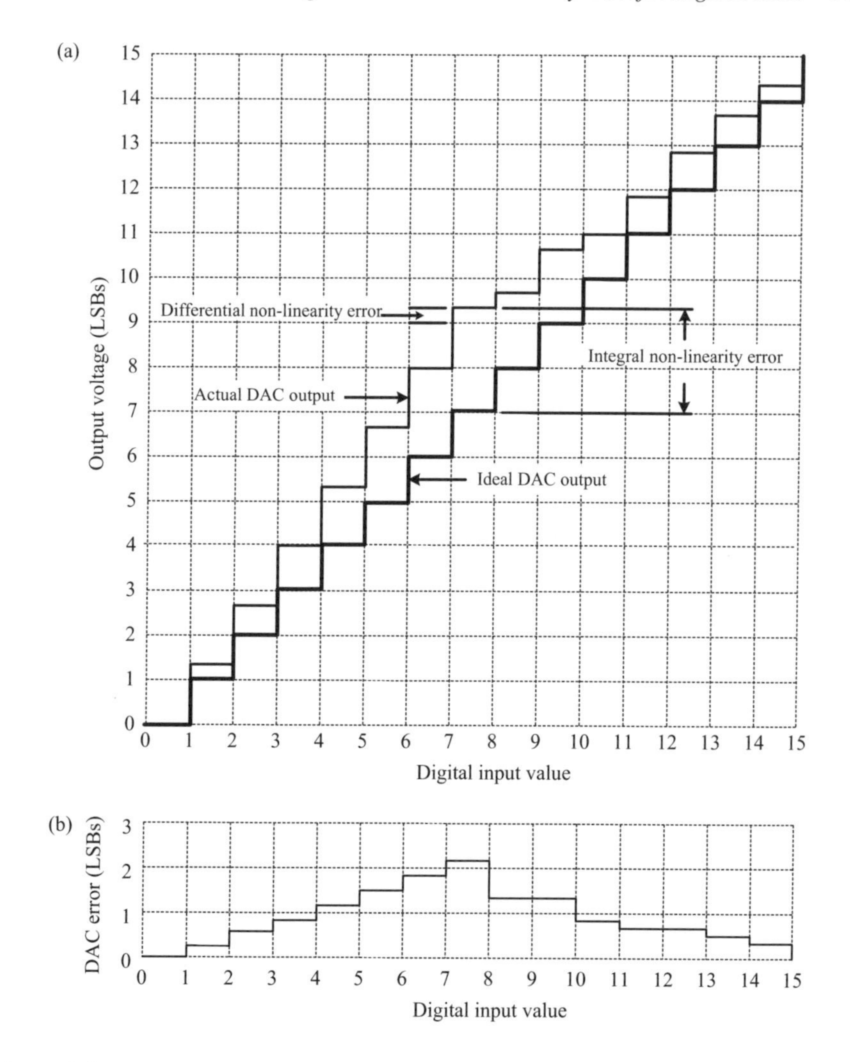

Figure 8.24 DAC linearity errors: (a) output voltage vs digital input; (b) error plot

8.9.2.2 Glitch energy

DAC accuracy errors are one source of unwanted spurious signals, but these errors are really static errors. In other words, these refer to the d.c. characteristics of the DAC and only apply at low frequencies where a.c. characteristics are not significant. For high speed arbitrary waveform generators the most troublesome error source is dynamic. The most meaningful specification of a DAC's a.c. characteristics as it relates to waveform generation is *glitch energy*. Glitch energy defines the error that occurs when the DAC output is transitioning from one discrete state to another. Not to be confused with settling time, glitch energy is dependent on the digital value and

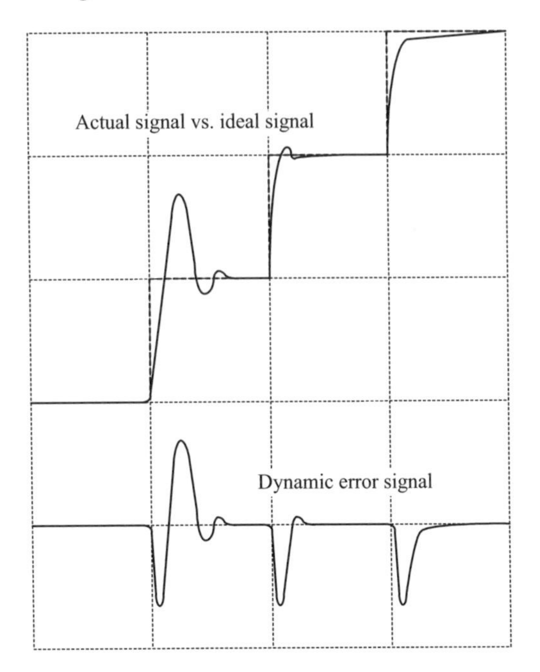

Figure 8.25 Glitch energy

the sample clock speed. Glitch energy can be thought of as the difference in energy between an ideal step transition and the actual step transition of the DAC. This energy difference will be unique for each transition. Figure 8.25 illustrates glitch energy related errors.

Notice that the dynamic error signal shown in Figure 8.25 always returns to zero error after the DAC output settles. This means that the effect of glitch errors becomes more and more significant as the sample clock rate is increased. In other words the relative power in the dynamic error signal increases as the sample clock frequency is increased. Glitch energy is typically specified in units of *LSB nanoseconds* or *PicoV-seconds*. Specifying glitch energy in this way provides an easy way to relate the dynamic error in terms of LSBs to a specific sample clock frequency. For example, a 12-bit DAC operating at 10 MHz with a glitch energy specification of 250 LSB-nanoseconds will have the following dynamic error:

$$\begin{aligned}\text{dynamic error at 10 MHz} &= 250\ \text{LSB-ns}/100\ \text{ns}\\ &= 2.5\ \text{LSBs}.\end{aligned}$$

Knowing that the dynamic error is 2.5 LSBs allows the upper bound on distortion for this 12-bit DAC to be computed as follows:

$$\text{distortion} = 20 \times \log(2.5/4096) = 64.3\ \text{dBc}.$$

If the same DAC is now operated at 30 MHz, then dynamic error will increase to 7.5 LSBs and result in a predicted upper bound on distortion of −54.7 dBc.

Similar computations can be made for glitch energy in units of PicoV-s, as long as the full scale output voltage of the DAC is known. As a point of reference, 250 LSB-ns is equivalent to 61 PicoV-s for a DAC with a 1 V full scale output.

8.9.2.3 Subharmonic spurs

Subharmonic spurious signals are the most difficult to deal with in an arbitrary waveform generator. A subharmonic spur is a signal that occurs at a 'submultiple' of the fundamental output frequency (such as $1/8F_0$ or $1/4F_0$, and so on). To illustrate a subharmonic signal and cause, we will use a DDS system generating a 20 MHz sinewave with 50 MHz clock. This situation will really push the limits of any design. The sinewave under these circumstances will consist of 2.5 discrete points per cycle. To actually build this waveform requires two cycles of a sinewave using five points. The DAC output for this waveform is illustrated in Figure 8.26. As improbable as it may seem, when properly filtered the waveform of Figure 8.26 will produce an acceptable sinewave.

Now consider the issues of glitch energy and code dependent errors discussed in the previous section, considering that the glitch energy is unique for each transition. Assume that the transition from point 3 to 4 has a high level of glitch energy compared to other transitions, as indicated in Figure 8.26 (highly exaggerated for the purpose of illustration). If the sinewave output is occurring at 20 MHz, then the large glitch is occurring at a 10 MHz rate or one half the fundamental. This energy will show up in the spectrum of the waveform at 10 MHz. Because the 10 MHz subharmonic is lower in frequency than the 20 MHz fundamental signal, it will not be removed from the spectrum by low pass filtering.

Other waveform configurations can be found that will also cause subharmonic spurious signals. Consider a sinewave that requires $3\frac{1}{3}$ points. In this case it would be constructed by programming 3 cycles of a sinewave into 10 points. With a clock rate of 50 MHz the fundamental output frequency is 50 MHz $\div\ 3\frac{1}{3} =$ 15.015 MHz.

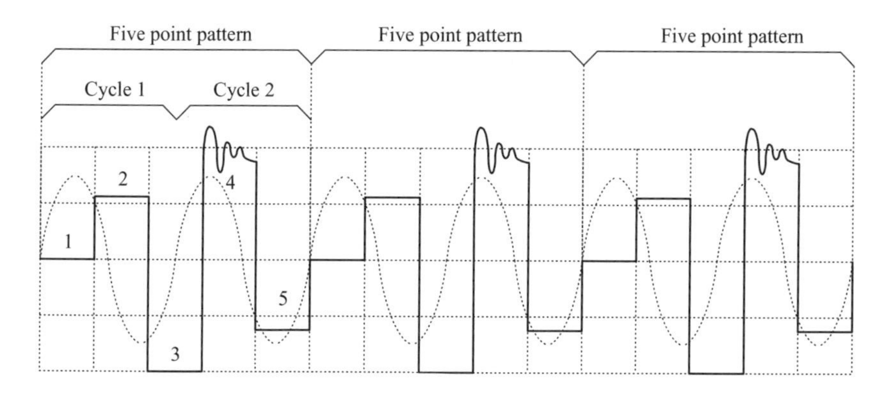

Figure 8.26 A 2.5 point sinewave

In this case the subharmonic due to code dependent glitch energy of the DAC could be found at $\frac{1}{3}$ and $\frac{2}{3}$ of the fundamental output frequency (approximately 5 MHz and 10 MHz).

8.10 References

1 KULARATNA, N.: *Modern Electronic Test & Measuring Instruments*, IEE, London, 1996.
2 MAXIM INTEGRATED PRODUCTS: Versatile Waveform Generator operates from 0.1 Hz to 20 MHz, *Maxim Engineering Journal*, **19**, pp. 3–11.
3 McLEOD, J.: 'One chip waveform generator cuts purchase price in half', *Electronics*, 3 September 1987, pp. 143–144.
4 NOVELLINO, J.: 'Pulse generators evolve as needs change', *Electronics Design*, April 1, 1993, pp. 69–70.
5 *Frequency Synthesis Handbook, RF Design Series*, Cardiff Publishing, 1993.
6 BREED, G.A.: *Oscillator Design Handbook*, Cardiff Publishing, 1991, 2nd edn.
7 MANASSEWITCH, V.: *Frequency Synthesisers: Theory and Design*, John Wiley & Sons, New York, 1987.
8 HELSZAJN, J.: *YIG Resonators and Filters*, John Wiley & Sons, New York, 1985.
9 Analog Devices: 'System Applications Guide', 1993.
10 HICKMAN, I.: 'Direct digital synthesis (Part I)', *Electronics and Wireless World*, August 1992, pp. 630–634.
11 HICKMAN, I.: 'Direct digital synthesis (Part II)', *Electronics and Wireless World*, August 1992, pp. 746–748.
12 HICKMAN, I.: 'Genuine solutions to spurious arguments', *Electronics and Wireless World*, October 1992, pp. 842–845.
13 SCHWEBER, B.: 'Give DACs due diligence for superior DDS performance', *EDN*, July 17, 1997, pp. 59–69.
14 CLARKE, R.M.: 'Signal generators meet waveform demands', *Electronic Test*, April 1990, pp. 30–35.
15 ROWE, M.: 'Don't let AWG specs confuse you', *Test & Measurement World*, May 1995, pp. 57–62.
16 GOLDBERG, J.M.: 'Don't overspecify microwave sources', *Test & Measurement World*, June 1997, pp. 42–46.
17 GOLD, B.: *Arbitrary Waveform Generators A Primer*, Wavetek Instruments Division, 1992.
18 BARKER, D.: 'Function generators test digital devices', *Electronic Design*, December 2, 1993, pp. 89–104.
19 ROWE, M.: 'Equations shape AWG waveforms', *Test & Measurement World*, May 2001, pp. 59–62.

Chapter 9
Spectrum analysis

Nihal Kularatna and Sujeewa Hettiwatte

9.1 Introduction

All electrical signals can be described either as a function of time or of frequency. When we observe signals as a function of time they are called the time domain measurements. Sometimes, we observe the frequencies present in signals, in which case they are called the frequency domain measurements. The word spectrum refers to the frequency content of any signal.

When signals are periodic, time and frequency are simply related; namely, one is the inverse of the other. Then we can use the Fourier series to find the spectrum of the signal. For non-periodic signals, a Fourier transform is used to get the spectrum. However, performing a Fourier transformation involves integration over all time, i.e. from $-\infty$ to $+\infty$. Because this is not practically desired, we approximate the Fourier transform by a discrete Fourier transform (DFT), which is performed on a sampled version of the signal. The computational load for direct DFT, which is usually computed on a personal computer (PC), increases rapidly with the number of samples and sampling rate. As a way out of this problem, Cooley and Tukey invented the fast Fourier transform (FFT) algorithm in 1954. With FFT, the computational loads are significantly reduced.

In chapter 3 we discussed the basis of converting analogue signals to digital samples and vice versa for accurate digital processing of a time varying signal. Following from there, this chapter provides an overview of FFT techniques, as applied to dynamic signal analysers (or FFT analysers) or DSOs where spectrum components of a time varying signal are to be displayed. In addition, the essential principles and applications of swept-tuned spectrum analysers are discussed, because spectrum observations of higher frequency signals, such as those used in communications systems, are still beyond the capability of FFT analysers.

9.2 Why spectrum analysis?

Most common test and measuring instruments such as multimeters, oscilloscopes and digitisers look at signals in the time domain. Therefore, one might raise the question 'why bother with spectrum analysis?'. All electronic components and modules, which are the building blocks of these instruments, perform according to design specifications only within a limited frequency range. Therefore, we can define a bandwidth for each building block of the instrument and an overall bandwidth of operation for the instrument. In order to make accurate measurements, our signal under observation or measurement should have significant frequency components only within the bandwidth of the instrument. It is therefore necessary for us to be able to analyse real life signals for their frequency content or the spectrum. In transmission systems, particularly the links with long paths, frequency behaviour is most important to consider in design, analysis, installation and commissioning, etc. Another common reason is to filter out signals buried in noise. This can be performed easily by analysing the frequency spectrum.

9.3 Continuous signals and Fourier analysis

Let $f(t)$ be an arbitrary function[1] of time. If $f(t)$ is also periodic, with a period T, then $f(t)$ can be expanded into an infinite sum of sine and cosine terms. This expansion, which is called the Fourier series of $f(t)$, may be expressed in the form:

$$f(t) = \frac{a_0}{2} + \sum_{n=1}^{\infty}\left[a_n \cos\left(\frac{2\pi nt}{T}\right) + b_n \sin\left(\frac{2\pi nt}{T}\right)\right], \tag{9.1}$$

where Fourier coefficients a_n and b_n are real numbers independent of t and which may be obtained from the following expressions:

$$a_n = \frac{2}{T}\int_0^T f(t)\cos\left(\frac{2\pi nt}{T}\right)\mathrm{d}t, \quad \text{where } n = 0, 1, 2, \ldots, \tag{9.2a}$$

$$b_n = \frac{2}{T}\int_0^T f(t)\sin\left(\frac{2\pi nt}{T}\right)\mathrm{d}t, \quad \text{where } n = 1, 2, \ldots. \tag{9.2b}$$

Further, $n = 0$ gives us the d.c. component, $n = 1$ gives us the fundamental, $n = 2$ gives us the second harmonic, and so on.

Another way of representing the same time function is,

$$f(t) = \frac{a_0}{2} + \sum_{n=1}^{\infty} d_n \cos\left(\frac{2\pi nt}{T} + \phi_n\right), \tag{9.3}$$

[1] Function $f(t)$ must satisfy Dirichlet's conditions [1].

where $d_n = \sqrt{a_n^2 + b_n^2}$ and $\tan \phi_n = -b_n/a_n$. The parameter d_n is the magnitude and ϕ_n is the phase angle.

Equation (9.3) can be re-written as

$$f(t) = \sum_{n=-\infty}^{+\infty} c_n \mathrm{e}^{\mathrm{j}2\pi nt/T}, \tag{9.4}$$

where

$$c_n = \frac{1}{T} \int_0^T f(t)\mathrm{e}^{-\mathrm{j}2\pi nt/T}\,\mathrm{d}t, \quad n = 0, \pm 1, \pm 2, \ldots.$$

The series expansion of eq. (9.4) is referred to as the complex exponential Fourier series. The c_n are called the complex Fourier coefficients.

According to this representation, a periodic signal contains all frequencies (both positive and negative) that are harmonically related to the fundamental. The presence of negative frequencies is simply a result of the fact that the mathematical model of the signal given by eq. (9.4) requires the use of negative frequencies. Indeed this representation also requires the use of complex exponential functions, namely $\mathrm{e}^{\mathrm{j}2\pi nt/T}$, which have no physical meaning either. The reason for using complex exponential functions and negative frequency components is merely to provide a complete mathematical description of a periodic signal, which is well suited for both theoretical and practical work.

9.3.1 The Fourier transform

The transformation from the time domain to the frequency domain and back again is based on the Fourier transform and its inverse. When the arbitrary function $f(t)$ is not necessarily periodic, we can define the Fourier transform of $f(t)$ as

$$F(f) = \int_{-\infty}^{\infty} f(t)\mathrm{e}^{-\mathrm{j}2\pi ft}\,\mathrm{d}t. \tag{9.5}$$

The time function $f(t)$ is obtained from $F(f)$ by performing the inverse Fourier transform:

$$f(t) = \int_{-\infty}^{\infty} F(f)\mathrm{e}^{\mathrm{j}2\pi ft}\,\mathrm{d}f. \tag{9.6}$$

Thus $f(t)$ and $F(f)$ form a Fourier transform pair. The Fourier transform is valid for both periodic and non-periodic functions that satisfy certain minimum conditions. All signals encountered in the real world easily satisfy these conditions. These conditions, known as the Dirichlet conditions, determine whether a function is Fourier expandable [1].

An oscilloscope displays the amplitude of a signal as a function of time whereas a spectrum analyser represents the same signal in the frequency domain. The two types

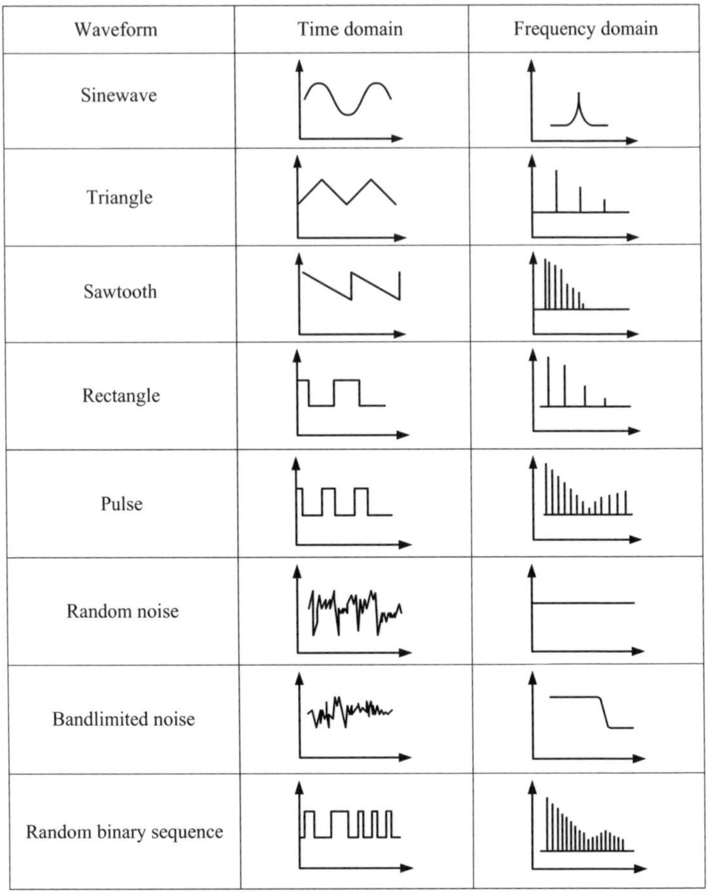

Waveform	Time domain	Frequency domain
Sinewave		
Triangle		
Sawtooth		
Rectangle		
Pulse		
Random noise		
Bandlimited noise		
Random binary sequence		

Figure 9.1 Signal analysis with oscilloscope and spectrum analyser

of representations (which form Fourier transform pairs) are shown in Figure 9.1 for some common signals encountered in practice [2].

9.4 Discrete and digital signals

A discrete signal is either discrete in time and continuous in amplitude, or discrete in amplitude and continuous in time. Discrete signals, for example, can occur in CCD arrays or switched capacitor filters. A digital signal, however, is discrete in both time and amplitude, such as those encountered in DSP applications.

9.4.1 Discrete time Fourier transform (DTFT)

The DTFT maps a discrete time function $h[k]$ into a complex function $H(e^{j\omega})$, where ω is a normalised real variable in $[-\pi, \pi]$. The formal definitions for the forward and inverse transforms are

$$H(\mathrm{e}^{\mathrm{j}\omega}) = \sum_{k=-\infty}^{\infty} h[k]\mathrm{e}^{-\mathrm{j}k\omega}, \tag{9.7}$$

$$h[k] = \frac{1}{2\pi}\int_{-\pi}^{\pi} H(\mathrm{e}^{\mathrm{j}\omega})\,\mathrm{d}\omega. \tag{9.8}$$

Note that the inverse transform, as in the Fourier case, is an integral over the real frequency variable. A complex inversion integral is not required.

The DTFT is useful for manual calculations, but not for computer calculation because of continuous variable ω [3]. The role of the DTFT in discrete time system analysis is very much the same as the role the Fourier transform plays for continuous time systems.

9.4.2 Discrete Fourier transform (DFT)

An approximation to the DTFT is the DFT. The DFT maps a discrete time sequence of N-point duration into an N-point sequence of frequency domain values. Formally, the defining equations of the DFT are [3]:

$$H\left(\mathrm{e}^{\mathrm{j}\frac{2\pi n}{N}}\right) = H[n] = \sum_{k=0}^{N-1} h[k]\mathrm{e}^{-\mathrm{j}\frac{2\pi nk}{N}} \quad \text{for } k, n = 0, 1, \ldots, N-1, \tag{9.9}$$

$$h[k] = \frac{1}{N}\sum_{n=0}^{N-1} H[n]\mathrm{e}^{\mathrm{j}\frac{2\pi nk}{N}} \quad \text{for } k, n = 0, 1, \ldots, N-1. \tag{9.10}$$

These equations are the DTFT relationships with the frequency discretised to N points spaced $2\pi/N$ radians apart around the unit circle, as in Figure 9.2. In addition, the time duration of the sequence is limited to N points.

The DFT uses two finite series of N points in each of the definitions. Together they form the basis for the N-point DFT. The restriction to N points and reduction from integral to summation occurs because we are only evaluating the DTFT at N points in the Z domain. The value of N is usually determined by constraints in the problem, or resources for analysis. The most popular values of N are in powers of 2. However, many algorithms do not require this. N can be almost any integer value [3].

9.4.3 Fast Fourier transform (FFT)

The FFT is not a transform but an efficient algorithm for calculating the DFT. The FFT algorithms remove the redundant computations involved in computing the DFT of a sequence by exploiting the symmetry and periodical nature of the summation kernel, the exponential term, and the signal. For large values of N, FFTs are much faster than performing direct computation DFTs. Table 9.1 shows the approximate computational

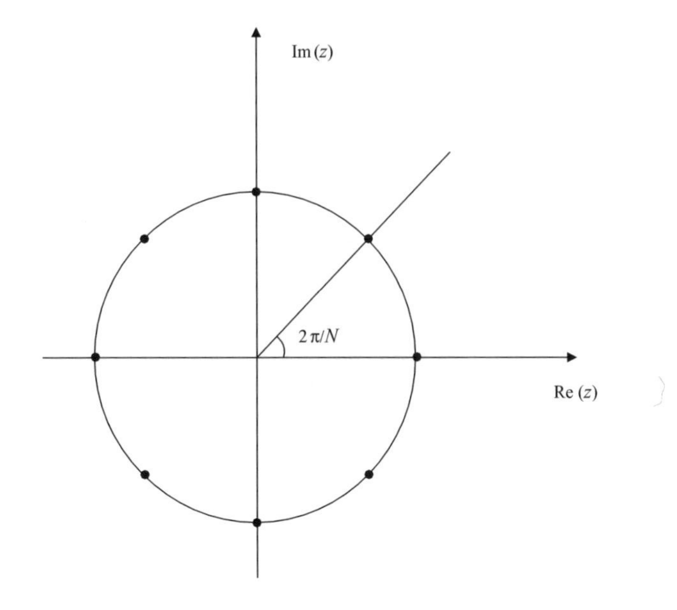

Figure 9.2 Unit circle with $N = 8$

Table 9.1 Comparison of approximate computational loads for direct DFT and decimation-in-time FFT

Number of points N	Multiplications		Additions	
	DFT	FFT	DFT	FFT
4	16	4	12	8
8	64	12	56	24
16	256	32	240	64
32	1024	80	992	160
64	4096	192	4032	384
128	16 384	448	16 256	896
256	65 536	1024	65 280	2048
512	262 144	2304	261 632	4608
1024	1 048 576	5120	1 047 552	10 240
2048	4 194 304	11 264	4 192 256	22 528
4096	16 777 216	24 576	16 773 120	49 152

loads for various sizes of the direct calculated DFT, and the DFT computed using the FFT decimation-in-time algorithm. Note that as N grows beyond 16, the savings are quite dramatic [3].

For more details on practical FFT algorithms, in particular on decimation in time algorithm, refer to Reference 4. A more detailed treatment can be found in Reference 5.

9.5 Dynamic signal analysis

Dynamic signal analysis involves transforming a time domain signal to the frequency domain. It is dynamic because the signal varies with time as we perform the transformation. The fast Fourier transform (FFT) algorithm is employed in dynamic signal analysis. The basic block diagram of a dynamic signal analyser is shown in Figure 9.3.

9.5.1 Sampling and digitising

Our time domain signal must be sampled before digitising and feeding to the FFT processor. The samples should be a replica of the original signal and hence the rate of sampling is important. It should confirm to the Shannon sampling theorem and Nyquist criteria (chapter 3). After sampling the signal, digitisation is done using the analogue-to-digital converter (ADC). The sampler must sample the input at exactly the correct time and must accurately hold the input voltage measured at this time until the ADC has finished its conversion. The ADC must have high resolution and linearity. For 70 dB of dynamic range the ADC must have at least 12 bits of resolution and one half least significant bit linearity [6].

Having digitised each sample, we take N consecutive, equally spaced samples, which is called a *time record*, as our input to the FFT processor. For ease of simplification we take N as a multiple of 2, for instance 1024. The input time record is transformed into a complete block of frequency lines by the FFT processor. This is depicted in Figure 9.4. It is important to note that all the samples of the time record are needed to compute each and every line in the frequency domain.

As the FFT transforms the entire block of time record into a block of frequency domain lines, there cannot be a valid frequency domain result until the time domain record is complete. However, once the time record is initially filled, at the next instance of sampling, the oldest sample could be discarded and the new sample could be added. This in turn enables all the samples to be shifted in the time record, as shown in Figure 9.5.

Therefore, there is a new time record at each instance of sampling and consequently a new block of frequency domain lines at each instance of sampling.

It should be noted here that calculating a new spectrum every sample is usually too much information, too fast. This would often give us thousands of transforms per second. Therefore, how fast should a dynamic signal analyser transform the time record? This issue is addressed in section 9.5.5.

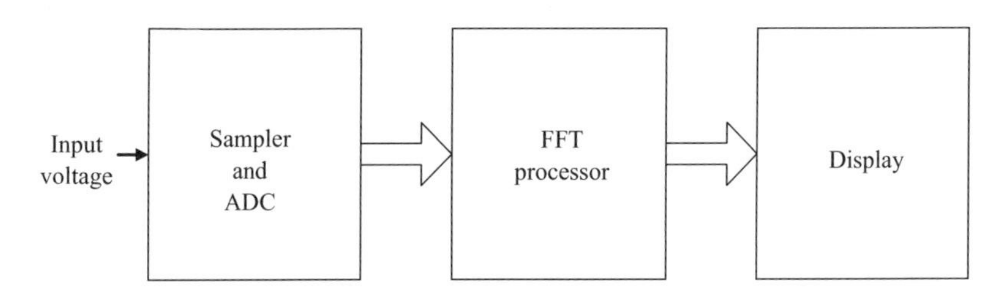

Figure 9.3 Block diagram of dynamic signal analyser

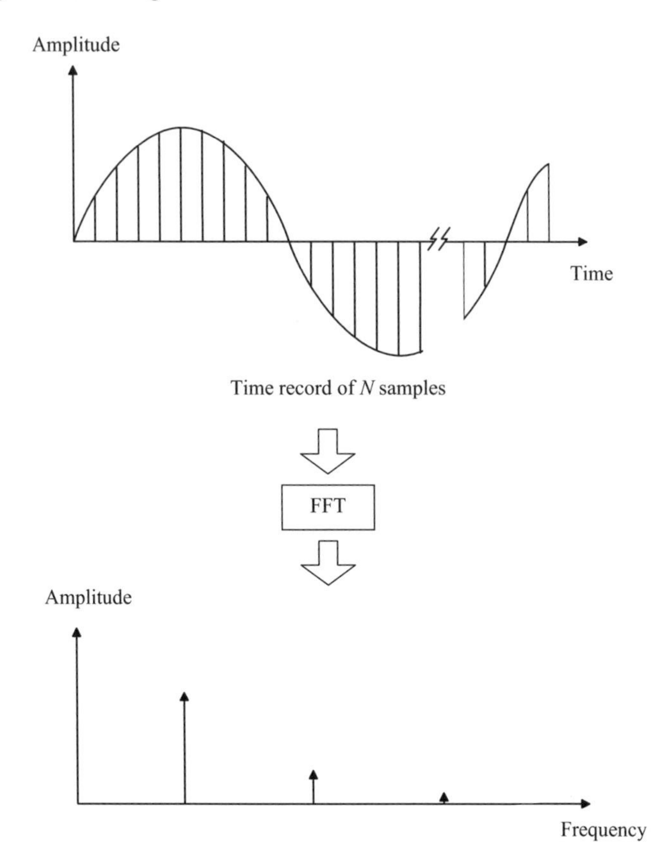

Figure 9.4 The FFT works on blocks of data

9.5.2 Aliasing

The reason as to why an FFT spectrum analyser needs so many samples per second is to avoid a problem called aliasing. Aliasing occurs if our sampling rate falls below the minimum required rate given by the Nyquist criteria. This criteria states that 'any band limited signal is adequately represented by its samples, if the sampling frequency is at least twice the maximum frequency of the signal'. More information on aliasing in a practical sense is described in chapter 3.

9.5.2.1 Antialias filter

To avoid aliasing the signal to be sampled must be low pass filtered before sampling. The filter employed for this purpose is referred to as the *antialias filter*. An ideal antialias filter would have a frequency response as in Figure 9.6(a). This filter would pass all frequencies up to F_c, and completely reject any higher frequencies.

However, it is not practically possible to build a filter having such a sharp transition from the pass band to the stop band. Instead, a more practical filter would have

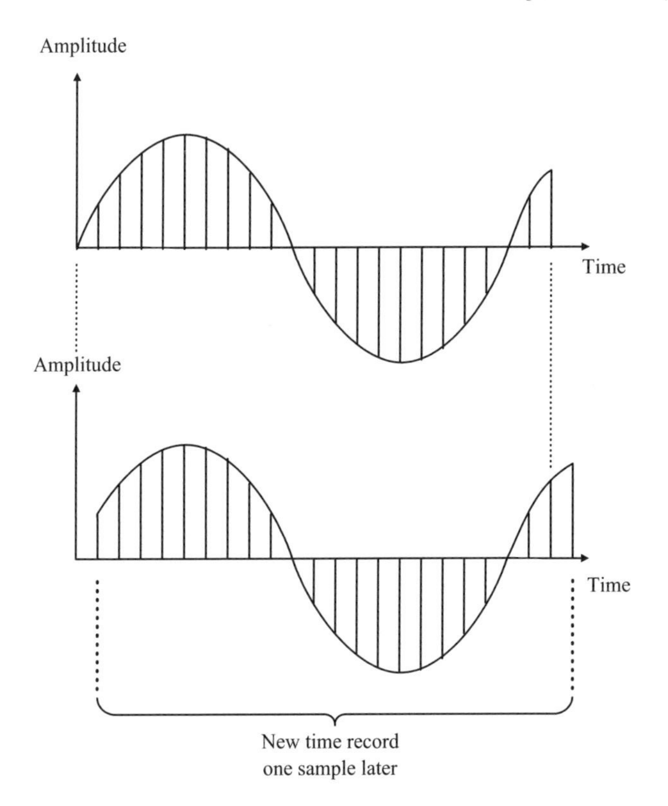

Figure 9.5 A new time record every sample after the time record is filled

a frequency response as in Figure 9.6(b). It has a gradual roll-off from the pass band to stop band with a finite transition band. With this real filter, large signals that are not well attenuated in the transition band could still alias into the desired input frequency range. To avoid this problem, sampling frequency is further raised, typically to twice the highest frequency in the transition band. This means that now the sampling rate is two and a half to four times the maximum desired input frequency. Therefore, a 25 kHz FFT spectrum analyser can require an ADC that runs at 100 kHz [6].

9.5.2.2 Digital filtering

We must vary the sampling rate to vary the frequency span of our dynamic signal analyser in Figure 9.3. To reduce the frequency span, we must reduce the sampling rate. This also leads to the reduction in the antialias filter frequency by the same amount.

A versatile instrument like a dynamic signal analyser should have a wide range of frequency spans. A typical instrument might have a minimum span of the order of 1 Hz and a maximum span of hundreds of kilohertz. This would lead to a bank of antialias filters as shown in Figure 9.7(a). This is not desirable when we consider the accuracy and the price of such an analyser.

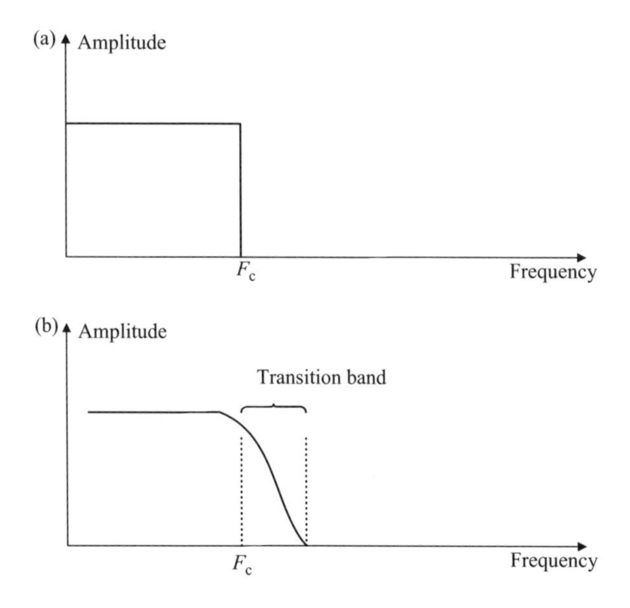

Figure 9.6 Antialias filter: (a) ideal; (b) practical

A more versatile and cheaper method is to employ digital filtering in our dynamic signal analyser. In this method we use a single antialias filter and a digital filter, as shown in Figure 9.7(b). It is called digital filtering because it filters the input signal after we have sampled and digitised it. In analogue filtering it is necessary to use a new filter every time we change the sample rate of the analogue-to-digital converter (ADC). However, with digital filtering we keep the sample rate of the ADC constant but vary the sample rate of the digital filter to match the frequency span used.

There are a few advantages of digital filtering compared with analogue filtering. There are no manufacturing variations, ageing or drift in the filter. Therefore, in a two channel analyser the filters in each channel are identical. A single digital filter can be designed to work on many frequency spans avoiding the need for multiple filters per channel. All these factors lead to digital filtering being much more cost effective than analogue filtering.

9.5.3 *Windowing*

The FFT computes the frequency spectrum of a signal from a block of samples of the time record. In addition, the FFT algorithm is based on the assumption that this time record is repeated throughout time, as shown in Figure 9.8.

However, this might lead to some distortion of the frequency spectrum even when the input signal is periodic in time but aperiodic in time record, as shown in Figure 9.9. For such a case when we construct a waveform using our time record we get an aperiodic waveform in time and hence a distorted frequency spectrum. This is

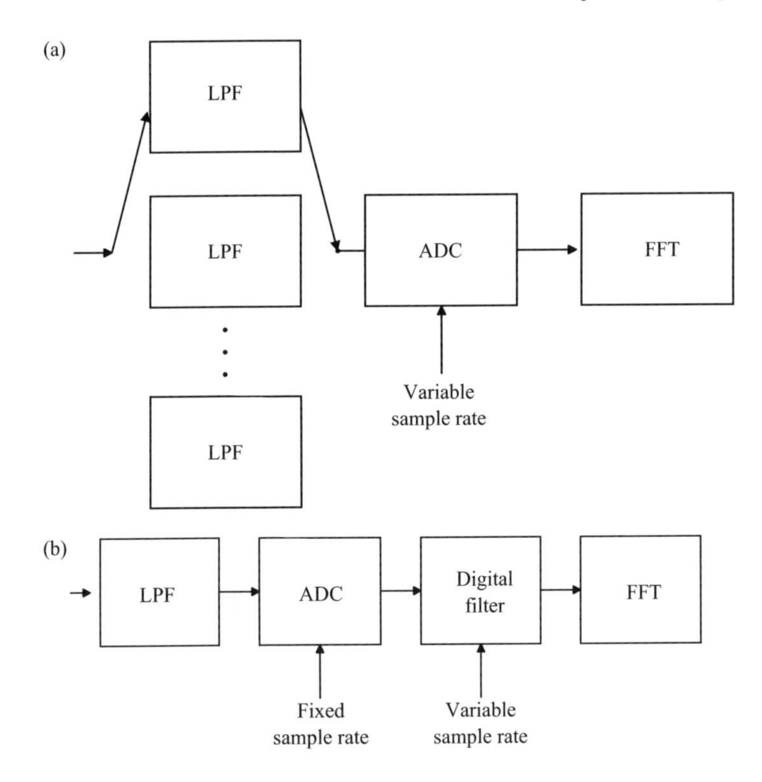

Figure 9.7 Use of filters in a dynamic analyser: (a) use of bank of antialias filters; (b) use of digital filtering

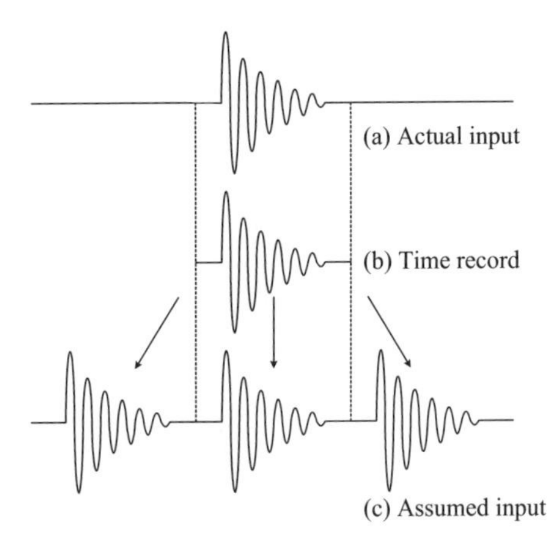

Figure 9.8 FFT assumption time record repeated throughout all time

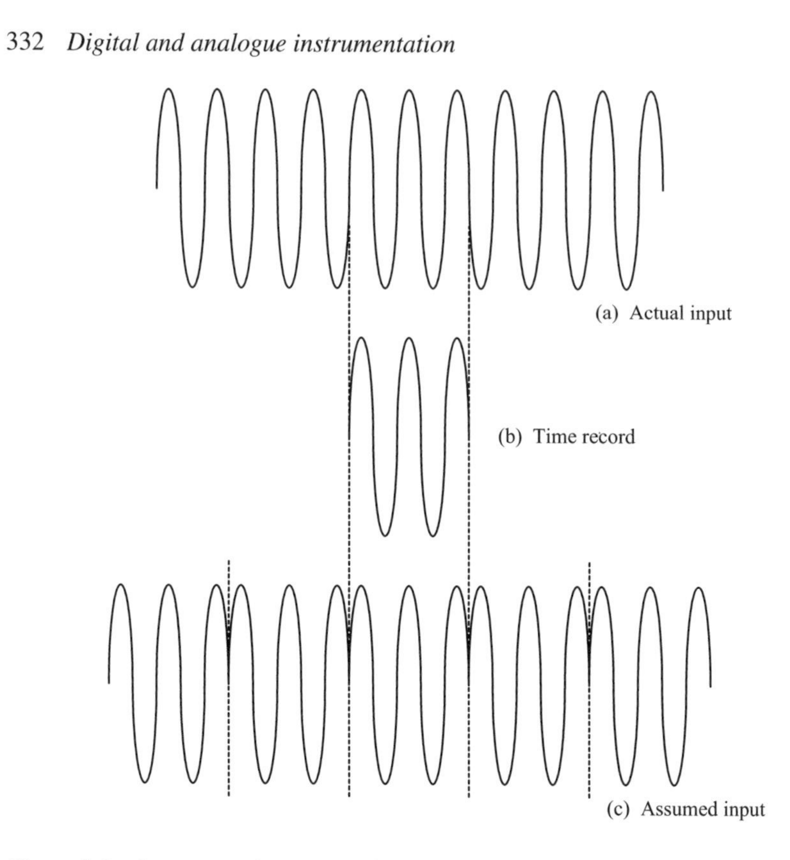

Figure 9.9 Input signal not periodic in time record: (a) actual input; (b) time record; (c) assumed input

mainly a result of the edges of our time record function not fitting together smoothly to give a periodic function in time.

The solution to this problem is windowing the time record. In windowing, we force our time record to have zero value at the edges as illustrated in Figure 9.10. However, windowing might lead to loss of information in some cases as shown in Figure 9.11.

9.5.4 *Types of window functions*

The FFT takes the captured waveform and assumes it repeats indefinitely. This works well if the captured waveform contains an integer multiple of the signal's period. If you append a captured waveform containing an integer multiple of cycles to a duplicate of itself, you will get perfect continuity from one waveform transition to the next. Discontinuities occur when a processor system captures a non-integer multiple of a signal resulting in '*leakage*'. Leakage amounts to spectral information from an FFT showing up at the wrong frequencies. Windowing functions act on raw data to reduce the effect of the leakage that occurs during the FFT. Once the user selects

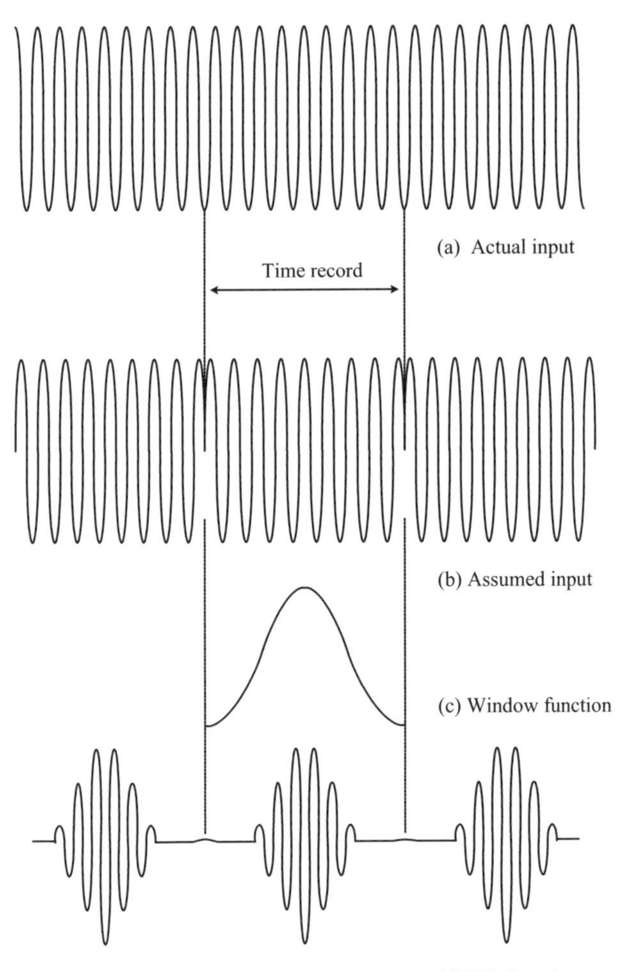

Figure 9.10 *The effect of windowing in the time domain: (a) actual input; (b) assumed input; (c) windowing function; (d) windowed input*

the windowing function, the processor system in the DSO or the FFT analyser will multiply the captured waveform by the curve of the window.

There are several types of window functions. Table 9.2 gives some specifications for the most commonly used FFT window functions [7]. The following sections describe the essential terminology related to FFT processing. For more details and practical implications of using filters and windows, References 8 and 9 are suggested. Applications of FFT routines and window functions in DSOs are described in Reference 10. The following sections provide terminology associated with FFT and windowing.

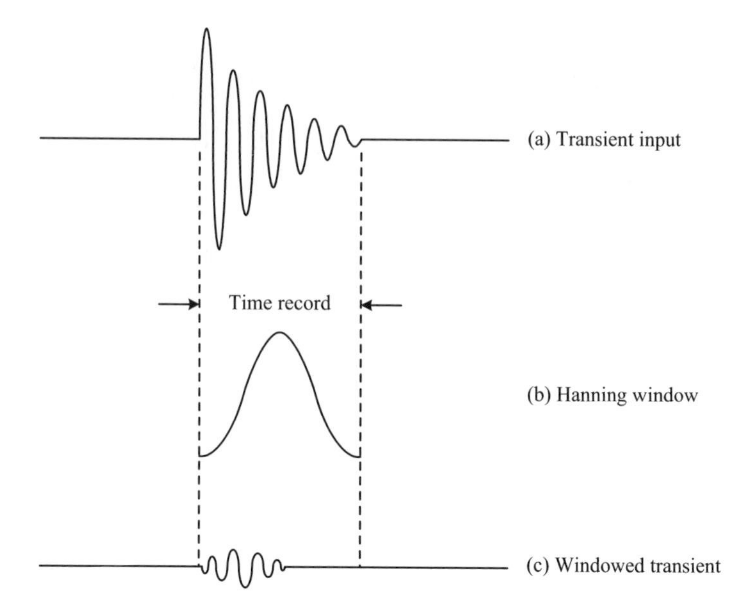

Figure 9.11 Loss of information due to windowing in a transient event

9.5.4.1 Bin

The FFT algorithm in a processor subsystem converts the selected points into a frequency domain record. Each point in a frequency domain record is called a bin.

9.5.4.2 Side lobes

When a record is processed using a FFT algorithm, owing to the effect of windowing, in addition to main lobes (bins), created side lobes also occur. The side lobe value, given in decibels, tells you how far below the main lobe you will find the adjacent lobes.

When sampling a sinewave with an ADC processor system it acquires discrete samples. Saving these samples in effect multiplies the infinitely long sequence of sinewave values by a sequence of discrete values of 1s. It means that the frequency spectrum of the limited element sinewave values is the convolution of the Fourier transform of the infinite duration sine wave and the Fourier transform of a set of 1s. The Fourier transform of the infinite sinewave yields a single value at the sinewave frequency, while the Fourier transform of a set of 1s yield the function $\sin(x)/x$. This process creates side lobes and reducing the high side lobes of the $\sin(x)/x$ curve would minimise leakage into the adjacent FFT bins.

9.5.4.3 Scalloping error

The FFT produces a discrete, rather than continuous, frequency spectrum. Suppose a frequency falls halfway between two of these points. In that case, the frequency

Table 9.2 Specifications for popular FFT window functions

Window	Shape of the window	Highest side lobe level (dB)	3 dB bandwidth (bins)	Scallop loss (dB)	Noise equivalent bandwidth (bins)
Rectangular	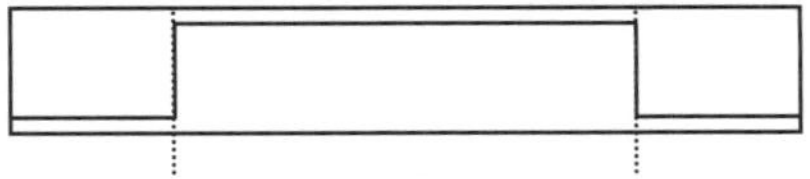	−13	0.89	3.92	1.00
Hanning	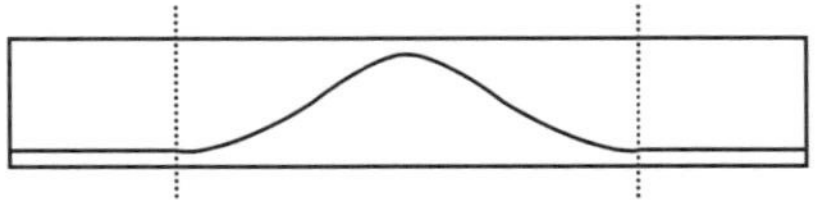	−32	1.44	1.42	1.50
Hamming	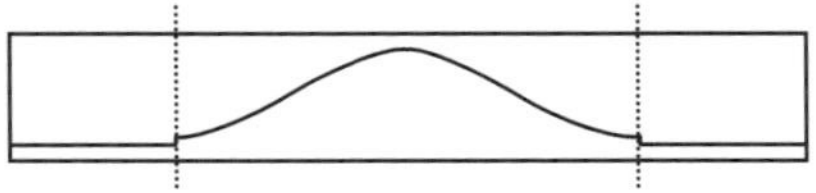	−43	1.30	1.78	1.36
Three-sample Blackman–Harris	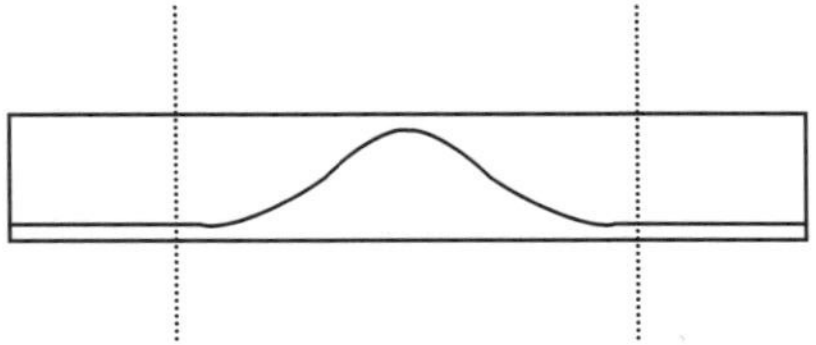	−67	1.66	1.13	1.71
Four-sample Blackman–Harris		−92	1.90	0.83	2.00

will appear at the two nearest points with equal amplitude. Both, however, will suffer from attenuation. This loss is called scalloping loss or error. An FFT's resolution (the difference in frequency, ΔF between two adjacent bins) depends on the sampling rate and record length:

$$\Delta F = \text{sample rate in samples per second/FFT record length.} \tag{9.11}$$

9.5.4.4 Equivalent noise bandwidth

Equivalent noise bandwidth is the ratio of the input noise power to the noise power in the output of an FFT filter times the input data sampling rate. Every signal contains some noise. That noise is generally spread over the frequency spectrum of interest, and each narrow band filter passes a certain amount of that noise through its main lobe and side lobes. White noise is used as the input signal and the noise power out of each filter is compared with the noise power into the filter to determine the equivalent noise bandwidth of each passband filter. In other words, equivalent noise bandwidth represents how much noise would come through the filter if it had an absolutely flat pass band gain and no side lobes.

9.5.5 Real time operation

It is important to realise that it takes a finite time interval to compute the FFT of a time record. If we can calculate the FFT of our time record while acquiring the current time record then we are said to be *operating in real time*. This can be performed by inserting a time buffer in our system as shown in Figure 9.12.

Real time operation of a system can be depicted by the timing diagram shown in Figure 9.13. However, if the time taken to compute the FFT of the previous time record is more than the time required to acquire the current time record, then we are operating in *non-real time*, as depicted by Figure 9.14.

The time record is not a constant for a spectrum analyser. It varies with the frequency span used. For wide frequency spans the time record is shorter. Therefore, as we increase the frequency span of the analyser, we eventually reach a span where the time record is equal to the FFT computation time. This frequency span is called the *real time bandwidth*. For frequency spans at and below the real time bandwidth, the analyser does not miss any data.

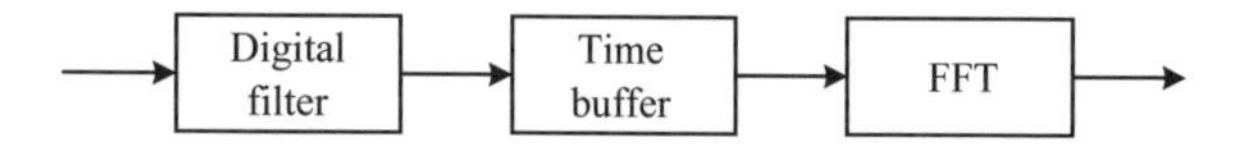

Figure 9.12 Time buffer added to block diagram of a FFT analyser

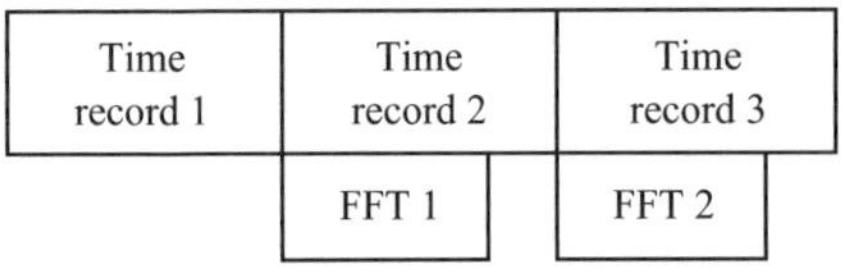

Time record 1 | Time record 2 | Time record 3

FFT 1 | FFT 2

Figure 9.13 Real time operation

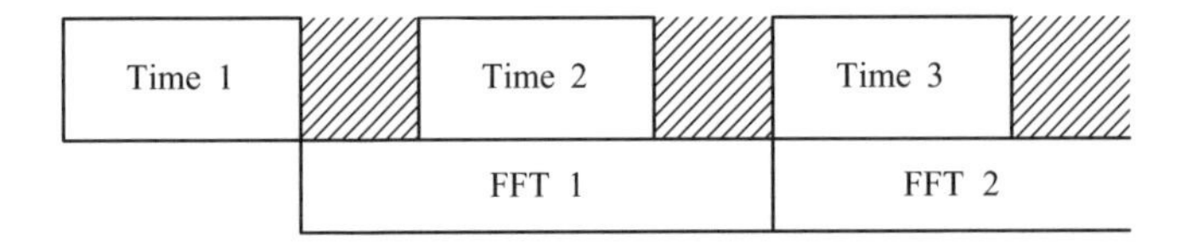

Figure 9.14 Non-real time operation

9.6 Types of spectrum analyser

There are three ways to make frequency domain measurements. These are: (i) the real time technique, (ii) Fourier transform analysis, and (iii) the swept tuned technique.

9.6.1 *Real time technique*

The real time technique, which uses a group of fixed band pass filters side by side, is useful for a range from d.c. to above audio frequencies. This type of spectrum analyser may be constructed as shown in Figure 9.15. In this case, the input signal is applied to a multi-coupler, which distributes the signal equally among several filters. The output of each is then detected and recorded as shown. Even though the capability is restricted within the audio band, one advantage of this type of analyser is that it measures signals as they occur. For this reason the '*multiple filter analyser*' is often called a '*real time analyser*'.

9.6.2 *Fourier transform analyser*

The Fourier transform analyser digitally processes a signal over a specific period of time to provide frequency, amplitude and phase data. This technique can also analyse periodic and non-periodic signals. Capabilities of most practical instruments are limited to about 100 kHz. A block diagram of an '*FFT spectrum analyser*', sometimes called a '*dynamic signal analyser*', is shown in Figure 9.16. During the past two decades, the development of digital storage scopes, making use of the processor techniques, has opened up a whole new world of FFT measurement opportunities. For example, the lower ends of DSOs have even incorporated the FFT techniques, where the oscilloscope works as a limited capability spectrum analyser. Reference 11 provides a detailed account of FFT applications in a commercial DSO family such as the TDS series from Tektronix. Fundamentals as applied to dynamic signal analysers are detailed in Reference 6.

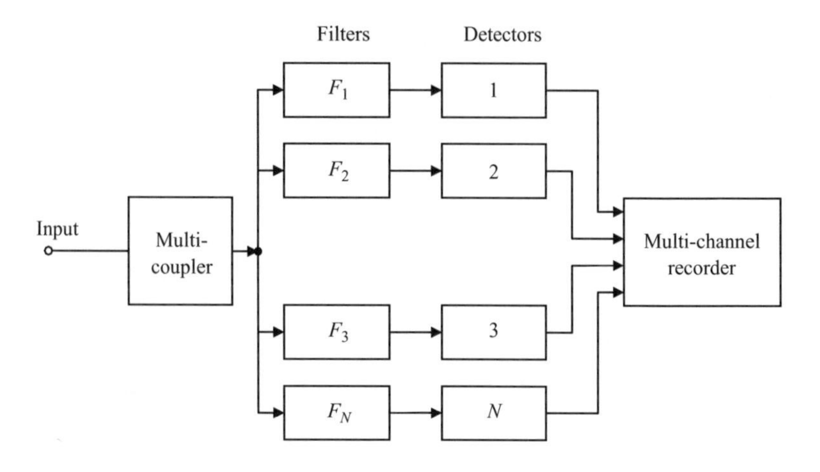

Figure 9.15 Multiple filter analyser

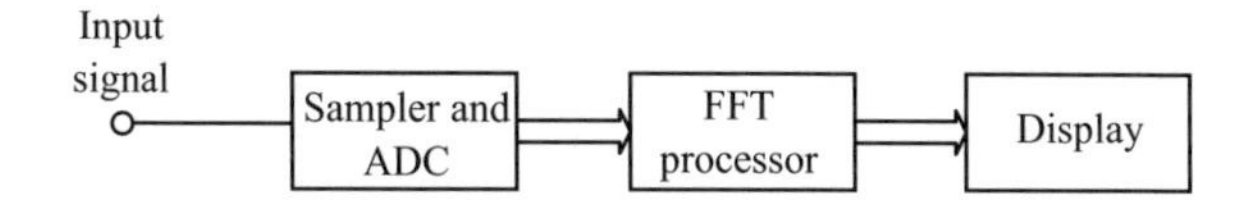

Figure 9.16 FFT analyser (dynamic signal analyser)

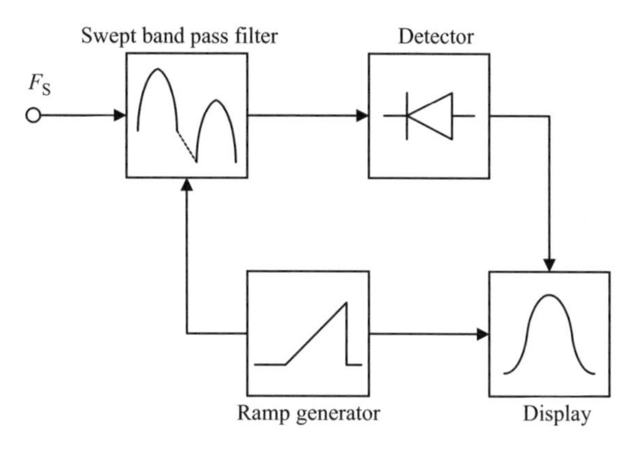

Figure 9.17 Tuned filter spectrum analyser

9.6.3 Swept tuned analyser

The '*swept tuned analyser*' can be either a tuned filter or a heterodyned receiver. The tuned filter, as shown in Figure 9.17, is cost effective, but does not provide the performance required for most signal analysis and measurements.

The '*superheterodyne spectrum analyser*' gives a balance of frequency resolution, speed and cost over the other techniques while substantially enhancing sensitivity and

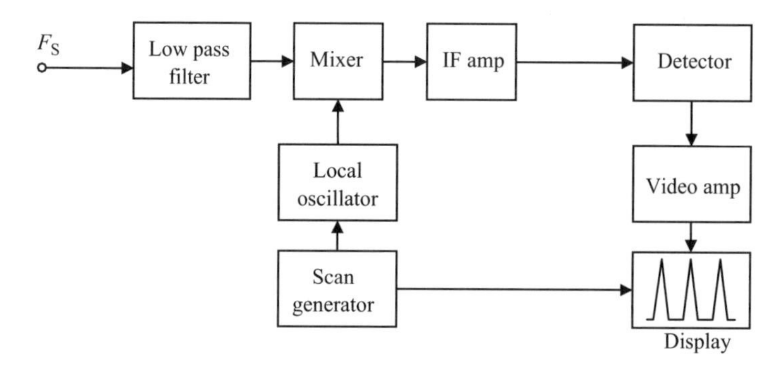

Figure 9.18 Simplified block diagram of superheterodyne spectrum analyser

frequency range. These types of instrument are very popular in industry because they have the capability of analysing signals beyond a few hundred GHz. A simplified block diagram of a superheterodyned spectrum analyser is shown in Figure 9.18. The heterodyned spectrum analyser, or simply '*spectrum analyser*', can show the individual frequency components that make up a complex signal. It does not, however, provide phase information about a signal.

9.7 Superheterodyne spectrum analyser

As mentioned briefly in the previous section, the superheterodyne spectrum analyser, sometimes called a scanning spectrum or sweeping signal spectrum analyser, operates on the principle of the relative movement in frequency between the signal and a filter. The important parameter is the relative frequency movement. It does not matter whether the signal is stationary and the filter changes or whether the filter is stationary and the signal is made to change the frequency.

To explain this consider Figure 9.19(a), which represents a spectrum composed of three discrete frequency continuous wave (CW) signals and a continuous dense spectrum in the middle. This spectrum is passed through a filter having the gain characteristics shown in Figure 9.19(b). The filter and spectrum have a relative frequency translation as indicated by arrows of opposite sense. The resultant display, shown in Figure 9.19(c), has the units of frequency, but takes a real time t to occur. Some of the fine details of the theoretical spectrum (Figure 9.19(a)), when translated into a Fourier transform $F(w)$, are lost in Figure 9.19(c) because of the finite frequency width and, hence, resolution of the filter. If the filter characteristics are made narrower, the ideal (cf. Figure 9.19(a)) and actual spectral representations become more alike. In other words, if the filter has zero bandwidth, the ideal and actual will have no difference at all.

Although the basic operation of the system is apparent from Figure 9.19, there are many ramifications, particularly with regard to the speed of relative frequency translation, which are not at all obvious. Now consider some of the details of the

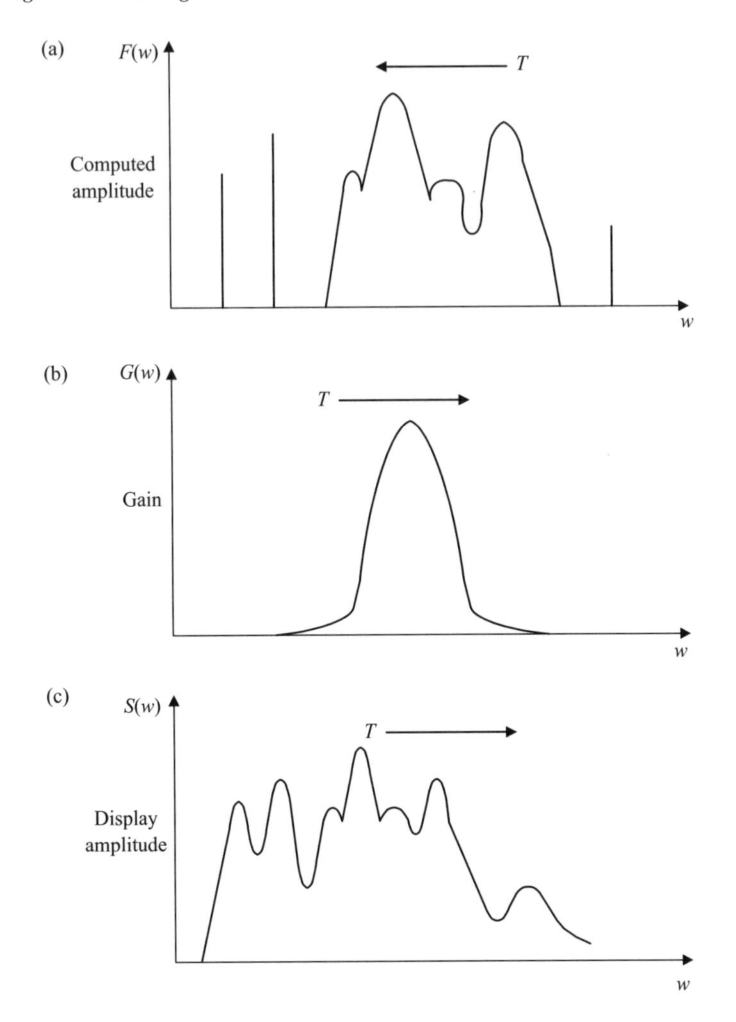

Figure 9.19 Sweeping spectrum analyser spectrum representation: (a) signal spectrum; (b) filter gain characteristics; (c) spectrum analyser display

sweeping signal spectrum analyser system. For the tuned filter analyser shown in Figure 9.17, the response to an input signal consisting of two sinewaves is illustrated in Figure 9.20. Note that the shape of the display is just the pass band of the filter as it sweeps by the incoming signals. This system is relatively simple and compact, yet there are practical difficulties.

In swept tuned spectrum analysers, the main problems stem from the tuneable filter construction. These generally have much wider bandwidths than is desired for most applications. If, for example, additional signals are added to those shown above, the CRT display would be as in Figure 9.21. The bandwidth of the filter is called the '*resolution*' of the spectrum analyser. By taking advantage of the latest digital filter

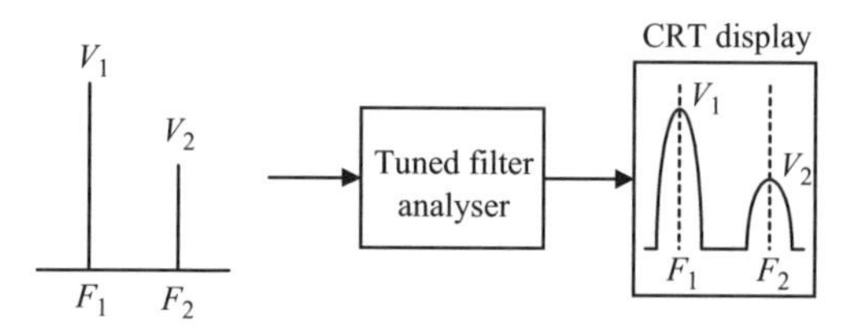

Figure 9.20 CRT display using a swept filter analyser

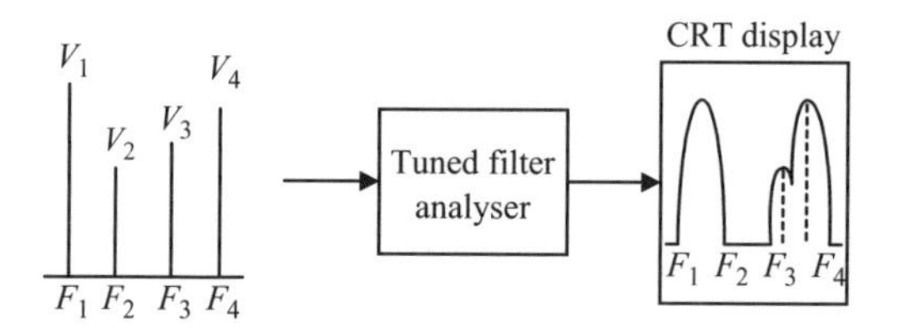

Figure 9.21 Problem due to filter bandwidth

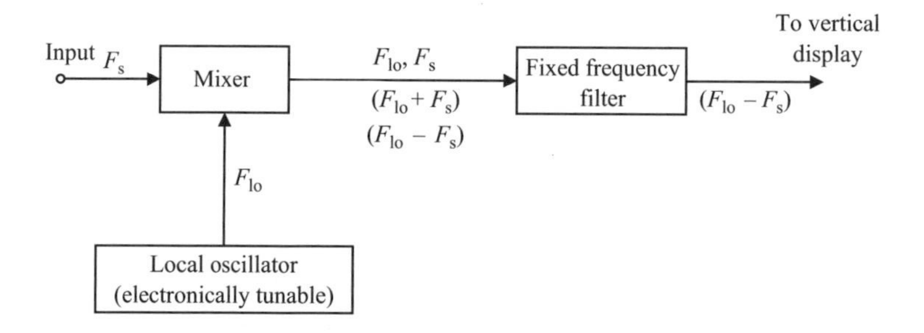

Figure 9.22 Mixer output frequencies

and FFT techniques some commercial designs have broken resolution speed barriers [12].

9.7.1 Block diagram of a swept tuned spectrum analyser

superheterodyne analyser is the most common type of instrument usable up to GHz order range of frequencies. Almost all modern spectrum analysers employ the superheterodyne principle. The fact that it provides better resolution and frequency coverage outweighs the fact that it is more complex than other types of analysers.

The superheterodyne system is based on the use of a mixer and a local oscillator. See Figure 9.22. The primary output signals from a mixer are shown where F_{lo} is the local oscillator frequency and F_s is the signal frequency.

The horizontal axis of the CRT can now be transformed from the time domain to the frequency domain by varying the local oscillator frequency in synchronisation

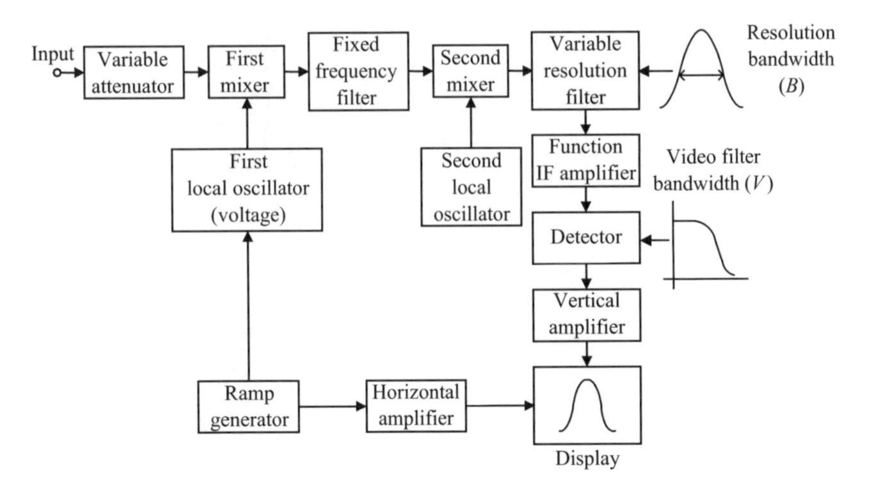

Figure 9.23 A basic superheterodyne spectrum analyser

with the horizontal position voltage. Compared with the tuned filter analyser performing the time-to-frequency domain transformation by varying the frequency of the filter with respect to the signal, the superheterodyne analyser performs this transformation by effectively varying the signal at the mixer output with respect to the filter frequency.

A basic superheterodyne spectrum analyser is shown in Figure 9.23. This system uses two mixers, a fixed frequency filter and a variable resolution filter, in addition to other basic components needed to display results on a CRT. Note that the variable resolution filter can be designed with a narrow bandwidth because the signal frequency has been substantially lowered by using two heterodyne stages. The filter is designed to have a bandwidth that can be varied manually. This is an important feature because it is normally desirable to use narrow resolution to observe signals that are close together and wide resolution to observe signals that are far apart.

For purposes of explanation in this chapter, the block diagram in Figure 9.23 shows a spectrum analyser with two local oscillator/IF stages. Practical instruments, however, can have as many as four stages.

9.7.2 Spurious and image responses

From the block diagram shown in Figure 9.24, applying the basic characteristics of a mixer, mixer output frequency (F_{mo}) can be described by

$$mF_{lo} \pm nF_{rf} = F_{mo}, \tag{9.12}$$

$$nF_{rf} \pm mF_{lo} = F_{mo}, \tag{9.13}$$

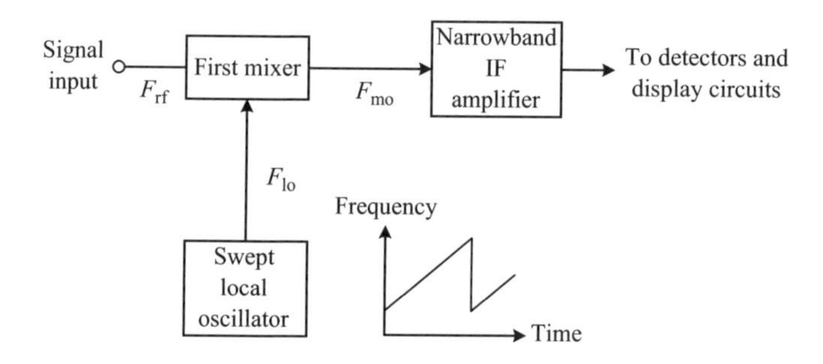

Figure 9.24 Swept front end of the heterodyne spectrum analyser

where,

$$\begin{aligned} F_{lo} &= \text{local oscillator frequency,} \\ F_{rf} &= \text{signal input frequency,} \\ F_{mo} &= \text{mixer output frequency, and} \\ m, n &= \text{positive integers including zero.} \end{aligned}$$

When the mixer output frequency, F_{mo} is equal to the IF centre frequency (F_0), there is a spectrum analyser response.

The above simple explanation indicates that in a spectrum analyser, you can have more than one possible output (which should truly represent the spectrum component of the input applied) from the mixing process. This causes components such as (i) IF feed through, and (ii) image response, in addition to the true response. Modern instruments are designed to minimize the effects of these image responses and IF feedthrough, etc., using appropriate circuitry. For details, Reference 13 is suggested.

9.7.3 Controls

Most modern spectrum analysers employ three primary controls. These are frequency, span per division and reference level. It is possible to make a variety of measurements using only the primary controls, although additional controls are provided. The added controls not only make the analyser more convenient to use, but also make the analyser more adaptable to measurement requirements. Many features of modern spectrum analysers are microprocessor controlled and selectable from on-screen menus. In modern designs microprocessors are used to provide selectable on-screen menus, etc.

9.7.3.1 Frequency control

Figure 9.25 shows a detailed block diagram of the input stages of a spectrum analyser. As shown in Figure 9.23, scanning spectrum analysers use a series of local oscillators and mixing circuits to measure the spectrum of the input signal. The first local oscillator (LO) determines the range of input frequencies analysed. For the details

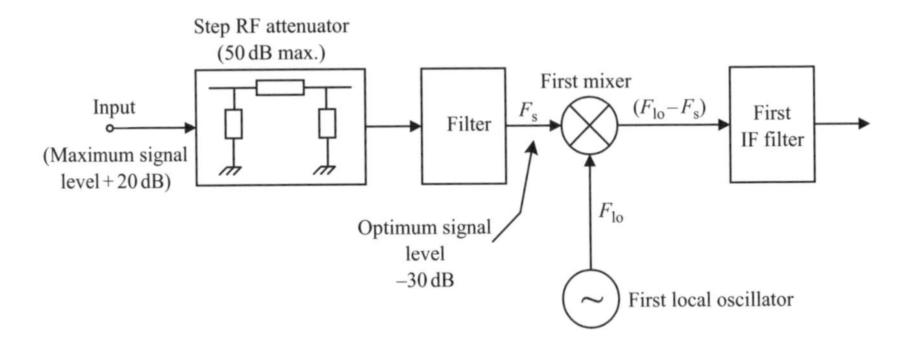

Figure 9.25 Spectrum analyser input stages indicating optimum first mixer level and frequency mixing

in Figure 9.25, by sweeping the frequency of the first LO over a specified frequency range or span, a corresponding range of input signal frequencies is swept past the resolution bandwidth (RBW) filter (discussed in section 9.7.4.1). The frequency control customarily determines the centre of the swept frequency range. In other words, the (centre) frequency control adjusts the average ramp voltage that is applied to the tunable oscillator.

9.7.3.2 Span control

The span control or 'span div^{-1}' control regulates the width of the frequency spectrum that is displayed by controlling the width of the local oscillator sweep. This control adjusts the amplitude of the ramp voltage. For example, if the user selects 20 MHz div^{-1} as the span div^{-1} for a 10 (horizontal) division CRT, the total sweep range will be 200 MHz. Setting the centre frequency to 175 MHz causes the analyser to sweep from 75 to 275 MHz. Most spectrum analysers have two special span control settings. They are maximum span and zero span. At maximum span the analyser sweeps across its maximum frequency range. For zero span, the analyser no longer sweeps across a spectrum; instead it behaves like a conventional (superheterodyne) radio receiver. The analyser is tuned to the centre frequency and the signal present in the RBW filter pass band is continuously displayed.

9.7.3.3 Reference level control

The reference level control varies the level of the signal necessary to produce a full screen deflection. For example, if the reference level is set for -10 dBm, a -10 dBm signal would rise just to the top graticule line. If a displayed signal is two divisions down from the top and the scale factor is 10 dB division^{-1} the actual signal is $-10\text{ dBm} - (2 \times 10)\text{ dB} = -30\text{ dBm}$.

The reference level is determined by the RF attenuation and the IF gain, but attenuation and gain are controlled by independent sections of the analyser. To avoid having to operate two controls, most analysers automatically select the proper amounts

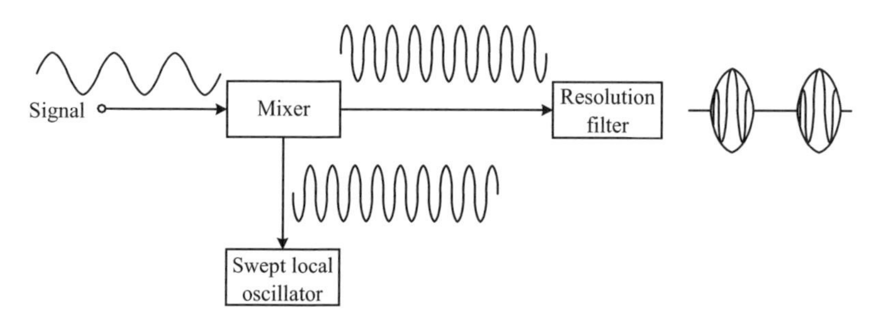

Figure 9.26 Simplified block diagram of spectrum analyser

of RF attenuation and IF gain. The RF attenuator determines the amount of attenuation the signal encounters just after it enters the analyser; see Figure 9.25. For optimum performance, the input signal reaching the first mixer must be attenuated to a level specified by the manufacturer. Exceeding the specified first mixer input level can result in distortion and spurious signal products, or, in extreme cases, damage to the mixer. All analysers have a maximum input level that must not be exceeded. Typically this level is +20 to +30 dBm (2.24 to 7.07 V across 50 Ω). Further details can be found in References 13–15.

9.7.4 Other controls

Other important controls are resolution bandwidth, sweep speed and dynamic range. To understand the importance of these parameters and controls consider the simplified spectrum analyser block diagram shown in Figure 9.26. The time/frequency diagram for this system is shown in Figure 9.27.

Here it is assumed that system operation is based on a mixer output composed of the difference frequency between local oscillator and signal. Figure 9.27 is for a signal composed of two discrete frequency components. The signal components at frequencies F_1 and F_2 are shown as straight lines having infinitesimal frequency width and infinite time duration.

A constant frequency signal is converted to a frequency sawtooth by combining it in a mixer with a frequency sawtooth from the swept local oscillator. In the example, it was assumed that the mixer output consists of the difference frequency between the local oscillator frequency sawtooth and the input. Other combinations, such as the sum of the frequencies, lead to similar diagrams. The display consists of pulses whose time position is determined by the time of intersection of the filter passband and the sweeping signal, and whose width is equal to the time interval during which the sweeping signal frequency is within the filter passband. The bursts or pulses generated by the relative translation of signal and filter are pseudo-impulses representing the frequency domain characteristics of the signal. Whereas the time position of these pulses represents the input signal frequency and is determined by the incoming signal, the width τ of these pulses is determined solely by the spectrum analyser parameters. The width τ is equal to the time that the sweeping signal frequency is within the

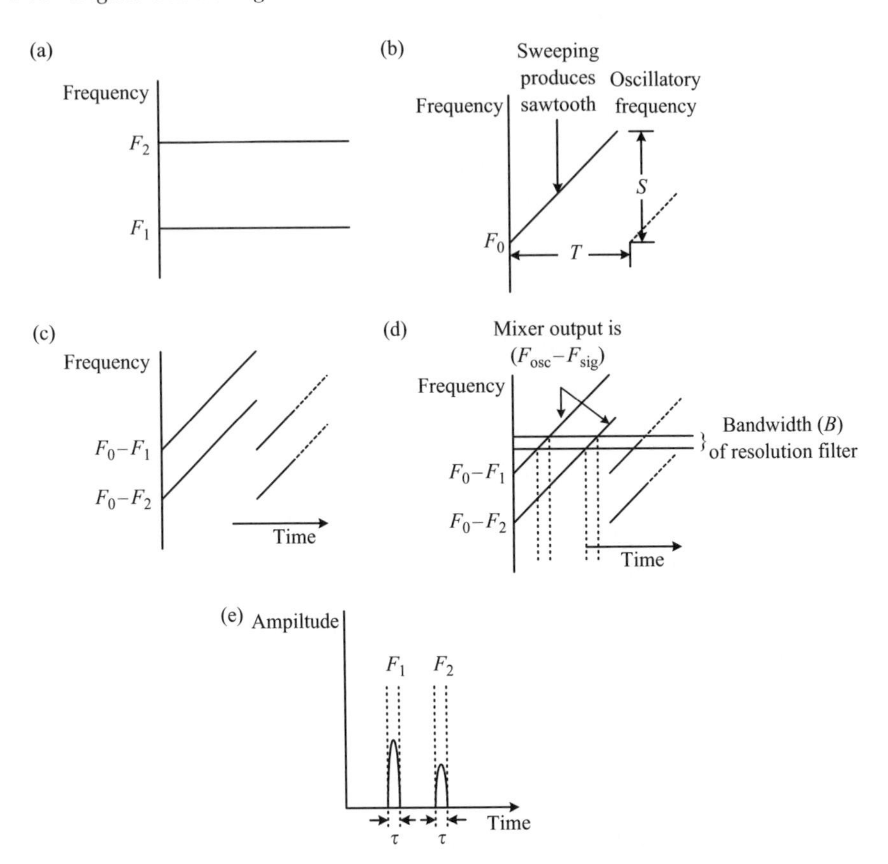

Figure 9.27 Time/frequency diagram for sweeping signal spectrum analyser

passband of the filter, and from simple geometrical considerations is

$$\tau = \frac{B}{S}T. \tag{9.14}$$

The sweep time T is the time it takes the electron beam to traverse the horizontal width of the CRT. Hence the physical width of τ in cm does not change with changing T. The actual time duration of τ, however, given by eq. (9.14), is directly determined by the sweep time T. At low sweep time T, or with narrow resolution filter bandwidth B, or with large span S, the pulse width τ can become quite small. For example, a full screen sweep time of 1 ms (100 μs div^{-1}), a resolution bandwidth of 10 kHz, and a span of 10 MHz result in a burst at the filter output that is only 1 μs wide. Such a narrow pulse cannot be passed by a 10 kHz wide filter without distortion. As with any pulse that is passed through a filter of insufficient bandwidth, the output is of lesser amplitude and greater time duration than the input, as illustrated in Figure 9.28. Because the distorted response is what appears on the CRT screen,

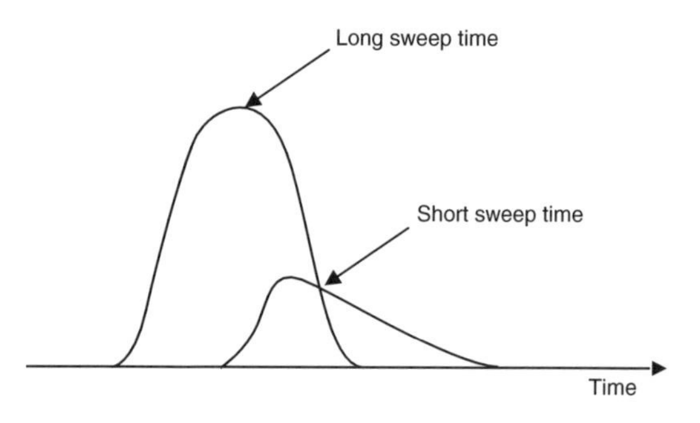

Figure 9.28 Resolution distortion for short sweep time

the loss in amplitude shows a loss in sensitivity, and the apparent widening of the resolution bandwidth shows a loss in resolution.

Equation (9.14) above indicates the limits of manipulations on the span and RBW controls of an instrument.

Analytical expressions relating to loss of sweep time to RBW and to span are discussed in Reference 13.

9.7.4.1 Resolution bandwidth selection

Resolution bandwidth (RBW) filters are bandpass filters located in the spectrum analyser's final IF stages (Figure 9.23). They determine how well closely spaced signals can be separated. The narrower the RBW filter the more clearly two close signals can be seen as separate signals. The RBW filters also determine the analyser response to pulse signals and to noise. The resolution bandwidth control selects which RBW filter is used.

The shape of a spectrum displayed on the analyser screen is a combination of the shape of the RBW filter and the shape of the true signal spectrum (Figure 9.19(c)). Thus, the measured analyser response to two equal amplitude sinewave signals that are one RBW apart in frequency resembles Figure 9.29.

Definitions for spectrum analyser resolutions are published by the IEEE, IEC, etc., as international standards. RBW filters are defined by their bandwidths and shape factors. Bandwidth is specified in Hz either at 3 dB or 6 dB down from the filter peak. The shape factor, which indicates the steepness of the filter, is the ratio of the RBW filter bandwidth 60 dB down from the peak to its normal (3 dB or 6 dB) bandwidth; see Figure 9.30.

The shape factor is important in determining the minimum separation between two signals which have equal amplitudes to be resolved. Ideally, RBW filters should be extremely narrow in order to trace out signal spectral shapes faithfully and to resolve very closely spaced signals. The smaller the ratio the sharper the filter. However, using a narrow RBW filter with a wide span results in a signal sweep that is too long. Therefore, to maintain reasonable speeds the resolution bandwidth must increase as the span div^{-1} increases.

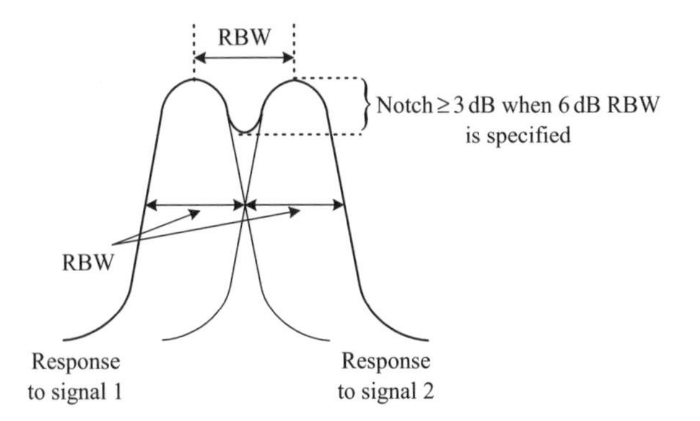

Figure 9.29 Two closely spaced sinewaves 'just resolved'

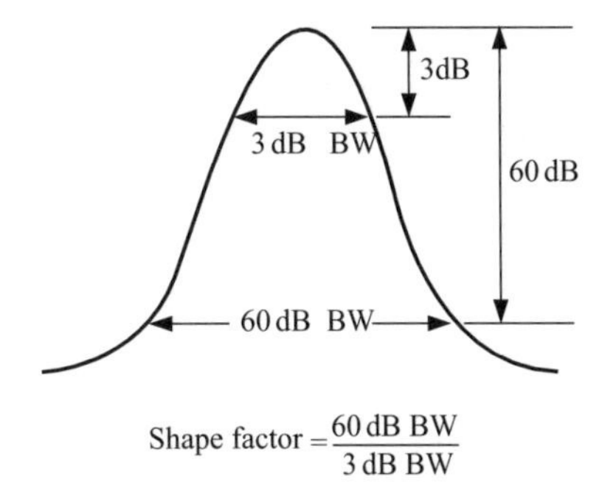

Figure 9.30 Filter bandwidth and shape factor

Another characteristic associated with RBW filters is the decrease in displayed noise floor as the bandwidth is narrowed. The '*noise floor*' is the baseline or lowest horizontal part of the trace. The noise floor decreases because noise power is proportional to bandwidth. A change in the bandwidth of the RBW filter by a factor of 10 (say, from 300 to 30 kHz) should decrease the noise floor by about 10 dB. The reduction in the noise floor works to advantage when we are looking for low level narrow band signals. Figure 9.31 is a composite photograph of the spectra produced by the same signal when different RBWs are used. In the upper trace the sidebands of the signal are buried in the noise. In the lower trace the RBW is reduced by a factor of ten between upper and lower traces which, in effect, reduces the noise by 10 dB to reveal the sidebands.

The limitations imposed on a spectrum analyser by the RBW filter are significant. Through the use of microprocessors, modern spectrum analysers automatically choose the best resolution bandwidth as a function of the span div^{-1} and sweep rate selected.

9.7.5 *Practical aspects in selecting span, sweep time and BW*

Referring to Figure 9.23, we can relate spectrum analyser operating functions to circuit blocks. The frequency sweep width of the first local oscillator determines the

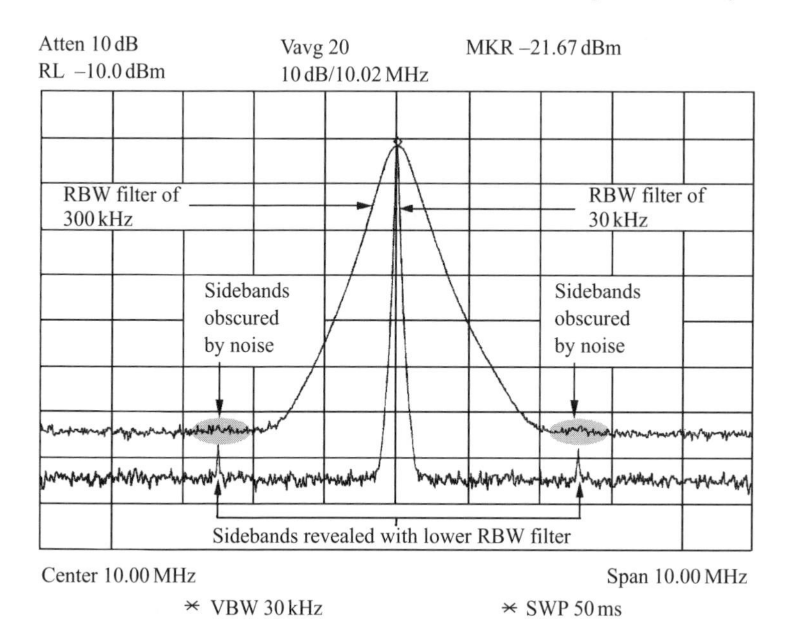

Figure 9.31 Composite photograph showing sidebands obscured by noise when using wider bandwidth

span. The swept mixed signal is translated after the bandpass filter that determines the resolution, or signal separation. After the resulting spectrum is detected, it is amplified in a low pass video filter amplifier prior to display on the CRT. The two most critical functions are the pre-detection resolution filter bandwidth (B) and post-detection video filter bandwidth (V).

Supported by the basic concepts discussed summarised by eq. (9.14), slower sweep speeds provide improved detail. In an overall sense, referring to Figure 9.23, a spectrum analyser's sweep time (T) depends on the frequency span to be displayed (S), the resolution bandwidth (B), the video filter bandwidth (V), the signal intercept or detector function (D), and the vertical display setting (VD). (Some secondary factors that also affect the sweep time, such as the digital storage sample rate, are not discussed here.) The sweep time increases directly with the span and inversely with the video filter bandwidth and the square of the resolution bandwidth. The peak detector function in the 10 dB div^{-1} logarithmic vertical mode usually provides the fastest sweep and the minimum measurement time. In addition, faster sweep times can be achieved by accepting reduced measurement accuracy or by using accessory items such as a preamplifier.

Reference 16 shows that for spectrum analysers whose bandwidth is specified at the −3 dB points (for example, those made by Hewlett Packard) the span (S), resolution bandwidth (B), and sweep time (T) are related by the formula $S/TB^2 = 0.5$. For spectrum analysers that specify resolution bandwidth at the −6 dB points (those made by Tektronix, for example), the relationship is $S/TB^2 = 0.22$. The two formulas are identical when you consider that the resolution bandwidth defined at the −6 dB points, $B_{6\,\text{dB}} = 1.5B_{3\,\text{dB}}$ (the resolution bandwidth defined at the −3 dB points), and $0.5/0.22 = 1.5^2$. This relationship holds for the normal spectrum analyser default setting where the video filter bandwidth is equal to or greater than the resolution bandwidth. Real spectrum analysers sweep about 25 per cent slower than the theory predicts.

9.7.6 Use of DSP techniques to improve on resolution–speed barriers

Traditional swept spectrum analyser techniques require designers to sacrifice measurement time for improved resolution. In fact, dropping the resolution bandwidth by a factor of three increases the sweep time by a factor of nine. By taking advantage of the latest in digital filter and fast Fourier transform technology, Hewlett Packard has broken that resolution–speed bottleneck.

The HP 3588A spectrum analyser offers a range of resolution bandwidths from 20 kHz to 0.0045 Hz, while increasing sweep speeds by up to 100 times or more, depending on the application. The analyser offers a wide variety of frequency spans from 10 Hz to 150 MHz (Figure 9.32). Previously, the best resolution bandwidth available in a swept analyser in the MHz range was about 3 Hz [12].

In this case the use of digital filter and FFT techniques allows the instrument to sweep the components about four times faster than it could sweep a conventional filter. 'Over-sweeping' the filter distorts the signal, but the distortion is predictable and can be corrected by an internal algorithm.

In the narrowband zoom mode, the HP 3588A uses an FFT to make spectrum measurements in bandwidths of up to 40 kHz anywhere within the instrument's 150 MHz range. At comparable resolution bandwidths, the unit makes these measurements 50–400 times faster than other swept tuned analysers do.

9.8 Sweep control/use of video filters and display storage

The sweep control selects the sweep speed at which the spectrum is swept and displayed. Sweep speed units are in time per division (div); a typical value might be 20 ms div^{-1}. The control can be either manually or auto-selected. Automatic selection is the normal setting for sweep control and, in this case, as with the automatic selection of RBW, most analysers can automatically select the optimum sweep speed, depending on the other parameter settings such as span, RBW and video filter BW. If manually selected, one should bear in mind that too fast a sweep speed may cause inaccurate measurements owing to the RBW filter not having sufficient time to charge. When swept too slowly the display accuracy is not affected but the display may flicker objectionably or fade out entirely before the start of the next sweep. Flicker and fade-out can be overcome using display storage. A video filter is a post-detection filter (sometimes referred to as a noise averaging filter) and it is used primarily to reduce noise in the displayed spectrum. A simplified block diagram in Figure 9.33 is illustrative of its function. The sensitivity of a spectrum analyser can be specified as that

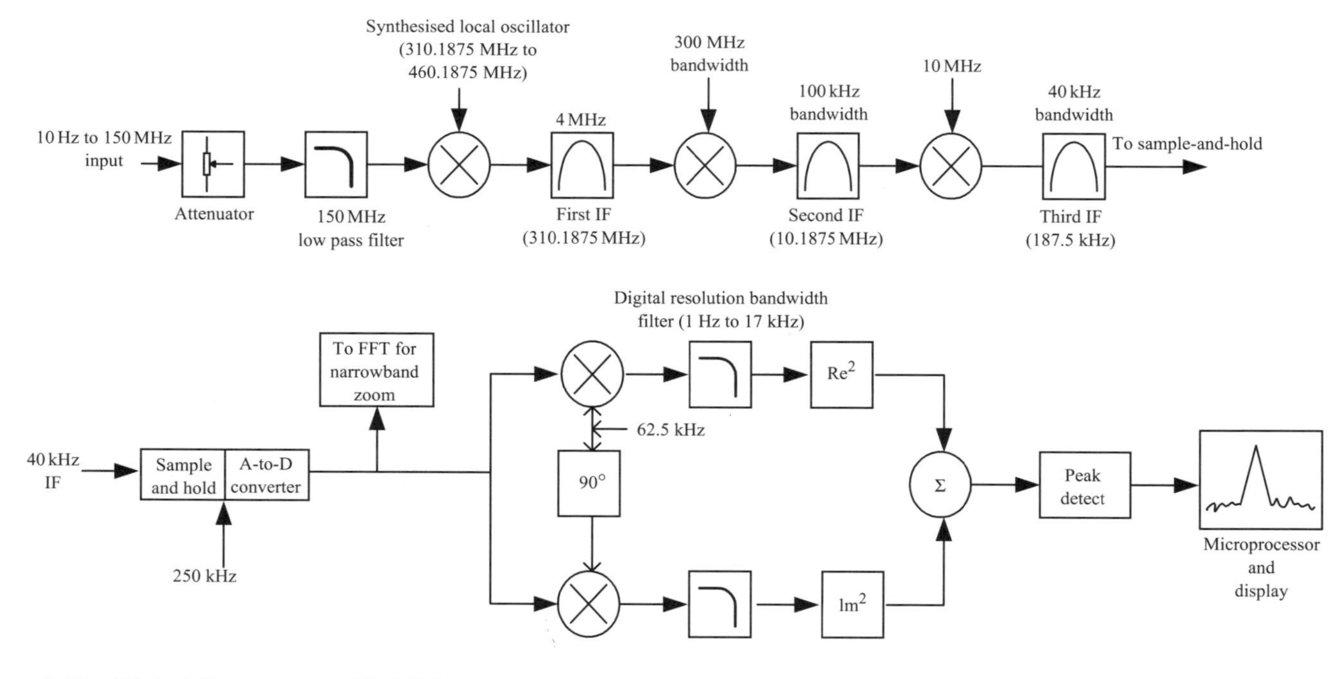

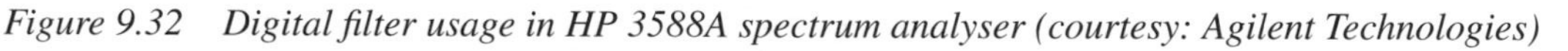

Figure 9.32 Digital filter usage in HP 3588A spectrum analyser (courtesy: Agilent Technologies)

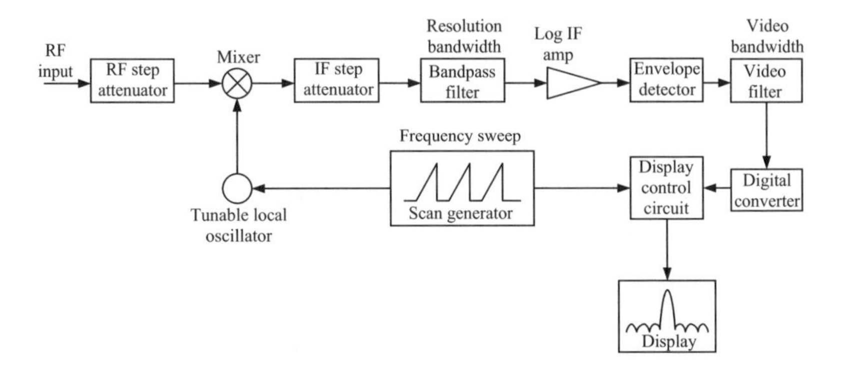

Figure 9.33 A spectrum analyser having video filter and digital storage

condition at which the signal level equals the displayed average noise level. This is the level where the signal appears to be approximately 3 dB above the average noise level.

In using video filters care should be taken as they may also reduce the signal amplitude in certain types of signals such as video modulation and short duration pulses. Most analysers provide several video filter bandwidths. The video filter control enables the user to turn the filter on and off and to select its bandwidth. As with the RBW and sweep controls many analysers can automatically select the video filter bandwidth. 'Auto' is the normal setting for this control.

Scanning spectrum analysers typically require from a few milliseconds to several seconds to complete a sweep of the signal spectrum. At high sweep speeds use of an analogue CRT display is practicable. However, at slower speeds the trace can flicker objectionably or fade out entirely before the next sweep can be started. To overcome this difficulty modern analysers use digital storage. In this case the spectrum is divided into small frequency increments and, as the analyser sweeps through each increment, the amplitude of the signal spectrum is sampled and digitised. The digitised data are then retrieved, reconverted to an analogue signal, and hence displayed on the screen. Even if the sweep speed is very slow, the trace can be refreshed from the stored data at a rate that gives it a constant and flicker-free appearance. For more details References 13, 15 and 17 are suggested.

9.9 Use of a tracking generator with a spectrum analyser

9.9.1 Tracking generator

A tracking generator is a signal generator whose output frequency is synchronised to, or tracks with, the frequency being analysed at any point in time. When used with a spectrum analyser, a tracking generator allows the frequency response of systems (such as filters, amplifiers or couplers) to be measured over a very wide dynamic range. The measurements are performed by connecting the output of the tracking generator to the input of the device being tested, and monitoring the output of the DUT with

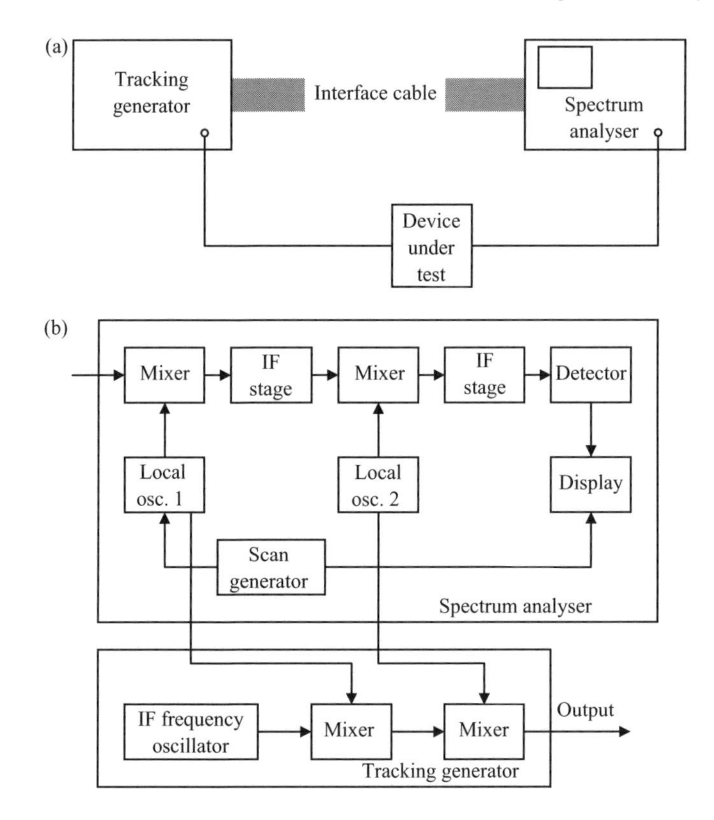

Figure 9.34 Spectrum analyser and tracking generator combination: (a) tracking generator/spectrum analyser test set up; (b) local oscillator connections

the spectrum analyser. A tracking generator is an oscillator/mixer combination that uses the local oscillator outputs of the spectrum analyser.

9.9.2 Spectrum analyser/tracking generator combination

Figure 9.34(a) shows the tracking generator/spectrum analyser test set up for use with a DUT. Figure 9.34(b) shows a simplified block diagram of a spectrum analyser/tracking generator combination showing the local oscillator connections. The spectrum analyser and tracking generator are designed to work as a pair. Some manufacturers have built-in tracking generators for their spectrum analyser products and some sell these separately.

The response displayed on the screen of the analyser is the combination of the '*unflatness*' of the tracking generator/spectrum analyser, pair and the DUT. In modern spectrum analysers, which have the digital storage facility, the flatness added due to the spectrum analyser and tracking generator pair can be easily compensated using the storage functions. In these situations it is necessary that the user is capable of using the same vertical display modes to reduce errors.

9.10 Alternative instruments for spectral analysis

When one wishes to observe the spectral content of a signal, there are several choices of test equipment. These include:

- spectrum analysers,
- Fourier analysers,
- vector signal analysers, and
- digitising oscilloscopes.

Each of these instruments has advantages and disadvantages, and one's choice will depend on its application. Table 9.3 lists a few benefits and limitations in using the above equipment for spectral analysis. For more details References 8 and 9 are suggested.

Table 9.3 Benefits and limitations of equipment for spectral analysis

Instrument	Benefits	Limitations	Applications
Spectrum analyser	Widest frequency range Widest frequency span	Requires a stable signal Scalar measurements Does not go down to d.c. Large minimum frequency resolution	RF spectral plots EMI scans
FFT analyser	Narrow frequency resolution Vector measurements Measurements down to d.c. Transient analysis Rich analysis capabilities	Narrow frequency range Narrow frequency span	Vibration analysis Acoustic analysis
Vector analyser	Benefits of FFT analyser plus wider frequency range	Does not go down to d.c.	RF spectral plots with phase information
Digitising oscilloscope	Narrow frequency resolution Widest bandwidth for transient analysis Measurement down to d.c.	Narrow frequency span Lower dynamic range Aliasing Spurious signals Limited analysis freatures	Spectral plots of transient signals Phase analysis of transient and continuous signals

9.11 Spectrum analyser performance factors

Modern spectrum analysers have frequency ranges varying from a few Hz to over 300 GHz. Spectrum analysers are classified in three categories: low frequency, communications and microwave analysers. Low frequency (or audio frequency) analysers employ DSP techniques to perform FFT on signals in the 0–1 MHz range.

Spectrum analyser designed for communications frequency range, 10 kHz to 2 GHz, best illustrate the advances made over the past two decades. These instruments have RF sections with low enough residual FM to permit bandwidth resolutions as low as 3 Hz. The resolutions bandwidth range of modern communications spectrum analysers are within a few Hz to about 5 MHz. Prices vary from US$8000 to over US$60 000. Reference oscillator stabilities may be 0.2–10 ppm with input amplitude ranges varying from +30 dBm to −150 dBm.

Most modern spectrum analysers have digital storage and processing capabilities. They have automatic adjustment capabilities for RBW, video bandwidth, sweep time, etc. In addition, digital signal processing allows built-in diagnostics, direct plot capabilities, programming custom measurement routines and frequency settable markers, etc. The IEEE-488 interface comes as a standard or option with most expensive spectrum analysers.

9.12 References

1 KULARATNA, N.: *Modern Electronic Test and Measuring Instruments*, IEE, London, 1996.
2 ROHDE & SCHWARZ: 'Fundamentals of spectrum analysis using spectrum analyzer FSA', Application Note, Info 001 138.
3 BEADLE, E.R.: 'Unifying overview of applied transform theory', *Electronic Design*, May 15, 1995, pp. 107–128.
4 LYNN, P.A. and FUERST, W.: *Introducing Digital Signal Processing*, John Wiley & Sons, New York, 1990.
5 BERGLAND, G.D.: 'A guided tour of the fast Fourier transform', *IEEE Spectrum*, July 1969, pp. 228–239.
6 HEWLETT PACKARD: *The Fundamentals of Signal Analysers*, Application Note 243, 11/89.
7 MARTIN, R.: 'Shift your domain through FFTs', *Test & Measurement World*, September 1993, pp. 63–68.
8 ALLEN, M.: 'Weigh the alternatives for spectral analysis', *Test & Measurement World*, November 1999, pp. 39–44.
9 LYONS, R.: 'Windowing functions improve FFT results – Part I', *Test & Measurement World*, June 1998, pp. 37–44.
10 LYONS, R.: 'Windowing functions improve FFT results – Part II', *Test & Measurement World*, September 1998, pp. 53–60.
11 TEKTRONIX, INC.: 'FFT applications for TDS', Application note-55W-8815-0, February 1993.

12 NOVELLINO, J.: 'Spectrum analyser breaks resolution–speed barriers', *Electronic Design*', 11 January 1990, pp. 141–142.
13 ENGELSON, M.: *Modern spectrum analyser theory and applications*, Artech House Publishers, 1984.
14 ENGELSON, M.: *Modern Spectrum Analyser Measurements*, JMS/Artech House Publishers, 1991.
15 TEKTRONIX, INC.: 'Spectrum analyser fundamentals', Application Note 26W-7037-1, Tektronix, USA, 1989.
16 TEKTRONIX, INC.: 'Sweep-time results in spectrum analysis', Tektronix Application Note 2EW-8710-0.
17 GALLANT, J.: 'Spectrum analysers: Modern instruments ease frequency analysis', *EDN*, 9 May 1991, pp. 95–102.
18 STRASSBERG, D. and MINK, J.: 'Signal and network analysers span the spectrum from audio to light', *EDN*, January 1997.

Further reading

ENGELSON, M.: 'To see spectrum analysis in detail fast, match sweep time to measurement', *EDN*, 7 July 1994, pp. 125–130.

SMITH, W.W., and SMITH, J.M.: *Handbook of Real-time Fast Fourier Transforms*, ISBN 0-7803-1091-8, IEEE Press, USA, 1995.

Chapter 10
Logic analysers

Dileeka Dias

10.1 Introduction

In troubleshooting analogue circuits, the oscilloscope is generally the instrument of choice. The oscilloscope also has its uses in digital circuits although it is inadequate for effective troubleshooting. Digital circuits, such as those found in microprocessor based systems, possess a very large number of non-periodic signals, which, most importantly, carry binary information. What is required here is an instrument that extends the capabilities of the oscilloscope such that the user will be able to lock into the system and trace the flow of digital information. This is how the logic analyser has been created. It has the ability to acquire and store digital data over a very large number of channels. It can be very selective about the data it acquires such that only the wanted section of data can be selected out of the vast amount of data available. Finally, it can process the stored data, and present them in a variety of useful formats, from signals to source code [1].

Increased product requirements, complex software and innovative hardware technologies make it difficult to meet time-to-market goals in today's digital systems. Meeting that challenge depends on the designer's ability to debug and characterise hardware, design and test software, perform system integration, and troubleshoot tough problems. The logic analyser is designed to enable the engineer to tackle all these design aspects that could be very time consuming without a proper tool.

The chapter will start with a survey of special requirements for troubleshooting digital circuits as opposed to analogue circuits. It will be shown how the oscilloscope fails to meet these requirements. The logic analyser will then be presented as a tool developed as an extension of the oscilloscope to handle digital systems testing. The basics of data acquisition in a logic analyser, which include probing, clocking, triggering and display will be described. This will be followed by the modern techniques developed by major equipment manufacturers in each of the above areas.

Several families of modern logic analysers are also presented towards the latter part of the chapter, with a summary of their performance features. The chapter ends with some application examples.

10.2 Digital circuit testing and logic analysis

10.2.1 Analogue and digital circuits

Even the most complex of analogue circuits tend to be sequential in nature with regard to the transmission of signals. Signal flow tends to be confined to a few major paths in the circuit. Information is carried by the frequency or the phase of signals. The most useful instrument for troubleshooting in this situation is a multichannel oscilloscope with sufficiently high bandwidth. By tracing the progress of signals along these paths and checking them against reference values, it is generally possible to isolate malfunctioning areas of the circuit [1].

By contrast, signals in digital circuits tend to be parallel in nature. A large number of input signals can be present, which can propagate to several outputs through an equally large number of paths. In addition, memory components such as flip-flops and registers complicate the situation because they can introduce unpredictable initial values. In the case of microprocessor circuits, hundreds of different signals can be present. Microprocessor signals, however, can be grouped into several classes such as data, address and control. A typical microprocessor found in a personal computer today has a large number of such signals. Most microprocessor signals are governed by complex and very tight timing requirements. An example of a read cycle timing diagram of a microprocessor is shown in Figure 10.1 [1]. Whereas a few tens of nanoseconds error in timing of these signals could cause a malfunction at clock speeds of tens of MHz, at today's processor speeds of several hundreds of MHz, a few nanoseconds error can be critical. The ability to look at these signals simultaneously and at the right time is essential in troubleshooting these circuits.

Interest in microprocessor circuits does not depend solely on the occurrence of a certain set of signals, but in their higher level effect, that is, the execution of the machine code instruction, and consequently, the machine code program. It would be very useful if execution of the program could be traced in terms of data, address and the instructions executed. It would be possible then to check if the program has, for example, got stuck in one place or is going into an endless loop. Another characteristic of computer circuits is that devices external to the CPU can interrupt the microprocessor. Interruptions cause the current program to be suspended, and an interrupt service routine (ISR) to be executed. Once the ISR is complete, the previous program is restored. Problems can creep in here either because of faults in the interrupt controller circuits, or bugs in the interrupt service routine. Being able to look at the program flow in such instances would certainly help in determining what went wrong [1].

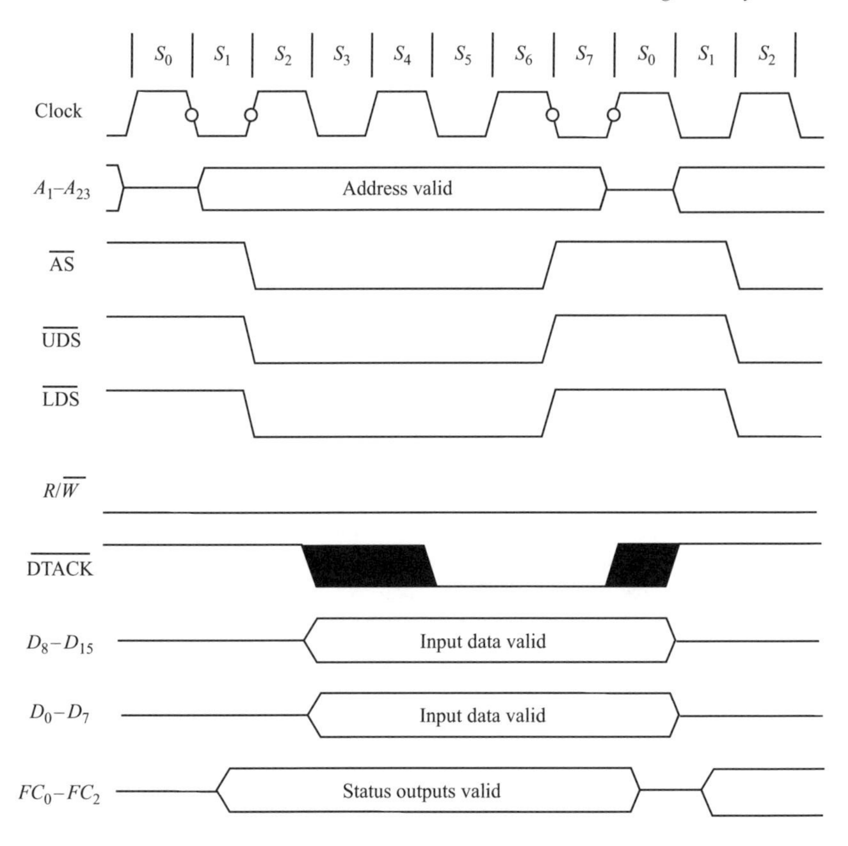

Figure 10.1 A read cycle timing diagram of a microprocessor

10.2.2 Types of digital faults

The types of fault that can occur in a microprocessor system may be divided into software and hardware faults. Software faults are caused by errors in programming. These can simply be wrong, or have missing instructions, wrong addresses, etc. Some software errors can be hard to detect because they may not occur each time the program is run. An example may be a conditional jump instruction for which the conditions are not properly defined. To detect software faults, it would be helpful to capture and list data flow on the address and data buses. It would be even more helpful if this information could be represented in disassembled format.

Hardware faults can arise in several forms. They may be the result of some signals permanently being high or low (stuck-at faults). They may be caused by timing errors or glitches (spurious narrow pulses), or they may be caused by races between signals. If the signals of interest can be displayed in timing diagram form, these faults can be detected. In addition, it is helpful if the actual signal voltages can be monitored for a given period to verify that it stays within the accepted logic levels [1].

An analogue oscilloscope would not help in the situations mentioned above as most digital signals are non-periodic. A digital oscilloscope will fare a little better. Even so, most of the conditions described above will go undetected for the following reasons [1]:

(i) the large number of channels that need to be monitored,
(ii) the complex triggering needed, which is beyond most medium range oscilloscopes, and
(iii) the high level of data processing and display required.

10.2.3 The logic analyser for troubleshooting digital circuits

The logic analyser is specially tailored to troubleshoot digital and, especially, processor based hardware. It is usually equipped with many channels (sometimes numbering in the hundreds) and has the ability to sample at a high rate, using either an internal or an external clock. It is provided with a complex triggering facility so that one particular event that is of interest can be captured. After the storage of data they can be displayed in an easily digested form, such as timing waveforms, state tables or disassembled listings [1].

At this stage, it is helpful to contrast the logic analyser with the oscilloscope. An oscilloscope plots signals against time and usually has up to four channels. It shows signals with a high vertical resolution and its accuracy will be determined by its bandwidth. A logic analyser, by contrast, shows a large number of channels. Its vertical resolution is only one bit, that is, it shows the signal only as zero or one and the transitions between the two levels. The accuracy of the logic analyser is determined by its sampling rate. Table 10.1 compares the generalised characteristics of analogue scopes, digital scopes and logic analysers [1].

Table 10.1 Comparison of oscilloscopes and logic analysers

Analogue oscilloscope	Digital oscilloscope	Logic analyser
2 to 8 channels	2 to 8 channels	8 to 64 or more channels
Single channel/probe	Single channel/probe	Multiple channels/probe
Displays actual waveform	Displays actual waveform	Idealised timing diagram State table Disassembly display
Trigger starts trace	Trigger stops acquisition	Trigger stops acquisition
Single trigger	Single trigger	Multiple trigger levels
Displays repetitive signals easily	Displays single or repetitive signal	Displays single events
No memory for waveform storage	Memory requirements moderate	Moderate to high memory requirements

Table 10.2 Selecting between an oscilloscope and a logic analyser

When to use an oscilloscope	When to use a logic analyser
• When you need a high vertical (voltage) resolution. e.g. to see small voltage excursions (such as noise spikes) in a signal • When you need high time interval accuracy, e.g. rise times, fall times of pulses	• When you need to see the relationship of many signals at once • When you need to examine the functional behaviour of a circuit • When you need to trigger on a data pattern appearing on a combination of several lines

10.2.4 Guidelines on selecting between an oscilloscope and a logic analyser for testing

Because of some overlapping capabilities between an oscilloscope and a logic analyser, either may be used in some cases. When given the choice between using an oscilloscope and using a logic analyser, most people will choose an oscilloscope. This is because the latter is more familiar to most users. It is one of the most 'general purpose' of all electronic instruments [2]. However, because a logic analyser is tuned to the digital world, it does not have as broad a use as an oscilloscope; see Table 10.2.

In the initial stage of an electronic design, the oscilloscope is more useful, as the designer is more interested in the parametric information of the signals. However, at later stages, the relationships between signals and the activities performed by the signals are of more interest. Thus, at this stage, the logic analyser is more useful.

Mixed signal oscilloscopes (MSOs) have both logic analyser and digitising oscilloscope channels. This helps the designer to observe exact signal voltages and/or timing information from the oscilloscope channels simultaneously with information from the logic analyser channels.

10.3 Logic analyser fundamentals

10.3.1 Basic operation

A logic analyser works by sampling input signals and saving the samples in its internal memory. The selection of data to be saved is done by triggering the analyser, and the saved data are displayed in a variety of ways. A large number of input channels, as well as complex triggering and display mechanisms, are incorporated into the logic analyser. In brief, logic analysers perform the broad functions of multichannel, high speed, data acquisition and display.

The main functional units of a logic analyser are: the memory, the sequence controller, a clock generator, a processor, and a display (screen) connected to a CPU. Data and external clock signals are acquired from the device under test via probes. This configuration is illustrated in Figure 10.2.

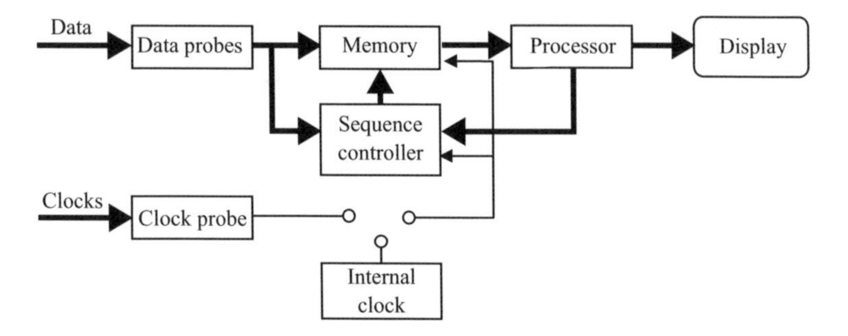

Figure 10.2 Main functional units of a logic analyser

The probes are comparators programmed to the thresholds of the logic family to be analysed and therefore only output either HIGH or LOW levels. Data are supplied from the data probes to the memory if a clock signal occurs and if the sequence controller requests this. The clock may come from the device under test itself (external clock) or from a clock generator in the logic analyser (internal clock). The sequence controller is a circuit that can be programmed by the user to store data to the memory selectively. At the end of acquisition, the data collected in the memory are processed by the CPU and displayed [1].

The two main types of analysis performed by a logic analyser are *timing analysis* and *state analysis*. In timing analysis, timing relationships between multiple signals are considered, and this is used mainly in hardware testing and debugging. State analysis considers the data transfer inside the circuit, and is used in software testing and debugging of systems. In timing analysis, a relatively small number of signals are analysed in parallel, with each being sampled at a very high speed. In state analysis, a larger number of signals, most commonly the system buses, are analysed for the activities they perform at lower sampling speeds.

The display presents the data stored in the memory in a form convenient to the user. Some of these forms require very little processing from the stored state (binary, hex, octal displays). Others require a considerable amount of processing (disassembler, event timing forms).

In timing analysis, the display shows the sampled waveforms in a manner analogous to an oscilloscope. In state analysis, the display can show the data on the system buses in binary, hex, octal or ASCII forms, or the data can be analysed and disassembled to show the bus activity as machine language mnemonics.

Detailed operation of timing and state analysis as well as the corresponding display formats are provided in section 10.4. Sections 10.5–10.7 describe probing, clocking and triggering of logic analysers.

10.3.2 Requirements of a logic analyser

A general purpose logic analyser should be able to selectively store and display the needed information in a useful manner. More specifically it should have the capability to trace, store and display a large number of signals at the same time. It should also be possible to select the signals that will be stored and displayed. A good quality logic analyser will allow the labelling of signals (data, address, etc.) while displaying.

The sampling rate should be high enough to meet the required application and the input voltage level should be adaptable to all the common logic families. Extensive trigger facilities should be available to capture data selectively. A variety of different display modes help to make data interpretation and understanding easier.

To detect hardware errors (e.g. to capture glitches) a high clock rate is required. The clock is usually generated inside the logic analyser (asynchronous clock). The ability to make accurate time measurements is also important. Logic analysers specialising in this type of work are known as *timing analysers*. They have fewer channels, less sophisticated trigger facilities and a large amount of fast (and expensive) memory. The type of logic analyser specialising in detecting software faults is known as a *state analyser*. A large number of channels are available, which are sampled using a system clock (synchronous clock). Extensive trigger facilities and disassembled listings are standard in this type of analyser. Many modern analysers provide both functions within a single instrument.

10.3.3 Uses of logic analysers

The application of logic analysers is largely to be found in the hardware verification of a circuit in the initial design stages, in the integration of hardware and software in later design stages, and in the fine tuning of a working system to obtain the best effect. Logic analysers are also used in the area of fault diagnosis of digital systems. The other area of application is in automated testing. The remote control facility of the instrument is used in such applications.

However, the logic analyser is still far less frequently used than the oscilloscope, probably owing to cost and complexity. Low cost PC based logic analysers, as well as the development of user friendly Windows based interfaces that have appeared in recent years, might well change this picture.

10.4 Types of analysis

There are two basic types of logic analysers, state and timing [3]. Most logic analysers provide both in one instrument. In this section we will look at these two analysis techniques, and also continue on to other specialised types of analysis that logic analysers can perform. Detailed descriptions of these analysis techniques are found in References 4–6.

10.4.1 Timing analysis

Timing analysis is analogous to the functioning of an oscilloscope. A timing analyser displays information in the same general form as a scope, with the horizontal axis representing time, and the vertical axis showing the logic levels of the signal.

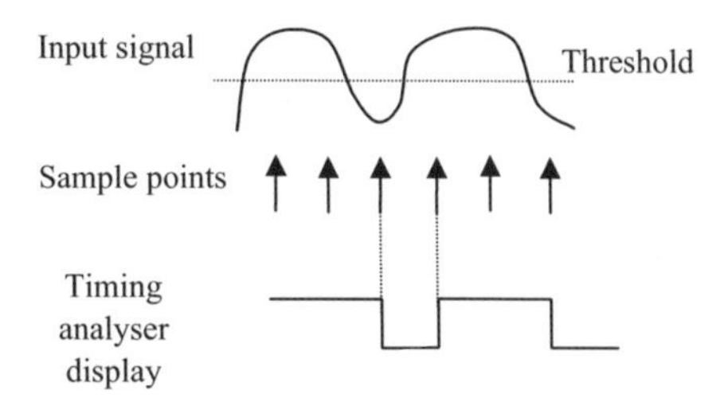

Figure 10.3 Basic timing analysis

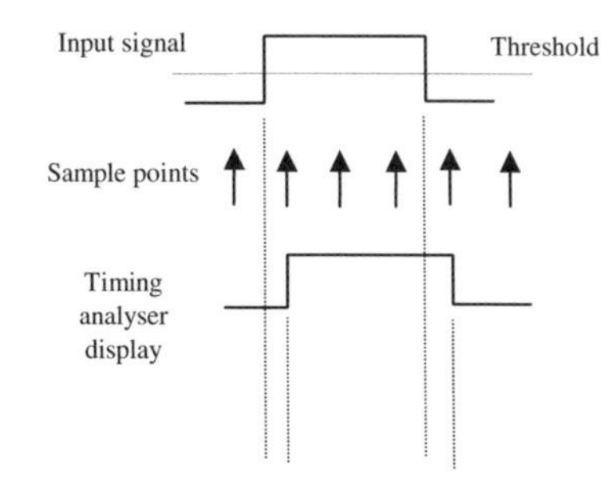

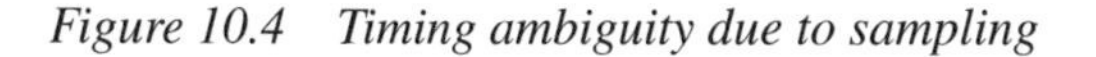

Figure 10.4 Timing ambiguity due to sampling

A timing analyser works by sampling the input waveforms. Clock signals used for sampling are usually asynchronous or internal for timing analysis. By comparison of the samples with a threshold level, a series of ones and zeros is generated that represents a one-bit picture of the input waveform. This sequence is stored in memory, and is also used to construct a display as shown in Figure 10.3.

Timing analysis is used to detect hardware faults in a circuit. As an example, if it is needed to make certain that an entire dynamic memory is refreshed within a particular period, a timing analyser can be set to trigger when the refresh counter starts, and display all the counts that it goes through before starting over.

Owing to the periodic sampling in timing analysis, it is not possible to detect the exact time of level transitions of the input signal. This is illustrated in Figure 10.4. The worst case for this ambiguity is one sample period, assuming that the transition occurred immediately after the previous sample point.

The resolution can be increased (ambiguity can be reduced) by increasing the sampling rate. However, there is a trade-off between resolution and the total acquisition time as the amount of acquisition memory available in the analyser is fixed. The higher the sampling rate (resolution), the shorter the acquisition window. Glitches (extremely narrow, spurious pulses) also can go undetected in the above sampling process.

10.4.1.1 Display of information in timing analysis

The display in timing analysis can be in sample, latch or glitch modes. An example of these modes is shown in Figure 10.5. In the sample mode, the logic value obtained

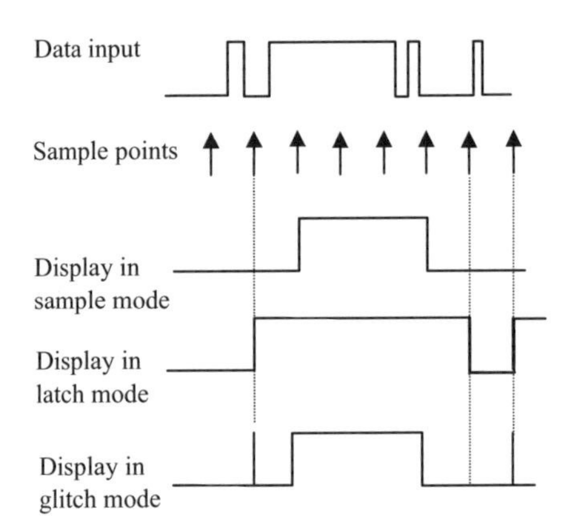

Figure 10.5 Display modes for timing analysis

in the current clock is held till the next clock. This mode, however, will not show spikes or glitches occurring between the clock pulses. In the latch mode, any such glitch is expanded and displayed in the next clock cycle. In the glitch mode, instead of the glitch being expanded to fill the clock cycle, it is shown on a vertical line.

Modern logic analysers have glitch trigger capabilities that makes it easy to track down elusive glitch problems. Figure 10.6 shows a timing analysis display.

10.4.2 State analysis

During state analysis, a hardware or software engineer studies the state of the signals to study how the circuit is functioning. The main purpose of state analysis is to detect software faults. For state analysis, the clock is derived from the circuit itself (a synchronous clock), unlike in the case of timing analysis. The logic analyser therefore runs synchronously with the circuit under test, capturing samples at the rising or falling edge of the clock.

The system clock is usually not the best choice for this. Signals such as MEM, which indicates that a memory access cycle is active, READ or WRITE can be used for state analysis. The best clock signal is that which allows you to select exactly those samples which you need to monitor the functioning of the system. This way the data stored will be more selective and the record stored in acquisition memory will be longer. The sampling speed required is lower for state analysis than for timing analysis.

The number of signals that a state analyser has to trace varies greatly from application to application. Most designs are built around a microprocessor. To trace what happens, the logic analyser must trace the input and output lines. It must therefore trace the address, data and many control lines, as well as a number of additional lines from peripheral devices. Therefore, the number of signals monitored simultaneously is larger than in timing analysis.

Data acquisition memories for state analysers must be fast and wide. A state analyser with 100 state channels will need a state memory that is 100 bits wide. The

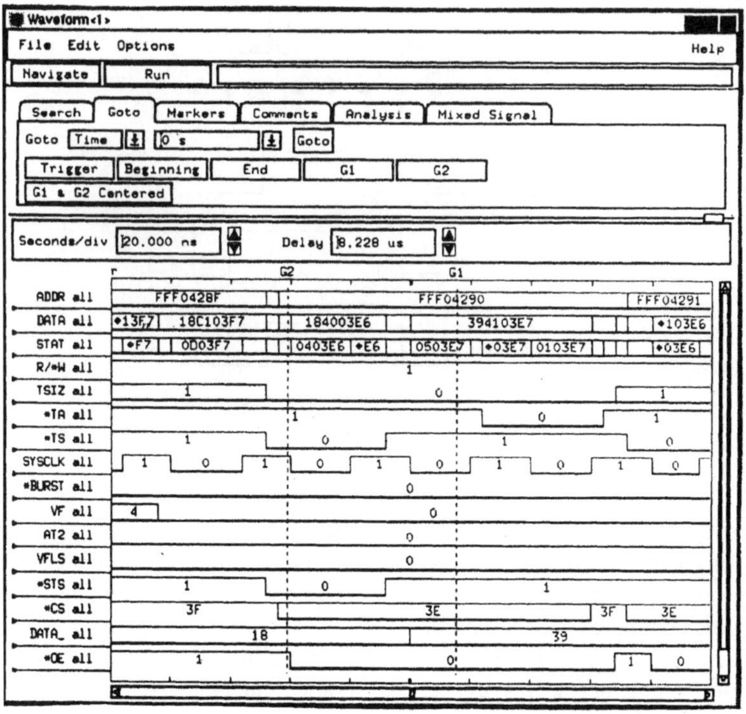

Figure 10.6 Timing analysis display in a modern logic analyser (reproduced from Reference 7, with permission of Hewlett-Packard)

information stored in each memory location will correspond to a bus transfer such as instruction fetches, memory read/writes and I/O actions. Each instruction will typically consist of three or four such bus cycles.

Although a deeper memory would be useful for showing more of the program flow, longer traces are difficult to interpret. It is instead better for an analyser to incorporate powerful trigger facilities and selective data acquisition facilities, as discussed later in sections 10.6 and 10.7.

10.4.2.1 Display of information in state analysis

Samples stored in the state analyser's memory can be displayed as a state listing (state table) or as a disassembly list.

In a state list, the data can be displayed in binary, hexadecimal or octal form, with the individual rows on the display showing one sample taken from each of the channels during a qualified clock pulse. For ease of interpretation, the groups of lines can also be translated into decimal or ASCII form. Although this can be useful for interpreting the events of, for instance, a state machine, it does not show the actual times at which events occurred. Instead, it assumes that the clock is a regular pulse train.

```
Listing<1>
File  Edit  Options  Invasm  Source                                    Help
Navigate    Run
Search  Goto  Markers  Comments  Analysis  Mixed Signal
G1:  DATA    = 394103E7   Time  from  Trigger  = 5.544 us
G2:  ADDR    = FFF04290   Time  from  Trigger  = 8.192 us

PC                     MPC821/860 Inverse Assembler             ADDR      DATA      STAT
Symbols                10=hex,  10.=decimal, X10=binary         Hex       Hex       Hex
    proc_specifi+01A8     lbz     r9 41AD(r12)                  FFF04274  894103E7  0103E7
    q.el:current_temp             read  5D                      000041AD  5D4123D7  0923D7
    proc_specifi+01AC     stb     r9 0000(r11)                  FFF04278  994103E7  0103E7
    proc_specifi+01B0     addi    r7 r1  0018                   FFF0427C  384103E7  0103E7
G1. proc_specifi+01B4     addi    r8 r7  0002                   FFF04280  394103E7  0103E7
    proc_specifi+01B8     li      r0 00000022                   FFF04284  384103E7  0103E7
    proc_specifi+01BC     stb     r0 0000(r8)                   FFF04288  984103E7  0103E7
    proc_specifi+01C0     addi    r10 r1 0018                   FFF0428C  394103E7  0103E7
G2. proc_specifi+01C4     addi    r11 r10 0003                  FFF04290  394103E7  0103E7
    proc_specifi+01C8     lis     r12 0000                      FFF04294  3D4103E7  0103E7
    proc_specifi+01CC     lbz     r9 41AC(r12)                  FFF04298  894103E7  0103E7
    q.el:outside_temp             read  5E                      000041AC  5E4123D7  0923D7
    proc_specifi+01D0     stb     r9 0000(r11)                  FFF0429C  994103E7  0103E7
    proc_specifi+01D4     addi    r7 r1  0018                   FFF042A0  384103E7  0103E7
    proc_specifi+01D8     addi    r8 r7  0004                   FFF042A4  394103E7  0103E7
    proc_specifi+01DC     li      r0 00000033                   FFF042A8  384103E7  0103E7
    proc_specifi+01E0     stb     r0 0000(r8)                   FFF042AC  984103E7  0103E7
    proc_specifi+01E4     addi    r10 r1 0018                   FFF042B0  394103E7  0103E7
    proc_specifi+01E8     addi    r11 r10 0005                  FFF042B4  394103E7  0103E7
    proc_specifi+01EC     lis     r12 0000                      FFF042B8  3D4103E7  0103E7
    proc_specifi+01F0     lbz     r9 4088(r12)                  FFF042BC  894103E7  0103E7
    q.elf:target_temp             read  5E                      00004088  5E4123D7  0923D7
    proc_specifi+01F4     stb     r9 0000(r11)                  FFF042C0  994103E7  0103E7
    proc_specifi+01F8     addi    r7 r1  0018                   FFF042C4  384103E7  0103E7
```

Figure 10.7 *State analysis display of a modern logic analyser combining the state list and the disassembly list (reproduced from Reference 7, with permission of Hewlett-Packard)*

Most state analysers incorporate a disassembler to make the program flow of a target microprocessor simpler to interpret. The bus cycles captured during state analysis are converted into processor mnemonics or bus transactions and displayed in the disassembly display. This allows a direct comparison between the circuit activities and the program listing.

Scanning a disassembly list will reveal errors much more quickly than a state list, particularly since any faults on the buses will usually generate illegal opcodes, which will be detected and reported by a disassembler. However, disassembly requires considerably more processing and interpretation of the acquired data inside the analyser, compared to a simple state table display. Disassembly modules specific to different processors are available as options in logic analysers.

Figure 10.7 shows a modern state analyser display combining both the state list and the disassembly list.

In state analysis, it is also important to know the time intervals that have elapsed between the consecutive states of the system under test. It is possible to measure these times by time stamping the required states in the state list. Time stamping can be absolute, i.e. starting from a defined starting point such as a trigger word, or relative, in which the elapsed time of each state is known individually.

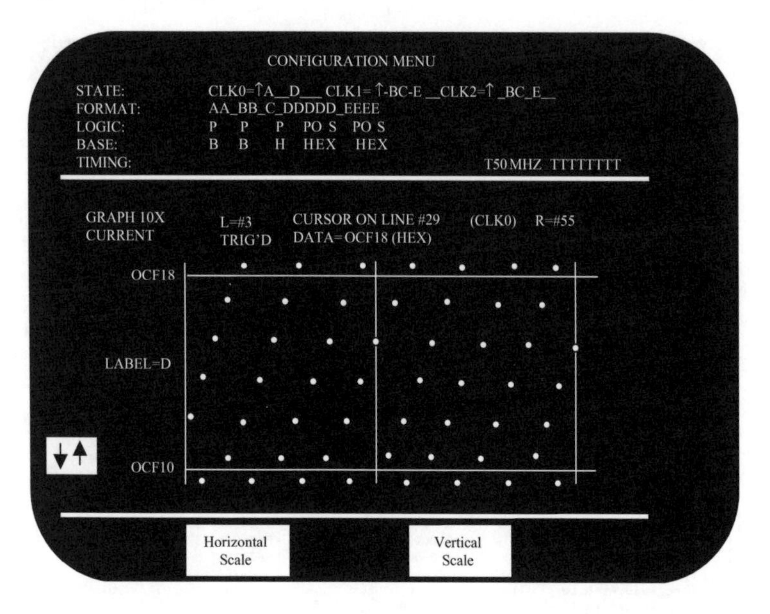

Figure 10.8 A graph display (reproduced from Reference 5, with permission)

Another method of displaying state analysis information is the *graph display* or *the map display*. This shows the captured data in a 'geographical' way. For this purpose, each data word in the memory is split into two halves: the least significant byte and the most significant byte. Each byte is separately D/A converted, and applied as X and Y axis data to be plotted on the display. Each data word is thus converted into a point on the display.

If the address lines of the system under test are taken as data inputs, the graph display shows the flow of the captured part of the program. It will clearly show subroutine calls, executions and returns. When subroutines are executed several times, the points corresponding to the relevant addresses will be more pronounced. Reference 6 gives a detailed description of the graph display and its applications. Figure 10.8 shows a graph display.

Recognition of repetitive loops or illegal memory accesses is much easier in a graph display than in a state display. State graph display and the state list are tightly coupled in logic analysers. Therefore, placing the cursor on a specific point of interest in the graph display, and then switching-over to the state list directly reveals the problem in terms of system state lines.

10.4.3 Simultaneous state and timing analysis

Regardless of the way in which state and timing channels are implemented, an analyser must allow the information from both state and timing channels to be studied

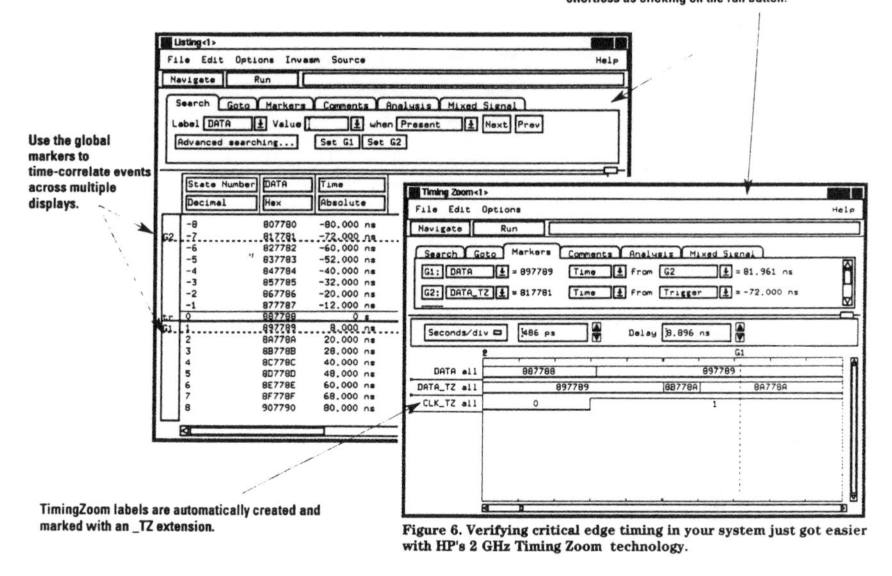

Figure 10.9 Time correlated timing and state displays (reproduced from Reference 7, with permission)

in parallel. This is important because a software error may be caused by a hardware fault, or vice versa. The facility of simultaneous state and timing analysis is therefore very useful in the hardware/software integration stage of a development project.

In one form of simultaneous analysis, the state analyser can trigger the timing section, and vice versa. It is also possible for one analyser section to wait in its trigger sequence for the other analyser to proceed. By coupling the state analyser to the timing analyser in this manner, the sample can be time stamped, to show how much time has elapsed between two samples, or since the trigger occurred.

In another form of analysis, time correlated measurements can be done by running the two analysers in synchronism with each other. In this situation, state and timing displays are fully cross referenced, so both state and timing conditions can be examined simultaneously at any specific moment. Global markers allow cross referencing of displays in this manner as shown in Figure 10.9.

10.4.4 Compare facilities

Logic analysers provide sophisticated compare facilities for the fast location of system faults such as wrongly executed branches or incorrect data transfers. Data patterns of interest can be stored in reference memory, after which acquisitions of new data can be compared with the reference pattern.

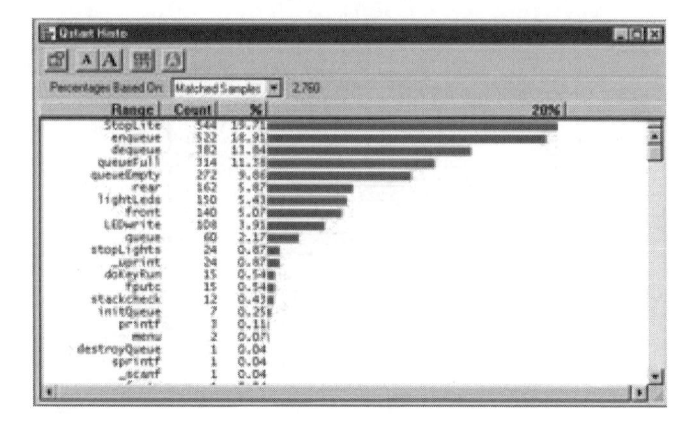

Figure 10.10 A display showing results of performance analysis (reproduced from Reference 8)

10.4.5 Performance analysis

Performance analysis generates statistical representations by post-processing real time data acquired and displaying the results in a histogram window. Figure 10.10 shows a performance analysis display.

This type of analysis is a vital step in the development route, taking over after completion of functional system tests. It allows system debugging and optimisation before production commitments are made.

As well as identifying and correcting bugs in hardware/software interaction, performance analysis is a powerful tool in checking the utilisation of specific sectors and optimising overall operational efficiency of the target system. This lets you locate system bottlenecks and determine the necessary corrective measures. For example, with performance analysis, you can measure the average execution time of a sub-routine, or how much time is available for it.

10.4.6 Signature analysis

Signature analysis is a technique used in component level troubleshooting of microprocessor based systems. The pulse stream derived from a test point in the circuit is converted to a unique 16-bit number known as its signature. The pulse stream is obtained under controlled conditions. The signature of a node in a faulty board can be compared with a reference value given by the manufacturer or derived from the corresponding point from a good board. By tracing signatures along a path, it is possible to isolate the faulty component.

Signature analysis requires four inputs: the data line, a clock input, a start and a stop input. The signature is computed only for the samples taken at the clock pulses occurring between start and stop signals. That period is known as the 'gate'. Figure 10.11 illustrates typical waveforms. The clock, start and stop signals are taken from the circuit itself and can be defined for the rising edge or the falling edge.

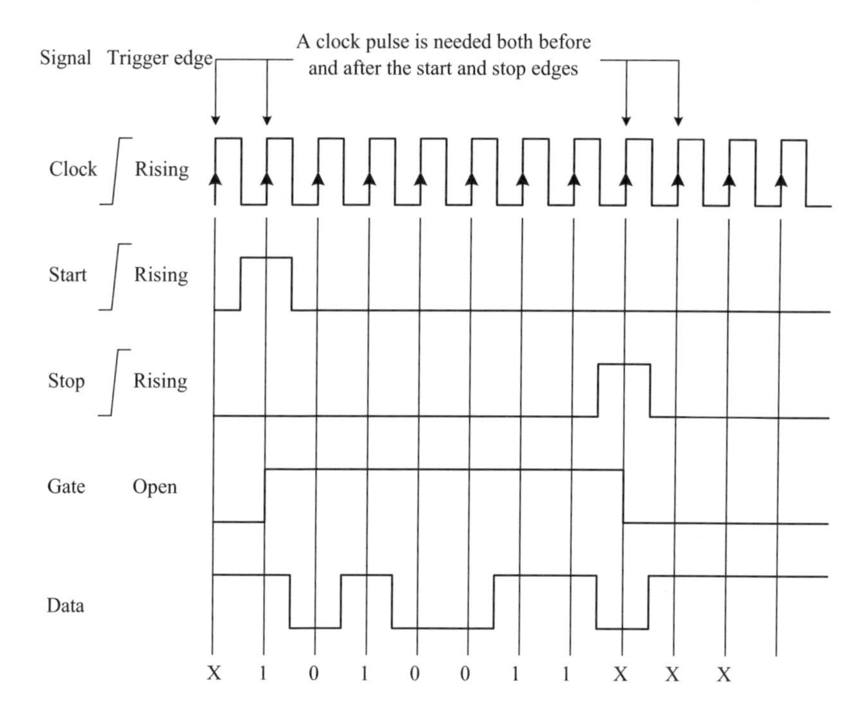

Figure 10.11 Typical signature waveforms

A stimulation data pattern is fed to the circuit during the measurement. Some logic analysers incorporate the signature analysis facility.

10.5 Probing

Data are collected from the system under test by using colour-coded probes to connect the logic analyser inputs to the required signals. Probes are comparators programmed to the thresholds of the logic family to be analysed. Probes can be programmed for different threshold values.

Accuracy of measurements depends on reliable and non-intrusive connections to the target system. Connections from the logic analyser to the target circuit must not load the target. Therefore, probes need to be short and have high impedance. Resistive loading affects the amplitude of the signal under test, and capacitive loading affects the timing of the signal by rounding and slewing the edges. Probing effects can be more troublesome as clock frequencies increase.

Many logic analysers use passive or active pods to get close to the test points, with flat cables connecting the pods to the analyser. Passive pods reduce the load on signals by using a matched resistor network. This type of pod is inexpensive, but gives a relatively low impedance of about $100\,\mathrm{k}\Omega/10\,\mathrm{pF}$. It therefore demands sensitive threshold detectors in the logic analyser. However, it is smaller in size,

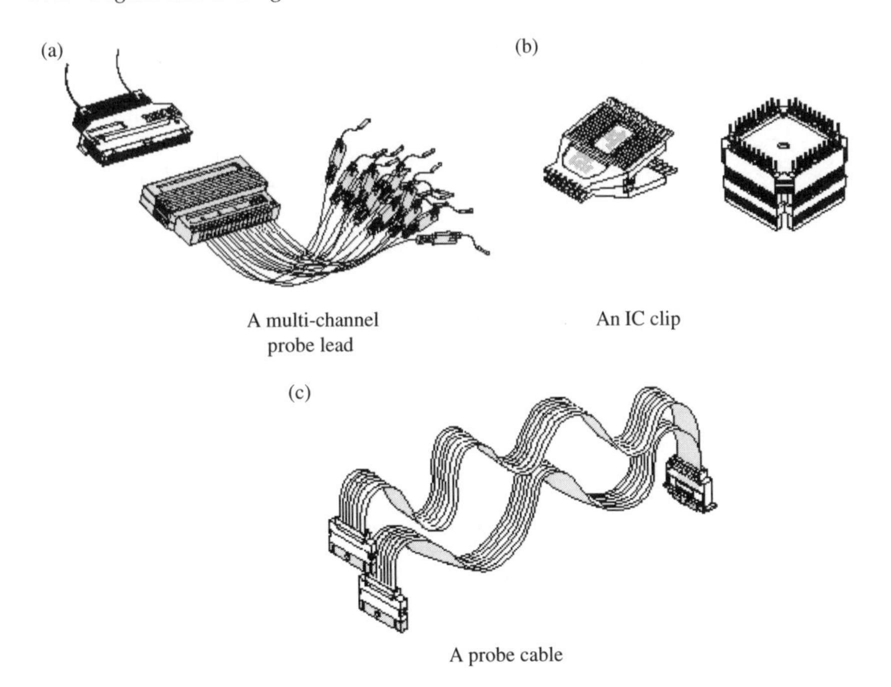

Figure 10.12 Components of a standard probing system (reproduced from Reference 9, with permission)

and is more reliable compared with active probes. Passive probing is similar to the probing system used on high frequency oscilloscopes. Active pods have a higher impedance of about $4\,M\Omega/6\,pF$, and therefore isolate the signals more effectively from the analyser. Another important consideration in probes is the immunity to noise. Electromagentic noise can corrupt data captured by the logic analyser. Active probing can be particularly susceptible to noise effects.

A standard probing system consists of IC clips, probe leads, probe housing and probe cable. Probe leads are configured into lead sets, which can probe multiple data channels with ground, one clock channel, and a common ground. An IC clip is a single hook that fits around IC pins and component leads. Figure 10.12 shows the different parts of a standard probing system.

Specialised probes (microprocessor probes) group signals as data, address and control, and connect these groups to the system. More information on modern probes is found in Reference 9.

10.6 Clocking

The logic analyser's sampling process is controlled by a clock. This inevitably causes some inaccuracies, depending on the sampling interval. Clock signals are usually

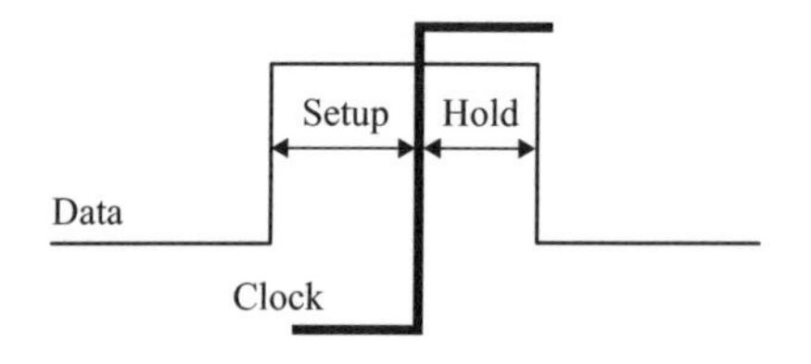

Figure 10.13 Setup and hold times

asynchronous or internal for timing analysis. This will provide the best resolution. For state analysis, the clock is derived from the circuit itself (synchronous clock).

More than one clock may be used by the analyser to track different types of activity. Some logic analysers can assign groups of channels to different clock signals, effectively creating two or more analysers in a single instrument. This allows a designer, for example, to monitor two processors in parallel.

10.6.1 Setup and hold times

Two important parameters connected with data sampling in state analysis are the *setup time* and the *hold time*. The data are sampled on the rising or falling edge (usually user selectable) of the sampling signal. The data have to be stable for some time before the sampling edge comes along. This time period is what is known as the setup time (usually in the order of a few nanoseconds). Similarly, data will have to be stable for a short time after the sampling edge. This is known as the hold time and is usually very close to zero. Figure 10.13 illustrates both setup and hold times.

10.6.2 Memory requirements and timing resolution

In timing analysis, the important considerations are resolution and glitch capture. Increasing the sampling frequency results is improving both. However, this increases the memory requirements, in both capacity and speed of the logic analyser, and thus reduces the record length. This is illustrated in Figure 10.14.

10.6.3 Making the best use of the available memory

10.6.3.1 Clock qualifiers

To prevent unnecessary data from being saved in the limited memory of the logic analyser, a clock qualifier is used to selectively store data. Usually two or more clock qualifier inputs are available in a logic analyser in addition to the clock input. When a clock pulse occurs, if the clock qualifier inputs from the circuit match the corresponding settings in the logic analyser, the sampled data will be stored in memory. For example, if we want to read in instructions that perform memory writes, we can use the system read/write signal as a clock qualifier signal to selectively sample data.

Although these qualifiers make the best use of available memory, their use may substantially increase the minimum setup time, because the clock signal must pass through additional gates. Clocking mechanisms are highlighted in Figure 10.15.

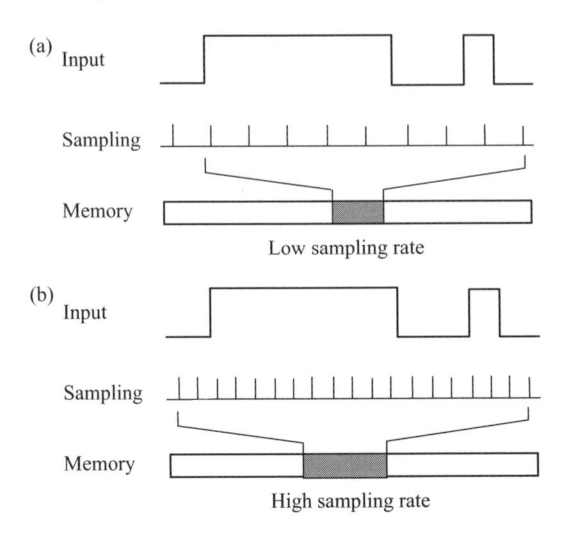

Figure 10.14 Relationship between sampling rate and memory requirement

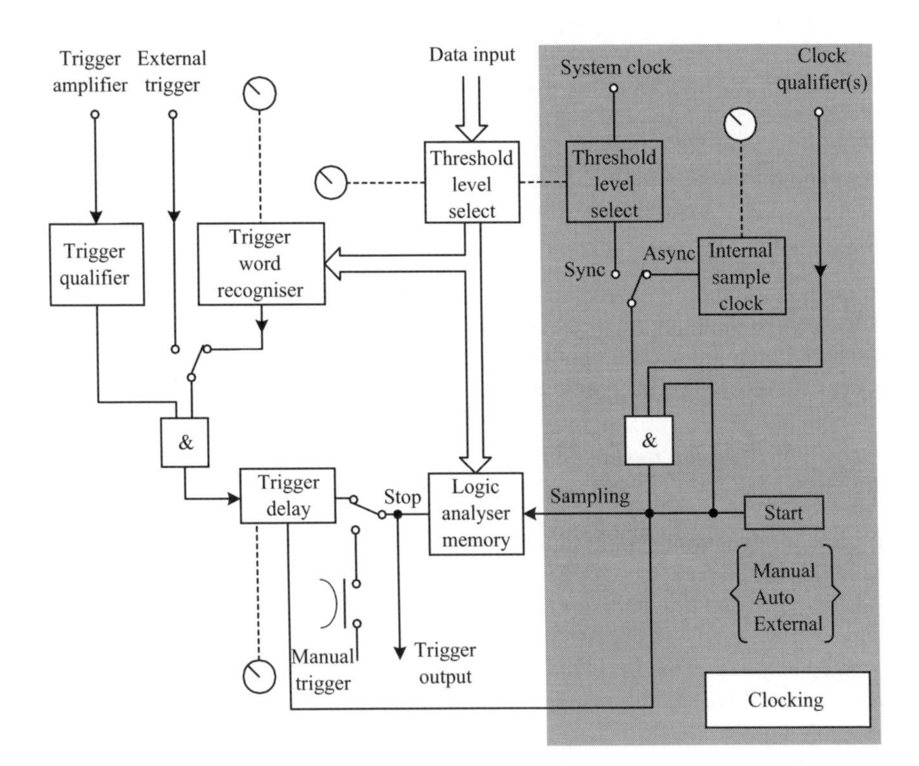

Figure 10.15 Clocking mechanisms

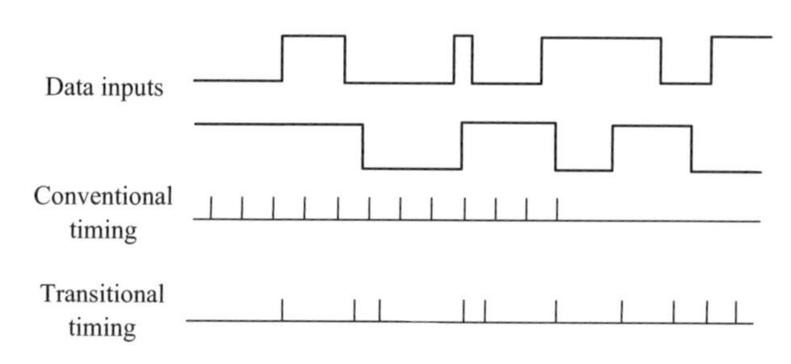

Figure 10.16 Transitional timing

10.6.3.2 Demultiplexing channels

For analysers that use conventional timing schemes, the recording length is given by dividing the memory length by the sampling frequency. Longer recording length can be achieved by lowering the sampling frequency (thus reducing the resolution). An alternative method to improving resolution is to demultiplex the input channels. For example, four 100 MHz channels with 1 k word of memory each can be combined into two 200 MHz channels with 2 k words each, or into one 400 MHz channel with 4 k word of memory. This trades off the number of channels for speed (resolution).

10.6.3.3 Transitional timing

Another alternative is transitional timing. Conventional timing analysers store a sample on each clock pulse, whether or not the incoming information has changed. This usually means that most of the memory is filled with repeated, redundant information. To avoid this, transitional timing analysers store samples only when there is a signal transition. Figure 10.16 illustrates transitional timing.

With this technique, only those events that you want to see are captured, without wasting memory. For example, two 5 ns pulses, separated in time by 20 min can be captured in a single acquisition. Using conventional storage techniques, achieving this capability would require a memory capacity of more than 130 Gbits.

The transitional timing technique uses two memories: one for data and one for time. If there is any change in one of the channels being examined between one sample and the next, the old data are loaded into the data memory, while the duration for which these data were valid is loaded into the time memory. Therefore, although only the transitions are captured, the complete waveform can be reproduced and displayed from the data in these two memories.

10.7 Triggering

It is very important that, among the millions of samples that are received every second, only the data of interest be captured and displayed. This is achieved by triggering the logic analyser only at a selected point. The trigger is an event (defined by the user) in the data stream that initiates the final phase of the measurement. This is different to the oscilloscope, where the trigger starts data acquisition.

Triggering is an important concept for logic analysers, because it is an acquisition instrument with a built-in circular memory buffer. The circular buffer stores data continuously, storing each new sample in the next available memory location. When the buffer is full, it simply wraps around and stores each subsequent sample over the 'oldest' sample in the memory. The primary responsibility of the trigger mechanism is to stop the acquisition at the right time so that the sampled information left in the memory represents the slice in time the user needs to view.

Measurement is immediately stopped (or stopped after a pre-settable number of clocks) when the trigger occurs. An immediate stop of the measurements is referred to as 'pre-trigger', because all the data contained in the memory have been read in before the occurrence of the trigger. If the user has specified a trigger delay, the memory is filled by data following the trigger. This is referred to as 'post-trigger', because only data following the trigger can be displayed. The trigger delay can also be selected by setting the post-trigger counter accordingly. Pre-trigger is useful if a faulty condition is defined as the trigger and the events leading up to this fault are to be examined. Post-trigger enables the response of the device under test to the trigger event to be examined. Trigger delay allows sampling to continue for a certain (user variable) number of clock pulses after the trigger signal is enabled.

Internal triggering is done in conjunction with a trigger word. The trigger word can be set by the user. This is the most important trigger facility and is used with most measurements. A trigger word recogniser is connected to the data path. It compares the incoming data word with the pre-set trigger word. If there is a match, triggering occurs and the sampling will be stopped after a delay of a certain number of samples pre-selected by the user.

In most cases a single trigger word is not sufficient as this may occur many times before the actual point of interest. To overcome this problem, some logic analysers provide sequential or nested triggering. In this mode, the trigger is enabled if all the trigger words are received in the proper sequence. This facility is useful when we want to check the execution of sub-routines in machine code. When the number of sequential trigger words is two, the first one is known as the 'arm' word. Once that word is received the logic analyser is armed and ready for triggering.

The trigger qualifier allows extra conditions to be added to the trigger word. We can feed in an external condition or use this facility to increase the width of the trigger word. Trigger qualifier inputs are not stored and displayed.

External triggering can be used when sampling is to be stopped on a condition on an external line such as an interrupt, status line, etc. External trigger inputs are usually provided to logic analysers. Manual triggering is done by a push button in the front panel to freeze the data and see what is happening. A range recogniser allows triggering when data fall within or outside (as set by the user) a specified range. This is useful to check if certain areas of the program are being executed.

Figure 10.17 highlights triggering mechanisms typically found in logic analysers.

In many applications, state and timing analysis must be performed simultaneously, and the triggering of the state and timing signals must therefore be synchronised.

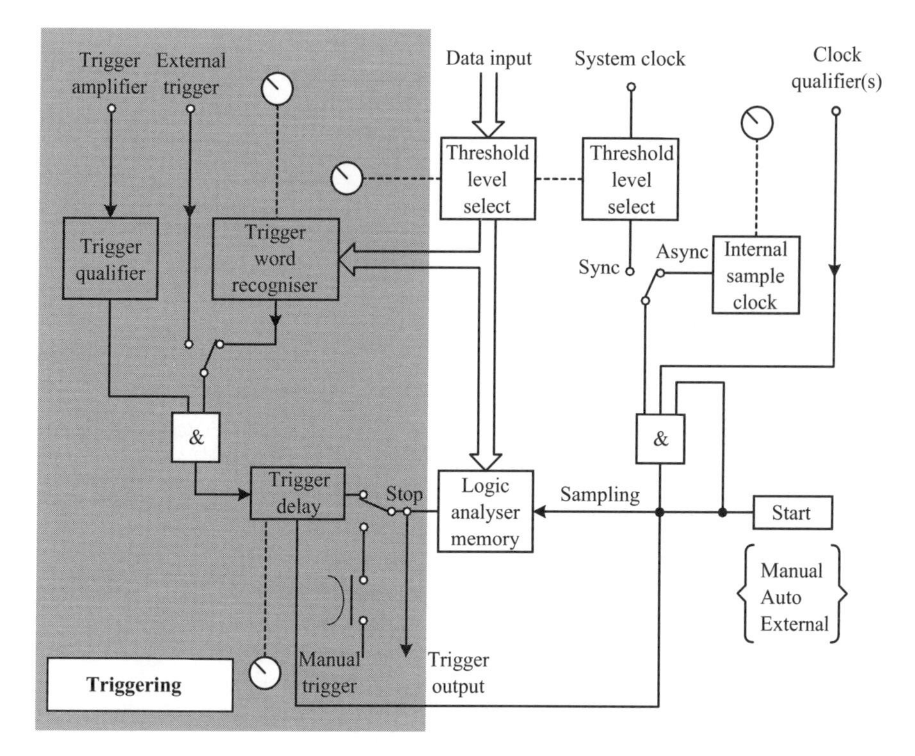

Figure 10.17 Triggering of logic analysers

Cross-triggering mechanisms in which a trigger sequence for one analyser triggers the other to freeze both acquisition memories simultaneously are available for this purpose. A logic analyser can also send a trigger signal to a second instrument such as a digital storage oscilloscope (DSO). Such advanced techniques are described in section 10.8.

10.8 Advanced features and measurement techniques

Traditionally, digital circuit designers typically employ multi-channel general purpose logic analyser for state analysis, and either specialised high speed logic analyser modules or oscilloscopes for timing analysis. When a general purpose logic analyser's speed is insufficient for timing analysis, equipment manufacturers trade-off channel count in specialised logic analyser modules. Trading-off channels for speed compromises the instrument's triggering capabilities as well.

However, innovative design techniques have been developed recently for logic analysers as well as other equipment for digital circuit testing, which improve the sampling rate as well other features, without compromising the channel count, or other capabilities. This section describes some of these techniques briefly.

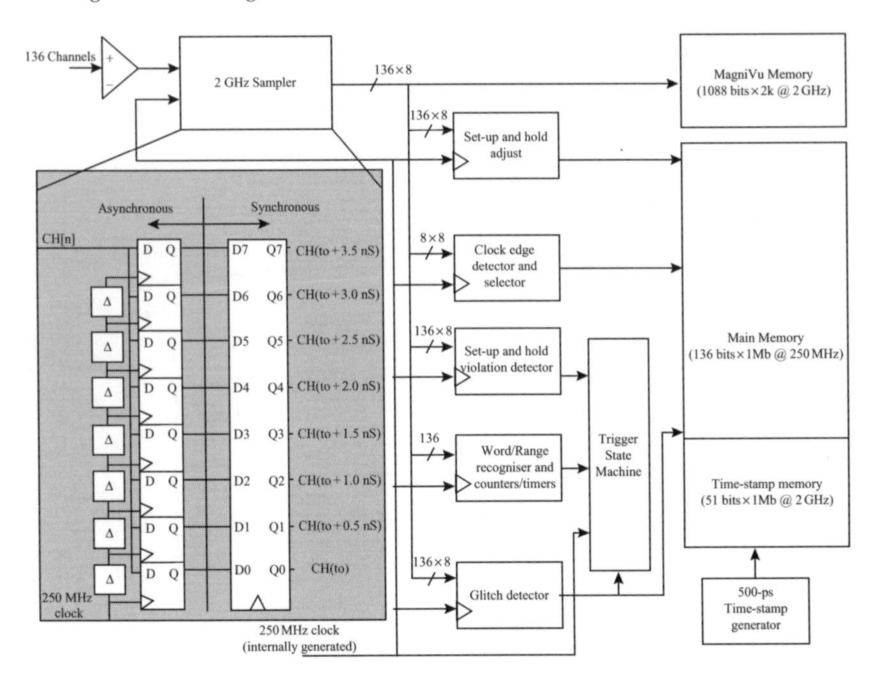

Figure 10.18 Asynchronous digital over-sampling (reproduced from Reference 8, with permission from Tektronix, Inc.)

10.8.1 Asynchronous digital over-sampling

This is a high speed sampling architecture developed by Tektronix, Inc., [8] that provides 500 ps timing resolution (2 GHz sampling frequency) on a large number of channels, along with 200 Msamples s^{-1} synchronous state acquisition.

The 2 GHz sampling frequency is achieved using circuitry driven by a 250 MHz clock. This clock is used with a chain of eight 500 ps delays to clock a set of eight sampling circuits, as shown in Figure 10.18. For every 4 ns tick of the 250 MHz clock, eight samples are acquired at 500 ps intervals for each channel. Each consecutive eight bits of sampled data are loaded in parallel every 4 ns into a shift register that feeds out selected pieces of the data to a 1 M memory based on user-defined clocking, triggering and storage parameters. This large memory, implemented using common RAM devices, can store data asynchronously acquired at up to 250 Msamples s^{-1} or synchronously acquired state data at up to 200 Msamples s^{-1} for all channels at all times.

At the same time, another high speed memory 2 k samples deep directly stores the unconditioned stream of data straight from the sampler for each channel. This high speed memory is large enough to store 1 μs worth of data for each channel. So, as the overall state activity is stored in the larger, slower memory, complete timing information is simultaneously captured in this faster memory. This dual memory architecture ensures that the logic analyser gathers a complete picture of the design's operation during state analysis, providing simultaneous 2 GHz timing and 200 MHz state analysis.

Because both the state and timing information are derived from the same stream of over-sampled data, only one set of probes need to be attached to the circuit under test.

New, high speed probes with only 2 pF loading for the entire probe have been developed to be used with the over-sampling techniques, answering the desire for fewer and simpler connections to the system under test.

Another advantage of this over-sampling technique is that, as the state and timing information are derived from the same 2 GHz sampler, there is absolute timing correlation between the two.

The TLA600 Series Logic Analyser implements the over-sampling technique to provide 2 GHz sampling frequency on 36 channels [8].

10.8.2 Advanced triggering techniques

In today's embedded microprocessor software debug applications, the system often does not really show signs of problems until long after they originate. Hence, the real challenge in triggering a logic analyser in such applications is recognising very complex events that characterise the symptom, requiring sophisticated triggering mechanisms.

Most logic analysers allow the user to specify a list of events that can be sequentially evaluated, usually culminating in the triggering of the analyser. Most implement these as sequential conditions with fixed logic choices that allow a very limited set of actions.

A new development from Tektronix uses sophisticated trigger circuitry called a trigger state machine. It is implemented as a true random access, fully programmable state machine rather than one with hardwired or fixed sequential logic. This technique can therefore evaluate several conditions simultaneously, can evaluate more complex combinations of events, and execute more extensive combinations of actions. It allows the user to draw a state diagram of the circuit behaviour and enter that diagram directly into the trigger program [10].

10.8.3 Advanced measurement techniques

10.8.3.1 Stimulus generation

In developing a digital system with many sub-components, the ability to test some of the sub-components without waiting for all to be developed is important in reducing the overall design time. Instrument manufacturers offer stimulus generator modules for logic analysis systems to provide stimulus to circuits in the event of missing boards, ICs or buses. Patterns needed to put the circuit in the desired state, to operate the circuit at full speed, or step the circuit through a series of states can be generated through these modules.

10.8.3.2 Cross-triggering and inter-module triggering

In applications where state and timing analyses must be performed simultaneously, with synchronised triggering of state and timing analysers, cross-triggering is a powerful technique. In this technique, a trigger sequence for one of the analysers triggers the other to freeze both acquisition memories simultaneously. The state trigger sequencer can be used to zoom in on a specific software problem, and then 'arm' the timing trigger logic so that the timing analyser triggers on a glitch or a pulse of a specified length. The timing trigger sequencer could alternatively be used to detect timing problems, and then arm the state analyser to freeze the state acquisition memory at the relevant instructions in the program. This could reveal the sequence of events that led up to the error condition: the program may have reacted too late to a hardware request, for example, or a gate may have been switched by a spurious spike.

Logic analyser modules can also be used to trigger other modules such as DSOs with complete time correlation. This way, the DSO can be synchronised with the logic analyser to show the exact waveform of the digital signals at the time they were sampled by the timing analyser, allowing, for instance, the rise and fall times to be studied. Alternatively, the logic analyser can be triggered by a DSO if the scope detects an incorrect rise or fall time.

10.8.3.3 Remote control

A remote control feature allows the logic analyser to be controlled by a computer via an RS-232 or GPIB interface. Measurement setups can be loaded by the computer and changed at will. Data memory can be downloaded to the computer for further analysis. The remote control feature is useful when the logic analyser is used in automated measurements.

10.9 Current logic analysers and related instruments

Logic analysers are available in a wide range of capabilities and prices for the facilities offered, to meet real world design problems varying from 8-bit designs to 64-bit multiprocessor systems. High end products provide a set of scalable debug tools having the capability to hold combinations of modules for timing and state analysis, digitising oscilloscope modules, pattern generator modules, emulation modules, etc. Mid-range, stand alone logic analysers have, in addition to their basic functions, the options of including additional measurement modules. The lowest cost logic analysers are those that are PC based.

Agilent Technologies, for example, offers logic analysis system mainframes [11], benchtop logic analysers [12] and PC hosted logic analysers [13].

The TLA family of logic analysers from Tektronix consists of the TLA600 series and the TLA700 series. The TLA600 series offers a selection of standalone logic analyser instruments. The TLA700 series consists of portable and bench-top modular mainframes with expansion mainframe capability [14].

Most logic analysers available today offer wide ranging processor and bus support, as well as a full line of support products.

10.9.1 Modular logic analysis systems

Logic analysis systems are integrated systems of logic analysis, emulation and software tools. They provide comprehensive system level debugging optimised for

Table 10.3 Comparison of the HP16700A and 16600A series (source: Reference 11)

Model	16700A	16702A	16600A/16601A/ 16602A/16603A
Slots for measurement modules	5 (10 total with 16701A expansion frame)		1
Built-in state/timing channels	None	None	204, 136, 102, 68
Built-in display	No	800 × 600, 10.3	No

Table 10.4 Built-in logic analysis capability of the HP 16600A series analysers (source: Reference 11)

Models	16600A	16601A	16602A	16603A
Maximum state clock	100 MHz	100 MHz	100 MHz	100 MHz
Maximum timing sampling rate (full/half channels)	125/250 MHz	125/250 MHz	125/250 MHz	125/250 MHz
Memory depth (full/half channels)	64/128 k[1]	64/128 k[1]	64/128 k[1]	64/128 k[1]
Channels supported	204	136	102	68
Setup/hold time	0/4.5 ns to 4.5/ 0 ns adjustable in 500 ps increments			

[1] Increased memory depth in half channel timing mode only.

multiple processor/bus designs. Most of these systems are scalable, with modules for state/timing analysis, as well as oscilloscope, function generator and emulation modules.

10.9.1.1 Agilent Technologies 16600 and 16700 series logic analysis systems

The HP 16700A series platform is a modular system that can change with the needs. The HP16600A series logic analysis system offers built-in state and timing capabilities from 68 to 204 channels. Though these channel configurations are fixed, each frame contains one expansion slot for additional measurement capability, such as an oscilloscope, pattern generator or analysis module.

Table 10.3 compares the HP16700A and 16600A series, and Table 10.4 summarises the built-in capability of the HP16600 series analysers. Figure 10.19 shows a logic analysis system.

Figure 10.19 Logic analysis system

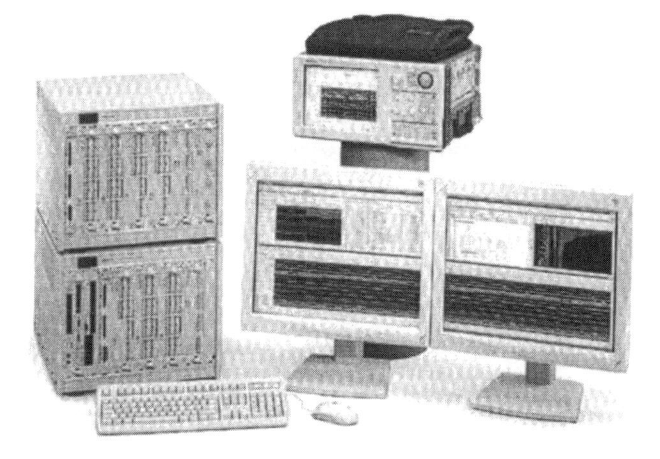

Figure 10.20 Logic analysis modules for the TLA700 series (reproduced with permission from Tektronix – source: [11])

10.9.1.2 Tektronix TLA700 series

This series of logic analysers consist of portable (TLA714) and bench-top (TLA720) modular mainframes. These logic analysis modules illustrated in Figure 10.20 are available in 34-, 68-, 102- and 136-channel modules.

Typical applications for this series include in addition to the basic functions are, multi-processor/bus analysis, performance analysis, digital stimulus and control, digital signal quality analysis and system validation. Table 10.5 shows key characteristics of the TLA700 series.

Measurement modules for this series include pattern generators with up to 268 MHz pattern speeds and 2 Mb memory depth, and 4-channel, 1 GHz,

Table 10.5 Key characteristics of the TLA700 series. (source: Reference 10)

	TLA714	TLA720
Display	Built-in	External
Number of channels per module	34, 68, 102, 136	
Maximum channels per mainframe	272	680
Maximum channels per system	3264	8160
Timing resolution	500 ps	
State clock rate	100 MHz standard/ 200 MHz optional	
State data rate	400 MHz (half)/200 MHz (full)	
Memory depth	128 (half)/64 (full) kb to 128/64 Mb	

5 Gsamples s^{-1} digitising oscilloscopes. The expansion mainframe can support up to 60 modules per system.

10.9.2 Stand-alone logic analysers

These bench-top logic analysers are all-in-one instruments, with a wide variety of models and options.

10.9.2.1 Agilent Technologies 1670G series bench-top logic analysers

The 1670G series offers 150 MHz state analysis and timing analysis at speeds up to 500 MHz. The high speed acquisition memory can be configured for depths up to 2 M samples over up to 136 acquisition channels. An optional 2-channel, digitising oscilloscope or a pattern generator can be added to the unit. The oscilloscope can be triggered by the logic analyser and vice versa. The 1664A is a low cost version of the 1670G series logic analyser family.

Another trend is towards the additional feature of test pattern generators (TPGs) which can provide reference signals to the system under test.

Table 10.6 summarises the specifications of this family of logic analysers. Figure 10.21 shows this series.

10.9.2.2 Tektronix TLA 600 series

This is an entry-level logic analyser, typically suitable for general purpose state/timing analysis and single processor/bus applications (see Table 10.7). These also include real time instruction trace analysis, source code debug and performance analysis. State and high speed timing analysis can be done on all channels using the same probes.

10.9.3 PC-hosted logic analysers

10.9.3.1 Agilent Technolgies Logic*Wave* E9340A PC-hosted logic analyser

This is a transportable, low cost logic analyser, which runs as a Microsoft Windows 95/98/NT application. This connects to a PC or a laptop via the parallel port. Figure 10.22 shows this logic analyser.

Table 10.6 *Key specifications for the 1670G series bench-top logic analysers (source: Reference 12)*

Model no.	1670G	1671G	1672G	1673G	1664A
Channel count	136	102	68	34	34
Timing analysis speed	250/500 MHz(full/half channels)				
State analysis speed	150 MHz				50 MHz
State/clock qualifiers	4	4	4	2	
Memory depth/channel	64/128 k (full/half channels) Optional 256/512 k or 2/4 M				4/8 k
Optional oscilloscope	2-channel, 500 MHz, 2 Gsamples s^{-1}, 32 ksample				N/A
Optional pattern generator	32-channel, 100/200 MHz, 256 k, vector pattern generator				N/A

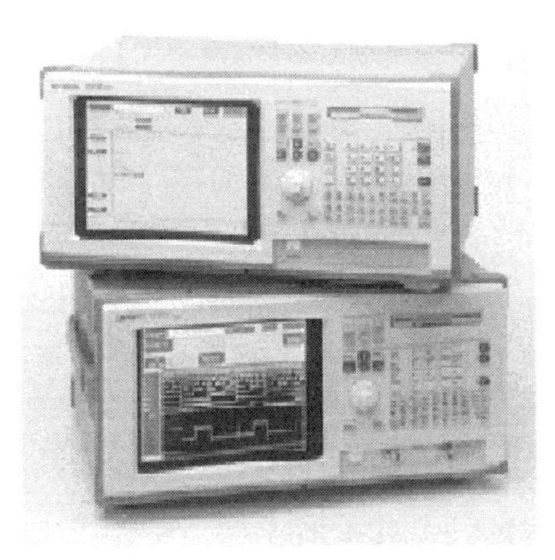

Figure 10.21 *Bench-top logic analysers*

The key features of the logic analyser are: 34 channels, memory depth of 128 k samples on each channel in timing mode, 64 k in state mode, 100 MHz state analysis and 250 MHz timing analysis.

This type of logic analysers find applications in debugging digital circuits that use 8- or 16-bit microprocessors, microcontrollers, or FPGAs.

10.9.4 Logic scopes

Logic scopes are instruments that seamlessly combine in a single instrument, the analogue acquisition system of a digital storage oscilloscope (DSO), with the triggering and display systems of a logic analyser. These instruments, usually having fewer

Table 10.7 Key characteristics of the TLA600 series (source: Reference 8)

	TLA60X	TLA61X	TLA62X
Display	External	Built-in	
Number of channels	34, 68, 102, 136		
Timing resolution	500 ps		
State clock rate	100 MHz standard/200 MHz optional		
State data rate (half/full channels)	400 MHz/200 MHz		
Memory depth (half/full channels)	128/64 kb to 2/1 Mb		

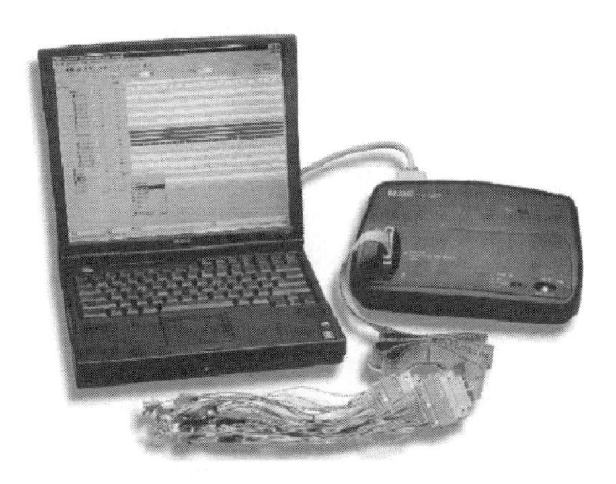

Figure 10.22 PC-hosted logic analyser

channels than a general purpose logic analyser, are low cost instruments suitable for applications such as multichannel data acquisition, mixed-signal analysis, and A/D/A analysis.

The TLS216 logic scope from Tektronix, Inc., as shown in Figure 10.23, has a bandwidth of 500 MHz and samples all 16 input channels simultaneously at 2 Gsamples s^{-1}. The timing resolution is less than 100 ps [15].

10.9.5 Mixed signal oscilloscopes

Another feature that has developed recently is the integration of the logic analyser with the digital storage oscilloscope (DSO). Although this is an excellent instrument for looking at digital signals, the logic analyser will not provide all the information required in some instances. If the actual waveforms need to be examined (for example, to check the rise and fall times of the system clock), then a DSO is required. Having both functions together offers a very powerful tool although the task of designing such an instrument is quite formidable. Both Agilent Technologies and Tektronix

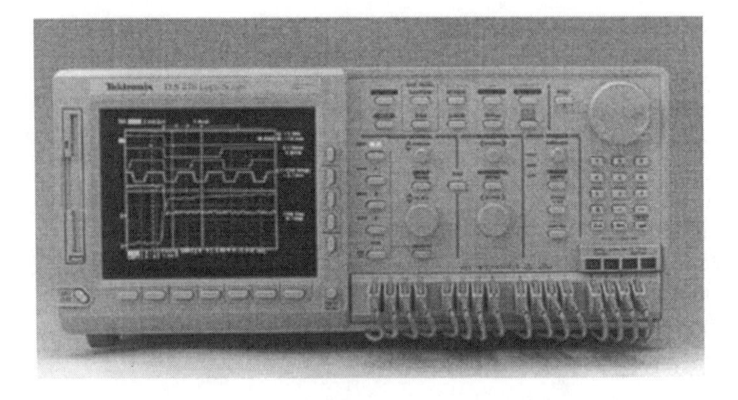

Figure 10.23 The TLS216 logic scope (reproduced with permission from Tektronix – source: Reference 15)

have developed logic analyser and digital storage oscilloscope combinations, named the mixed signal oscilloscope (MSO).

Mixed signal oscilloscopes have both logic analyser and digitising oscilloscope channels. However, these are more advanced instruments than the logic scope, having higher sampling rates and a larger number of channels.

The display of a MSO is shown in Figure 10.24. The upper trace is what you would see typically in an oscilloscope, and the lower traces are what you would see in a logic analyser. In certain troubleshooting instances, it is advantageous to look at exact signal levels or exact rise/fall times in order to isolate the problem faster. MSOs prove useful in such cases.

Given the cheaper computing power and memory it is likely that this instrument will be more popular in the future.

10.9.6 Present and future trends

Cheaper computing power and memory have influenced the design of logic analysers in recent years and will probably continue to do so in the future. With the instrument developing higher intelligence, the user interface has become simpler. This development is obvious when comparing, say, a top-end logic analyser of the early 1980s with that of the late 1990s. The number of knobs and buttons in the front panel has dramatically declined. The instrument has also become more user friendly. Menu-driven systems with online help remove some of the difficulties of using logic analysers. This is especially true for PC-based logic analysers. The cost of such an instrument is relatively low, since most of the hardware is needed for data acquisition. Excellent data evaluation and display capacity are available in the host processor. PC-based analysers also have the added advantage of being able to store data and instrument setups within the PC's hard disk. Programming the analyser for automated measurement is also very easy [1].

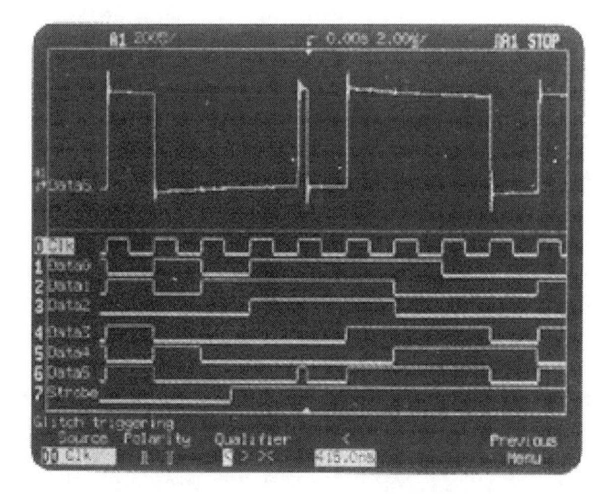

Figure 10.24 The display of a mixed signal oscilloscope showing the digital storage oscilloscope channel (top) and several logic analyser channels (reproduced from Reference 2, with permission)

10.10 Application examples of logic analysers

Logic analysers are used in hardware or software verification or in troubleshooting. This section illustrates two examples of such applications.

The first example deals with a hardware error. On testing a static RAM circuit, it is found that the data written out differ from those written in. A timing violation is suspected. The timing diagram for a write cycle is shown in Figure 10.25. This is obtained by setting to the asynchronous mode (10 ns resolution) and triggering when the chip select (CS) line goes low and the RAM address range is accessed. Examination of the timing diagram shows that the write enable (WE) signal is delayed (probably because of gate propagation delays) and the microprocessor data change before they are written in. A solution might be to reduce the delay by using faster gates [1].

The second example deals with software. A parallel printer interface detects the ready (online) status of the printer and sets bit 0 of the status register and at the same time generates an interrupt. The CPU, responding to this interrupt, checks that the appropriate status bit is set and, if so, sends a character to the printer. However, on operation, although the printer stays online, no characters are received. It is possible that the program may have crashed or it may be hung up on a loop. The logic analyser is first used to check if the interrupt is called. This can be detected by triggering the logic analyser on the start address of the printer interrupt service routine. The logic analyser is thus triggered and the display is put to the control flow mode. A single conditional branch instruction is found to be repeated endlessly. Therefore, the interrupt is called but the program is hung in a loop thereafter. By switching the display on to disassembly mode and displaying the data bus in binary, it is soon obvious what the fault is: namely, the program is checking bit 1 rather that 0 for busy/ready status [1].

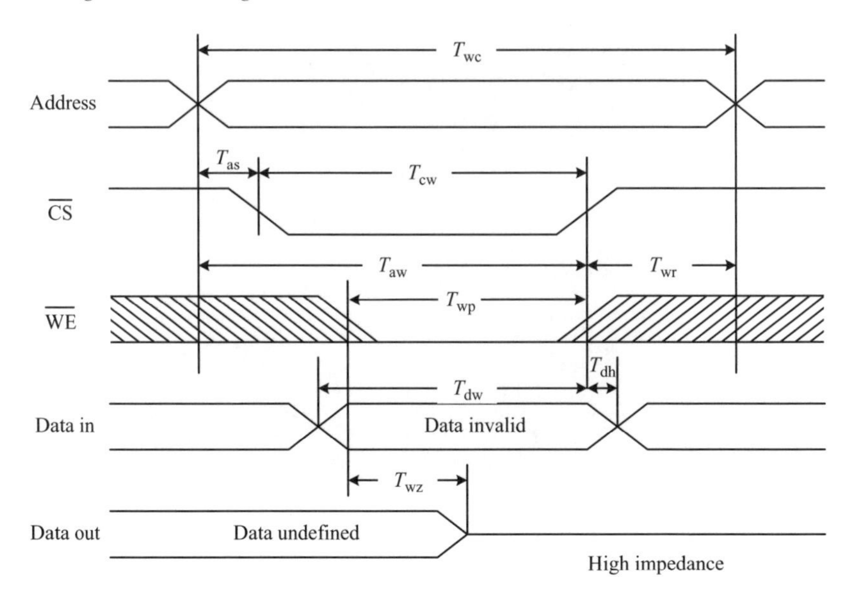

Figure 10.25 Static memory write cycle timing diagram

10.11 References

1 KULARATNA, N.: 'Modern Electronic Test & Measuring Instruments', IEE Press, 1996.
2 'Oscilloscope or Logic Analyser?', http://contact.tm.agilent.com/tmo/technifo/English/ BI_Lgic_a.html, Agilent Technologies, 2000.
3 'What's a Logic Analyser ? What's a Timing Analyser', http://contact.tm.agilent.com/tmo/technifo/English/BI_Lgic_b.html, Agilent Technologies, 2000.
4 'An ABC of Logic Analysis', Philips Eindhoven, The Netherlands.
5 'Philips PM3565 and 3570: Logic Analysers for Tomorrow's Systems', Philips Eindhoven, The Netherlands.
6 'Digital Instruments Course Part 5 – Logic Analysers', Philips Eindhoven, The Netherlands.
7 'State and Timing Modules for HP Logic Analysis Systems', Hewlett Packard Product Overview, 5966-3367E, 5/99.
8 'A New Breed of Affordable Logic Analysers', http://www.tektronix.com/Measurement/Products/backgrounders/jedi/index.html, Tektronix Inc., 2000.
9 'Probing Solutions for HP Logic Analysis Systems', Hewlett Packard Product Overview, 5968-4632E, 8/99.
10 'Advanced Triggering with the TLA700 Series Logic Analyzers', http://www.tektronix.com/Measurement/App_Notes/tla700/advtrig, Tektronix Inc., 2000.
11 'HP16600A and 16700A Series Logic Analysis System Mainframes', Hewlett Packard Product Overview, 5966-3107E, 5/99.

12 'Agilent Technologies 1670G Series Benchtop Logic Analysers', Agilent Technologies Technical Data, 5968-6421E, 11/99.
13 'Agilent Technologies Logic *Wave* E9340A PC-Hosted Logic Analyser', Agilent Technologies Technical Data, 5968-5560E, 11/99.
14 'Logic Analysers – TLA Family, Features and Benefits', http://www.tek.com/Measurement/Products/catalog/52_14255/eng/index.html, Tektronix Inc., 52W-14255-Op1, 08/2000, 09/15/2000.
15 'Logic Scope TLS216', http://www.tektronix.com/Measurement/Products/catalog/tls216/index.html, Tektronix Inc., 49A-10733-4p154, 06/1997, 08/10/2000.

Further reading

CONNER, D.: 'Consider logic analysers for more than microprocessor applications', *EDN*, 24 December 1987, pp. 55–56.
CONNER, D.: 'Combined instruments find logic problems', *EDN*, 5 August, 1991, pp. 77–84.
'Logic Analysis – Fundamentals, Functions, Applications', Rhode and Schwartz, 1986.
MCCULLOUGH, R.: 'Low-cost logic analysers: combining capabilities', *Electron. Test.*, August 1990, pp. 56–57.
RISHAVY, D.: 'Logic analyser triggering techniques tame complex computer buses', *Electronic Design*, January 24, 2000.
STRASSBERG, D.: 'PC based logic analyzers', *EDN*, 13 October 1998, pp. 134–142.
TEKTRONIX, INC.: 'Basic Concepts of Logic Analysis', Tektronix, Inc., 1989.
ZIVNY, P.: 'Scope/logic analyser team measures timing compliance of high-speed, synchronous designs', *EDN*, June 5, 2000.

Chapter 11
An introduction to instrument buses and VLSI testing

Nihal Kularatna and Shantilal Atukorala

11.1 Introduction

This chapter provides an introduction to common instrument interfaces and VLSI testing. Technical details of two matured standards such as the IEEE-488 interface system and the VME bus extension for instrumentation (VXIbus) are given, along with an insight into the evolution of the standards and the future trends.

The IEEE-488 interface system, more commonly known as the general purpose interface bus (GPIB), has been carefully designed to integrate several instruments, together with a computer or a dedicated instrument controller, to simplify the automated measurement process. The bus allows a two-way communication between the controller or the computer with the instruments on the test setup and assists the test engineer in the design and implementation of automatic test equipment (ATE). The particular design philosophy and the interface system allow the necessary flexibility to accommodate different types of electronic instruments manufactured by different manufacturers. The IEEE-488 interface is a digital system, which allows up to 15 instruments or devices to communicate with each other under the control of a master unit or a controller. The devices and instruments are connected together in parallel using specially designed connectors and a cable system referred to as a 'bus'. The supervision of the integrated test setup is provided by the master unit called the controller which is usually a personal computer or a dedicated bus controller. The software needed to control the test setup can be easily written using the popular computer languages. In the past, languages like BASIC or FORTRAN were very commonly used and the present trend is towards languages such as C or its variations.

With wide experience gained over two decades on processor-based systems, computers, work stations, software packages, etc., we have now entered a new era of modular instrumentation based on the VXIbus or the instrument on a card (IAC) technology.

11.2 IEEE-488 bus

11.2.1 Standardisation and evolution of IEEE-488

Prior to the 1970s there were few attempts at standardisation. Each manufacturer created its own special interface to connect a particular computer to a specific piece of test equipment. One of the earliest attempts at standardisation was a CAMAC (computer automated measurement and control) initiated by the European Scientific Community for nuclear instrumentation. A large controller was required and each piece of test equipment required a plug-in interface adaptor. This was eventually accepted as IEEE standard 583/1973; however, it has seen little use outside the scientific community. Another early interface standard, still widely used today, is the popular RS-232 interface developed by the Electronic Industries Association (EIA). This is a serial interface designed for connecting terminals, printers and modems to mainframe computers. Because it works on serial mode and provides no easy means of addressing instruments, it is very inefficient at transferring data across various instruments. Therefore, it is not ideally suited for ATE applications. However, most manufacturers still provide optional RS-232 interfaces for test instruments to allow access to remote test equipment through telephone lines and modems.

The Hewlett-Packard Company was one of the instrument manufacturers who investigated the possibility of interfacing any and all future instruments with each other. After commencing its initial investigations in 1965, the company had, by 1972, introduced a bus to interface its programmable test instruments with an instrument controller. This bus was a byte serial, a bit parallel data bus that could support a controller and one or several instruments configured as 'talkers' and 'listeners'. The concept of this Hewlett-Packard interface bus (HP-IB) was soon endorsed by the electronics industry and, after some modifications, was adopted by Institution of Electrical and Electronics Engineers (IEEE) in 1975 and was entitled *Digital Interface for Programmable Instrumentation*. This contained the electrical, mechanical and functional specifications of an American standard interface system. In January 1976 the American National Standards Institute (ANSI) published an identical standard called MC1.1.

The IEEE-488/1975 was later revised into the version IEEE-488/1978 and today it is commonly referred to as either IEEE-488 bus, HP-IB or GPIB (general purpose interface bus). The European IEC-625 bus is identical except for the specifications of bus connectors. Four major standards based on IEEE-488 are currently available. These are:

(i) IEEE-488/1978 (now referred to as IEEE-488.1),
(ii) ANSI MC1.1 (identical to IEEE-488.1),
(iii) IEC-625.1 (identical except connector),
(iv) BS 6146 (identical to IEC-625.1).

The IEEE-488/1978 was a common interest among test engineers for over a decade; instrument designers, as well as users, became familiar with drawbacks and possible improvements. Because the IEEE-488 document had purposely left some problems unsolved, it was up to the users to handle the message protocol and data differently. Experience gained during the first half of the1980s clearly indicated that the software was not adequately standardised. Users were forced to spend long periods in software tune-ups rather than in configuring instruments. During the latter half of the 1980s revisions to the standard came in the wake of the US Air Force attempting to have its 'control intermediate interface language' (CIIL) adopted as a commercial standard. Manufacturers, however, disliked the idea and so a new set of IEEE standards were proposed. The first attempt to standardise the data formats which were recommended here were acceptable to the test engineers and helped in creating a new series of revisions to IEEE-488/1978.

In June 1987 the IEEE approved a new standard for programmable instruments and devices. This was IEEE-488.2/1987 entitled *IEEE Standard Codes, Formats, Protocols and Common Commands*. The original document, IEEE-488/1978 was revised as IEEE-488.1 with a few variations. The major changes introduced were three new appendices offering guidelines on address switch labels, status indicators and reducing electromagnetic interference. Further, the 488.1 version incorporated several minor revisions to match it with the software version IEEE-488.2. The IEEE-488.1 and 488.2 were approved in 1987 and were published in June 1988. At the same time the International Electro-technical Commission (IEC) was also considering its own new revisions, IEC-625.1 and IEC-625.2.

IEEE-488.2 defined data formats, status reporting, a message exchange protocol, IEEE-488.2 controller requirements and common configuration commands to which all IEEE-488.2 instruments must respond in a precise manner. While IEEE-488.2 incorporated 'precise talking', in order to be compatible with non-IEEE-488.2 instruments, 'forgiving listening' was also included. IEEE-488.2 simplified programming and created more compatible and more reliable systems; it did not, however, address the problem of different command sets that were used by different instruments. In 1990, a consortium of major IEEE-488 product manufacturers developed a specification known as SCPI (standard commands for programmable instruments). This was based on the IEEE-488.2, the Hewlett-Packard 'test and measurement systems language' (TMSL), and the Tektronix 'analogue data interchange format' (ADIF). SCPI defines a single comprehensive command set suitable for all instruments. It simplifies programming and allows interchange of different instruments within the same product line with minimum changes to the program. Figure 11.1 shows the relations between 488.1 and 488.2 on various device commands and interface levels.

In summary, the primary reasons for the proposal of 488.1, 488.2 and SCPI were that although the original IEEE-488 specifications described very well how to exchange data bytes, it did not say what those bytes should be. The result was a set of standards that tell the designers exactly how to format messages and how device intended messages should appear on the bus. Most concepts were originally presented in the recommended practice ANSI/IEEE standard 728. By the 1990s the IEEE-488 standard had been published in over nine languages and has now been used by more

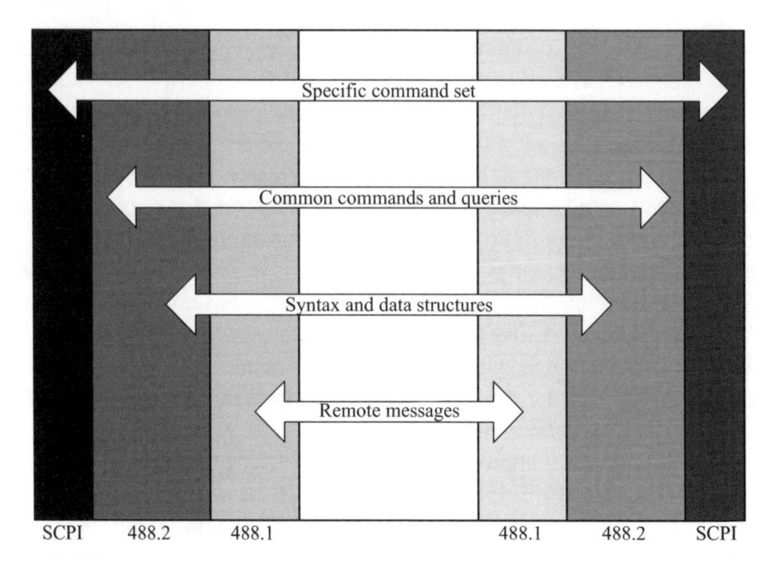

Figure 11.1 IEEE-488.1, 488.2 and SCPI functional layers

than 250 manufacturers in over 14 countries. It is one of the most carefully designed, consistent and highly used instrument interface systems in the world.

11.2.2 IEEE-488 major specifications

The IEEE-488 is a digital, 8-bit parallel communication interface with data transfer rates up to a maximum of 1 Mbyte s^{-1}. The bus supports one system controller (usually a computer) and up to 14 additional instruments. Because of its fairly high data transfer rate and the 8-bit parallel interface, it has gained popularity in other applications such as inter-computer communication and peripheral control.

Devices are connected with a cable assembly of 24 conductor cables (shielded) with both a plug and a receptacle connector at each end. This design allows the devices to be linked in linear (bus) configuration or star configuration, as illustrated in Figure 11.2.

The GPIB uses negative logic (with standard TTL voltage levels) where a zero logic level on any line represents a true condition. One important reason for adopting the negative logic convention (active-low condition) is to allow the devices to be designed with transistor open collector output circuits, which can pull the lines to a zero voltage level to indicate a true condition. Furthermore, the negative logic convention improves noise immunity. The GPIB connector and pin assignments are given in Figure 11.3.

11.2.3 GPIB operation

GPIB equipment communicates using a byte serial/bit parallel protocol: 'bit parallel' means that eight bits are sent over the interface at one time and 'byte serial' means that each byte is sent one after another.

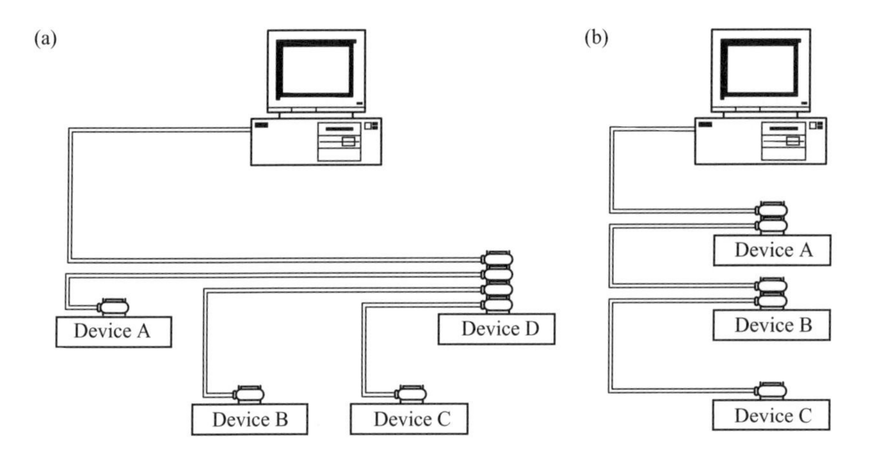

Figure 11.2 Different GPIB configurations: (a) star configuration; (b) linear configuration

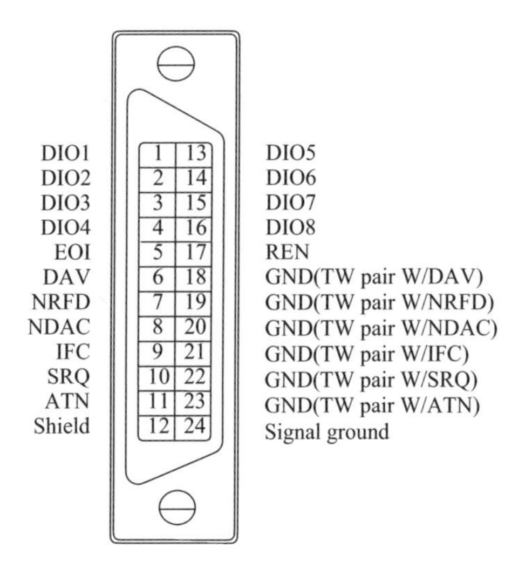

Figure 11.3 GPIB connector pin assignments

11.2.3.1 Device communication

Among GPIB devices, communication is achieved by passing messages on the interface bus. There are two types of message: namely, device dependent messages and interface messages. Device dependent messages are often called data or data messages and contain device specific information such as programming instructions, results of measurement, machine status or data files. These device dependent messages are sent with GPIB ATN line unasserted (discussed later).

Interface messages manage the bus itself. They are usually called commands or command messages. These perform tasks such as initialising the bus, addressing/un-addressing devices and setting device modes (local/remote). These messages are sent with the ATN line asserted. (The term 'command' here does not mean device specific instructions. Device specific instructions are data messages.)

Data are transferred from device to device over the bus using the eight bi-directional data lines. Normally the 7-bit American Standard Code for Information Interchange (ASCII) is used. The equivalent international standard is the International Standards Organisation (ISO) code. The eight data lines from DIO-1 to DIO-8 carry both data and command messages. All commands and most data use the 7-bit ASCII or ISO code set. The eighth bit, DIO-8, is either unused or used for parity. The ATN line is used to identify whether data or command messages are transferred across the line.

Figure 11.4 indicates a case where the transfer of the 3-byte sequence 'bus' would occur over the data lines. It is important to note the convention of negative logic signals on these data lines.

11.2.4 GPIB devices and interface functions

Devices are categorised as controllers, talkers and listeners. The operation of a GPIB based test setup could be equated to a committee where the chairperson is like the controller. The chairperson dictates who has to talk and who has to listen.

11.2.4.1 Controller

Most GPIB systems are configured around a computer and in these cases the computer becomes the system controller. When multiple computers are used anyone could become a controller. However, only one of these could become the active controller or the controller-in-charge. The system controller must be defined for each GPIB system. This is done either by jumper settings on the interface boards or by configuring the software.

11.2.4.2 Talker/listener

Most GPIB devices and instruments fit into the category of talkers or listeners. Some devices can act only as a talker or a listener. A personal computer used as a controller could act as all three devices (controller/talker/listener). Table 11.1 gives the characteristics of talkers and listeners. Examples of talkers are DVMs, oscilloscopes, etc. Listeners are devices like printers or power supplies. The controller is usually an interface card mounted on a personal computer.

The IEEE-488 standard specifies a total of 11 interface functions, which can be implemented in any GPIB device. Equipment manufacturers are free to use any of these functions to perform various functions. Each interface function is identified by a mnemonic that is a one, two or three letter word used to describe a particular capability. A brief description of these interface functions is given in Table 11.2. In addition to these basic interface functions the IEEE-488 standards describe subsets of all functions in detail. Each subset is identified by assigning a number after the mnemonic. For these details the reader may consult References 1 and 2.

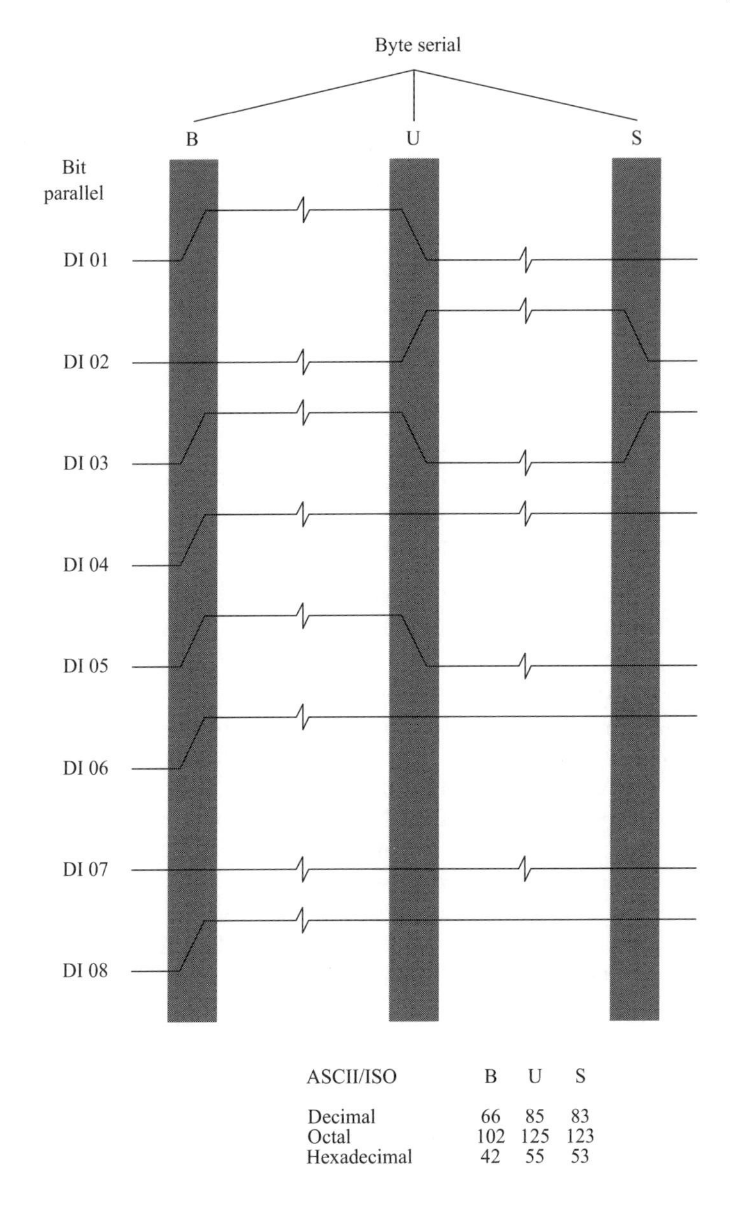

Figure 11.4 Data bus format

11.2.5 GPIB signals and lines

The GPIB has 16 signals and eight ground return (or shield drain) lines. All devices on the GPIB share the same 24 bus lines. Figure 11.5 shows the bus arrangements.

The 16 signal lines are divided into three groups: that is, eight data lines, five interface management lines and three handshake lines.

Table 11.1 Characteristics of GPIB talkers and listeners

Talker	Listener
Instructed by the controller to talk	Instructed by the controller to listen
Places data on the GPIB	Reads data placed on the GPIB by the talker
Only one device can be addressed to talk at one time	Multiple devices can be addressed to listen at one time

11.2.5.1 Interface management lines

The five interface management lines are used to manage the flow of information across the GPIB. These five lines are: (i) 'interface clear' (IFC), (ii) 'attention' (ATN), (iii) 'remote enable' (REN), (iv) 'end or identify' (EOI), and (v) 'service request' (SRQ) as shown in Figure 11.5. They are explained as follows.

(i) Interface clear (IFC) can only be controlled by the system controller. The system controller uses IFC to take control of the bus asynchronously. The IFC line is the master reset of the GPIB.
(ii) Attention (ATN) is asserted (or unasserted) by the controller-in-charge and is used to notify the devices of the current data type. When unasserted, information on the bus is interpreted as a data message, and when asserted, information on the bus is interpreted as a command message.
(iii) Remote enable (REN) is used by the controller to put the devices into a remote state. This is asserted by the system controller.
(iv) End or identify (EOI) is used by some devices to terminate their data output. A talker would assert EOI along with the last byte of data. A listener stops reading data when EOI along with the last byte of data. A listener stops reading data when EOI is asserted. This line is also used in parallel polling (discussed later).
(v) Service request (SRQ) is asserted at any time by a device to notify the controller-in-charge that it needs service. It is the responsibility of the controller to monitor the SRQ line.

11.2.5.2 Handshake lines

Three lines asynchronously control the transfer of message bytes among devices. This three wire interlocking handshake scheme guarantees that the message bytes on the data line are sent and received without errors. These handshake lines are (i) 'not ready for data' (NRFD), (ii) 'not data accepted' (NDAC), and (iii) 'data valid' (DAV). These are explained as follows.

(i) NRFD indicates whether a device is ready (or not ready) to receive a data byte. This line is driven by all devices when receiving commands and only by listeners when receiving data messages.

Table 11.2 Available interface functions

Interface functions that may be included in a GPIB device	Mnemonic	Comments
Talker or extended talker	T,TE	Capability required for a device to be a 'talker'
Listener or extended listener	L,LE	Capability required for a device to be a 'listener'
Source handshake	SH	Provides a device with the capability to properly transfer a multi-line message
Acceptor handshake	AH	Provides a device with the capability to guarantee proper reception of remote multi-line message
Remote/local	AL	Allows device to select between two sources of input information. 'Local' corresponds to front panel controls and 'remote' to the input information from the bus
Service request	SR	Permits a device to asynchronously request service from the controller
Parallel poll	PP	Allows a device to uniquely identify itself if it requires service when the controller is requesting a response
Device clear	DC	This differs from 'service request' in that it requires a commitment of the controller to periodically conduct a parallel poll
Device trigger	DT	Permits a device to have its basic operation initiated by the talker on the bus
Controller	C	Permits a device to send addresses, universal commands and addressed commands to other devices on the GPIB. It may also include the ability to conduct polling to determine devices requiring service
Drivers	E	This code describes the type of electrical drivers used in the device

(ii) NDAC indicates whether a device has (or has not) accepted a data byte. The line is driven by all devices when receiving commands and only by listeners when receiving data messages. By this handshake scheme the transfer rate will be limited by the slowest active listener. This is because a talker waits until

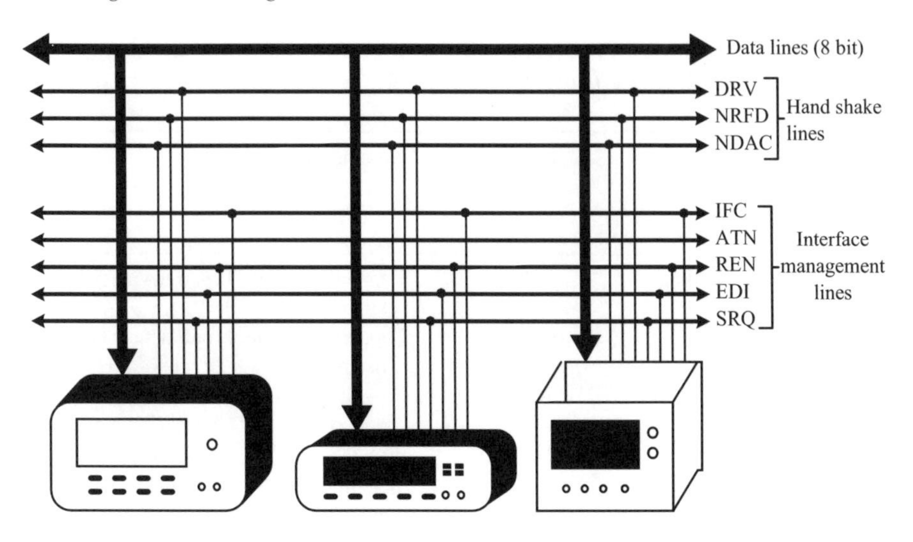

Figure 11.5 GPIB bus arrangement

all listeners are ready. NRFD is driven false before sending data and waits for listeners to accept data.

(iii) DAV indicates whether the signals on the data line are stable and valid, and can be accepted by devices. The controller asserts DAV when sending commands and the talker asserts it when sending data messages. Figure 11.6 shows the timing sequence of the three wire handshake process.

11.2.6 Bus commands

All devices must monitor the ATN line and respond within 200 ns. When the ATN is true all devices accept data on the data lines and treat them as commands or addresses. Commands accepted by all devices are known as universal demands. These could be uni-line such as ATN or others such as IFC, REN and EOI. Universal commands also can be multi-line, in which the command is a code word on the data lines. Some of the commands are addressed; that is, they are meant for addressed devices only. By using these commands a controller can designate talkers and listeners (send talk and listen addresses), disable talkers or listeners (send untalk and unlisten commands), set a device to a predefined device dependent state (device clear command), and enable polling of devices to determine which device has requested attention (parallel poll configure, serial poll enable commands). The five universal commands are shown in Table 11.3 and addressed commands are given in Table 11.4.

11.2.7 Polling

In IEEE-488 systems, the controller initiates each data transfer. (One exception may be systems that have a single dedicated talker, which can function without a controller.)

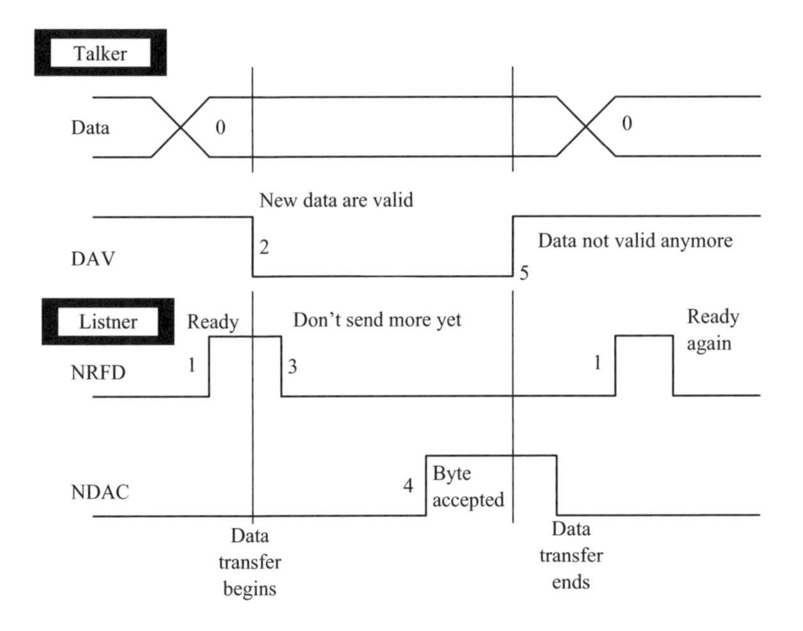

Figure 11.6 Three wire handshake process

Table 11.3 Universal commands

Multi-line command	Mnemonic	Octal code
Device clear	DCL	24
Local lockout	LLO	21
Serial poll enable	SPE	30
Serial poll disable	SPD	31
Parallel poll unconfigure	PPU	25

Table 11.4 Addressed commands

Addressed command	Mnemonic	Octal code
Group execute trigger	GET	10
Selected device clear	SDC	04
Go to local	GTL	01
Parallel poll configure	PPC	05
Take control	TCT	11

In this sense the controller is somewhat in the position of the exchange in a public telephone system, where as soon as the sender and the receiver are known, the exchange sets up the path. If any device desires to become a talker (i.e. to send data or report an error) it will have to indicate that fact to the controller. This is done by asserting the SRQ line that acts as an interrupt to the controller. The controller must then poll each device to find out which one or more of them have asserted the SRQ. Each device would be addressed as a talker and will then send a status byte to the controller, which will indicate whether it requires attention or not. This method of interrogation is known as a serial poll. It is also possible for the controller to initiate a parallel poll, in which selected devices may send a status bit on a predefined data line. This method allows faster checking of the status of certain actions. A parallel poll is initiated by asserting the ATN and EOI lines together. A serial poll is carried out by a multi-line command.

11.2.8 Physical and electrical characteristics

The GPIB bus uses a transmission line system. Hence, cable impedances and terminations control the maximum data speeds. The following rules are dictated by the IEEE-488 specifications.

(i) The total length of all cables used should be less than 2 m per device, with a maximum limit of 20 m.
(ii) At least two-thirds of the devices must be powered on.

When these limits are exceeded bus extenders and expanders should be used. Even though the standard specifies a maximum data speed of 1 Mbyte s^{-1}, when the cables are at the maximum length of 20 m, the maximum speed drops to about 250 kbytes s^{-1}. National Instruments (USA) have developed a high speed data protocol (HS488) that is compatible with the normal protocol. In the high speed mode, the data transfer rate can go up to 8 Mbyte s^{-1}. For a detailed account of the restrictions on IEEE-488 bus usage the reader may consult References 1 and 2.

11.2.9 IEEE-488.2

The ANSI/IEEE standard 488/1975 (now called IEEE-488.1) had several shortcomings. It defined the electrical and mechanical interface and setting-up communication. It did not, however, define data formats of messages, status reporting, message exchange protocol, common configuration commands or device specific commands. Different manufacturers have interpreted these areas differently leaving a heavy programming burden on the user. The IEEE-488.2 standard addresses these problems. Communication was defined such that the system may be more reliable and efficient compared with IEEE-488.1 systems. However, the IEEE-488.2 controllers also possess the 'forgiving listening' capability which makes them backward compatible with IEEE-488.1 instruments. On the controller side, the new standard puts down precise requirements such as the method of pulsing the IFC line for 100 pulses s^{-1}, asserting the REN line, sensing the state and the transitions of the SRQ line and timing out of any I/O transaction. The IEEE-488.2 also defines the control sequences that specify the exact messages sent by the controller to initiate certain actions. To ease programming, the new standard also defined high level controller protocols, which make use of several control sequences. For example, the protocol ALLSPOL carries a serial poll and returns a status byte for each device. The IEEE-488.2 control sequences and controller protocols are given in Tables 11.5 and 11.6, respectively.

On the instrument side, the IEEE-488.2 defines a set of mandatory common commands that all devices must possess. Table 11.7 lists these. The SCPI specification uses these common commands as the foundation. The new standard also defines the status reporting, expanding on the status byte used by IEEE-488.1 instruments. Four registers are used with status reporting. The common command set allows the controller to access these registers. Figure 11.7 illustrates the status report model. The status byte register is returned by the serial poll. The service request enable register (SRE) will determine the conditions under which service requests are generated and can be set by the controller. More details about the condition that generated the service request for a standard event can be obtained by reading the 'standard events status

Table 11.5 IEEE-488.2 required and optional control sequences

Description	Control sequence	Compliance
Send ATN true commands	SEND COMMAND	Mandatory
Set address to send data	SEND SETUP	Mandatory
Send ATN false data	SEND DATA BYTES	Mandatory
Send a program message	SEND	Mandatory
Set address to receive data	RECEIVE SETUP	Mandatory
Receive ATN false data	RECEIVE RESPONSE MESSAGE	Mandatory
Receive a response message	RECEIVE	Mandatory
Pulse IFC line	SEND IFC	Mandatory
Place devices in DCAS	DEVICE CLEAR	Mandatory
Place devices in local state	ENABLE LOCAL CONTROLS	Mandatory
Place devices in remote state	ENABLE REMOTE	Mandatory
Place devices in remote with local lock out state	SET RWLS	Mandatory
Place devices in local lockout state	SEND LLO	Mandatory
Read IEEE-488.1 status byte	READ STATUS BYTE	Mandatory
Send group execution trigger (GET) message	TRIGGER	Mandatory
Give control to another device	PASS CONTROL	Optional
Conduct a parallel poll	PERFORM PARALLEL POLL	Optional
Configure devices parallel poll responses	PARALLEL POLL CONFIGURE	Optional
Disable devices parallel poll capability	PARALLEL POLL UNCONFIGURE	Optional

Table 11.6 IEEE-488.2

Keyword	Name	Compliance
RESET	Reset system	Mandatory
FINDRQS	Find device requesting service	Optional
ALLSPOL	Serial poll all devices	Mandatory
PASSCTL	Pass control	Optional
REQESTCTL	Request control	Optional
FINDLSTN	Find listeners	Optional
SETADD	Set address	Optional but requires FINDLSTN
TESTSYS	Self-test system	Optional

Table 11.7 IEEE-488.2 mandatory common commands

Mnemonic	Group	Description
*IDN?	System data	Identification query
*RST	Internal operations	Reset
*TST?	Internal operations	Self-test query
*OPC	Synchronisation	Operation complete
*OPC?	Synchronisation	Operation complete query
*WAI	Synchronisation	Wait to complete
*CLS	Status and event	Clear status
*ESE	Status and event	Event status enable
*ESE?	Status and event	Event status enable query
*ESR?	Status and event	Event status register query
*SRE	Status and event	Service request enable
*SRE?	Status and event	Service request enable query
*STB?	Status and event	Read status byte query

register' (ESR). Setting the standard events status enable register (ESE) will determine which of these events are allowed to generate a service request.

Although the IEEE-488.2 has improved compatibility and reliability of GPIB systems, it does not specify device specific commands. For example, to program an IEEE-488.2 oscilloscope to select channel 2 with the attenuator range set to 5 V, may require one of the following commands, depending on the model being used.

(i) CH2 VOLTS:5.0
(ii) CHAN2:RANGE 5.0
(iii) VERT A
(iv) ATT 5.0

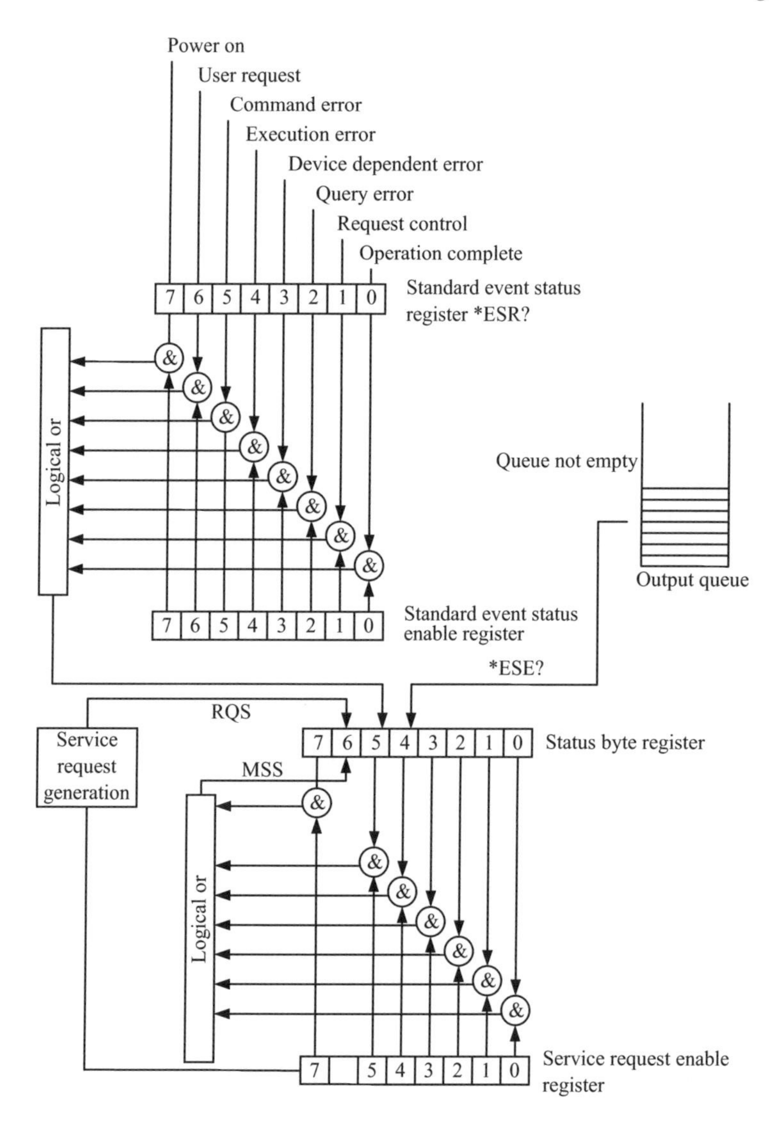

Figure 11.7 IEEE-488.2 status reporting model

11.2.10 SCPI

In 1990, a consortium of GPIB manufacturers announced a specification known as the 'standard commands for programmable instruments' (SCPI) to address this need. SCPI defines a standard command set, which can be extended, applicable to all instruments that simplifies or eliminates software updating when instruments are interchanged. SCPI specification is organised into three sections: (i) syntax and style, (ii) command reference, and (iii) data interchange format. The syntax and style section

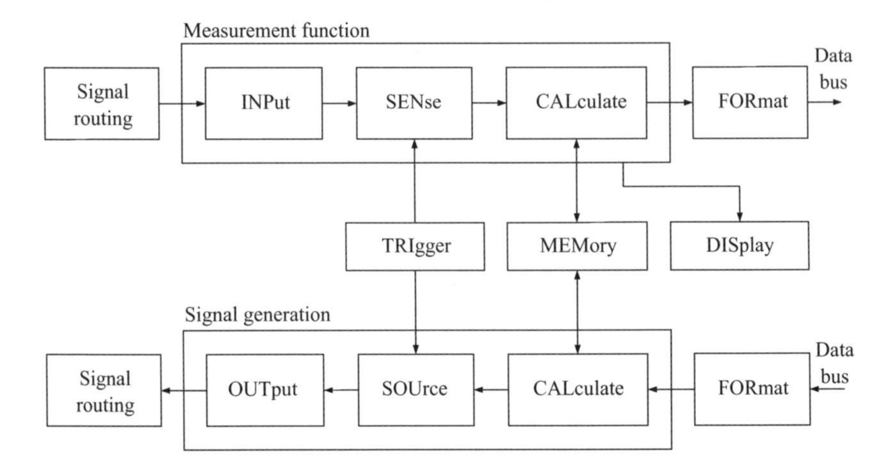

Figure 11.8 SCPI instrument model

covers rules for generating commands. This includes rules for mnemonics, command structures, command parameters, expressions, status reporting and reset conditions. The command reference section builds on the IEEE-488.2 common commands with 10 more mandatory commands plus over 400 optional commands. The final section defines the syntax, grammar, data format extensions and block descriptions necessary to build interchangeable data. This will allow different instruments to interchange data. An oscilloscope sending an oscillogram to a plotter is an example of such an interchange.

The SCPI command set is generated according to the SCPI instrument model illustrated in Figure 11.8. The signal routing component defines the connection of signal to the instrument. The measurement function converts real world signals to preprocessed form. The format converts it to IEEE-488.2 message format. Similarly messages from the bus are converted through the signal generating function to real world signals. Each block represents a class of functions or settings available in an instrument. A single instrument may not have all the components shown in the model. An SCPI command to program a digital multimeter to make an a.c. voltage measurement and send back the results to the controller on a signal of maximum 20 V with a resolution of 0.001 V resolution is as follows.

: MEASURE:VOLTage:AC? 20, 0.001

This would remain the same for all DVMs that are SCPI compatible.

11.2.11 Software support for GPIB

Software support for IEEE-488 systems is available at various levels. Normally, a manufacturer provides a library of routines which can be linked with a program written in a high level language such as C. National Instruments (USA), for example, provides a library called NI-488.2 routines, some of which are shown in Table 11.8.

Table 11.8 NI-488.2 bus management routines

NI-488.2 routine	Description
ReseTSys	Initialise an IEEE-488 system
SendFIC	Clear GPIB interface functions with IPC
FindLstn	Find all listeners
TestSRQ	Determine the current state of the SRQ line
WaitSRQ	Wait until a device asserts service request
RestSys	Cause device to conduct self-test
SendLLO	Send the local lockout message to all devices
SetRWLS	Place devices in the remote lockout state.

```
Char id [20];
SendIFC(0);
Send(0,6,"IDN?",5,NLEnd);
Receive(0,6,id,20,STOPend);
```

Figure 11.9 Use of NI-488.2 routines

The routine SendIFC, for instance, clears the devices by asserting the IFC line and the routine Send transmits data types to a device. Figure 11.9 shows a part of a program where the bus is cleared and then the device at address 6 is sent the 'identification query' (*IDN?). The response is then received.

These routines are designed to be called from standard compilers running under commonly used environments such as DOS or Windows. Other software helps are also available such as interactive control programs, which allow the user to control devices directly rather than via a program. Such a program is a powerful debugging and development tool. Companies such as National Instruments (USA) also provide other environments such as LabWindows and LabView. The former is an environment for developing programs in languages such as C and BASIC. LabWindows provides a powerful graphic library such that the users can easily create realistic graphical interfaces. It also provides GPIB, RS-232, VXI software drivers for over 300 instruments, simplifying device interaction. LabView is a package for building up virtual instruments without writing programs. You can specify the inputs by graphically creating the front panel of your virtual instrument and create a data flow block diagram to indicate the operations. LabView also has software driver support for a large number of instruments, which use GPIB, RS-232 or VXI. The results of measurement can be analysed and presented through 'graphical user interfaces' (GUIs).

11.2.12 GPIB applications

A simple ATE application using GPIB is illustrated in Figure 11.10. This setup measures the frequency response of a device (such as an amplifier or a filter) and displays

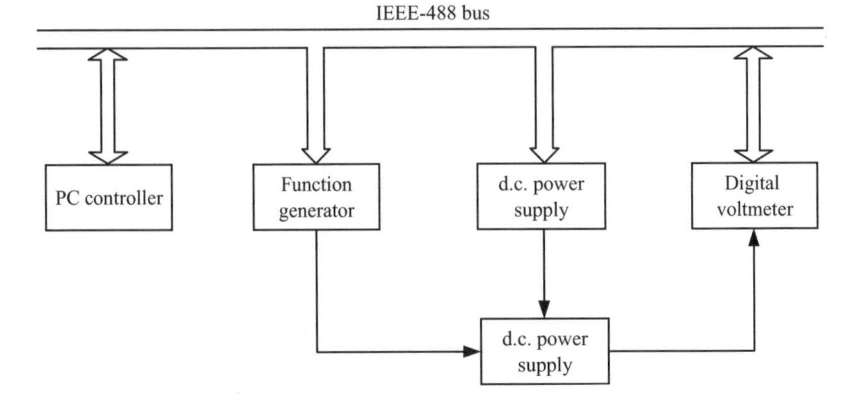

Figure 11.10 Test setup for automated frequency response

or prints out a plot of the results. The 'unit under test' (UUT) receives input from a function generator. A wideband digital multimeter is connected to the output. The manual method of obtaining the frequency response is to feed a.c. signals of constant amplitude but varying frequency over the desired range and record the output. The response then can be plotted, for example, as gain in decibels against log frequency. This procedure can be easily automated with instruments that can be remotely controlled via GPIB. The two instruments will be connected via GPIB cables to the controller, which is an IEEE-488 controller card connected to a personal computer. Both the multimeter and the function generator are primarily listeners. Automating measurement removes the tedium of the process, can be quickly repeated and shortens the design cycle. Figure 11.11 shows a flow chart of the program. This flow chart was implemented in C language using National Instruments NI-488 high level routines in the LabWindows environment.

Commands were sent using the 'ibwrt' routine, and data were read using the 'ibrd' routine[1]. The HP34401A DVM accepted SCPI compatible commands whereas the HP8116A function generator did not. To configure the function generator, the ASCII command strings shown in Table 11.9, were sent.

To set the frequency, the command string 'FRQfHz' was sent, which sets the signal frequency to f Hz.

To set the DVM to measure output and send the signal to the output buffer of the device, the following SCPI compatible command string was sent.

MEAS: VOLT: AC? 10, 0.003

This sets the DVM on the 10 V range with a resolution of 0.003 V for a.c. measurement. The 'ibrd' routine used after this command string will read measured data to the computer.

[1] NI-488 is an IEEE-488 based library of software routines from National Instruments Corp., USA; 'ibwrt' means 'write data from string' and 'ibrd' means 'read data to string'.

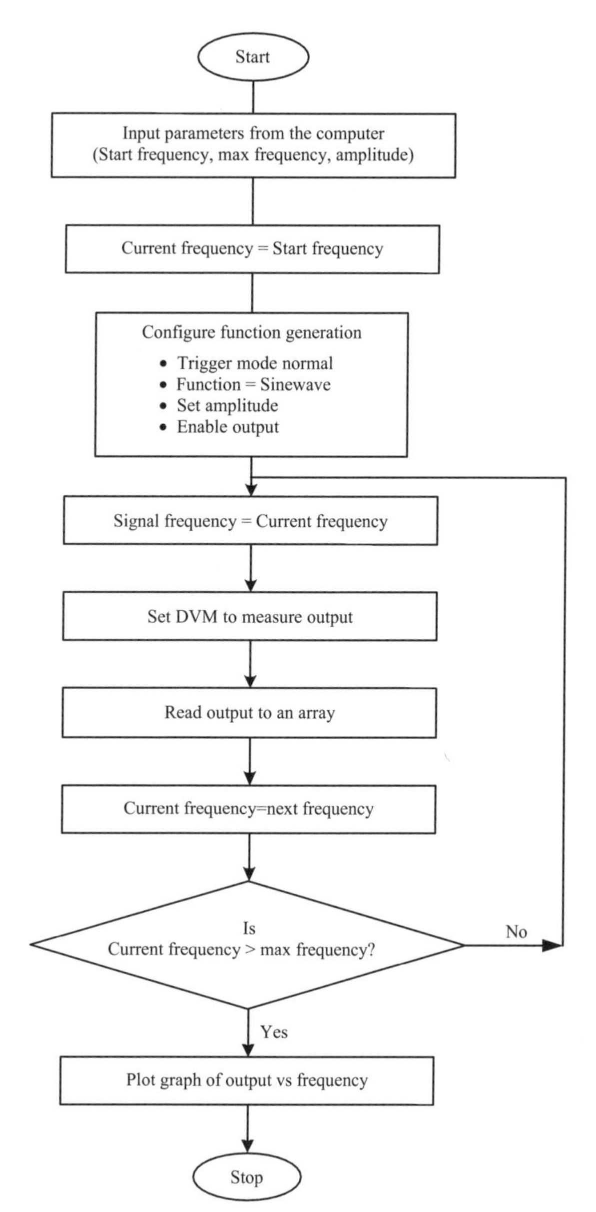

Figure 11.11 Flow chart for frequency response measurement

GPIB can be used in many situations where several instruments are connected to a test setup and repetitive measurements are taken. Another example is in automated testing of digital printed circuit boards (PCBs). The UUT in this case would be the PCB, which will have a test socket built into it. A 'test pattern generator' (TPG) will

Table 11.9 Command strings to configure function generator

Command	Comment
M1	Set normal trigger mode
AMP c V	Set amplitude to 'c' volts
Do	Enable signal output
W1	Function is sine wave

be connected to the socket and will generate the test stimuli. A logic analyser will be connected to another socket, which contains the test points. The system can be remotely controlled to generate the test patterns and receive the output to the logic analyser. There data can then be sampled with data in a reference memory to see if the board is good.

For detailed applications of IEEE-488 standards refer to texts listed in Further Reading, at the end of the chapter.

11.3 VXIbus

11.3.1 History of VXIbus

VMEbus extension for instrumentation (VXIbus) describes the technology of VMEbus compatible modular instrumentation. The architectural concepts of the VMEbus date back to Motorola's development of the 68000 microprocessor in the late 1970s. In 1979, Motorola published a brief description of a 68000 oriented bus known as VERSAbus [3]. Several revisions followed, the last being in July 1981. At the same time, a new printed circuit board standard known as the 'Eurocard' standard (IEC 297–3) was developed. In October 1981 several US companies announced their agreement to support a line of cards based on VERSAbus with Eurocard module dimensions, which was renamed VMEbus. After several revisions of the VMEbus specifications it was standardised under the title '*IEEE Standard for a Versatile Backplane Bus: VMEbus* (IEEE standard 1014/1987)'.

The marketplace has demonstrated the open system nature of VMEbus. There are thousands of VMEbus cards from over 170 vendors. With the development of these cards, which were primarily computer oriented, there have been numerous requests for instrumentation modules for VMEbus, particularly from the US Department of Defense. Reduction of the size of automated test equipment (ATE) was a major factor driving users towards VMEbus. The most serious impediment to VMEbus-based instrumentation was the lack of standards beyond that of VMEbus. In June 1987, five US instrument companies (Colorado Data Systems, Hewlett-Packard, Racal Dana, Tektronix and Wavetek) formed an *ad hoc* committee to engineer the additional standards necessary for an open architecture instrumentation bus based on VMEbus, the Eurocard standard, and other instrumentation standards such as IEEE-488.1/488.2.

In July 1987 they announced their agreement to support a common architecture of a VMEbus extension for instrumentation, named VXIbus.

11.3.2 VXIbus objectives

The idea of the VXIbus is to define a technically sound modular instrument standard that is based on the VMEbus. This is available to all manufacturers and it is compatible with present industry standards. The VXIbus specification is intended to be used by designers interested in generating compatible components for a system.

The VXIbus specification details the technical requirements of VXIbus compatible components, such as mainframes, back planes, power supplies and modules. The specification has the following objectives.

(i) To allow communication among devices in an unambiguous fashion.
(ii) To allow for physical size reduction of standard rack-and-stack instrumentation systems.
(iii) To provide higher system throughput for test systems through the use of higher bandwidth channels for inter-device communication, and the use of new protocols specifically designed to enhance throughput.
(iv) To provide test equipment which could be used in military 'instrument on a card' (IAC) systems.
(v) To provide the ability to implement new functionality in test systems through the use of virtual instruments.
(vi) To define how to implement multimode instruments within the framework of this standard.

11.3.3 VXIbus specifications

The VXIbus specification is built on top of the VMEbus specification [4, 5] and the VXIbus specification [6] covers the following details: VMEbus implementation, electrical and mechanical, electromagnetic compatibility, system power, VXIbus device operation, VXIbus device communication protocols, system resources, VXIbus instruments, IEEE-488 VXIbus interface, command and event formats, dynamic configurations and shared memory protocols.

11.3.3.1 VXIbus mechanical specifications

The VXIbus mechanical specification allows VXI modules in four sizes based on Eurocard dimensions. A module can be a single printed circuit board or an enclosed housing that contains multiple boards. If an instrument needs more than the width of a single slot, it can take up multiple slots on a VXIbus mainframe (Figure 11.12(a)).

11.3.3.2 VXIbus electrical specifications

The electrical architecture defines the structure and functions of the back plane and the electrical connections between VXIbus modules within a VXI subsystem. A VXI subsystem has the capacity up to 13 slots or modules. A VXI system can have a

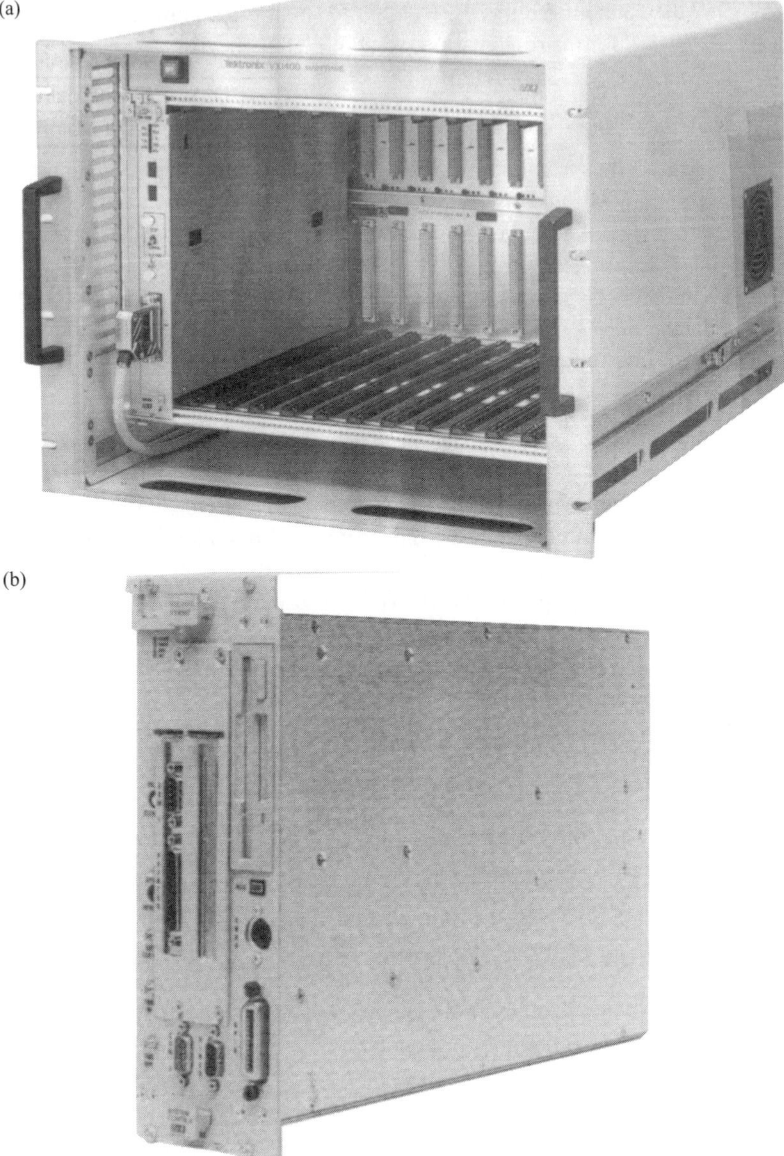

Figure 11.12 VXI system components: (a) VXI mainframe; (b) system controller (reproduced by permission of Tektronix, Inc., USA)

maximum of 256 devices. Normally a device will consist of one VXI board assembly. However, multiple slot devices and multiple device modules are permitted. Examples of the device include computers, multimeters, multiplexers, oscillators and operator interfaces. A module is required to connect to the back plane through the P1 connector

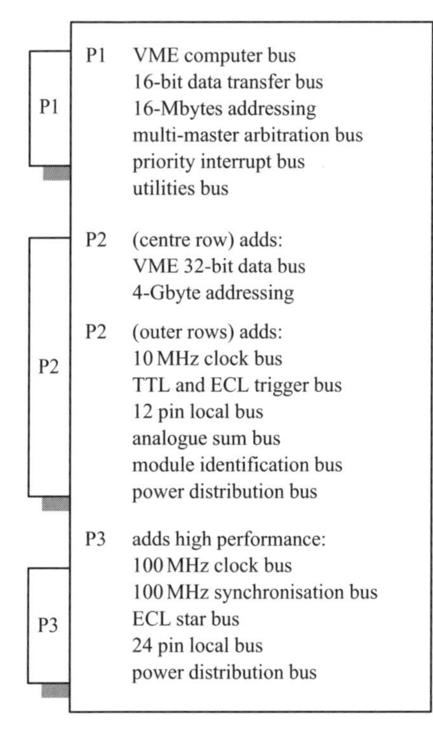

Figure 11.13 VXIbus electrical functions

and optionally through P2 and P3 connectors. Only connector P1 is required by the VXIbus regardless of the instrument module size.

The VXIbus encompasses the VMEbus. VMEbus defines all the pins of P1 and the centre row pins of P2. The P2 outer rows pins are user defined by the VMEbus specification. VXI adds its signals by defining these outer rows of P2, and completely defining P3 (see Figure 11.13).

The VXIbus can be logically grouped into eight buses and a few reserved pins. These buses are indicated in Table 11.10.

The VME computer bus system comprises four buses: namely, data transfer, bus arbitration, priority interrupt and utilities. The VME computer bus allows multiple devices to take control of the back plane. This (multimaster) capability provides a tiered hierarchy with one top level master and its slaves. Each slave may also be the master of other slaves. The multimaster capability of VME allows true parallel processing. In VXI, these capabilities are enhanced to commander/servant relationship with associated communication and control protocols.

The 'trigger bus' defines the facilities for several trigger protocols which ensures compatibility between manufacturers. Trigger bus can be subdivided into eight TTL trigger lines and six ECL trigger lines. These triggers and related protocols include synchronous and asynchronous (to VXI clock signals) as well as simple start/stop

Table 11.10 Buses within VXI system and their types

Bus	Type
VME computer bus	Global
Trigger bus	Global
Analogue sum bus	Global
Power distribution bus	Global
Clock and sync. bus	Unique
Star bus	Unique
Module identification bus	Unique
Local bus	Private

triggers. Because there are multiple trigger lines, it is possible to have multiple groups of instruments in a VXIbus subsystem each with separate and simultaneous synchronisation.

The 'analogue sum bus' is an analogue-summing node that is bussed the length of the VXI subsystem back plane. This bus is terminated to signal ground through a 50 Ω resistor at each end of the back plane. An example of using the analogue sum bus is in the generation of complex waveforms.

The 'power distribution bus' provides the power distribution to the modules containing +5, +12 and −12 V (from VME) and +24 V (for analogue circuits) and −5.2 and −2 V for high-speed ECL circuits.

The 'clock bus' provides two clocks and a clock synchronisation signal. A 10 MHz clock is located on P2 and a 100 MHz clock with synchronisation signal (SYNC 100) is located on P3. The 10 MHz and 100 MHz clocks are synchronous. Both clocks and SYNC 100 are obtained from slot 0 (master controller) and are buffered on the back plane.

The 'star bus' is a high speed two wire bus for connecting devices in a star using slot 0 of a VXI subsystem as the centre. The star bus provides high speed communication between modules. This communication may be a clock with a synchronisation signal or various forms of triggering. The star bus is located on P3. The 'module identification bus' allows a logical device to be identified with a particular physical location or slot. These lines are sourced from the VXIbus slot 0 module to each of the slots from 1 to 12. Using the module identification (MODID) lines, the slot 0 module can detect the presence of a module in a slot even if that module has failed.

The 'local bus' is a private module-to-adjacent module communication bus. The local bus back plane layout only provides connections from one side of a module to the adjacent side of the next module. The purpose of the local bus is to decrease the need for cable jumpers between modules. There are 12 local bus lines on P2 and 24 local bus lines on P3.

In order to extend the number of devices beyond 256, the Multisystem extension interface was proposed in 1989 [7].

Further details on the VXIbus system specifications and usage is supplied in Reference 8.

11.3.4 VXIbus standards and product ranges

By 1993 the VXI bus system reached its mature development stages based on field experiences by both vendors and end users. The VXI bus was picked up by the IEEE as IEEE-P1155, where the IEEE version is equivalent to the consortium-developed specification, version 1.3. By September 1993, the VXI plug and play systems alliance was formed, with a view to coming up with a set of software standards.

The VXIbus consortium did an excellent job with the hardware standard, ensuring mechanical and electrical inter-operability between mainframes and modules from different manufacturers. But as an architecture with no physical front panels, VXI required system designers to come up with 'soft' front panels, and to make sure software drivers were compatible with the operating system.

In fact, a VXI system needs at least three software interfaces: an instrument interface, a communication interface, and an instrument-driver interface (see Figure 11.14). The instrument interface transfers information between the instrument and the controller. Examples include the VXI word serial and register based protocols, IEEE-488 and RS-232. The communication interface connects each instrument interface in a standardised way. The instrument-driver interface links a communication

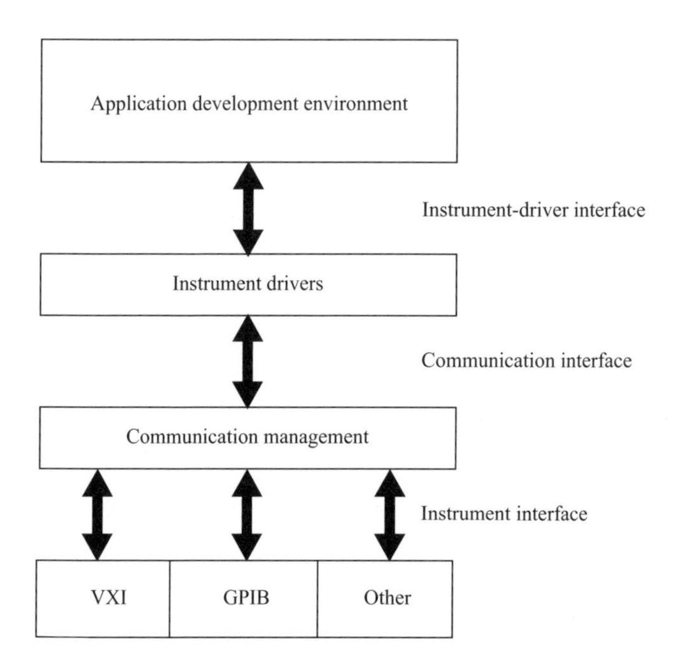

Figure 11.14 Three interfaces needed in an instrument system (source: Reference 9)

interface and an application development environment. The instrument drivers themselves add another level of complexity to the mix. This kind of software interfacing in a system makes the interoperability of items difficult.

During 1994 and 1995, the VXI plug and play system alliance addressed that problem by establishing standard software frameworks. A test system designer who specifies a VXI plug and play compliant module could expect a disk with the necessary drivers and automatically provides a soft front panel. VXI plug and play (Nath, 1995) saves a lot of time in system development and hence will make VXI systems less costly and more competitive with other architectures, like IEEE-488.

For details of VXI plug and play, References 10 and 11 are suggested. By mid-1996, system vendors had introduced over 200 different VXI plug and play compliant products [12]. In addition to the software framework work (i) a fast data channel and (ii) TCP/IP networking capability also were added to the system during the mid-1990s [12]. The effect of these developments were to cut the system integration time as well as costs.

Another recent development related to the range of products was the introduction of mezzanine modules [13]. These modules are used to enhance the performance factors of instrument functions on a given board. For example, if a VXI module carries the functions to test a cell phone, via an RS-232 port on the VXIbus card, one could use a mezzanine module to test several phones simultaneously using multiple RS-232 ports. By the end of 2000, there were over 350 varieties of industry pack (IP) mezzanine modules from more than 130 manufacturers. For details, Reference 13 is suggested.

11.4 The RS-232 interface

The RS-232 interface is a standard parallel-to-serial interface most commonly supplied with personal computers and low speed peripherals of all types. It was initially developed by the Electronic Industries Association (EIA) as RS-232C to standardise the connection of computers with telephone line modems. The standard allows as many as 20 signals to be defined, but gives complete freedom to the user. In practice three wires are sufficient: send data, receive data and signal ground. The remaining lines can be hardwired 'on' or 'off' permanently. The signal transmission is bipolar, requiring two voltages, from 3 to 25 V, of opposite polarity. This standard has gone through several revisions and in 1991 this standard was revised to ANSI EIA/TIA-232-E from its 1986 version EIA-232-D. RS-232 standards support operations for full or half duplex synchronous/asynchronous transmissions at speeds up to 20 kbps. Even though this standard is not primarily aimed at instruments, many instrument manufacturers supply their products with this very popular serial interface to a computer.

11.5 VLSI testing and automatic test equipment

The testing of logic devices involves simulating the voltage, current and timing environment that might be expected in a real system, sequencing the device through a series

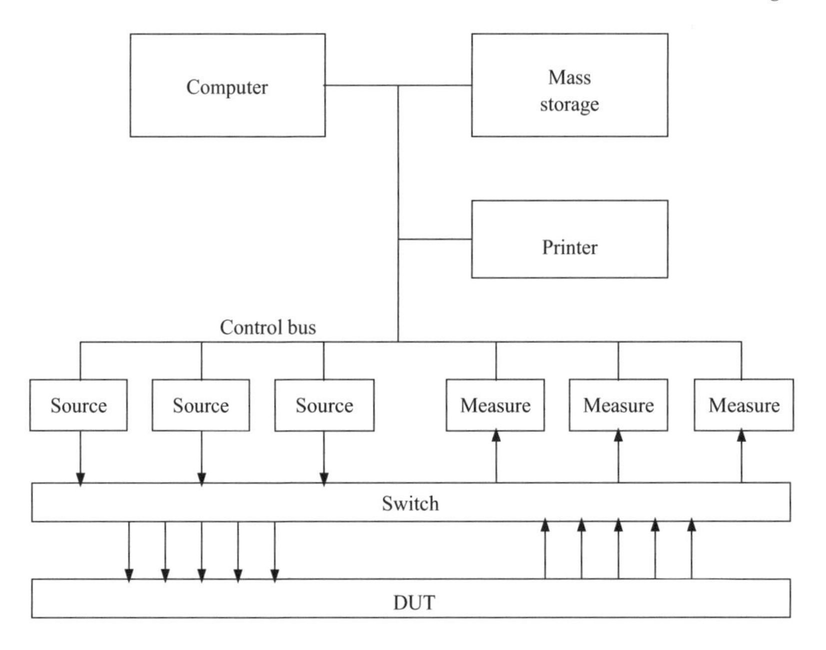

Figure 11.15 Configuration of a test system where the DUT is excited with a stimulus using various signal sources and the resultant output is measured using various instruments

of states and checking its actual response against its expected response. When the logic devices need to be tested at the end of a production line, manual testing becomes too slow and can be inconsistent, particularly when the devices are complex in architecture and pincount. Today, VLSI components have many pins, usually more than 200 in complex ICs, and they come in many different packages. These devices may have clock speeds well beyond 200 MHz. Testing these devices at the end of a production line has created the need for very sophisticated families of testers which are in the category of automatic test equipment (ATE). In this area of application computer controlled techniques such as IEEE-488, card cage based VXI systems or expensive VLSI testers based on proprietary architectures are used.

When the different kinds of devices (usually called 'device under test', DUT) need to be tested a flexible ATE may be required. Figure 11.15 shows the configuration of a test system where the DUT needs to be excited with a stimulus using various signal sources and then the resultant output is measured using various instruments. A switch is used to provide the flexibility of coupling signal sources and the outputs from the DUT. A typical system always uses a computer system as the controlling unit.

For low volumes of production, field service or production rework requirements, an economical ATE alternative using the IEEE-488 is shown in Figure 11.16. This flexible ATE system comprises off-the-shelf instruments coupled by the IEEE-488 interface bus to relieve the incompatibility problem that can occur in linking different

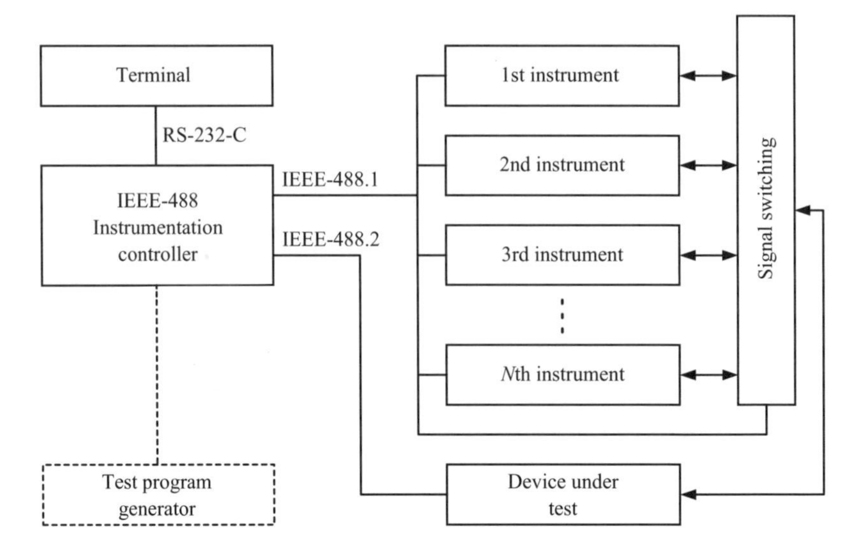

Figure 11.16 Economical ATE using IEEE-488 interface bus

instruments from different suppliers. With the most modern families of test and measuring instruments, which are compatible with the SCPI command set, the situation becomes much easier. The concept of a flexible ATE system implies reconfigurability for new requirements as new test applications arise. Because writing test programs can be the largest single task in such a system, and can also become the largest cost component, a test program language or test program generator is highly desirable. A typical test program generator is a menu driven selection guide for entering the required test procedures.

Such a low cost flexible system with stand-alone instruments and IEE-488 interfaces may, however, reach its limitations when very complex digital components are tested. For situations where complex VLSI components, using 100 MHz or higher clock rates with pin counts over 100, need to be tested, special testers are available. These systems use the technique called ‘tester-per-pin’ architecture.

The basic architecture of such a system is shown in Figure 11.17. This system has four subsystems: namely, per-pin resources, tester-control resources, a computer and the system peripherals. Three board types make up the per-pin resources that account for over 70 per cent of the total system. These are (i) the test vector memory board, which allocates over one million elements of test vectors for each pin, (ii) the timing generators, which generate timing with picosecond accuracies, and (iii) the pin electronics board, which turns the timing and command data from the timing generator into complete waveforms. The per-pin resources controlled under the tester-control resources result in high throughput for most VLSI circuits. For example, the test time for a typical VLSI device like 8086 is under 200 ms, compared with older generations of testers which take from 5 to 10 s. However, these very sophisticated VLSI testers, controlled by 32-bit multiprocessor computer systems, are very expensive and are used

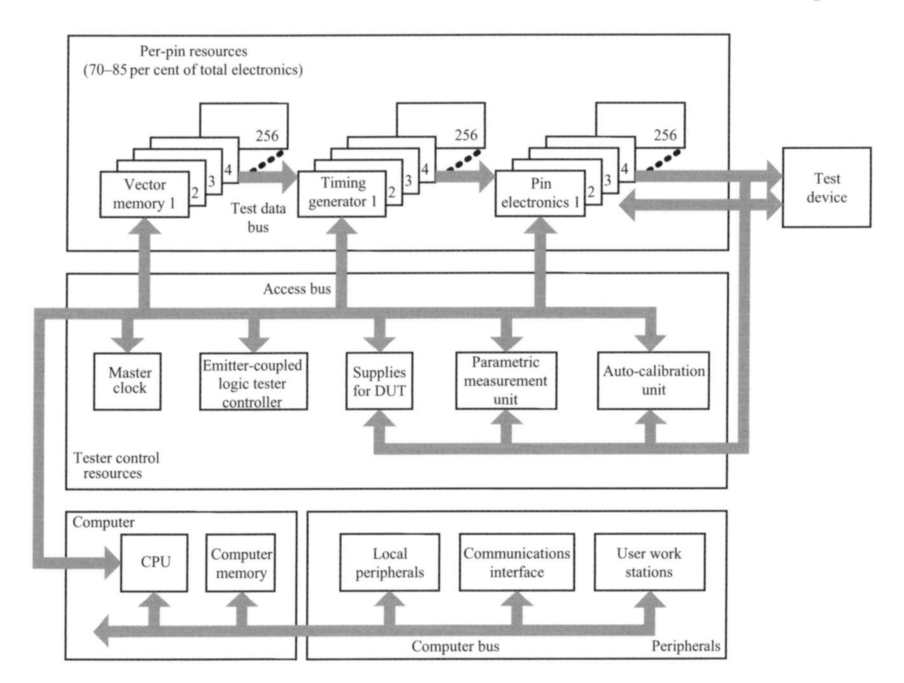

Figure 11.17 Basic tester-per-pin architecture (reproduced by permission of Magatest Corporation, USA)

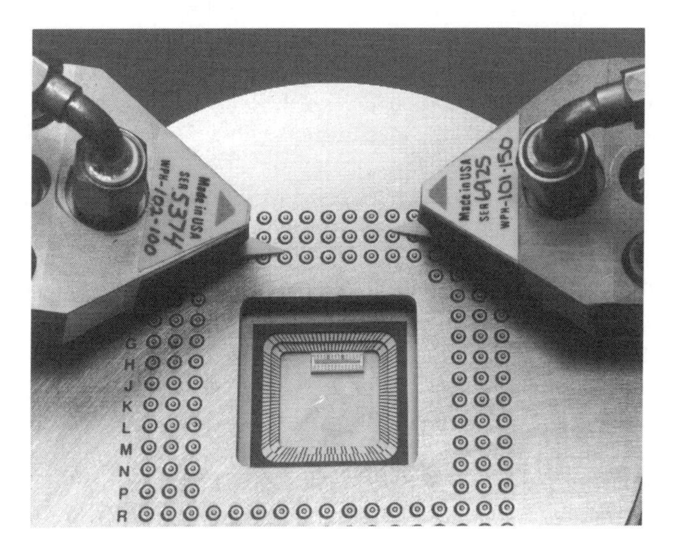

Figure 11.18 Package test fixture with a PGA attached (reproduced by permission of Cascade Microtech, Inc., USA)

only in very sophisticated production lines. Such systems are very expensive, costing a few million dollars, depending on the configuration; they also have auto-calibration and self-diagnostics built into the system.

In VLSI testing the other complex problem is to couple the DUT placed on the test heads to the test system. Digital LSI and VLSI packages are now required to operate at clock rates of 50–250 MHz and even beyond this. This requires analogue bandwidths of 500–2500 MHz for good edge definition [14]. At these frequencies, typical digital packages have metal pattern dimensions that resonate and/or couple to the outside environment. These elements can act as antennas and resonators. This is the microwave domain, where open lines may appear as short circuits and shorted lines as opens, depending on the frequency. Microwave designers have been successful working with these issues for years, and now these techniques have been adapted to the high speed digital design environment.

For the purpose of coupling a DUT to the tester there are very special test fixtures. Figure 11.18 shows a package of a test fixture together with attached probes.

A detailed description of these advanced techniques is beyond the scope of this brief chapter, and for more detailed descriptions References 15–19 are suggested.

11.6 References

1 IEEE standard digital interface for programmable instrumentation. ANSI/IEEE-488.1-1987.

2 HEWLETT PACKARD: 'Tutorial description of the Hewlett Packard interface bus', 1984.

3 HILF, W.: 'The M68000 family', Vol. 2, 'Applications and the M68000 devices', Vol. 2, Prentice-Hall, Englewood Cliffs, NJ, 1990.

4 MOTOROLA: 'The VMEbus specification', rev c.1 – Motorola series in Solid State Electronics.

5 DI GIACOMO, J.: *Digital Bus Handbook*, McGraw-Hill, New York, 1990.

6 TEKTRONIX, INC. (USA): *VXIBus System Specification*, revision 1.4, Tektronix Publication, 1992.

7 WOLFE, R.: 'MXIbus extends the VXIbus to multiple mainframes and PCs', *Electronics Test*, April 1989, pp. 31–35.

8 HAWORTH, D.: *Using VXIbus: A Guide to VXIbus System*, Tektronix Publication, 1991.

9 NOVELLINO, J.: 'VXI becomes easier to use, less expensive', *Electronic Design*, 14 October 1994, pp. 109–111.

10 HOWARTH, D. and HAGEN, M.: 'Step-by-Step Process eases VXI integration', *Electronic Design*, 14 October 1994, pp. 113–118.

11 TITUS, J.: 'VXIplug & play cuts system integration time', *Test & Measurement World*, June 1998, pp. 55–60.

12 NOVELLINO, J.: 'Plug and Play should solve VXI software woes', *Electronic Design*, 16 September 1996, pp. 137–140.

13 TITUS, J.: 'VXIbus Mezzanine modules maximize I/O and function choices', *Test & Measurement World*, August 2000, pp. 13–18.
14 CASCADE MICROTECH, INC. (USA): 'High speed digital micro probing – principles and applications', pp. 15–39, 1990.
15 SWAN, R.: 'General purpose tester prints a separate set of resources behind each VLSI device pin', *Electronics*, 8 September 1983, pp. 101–106.
16 ADAMS, R.: 'Modules cut ATE costs', *Electronics Week*, 15 April 1985, pp. 49–51.
17 'VLSI chip test system tests itself at board level', *Electronics*, 5 August 1985, pp. 40–49.
18 KIRSOP, D.: 'Switching systems speed test and add flexibility', *Electronics Test*, May 1989, pp. 40–46.
19 SONTONI, A.: 'Fast accurate calibration improves device test systems', *Electronic Test*, May 1989, pp. 47–51.

Further reading

CARISTI, A.J.: 'IEEE-488 general purpose instrumentation bus manual', Academic Press, London, 1989.
CORNER, D.: 'IEEE-488 test and measurement software' *EDN*, 13 April 1989, p. 142–150.
DEHNE, T.: 'Software for PC-based instrument control', Part I, *Electron Test*, September 1989, pp. 52–57; Part II, October 1989, pp. 58–60.
DEHNE, T.: 'Programming with NI-488 software'. National Instruments (USA), Application note PN 320191-01, June 1989.
HAWARTH, D.: 'SCPI: the next step in the evolution of ATE systems', *Electron. Test*, August 1990, pp. 14–18.
HOLLISTER, A.: 'A comparison of IEEE-488 to VXI', *Electron. Test*, April 1989, p. 1821.
IEEE standard codes, formats, protocols and common commands. ANSI/IEEE-488.2-1987.
LEIBSON, S.H.: 'IEEE-488.2 products are just now appearing', *EDN*, 25 April 1991, pp. 91–99.
MOSSLEY, J.D.: 'PC based control and data acquisition products', *EDN*, 6 August 1987, pp. 94–104.
NATH, M.: 'Plug-&-Play VXI helps ATE users', *EDN*, June 8, 1995, pp. 95–99.
NATIONAL INSTRUMENTS CORPORATION: 'GPIB training manual'.
NOVELLINO, J.: 'New standard simplifies GPIB programming', *Electron. Des. Int.*, November 1988, pp. 47–52.
WRIGHT, M.: 'SCPI instruments will ease ATE development', *EDN*, 1 October 1991, pp. 39–49.

Chapter 12
Transmission measurements

12.1 Introduction

In modern electronic and telecommunication systems, transmission media such as cables and wave guides, free space radio links and fibre optic cable systems are used. These systems today carry a wide range of frequencies from a few kilohertz to gigahertz and optical range signals. These systems, such as cables or wave guides, are characterised by distributed parameters and behave in a slightly different manner than the systems that are designed by using the lumped parameter systems. For this reason transmission measurements need be carried out using specially designed instruments. In practical environments, engineers and technicians use different types of instruments for the measurement of power, field strength and relative amplitude of signals transmitted, as well as other important parameters of the medium such as signal reflections and loss. As the length of a transmission medium becomes long compared with the wavelengths of the signals transmitted, it becomes very important to measure these parameters accurately. At microwave frequencies the behaviour of transmission media and accessories becomes very critical, thus the accurate measurement of parameters related to transmission becomes quite crucial to the operation of a system.

In the digital telecommunication networks, the present industry uses technologies such as digital subscribe loop (DSL), plesiochronous digital hierarchy (PDH), synchronous digital hierarchy (SDH) and synchronous optical networks (SONET). In these systems digital instruments are utilised for nanosecond and picosecond order timing measurements to guarantee the quality of transport. These have created a need for a very sophisticated digital transmission measurement test system.

With the clock speeds of processors and peripherals crossing a few hundred MHz, designers of circuits and systems are forced to apply transmission line theory and techniques in the design of printed circuit boards.

This chapter is a review of basic transmission measurements and an introduction to measurements in the digital telecommunications domain and the packaging of high speed circuitry.

12.2 Transmission line behaviour

12.2.1 Lumped and distributed circuits

When the physical dimensions of electrical components are insignificant for the operation of an electrical circuit, the elements are 'lumped'. When the physical dimensions of an element significantly affect the propagation of signal information, the element is 'distributed'. In electrical circuits, lumped elements permit the use of Kirchhoff's circuit laws and the analysis of circuits is quite straightforward. However, with distributed elements, Kirchhoff's laws fail and the mathematics of the system becomes more complex. For example, consider applying Kirchhoff's current law to a transmitting antenna where the current enters at the feeder but does not exit in the classical sense.

In the past most engineers had to apply the distributed parameters to real transmission lines, such as feeder cables or wave guides, in the true sense of transporting RF energy from one location to the other. However, with the usage of high speed digital circuits on printed circuit boards, the designers are forced to consider the metal interconnections as transmission lines. In these situations, even though it is hard to distinguish the transmission lines from simple interconnections, some rules can be effectively utilised for determining whether a signal path is lumped or distributed [1–3]. A practical criterion by which a system can be judged as lumped or distributed for the purposes of digital electronics is discussed in Reference 4.

12.2.2 Propagation on a transmission line

The classical transmission line is assumed to consist of a continuous structure of R, L, C and G as shown in Figure 12.1. By studying this equivalent circuit, several characteristics of the transmission line can be determined. If the line is infinitely long and R, L, G and C are defined per unit length, several useful parameters, as given in Table 12.1, can be defined and used to characterise the lines.

The coefficients ρ and σ can be measured with presently available transmission test equipment. But the value of the VSWR measurement is limited. Again, if a system consists of a connector, a short transmission line and a load, the measured standing wave ratio indicates only the overall quality of the system. It does not tell us which of the system components is causing the reflection. It does not tell us if the reflection from one component is of such a phase as to cancel the reflection from

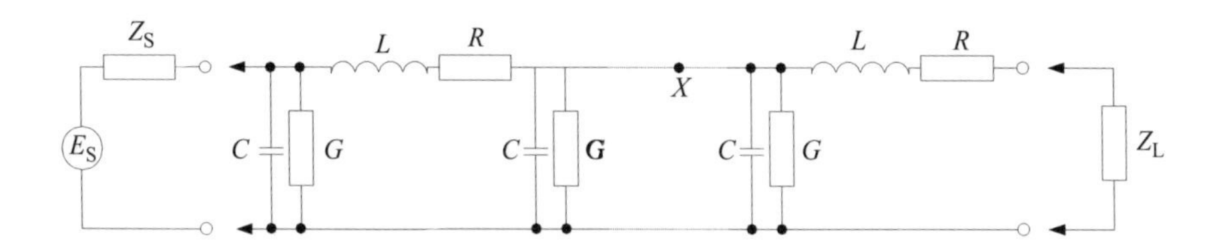

Figure 12.1 Classical model of a transmission line

Table 12.1 A summary of transmission line parameters and their significance

Parameter	Symbol	Expression	Significance
Characteristic impedance	Z_0	$Z_0 = \sqrt{\dfrac{R + \mathrm{j}wl}{G + \mathrm{j}wC}}$	This characteristic parameter is used to match input and output transmissions
Attenuation per unit length	α		Indicates the attenuation of signal per unit length
Phase shift per unit length	β	$\beta = \dfrac{w}{V_\mathrm{p}}$	Indicates phase shift per unit length
Velocity of propagation	V_p	$V_\mathrm{p} = \dfrac{Vc}{\sqrt{E_\mathrm{r}}}$	Propagation velocity along a transmission media is a fraction of velocity of light and related to permittivity of medium
Propagation constant	γ	$\gamma = \alpha + \mathrm{j}\beta = \sqrt{(R + \mathrm{j}wl)(G + \mathrm{j}wC)}$	Indicates the level of attenuation and the phase change along the line
Voltage at a distance (x)	E_x	$E_x = E_\mathrm{in}\mathrm{e}^{-\gamma x}$	Provides the actual voltage at a given distance x from the feeding point
Current at a distance (x)	I_x	$I_x = I_\mathrm{in}\mathrm{e}^{-\gamma x}$	Provides the actual current at a given distance x from the feeding point
Voltage reflection coefficient	ρ	$\rho = \dfrac{Z_\mathrm{L} - Z_0}{Z_\mathrm{L} + Z_0} = \dfrac{e_\mathrm{r}}{e_\mathrm{i}}$	Indicates the level of reflection along the transmission line
Voltage standing wave ratio (VSWR)	σ	$\sigma = \dfrac{1 + \lvert\rho\rvert}{1 - \lvert\rho\rvert}$	The magnitude of the steady state sinusoidal voltage along a line terminated in a load other than Z_0 varies periodically as a function of distance between a maximum and minimum value. This variation is caused by the phase relationship between incident and reflected waves and provides a picture of voltage maximum and minimum along the line

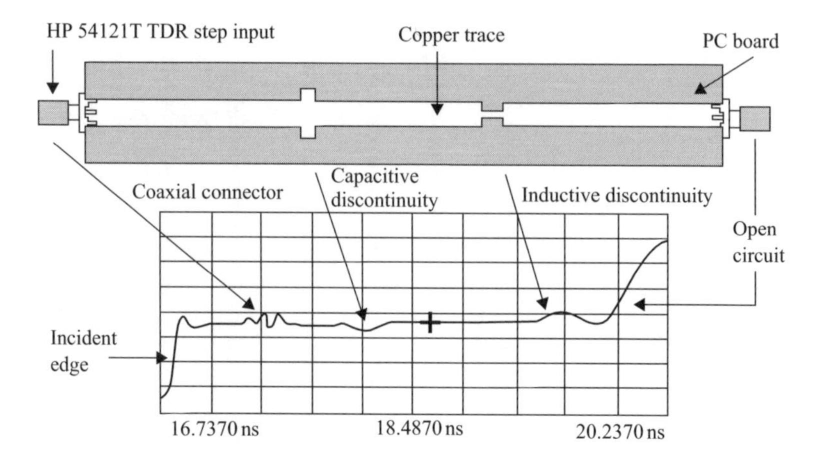

Figure 12.2 Effect of reflections along a transmission line (courtesy of Agilent Technologies Inc.)

another. The engineer must make detailed measurements at many frequencies before it is known what must be done to improve the broadband transmission quality of the system. Figure 12.2 provides a pictorial view of the effect of voltage reflections along a transmission media on a printed circuit board path with some discontinuities.

12.3 Decibel measurements

The language of decibels originated in the telephone industry. Because the response of human senses to stimuli, such as sound and light, is closely proportional to the logarithm of the power level of the stimuli, the use of decibels in telecommunications is justified. Moreover, logarithmic measures simplify many calculations that are commonly needed in communication systems. Such systems consist of distinct units such as transmission lines, amplifiers, attenuators and filters. Each of these units changes the signal power level by some factor and the ratio of power delivered to the terminal load to the power supplied by the signal source is the product of all these factors. If the effect of each of these units is expressed by the logarithm of the factor by which it modifies the signal power, then the logarithm of the effect of all the units combined is the sum of the logarithms of the effects for each of the separate units.

12.3.1 Bel and decibel

The bel (B), named in honour of Alexander Graham Bell, is defined as the common logarithm of the ratio of two powers, P_1 and P_2. Thus, the number of bels N_B is given by

$$N_B = \log(P_2/P_1). \tag{12.1}$$

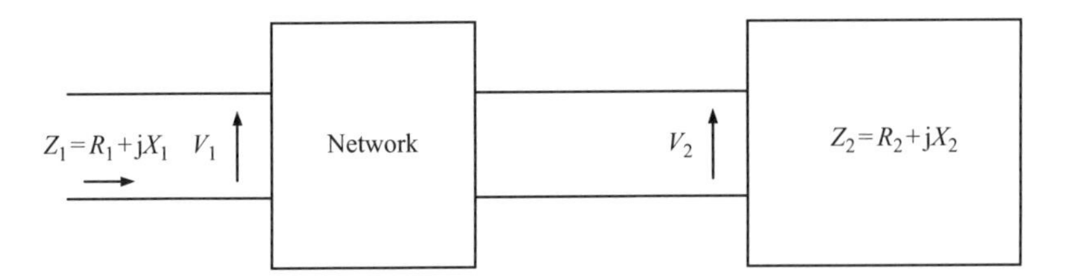

Figure 12.3 Two port linear passive network terminated with an arbitrary impedance

If P_2 is greater than P_1, N_B is positive, representing a gain in power; if $P_2 = P_1$, N_B is zero and the power level of the system remains unchanged; if P_2 is less than P_1, N_B is negative and the power level is diminished.

A smaller and more convenient unit for engineering purposes is the decibel, whose magnitude is 0.1 B. Thus,

$$N_{dB} = 10N_B = 10\log(P_2/P_1). \tag{12.2}$$

An absolute meaning can also be given to the decibel value of a power level by making P_2 a standard reference value. If $P_2 = 1\,W$, the decibel value of a power level P_1 is designated by $\pm N_{dB}$ dBW (e.g. 0 dBW $\equiv$ 1 W, 10 dBW $\equiv$ 10 W). When 1 mW is used as a reference value (i.e. $P_2 = 1\,mW$) then the power level is designated by $\pm N_{dB}$ dBm.

The power level change, expressed in decibels, is correctly given in terms of voltage and current ratios alone only for the special case for which impedances across the circuit are equal, as in Figure 12.3. As shown in Figure 12.3, for the special case where $Z_1 = Z_2$ and hence $R_1 = R_2$,

$$N_{dB} = 20\log\left|\frac{V_1}{V_2}\right|. \tag{12.3}$$

When this condition of equal impedances does not exist, caution should be taken to interpret values accurately. For example, it is not correct to state a voltage amplifier with voltage amplification of 1000 as having 60 dB in a case where input impedance and the load impedance have different values. For details, Reference 5 is suggested.

12.3.2 Measurement of power levels in telecommunications circuits

12.3.2.1 System level and zero transmission level point

A multichannel telephone system may be long and contain many stages of gain and attenuation. To state the power level of a signal measured in such a system, a reference point is required and this reference point is designated as the zero transmission level point (zero TLP). It is the point in the system at which the standard test zone has an absolute power of 1 mW or 0 dBm. The level of any point in the system, expressed in decibels relative to the zero TLP, is referred to as the relative transmission level of that point. It is commonly designated by dBr or TL. For example a −33 dBr (or −33 TL) point in a system will be 33 dB below the zero TLP. Because the standard test tone level is 0 dBM (1 mW) at the zero transmission level point, the actual power of the standard test tone at any other point in the system when specified in dBm should be numerically equal to the transmission level of that point: dBm0 indicates the level of a signal referred to the circuit's relative zero level.

12.3.2.2 Noise level measurements

In measuring noise level in telephone circuits emphasis should be placed in determining the point at which noise becomes troublesome to the subscriber rather than on simply determining the absolute level. For this reason the CCITT has established a measure of evaluation for noise in telephone circuits, and the instrument which measures such noise under the special weighting characteristics is called a psophometer [6]. A psophometer indicates channel noise more in accord with the actual interfering effect of noise than would a set with a flat frequency response. Noise measured with a psophometer is designated by the suffix 'p'. Thus dBm becomes dBmp. DBm0p expresses the psophometric noise value referred to the circuit's relative zero level. The psophometric measurement is rarely used in the USA. The common system is C-message weighting, and the term used is dBrn [6].

12.3.2.3 Power level measurements and terminating resistance

In telecommunication systems the power level of a given point in a circuit is usually determined by an indirect measurement. The circuit is terminated at the point in question by a resistance and the voltage developed across this resistance is measured by a level measuring set, effectively a voltmeter calibrated in voltage level. The value of the terminating resistance is chosen according to the nominal impedance of the circuit at the measurement point. The standardised value for speech frequencies is 600 Ω. Voltage level is defined for convenience according to the following principle. For a circuit impedance of 600 Ω the values of voltage level and power level are identical. Therefore, when defining voltage level, we choose the reference voltage to be just that which produces 1 mW in 600 Ω. That is,

$$V_{\text{ref}} = \sqrt{1 \times 10^{-3} \times 600} = 775\,\text{mV}. \tag{12.4}$$

When L_1 is the measured voltage across 600 Ω (measured in millivolts), then

$$\text{voltage level} = 20 \log_{10} \frac{L_1}{775}. \tag{12.5}$$

If the terminated impedance differs from 600 Ω a correction, K, is added. This is given by

$$K = 10 \log \frac{(600\,\Omega)}{Z}, \tag{12.6}$$

where K is measured in dB. So, the power level is expressed as

$$\text{power level} = \text{voltage level} + K. \tag{12.7}$$

Table 12.2 Approximate values of K for some common values of impedance

$Z(\Omega)$	K
600	0
300	3
150	6
75	9
50	10.8

The value of K for some common values of impedance can be found from Table 12.2. Measurements on digital subscriber loops are discussed in Reference 7.

12.4 Practical high frequency measurements

12.4.1 High frequency power measurements

Instruments used to measure power at RF and microwave frequencies are of two types: absorption power meters and throughline power meters. Absorption power meters contain their own load. Throughline meters use the (actual) load, which may be remote from the meter. Absorption types are more accurate and usually contain a 50 or 75 Ω load at RF. The main instruments currently used for laboratory work are based on thermistor, thermocouple or diode sensors. These instruments cover a power range from 70 dBm to a few watts and are similar in use, although based on different detecting principles.

A power sensor should accept all the RF power input, because the meter connected to the sensor measures the power which has been absorbed. However, owing to mismatch between impedances of the RF source and sensor, some of the power is reflected back. Owing to non-ideal circumstances (such as reflected power) in RF detection elements, calibration factors are used. For details, Reference 5 is suggested.

12.4.1.1 Thermistor power meters

Elements that respond to power absorption by changing their resistance are known as bolometers. There are two well known types: barretters and thermistors. The barretters consist of a thin piece of wire with a positive temperature coefficient of resistance, whereas the thermistors are semiconductors with a negative coefficient. Because of their low coefficients, small power changes are detectable with barretters only if they have very thin wires, with the result that at high power levels they operate very close to burn out and are subject to accidental destruction. Even so, they are easily made and have the advantage of greater immunity to ambient temperature changes.

However, they are not mechanically robust and, with their fast thermal response, are not well suited to measuring average power. For these reasons they are no longer in common use, although they are still to be found in some laboratories for measuring low milliwatt order powers in wave guides. In thermistors the disadvantage of high drift with temperature change, and consequent increase in compensating circuitry, is balanced by a wide power range and better resistance to damage.

Thermistor based systems are rugged and difficult to damage with RF overload. They are not, however, suitable for direct reading instruments because of the large variation in characteristics from element to element. For these reasons substitution methods are used where d.c. power is exchanged with RF power to maintain a fixed operating point to avoid non-linear effects. Bridge circuits are used in these situations, and for further details References 8–10 are suggested. The power range of thermistor based instruments is from 1 μW to 10 mW with typical SWR from 1.5 to 2.0. Typical operating ranges for coaxial mounts are from 10 MHz to 18 GHz. The reflection coefficient is typically 0.2 over most of the range and the calibration factors vary from 0.70 to 0.99 depending on the types and the manufacturers. Wave guide thermistor mounts cover frequencies up to about 140 GHz.

12.4.1.2 Thermocouple power meters

Thermocouples deposited on silicon chips are used to achieve much better sensors in modern power meters. These sensors, which have better reflection coefficients, typically 0.1, are capable of measuring RF power from 100 nW to 1 W. For range extension calibrated attenuators can be used. Thermocouple power meters have the following important characteristics:

- very wide dynamic range,
- high thermal stability,
- frequency range from d.c. to 40 GHz,
- high accuracy (typically 1 per cent).
- fast response time, and
- short term overload up to 300 per cent.

Power head design depends on the frequency and power range. For instance, lower frequencies require a larger blocking capacitor, which can compromise high frequency performance. The mass and shape of the resistive load (Figure 12.4) has an important bearing on both dynamic range and matched bandwidth. Drifting due to ambient temperature variations is reduced to very low levels because hot and cold junctions generate e.m.f.s of opposite polarity, thus giving no net output when the ambient temperature changes. Power sensor d.c. output is fed to a power meter which consists of a very low noise, high gain d.c. amplifier and a meter. A thermocouple based power meter is shown in Figure 12.5. In use, the very low minimum input, close to the noise level in the d.c. output of less than 1.5 μV on the most sensitive range, necessitates the use of a chopper amplifier system to avoid drift. Also with thermocouple based meters, a built-in calibration source is used to calibrate the system. This is mainly to take care of the drift in the thermocouple and associated electronics.

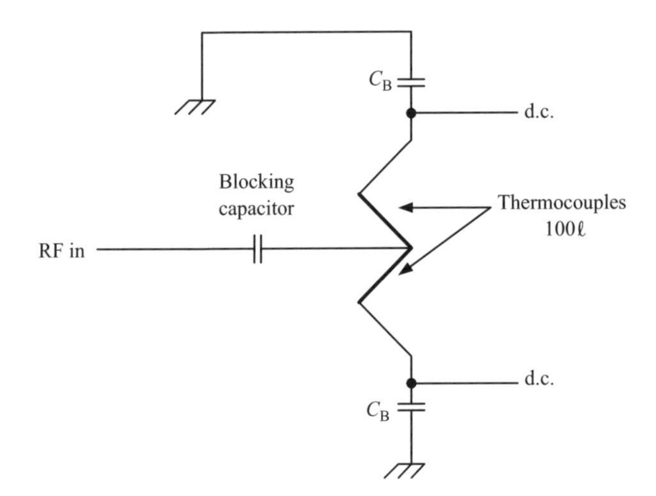

Figure 12.4 Thermocouple arrangement showing decoupling of d.c. from RF

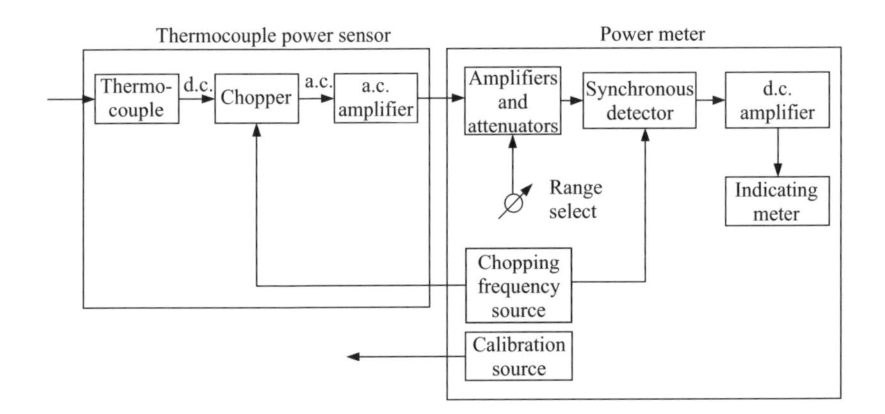

Figure 12.5 Block diagram of a thermocouple based power meter for high frequencies.

Recent advances in thermocouple based power sensors have eliminated the need for blocking capacitors [9] at the input of the RF circuit. The photograph in Figure 12.6 shows such a power sensor from Rohde and Schwartz. This offers the following clear advantages.

(i) A single sensor can cover a wide frequency from d.c. to over 18 GHz.
(ii) There is no lower limit frequency below which matching and measurement accuracy deteriorate.
(iii) Low SWR due to no blocking capacitor to affect the circuit at higher frequencies.

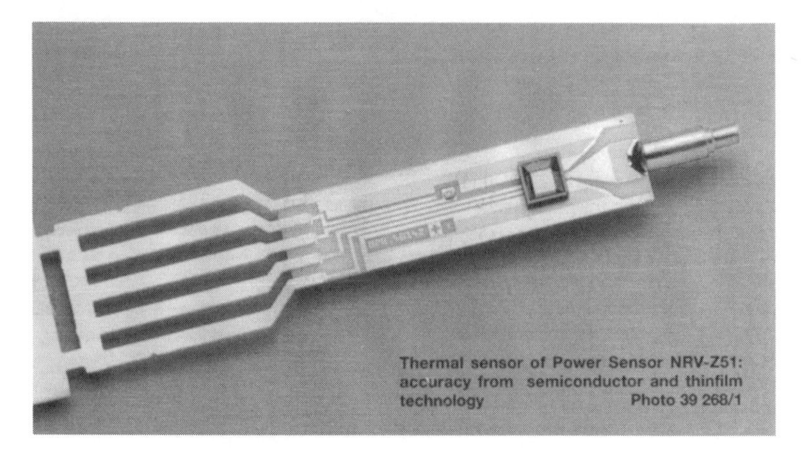

Figure 12.6 A thermal power sensor for d.c. to 18 GHz (reproduced by permission of Rohde and Schwartz)

Some of these modern sensors carry a data memory (an EPROM), which carries all calibration data such as absolute sensitivity, frequency response, linearity and temperature response on each individual sensor. This eliminates the need for calibration sources and allows exchange of measuring heads. Further details are to be found in Reference 11.

12.4.1.3 Diode sensors

A diode detector can be used to measure RF power levels down to 100 pW, over the frequency range 10 MHz to 20 GHz. It is faster than a thermistor or thermocouple probe, but less accurate. In these detectors, Schottky barrier diodes are used and they generally operate in a range from −50 dBm to +20 dBm. The square law range is from −70 dBm to −20 dBm, above which they become linear. Diode detectors conveniently interchange with thermocouple heads, using the same power meters to extend the dynamic range. For details on these sensors, Reference 8 is suggested.

12.4.1.4 Throughline watt meters

Most of the above power measurement techniques are generally limited to a few watts. For higher power measurements up to about 1 kW in the frequency range from about 25 to 1000 MHz, throughline watt meters become very practical. For the lower frequency ranges (from 200 kHz to 30 MHz) measurements up to 10 kW are possible.

The arrangement of such an instrument is shown in Figure 12.7. The main RF circuit of the instrument is a short section of a uniform air-type line section whose characteristic impedance is matched to the transmission line (usually 50 Ω). The coupling circuit that samples the travelling wave is in the plug-in element. The coupling element absorbs energy by the mutual inductance and the capacitive coupling. Inductive currents within the line section will follow according to the direction of the travelling wave and the capacitive portion of the line is independent of the travelling

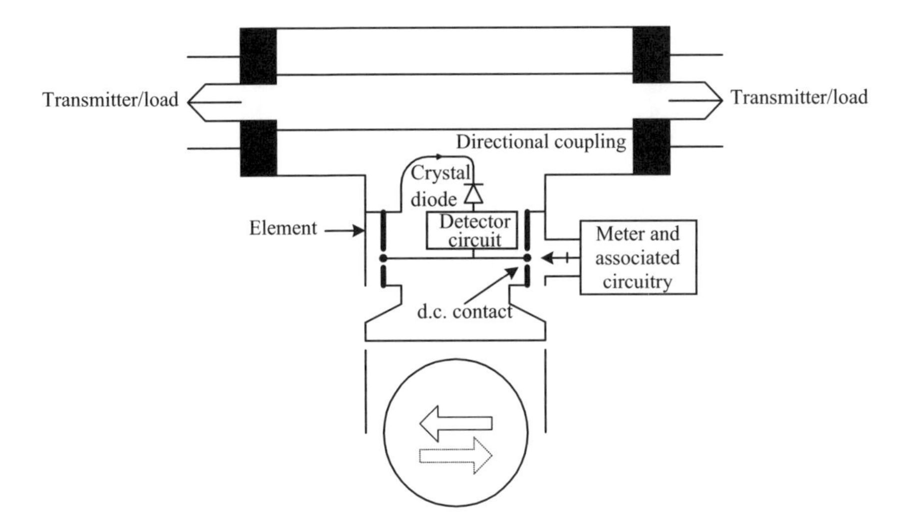

Figure 12.7 Construction of a thruline[1] wattmeter (reproduced by permission of Bird Electronic Corporation, USA)

wave. The element is therefore designed to respond directly to the forward or reverse wave only and the meter could be used to measure forward and reflected powers separately. The elements could be designed for different ranges of forward or reflected power usually varying from a few milliwatts to over 1 kW. The plug-in element is so designed that it can be rotated by 180° in its socket in the line section allowing the measurement of both forward and reverse power.

12.5 RF insertion units for SWR and reflection coefficient measurements

To measure transmission parameters (VSWR, reflection coefficients, etc.) many insertion units are available for use with RF millivoltmeters and RF power meters. Usually these RF insertion units are available for 50 and 75 Ω coaxial systems for use up to about 2 GHz from less than 10 kHz. RF insertion units for measuring the voltages have the ranges from a few hundred microvolts to about 100 V. These units can be inserted in the lines at suitable places for measuring the voltage along the line without affecting the VSWR on the line. Dual directional couplers for measuring the forward and reflected power have their range from a few microwatts to about 2 kW. The frequency ranges for these are typically from a few hundred kHz to about 100 MHz. In modern units the calibration parameters may be stored on the individual units and allow elimination of the errors encountered due to (i) non-linearities, (ii) level dependent temperature effects, and (iii) frequency response.

[1] 'thruline' is a registered trade mark of Bird Electronic Corporation, USA.

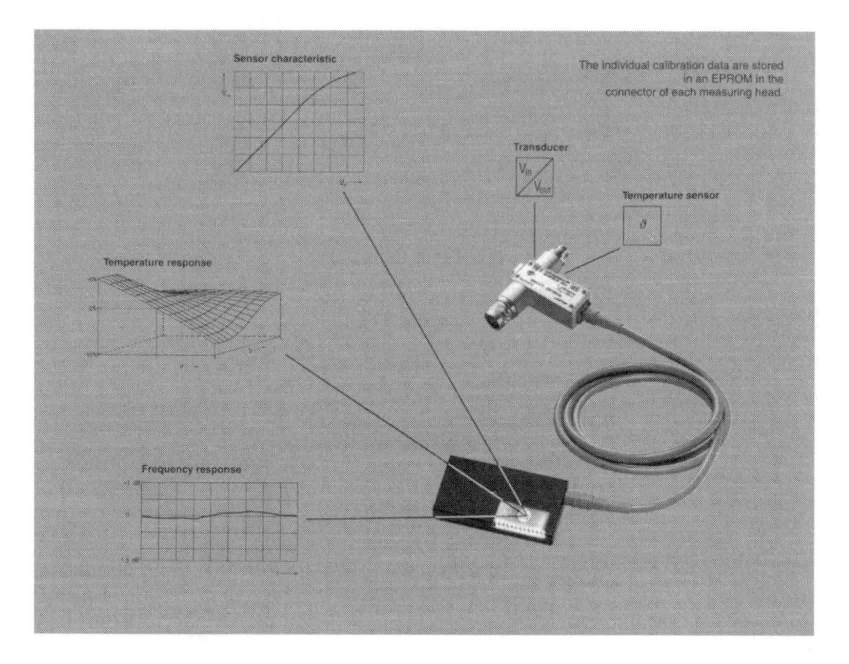

Figure 12.8 A modern sensor carrying calibration data on an EPROM in the connector of the measuring head (reproduced by permission of Rhode and Schwartz)

Sensor characteristics, temperature response and the frequency response are all stored for each sensor at the factory (measured parameters for each individual head) in an EPROM device; see Figure 12.8. The level-dependent temperature effects are stored as two-dimensional characteristics with a great number of measuring points. Each head comprises a temperature sensor, the signal of which is evaluated in the meter at regular intervals. From the measured temperature and the level values, the respective characteristic yields the correction values for the output voltage of the measuring head. The input parameter is then calculated from this corrected value with the aid of the transmission characteristics stored in the head. Subsequently, a frequency response correction is carried out by using a stored correction factor for the signal frequency. This comprehensive correction technique offers the following advantages:

- unrestricted exchange of measuring heads,
- optimum measuring accuracy,
- traceability of calibration of measuring heads, and
- fast and convenient operation.

12.6 Field strength measurements

Radio communication services (such as telecommunication service providers, broadcasting corporations, military, traffic and security services) use field strength meters for propagation measurements in the planning stages and for coverage measurements during operation of communication networks.

Field strength is determined by placing an antenna of known effective length in the electric field under measurement. Then, using a calibrated radio receiver, the induced voltage in the antenna is measured. A loop antenna is usually used in the HF band, a half wave doublet antenna in the VHF or UHF bands. For wideband measurements more sophisticated antenna types (log periodic, biconical, conical log spiral, etc.) are used. The calibrated radio receiver must have the following characteristics.

(i) Selective measurement of a particular radio signal without being affected by other radio signals.
(ii) Capability of measuring the field strength over a wide range from $\mu V\,m^{-1}$ to $mV\,m^{-1}$.
(iii) Capability of accurate measurement of the induced e.m.f. of an antenna.
(iv) Ability of avoiding errors in measurements due to shunting of the RF sections by strong fields.

For these reasons a calibrated radio receiver with a tunable amplifier in the front with excellent linearity, selectivity, stability and sensitivity is configured as a field strength meter. Field strength is obtained by measuring the induced e.m.f. of the antenna and dividing the value by the effective length of the antenna.

A block diagram of a field strength meter is shown in Figure 12.9. At the RF input stage, the signal is amplified and mixed with suitable local oscillator frequencies to down convert the input frequency. A calibration oscillator is used to calibrate the gain, and the control section interacts with the frequency dividers and mixers in this stage to suitably down convert the incoming frequency to the IF frequency. At the IF stage a bandwidth switch is used to select the bandwidth of the signal under measurement. In addition, step amplifiers with gain control (in inverse proportion to the input signal level) are used to keep the detector output within a fixed range. A detector stage in the IF section and the log amplifiers output the detected signal for displaying the level on a digital display. In modern instruments, IEEE-488 interfaces are used for automatic control. Field strength measurements, in particular propagation

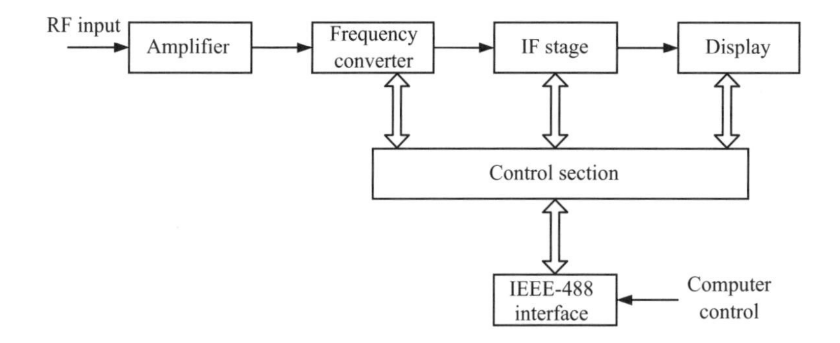

Figure 12.9 Block diagram of a field strength meter

and coverage measurements, are usually made in mobile mode. Therefore, portability and battery operation are important criteria in the choice of a test receiver.

For field strength measurement a calibrated antenna with known effective length is used and the instruments store this in a memory and use it for direct readout of the field strength. The instrument manufacturers supply calibrated antennas in general; however, other antenna elements could also be used by storing the effective length in the instrument memory. In general, an antenna factor is used for this purpose which takes into account the effective length of the antenna as well as other anomalies due to mismatch and loss. For more details Reference 5 is suggested.

12.7 Digital telecommunication transmission systems and associated measurements

In the previous sections we have discussed the analogue and high frequency aspects of transmission measurements. These are applicable from the analogue subscriber loop to radio transmission systems. During the past two decades many terrestrial microwave and fibre optic systems were introduced to carry digitally multiplexed high speed bit steams, using complex time division multiplex (TDM) techniques. In addition, developments related to digital subscriber loop (DSL) technology also have generated many new domains for measurements.

The following sections give a summary to review essentials of such systems and measurements. Because of space limitations, measurements on analogue subscriber loops, etc., are not discussed here. For details, Reference 12 is suggested.

12.7.1 An overview of DSL techniques

Bell Communications Research, Inc., developed the first DSL in 1987 to deliver video on demand and interactive TV over the copper wires. Although the effort initially stalled, interest in DSL regained momentum in the mid-1990s, for several reasons [13]. All DSL technologies run on existing copper phone lines and use advanced modulation techniques (rather than voice grade modems) to boost transmission rates. But different variations, as shown in Table 12.3, are available for different applications. These are:

- asymmetric DSL (ADSL),
- rate adaptive DSL (RADSL),
- high bit rate DSL (HDSL),
- single line DSL (SDSL),
- very high bit rate DSL (VDSL).

ADSL is a technology for providing megabit data rates over existing twisted pair copper wires that make up the last 12 000 to 18 000 feet of wire from the customer's premises to the local phone company's central office. The megabit per second data rates for ADSL modems are for the downstream direction only. For upstream communications, ADSL modems operate up to 640 kbps.

Table 12.3 Characteristics of digital subscriber line (DSL) systems (source: Reference 13)

Acronym	Standard or document[1]	No. of wire pairs	Modulation	Payload data rate (Mb s^{-1})	Mode	Distance[3]	Applications	Splitter
HDSL	G.991.1	1–3	2B1Q/CAP	1.544–2.048	Symmetric	$\leq$ 5 km, $\leq$ 12 km with repeaters	T1 or E1 service access	None
HDSL	T1E1.4 Tech report 28	2	2B1Q/CAP	1.544–2.048	Symmetric	$\leq$ 5 km, $\leq$ 12 km with repeaters	T1 or E1 service access	None
SDSL	G.shds1 (working title)	1[4]	TC-PAM	0.192–2.32	Symmetric	2 km at maximum data rate	Feeder plant, LAN, WAN and server access	None
SDSL	T1E1.4 HDSL2[2]	1[4]	TC-PAM	1.544–2.048	Symmetric	$\leq$ 5 km	Feeder plant, LAN, WAN, and server access	None
ADSL	G.992.1	1	DMT	Downstream: $\leq$ 6.144 Upstream: $\leq$ 0.640	asymmetric	3.6 km at maximum data rate	Internet access, video on demand, simplex video, LAN access, interactive multimedia	At entrance
ADSL	T1.413 Issue 2	1	DMT	Downstream: $\leq$ 6.144 Upstream: $\leq$ 0.640	Asymmetric	3.6 km at maximum data rate	Internet access, video on demand, simplex video, LAN access, interactive multimedia	At entrance

Table 12.3 Continued

Acronym	Standard or document[1]	No. of wire pairs	Modulation	Payload data rate (Mb s^{-1})	Mode	Distance[3]	Applications	Splitter
ADSL Lite	G.992.2	1	DMT	Downstream: ≤ 1.5 Upstream: ≤ 0.512	Asymmetric	Best-effort service	Internet access	No entrance splitter, but micro filter is used
VDSL[1]	G.vdsl (working title)	1	Modulation scheme not agreed	≤ 26 or 52	Symmetric or asymmetric	≤ 300 m at maximum data rate	Same as ADSL plus HDTV	Not decided

2B1Q/CAP: 2 binary, 1 quaternary, carrierless amplitude modulation
DMT: Discrete multitone TC-PAM: Trellis-coded pulse-amplitude modulation
1. G numbers are for ITU documents; T1 numbers are from ANSI T1.413 Committee.
2. Standard still under development; parameters may change.
3. Loop reach for 0.5 mm (AWG 26) wire with no bridged taps.
4. Two-wire pairs may be used for longer loops.

HDSL – high bit rate DSL
ADSL – asymmetric DSL
SDSL – single line DSL
VDSL – very high bit rate DSL

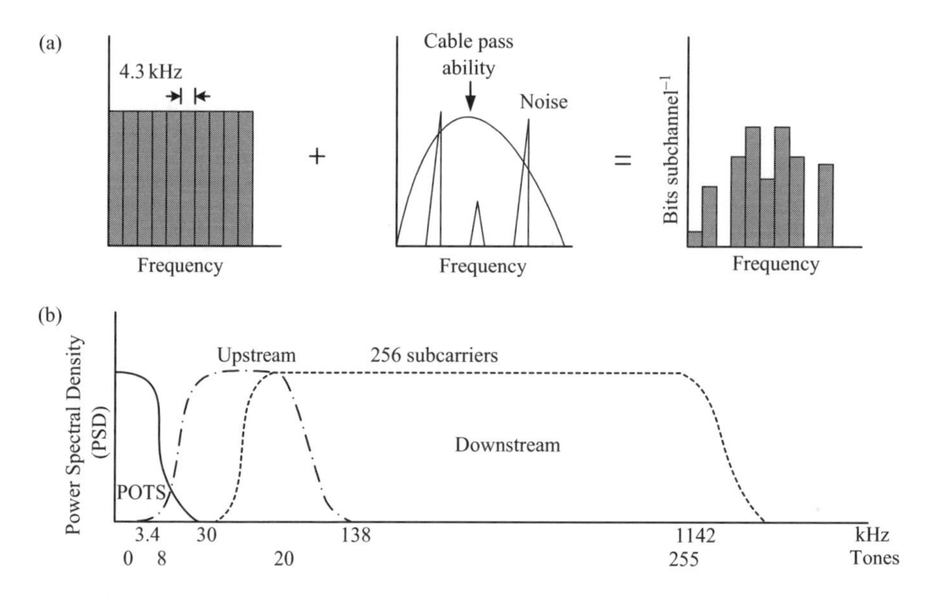

Figure 12.10 DMT technique used in ADSL: (a) use of bins (or tones); (b) spectrum utilisation

As Figure 12.10 shows, ADSL uses frequencies up to 1 MHz and this is very much higher than those used by the ordinary phone equipment (normally 300–3400 Hz range). Those frequencies higher than 4 kHz were unusable until recently because ICs did not have enough processing power to provide enough dynamic range and compensate for line impedances at the higher frequencies.

Figure 12.10 shows the frequency spectrum used by ADSL modems. The lower 4 kHz is still available for plain old telephony service (POTS) equipment. ADSL then divides the higher frequencies into 256 tones, where each tone occupies 4.3 kHz. This technique is called discrete multitone (DMT). The lower frequency tones are used for upstream transmission and most of the tones are used for downstream transmission.

Each tone can carry up to 16 bits, depending on the frequency and delay response of the line at that tone's frequency. If, for example, the line has a higher noise level at one frequency than at another, the number of bits (amplitudes) available at that frequency will be fewer compared to other frequencies (as shown by Figure 12.10(a), extreme right). An ADSL receiver performs an inverse fast Fourier transform (FFT) on the tones, transforming them back into a single time domain waveform from which the modem can recreate a serial bit stream. ADSL modems use cyclic redundancy check (CRC) and forward error correction (FEC) techniques to correct bit errors.

ADSL modems establish the data rates by transmitting at the highest data rate at the beginning, lowering the data rates until an error free connection is made. This process is called training. During cross-talk testing, for example, two ADSL modems must first train each other at a reference noise level before you commence increasing that noise to complete the test.

In the following sections of ADSL measurements ANSI standard T1.413 based DMT technology is discussed. For details of other techniques References 14 and 15 are suggested.

12.7.1.1 ADSL standard – ANSI T1.413

The ANSI T1.413 standard, drafted by the ANSI T1E1.4 committee, defines ADSL modems that use the discrete multitone (DMT) technique for data transmission. DMT is one of the two modulation techniques used by ADSL modem manufacturers. The other is called carrier amplitude phase (CAP). The ANSI T1.413 document also defines conformance tests for DMT ADSL modems. In tests, the test engineer must simulate a line of twisted pair wires (called a test loop), add electrical impairments to the lines, and measure the bit error rate (BER).

ANSI T1.413 defines several tests for ADSL modems. These tests are:

- cross-talk tests;
- impulse-noise tests; and
- POTS interference tests.

12.7.2 ADSL measurements

ADSL modems designed with the ADSL chip sets have built-in test features, similar to the analogue modems. Chip sets have loop-back features that allow the testing of the devices' transmitter and receiver circuits. The loop-back tests can take place within the chips or over a simulated local loop to a remote transmitter. Figure 12.11 shows a typical setup for remote loop back testing. These are the basic tests performed at the manufacturers' locations for testing and confirming the functioning of the chip sets.

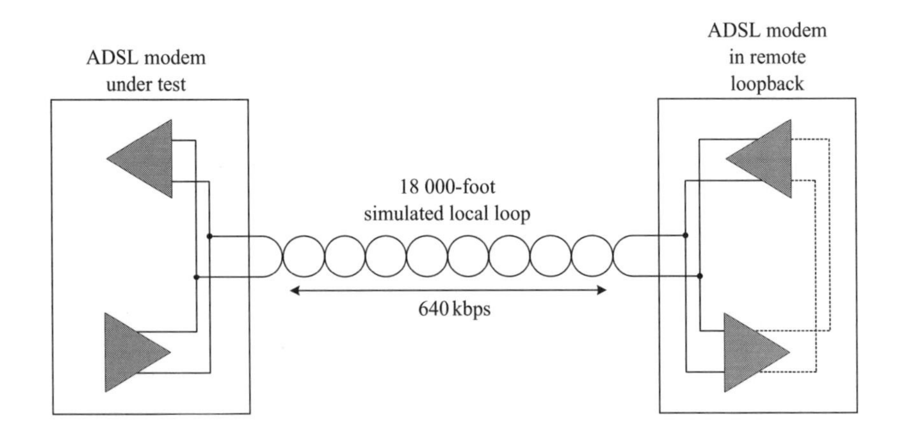

Figure 12.11 Remote loop-back testing of ADSL loops

12.7.2.1 Lab test setup and measurements

Figure 12.12 shows the lab test setup [16] for testing an ADSL modem's conformance to ANSI T1.413. The ADSL transceiver unit (ATU)-central office [ATU-C] is the ADSL phone company's central office (CO) ADSL modem, and the ATU-remote [ATU-R] is the modem at the customer's premises. The ATU-C is sometimes called the transmitter because it transmits at a rate of megabits per second. The ATU-R is called the receiver.

Simulation of the test loop is carried out by using a line simulator, or a spool of wire. The splitters separate POTS signals (normal voice signals) from ADSL signals. During the tests, a POTS phone or modem is connected to the ATU-R splitter and a central office simulator to the ATU-C splitter. It is also necessary to use a high impedance network to introduce impairments into the test loop without altering the characteristics of the test loop. Conformance testing must be done with any error correction enabled.

The first conformance test is the cross-talk test. Cross-talk testing is necessary because phone lines are typically bundled between the CO and their destinations. Bundles use either 25 or 50 pairs of wires. The tests are based on interference from high speed DSL (HDSL), T1, ISDN (called DSL in ANSI T1.413), and ADSL transmissions on other lines in the same bundle.

ANSI T1.413-Annex B defines the characteristics of each type of signal. Each interference signal (called a disturber) has its own power spectral density (PSD).

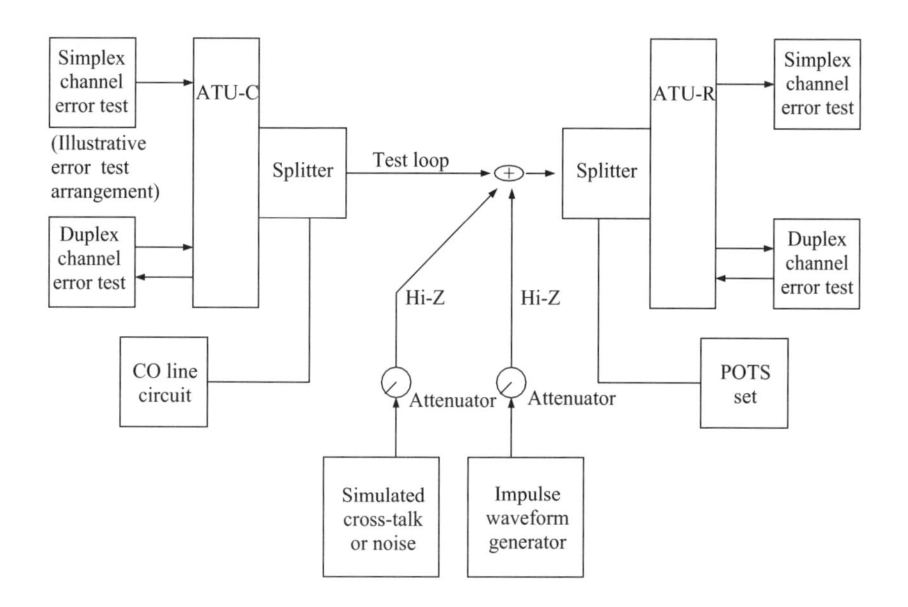

Figure 12.12 Conformance testing on ADSL modems by injection of signals into a simulated test loop (courtesy of Test & Measurement World Magazine, Feb. 1997)

Therefore, the noise generated by each form of cross-talk will differ. A test engineer must also make sure that the PSD of the test system's background noise is flat.

One of the tests to be performed is the 24 disturber DSL near-end cross-talk (NEXT) test. In this test, the ADSL signals are fed over one pair of a bundle of 25 wire pairs. Then the PSD level of the noise is varied to simulate the number of disturbers.

The noise used to simulate the disturbers must be Gaussian white noise that has a crest factor of 5. (The T1E1.4 committee is investigating this specification, and may change it in the future.) Connection of the noise source to the test loop must have an impedance greater than 4 kΩ.

12.7.2.2 Simulation of wire loops

In addition to simulating disturbers, one must also simulate several configurations of twisted pair lines. These lines are called revised-resistance design (RRD) lines (for 1.544 Mbps data rates) and carrier service area (CSA) lines (for 6 Mbps data rates). In the USA, the Bellcore SR-TSV-002275 and ANSI T.601 specifications define the lengths and gauges of the test loops. They also define the locations of bridge taps in the loops. Bridge taps are unterminated connections to a subscriber loop. Those unterminated wires cause reflections and can distort signals. Figure 12.13 shows two subscriber loops with bridge taps used for testing. In this example, the T1.601 loop number 13 has both 24 AWG and 26 AWG wires; the CSA loop number 4 is entirely of 26 AWG wire.

While performing a cross-talk test, noise source should be set to the appropriate level for the type of cross-talk you are simulating. That becomes the reference level, 0 dB. Establishment of communication between the ATU-C and ATU-R is necessary

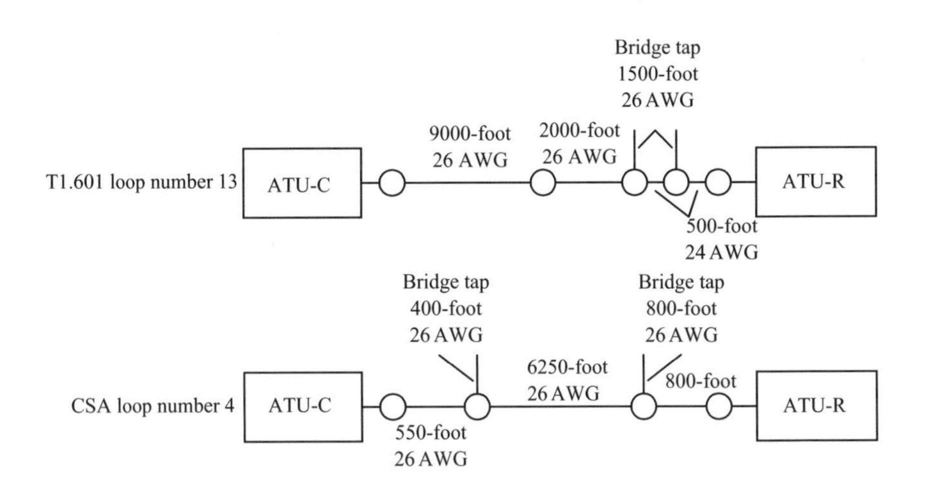

Figure 12.13 Simulations of subscriber loops with bridge taps (courtesy of Test & Measurement World Magazine, Feb. 1997)

Table 12.4 Minimum test times for cross-talk tests

Bit rate	Minimum test time
Above 6 Mbps	100 s
1.544 Mbps–6 Mbps	500 s
Below 1.544 Mbps	20 min

at the reference noise level (assuming that the ATU-R is the unit under test). The process of establishing the connection between the two ADSL modems is called training.

Once training is complete, measurement of the bit error rate (BER) with a BER tester is necessary while using a pseudo-random test pattern of $2^{23} - 1$ bits. By increasing the noise level in increments of 1 dB test engineers can find the highest noise level that produces a BER that is less than 10^{-7}. The difference between the noise level and the reference level is called a margin. To pass cross-talk conformance tests, recommendations specify a margin of 6 dB. Because the noise signals are random, the tests must be performed long enough to accurately measure BER. Table 12.4 shows the minimum test times for cross-talk conformance testing.

12.7.2.3 Impulse-noise test

After the cross-talk test, impulse-noise testing can be performed. Impulse-noise testing subjects the ADSL products to wideband noise delivered in short bursts. The impulses specified in ANSI T1.413-Annex C are reconstructions of pulses recorded at field sites. There are two impulse test waveforms in ANSI T1.413 as shown in Figure 12.14. To generate those waveforms, an arbitrary waveform generator or a line simulator that generates impulse impairments can be used.

The impulse tests could begin after the training of the ADSL modems using the 0 dB noise level that was used for the cross-talk tests. After communication is established between the two modems, each impulse is applied 15 times at intervals of at least 1 s. The tests need be repeated by adjusting the amplitudes of the impulses until you measure errored seconds on the BER tester. What is then measured is the amplitude that produces errors in half of the number of seconds in a test. Test engineers can then follow a formula in ANSI T1.413 to calculate the probability that a given amplitude will produce an errored second. To pass the test, that probability must be less than 0.14 per cent.

12.7.2.4 Errors created by POTS equipment

The above tests do not account for errors created by POTS equipment. ADSL and POTS equipment must share the same wires, so testing must be carried out on the

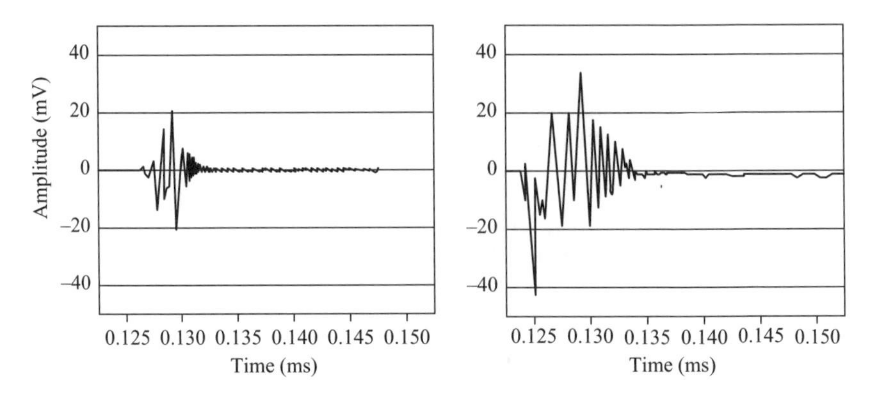

Figure 12.14 Use of two waveforms for impulse testing on ADSL equipment (courtesy of Test & Measurement World Magazine, Feb. 1997)

ADSL equipment with interference from POTS equipment on the same line. Connection of a CO simulator to the ATU-C splitter (Figure 12.12), and connecting a POTS phone to the ATU-R splitter allows this. The POTS equipment should not interfere with ADSL operation. During the POTS interference test, simulations are necessary on the common conditions that POTS equipment produces such as signalling, alternating, ringing, answering, and going off hook and on hook.

12.7.3 G-Lite ADSL

G-Lite ADSL (or simply G-Lite) is a medium bandwidth version of ADSL developed for the consumer market. That allows Internet access up to 1.5 Mbps downstream and up to 500 kbps upstream. In most cases, G-Lite will operate over existing telephone wiring, providing a means of simpler installation. The standard specification applicable is ITU Rec. G 992.2.

G-lite modems use only 96 tones for data rather than the 256 tones used by full rate ADSL modems (figure 12.15). The fewer tones limit the bandwidth to 420 kHz, a data rate of about 1 Mbps downstream and 512 kpbs upstream. G-Lite modems are designed to be compatible with the full rate ATU-C systems at the CO side, but without letting the ATU-C assign data channels to tones greater than 96. G-Lite modems use less bandwidth, and hence less power, than full rate ADSL modems. Because the bandwidth of G-Lite modems is about half that of full rate ADSL, their total power usage drops by about 3 dB. For testing of these, Reference 17 is suggested.

12.7.4 Digital transmission hierarchy: an overview

With voice telephony entering the digital domain in the late 1970s, worldwide switching and transmission systems gradually commenced the conversion process from analogue transmission systems to digital systems. The basic voice digitisation process used is known as pulse code modulation (PCM), and systems for the multiplexing of

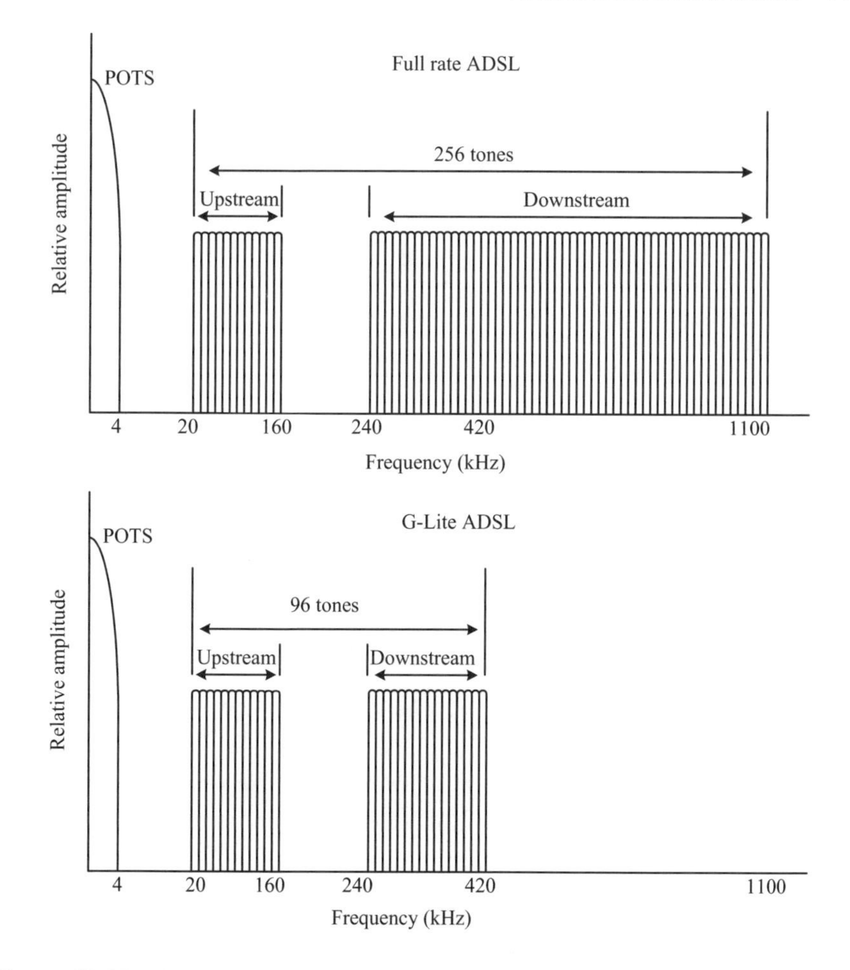

Figure 12.15 G-Lite ADSL bandwidth compared with full rate ADSL

many voice channels to one single data stream are known as time division multiplex (TDM) systems.

12.7.4.1 Pulse code modulation (PCM)

PCM is a sampling, quantisation and coding process that converts a voice conversation into a 64 kbps standard rate. This basic transmission level is known as digital signal level zero (DS0).

As shown in Figure 12.16, a voice signal is time sampled at a rate of 8000 Hz (where the incoming signal is expected to have frequencies only up to 4000 Hz). These samples are then converted to pulses by using a process known as pulse amplitude modulation (PAM) as in Figure 12.16(a). In the next stage the highest of each pulse is assigned an equivalent 8-bit binary value, going through the process known as

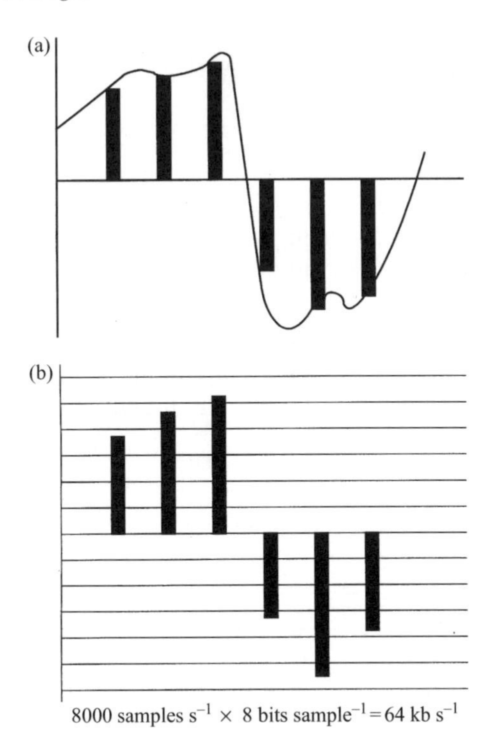

Figure 12.16 The PCM process: (a) sampling and pulse amplitude modulation (PAM); (b) assigning binary values to pulses

quantising and coding. The resulting output is a digital representation of the pulse and, by extension, of the sampled analogue waveform.

The above is a very simple explanation and in the actual process in telecom applications use is made of a non-uniform coding process based on the A-law and μ-law prescribed by telecom standardisation bodies such as CCITT. For details of this process, known as companding, Reference 18 is suggested.

12.7.4.2 E1 and T1 systems

Once digitised, voice and/or data signals from many sources can be combined (i.e. multiplexed) and transmitted over a single high speed link. This process is made possible by the technique called TDM. TDM divides the link into 24 or 30 discrete 64 kbps timeslots. An identical number of DS0 signals (representing multiple separate voice and/or data calls) is assigned to each timeslot for transmission within the link, as shown in Figure 12.17.

In the USA and Japan, 24-channel systems known as T1 links are used, while 30-channel systems called E1 links are used in CCITT standards dominant countries such as Europe and elsewhere.

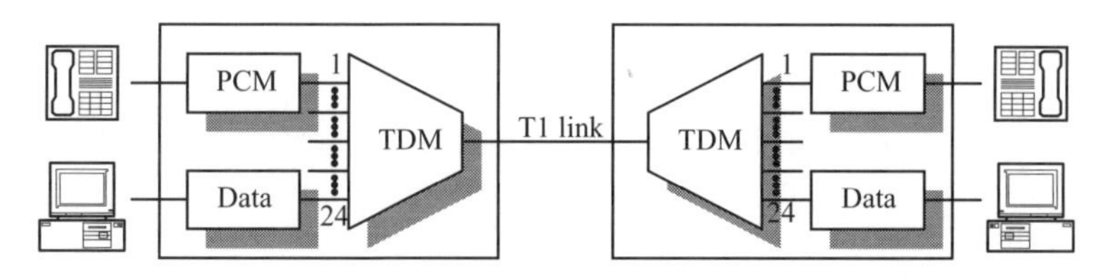

Figure 12.17 Time division multiplexing

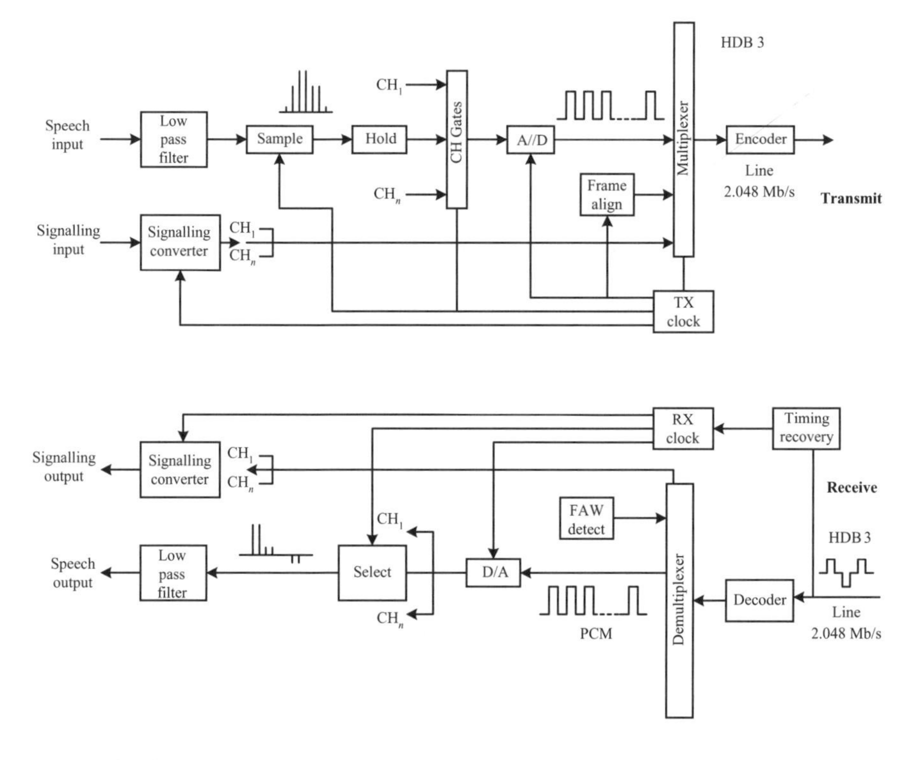

Figure 12.18 The basic asynchronous transmission concept in E1 systems

12.7.4.2.1 E1 systems

A typical E1 based PCM transmission concept is shown in Figure 12.18.

Transmission path The speech signal is first band limited by the low pass filter so that only the frequency band 300–3400 Hz is transmitted. The speech signal is then sampled at the rate of 8 kHz to produce the PAM signal. The PAM signal is temporarily stored by a hold circuit so that the PAM signal can be quantised and encoded in the A/D converter. Samples from a number of telephone channels (typically 24 or 30) can be processed by the A/D converter within one sampling period of 125 μs. These samples are applied to the A/D converter via their respective gates selected by the transmit timing pulses. At the output of the A/D converter, the speech samples exit as 8-bit PCM code words. These code words from the speech path are combined with the frame alignment word, service bits, and the signalling bits in the multiplexer to form frames and multiframes. They are then passed on to the high density bipolar 3 (HDB 3) line encoder, which converts the binary signals into bipolar (pseudo-ternary) signals for transmission over the wire line, the digital microwave radio (DMR), or the optical fibre cable.

In the European CCITT systems, each frame contains 32 time slots of approximately 3.9 μs duration. The time slots are numbered from 0 to 31. Time slot 0 is reserved for the frame alignment signal and service bits. Time slot number 16 is reserved for multiframe alignment signals and service bits and also for the signalling information of each of the 30 voice channels. Each multiframe consists of 16 frames, so the time duration of one multiframe is 2 ms. The purpose of the formation of multiframes is to allow the transmission of signalling information for all 30 channels during one complete multiframe.

The signalling information for each telephone channel is processed in the signalling converter, which converts the signalling information into a maximum of 4-bit codes per channel. These bits are inserted into time slot 16 of each PCM frame except frame number 0. The 16 frames in each multiframe are numbered 0 to 15. Because in each frame signalling information from two telephone channels is inserted into time slot 16, signalling information from the 30 telephone channels can be transmitted within one multiframe.

The transmission rate of the PCM signals is 2048 kbps (2.048 Mbps). This is controlled by the timing clocks in the transmission end that control the processing of the speech, signalling, synchronising, and service information.

Reception path The 2048 kbps pseudo-ternary signal that comes from the line is first decoded by the HDB 3 decoder into a binary signal. This signal is then separated by the input demultiplexer or separator into the respective speech channels, together with supervisory information (signalling, etc). The speech codes are sent to the D/A converter, the signalling bits are sent to the signalling converter, and the frame alignment bits and service bits for alarms, etc., are sent to the frame alignment detector and alarm unit. The timing signals for the receiver are recovered from the line codes and processed in the receiver timing unit to generate the clock signals for processing the received signals. In this manner the receiver is kept synchronised to the transmitter. Synchronisation between the transmitter and receiver is vital for TDM systems. The codes belonging to the speech signal are then converted to PAM signals by the D/A converter. Next, they are selected by their respective gates and sent to their own channels via the respective low pass filters, which reconstruct the original analog speech patterns. The bits belonging to signalling are converted into signalling information by the receive signalling converter and sent to the respective telephone channels. The frame alignment word and service bits are processed in the frame alignment and alarm units. Frame alignment word (FAW) detection is done here, and if a FAW error is detected in four consecutive frames, the frame alignment loss (FAL) alarm is initiated. Some of the service bits are used to transmit and receive alarm conditions.

12.7.4.2.2 Frame alignment word (FAW) formats for 30-channel PCM systems

The construction of the frame and multiframe is shown in Figure 12.19. The details of a multiframe in time slot 0 and time slot 16 are not discussed here, and can be found in Reference 17. In time slot 0 of each frame number, the FAW (0011011) is sent on every even frame, and the service bits (Y1ZXXXXX) are sent on every odd frame.

In time slot 16 of frame number 0 only, the multiframe alignment word (0000) is sent. In time slot 16 of frame numbers 1 to 15 the signalling information of channel pairs 1/16, 2/17, etc., are sent in the respective frame order.

Frame alignment As mentioned earlier, a time slot of 8 bits per frame is available for frame alignment. This means that 64 kbps are reserved for this purpose. The basic principle of frame alignment is that the receiver identifies a fixed word and then checks its location at regular intervals. This makes it possible for the receiver to organise itself to the incoming bit flow and to distribute the correct bits to the respective channels. In addition to frame alignment, the assigned time slot is also used for transmission of information concerning the alarm states in the near end terminal to the remote-end terminal. Spare capacity is also available for both national and international use. The 16 frames are numbered 0 to 15. The words in time slot 0 in frames with even numbers are often called frame alignment word 1, while those in odd frames are called frame alignment word 2. For details, Reference 18 is suggested.

12.7.4.3 Format of T1 based 24-channel PCM systems

The 30-channel primary systems used in Europe and most of the developing world are not used in North America and Japan, where they have designed a system which has 24 channels for the primary PCM system, and the basic concept of A/D conversion is the same. The companding[1] for each system achieves the same objective but uses slightly different companding curves. The frame structure is the major difference, as indicated in Figure 12.20. The 24-channel frame is 125 μs long, just the same as the 30-channel frame. However, the 24-channel frame contains only 24 times slots, each having 8 bits. The first seven bits are always used for encoding, and the eighth bit is for encoding in all frames except the sixth frame, where it is used for signalling. At the start of every frame, 1 bit is included for frame and multiframe alignment purposes. Each frame therefore contains $(24 \times 8) + 1 = 193$ bits. Since the sampling rate is 8 kHz, there are 8000 frames per second, giving $193 \times 8000 = 1.544\,\text{Mb}\,\text{s}^{-1}$. The signalling bit rate is $(8000/6) \times 24 = 3200$ bps.

The 24-channel system also has a 1.5 ms multiframe consisting of 12 frames. The frame and multiframe alignment words are transmitted sequentially, by transmitting 1 bit at the beginning of each frame. They are sent bit by bit on the odd and even

[1] The process of first compressing and then expanding a signal is referred to as compounding.

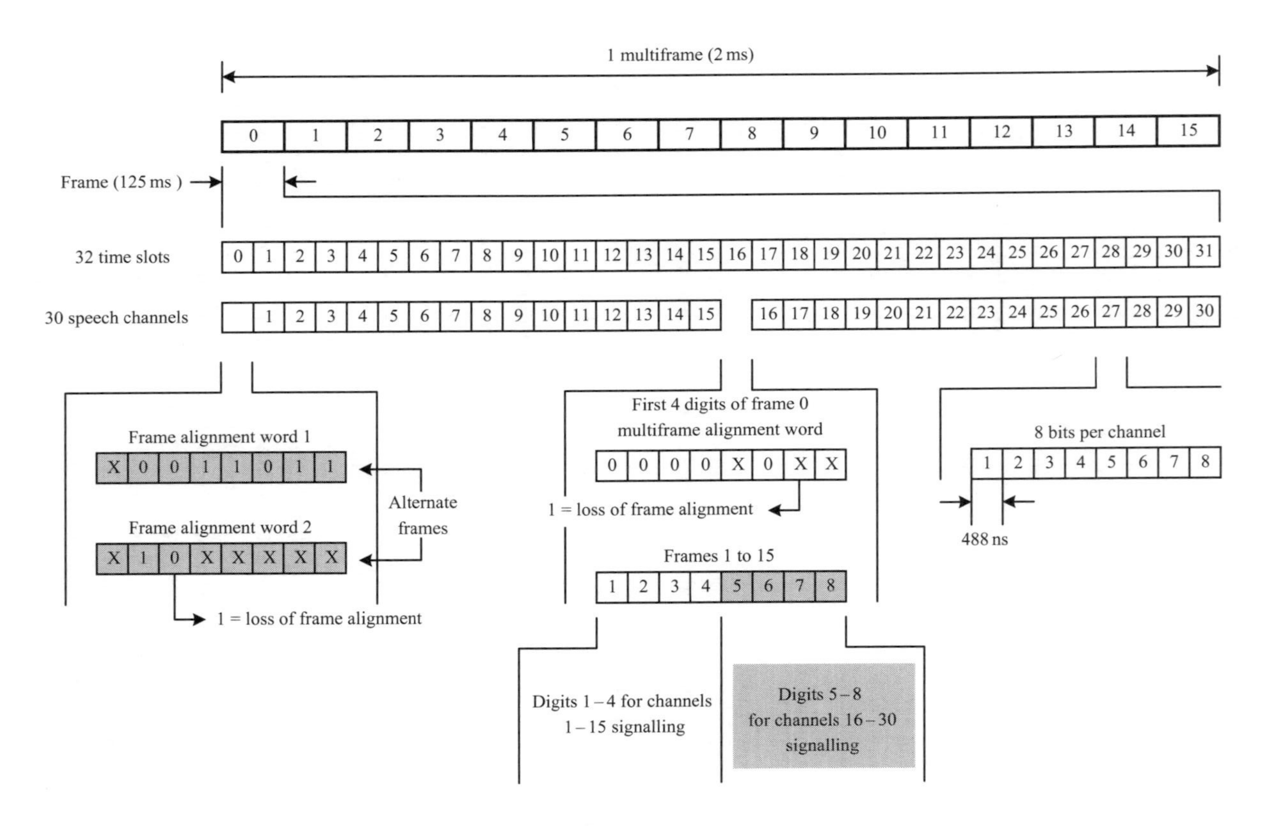

Figure 12.19 A 30-channel PCM frame and multiframe details

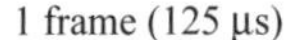

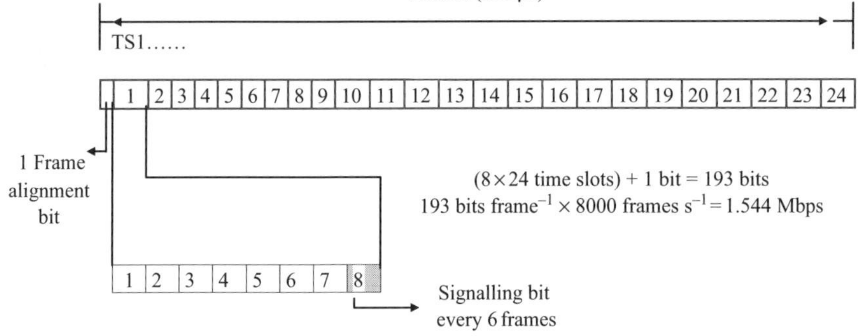

Figure 12.20 The 24-channel T1 system

Table 12.5 Comparison of 24- and 30-channel systems

	24-channel systems	30-channel systems
Sampling frequency (kHz)	8	8
Duration of time slot (μs)	5.2	3.9
Bit width (μs)	0.65	0.49
Bit transfer rate (Mb s^{-1})	1.544	2.048
Frame period (μs)	125	125
No. of bits per word	8	8
No. of frames per multiframe	12	16
Multiframe period (ms)	1.5	2
Frame alignment signal in	Odd frames	Even frames
Frame alignment word	101010	0011011
Multiframe alignment word	001110	0000

frame cycles, and their transmission is completed only after each multiframe has been transmitted. The frame and multiframe alignment words are both 6-bit words (101010 and 001110, respectively).

A comparison of the construction of the frame and multiframe for the CCITT 30-channel and US/Japan 24-channel primary PCM systems is summarised in Table 12.5. For details, Reference 19 is recommended.

12.7.4.4 HDB 3 code

The purpose of the HDB 3 code is to limit the number of zeros in a long sequence of zeros to three. This assures clock extraction in the regenerator of the receiver. This code is recommended by the CCITT (G.703) for the 2, 8, and 34 Mb s^{-1} systems. An example is shown in Figure 12.21. Longer sequences of more than three zeros are avoided by the replacement of one or two zeros by pulses according to specified rules. These rules ensure that the receiver recognises that these pulses are replacements for

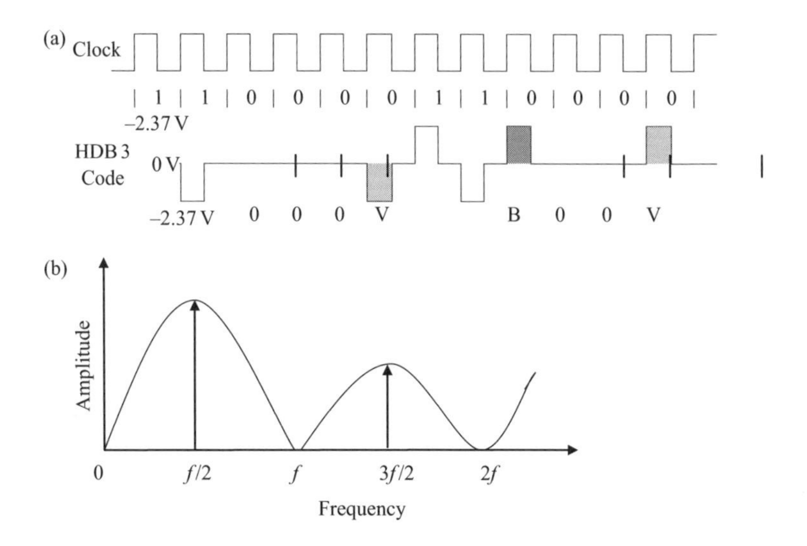

Figure 12.21 The HDB 3 code: (a) bit stream; (b) spectrum

zeros and does not confuse them with code pulses. This is achieved by selecting the polarity of the alternative mark inversion (AMI) code [18]. Also, the replacement pulses themselves must not introduce an appreciable d.c. component. The rules for HDB 3 coding can be summarised as follows.

(i) Invert every second '1' for as long as a maximum of three consecutive *zeros* appear.
(ii) If the number of consecutive *zeros* exceeds three, set *the violation pulse* in the fourth position. The violation pulse purposely violates the AMI rule.
(iii) Every alternate violation pulse shall change polarity. If this rule cannot be applied, set a '1' according to the AMI rule in the position of the first *zero* in the sequence.

For more details, Reference 18 is suggested.

12.7.5 Plesiochronous digital hierarchy

Until the mid-1990s most countries used digital hierarchical transmission systems known as plesiochronous digital hierarchy. In the following sections some details of these systems and the reasoning behind the conversion to newer and more flexible systems are discussed.

12.7.5.1 Asynchronous higher order digital multiplexing

The 30- or 24-channel PCM systems are only the first, or primary, order of digital multiplexing as designated by standardisation bodies such as CCITT or ANSI, etc. If it is necessary to transmit more than 30 or 24 channels, the system is built up as in the heirarchy diagram of Figure 12.22, comparing the use in different parts of the world.

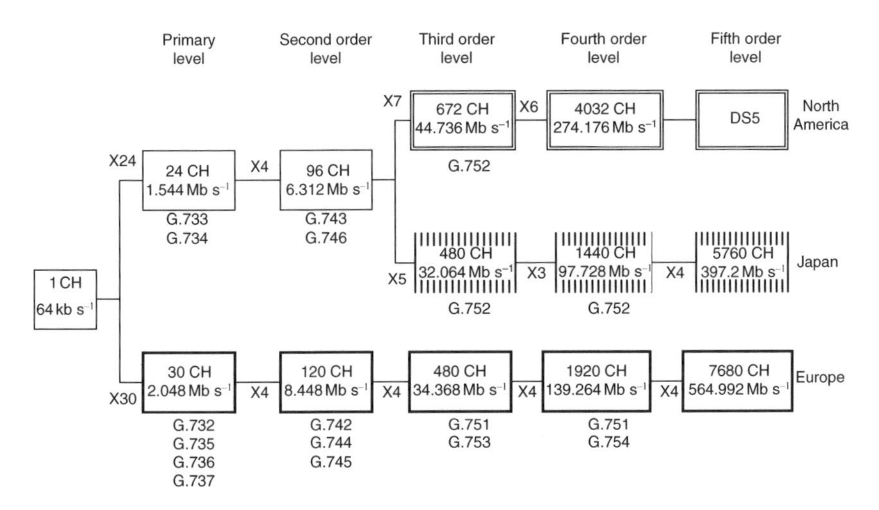

Figure 12.22 The asynchronous digital hierarchy

Table 12.6 Asynchronous multiplexer levels and bit rates for the CCITT system

Level	No. of channels	Bit rate (Mbps)
First	30	2.048
Second	120	8.448
Third	480	34.368
Fourth	1920	139.264
Fifth	7680	565.992

In CCITT based systems four primary systems are multiplexed to form an output having 120 channels. This is called the second order of multiplexing. Similarly, four 120-channel systems can be multiplexed to give an output of 480 channels in third order multiplexing. Table 12.6 indicates these levels and the corresponding bit rates for each asynchronous multiplexer levels for CCITT systems. Similarly for 24-channel T1 systems, Table 12.7 depicts the levels and bit rates.

For details of these systems, Reference 18 is suggested.

12.7.6 Synchronous digital multiplexing and SDH/SONET systems

The asynchronous multiplexer systems described earlier have the benefit of operating independently without a master clock to control them. Each lower rate multiplexer (such as a 2 Mbps E1 or a 1.5 Mbps T1 link) has its own independent clock.

Table 12.7 Asynchronous multiplexer details for the North American system

Level	No. of channels	Bit rate (Mbps)
DS-1	24	1.544
DS-1C	48	3.152
DS-2	96	6.312
DS-3	672	44.736
DS-4	4032	274.176

These plesiochronous transmission systems could have minute differences in frequency from one multiplexer to another, so when each provides a bit stream for the next hierarchical level, bit stuffing (justification) is necessary to adjust for these frequency differences.

Despite the attractive aspects of asynchronous multiplexing, there is one major drawback. If, for example, a 140 Mb s^{-1} system is operating between two major cities, it is not possible to identify and gain access to individual channels at towns *en route*. In other words, drop and insert capability requires a complete demultiplexing procedure.

Recognising these disadvantageous reasons in PDH systems, in the latter part of the 1980s industry and standardisation bodies commenced working on synchronous systems leading to synchronous digital hierarchy (SDH) and synchronous optical network (SONET) systems. In 1988, the CCITT reached an agreement on a worldwide standard for the synchronous digital hierarchy (SDH) in the form of Recommendations G.707, 708 and 709. In addition to being a technical milestone, this agreement also unified the bit rates so that this new synchronous system does not have the existing interface problems between North America and Japan and the rest of the world. The resulting recommendations were intended for application to optical fibre transmission systems and they were originally called the synchronous optical network (SONET) standard. Although SDH now supersedes the SONET description, they both refer to the same subject matter and are sometimes used interchangeably in the literature.

12.7.6.1 Synchronous digital hierarchy

SDH standards approved in 1988 define transmission rates, signal format, multiplexing structures and tributary mapping for the network mode interface (NNI) – the international standard for the SDH. In addition to defining standards covering the NNI, the CCITT also embarked on a series of standards governing the operation of synchronous multiplexers (G.781, G783 etc.) and SDH network management (G.784). It is the standardisation of these aspects of SDH equipment that will deliver the flexibility required by network operators to manage in a cost effective manner, the growth in bandwidth and provisioning of new customer services.

The SDH specifications define optical interfaces that allow transmission of lower rate (e.g. PDH) signals at a common synchronous rate. A benefit of SDH is that it allows multiple vendors' optical transmission equipment to be compatible in the same span. SDH also enables dynamic drop-and-insert capabilities on the payload; PDH operators would have to demultiplex and remultiplex the higher rate signal, causing delays and requiring additional hardware. Because the overhead is relatively independent of the payload, SDH easily integrates new services, such as asynchronous transfer mode (ATM) and fibre distributed data interface (FDDI), along with existing European 2, 34 and 140 Mbit s^{-1} PDH signals, and North American 1.5, 6.3 and 45 Mbit s^{-1} signals.

12.7.6.2 Synchronous optical network (SONET)

In 1985, Bellcore proposed the idea of an optical carrier-to-carrier interface that would allow the interconnection of different manufacturers' optical equipment. This was based on a hierarchy of digital rates, all formed by the interleaving of a basic rate signal. The idea of SONET attracted the interest of carriers, Regional Bell Operating Companies (RBOCs), and manufacturers alike and quickly gained momentum. Interest in SONET by the CCITT (now ITU-T) expanded its scope from a domestic to an international standard, and by 1988 the ANSI committee had successfully integrated changes requested by the ITU-T, and were well on their way towards issuing the new standard. Today, the SONET standard is contained in the ANSI specification T1.105 *Digital Hierarchy – Optical Interface Rates and Formats Specifications (SONET)*, and technical recommendations are found in Bellcore TR-NWT-000253 *Synchronous Optical Network (SONET) Transport System: Common Generic Criteria*. The SONET specifications define optical carrier (OC) interfaces and their electrical equivalents to allow transmission of lower rate signals at a common synchronous rate.

As the overhead is relatively independent of the payload, SONET is able to integrate new services, such as ATM and FDDI, in addition to existing DS3 and DS1 services. Another major advantage of SONET is that the operations, administrations, maintenance, and provisioning (OAM&P) capabilities are built directly into the signal overhead to allow maintenance of the network from one central location.

12.7.6.3 SDH/SONET multiplexing

12.7.6.3.1 SDH multiplexing and frame formats

SDH multiplexing combines low speed digital signals such as 2, 34 and 140 Mbit s^{-1} signals with required overhead to form a frame called synchronous transport module at level one (STM-1). Figure 12.23 shows the STM-1 frame, which contains 9 segments of 270 bytes each. The first 9 bytes of each segment carry overhead information; the remaining 261 bytes carry payload. When visualised as a block, the STM-1 frame appears as 9 rows by 270 columns of bytes. The STM-1 frame is transmitted with row number 1 first, with the most significant bit (MSB) of each byte transmitted first.

In order for SDH easily to integrate existing digital services into its hierarchy, it operates at the basic rate of 8 kHz or 125 μs per frame, so the frame rate is 8000 frames s^{-1}. The frame capacity of the signal is the number of bits contained within

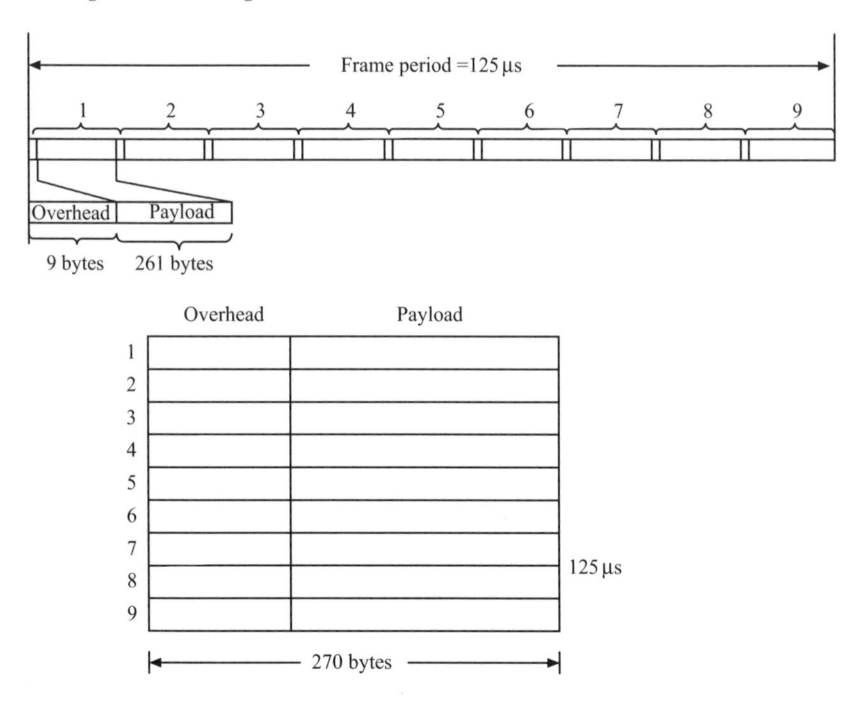

Figure 12.23 SDH frame format

a signal frame. Figure 12.23 shows that

$$\begin{aligned}\text{frame capacity} &= 270\ \text{bytes row}^{-1} \times 9\ \text{rows frame}^{-1} \times 8\ \text{bits byte}^{-1}\\ &= 19\,440\ \text{bits frame}^{-1}.\end{aligned}$$

The bit rate of the STM-1 signal is calculated as follows using the formula,

$$\begin{aligned}&\text{bit rate} = \text{frame rate} \times \text{frame capacity}\\ &\text{bit rate} = 8000\ \text{frames s}^{-1} \times 19\,440\ \text{bits frame}^{-1} = 155.52\ \text{Mbps}.\end{aligned} \tag{12.8}$$

Three transmission levels (STM-1, STM-4 and STM-16) have been defined for the SDH hierarchy. As figure 12.24 shows, the ITU has specified that an STM-4 signal should be created by byte interleaving four STM-1 signals. The basic frame rate remains 8000 frames s^{-1}, but the capacity is quadrupled, resulting in a bit rate of 4 × 155.52 Mbit s^{-1}, or 622.08 Mbps. The STM-4 signal can then be further multiplexed with three additional STM-4 tributaries to form an STM-16 signal. Table 12.8 lists the defined SDH frame formats, their bit rates, and the maximum number of 64 kbps telephony channels that can be carried at each rate.

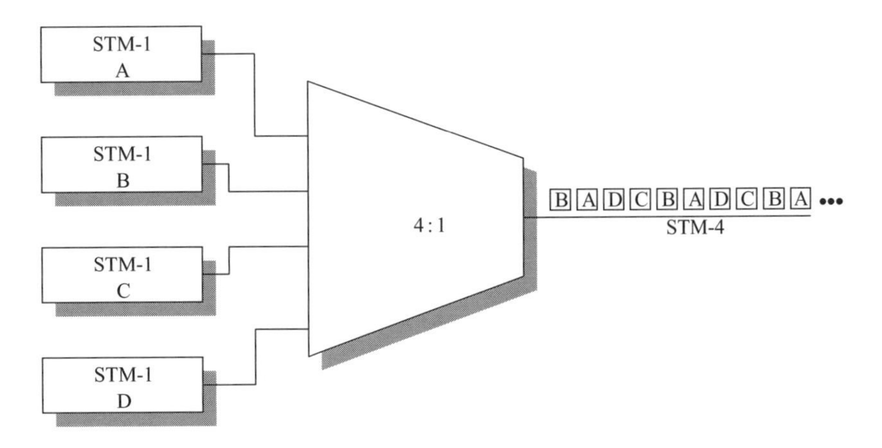

Figure 12.24 SDH multiplexing

Table 12.8 SDH levels

Frame format	Bit rate	Max. number of telephony channels
STM-1	155.52 Mbps	1920
STM-4	622.08 Mbps	7680
STM-16	2.488 Gbps	30 720

12.7.6.3.2 SONET multiplexing and frame formats

SONET multiplexing combines low speed digital signals such as DS1, DS1C, E1, DS2 and DS3 with required overhead to form a building block called synchronous transport signal level one (STS-1). Figure 12.25 shows the STS-1 frame, which is organised as 9 rows by 90 columns of bytes. It is transmitted row first, with the MSB of each byte transmitted first.

In order for SONET easily to integrate existing digital services into its hierarchy, it was defined to operate at the basic rate of 8 kHz or 125 μs frame^{-1}, so the frame rate is 8000 frames s^{-1}. This is similar to the case of SDH. Figure 12.25 shows that

$$\begin{aligned}\text{frame capacity} &= 90\ \text{bytes row}^{-1} \times 9\ \text{rows frame}^{-1} \times 8\ \text{bits byte}^{-1}\\ &= 6480\ \text{bits frame}^{-1}.\end{aligned}$$

Now the bit rate of the STS-1 signal is calculated as follows:

$$\text{bit rate} = 8000\ \text{frames s}^{-1} \times 6480\ \text{bits frame}^{-1} = 51.840\ \text{Mb s}^{-1}.$$

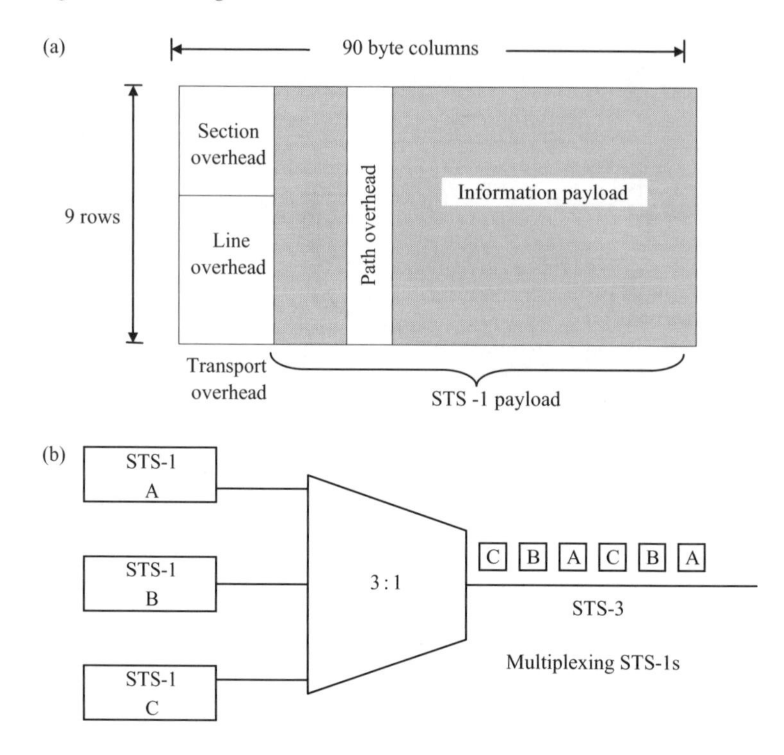

Figure 12.25 SONET frame format and multiplexing: (a) frame format; (b) multiplexing

Higher rate signals are formed by combining multiples of the STS-1 block by interleaving a byte from each STS-1 to form an STS-3, as shown in Figure 12.25(b). The basic frame rate remains 8000 frames s^{-1}, but the capacity is tripled to result in a bit rate of 155.52 Mb s^{-1}. The STS-3 may then be converted to an optical signal (OC-3) for transport, or further multiplexed with three additional STS-3 tributaries to form an STS-12 signal, and so on. Table 12.9 defines common SONET optical rates, their equivalent electrical rates, and the maximum number of DS0 voice channels which can be carried at that rate. Comparing the contents of Table 12.6 and Table 12.7, STS-1 carries the same number of DS0s as a DS-3 signal. OC-3, OC-12 and OC-48 are the most popular transport interfaces today.

Table 12.10 indicates relationship between SONET & SDH transmission rates

12.7.6.4 Transport capabilities of SDH/SONET

All of the tributary signals that appear in today's PDH networks can be transported over SDH. The list includes: CEPT 2, 34 and 140 Mb s^{-1} tributary signals, plus North American DS1, DS2 and DS3 signals. This means that SDH is completely backwards compatible with the existing networks. SDH can be deployed, therefore, as an overlay network supporting the existing network with greater network flexibility whilst the

Table 12.9 SONET rates

Frame format	Optical	Bit rate	Maximum DS0s
STS-1	OC-1	51.84 Mbps	672
STS-3	OC-3	155.52 Mbps	2016
STS-12	OC-12	622.08 Mbps	8064
STS-24	OC-24	1.244 Gbps	16 128
STS-48	OC-48	2.488 Gbps	32 256
STS-192	OC-192	9.953 Gbps	129 024

Table 12.10 SONET/SDH transmission rate comparison

SONET signal	SDH signal	Transmission rate
STS-1		51.84 Mbps
STS-3	STM-1	155.52 Mbps
STS-12	STM-4	622.08 Mbps
STS-24		1244.16 Mbps
STS-48	STM-16	2488.32 Mbps

transfer to SDH takes place. In addition SDH/SONET transport capabilities have flexibility to accommodate the more advanced network signals such as:

- asynchronous transfer mode (ATM – the standard for broadband ISDN),
- fibre distributed data interface (FDDI – a high speed local area network standard),
- distributed queue dual bus (DQDB – a metropolitan area network standard).

It is beyond the limitations of this chapter to describe more details of SDH/SONET, and the interested reader is advised to read References 18 and 20–23, and the relevant ITU or North American standards before the next section on testing.

12.7.7 Testing of digital hierarchy

12.7.7.1 Timing and synchronisation in digital networks

Based on the discussion of the previous sections, modern SDH/PDH telecommunication networks transport two basic entities – data and timing – as part of a service. As a result, timing has always been very carefully specified, controlled and distributed within networks, across network interfaces and between customers. To deliver the timing part of the service, the network must be properly synchronised. Compared with PDH technology, the new SDH technology in public networks around the world represents a quantum leap in performance, management and flexibility.

As a consequence, timing and synchronisation are of strategic importance to network operators as they work in the new de-regulated environment of the 1990s.

In the next few sections, essential elements of timing measurements are discussed.

12.7.7.2 Jitter and wander – definitions and related standards

Jitter and wander are defined respectively as '*the* ***short term*** *and the* ***long term*** *variations of the significant instants of a digital signal from their ideal positions in time*'. One way to think of this is to imagine a digital signal continually varying its position in time by moving backwards and forwards with respect to an ideal clock source. Most engineers' first introduction to jitter is as viewed on an oscilloscope (Figure 12.26). When triggered from a stable reference clock, jittered data are clearly seen to be moving in relation to a reference clock.

In fact, jitter and wander on a data signal are equivalent to a phase modulation of the clock signal used to generate the data (Figure 12.27). Naturally, in a practical situation, jitter will be composed of a broad range of frequencies at different amplitudes.

Jitter and wander have both an amplitude and a frequency. Amplitude indicates how much the signal is shifting in phase, while the frequency indicates how quickly the signal is shifting in phase. Jitter is defined in G.810 as phase variation with

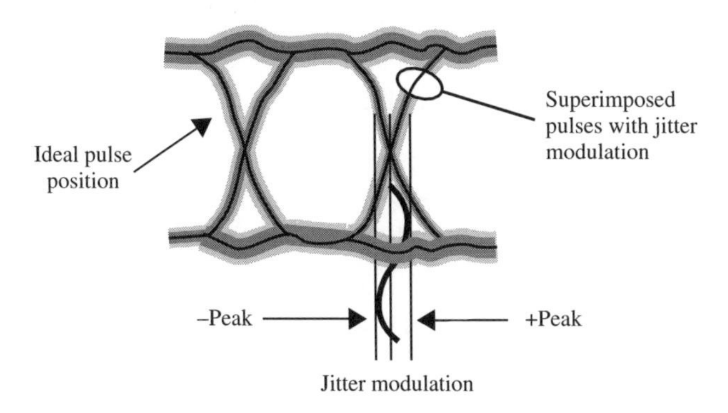

Figure 12.26 Jitter as viewed on an oscilloscope

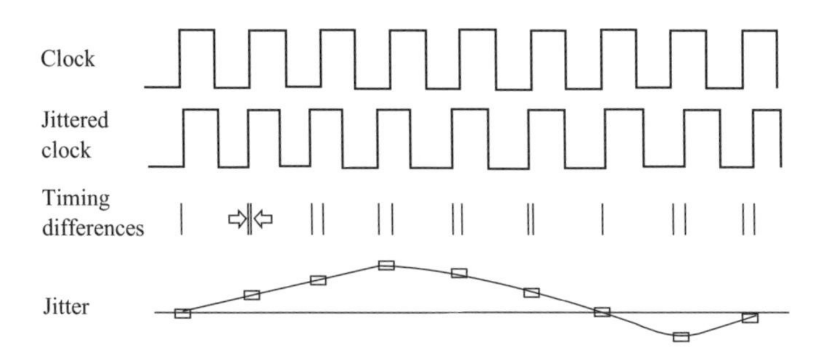

Figure 12.27 Phase variation between two signals

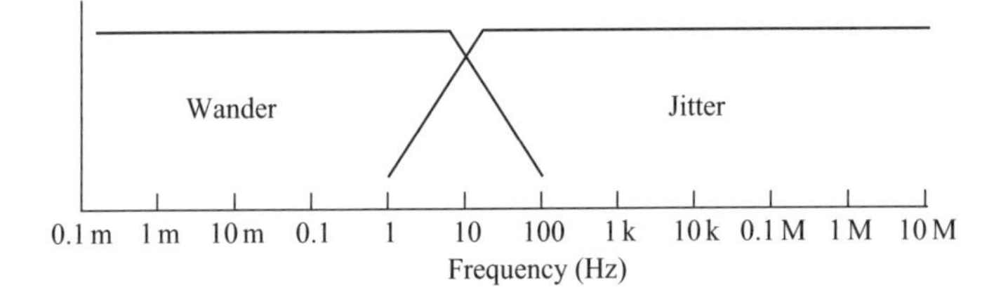

Figure 12.28 Frequency ranges of jitter and wander according to the ITU G.810 recommendation

frequency components greater than or equal to 10 Hz, whereas wander is defined as phase variations at a rate less than 10 Hz (Figure 12.28).

12.7.7.3 Metrics of jitter

Jitter is normally specified and measured as a maximum phase amplitude within one or more measurement bandwidths. A single interface may be specified by using several different bandwidths because the effect of jitter varies depending on its frequency, as well as its amplitude.

12.7.7.3.1 Unit intervals (UI)

Jitter amplitudes are specified in unit intervals (UI), such that one UI of jitter is one data bit-width, irrespective of the data rate. For example, at a data rate of 2048 kbit s^{-1}, one UI is equivalent to 488 ns, whereas at a data rate of 155.52 Mbit s^{-1}, one UI is equivalent to 6.4 ns.

Jitter amplitude is normally quantified as a peak-to-peak value rather than an RMS value, since it is the peak jitter that would cause a bit error to be made in network equipment. However, RMS values are also useful for characterising or modelling jitter accumulation in long line systems using SDH regenerators, for example, and the appropriate specifications use this metric instead of the peak-to-peak value.

12.7.7.4 Metrics of wander

A wander measurement requires a ‘wander-free’ reference, relative to which the wander of another signal is measured. Any primary reference clock (PRC) can serve as a reference because of its long term accuracy (10^{-11} or better) and good short term stability. A PRC is usually realised with a caesium-based clock, although it may also be realised with GPS technology.

Because it involves low frequencies with long periods, measured wander data can consist of hours of phase information. However, because phase transients are of importance, high temporal resolution is also needed. So to provide a concise measure of synchronisation quality, three wander parameters have been defined and are used

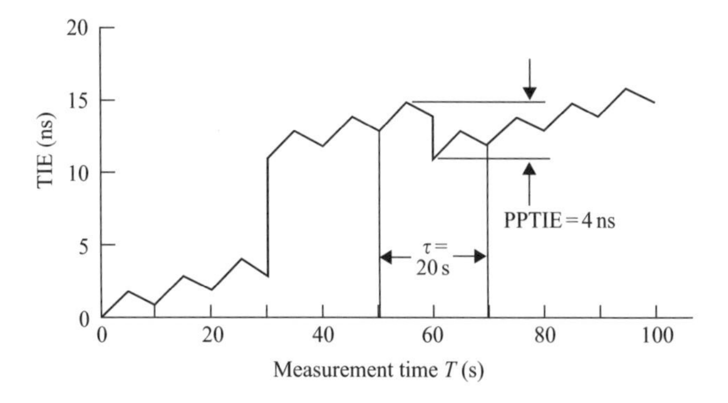

Figure 12.29 An example of TIE wander measurement (courtesy of Tektronix, Inc., USA)

to specify performance limits:

- **TIE:** time interval error (wander in ns),
- **MTIE:** maximum time interval error (related to peak-to-peak wander), and
- **TDEV:** time deviation (related to r.m.s. wander).

Formal mathematical definitions of these and other parameters can be found in G.810.

12.7.7.4.1 Time interval error (TIE)

TIE is defined as the phase difference between the signal being measured and the reference clock, typically measured in ns. TIE is conventionally set to zero at the start of the total measurement period T. Therefore TIE gives the phase change since the measurement began. An example is given in Figure 12.29. The increasing trend shown results from a frequency offset – about 1 ns per 10 s, or 10^{-10} in this case.

12.7.7.4.2 Maximum time interval error [MTIE]

MTIE is a measure of wander that characterises frequency offsets and phase transients. It is a function of a parameter τ called the *observation interval*. The definition of MTIE (τ) is the largest peak-to-peak TIE (i.e. wander) in any observation interval of length τ; see Figure 12.30.

In order to calculate MTIE at a certain observation interval τ from the measurement of TIE, a time window of length τ is moved across the entire duration of TIE data, storing the peak value. The peak value is the MTIE (τ) at that particular τ. This process is repeated for each value of τ desired. For example, Figure 12.29 shows a window of length $\tau = 20$ s at a particular position. The peak-to-peak TIE for that window is 4 ns. However, as the 20 s window slides though the entire measurement period, the largest value of PPTIE is actually 11 ns (at about 30 s into the measurement). Therefore MTIE (20 s) $= 11$ ns.

Figure 12.30(b) shows the complete plot of MTIE (τ) corresponding to the plot of TIE in Figure 12.29. The rapid 8 ns transient at $T = 30$ s is reflected in the value MTIE (τ) $= 8$ ns for very small τ.

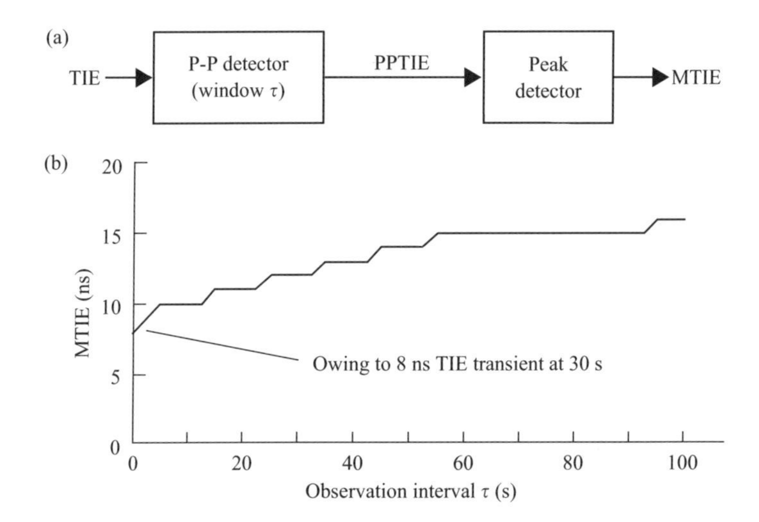

Figure 12.30 MTIE definition and measurement (courtesy of Tektronix, Inc., USA): (a) functional definition of MTIE; (b) plot of MTIE

It should be noted that the MTIE plot monotonically increases with observation interval and that the largest transient masks events of lesser amplitude.

12.7.7.4.3 Time deviation (TDEV)

TDEV is a measure of wander that characterises its spectral content. It is also a function of the parameter τ called *observation interval.* The definition (Figure 12.31(a)) is: TDEV (τ) is the r.m.s. of filtered TIE, where the band pass filter (BPF) is centred on a frequency of $0.42/\tau$.

Figure 12.31(b) shows two plots of TDEV (τ). The first plot (for $T = 100$ s), corresponding to the TIE data of Figure 12.29, shows TDEV rising with τ. This is because, for the short measurement period $T = 100$ s, the two transients in Figure 12.26 dominate.

If we were to make a longer TIE measurement out to $T = 250$ s, the effect of the two transients on TDEV would become less, assuming there are no more transients. The TDEV characteristic labelled $T = 250$ s would be the result. It should also be noted that TDEV is insensitive to constant phase slope (frequency offset).To calculate TDEV for a particular τ, the overall measurement period T must be at least 3τ. For an accurate measure of TDEV, a measurement period of at least 12τ is required. This is because the r.m.s. part of the TDEV calculation requires sufficient time to get a good statistical average.

12.7.7.5 Bit error rate testers (BERT)

A critical element in a digital transmission system is how error free its transmissions are. This measurement is made by a bit error rate tester (BERT), which replaces one or more of the system's components during a test transmission.

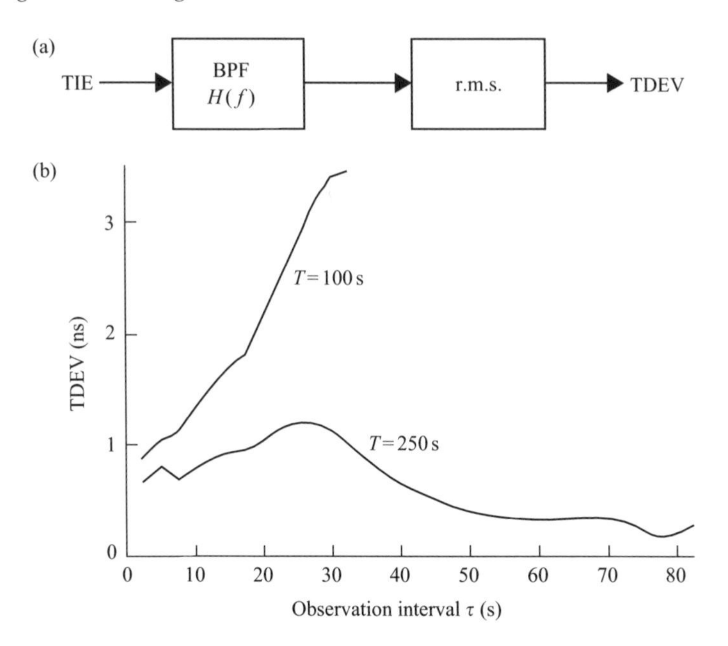

Figure 12.31 TDEV measurement: (a) functional definition; (b) example of TDEV wander measurement for case of Figure 12.26 (courtesy of Tektronix, Inc., USA)

A digital transmission system (see Figure 12.32) includes a data source – such as computer memory, a voice digitiser, or a multiplexer – that is the origin of a digital signal, D.

A clock source produces a clock signal, C, that times the occurrence of each bit in the digital signal. A driver, which may be a power amplifier, a laser diode, an RF modulator, or a tape head, prepares the signal for the system under test. The system under test can be a transmission line with repeaters, an optical fibre link, microwave radio link, or digital tape recorder. The received signal, F, exhibits the noise and pulse dispersion that the transmission system adds to the digital signal.

If the noise and distortion are within limits, the decision circuit can correctly decide whether the original bit was a 1 or a 0. The circuit does this by comparing F (at sampling instants determined by clock signal G) with a threshold halfway between the two levels (see Figure 12.30). If no errors are made in the decision process, H is a delayed replica of the original data signal D. A clock recovery circuit generates G from information in data signal F.

A malfunction in any system component can cause the recovered data to differ from the original data. The primary job of a BERT is to determine the system's error rate rather than to isolate the faulty component. But for the sake of convenience, the BERT may replace the clock source in the transmitter or receiver. In this case, some fault isolation may be possible by comparing the performance of the system clock

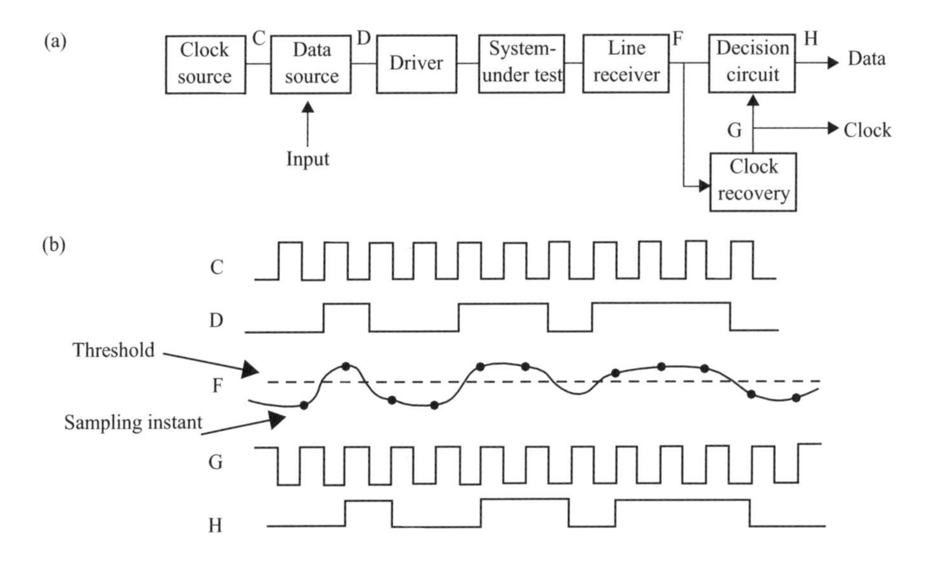

Figure 12.32 A typical digital transmission system and its signals: (a) block representation of system; (b) interrelations or signals (courtesy of Tektronix, Inc., USA)

sources with that of the BERT. But for the comparison to be meaningful, users must understand the timing jitter specifications of both units. To measure the system's error rate, the test set performs one or more of the following pairs of functions:

- data pattern generation and error monitoring;
- clock generation and recovery; and
- jitter generation and measurement.

Which functions are used depends on how the BERT is connected in the system.

12.7.7.5.1 Detection of data errors

The simplest measuring technique (see Figure 12.33) is to replace the system's data source with the BERT's data pattern generator and have the BERT receiver monitor the recovered signal for errors.

The data signal D then becomes D′. The data pattern generator can mimic typical traffic by creating pseudorandom patterns, or it can stress the system by outputting fixed patterns stored in memory. To monitor the transmission, the BERT receiver generates its own data pattern H′, which is the same as the desired data D′. The BERT receiver compares the received signal H, with H′, and looks for errors. The tester records the total number of errors, the ratio of errors to bits (the bit error rate), the number of '*errored seconds* (ES)', and the ratio of ES to total seconds. To make a valid comparison, the BERT receiver must synchronise H′ with H. Accomplishing synchronisation depends on whether the data are a fixed or pseudo-random pattern.

Supplying clock Sometimes it is convenient for the BERT to supply its own clock signals for its transmitter and/or receiver. For instance, the system clock may be unavailable in a field situation, or the test engineer may want to avoid the trouble of providing and phasing the clock at the BERT receiver. In this case, the BERT's transmitter clock is C′, and its receiver is G′ (see Figure 12.33(b)). In laboratory applications, it is common for the BERT to provide a wide range of clock frequencies.

The BERT's clock source and clock recovery circuit must be as good as their counterparts in the system under test. The source must introduce negligible timing jitter, because phase jitter in the recovered clock signal, G, is relative to the received data signal, F. Likewise, the BERT's clock recovery circuit must tolerate at least as much jitter as the system's recovery circuit without causing errors.

Stressing with jitter Although the BERT clock source should be essentially jitter free to test the digital transmission system under normal conditions, users may wish to stress the system at times. In that case, the BERT must generate controlled jitter. To do so, some BERTs have a jitter generator that can sinusoidally modulate the phase of the clock source (see Figure 12.33(b)).

On the receive end, the BERT monitors the effect of the controlled jitter in two ways. First it looks for an increased error rate. Second, it measures the jitter remaining in the recovered data, which yields the system's jitter transfer function. The jitter

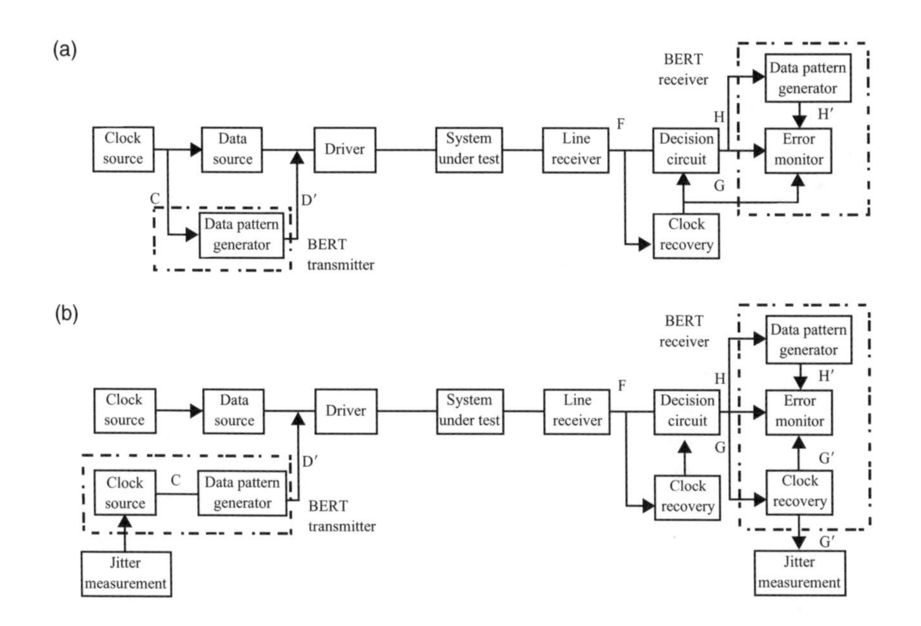

Figure 12.33 *A simple application of a BERT: (a) BERT generating a known data pattern D′; (b) BERT supplying both clock and data D′ (courtesy of Tektronix, Inc., USA)*

measurement circuit can also be used without the jitter generator to measure the system's own jitter.

For more details of BERT applications, Reference 24 is suggested.

12.7.7.6 Broadband network measurements

Figure 12.34 illustrates a simple SDH/PDH network model. A PDH circuit is transported over an SDH path, whilst being multiplexed with other PDH circuits, cross-connected with other SDH payloads and regenerated. This model network is synchronised from a logically separate sync network, although it is likely that the sync signals will be physically carried on parts of the SDH network. In such synchronous networks the incoming STM-N channels (of higher data rates) signals can have slow phase movement (defined as wander) relative to the reference.

The causes of such wander in a synchronous system are as follows.

- Changes in cable delay due to temperature changes.
- Drift due to d.c. offsets in the phase-locked loops of the synchronisation clocks.
- Random phase walk or phase transients as the synchronisation distribution chain is reconfigured, either manually or as a result of protection switching.
- Frequency differences due to loss of synchronisation at a node within the network (as much as 4.6 parts in 10^6).

12.7.7.6.1 Synchronisation of distribution networks

The timing distribution chain in an SDH network is designed to minimise the phase movement between synchronisation references. A typical timing distribution chain is shown in Figure 12.35 according to the relevant standards [24] such as ETS 300 462-2.

The primary reference clock (PRC) is very stable; its frequency is generally accurate to one part in 10^{11}, and it has very little short term wander. The timing signal is

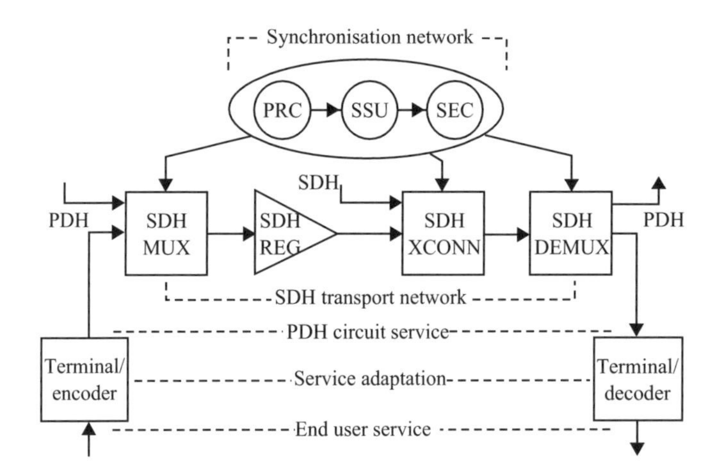

Figure 12.34 A model broadband network

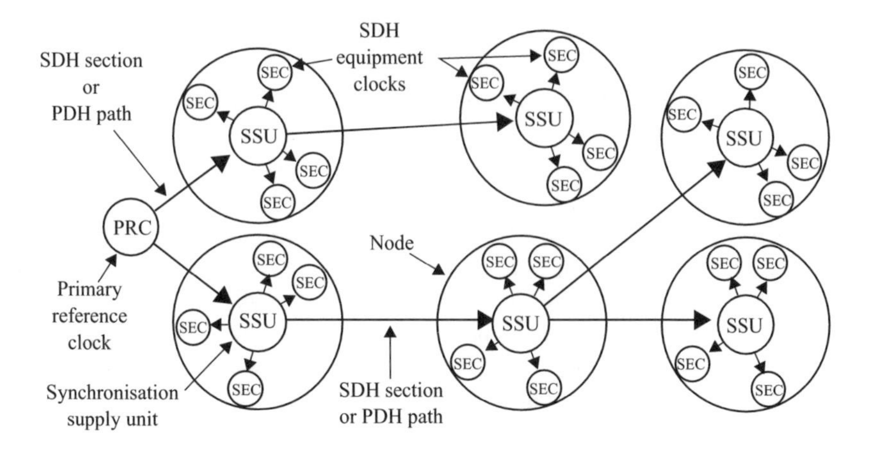

Figure 12.35 Synchronisation network architecture for SDH systems (courtesy of Tektronix, Inc., USA)

distributed (by either STM-N sections or PDH paths) to slave clocks called *synchronisation supply units* (SSUs). There is usually one SSU at a physical site, called a 'node', that supplies timing to all the *SDH equipment clocks* (SECs) at that node.

Time for new clocks An SSU selects one of several available synchronisation sources (e.g. from a PRC or GPS receiver) and distributes that clock to locally situated equipment. If that source fails, it selects another, or may use its own internal clock, a mode known as holdover. An SEC selects one of several available synchronisation sources (e.g. an SSU) and distributes that clock within its own equipment. If that source fails, it selects another, or may switch to holdover mode using its own internal clock.

An SSU can derive its timing from a number of sources, as shown in Figure 12.36 (this is based on the ETSI architecture described in ETS 300 462-2). The phase locked loop in the SSU provides filtering to smooth the outgoing phase (using a low pass filter with bandwidth less than 3 mHz). In the event that it loses all its timing inputs, the SSU can enter holdover mode in which it tries to maintain phase and frequency based on memory of past timing input.

The above details should give the reader the complexity of clock system for the need for synchronisation. A comprehensive description of these techniques and their measurement related parameters are described in Reference 25.

12.7.7.6.2 Jitter and wander testing on SDH systems

Traditional jitter tests Jitter testing has been an established part of the telecom industry for a long time, and several tests are popular and well known from their previous use in PDH systems:

- output jitter,
- input jitter tolerance, and
- jitter transfer function.

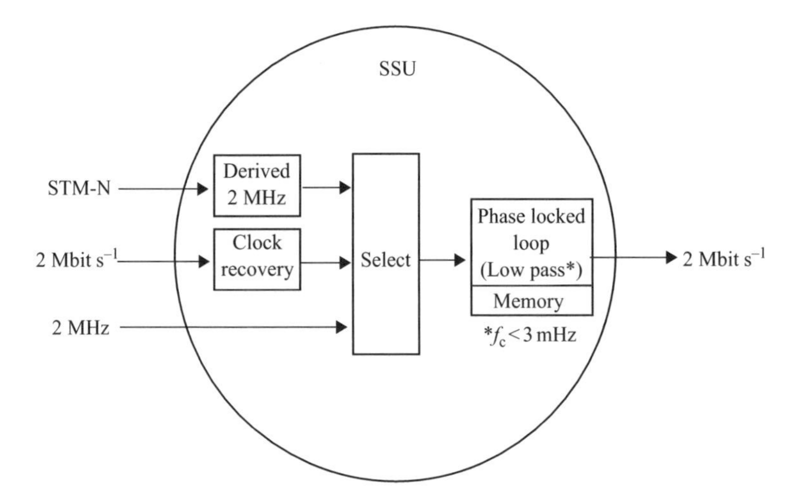

Figure 12.36 The SSU clock function

New jitter tests Several new jitter tests required for SDH systems are providing new challenges for test equipment. These new sources of jitter are so significant that 50 per cent of the allowed jitter at a demultiplexer PDH output is allocated to them:

- pointer jitter, and
- mapping jitter.

Pointer jitter is considered to be a more significant source of timing impairment than mapping jitter.

New wander tests New wander tests are rapidly increasing in importance as SDH is more widely deployed:

- output wander,
- input wander tolerance,
- wander noise transfer,
- phase transient response, and
- holdover performance.

ITU-T test equipment recommendations ITU-T Recommendations 0.171 and 0.172 are two specifications for test equipment for PDH and SDH, as follows.

- 0.171 – Timing jitter and wander measuring equipment for digital systems which are based on the PDH.
- 0.172 – Jitter and wander measuring equipment for digital systems which are based on the SDH.

Recommendation 0.172 has been developed in order to specify measurement requirements for the whole SDH network – that is, both SDH line interfaces operating at PDH bit rates (simply referred to as 'tributaries'); Figure 12.37 illustrates this concept.

For details, Reference 26 is suggested.

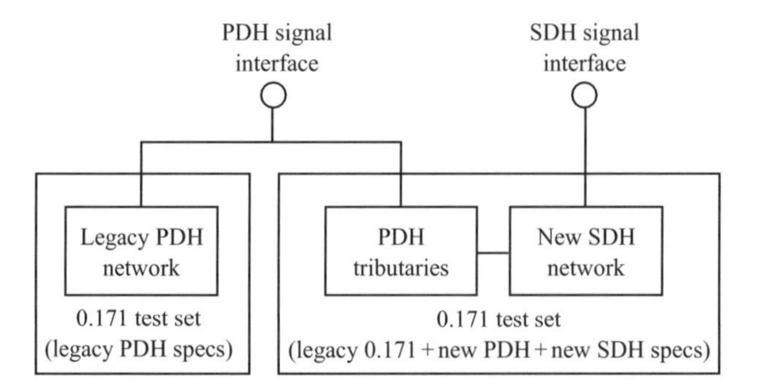

Figure 12.37 Relationship of 0.172 to 0.171 and application to SDH/PDH (courtesy of Tektronix, Inc., USA)

12.7.7.6.3 Use of sampling oscilloscopes for physical layer testing or tributary signals

Compared with BERTs, digital oscilloscopes with high sampling rates such as the DPO series from Tektronix could be used in debugging communications systems. BERTs can be very effective in finding errors that occur very infrequently, but only if the communication device under test is working well enough to accept inputs and/or transmit data. If the design is not working, a BERT cannot help a designer determine the source of the problem. When a transmitter is not working or partially working, an oscilloscope is invaluable for debugging the design. The DPO's graphical display of the digital data makes finding the problem easier. By contrast, the BERT receiver typically only has a numeric display that shows bit error rate.

In many cases, the advanced triggering available in a DPO allows quick location of signal errors. The TDS Series DPOs offer five different categories of advanced triggers including communications, logic, and pulse width triggering. Many of these advanced triggers can be used to quickly debug a communications problem. For example, Figure 12.38 shows an alternate mark inversion (AMI) encoded signal that has a polarity violation on the eighth pulse in the pulse sequence. This polarity violation could be found using the TDS oscilloscope's timeout trigger function. In timeout trigger operation, the TDS triggers if it has not seen a high-to-low transition in the specified period of time.

For more details on use of digital oscilloscopes for telecom tributary signals, Reference 27 is suggested.

12.8 PCB designs and measurements at high frequencies

In electrical circuits, lumped elements permit the use of Kirchhoff's circuit laws; with distributed elements, Kirchhoff's laws fail and the mathematics of the system become more complex. Consider, for example, applying Kirchhoff's current law to a transmitter antenna where current enters but does not exit – at least in the classical sense. Digital system speeds in networking and telecommunications are continuing

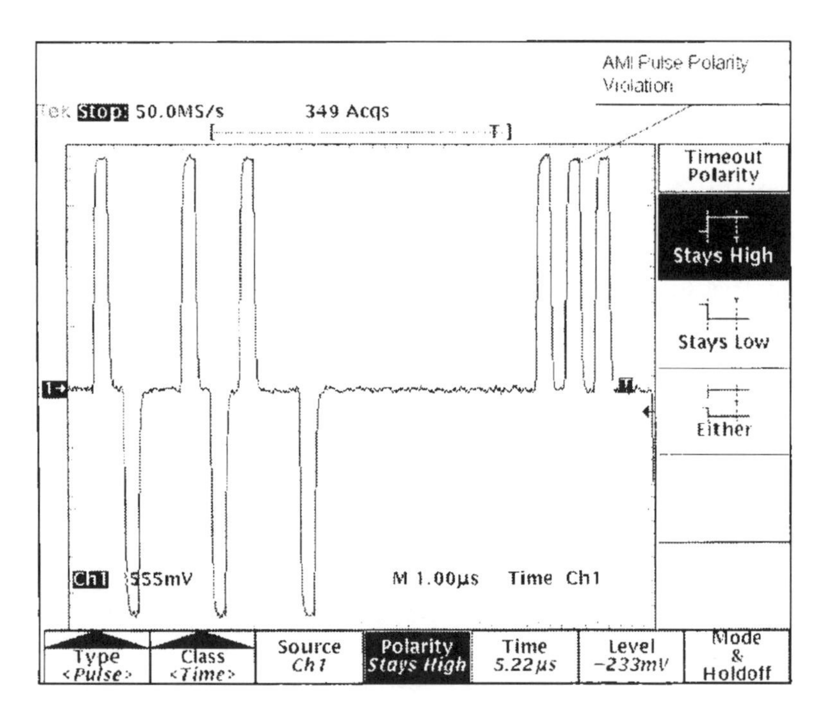

Figure 12.38 AMI pulse violation capturing by a DPO series oscilloscope (courtesy of Tektronix, Inc., USA)

to increase, and the rate at which these speed increases occur is also accelerating. In a large measure, this trend is occurring because of the movement from parallel buses, clocked at less that 100 MHz, to serial signal paths that operate at multiple gigabits per second. Currently, some systems operate at 1.2 Gbps, and several companies are working to push speeds to 10 Gbps.

At these speeds, interconnection structures, such as printed-circuit boards and connectors, can negatively affect data signal integrity. To successfully transmit data at multi-gigabit speeds, designers must simulate their systems and often must use unorthodox construction techniques.

Whether a designer should treat a system as a transmission line or a lumped parameter system in analysis the systems can be considered in either a frequency domain or a time domain perspective.

When using frequency domain analysis, if the length of the interconnection is greater than 1/15th of the wavelength of the signal it carries, consider the interconnection to be a transmission line. The frequency domain relationship applies to wideband signals whose wavelength is that of the highest frequency present, f_{max}.

When using time domain analysis, if the time required for the signal to travel the length of the interconnection is greater than 1/8th of the signal transition time, consider the interconnection to be a transmission line. In this domain, transition time refers to either the rise or fall time, whichever is smaller.

For the details of this treatment and analysis, References 1–3 and 28 are suggested; Reference 28 is an excellent text for high speed digital design.

12.8.1 Signal integrity and system design

In electronics, signal integrity means that the signal is unimpaired with regard to functionality. For example, a system with good signal integrity has data that arrive early enough to guarantee that setup timing requirements are met and that clocks make only one logic transition per physical transition. Digital systems are tolerant of many signal integrity effects, including delay mismatches and signal ringing. However, sufficiently large signal integrity problems can cause systems to fail or, worse yet, work only intermittently.

One common signal integrity problem is false clocking, where a clock crosses a logic threshold more than once on a transition, owing to ringing on the line. Some parameters related to product design and signal integrity are described in Table 12.11.

12.8.1.1 Clock skew and transmission line effects

The most difficult problem in high speed design is the clock skew. In a digital system, the clock must be distributed to every IC that is operating at the processor speed. Because these ICs can number 15 or more in a typical system, a designer is expected to use buffers with the crystal clock reference. Clock skew is the time differential between edge transitions for different buffered outputs.

Buffers generate three types of clock skew:

- intrinsic,
- pulse, and
- extrinsic.

Table 12.11 Definitions of signal integrity and related terms

Term	Description
Signal integrity	Signal integrity is the ability of a signal to generate correct responses in a circuit. A signal with good signal integrity has digital levels at required voltage levels at required times.
Cross-talk	Cross-talk is the interaction between signals on two different electrical nets. The one creating cross-talk is called an aggressor, and the one receiving it is called a victim. Often, a net is both an aggressor and a victim.
Overshoot	Overshoot is the first peak or valley past the settling voltage – the highest voltage for a rising edge and the lowest voltage for a falling edge.
Undershoot	Undershoot is the next valley or peak. Undershoot is the second peak or valley past the settling voltage – the deepest valley for a rising edge and the highest peak for a falling edge.
Skew	Signal skew is the difference in time in arrival of one signal to different receivers on the same net. Skew is also used to describe the difference in arrival time between clock and data at a logic gate.

Intrinsic skew is the differential between outputs on the same IC. Clock buffers also generate *pulse skew*, which is the difference in propagation delay between a low-to-high transition and a high-to-low transition for a single buffer output. *Extrinsic skew* is external to the buffers and it is the time differential that occurs when you send a fast-rise-time signal on board traces that have different lengths and capacitive loads.

To compensate for various types of clock skew related delays, designers have to use transmission line principles related design approaches. PC board transmission lines come in two flavours, namely strip lines and micro-strip lines. The clock driver must drive a line's characteristic impedance (Z_0), which is a function of trace geometry and board's dielectric constant (ε_r). For details, References 28–32 are suggested.

12.8.2 PCB interconnection characterisation

As the performance requirements for modern computer and communications systems grow, the demand for high speed printed circuit boards also increases. Speeds as fast as 1 Gbit s^{-1} are expected to be supported by standard PCB technologies, with the rise times of these signals being as fast as 100 ps.

At these speeds, interconnections on PCBs behave as distributed elements, or transmission lines, and reflections due to impedance mismatch are typical signal integrity problems that board designers encounter in their work. Vias between layers and connectors on a board create discontinuities that distort the signals even further. To predict accurately the propagation of the signals on board, designers need to determine the impedance of their traces of different layers and extract the models for board discontinuities.

Even at relatively low speed CMOS processor board designs, beyond 40 MHz transmission line effects and clock skews, etc., come into play [31]. In these situations, use of TDR techniques with high speed probing could assist in characterisation. For details of high speed probing for digital designs, Reference 33 is suggested.

In a case where one uses a 500 MHz probe and a 500 MHz scope for a measurement of 1 ns pulse the rise time, owing to probe output feeding the scope, the rise time at the probe output could be degraded to about 1.21 ns. The same measured on the 500 MHz scope can be 1.38 ns. For details, refer to chapter 4.

For details of planning for signal integrity and identifying the bandwidth limitations of PCB traces References 30 and 31 are proposed.

12.8.2.1 TDR techniques

Time domain reflectometry (TDR) measurements have always been the measurement approach of choice for board characterisation work. Based on TDR measurements, a circuit board designer can determine characteristic impedances of board traces, compute accurate models for board components and predict board performance more accurately.

For example, to compute the impedance profile for a PCB trace, a TDR technique can be used as shown in Figure 12.39.

In Figure 12.39, the incident waveform amplitude at the device under test (DUT) is typically half the original stimulus amplitude V at the TDR source. The smaller

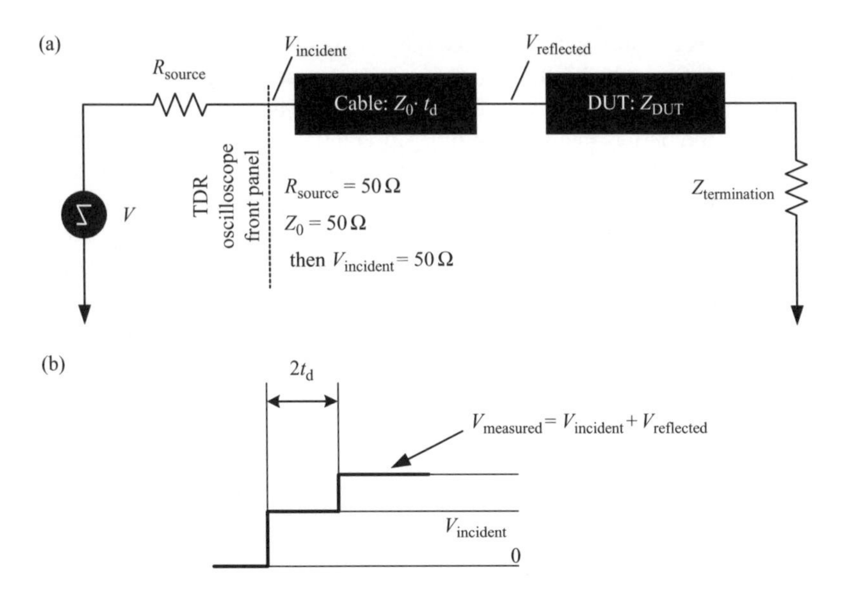

Figure 12.39 TDR oscilloscope equivalent circuit and waveform: (a) basic setup; (b) waveform

DUT incident waveform amplitude results from the resistive divider effect between the 50 Ω resistance of the source and 50 Ω impedance of the coaxial cables connecting the TDR sampling head and the DUT.

The impedance of the board trace can be determined from the waveform measured by the TDR oscilloscope, V_{measured}, which is the superposition of the incident waveform at the DUT and the reflected one, offset by two electrical lengths of the cable interconnecting the oscilloscope TDR sampling head to the DUT.

Traditionally, the impedance of the board is computed from TDR measurements by using an equation that relates the DUT impedance to the impedance of the cable interconnecting the TDR oscilloscope sampling head to the DUT.

For details, References 34 and 35 are suggested.

12.8.2.2 Quality measurements of PCB material

Transmission measurements can also be used as a means of PCB characterisation. For example, by constructing 50 and 75 Ω transmission lines of different geometries one could determine the suitability of various PCB materials for gigabits per second rate signal transmission.

For example, in an experiment carried out by Amp, Inc. [36], several sets of PCB samples using different dielectric material is summarised below. Table 12.12 depicts the material properties based on manufacturers' data.

In this example, using different dielectric materials allowed the designers to quantify how materials other than the traditional FR4 affect the behaviour of high

Table 12.12 Material properties based on manufacturers' data (source: EDN, Nov 11, 1999; Reproduced with permission of Cahners Business Information)

Material	Dielectric constant	Loss tangent	Transition temperature (°C)	Relative cost
Nelco 4000-6	4.4	0.018	180	1
Getek ML200	3.9	0.012	180	1.1
Rogers 4350	3.5	0.004	280	2.1
Arlon CLTE	2.9	0.0025	288	6.8

speed signals. After investigation, the designers selected woven-glass-epoxy-resin, polyphenylene-oxide (PPO), ceramic, and glass-reinforced-polytetrafluoroethelene (PTFE) materials.

The first set of boards used Nelco 4000-6, a high performance type of FR4 that belongs to the woven-glass-epoxy-resin material family. The second set used Getek ML200, a PPO-type material. The third set used Rogers 4350 ceramic material. The final set of boards used Arlon CLTE, a glass-reinforced PTFE material.

12.8.2.2.1 Time vs frequency data

In this case the initial approach was to measure system output eye patterns in the time domain. Eye pattern measurements in the time domain, however, may not provide all the design information. Also extracting material and transmission line parameters is easier when you use frequency domain data than time domain data.

In this case where time domain testing was limited to about 3 Gbps, a network analyser was used for frequency domain measurements up to 6 GHz (which can correspond to digital data as high as 10 Gbps). This enabled the calculation of key dielectric – material parameters such as dielectric constant and loss tangent.

Fourier transform algorithms can easily transform frequency data into better understood time-domain data. With this approach, the engineers could easily separate, or 'de-embed', the effects of the test structures.

12.8.2.2.2 De-embedding test points

In order to remove the effects of the test points, which contributed a significant error on the measurements due to impedance mismatches, a through reflect line (TRL) de-embedding technique was used. This technique characterises and removes from the overall measurements the repeatable measurement error that the test points introduce.

Figure 12.40(a) depicts the PCB traces on the test board that included a 16 inch section of a 0.012 inch wide trace that had a characteristic impedance of 50 V. This trace passed through HS3 connectors at each end and was joined to a test point at each end by a 1 inch section of a 0.012 inch wide trace. Figure 12.40(b) shows the measured

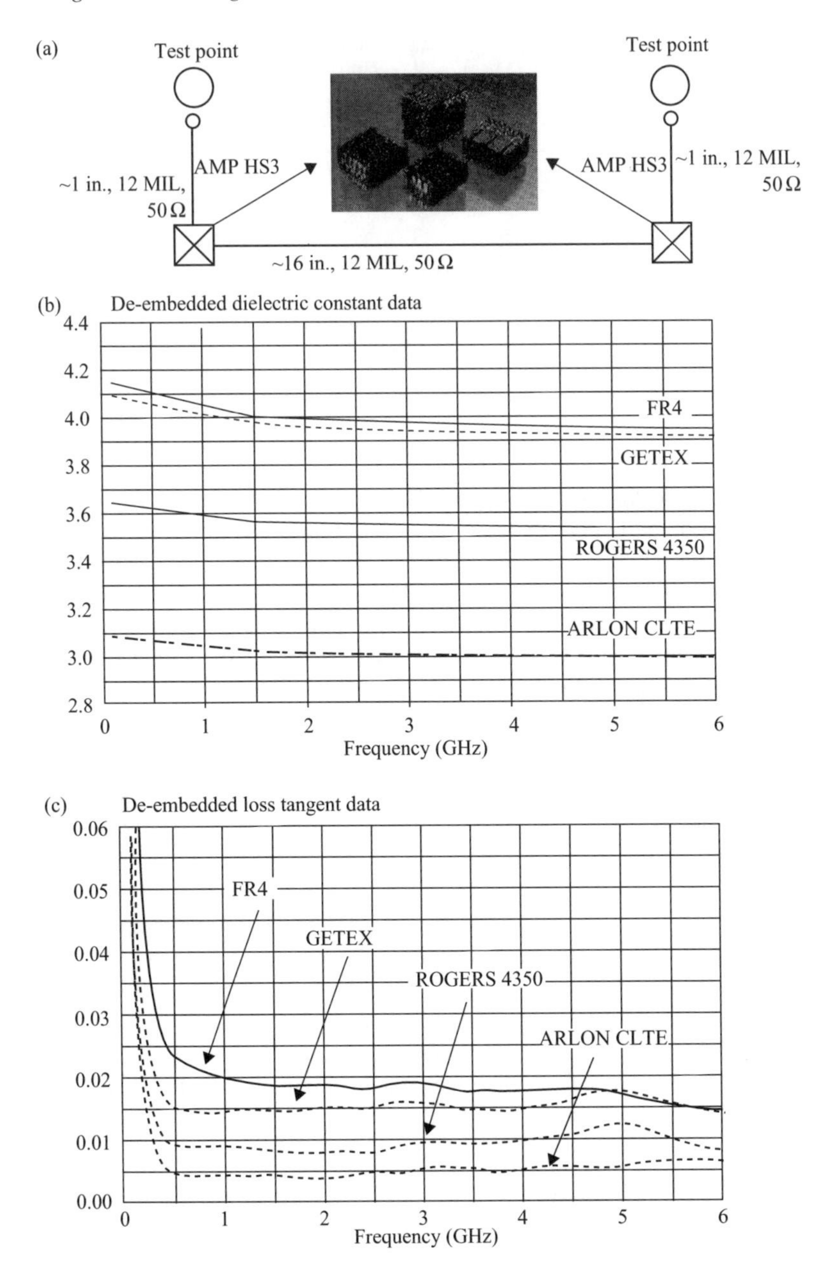

Figure 12.40 Test system and measured results with de-embedding: (a) test board; (b) dielectric constant; (c) loss tangent (source: EDN, Nov 11, 1999; Reproduced with permission of Cahners Business Information)

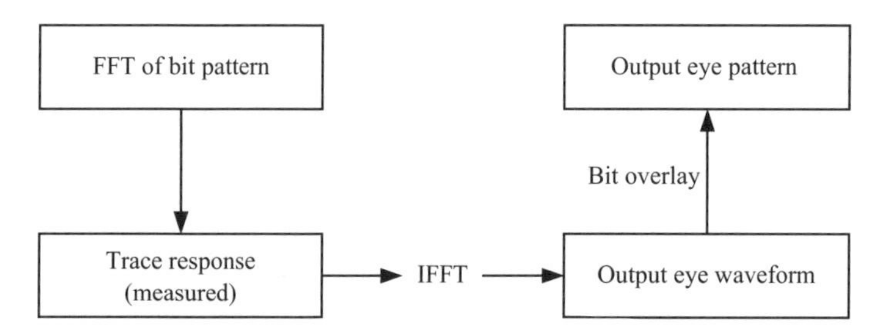

Figure 12.41 *Technique of eye pattern derivation from frequency response data to time domain (source: EDN, Nov 11, 1999; Reproduced with permission of Cahners Business Information)*

de-embedded dielectric constant for each of the board materials. Figure 12.40(c) shows the corresponding loss tangents.

Note how each of the basic material parameters varies with frequency. One can easily overlook this frequency dependence because most material manufacturers provide only one pair of supposedly constant values of dielectric constant and loss tangent for a material.

The values in Figures 12.40(b) and (c) are slightly higher than the specifications by the material manufacturers. This difference between predicted and measured values probably results from the changes that take place in a dielectric material during board manufacturing. The data in Figures 12.40(b) and (c) represent the most fundamental properties of each dielectric material. Knowing the material's dielectric constant and loss tangent allows accurate modelling of a lossy transmission line structure.

12.8.2.2.3 Time domain transform techniques

Since time domain representation is more comfortable for designers, conversion techniques from frequency to time domain are useable in these types of experiment.

Figure 12.41 shows the technique that transforms the frequency domain data into time domain eye patterns. The process involves several steps. The input bit pattern is transformed to the frequency domain and then multiplied by the circuit's frequency response. The product of this multiplication is then (inverse) transformed back to the time domain. The result is the output bit pattern that you would see if the input bit pattern had travelled through the interconnect. Overlaying this bit pattern's individual bits produces the output eye pattern.

Comparing a measured 3 Gbps eye pattern with a transformed 3 Gbps eye pattern validated this method (Figures 12.42(a) and (b)). Figure 12.42(a) shows the directly measured eye pattern of an 18 inch path in FR4. Figure 12.42(b) shows an eye pattern created for the same path by transforming frequency domain measurements into the time domain. The almost exact correlation provides confidence in the transform technique.

For more details on this, Reference 36 is suggested.

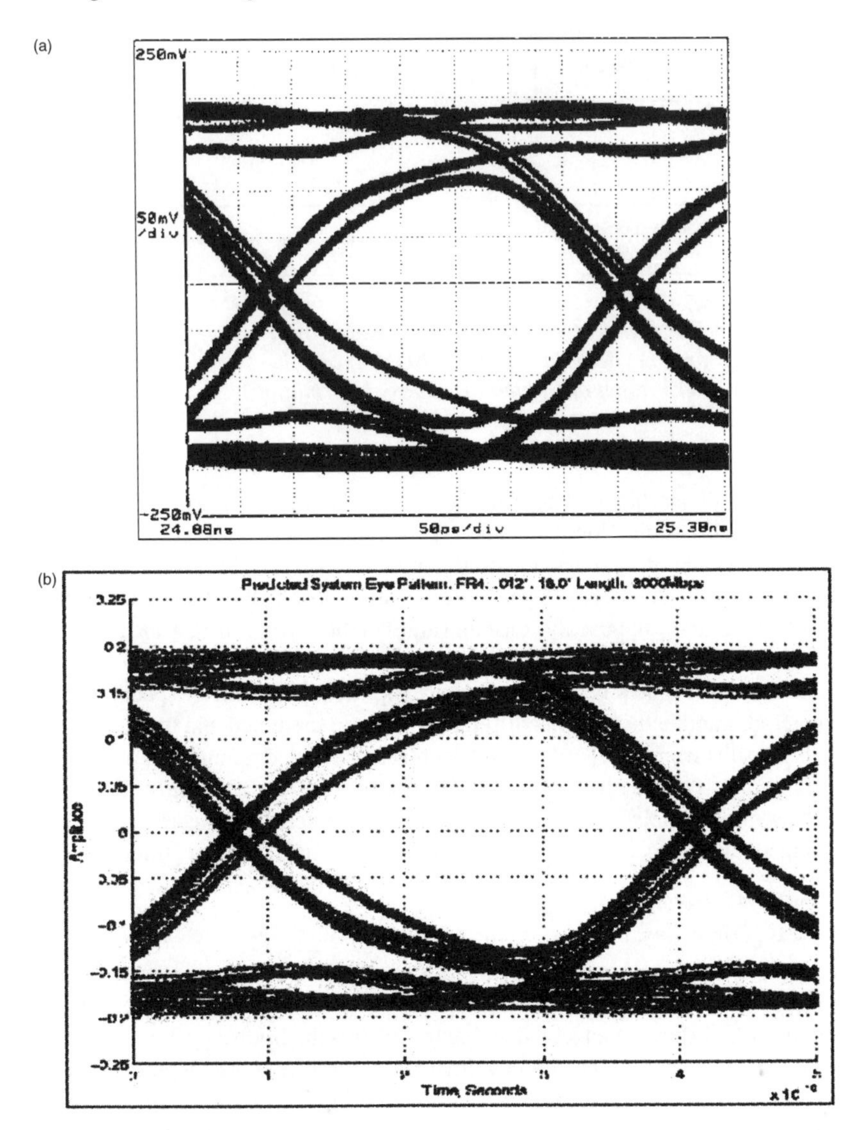

Figure 12.42 *Comparison of eye patterns: (a) measured; (b) transformed (source: EDN, Nov 11, 1999; Reproduced with permission of Cahners Business Information)*

12.9 References

1 ROYLE, D.: 'Rules tell whether interconnections act like transmission lines', *EDN*, 23 June 1988, pp. 131–136.

2 ROYLE, D.: 'Correct signal faults by implementing line-analysis theory', *EDN*, 23 June 1988, pp. 143–144.

3 ROYLE, D.: 'Quiz answers show how to handle connection problems', *EDN*, 23 June 1988, pp. 155–159.
4 HART, B.L.: *Digital Signal Transmission Line Technology*, Chapman & Hall, London, 1988.
5 KULARATNA, N.: *Modern Electronic Test & Measuring Instruments*, IEE, London, 1996.
6 TANT, M.J.: *The White Noise Book*, Marconi Instruments, UK, 1974.
7 REEVE, W.: *Subscriber Loop Signaling Handbook–Digital*, IEEE Press, 1995.
8 BRYANT, G.H.: *Principles of Microwave Measurements*, Peter Perigrinus, London, 1993, 2nd edn.
9 MAZDA, F.F.: *Electronic Instruments and Measurement Techniques*, Cambridge University Press, Cambridge, UK 1987.
10 LOSER, A.: 'New thermocouple power sensors NTV-Z51 and NRV-Z52 based on semiconductor technology', News from Rohde & Schwartz, No.139, 1992/IV, Volume 32, p. 34.
11 ROHDE & SCHWARTZ: 'Power meter NRVS', Tech. brochure/application note.
12 REEVE, W.: *Subscriber Loop Signaling Handbook – Analog*, IEEE Press, 1992.
13 DUTTA-ROY, AMITAVA: 'A second wind for wiring', *IEEE Spectrum*, 1999, pp. 52–60.
14 KEMPAINEN, S.: 'ADSL; The end of the wait for home internet', *EDN*, October 10, 1996, pp. 53–70.
15 CZAJKOWSKI, I.K.: 'High speed copper access: A tutorial overview', *Electronics and Communication Engineering, IEE*, June 1999, pp. 125–48.
16 ROWE, M.: 'ADSL products must confirm and perform', *Test & Measurement World*, 1997, pp. 39–48.
17 ROWE, M.: 'ADSL testing moves out of the lab', *Test & Measurement World*, April 1997, pp. 46–54.
18 WINCH, R. G.: *Telecommunication Transmission Systems*, McGraw Hill, 1993.
19 TELECOMMUNICATIONS TECHNIQUES CORPORATION: 'T1 Basics'. Technical Note: TIBTN – 11/94.
20 HEWLETT PACKARD: 'Introduction to SDH'. HP application note 5091-3935E (2/92).
21 TELECOMMUNICATIONS TECHNIQUES CORPORATION: 'The fundamentals of SDH'. TTC Application note I-SDH TN, 5/95.
22 TELECOMMUNICATIONS TECHNIQUES CORPORATION: 'The fundamentals of SONET'. TTC Application note SONET TN, 7/95.
23 BELLAMY, J.: *Digital Telephony*, 2nd Edition, John Wiley, 1991.
24 TEKTRONIX, INC.: 'Ensure accuracy of bit-error rate testers', Technical brief [.........].
25 TEKTRONIX, INC.: 'Performance assessment of timing and synchronization in broadband networks', Application note – FL 5336/XBS 2GW-11145-0, 4/97.
26 TEKTRONIX, INC.: 'New ITU-T 0.172 defines jitter and wander test equipment for SDH systems', Technical brief – TD/XBS 2GW-13178-0, 6/99.
27 TEKTRONIX, INC.: 'Testing telecommunications tributary signals', Application note – TD/XBS 55W – 12045-0, 5.98.

28 JOHNSON, H. and GRAHAM, M.: *High Speed Digital Design*, Prentice Hall, USA, 1993, ISBN 0-13-395724-1.
29 GALLANT, J.: '40 MHz CMOS circuits send designs back to school', *EDN*, March 2, 1992, pp. 67–75.
30 JOHNSON, H.: 'Planning for signal integrity', *Electronic Design*, May 12 1997, p. 26.
31 JOHNSON, H.: 'Transmission-line scaling', *EDN Europe*, February 1999, p. 26.
32 SUTHERLAND, J.: 'As edge speeds increase, wires become transmission lines', *EDN*, 1999, pp. 75–94.
33 JOHNSON, H.: 'Probing high speed digital designs', *Electronic Design*, March 17, 1997, pp. 155–162.
34 SCHMITT, R.: 'Analyze transmission lines with almost no maths', *EDN*, March 18, 1999, pp. 143–150.
35 SMOKYANSKY, D. and COREY, S.: 'PCB interconnect characterization from TDR measurements', *Electronic Engineering*, July 1999, pp. 63–68.
36 MORGAN, C. and HELSTER, D.: 'New printed-wiring board material guard against garbled giga bits', *EDN*, November 11 1999, pp. 73–82.

Chapter 13
Digital signal processors[1]

13.1 Introduction

During the period 1975–1985 many instruments designers recognised the value of microprocessors in the design of instruments. Digital storage scopes, function generators, and frequency counters, etc., were the early families of instruments that made use of the microprocessor subsystems. High performance analogue oscilloscopes were another classic example of use of microprocessor subsystems for add-on features and performance. FFT analysers were yet another kind of instruments to use digital signal processors (DSPs).

Early generations of 4- and 8-bit microprocessors have evolved into 16-, 32- and 64-bit components with complex instruction set computer (CISC) or reduced instruction set computer (RISC) architectures. DSPs can be considered as special cases of RISC architectures or sometimes parallel developments of CISC systems to tackle real time signal processing needs. Over the past two decades, the field of digital signal processing has grown from a theoretical infancy to a powerful practical tool and matured into an economical yet successful technology. At the early stages, audio and the many other familiar signals in the same frequency band appeared as a magnet for DSP development. The 1970s saw the implementation of signal processing algorithms especially for filters and fast Fourier transforms (FFT) by means of digital hardware, developed for the purpose. Early sequential program DSPs are described in Reference 1. In the late 1990s the market for DSPs was mostly generated by wireless, multimedia and several other applications. According to industry estimates, by the year 2001, the market for DSPs had been expected to grow up to about US$ 9.1 billion [2]. Currently, communications represents more than half of the applications for DSPs [2].

[1] This chapter is an edited version of chapter 5 of the Butterworth/Newnes book 'Modern Component Families and Circuit Block Design', reproduced with permission of the publisher.

With the rapid developments related to data converter ICs and other mixed signal components allowing the conversion of analogue real time signals into digital, instrument designers saw the advantage of using DSP chips inside the instruments. FFT analysers, the latest families of DSOs and new spectrum analysers, etc., have made use of these low cost parts to advantage the real time processing of digitised signals. In many cases both DSP and microprocessor chips were used inside the instruments. This chapter provides an essential guide for designers to understand the DSPs, and briefly compares microprocessors and DSPs.

13.2 What is a DSP?

A digital signal processor (DSP) accepts one or more discrete time inputs, $x_i[n]$, and produces one or more items of output, $y_i[n]$, for $n = \ldots, -1, 0, 1, 2, \ldots,$ and $I = 1, \ldots, N$, as depicted in Figure 13.1(a). The input could represent appropriately sampled (and analogue-to-digital converted) values of continuous time signals of interest, which are then processed in the discrete time domain to produce outputs in discrete time that could then be converted to continuous time, if necessary. The operation of the digital signal processor on the input samples could be linear or non-linear, time invariant or time varying, depending on the application of interest. The samples of the signal are quantised to a finite number of bits, and this word length can be either fixed or variable within the processor. Signal processors operate on millions of samples per second, require large memory bandwidth, and are computationally very demanding, often requiring as many as a few hundred operations on each sample processed. These real time capabilities are beyond the capabilities of conventional

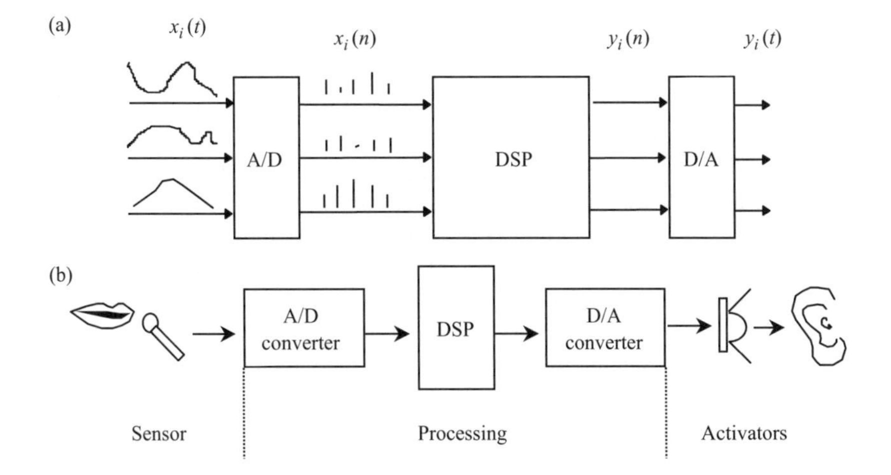

Figure 13.1 A digital signal processing system: (a) mathematical representation; (b) a practical example

microprocessors and mainframe computers. A practical example of voice processing by DSP is shown in Figure 13.1(b).

Signal processors can be either programmable or of a dedicated nature. Programmable signal processors allow flexibility of implementation of a variety of algorithms that can use the same computational kernel, whereas dedicated signal processors are hardwired to a specific algorithm or to a specific class of algorithms. Dedicated processors often are faster than, or dissipate less power than, general purpose programmable processors, although this is not always the case.

Digital signal processors have traditionally been optimised to compute finite impulse response convolutions (sum of products), infinite impulse response recursive filtering, and fast Fourier transform type (butterfly) operations that typically characterise most signal processing algorithms. They also include interfaces to external data ports for real time operation. It is interesting to note that one of the earliest digital computers, ENIAC, had characteristics of a DSP [3].

13.3 Comparison between a microprocessor and a DSP

General architectures for computers and single chip microcomputers fall into two categories. The architectures for the first significant electromechanical computer had separate memory spaces for the program and the data, so that each may be accessed simultaneously. This is known as *Harvard architecture*, having been developed in the late 1930s by Howard Aiken, a physicist at Harvard University. The Harvard Mark 1 computer became operational in 1944. A quick guide to basic microprocessor architecture is available in Reference 4.

The first general purpose electronic computer was probably the ENIAC (Electronic Numerical Integrator and Calculator) built from 1943 to 1946 at the University of Pennsylvania. The architecture was similar to that of the Harvard Mark 1 with separate program and data memories. Owing to the complexity of two separate memory systems, Harvard architecture has not proved popular in general purpose computer and microcomputer design.

One of the consultants to the ENIAC project was John von Neumann, a Hungarian-born mathematician, who is widely recognised as the creator of a different and very significant architecture, published by Burks, Goldstine and von Neumann (1946; reprinted in Bell and Newell, 1971) [20]. The so-called von Neumann architecture set the standard for developments in computer systems over the next 40 years and more. The idea was very simple and based on two main premises: that there is no intrinsic difference between instructions and data and that instructions can be partitioned into two major fields containing the operation command and the address of the operand (data to be operated upon); therefore, a single memory space could contain both instructions and data.

Common general purpose microprocessors such as the Motorola 68000 family and the Intel i86 family share what is now known as the von Neumann architecture. These and other general purpose microprocessors also have other characteristics typical of most computers over the past 40 years. The basic computational blocks

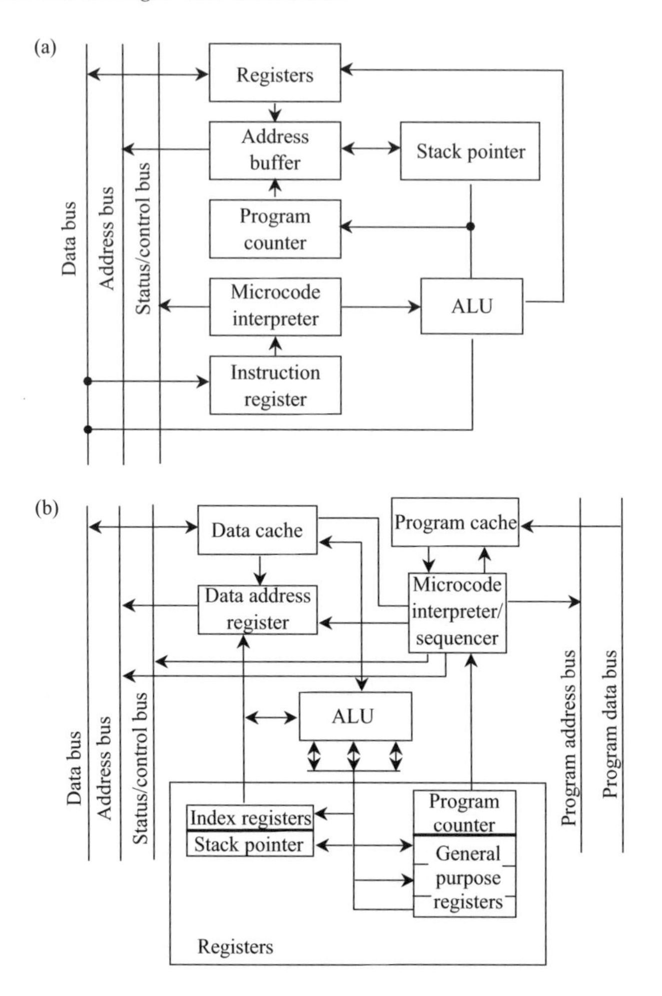

Figure 13.2 Comparison of microprocessor and DSP architectures: (a) traditional microprocessor architecture; (b) typical DSP architecture

are an arithmetic logic unit (ALU) and a shifter. Operations such as add, move and subtract are easily performed in a very few clock cycles. Complex instructions such as multiply and divide are built up from a series of simple shift, add or subtract operations. Devices of this type are known as *complex instruction set computers (CISC)*. CISC devices have multiply instructions, but this will simply execute a series of microcode instructions which are hard coded in on-chip ROM. The microcoded multiply operation therefore takes many clock cycles.

Figure 13.2 compares the basic differences between traditional microprocessor architecture and typical DSP architecture. Real time digital signal processing applications require many calculations of the form

$$A = BC + D. \tag{13.1}$$

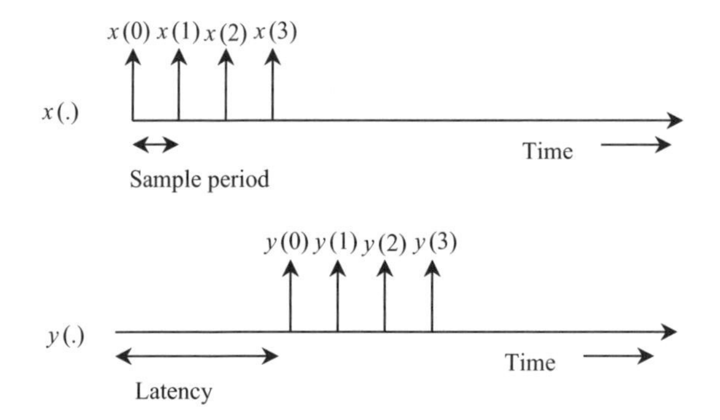

Figure 13.3 Sample period and latency

This simple little equation involves a multiply operation and an add operation. Because of its slow multiply instruction, a CISC microcomputer is not very efficient at calculating it. We need a machine that can multiply and add in just one clock cycle. For this, we need a different approach to computer architecture.

Many embedded applications are well defined in scope and require only a few calculations to be performed, but they require very fast processing. Examples of such applications are digital compression of images, compact disc players and digital telephones. In addition to these computation intensive functions demanding continuous processing, the processor has to perform comparatively simple functions such as menu control for satellite TV, selection of tracks for CD players, or number processing in a digital PBX, all of which require significantly less processing power.

In such applications, computation intensive functions such as digital filtering and data compression, etc., require continuous signal processing, which requires multiplication, addition, subtraction and other mathematical functions. While RISC processor architectures could be optimised to handle these situations by incorporating cache memory and direct access internal registers, and the like, DSP systems provide more computation intensive functions such as fast Fourier transforms, convolutions, and digital filters, etc. Particularly in a DSP based system, such tasks should be performed on a real time basis, as in Figure 13.3. This indicates that the sample period and computational latency are becoming key parameters.

13.3.1 The importance of sample period and latency in the DSP world

The sample period (the time interval between the arrival of successive samples of the input signal) depends on the technology employed in the processor. The time interval between the arrival of input and the departure of the corresponding output sample is the computational latency of the processor. To ensure the stability of the input ports, the output samples have to depart at the same sample period as the input samples. In signal processing applications, the minimum sample period that can be achieved often is more important than the latency of the circuit. Once the first output sample emerges, successive samples will also emerge at the sample period rate, hiding the effects of a large latency of circuit operation. This makes sense because typical signal processing applications deal with a few million samples of data in every second of operation. For details on the relationship between these two parameters, Reference 5 is suggested.

Other important measures are the area of the VLSI implementation and its power dissipation. These directly contribute to the cost of a DSP chip. One or more of these measures usually is optimised at the cost of others. These trade-offs again depend on the application. For instance, signal processors for portable communications require low power consumption combined with small size, usually at the cost of an increased sample period and latency.

13.3.2 The merging of microprocessors and DSPs

Diverse, high volume applications such as cell phones, disk drives, antilocking brakes, modems, and fax machines require both microprocessor and DSP capability. This requirement has led many microprocessor vendors to build in DSP functionality. In some cases, such as in Siemens' Tricore architecture [6], the functional merging is so complete that it is difficult to determine whether you should call the device a DSP or a microprocessor. At the other extreme, some vendors claim that their microprocessors have high performance DSP capability, when in fact they have added only a 'simple' 16×16-bit multiplication instruction.

13.4 Filtering applications and the evolution of DSP architecture

Digital signal processing techniques are based on mathematical concepts that are familiar to most engineers. From these basic ideas spring the myriad applications of DSP, including FFT, linear prediction, non-linear filtering, decimation and interpolation, and many more; see Figure 13.4. One of the most common signal processing functions is linear filtering. High pass, low pass, and band pass filters, which traditionally are analogue designs, can be constructed with DSP techniques. To build a linear filter with digital methods, a continuous time input signal, $x_c(t)$, is sampled to produce a sequence of numbers, $x[n] = x_c(nT)$. This sequence is transformed by a discrete time system – that is, a computational algorithm – into an output sequence of numbers, $y[n]$. Finally, a continuous time output signal, $y_c(t)$, is reconstructed from the sequence $y[n]$. Essentials of filtering and sampling concepts as applied to the world of DSP were discussed in chapter 3.

13.4.1 Digital filters

Digital filters have, for many years, been the most common application of digital signal processors. Digital design of any kind, ensures repeatability. Two other significant advantages accrue with respect to filters. First, it is possible to reprogram the DSP and drastically alter the filter's gain or phase response. For example, we can reprogram a system from low pass to high pass without throwing away the existing hardware.

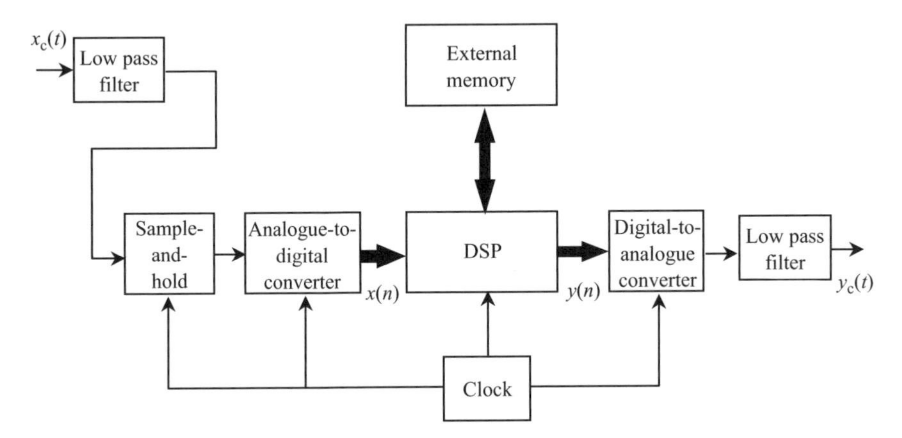

Figure 13.4 A DSP based filter implementation

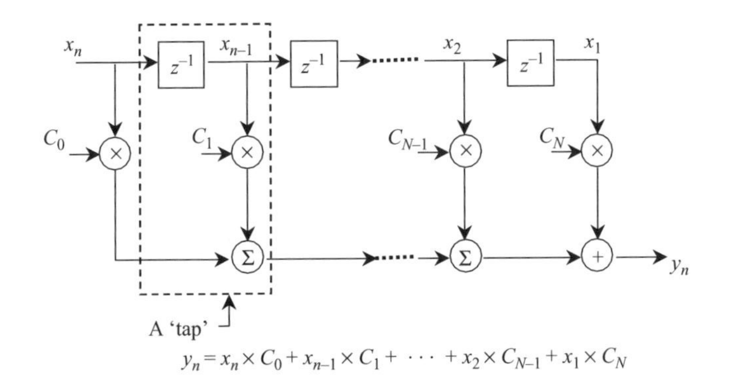

Figure 13.5 Finite impulse response filter

Second, we can update the filter coefficients whilst the program is running; that is, we can build 'adaptive' filters. The two basic forms of digital filter, the finite impulse response (FIR) filter and the infinite impulse response (IIR) filter, are explained next. The initial descriptions are based on a low pass filter. It is very easy to change low pass filters to any other types, high pass, band pass, etc. References 7 and 8 cover this in detail.

13.4.1.1 Finite impulse response (FIR) filter

The mechanics of the basic FIR filter algorithm are straightforward. The blocks labelled z^{-1} in Figure 13.5 are unit delay operators; their output is a copy of the input sample delayed by one sample period. A series of storage elements (usually memory locations) are used to simulate series of these delay elements (called a *delay line*). The FIR filter is constructed from a series of taps. Each tap includes a multiplication and an accumulation operation. At any given time, $n - 1$ of the most recent input

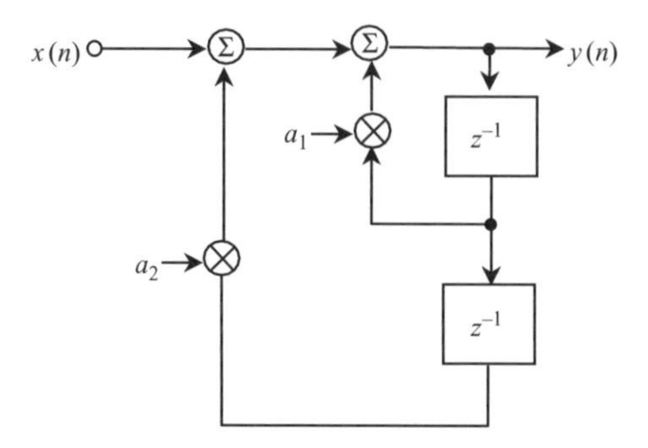

Figure 13.6 Simple IIR filter

samples resides in the delay line, where n is the number of taps in the filter. Input samples are designated x_k; the first input sample is x_1, the next is x_2, and so on. Each time a new input sample arrives, the previously stored samples are shifted one place to the right along the delay line and a new output sample is computed by multiplying the newly arrived sample and each of the previously stored input samples by the corresponding coefficient. In the figure, coefficients are represented as C_N, where N is the coefficient number. The results of each multiplication are summed together to form the new output sample, y_n. Later we discuss how DSPs are designed to help implement these.

13.4.1.2 Infinite impulse response (IIR) filter

The other basic form of digital filter is the infinite impulse response (IIR) filter. A simple form of this is shown in Figure 13.6. Using the same notations as we have just used for the FIR we can see that

$$y(n) = x(n) + a_1 y(n-1) + a_2 y(n-2) \tag{13.2}$$

$$= x(n) + \left[a_1 z^{-1} + a_2 z^{-2}\right] \cdot y(n)$$

$$= x(n) \frac{1}{1 - a_1 z^{-1} - a_2 z^{-2}}. \tag{13.3}$$

Take the maths for granted – it is just relatively simple substitution.

Therefore the transfer function is given by

$$H(n) = \frac{y(n)}{x(n)} = \frac{1}{1 - a_1 z^{-1} - a_2 z^{-2}}. \tag{13.4}$$

From eq. (13.2) we can see that each output, $y(n)$ is dependent on the input value, $x(n)$, and two previous outputs, $y(n-1)$ and $y(n-2)$. Taking this one step at a time, let us assume that there were no previous input samples before $n = 0$, then

$$y(0) = x(0).$$

At the next sample instant:

$$\begin{aligned} y(1) &= x(1) + a_1 y(0) \\ &= x(1) + a_1 x(0). \end{aligned}$$

Then at $n = 2$:

$$\begin{aligned} y(2) &= x(2) + a_1 y(1) + a_2 y(0) \\ &= x(2) + a_1[x(1) + a_1 x(0)] + a_2 x(0). \end{aligned}$$

Then at $n = 3$:

$$\begin{aligned} y(3) &= x(3) + a_1 y(2) + a_2 y(1) \\ &= x(3) + a_1[x(2) + a_1[x(1) + a_1 x(0)] + a_2 x(0)] + a_2[x(1) + a_1 x(0)]. \end{aligned}$$

We already can see that any output depends on all the previous inputs and we could go on, but the equation just gets longer. An alternative way of expressing this is to say that each output is dependent on an infinite number of inputs. This is why this filter type is called *infinite impulse response*.

If we look again at Figure 13.6, the filter is actually a series of feedback loops and, as with any such design, we know that, under certain conditions, it may become unstable. Although instability is possible with an IIR design, it has the advantage that, for the same roll-off rate, it requires fewer taps than FIR filters. This means that, if we are limited in the processor resources available to perform our desired function, we may find ourselves having to use an IIR. We just have to be careful to design a stable filter. More advanced forms of these filters are discussed with simple explanation in [3].

13.4.2 Filter implementation in DSPs

To explain the filter implementation, let us take the case of a first order recursive filter. A signal flow graph (SFG) or signal flow diagram is a convenient representation of a signal processing algorithm. Consider the first order recursive filter shown in Figure 13.7(a). The sequential computations involved are not clearly evident in the SFG, since it appears as if all the operations can be evaluated at the same time. However, operations have to follow a certain precedence to preserve correct operation. It is also not clear where the data operands and coefficients are stored prior to their utilisation in the computation. A more convenient mode of description would be the one in Figure 13.7(b), which shows the storage locations for each operand and the sequence of computations in terms of micro-operations at the register transfer level (RTL) ordered in time from left to right. We assume that the state variable $v(n-1)$ is stored in the data memory (DM) at location D_1, while the coefficient C_1 is stored in a coefficient memory (CM) at location C_1. Both these operands are fetched and multiplied, the result is added to the input sample, $x[n]$, and the sum is stored in a temporary location T_1. Then, another multiplication is performed using coefficient C_2, and the product is added to the contents of T_1. The final result is output as $y[n]$.

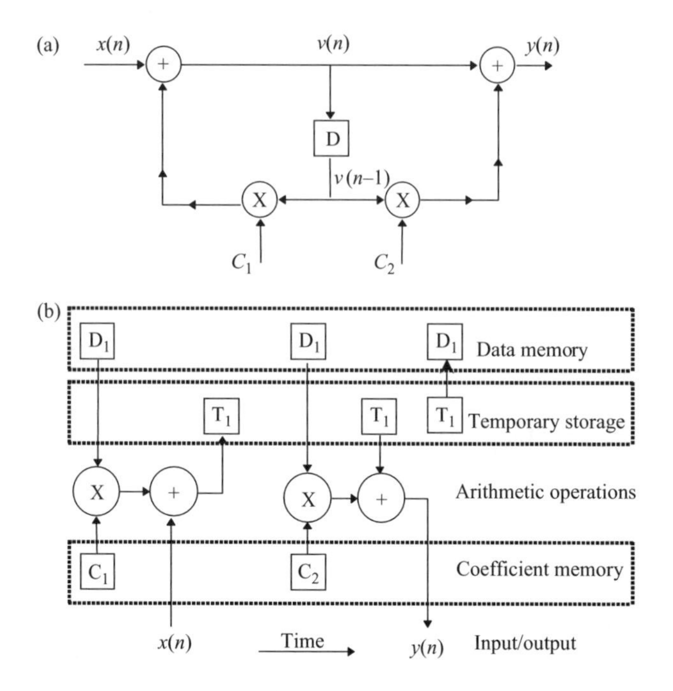

Figure 13.7 Filtering implementation by DSP techniques: (a) first order IIR filter; (b) assembler instructions at the register transfer level (RTL)

Then the new variable $v[n]$ is stored in memory location D_1. One may wonder why temporary location T_1 has been used. Temporary locations such as T_1 often provide a longer word length (or precision) than the word length of the memory. Repeated sums of products as required in this example can quickly exceed the dynamic range provided by the word length. Temporary locations provide the additional bits required to offset the deleterious effects of overflow. One also can observe that, in this example, the multiplier and adder operate in tandem, and the second coefficient multiplication can utilise the same multiplier when the input sample is being added. Thus, only one multiplier and one adder are required as arithmetic units. One data memory location, two coefficient memory locations, and one temporary storage register are required for correct operation of the filter. The specification of the sequence of micro-operations required to perform the computation is called *programming in assembler*.

From the preceding discussion, any candidate signal processor architecture for the IIR filter has to have a coefficient memory, a data memory, temporary registers for storage, a multiplier, an adder, and interconnection. In addition, addresses must be calculated for the memories as well as interpretation (or decoding) of the instruction (obtained from the program memory). The coefficient memory and the program memory can both be combined into one memory (the program memory). Nothing can be written into this read only memory (ROM). Data can be written and read from the random access data memory (RAM). The architecture shown in Figure 13.8 is

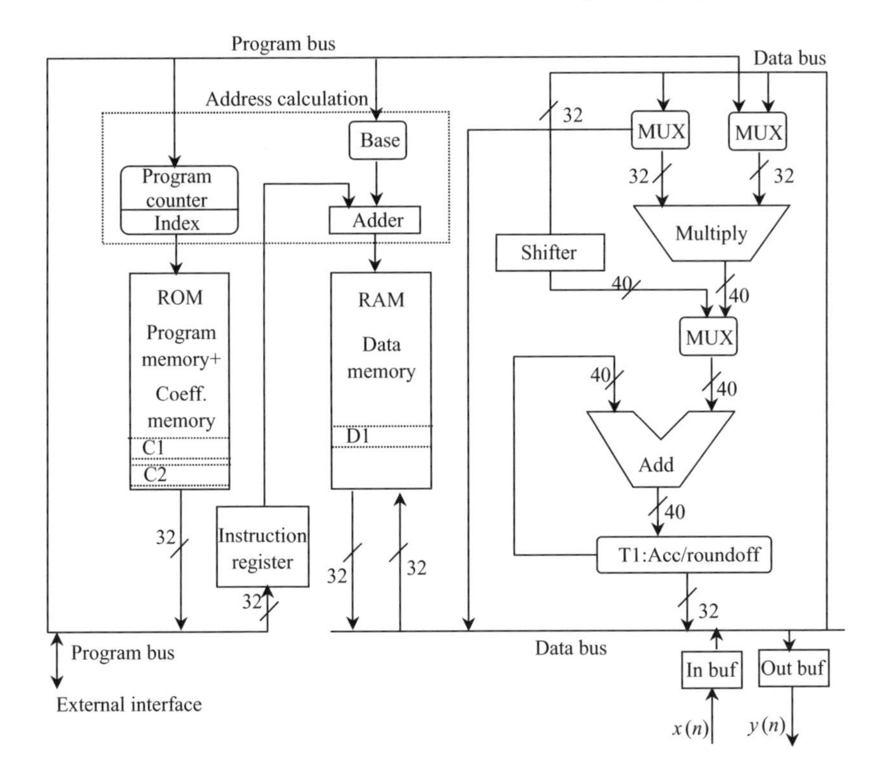

Figure 13.8 A candidate DSP architecture for IIR/FIR type filtering (source: Reference 8)

a suitable candidate architecture for this application. The program counter (PC) and the index registers are used in computing the addresses of the next instruction and the coefficients. The instruction is decoded by the instruction register (IR), where the address of the data is calculated using the adder and the base index register provided with the data memory. The program bus and the data bus are separate from each other, as are the program and data memories. This separation of data and program memories and buses characterises the so-called Harvard architecture for digital signal processors. The shifter is provided to allow incorporation of multiple word lengths within the data path (the multiplier and the adder) and the data and program buses. The T_1 register is configured as a higher precision accumulator. Input samples are read in from the input buffer and written into the output buffer. The DSP can interface to a host computer via the external interface. In Figure 13.8, the integers represent the number of bits carried on each bus. For a detailed account of digital filters chapter 7 of Reference 1 is suggested.

The inherent advantages of digital filters are these.

(a) They can be made to have no insertion loss.
(b) Linear phase characteristics are possible.

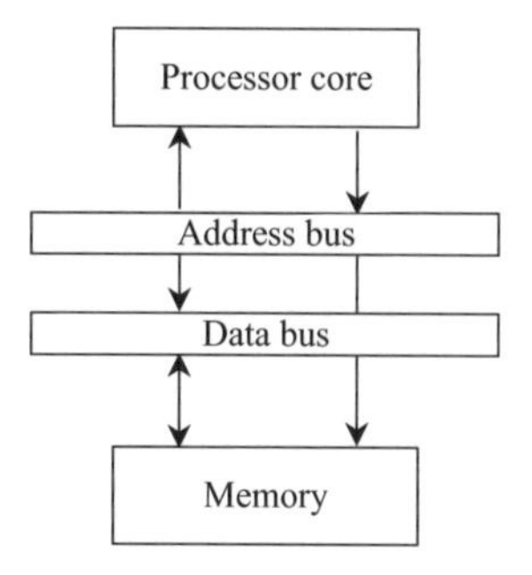

Figure 13.9 The von Neumann architecture for non-DSP processors

(c) Filter coefficients are easily changed to enable adaptive performance.
(d) Frequency response characteristics can be made to approximate closely to the ideal.
(e) They do not drift.
(f) Performance accuracy can be controlled by the designer.
(g) They can handle very low frequency signals.

13.4.3 DSP architecture

The simplest processor memory structure is a single bank of memory, which the processor accesses through a single set of address and data lines, as shown in Figure 13.9. This structure, which is common among non-DSP processors, is often considered a von Neumann architecture. Both program instructions and data are stored in the single memory. In the simplest (and most common) case, the processor can make one access (either read or a write) to memory during each instruction cycle.

If we consider programming a simple von Neumann architecture machine to implement our example FIR filter algorithm, the shortcomings of the architecture immediately become apparent. Even if the processor's data path is capable of completing a multiply–accumulate operation in one instruction cycle, it will take four instruction cycles for the processor to actually perform the multiply–accumulate operation, because the four memory accesses outlined above must proceed sequentially with each memory access taking one instruction cycle. This is one reason why conventional processors often do not perform well on DSP-intensive applications, and why designers of DSP processors have developed a wide range of alternatives to the von Neumann architecture, which we explore next.

The previous discussions indicate that parallel memories are preferred in DSP applications. In most DSPs, Harvard architectures coexist with pipelined data and instruction processors in a very efficient manner. These systems with specific addressing modes for signal processing applications could be best described as special instruction set computers (SISC). SISC architecture is characterised by their memory oriented special purpose instruction set.

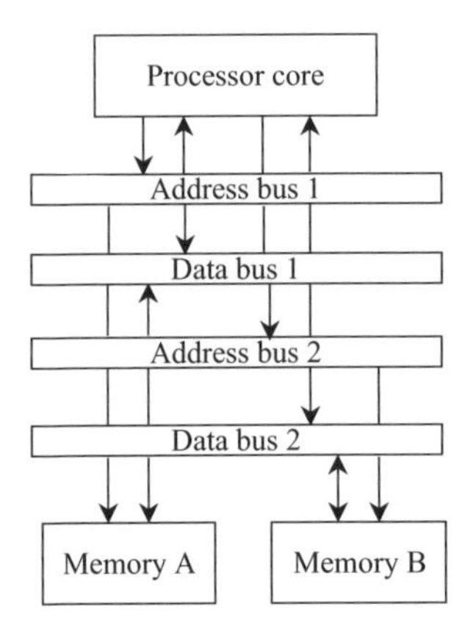

Figure 13.10 Harvard architecture

13.4.3.1 Basic Harvard architecture

Harvard architecture refers to a memory structure in which the processor is connected to two independent memory banks via two independent sets of buses. In the original Harvard architecture, one memory bank holds program instructions and the other holds data. Commonly, this concept is extended slightly to allow one bank to hold program instructions and data, while the other bank holds data only. This 'modified' Harvard architecture is shown in Figure 13.10. The key advantage of the Harvard architecture is that two memory accesses can be made during any one instruction cycle. Thus, the four memory accesses required for the example FIR filter can be completed in two instruction cycles. This type of memory architecture is used in many DSP families including the Analog Devices ADSP21xx.

13.4.3.2 SISC architectures

While microprocessors are based on register-oriented architectures, signal processors are memory oriented architectures. Multiple memories for both program and data have been present even in the first generation DSPs such as TMS320C10. Modern DSPs have as many as six parallel memories for the use of the instruction or the data processors. External memory is as easily accessible as internal memory. In addition, a rich set of addressing modes tailored for signal processing applications also are provided. We describe these architectures as representative of SISC computers and expect that future generations of SISC computers will have communication primitives as part of the standard instruction set. The basic instruction cycle is a unit of time measurement in the context of signal processing architectures, in some sense, the average time required to execute an ALU instruction. The basic instruction cycle is further divided into sub-cycles (usually two to four). The memory cycle time is that required to access one operand from the memory. The high memory bandwidth requirement in SISC computers can be met by either providing for memories with very low memory cycle times or multiple memories with relatively slow cycle times. Typically, an instruction cycle is twice as long as a memory cycle for on-chip memory (and equal to the memory cycle for external memory). Clearly, this facilitates the use of operand fetch and execution pipelines of two-operand instructions with on-chip data memories. If parallel data memories are provided, then the total number of memory cycles per instruction cycle is increased. The total number of memory cycles possible within a single basic instruction cycle is defined as the demand ratio [9] for a SISC machine. Higher demand ratios lead to a higher throughput of instructions.

$$\text{Demand ratio} = \frac{(\text{basic instruction cycle time}) \times (\text{number of memories})}{\text{memory cycle time}}. \quad (13.5)$$

13.4.3.3 Multiple access memory based architectures

As discussed, Harvard architecture achieves multiple memory accesses per instruction cycle by using multiple, independent memory banks connected to the processor data path via independent buses. While a number of DSP processors use this approach, there are also other ways to achieve multiple memory accesses per instruction cycle. These include using fast memories that support multiple, sequential accesses per instruction cycle over a single set of buses, and using '*multiported*' memories that allow multiple concurrent memory accesses over two or more independent sets of buses.

Achieving increased memory access capacity by use of multiported memory is becoming popular with the development of memory technology. A multiported memory has multiple independent sets of address and data connections, allowing multiple independent memory access to proceed in parallel. The most common type of multiported memory is the dual ported variety, which provides two simultaneous accesses. However, triple and even quadruple ported varieties are sometimes used. Multiported memories dispense with the need to arrange data among multiple, independent memory banks to achieve maximum performance. The key disadvantage of multiported memories is that they are much more costly (in terms of chip area) to implement than standard, single ported memories. The memory architecture shown in Figure 13.11, for example, includes a single ported program memory with a dual ported data memory. This arrangement provides one program memory access and two data memory accesses per instruction word and is used in the Motorola DSP561xx processors. For more detailed discussion on these techniques, see Reference 10.

13.4.4 Modifications to Harvard architecture

The basic Harvard architecture can be modified into six different types. This discussion is beyond the scope of the chapter and for details Reference 11 is suggested.

13.5 Special addressing modes

In addition to general addressing modes used in microprocessor systems, several special addressing modes are used in DSPs, including circular addressing and bit reversed

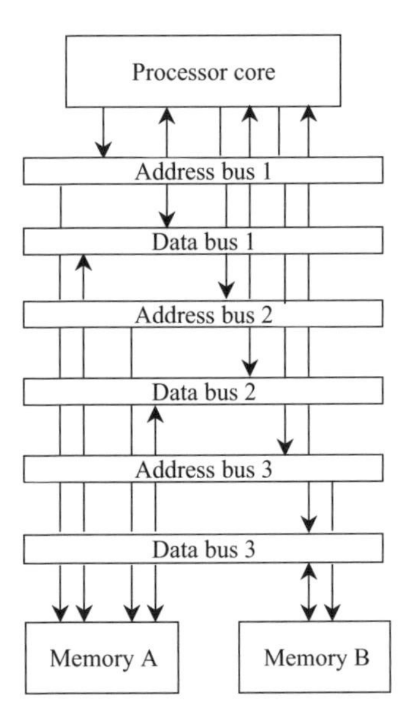

Figure 13.11 Modified Harvard architecture with dual ported memory

addressing. For a comprehensive discussion on addressing modes, Reference 10 is suggested, as only circular addressing and bit reversed addressing are discussed here.

13.5.1 Circular addressing

Many DSP applications need to manage data buffers. A data buffer is a section of memory that is used to store data that arrive from an off-chip source or a previous computation until the processor is ready to process the data. In real time systems, where dynamic memory allocation is prohibitively expensive, the programmer usually must determine the maximum amount of data that a given buffer will need to hold and then set aside a portion of memory for that buffer. The buffers generally use a first-in-first-out (FIFO) protocol, meaning that data values are read out of the buffer in the order in which they arrived.

In managing the movement of data into and out of the buffer, the programmer maintains two pointers, which are stored in registers or in memory: a read pointer and a write pointer. The read pointer points to (that is, contains the address of) the memory location containing the next data value to arrive, as illustrated in Figure 13.12. Each time a read or write operation is performed, the read or write pointer is advanced and the programmer must check to see whether the pointer has reached the last location in the buffer. The action of checking after each buffer operation whether the pointer has reached the end of the buffer, and resetting it if it has, is time consuming. For systems that use buffers extensively, this linear addressing can cause a significant performance bottleneck.

To address this bottleneck, many DSPs have a special addressing capability that allows them, after each buffer address calculation, to check automatically whether

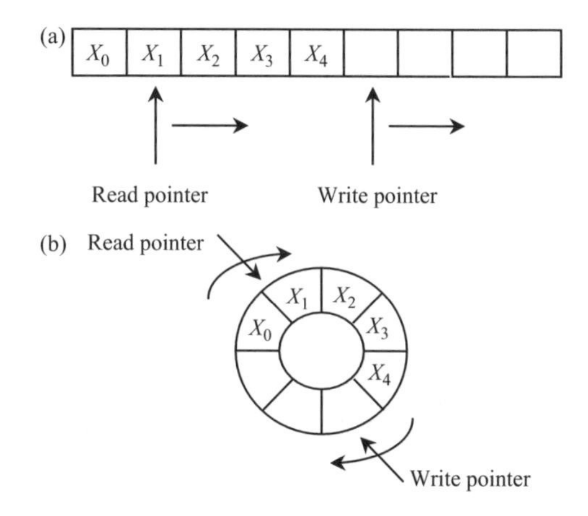

Figure 13.12 Comparison of linear and circular addressing: (a) FIFO buffer with linear addressing; (b) the same data in FIFO buffer with circular addressing

the pointer has reached the end of the buffer and reset it at the buffer start location if necessary. This capability is called *modulo addressing* or *circular addressing*.

The term modulo refers to modulo arithmetic, wherein numbers are limited to a specific range. This is similar to the arithmetic used in a clock, which is based on a 12-hour cycle. When the result of a calculation exceeds the maximum value, it is adjusted by repeatedly subtracting from it the maximum representable value until the result lies within the specified range. For example, 4 hours after 10 o'clock is 2 o'clock (14 modulo 12).

When modulo address arithmetic is in effect, read and write pointers (address registers) are updated using pre- or post-increment register indirect addressing [10]. The processor's address generation unit performs modulo arithmetic when new address values are computed, creating the appearance of a circular memory layout, as illustrated in Figure 13.12(b). Modulo address arithmetic eliminates the need for the programmer to check the read and write pointers to see whether they have reached the end of the buffer and reset them once they have reached the end. This results in much faster buffer operations and makes modulo addressing a valuable capability for many applications.

In most real time signal processing applications, such as those found in filtering, the input is an infinite stream of data samples. These samples are '*windowed*' and used in filtering applications. For instance, a sliding window of N data samples is used by an FIR filter with N taps. The data samples simulate a tapped delay line and the oldest sample is written over by the most recent sample. The filter coefficients and the data samples are written into two circular buffers. Then, they are multiplied and accumulated together to form the output sample result, which is stored. The address pointer for the data buffer is then updated and the samples appear shifted by one sample period, the oldest data being written out, and the most recent data are written into that location.

13.5.2 Bit reversed addressing

Perhaps the most unusual of addressing modes, bit reversed addressing, is used only in very specialised circumstances. Some DSP applications make heavy use of the fast Fourier transform (FFT) algorithm. The FFT is a fast algorithm for transforming a time domain signal into its frequency domain representation and vice versa [8, 12]. However, the FFT has the disadvantage that it either takes its input or leaves its output in a scrambled order. This dictates that the data be rearranged to or from natural order at some point.

The scrambling required depends on the particular variation of the FFT. The radix-2 implementation of an FFT, a very common form, requires reordering of a particularly simple nature, bit reversed ordering. The term 'bit reversed' refers to the observation that if the output values from a binary counter are written in reverse order (that is, least significant bit first), the resulting sequence of counter output values will match the scrambled sequence of the FFT output data. This phenomenon is illustrated in Figure 13.13.

Because the FFT is an important algorithm in many DSP applications, many DSP processors include special hardware in their address generation units to facilitate

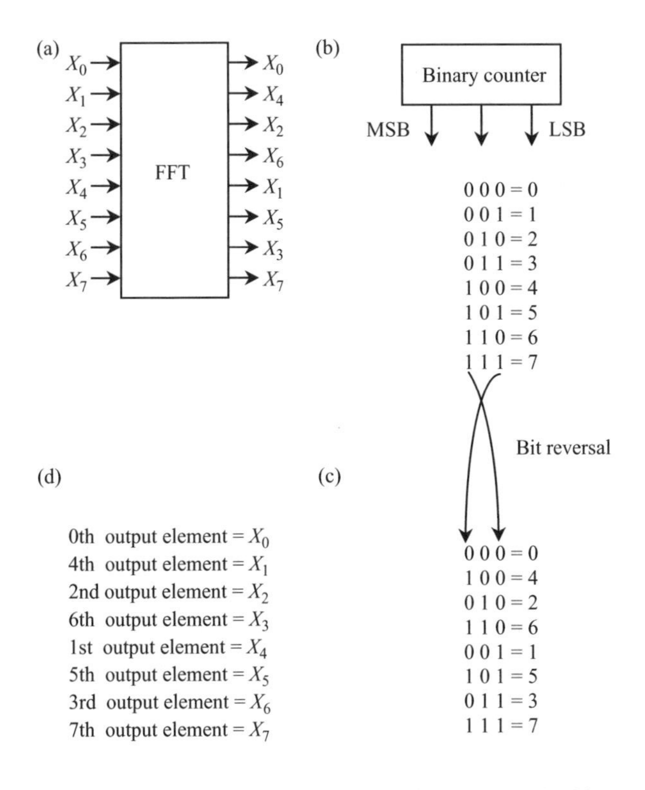

Figure 13.13 The output of an FFT algorithm and bit reversed addressing: (a) FFT output and input relations; (b) binary counter output; (c) bit reversal; (d) transformation of output into order

generating bit reversed address sequences for unscrambling FFT results. For example, the Analog Devices ADSP-210xx provides a bit reverse mode, which is enabled by setting a bit in a control register. When the processor is in bit reverse mode, the output of one of its address registers is bit reversed before being applied to the memory address bus.

An alternative approach to implementing bit reversed addressing is the use of reverse carry arithmetic. With reverse carry arithmetic, the address generation unit reverses the direction in which carry bits propagate when an increment is added to the value in an address register. If reverse carry arithmetic is enabled in the AGU, and the programmer supplies the base address and increment value in bit reversed order, then the resulting addresses will be in bit reversed order. Reverse carry arithmetic is provided in the AT&T DSP32xx, for example.

13.6 Important architectural elements in a DSP

Compared with architectural elements of a microprocessor (see Reference 12, chapter 4) it may be relevant for us to discuss special function blocks in a DSP chip. Performing efficient digital signal processing on a microprocessor is a tricky business. Although the ability to support single cycle multiply–accumulates (MACs) is the most important function a DSP performs, many other functions are critical for real time DSP applications. Executing a real time DSP application requires an architecture that supports high speed data flow to and from the computation units and memory through a multiport register file. This execution often involves the use of direct memory access (DMA) units and address generation units (AGU) that operate in parallel with other chip resources. AGUs, which perform address calculations, allow the DSP to bring two pieces of data per clock, which is a critical need for real time DSP algorithms.

It is important for DSPs to have an efficient looping mechanism, because most DSP code is highly repetitive. The architecture allows for zero overhead looping, in which you use no additional instructions to check the completion of loop iterations. Generally, DSPs take looping a step further by including the ability to handle nested loops.

DSPs typically handle extended precision and dynamic range to avoid overflow and minimise round-off errors. To accommodate this capability, DSPs typically include dedicated accumulators with registers wider than the nominal word size to preserve precision. DSPs also must support circular buffers to handle algorithmic functions, such as tapped delay lines and coefficient buffers. DSP hardware updates circular buffer pointers during every cycle in parallel with other chip resources. During each clock cycle, the circular buffer hardware performs an end of buffer comparison test and resets the pointer without overhead when it reaches the end of the buffer. FFTs and other DSP algorithms also require bit reversed addressing.

13.6.1 Multiplier/accumulator (MAC)

The multiplier/accumulator provides high speed multiplication, multiplication with cumulative addition, multiplication with cumulative subtraction, saturation and

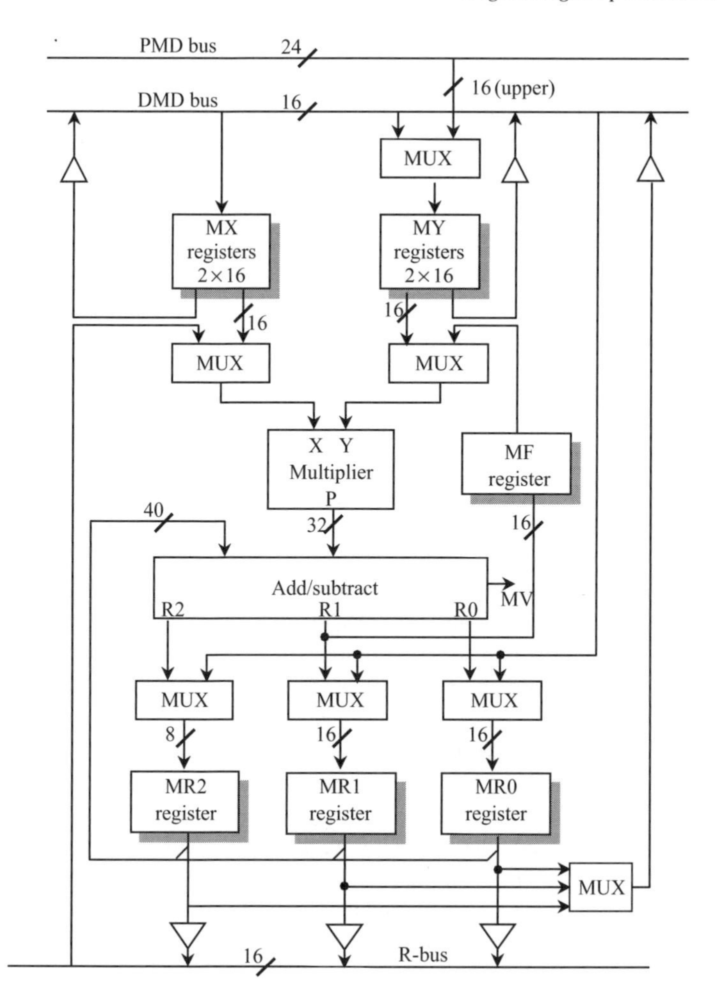

Figure 13.14 MAC block diagram of ADSP 2104 (reproduced by permission of Analog Devices, Inc.)

clear-to-zero functions. A feedback function allows part of the accumulator output to be directly used as one of the multiplicands of the next cycle. To explain MAC operation, let us take a real life example from the ADSP21XX family (see Figure 13.14).

The multiplier has two 16-bit input ports, X and Y, and a 32-bit product output port, P. The 32-bit product is passed to a 40-bit adder/subtractor, which adds or subtracts the new product from the content of the multiplier result (MR) register, or passes the new product directly to MR. The MR register is 40 bits wide. In this discussion, we refer to the entire register as MR, which actually consists of three smaller registers: MR0 and MR1, which are 16 bits wide, and MR2, which is 8 bits wide.

The adder/subtractor is greater than 32 bits to allow for intermediate overflow in a series of multiply/accumulate operations. The multiply overflow (MV) status bit is set when the accumulator has overflowed beyond the 32-bit boundary; that is, when there are significant (non-sign) bits in the top nine bits of the MR register (based on a two's-complement arithmetic). The input/output registers of the MAC section are similar to the ALU. The X input port can accept data from either the MX register file or from any register on the result (R) bus. The R bus connects the output registers of all the computational units, permitting them to be used directly as input operands. Two registers in the MX register file, MX0 and MX1, can be read and written from the DMD bus. The MX register file output is dual ported so that one register can provide input to the multiplier while either one drives the DMD bus.

The Y input port can accept data from either the MY register file or the MF register. The MY register file has two registers, MY0 and MY1, which can be read and written from the DMD bus and written from the program memory data (PMD) bus. The ADSP-2101 instruction set also provides for reading these registers over the PMD bus but with no direct connection; this operation uses the DMD-PMD bus exchange unit. The MY register file output also is dual ported so that one register can provide input to the multiplier while either one drives the DMD bus.

The output of the adder/subtractor goes to either the MF register or the MR register. The MF register is a feedback register which allows bits 16–31 of the result to be used directly as the multiplier Y input on a subsequent cycle. The 40-bit adder/subtractor register (MR) is divided into three sections: MR2, MR1, and MR0. Each of these registers can be loaded directly from the DMD bus and its output to either the DMD bus or the R bus.

Any register associated with the MAC can be both read and written in the same cycle. Registers are read at the beginning of the cycle and written at the end of the cycle. A register read instruction, therefore, reads the value loaded at the end of a previous cycle. A new value written to a register cannot be read out until the subsequent cycle. This allows an input register to provide an operand to the MAC at the beginning of the cycle and to be updated with the next operand from memory at the end of the same cycle. It also allows a result register to be stored in memory and updated with a new result in the same cycle.

The MAC contains a duplicate bank of registers, shown in Figure 13.14 behind the primary registers. There are actually two sets of MR, MF, MX, and MY register files. Only one bank is accessible at a time. The additional bank of registers can be activated for extremely fast context switching. A new task, such as an interrupt service routine, can be executed without transferring current states to storage. The selection of the primary or alternate bank of registers is controlled by bit 0 in the processor mode states register (MSTAT). If this bit is a 0, the primary bank is selected; if it is 1, the secondary bank is selected. For details, References 13 and 14 are suggested.

13.6.2 Address generation units (AGU)

Most DSP processors include one or more special address generation units (AGUs) that are dedicated to calculating addresses. Manufacturers refer to these units by various names. For example, Analog Devices calls its AGU a data address generator, and AT&T calls its a control arithmetic unit. An AGU can perform one or more

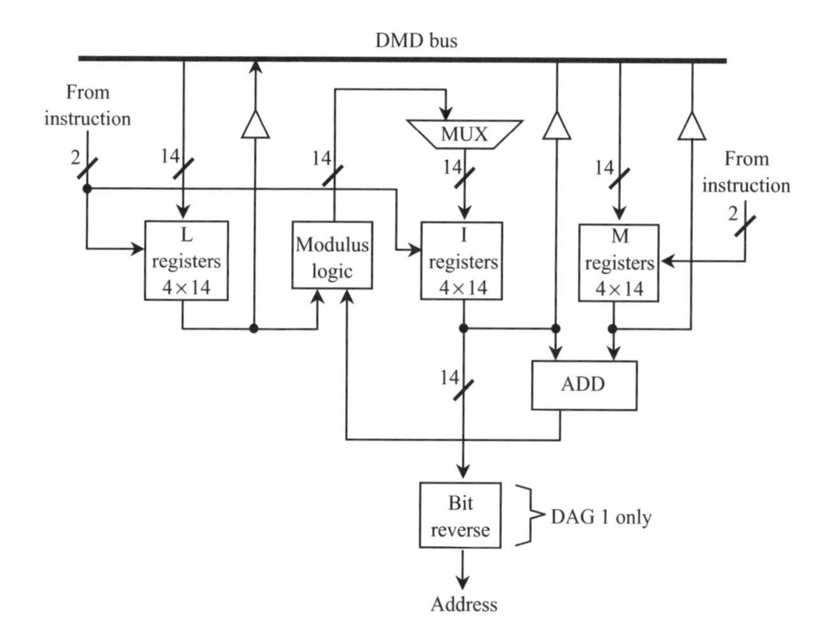

Figure 13.15 Data address generator block diagram of ADSP-2101 (courtesy of Analog Devices, Inc., USA)

complex address calculations per instruction cycle without using the processor's main data path. This allows address calculations to take place in parallel with arithmetic operations on data, improving processor performance. The differences among address generation units are manifested in the types of addressing modes provided and the capability and flexibility of each addressing mode. As an example let us take data address units in the ADSP-21xx family.

13.6.2.1 Data address units of ADSP-21xx family: an example

Data address generator (DAG) units contain two independent address generators so that program and data memories can be accessed simultaneously. Let us discuss the operation of the DAGs taking the ADSP-2101 as an example. The DAGs provide indirect addressing capabilities and perform automatic address modification. In the ADSP-2101, the two DAGs differ: DAG 1 generates data memory addresses and provides an optional bit reversal capability, DAG 2 can generate both data memory and program memory addresses but has no bit reversal.

Figure 13.15 shows a block diagram of a single DAG. There are three register files: the modify (M) register file, the index (I) register file, and the length (L) register file. Each file contains four 14-bit registers that can be read from and written to via the DMD bus. The I registers (I0-3 in DAG 1, I4-7 in DAG 2) contain the actual addresses used to access memory. When data are accessed in the indirect mode, the address stored in the selected I register becomes the memory address. With DAG 1, the output address can be bit reversed by setting the appropriate mode bit in the mode status register (MSTAT) as discussed next. Bit reversal facilitates FFT addressing.

The data address generator employs a post-modification scheme. After an indirect data access, the specified M register (M0-3 in DAG 2) is added to the specified I register to generate the new I value. The choice of the I and M registers are independent within each DAG. In other words, any register in the I0-3 set may be modified by any register in the M0-3 set in any combination, but not by those in DAG 2 (M4-7). The modification values stored in the M register are signed numbers so that the next address can be either higher or lower. The address generators support both linear and circular addressing. The value of the L register determines which addressing scheme is used. For circular buffer addressing, the L register is initialised with the length of the buffer. For linear addressing, the modulus logic is disabled by setting the corresponding L register to zero. L registers and I registers are paired and the selection of the L register (L0-3 in DAG 1, L4-7 in DAG 2) is determined by the I register used. Each time an I register is selected, the corresponding L register provides the modulus logic with the length information. If the sum of the M register content and the I register content crosses the buffer boundary, the modified I register value is calculated by the modulus logic using the L register value.

All data address generator registers (I, M, and L registers) are loadable and readable from the lower 14 bits of the DMD bus. Because the I and L register content is considered unsigned, the upper 2 bits of the DMD bus are padded with zeros when reading them. The M register content is signed; when reading an M register, the upper 2 bits of the DMD bus are sign extended. The modulus logic implements automatic pointer wraparound for accessing circular buffers. To calculate the next address, the modulus logic uses the following information:

- the current location, found in the I register (unsigned),
- the modify value, found in the M register (signed),
- the buffer length, found in the L register (unsigned), and
- the buffer base address.

From such an input, the next address is calculated with the formula:

$$\text{next address} = (\mathrm{I} + \mathrm{M} - \mathrm{B}) \text{ modulo } (\mathrm{L}) + \mathrm{B}, \tag{13.6}$$

where:

$$\begin{aligned} \mathrm{I} &= \text{current address;} \\ \mathrm{M} &= \text{modify value (signed);} \\ \mathrm{B} &= \text{base address (generated by the linker);} \\ \mathrm{L} &= \text{buffer length M+;} \\ \mathrm{I} &= \text{modified address;} \end{aligned}$$

and $\mathrm{M} < \mathrm{L}$ (which ensures that the next address cannot wrap around the buffer more than once in one operation).

13.6.3 Shifters

Shifting a binary number allows scaling. A shifter unit in a DSP provides a complete set of shifting functions, which can be divided into two categories: arithmetic and logical. A logical left shift by 1 bit inserts a 0 bit in the least significant bit, while a logical right shift by 1 bit inserts a 0 bit in the most significant bit. In contrast, an arithmetic right shift duplicates the sign bit (either a 1 or 0, depending on whether the number is negative or not) into the most significant bit. Although people use the term *arithmetic left shift*, arithmetic and logical left shifts are really identical: both shift the word left and insert a 0 in the least significant bit.

Arithmetic shifting provides a way of scaling data without using the processor's multiplier. Scaling is especially important in fixed point processors, where proper scaling is required to obtain accurate results from mathematical operations.

Virtually all DSPs provide shift instructions of one form or another. Some processors provide the minimum; that is, instructions to do arithmetic left or right shifting by one bit. Some processors may additionally provide instructions for 2- or 4-bit shifts. These can be combined with single-bit shifts to synthesise n-bit shifts, although at a cost of several instruction cycles.

Increasingly, many DSP processors feature a barrel shifter and instructions that use the barrel shifter to perform arithmetic or logical left or right shifts by any number of bits. Examples include the AT&T DSP16xx, the Analog Devices ADSP-21xx and ADSP-210xx, the DSP Group Oak DSP Core, the Motorola DSP563xx, the SGS-Thompson D950-CORE, and the Texas Instruments TMS320C5x and TMS320C54x. If you start with a 16-bit input, a complete set of shifting functions need a 32-bit output. These include arithmetic shift, logical shift, and normalisation. The shifter also performs derivation of exponent and derivation of common exponent for an entire block of numbers. These basic functions can be combined to efficiently implement any degree of numerical format control, including full floating point representation. Figure 13.16 shows a block diagram of the ADSP-2101 shifter.

A variable shifter section in ADSP2100 can be divided into a shifter array, an OR/PASS logic, an exponent detector, and the exponent compare logic.

The shifter array is a 16×32 barrel shifter. It accepts a 16-bit input and can place it anywhere in the 32-bit output field, from off-scale right to off-scale left, in a single cycle. This gives 49 possible placements within the 32-bit field. The placement of the 16 input bits is determined by a control code (C) and a HI/LO reference signal.

The shifter array and its associated logic are surrounded by a set of registers. The shifter input (SI) register provides input to the shifter array and the exponent detector. The SI register is 16 bits wide and is readable and writable from the DMD bus. The shifter array and the exponent detector also take as inputs arithmetic, shifter, or multiplier results via the R bus. The shifter result (SR) register is 32 bits wide and is divided into two 16-bit sections, SR0 and SR1. The SR0 and SR1 registers can be loaded from the DMD bus and sent to either the DMD bus or the R bus. The SR register is also fed back to the OR/PASS logic to allow double precision shift operations. The SE (shifter exponent) register is 8 bits wide and holds the exponent during the normalise and denormalise operations. The SE register is loadable and readable from the lower 8 bits of the DMD bus. It is a two's-complement, integer value.

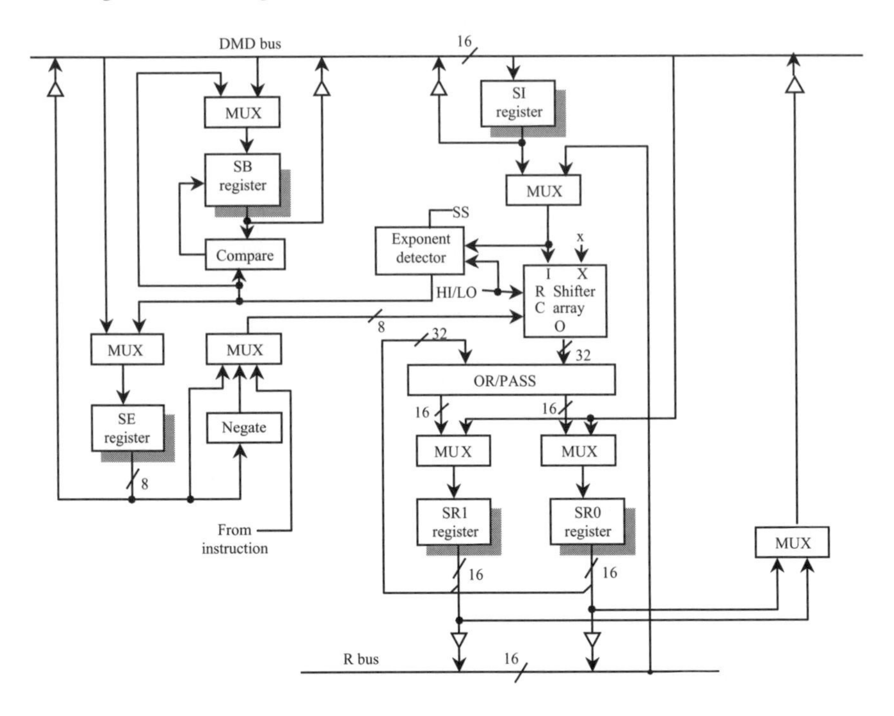

Figure 13.16 Block diagram of the ADSP-2101's shifter (courtesy of Analog Devices, Inc., USA)

The SB (shifter block) register is important in block floating point operations where it holds the block exponent value; that is, the value by which the block values must be shifted to normalise the largest value. The SB is 5 bits wide and holds the most recent block exponent value. The SB register is loadable and readable from the lower 5 bits of the DMD bus. It is a two's-complement, integer value.

Whenever the SE or SB registers are loaded onto the DMD bus, they are sign extended to form a 16-bit value. Any of the SI, SE or SR registers can be read and written in the same cycle. Registers are read at the beginning of the cycle and written at the end of the cycle. All register reads, therefore, read values loaded at the end of a previous cycle. A new value written to a register cannot be read out until a subsequent cycle. This allows an input register to provide an operand to the shifter at the beginning of the cycle and be updated with the next operand at the end of that cycle. It also allows a result register to be stored in memory and updated with a new result in the same cycle.

The shifter section contains a duplicate bank of registers, shown in Figure 13.16 behind the primary registers. There actually are two sets of SE, SB, SI, SR1, and SR0 registers; only one bank is accessible at a time. The additional bank of registers can be activated for extremely fast context switching. A new task, such as an interrupt service routine, can be executed without transferring current states to storage. The selection of the primary or alternative bank of registers is controlled by bit 0 in the processor mode status register (MSTAT). If this bit is a 0, the primary bank is selected; if it is a 1, the secondary bank is selected.

The shifting of the input is determined by a control code (C) and a HI/LO reference signal. The control code is an 8-bit signed value that indicates the direction and number of places the input is to be shifted. Positive codes indicate a left shift (upshift) and negative codes indicate a right shift (downshift). The control code can come from three sources: the content of the shifter exponent (SE) register, the negated content of the SE register, or an immediate value from the instruction.

The HI/LO signal determines the reference point for the shifting. In the HI state, all shifts are referenced to SR1 (the upper half of the output field); and in the LO state, all shifts are referenced to SR0 (the lower half). The HI/LO reference feature is useful when shifting 32-bit values because it allows both halves of the number to be shifted with the same control code. A HI/LO reference signal is selectable each time the shifter is used.

The shifter fills any bits to the right of the input value in the output field with zeros, and bits to the left are filled with the extension bit (X). The extension bit can be fed by three possible sources depending on the instruction being performed: the MSB of the input, the a.c. bit from the arithmetic status register (ASTAT), or a zero.

The OR/PASS logic allows the shifted sections of a multi-precision number to be combined into a single quantity. When PASS is selected, the shifter array output is passed through and loaded into the shifter result (SR) register unmodified. When OR is selected, the shifter array is bitwise ORed with the current contents of the SR register before being loaded there.

The exponent detector derives an exponent for the shifter input value. The exponent detector operates in one of three ways, which determine how the input value is interpreted. In the HI state, the input is interpreted as a single precision number or the upper half of a double precision number. The exponent detector determines the number of leading sign bits and produces a code that indicates how many places the input must be upshifted to eliminate all but one of the sign bits. The code is negative so that it can become the effective exponent for the mantissa formed by removing the redundant sign bits.

In the HI-extend state (HIX), the input is interpreted as the result of an add or subtract performed in the ALU section, which may have overflowed. Therefore, the exponent detector takes the arithmetic overflow (AV) status into consideration. If AV is set, then the $a + 1$ exponent becomes output to indicate that an extra bit is needed in the normalised mantissa (the ALU carry bit); if AV is not set, then HI-extend functions exactly like the HI state. When performing a derive exponent function in HI or HI-extend modes, the exponent detector also outputs a shifter sign (SS) bit, which is loaded into the arithmetic status register (ASTAT). The sign bit is the same as the MSB of the shifter input except when AV is set; when AV is set in the HI-extend state, the MSB is inverted to restore the sign bit of the overflow value. In the LO state, the input is interpreted as the lower half of a double precision number. In the LO state, the exponent detector interprets the SS bit in the arithmetic status register (ASTAT) as the sign bit of the number. The SE register is loaded with the output of the exponent detector only if SE contains P15. This occurs only when the upper half – which must be processed first – contains all sign bits. The exponent detector output is also offset by P16 to indicate that the input is actually the lower half of a 32-bit value.

The exponent compare logic is used to find the largest exponent value in an array of shifter input values. The exponent compare logic in conjunction with the exponent detector derives a block exponent. The comparator compares the exponent value derived by the exponent detector with the value stored in the shifter block exponent (SB) register and updates the SB register only when the derived exponent value is larger than the value in the SB register.

Shifters in different DSPs have different capabilities and architecture. For example, the TMS320C25 scaling shifter shifts to the left from none to 16 bits. Two other shifters can shift data coming from the multiplier left 1 bit or 4 bits or can shift data coming from the accumulator left from none to 7 bits. These two shifters add the advantage of being able to scale data during the data move instead of requiring an additional shifter operation.

13.6.4 Loop mechanisms

DSP algorithms frequently involve the repetitive execution of a small number of instructions, so-called inner loops or kernels. FIR and IIR filters, FFTs, matrix multiplications, and a host of other application kernels are performed by repeatedly executing the same instruction or sequence of instructions. DSPs have evolved to include features to handle efficiently this sort of repeated execution. To understand the evolution, we look at the problems associated with traditional approaches to related instruction execution. First, a natural approach to looping uses a branch instruction to jump back to the start of the loop.

Second, because most loops execute a fixed number of times, the processor must use a register to maintain the loop index; that is, the count of the number of times the processor has been through the loop. The processor's data path must be used to increment or decrement the index and test to see if the loop condition has been met. If not, a conditional branch brings the processor back to the top of the loop. All of these steps add overhead to the loop and use precious registers.

DSPs have evolved to avoid these problems via hardware looping, also known as *zero overhead looping*. Hardware loops are special hardware control constructs that repeat between hardware loops and software loops so that hardware loops lose no time incrementing or decrementing counters, checking to see if the loop is finished, or branching back to the top of the loop. This can result in considerable savings. In order to explain the way a loop mechanism improves the efficiency, we once again use ADSP-2101 as an example (Figure 13.17).

The ADSP-2100A program sequencer supports zero overhead DO UNTIL loops. Using the count stack, loop stack, and loop comparator, the processor can determine whether a loop should terminate and the address of the next instruction (either the top of the loop or the instruction after the loop) with no overhead cycle.

A DO UNTIL loop may be as large as program memory size permits. A loop may terminate when a 16-bit counter expires or when any other arithmetic condition occurs. The example below shows a three instruction loop that is to be repeated 100 times.

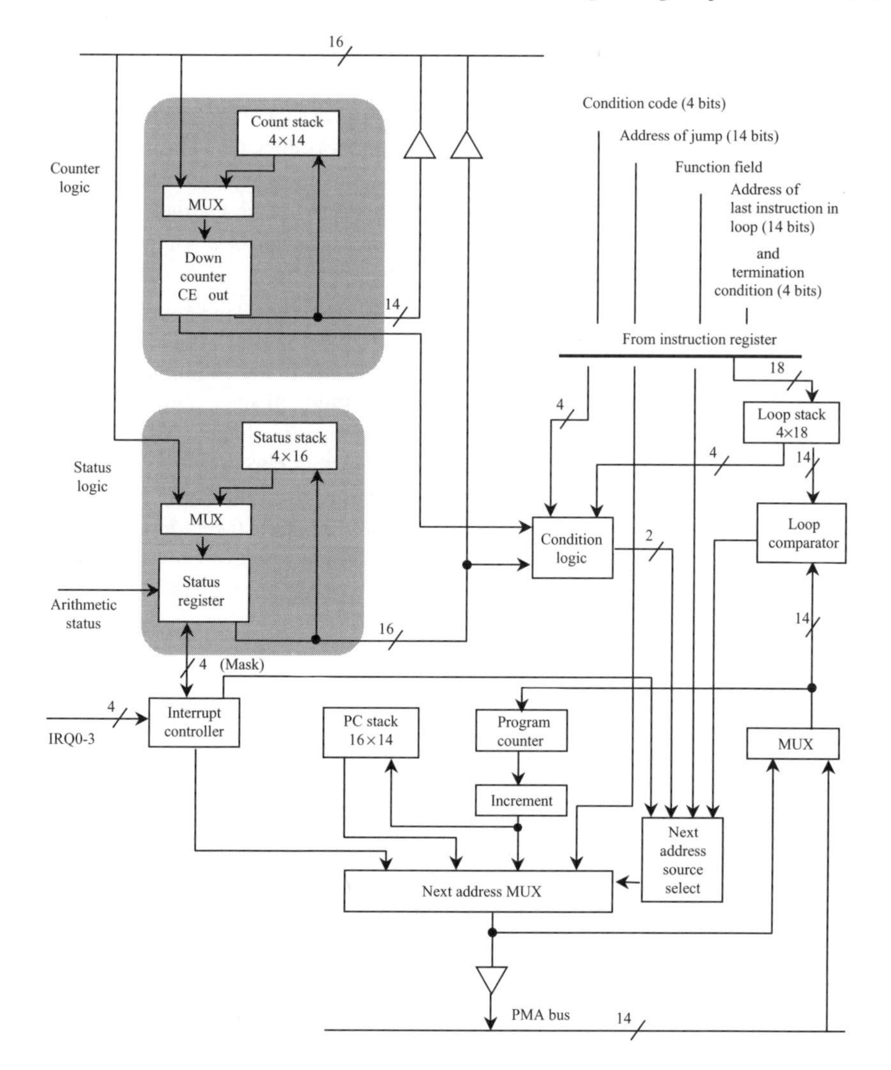

Figure 13.17 ADSP-2100A program sequencer architecture (reproduced by permission of Analog Devices, Inc., USA)

```
CNTR=100
Do Label UNTIL CE
        First instruction of loop
        Second instruction of loop
Label:  Last instruction of loop
        First instruction outside loop
```

The first instruction loads the counter with 100. The DO UNTIL instruction contains the address of the last instruction in the loop (in this case the address represented by the identifier label) and the termination condition (in this case the count expiring, CE). The execution of the DO UNTIL instruction causes the address of the first instruction of the loop to be pushed on the program counter stack and the address of the last instruction of the loop to be pushed on the loop stack (see Figure 13.17).

As instruction addresses are sent to the program memory address bus and the instruction is fetched, the loop comparator checks to see if the instruction is the last instruction of the loop. If it is, the program sequencer checks the status and condition logic to see if the termination condition is satisfied. The program sequencer then either takes the address from the program counter stack (to go back to the top of the loop) or simply increments the PC (to go to the first instruction outside the loop).

The looping mechanism of the ADSP-2100A is automatic and transparent to the user. As long as the DO UNTIL instruction is specified, all stack and counter maintenance and program flow will be handled by the sequencer logic with no overhead. This means that, in one cycle, the last instruction of the loop is being executed and, in the very next cycle, the first instruction of the loop is executed or the first instruction outside the loop is executed, depending upon whether the loop terminates or not. For further details of program sequencer and loop mechanisms of ADSP-2100A, References 13 and 15 are suggested.

13.7 Instruction set

Generally, a DSP instruction set is tailored to the computation-intensive algorithms common to DSP applications. This is possible because the instruction set allows data movement between various computational units with minimum overhead. For example, sustained single cycle multiplication/accumulation operations are possible.

Again, we use the ADSP-2101 as an example. The instruction set provides full control of the ADSP-2101's three computation units: the ALU, MAC and shifter. Arithmetic instructions can process single precision 16-bit operands directly with provisions for multi-precision operations. The ADSP-2101 assembly language uses an algebraic syntax for arithmetic operations and for data moves. The sources and destinations of computations and data moves are written explicitly, eliminating cryptic assembler mnemonics. There is no performance penalty for this; each program statement assembles into one 24-bit instruction, which executes in one cycle. There are no multi-cycle instructions in the ADSP-2101 instruction set. Some 50 registers surrounding the computational units are dual purpose, available for general purpose on-chip storage when not used in computation. This saves many memory access cycles and provides excellent freedom in coding. The control instructions provide conditional execution of most calculations and, in addition to the usual JUMP and CALL, support a DO UNTIL looping instruction. Return from interrupt (RTI) and the return from subroutine (RTS) are also provided. These services are made compact and speedy by the single cycle context save. The contents of the primary register set are held constant while the alternate set is enabled for subroutine and interrupt services. This eliminates the cluster of PUSHes and POPs of stacks common in general purpose microprocessors.

The ADSP-2101 also provides the IDLE instruction for idling the processor until an interrupt occurs. IDLE puts the processor into a low power state while waiting for interrupts. Two addressing modes are supported for memory fetches. Direct addressing uses immediate values; indirect addressing uses the two data addressing generators.

The 24-bit instruction word allows a high degree of parallelism in performing operations. The instruction set allows for a single cycle execution of any of the following combinations:

- any ALU, MAC or shifter operation (may be conditional),
- any register to register move,
- any data memory read or write,
- a computation with any data register/data register move,
- a computation with any memory read or write, and
- a computation with a read from two memories.

The instruction set provides moves from any register to any other register or from most registers to and from either memory. For combining operations, almost any ALU, MAC or shifter operation may be combined with any register-to-register moves or with a register move to or from either internal or external memory.

There are five basic categories of instruction: computational instructions, data move instructions, multifunction instruction, program flow control instructions, and miscellaneous instructions, all of which are described in the next several sections, with tables summarising the syntax of each instruction category. The notion used in an instruction is shown in Table 13.1.

As it is beyond the scope of a chapter of this kind to explain the whole group of instructions, the computation instructions of ADSP-2101 are described in a summary form below. A more detailed version instruction set overview can be found in Reference 13 and the ADSP literature.

13.7.1 Computation instructions: a summary of ADSP-21xx family

The computation group executes all ALU, MAC, and shifter instructions. There are two functional classes: standard instructions, which include the bulk of the computation operations, can be executed conditionally (IF condition ...), test the ALU status register, and may be combined with a data transfer in single cycle multifunction instructions; and special instructions, which are from a small subset and must be executed individually. Table 13.2 indicates permissible conditions for computation instructions, and Table 13.3 describes the computational input/output registers.

13.7.1.1 MAC functions

13.7.1.1.1 Standard functions

Standard MAC instructions include multiply, multiply/accumulate, multiply/subtract, transfer AR conditionally, and clear. As an example, consider a MAC instruction for

Table 13.1 Notation used in instruction set of ADSP 21xx family (courtesy of Analog Devices, Inc., USA)

Symbol	Meaning
+, −	Add, subtract.
*	Multiply.
a = b	Transfer into a the contents of b.
,	Separates multifunction instructions.
DM (addr)	The contents of data memory at location 'addr'.
PM (addr)	The contents of program memory at location 'addr'.
[option]	Anything within square brackets is an optional part of the instruction statement.
| option a |	List of parameters enclosed by parallel vertical lines require the choice of one parameter from among the available list.
CAPITAL LETTERS	Capital letters denote reserved words. These are instruction words, register names and operand selections.
Lower-case letters	Parameters are shown in small letters and denote an operand in the instruction for which there are numerous choices.
⟨data⟩	These angle brackets denote an immediate data value.
⟨addr⟩	These angle brackets denote an immediate value of an address to be coded in the instruction.
;	End of instruction.

multiply/accumulate in the form:

```
[IF Condition] MR = MR + xop * yop (SS) ;
MF                                   SU
                                     US
                                     UU
                                     RND
```

If the options 'MR' and 'UU' are chosen; if xop and yop are the contents of MXO and MYO respectively; and if the MAC overflow condition is chosen, then a conditional instruction would read:

```
IF NOT MV MR = MR + MXO * MYO (UU) ;
```

The conditional expression, IF NOT MV, tests the MAC overflow bit. If the condition is not true, a NOP is executed. The expression $MR = MR + MXO * MYO$ is the multiply/accumulate operation: the multiplier result register (MR) gets the value of itself plus the product of the X and Y input registers selected. The modifier selected in parentheses (UU) treats the operands as unsigned. Only one such modifier can be selected from the available set: (SS) means both are signed, (US) and (SU) mean that either the first or second operand is signed; (RND) means to round the (implicitly signed) result.

Table 13.2 Permissible conditions for computation instructions of ADSP2101 (reproduced by permission of Analog Devices, Inc.)

Condition	Keyword
ALU result is:	
equal to zero	EQ
not equal to zero	NE
greater than zero	GT
greater than or equal to zero	GE
less than zero	LT
less than or equal to zero	LE
ALU carry status:	
carry	AC
not carry	NOT AC
x-input sign:	
positive	POS
negative	NEG
ALU overflow status:	
overflow	AV
not overflow	NOT AV
MAC overflow status:	
overflow	MV
not overflow	NOT MV
Counter status:	
not expired	NOT CE

Table 13.3 Computational input/output registers (reproduced by permission of Analog Devices, Inc.)

Source for X input (xop)	Source for Y input (yop)	Destination*
ALU		
AX0, AX1, AR	AY0, AY1	AR
MR0, MR1, MR2	AF	AF
SR0, SR1		
MAC		
MX0, MX1, AR	MY0, MY1	MR(MR2, MR1, MR0)
MR0, MR1, MR2	MF	MF
SR0, SR1		
Shifter		
SI, SR0, SR1		SR (SR1, SR0)
AR		
MR0, MR1, MR2		

*Destination for output port R for ALU and MAC or destination for shifter output.

Accumulator saturation is the only MAC special function:

```
IF MV SAT MR ;
```

The instruction tests the MAC overflow bit (MV) and saturates the MR register (for only one cycle) if that bit is set.

13.7.1.2 ALU group functions

Standard ALU instructions include add, subtract, logic (AND, OR, NOT, exclusive-OR), pass, negate increment, decrement, clear, and absolute value. The – function does two's-complement subtraction while NOT obtains a one's-complement. The PASS function passes the listed operand but tests and stores status information for later sign/zero testing. As an example, consider an ALU addition instruction for add/add-with-carry in the form:

```
[IF Condition] AR = xop + yop ;
               AF       + c
                        + yop + c
```

Instructions are in similar form for subtraction and logical operations. If the options AR and + yop + C are chosen, and if xop and yop are the contents of AXO and AYO, respectively, the unconditional instruction would read:

```
AR = AXO + AYO + C;
```

This algebraic expression means that the ALU result register AR gets the value of the ALU x-input and y-input registers plus the value of the carry-in bit. This shortens code and speeds execution by eliminating many separate register move instructions.

When an optional IF condition is included, and if ALU carry bit status is chosen, then the conditional instruction would read:

```
IF AC AR = AXO + AYO + C ;
```

The conditional expression, IF AC, tests the ALU carry bit. If there is a carry from the previous instruction, this instruction executes; otherwise, an NOP occurs and execution continues with the next instruction.

Division is the only ALU special function. It is executed in two steps: DIVS computes the sign, then DIVQ computes the quotient. A full divide of a signed 16-bit divisor into a signed 32-bit quotient requires a DIVS followed by 15 DIVQs.

13.7.1.3 Shifter group functions

Shifter standard functions include arithmetic and logical shift as well as floating point and block floating point scaling operations, derive exponent, normalise, denormalise, and block exponent adjust. As an example, consider a shifter instruction for normalise:

```
IF NOT CE SR = SR OR NORM SI (HI) ;
```

The conditional expression, IF NOT CE, tests the 'not counter expired' condition. If the condition is false, an NOP is executed. The destination of all shifting operations is the shifter result register, SR. (The destination of the exponent detection instructions is SE or SB.) In this example, SI, the shifter input register, is the operand. The amount

Table 13.4 Instruction set groups (using the ADSP-21xx family as an example)

Instruction type	Purpose
Data move instructions	Move data to and from data registers and external memory.
Multifunction instructions	Exploits the inherent parallelism of a DSP by combinations of data moves and memory writes/reads in a single cycle.
Program flow control instructions	Directs the program sequence. In normal order, the sequence automatically fetches the next contiguous instruction for exertion. This flow can be altered by these.
Miscellaneous instructions	Such as NOP (no operation) and PUSH/POP, etc.

and direction of the shift is controlled by the signed value in the SE register in all shift operations except an immediate shift. Positive values cause left shifts; negative values cause right shifts.

The 'SR OR' modifier (which is optional) logically ORs the result with the current contents of the SR register; this allows the user to construct a 32-bit value in SR from two 16-bit pieces. NORM is the operator and (HI) is the modifier that determines whether the shift is relative to the HI or LO (16-bit) half of SR. If SR OR is omitted, the result is passed directly into SR.

Shift-immediate is the only shifter special function. The number of places (exponents) to shift is specified in the instruction word.

13.7.2 Other instructions

Other instructions in a DSP could be grouped as in Table 13.4. Details could be dependent on the DSP family and hence Table 13.4 should be considered only as a guideline.

13.8 Interface between DSPs and data converters

Advances in semiconductor technology have given DSPs fast processing capabilities and data converter ICs have the conversion speeds to match the faster processing speeds. This section considers the hardware aspects of practical design.

13.8.1 Interface between ADCs and DSPs

Precision sampling ADCs generally have either parallel data output or a single serial output data link. We consider these separately.

13.8.1.1 Parallel interfaces with ADCs

Many parallel output sampling ADCs offer three state output, which can be enabled or disabled by using an output enable pin on the IC. Although it may be tempting to connect these three state outputs directly to a back plane data bus, severe performance degrading noise problems will result. All ADCs have a small amount of internal stray capacitance between the digital output and the analogue input (typically 0.1–0.5 pF). Every attempt is made during the design and layout of the ADC to keep this capacitance to a minimum. However, if there is excessive overshoot and ringing and possibly other high frequency noise on the digital output lines (as would probably be the case if the digital output were connected directly to a back plane bus), this digital noise will couple back into the analogue input through the stray capacitance. The effect of this noise is to decrease the overall ADC SNR and ENOB. Any code dependent noise also will tend to increase the ADC harmonic distortion.

The best approach to eliminating this potential problem is provided as an intermediate three state output buffer latch which is located close to the ADC data outputs. This latch isolates the noisy signals on the data bus from the ADC data outputs, minimising any coupling back into the ADC analogue input. The ADC data sheet should be consulted regarding exactly how the ADC data should be clocked into the buffer latch. Usually, a signal called *conversion complete* or *busy* from the ADC is provided for this purpose.

It also is a good idea not to access the data in the intermediate latch during the actual conversion time of the ADC. This practice will further reduce the possibility of corrupting the ADC analogue input with noise. The manufacturer's data sheet timing information should indicate the most desirable time to access the output data.

Figure 13.18 shows a simplified parallel interface between the AD676 16-bit, 100 kSPS ADC (or the AD7884) and the ADSP-2101 microcomputer. (Note that the actual device pins shown have been relabelled to simplify the following general discussion.) In a real time DSP application (such as in digital filtering), the processor

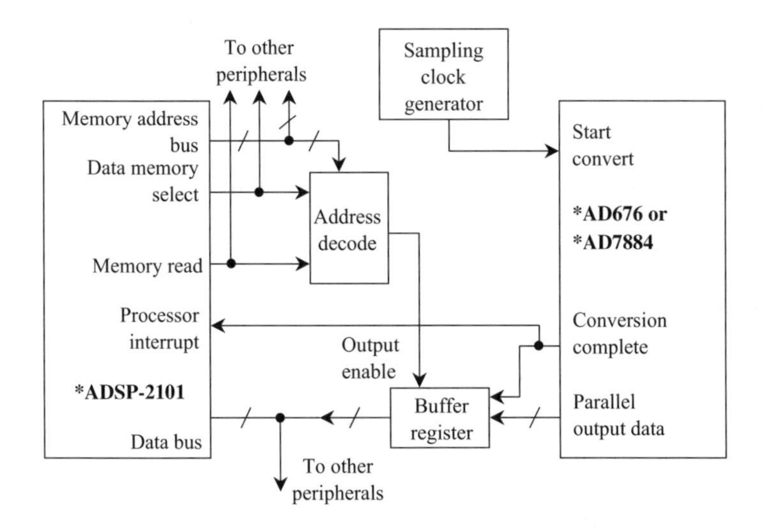

Figure 13.18 Generalised DSP to ADC parallel interface (reproduced by permission of Analog Devices, Inc., USA)

must complete its series of instructions within the ADC sampling interval. Note that the entire cycle is initiated by the sampling clock edge from the sampling clock generator. Even though some DSP chips offer the capability to generate lower frequency clocks from the DSP master clock, the use of these signals as precision sampling clock sources is not recommended owing to the probability of timing jitter. It is preferable to generate the ADC sampling clock from a well designed low noise crystal oscillator circuit as has been previously described.

The sampling clock edge initiates the ADC conversion cycle. After the conversion is completed, the ADC conversion complete line is asserted, which in turn interrupts the DSP. The DSP places the address of the ADC that generated the interrupt on the data memory address bus and asserts the data memory select line. The read line of the DSP is then asserted. This enables the external three state ADC buffer register outputs and places the ADC data on the data bus. The trailing edge of the read pulse latches the ADC data on the data bus into the DSP internal registers. At this time, the DSP is free to address other peripherals that may share the common data bus.

Because of the high speed internal DSP clock (50 MHz for the ADSP-2101), the width of the read pulse may be too narrow to access properly the data in the buffer latch. If this is the case, adding the appropriate number of programmable software wait states in the DSP will both increase the width of the read pulse and also cause the data memory select and the data memory address lines to remain asserted for a correspondingly longer period of time. In the case of the ADSP-2101, one wait state is one instruction cycle, or 80 ns.

13.8.1.2 Interface between serial output ADCs

ADCs that have a serial output (such as the AD677, AD776, and AD1879) have interfaces to the serial port of many DSP chips, as shown in Figure 13.19. The sampling clock is generated from the low noise oscillator. The ADC output data is presented on the serial data line one bit at a time. The serial clock signal from the ADC is used to latch the individual bits into the serial input shift register of the DSP serial port. After all the serial data are transferred into the serial input register, the serial port logic generates the required processor interrupt signal. The advantages of using serial output ADCs are the reduction in the number of interface connections as well as reduced noise because fewer noisy digital program counter tracks are close to the converter. In addition, SAR and sigma-delta ADCs are inherently serial output devices. The number of peripheral serial devices permitted is limited by the number of serial ports available on the DSP chip.

13.8.2 Interfaces with DACs

13.8.2.1 Parallel input DACs

Most of the principles previously discussed regarding interfaces with ADCs also apply to interfaces with DACs. A generalised block diagram of a parallel input DAC is shown in Figure 13.20(a). Most high performance DACs have an internal parallel DAC latch that drives the actual switches. This latch deskews the data to minimise the output glitch. Some DACs designed for real time sampling data DSP applications

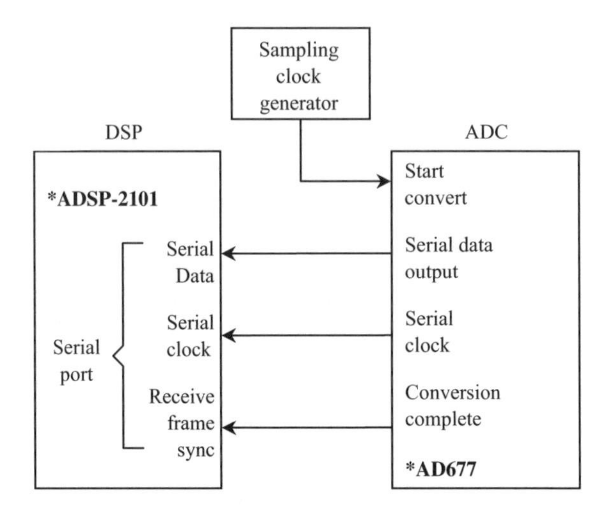

Figure 13.19 Generalised serial DSP to ADC interface (reproduced by permission of Analog Devices, Inc., USA)

have an additional input latch so that the input data can be loaded asynchronously with respect to the DAC latch strobe. Some DACs have an internal reference voltage that can be either used or bypassed with a better external reference. Other DACs require an external reference.

The output of a DAC may be a current or a voltage. Fast-video DACs generally are designed to supply sufficient output current to develop the required signal levels across resistive loads (generally 150 Ω, corresponding to a 75 Ω source and load terminated cable). Other DACs are designed to drive a current into a virtual ground and require a current-to-voltage converter (which may be internal or external). Some high impedance voltage output DACs require an external buffer to drive reasonable values of load impedance.

A generalised parallel DSP-to-DAC interface is shown in Figure 13.20(b). The operation is similar to that of a parallel DSP-to-ADC interface described earlier. In most DSP applications the DAC is operated continuously from a stable sampling clock generator external to the DSP. The DAC requires double buffering because of the asynchronous interface to the DSP. The sequence of events is as follows. Asserting the *sampling clock generator* line clocks the word contained in the DAC *input latch* into the DAC *latch* (the latch which drives the DAC switches). This causes the DAC output change to the new value. The sampling clock edge also interrupts the DSP, which then addresses the DAC, enables the DAC *chip select*, and writes the next data into the DAC *input latch* using the *memory write* and data bus lines. The DAC is now ready to accept the next sampling clock edge.

13.8.2.2 Serial input DACs

A block diagram of a typical serial input DAC is shown in Figure 13.21(a). The digital input circuitry consists of a serial-to-parallel converter driven by a serial data line and

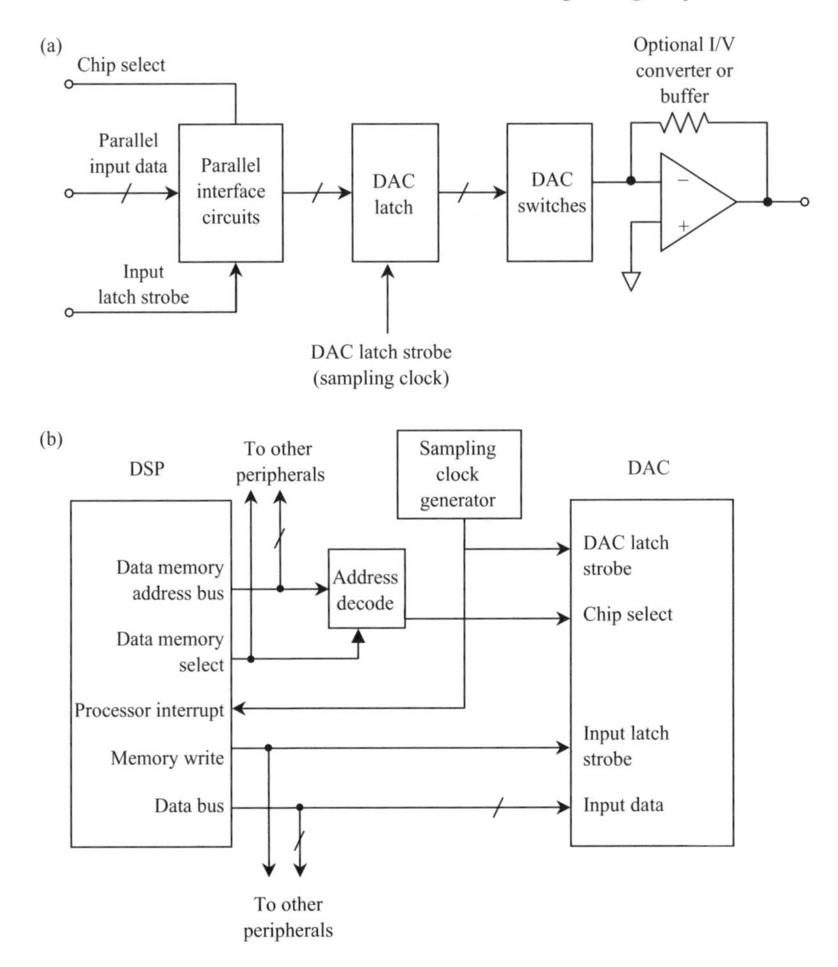

Figure 13.20 Interface between DSPs and parallel DACs: (a) parallel input DAC; (b) DSP and parallel DAC input

a serial clock. After the serial data is loaded, the DAC latch strobe clocks the parallel DAC latch and updates the DAC switches with a new word. Interface between DSPs and serial DACs is quite easy using the DSP serial port (Figure 3.21(b)). The serial data transfer process is initiated by the assertion of the sampling clock generator line. This updates the DAC latch and causes the serial port of the DSP to transmit the next word to the DAC using the serial clock and the serial data line.

13.9 A few simple examples of DSP applications

In most microprocessor or microcontroller based systems, a continuously running assembler program may be interrupted and deviated to subroutines as and when

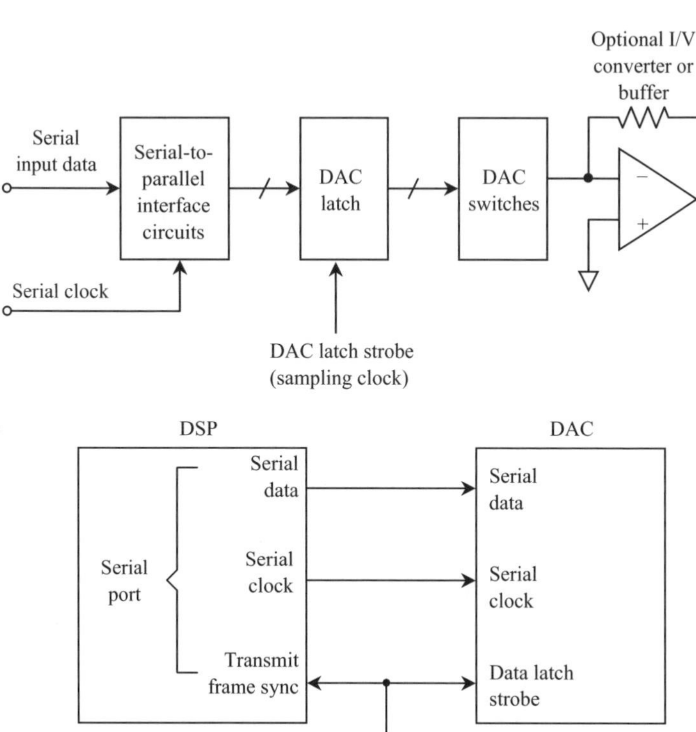

Figure 13.21 Interface between DSPs and serial DACs: (a) serial input DAC; (b) DSP and parallel DAC input

interrupts occur. Compared with this process, a DSP based system allows the processing of a continuously changing set of data samples, with no necessity to buffer and store the samples occurring.

One simple example we could discuss is to calculate the moving average of a set of data samples (see Figure 13.22). To calculate a moving average, say for temperature measurements taken at intervals of 1 s, several sequential measurements must first be saved. On reaching the current temperature, the one that occurred 10 s ago can be discarded, for a case of average of 10 samples. By moving the 10-reading span for each average, a moving average is performed.

A moving average provides a useful tool that can uncover a trend and smooth or sudden fluctuations. For example, in the case of temperature samples, one very high or very low temperature will not unduly influence the moving average calculated with ten values. Any effect an odd temperature has on the average lasts only for ten averages. The general formula for the temperature moving average contains one term

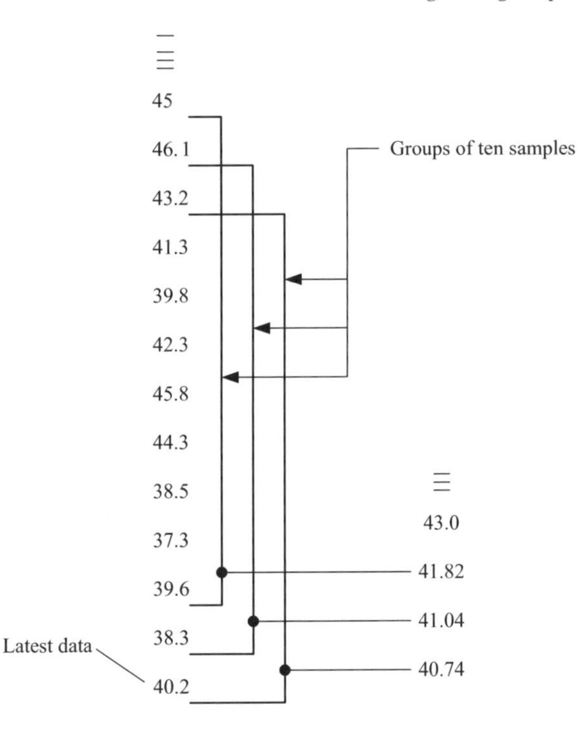

Figure 13.22 Moving average calculation as a simple example of DSP applications

per value:

$$\text{average} = \tfrac{1}{10}(T_{\text{now}}) + \tfrac{1}{10}(T_{\text{now}-1}) + \cdots + \tfrac{1}{10}(T_{\text{now}-9}).$$

You may expand or contract the equation to average more or fewer readings, depending on our specific averaging need. The general formula for a moving average incorporates as many values as you need:

$$\text{average} = \frac{1}{n}(X_n) + \frac{1}{n}(X_{n-1}) + \cdots + \frac{1}{n}(X_1).$$

Averaging makes up a small portion of a larger class of common DSP operations, namely filtering. Whereas analogue circuit engineers think of filter designs in terms of passive components and op amps, DSP system designers think of filters in terms of algorithms – ways to process information.

Most DSP filter algorithms fall into two general categories: infinite impulse response (IIR) and finite impulse response (FIR). You can use either type to develop DSP equivalents of analogue high pass, band pass, low pass, and band stop filters. Even if you have little DSP background, mastering the mathematics behind an FIR filter takes little time. The FIR filter equation below practically duplicates the equation

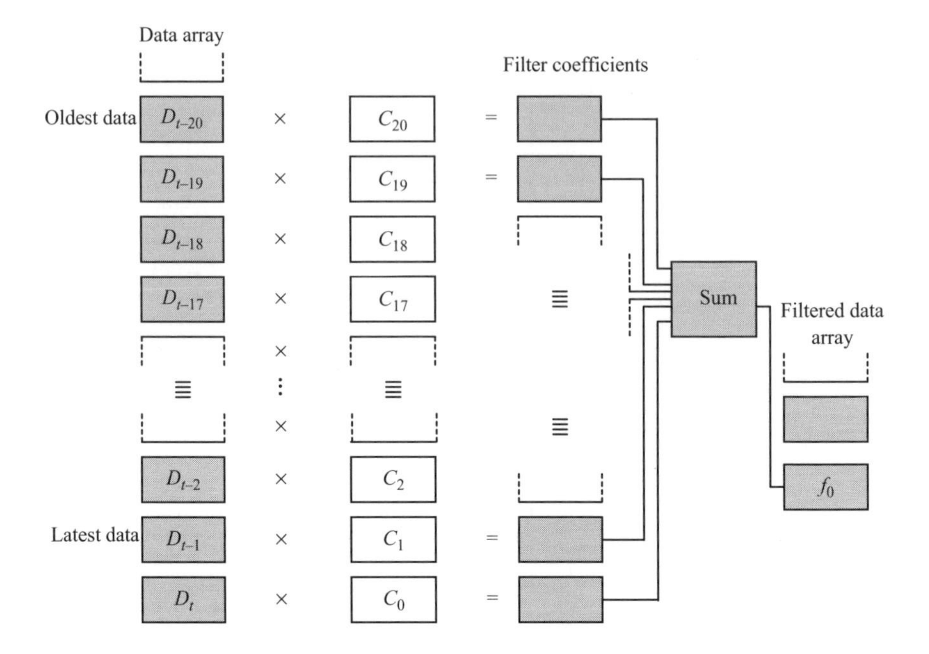

Figure 13.23 FIR filter implementation by multiplication and addition of data pointed

for a moving average:

$$F1 = (C_n * X_n) + (C_{n-1} * X_{n-1}) + \cdots + (C_1 * X_1).$$

Unlike the averaging equation, though, the FIR filter equation uses different coefficients for each multiplication term. Having discussed FIR and IIR filters earlier, let us discuss the case of an FIR filter compared to a moving average routine. Like a moving average routine, an FIR filter routine operates on the n most recent data values. (The number of coefficients, n, depends on a filter's characteristics.) Typically, a filter routine discards the oldest value as it acquires the newest value (Figure 13.23). After each new value arrives, the software multiplies each of the n values by its corresponding coefficient and sums the results.

The sum goes into a separate section of the system memory that holds the filtered information. If you use a 20-term FIR filter routine to process a signal, your computer must perform 20 multiplications and 20 additions for each value. Some FIR filter routines may require even more coefficients. The heavy use of mathematical operations differentiates DSP operations from other types of software tasks.

In spite of the amount of time a computer can spend working on maths for DSP, processing signals as discrete values has its advantages. To acquire analogue signals from 30 sensors for a strip chart recorder, for example, you may have to build, test, and adjust 30 analogue front-end circuits. You will also confront problems associated with thermal drift, voltage offsets, and component tolerances. If test conditions change, you'll have to modify and retest all 30 circuits.

When the signals are acquired with an analogue-to-digital converter (ADC) and processed digitally, though, many problems disappear. Instead of needing 30 individual filters, one DSP filter routine can operate on all the data, one channel at a time. If you plan to apply the same type of filtering to each channel, one set of coefficients will suffice. To perform different types of filtering on some channels, you simply set up an array of coefficients for each filter type. You still may have to provide antialiasing filters, though.

For details, Reference 16 is suggested.

13.10 Practical components and recent developments

During 1997 and 1998, incredible developments took place in the DSP components world. Vendors were focusing on several key aspects of the DSP architectures. The most obvious architectural improvements were in the increased 'parallelism': the number of operations the DSP can perform in an instruction cycle. An extreme example of parallelism is Texas Instruments' C6x very long instruction word (VLIW) DSP with eight parallel functional units. Although Analog Devices' super Harvard architecture (SHARC) could perform as many as seven operations per cycle, the company and other vendors were working feverishly to develop their own 'VLIW-ised' DSPs. In contrast to super scalar architectures, VLIW simplifies a DSP's control logic by providing independent control for each processing unit. During 1997, the following important developments were achieved [17].

- While announcing the first general purpose VLIW DSP, Texas Instruments also announced the end of the road for the C8x DSP family. The company emphasised the importance of the compilers for DSPs with the purchase of DSP compiler company Tartan.
- Analog Devices broke the $100 price barrier with its SHARC floating point architecture.
- Lucent Technologies discontinued new designs incorporating its 32-bit, floating point DSP. The company also focused its energy on application specific rather than general purpose DSPs. The application specific products target modems and other communication devices.
- Motorola's DSP Division recently became the Wireless Signal Processing Division, although the company still supports many general purpose DSP and audio applications.

Among the hottest architectural innovations during 1997 was the move to dual multiply-accumulate (MAC) units. The architectures of these MACs allow performing twice the digital signal processing as before. TI kicked off this evolution with its VLIW-based C6x. Meanwhile, engineers designing with DSPs need a simple method to compare processor performance. Unfortunately, as processor architectures diversify, traditional metrics such as MIPS and MOPS have become less relevant.

Alternatively, Berkeley Design Technology (BDTI, www.bdit.com) has become well known in the DSP industry for providing DSP benchmarks. Instead of using full application benchmarks, BDTI has adopted benchmark methodology based on DSP algorithm kernels, such as FFTs and FIR filters. BDTI implements its suite of 11 kernel based benchmarks (the BDTI benchmarks) on a variety of processors. The results of these benchmarks can be found in the company's *Buyer's Guide to DSP Processors* at Berkeley's web site.

To see the developments over the past 10 years, compare Reference 18 with References 17 and 19.

13.11 References

1 JONES, N.B. and WATSON J.D.M.: *Digital Signal Processing – Principles, Devices and Applications*, Peter Peregrinus/IEE 1990.
2 SCHNEIDERMAN, R.: 'Faster, more highly integrated DSPs – Designed for designers', *Wireless Systems Designs*, November 1996, pp. 12–13.
3 MARVEN, C. and EWERS. G.: *A Simple Approach to Digital Signal Processing*, Texas Instruments, 1994.
4 KULARATNA, N.: *Modern Component Families and Circuit Block Design*, Butterworth-Newnes, 2000, Chapter 4.
5 MADISETTI. V.K.: *VLSI Digital Signal Processors*, Butterworth Heinemann, 1995.
6 LEVY, M.: 'Microprocessors and DSP technologies unite for embedded applications', *EDN*, March 2, 1998, pp. 73–81.
7 PARKS, T.W. and BURRUS, C.S.: *Digital Filter Design*, Wiley & Sons, 1987.
8 OPENHEIM, A.V. and SCHAFER, R.W.: *Digital Signal Processing*, Prentice Hall, 1975 and 1988.
9 KOGGE, P.M.: *The Architecture of Pipe-Lined Computers*, Hemisphere Publishing Co., McGraw Hill, New York, 1981.
10 LAPSLEY, P., BIER, J., SHOHAM, A. and LEE, E.A.: *DSP Processor Fundamentals: Architecture and Features*, IEEE Press, 1997.
11 LEE, E. A.: 'Programmable DSP architectures : Parts I and II', *IEEE ASSP Magazine*, October 1988, pp. 4–19 and January 1989, pp. 4–14.
12 KULARATNA, N.: *Modern Electronic Test and Measuring Instruments*, IEE, 1996, Chapter 9.
13 INGLE, V.K. and PROAKIS, J.G.: *Digital Signal Processing Laboratory using the ADSP – 2101 Microcomputer*, Prentice Hall/Analog Devices 1991.
14 NEW, B.: 'A distributed arithmetic approach to designing scalable DSP chips', *EDN*, August 17, 1995, pp. 107–114.
15 FINE, B.: 'Considerations for selecting a DSP processor – ADSP 2100A Vs TMS 320C25', Analog Devices Application note ....
16 TITUS, J.: 'What is DSP all about?', *Test & Measurement World*, May 1996, pp. 49–52.

17 LEVY, M.: 'EDN's 1997 DSP – Architecture Directory', *EDN*, May 8, 1997, pp. 43–107.
18 CUSHMAN, R.H.: 'μP-like DSP chips', *EDN*, September 3, 1987, pp. 155–186.
19 LEVY, M.: 'EDN's 1998 DSP – Architecture Directory', *EDN*, April 23, 1998, pp. 40–111.
20 BELL, C.G. and NEWELL, A.: *Computer structures*, McGraw Hill, New York, USA, 1971.

Chapter 14
Sensors[1]

14.1 Introduction

Sensors convert information about the environment, such as temperature, pressure, force, or acceleration, etc., into an electrical signal. With the development of microelectronics technology with silicon as the base material in the 1970s, sensors using the properties of silicon entered the component market. Silicon's physical properties make it an ideal building material for mechanical devices. Silicon has the hardness of steel, the thermal conductivity of diamond, piezoresistive properties, a light weight, and low thermal expansion; also it is relatively inert. It is free of hysteresis and its crystalline structure is well suited to the fabrication of miniature precision products. Silicon micromechanical products have several advantages over their conventionally manufactured counterparts. They are generally much smaller. Their performance is higher because of the precise dimensional control in the fabrication and costs are lower owing to the possibility of mass scale production.

Silicon micromachining is a powerful outgrowth of semiconductor process technology whereby integrated circuit manufacturing techniques are supplemented by the silicon etching process to create very precise, miniature micromechanical structures. These silicon microstructures can have electronic features that allow conversion of physical input into electrical signals. Similarly, electrical signals can be applied to these devices to provide control functions. Initially developed in the 1950s and 1960s at leading semiconductor pioneers including Fairchild and National Semiconductor, the technology was further advanced in the 1970s at universities throughout the world. Commercial activities picked up in the early 1980s with a number of start-ups located in the Silicon Valley area of the United States.

By the beginning of 1980s, designers were able to incorporate integrated circuits on a single die with the sensor elements. Although this complicates the fabrication process and can limit the operating temperature range for the sensor, it often leads to

[1] This chapter is an edited version of chapter 7 of the Butterworth/Newnes book 'Modern Component Families and Circuit Block Design', reproduced with permission of the publisher.

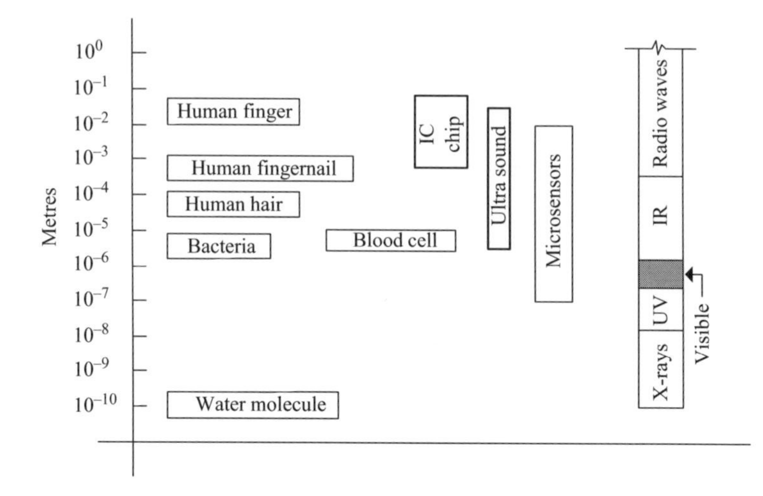

Figure 14.1 Comparative scale of microsensors

superior performance at an acceptable cost. These integrated microsensors can provide a more linear output than that of the sensor itself, or an output having a digital format that can readily be handled by associated data logging or display systems. By the late 1980s, microsensors for measuring pressure and temperature and the like were readily available, while silicon accelerometers and so forth were entering the market. Figure 14.1 shows the comparative scale of microsensors.

Nowadays, miniaturisation is the aim of many research laboratories and companies. As a part of microsystem technology sensors will also play a major role in the future and sensor interfaces and related standards are getting ready for this developing component sector. Many producers in Japan, Europe and the United States forecast growth rates for sensors above 10 per cent beyond the year 2000.

This chapter is a summary of modern semiconductor sensors, their characteristics and applications with some representative devices.

14.2 The properties of silicon and their effects on sensors

Silicon is a suitable material for sensor technologies as it manifests sufficient physical and chemical effects of an acceptable strength for use in uncomplicated structures across a wide range of temperatures. Table 14.1 presents the most important effects and their applications for sensor technology.

The use of silicon has a number of implications for sensors. Firstly, the physical properties of silicon can be used directly to measure the desired dimension, as indicated in Table 14.1. However, the range of possibilities is limited. Beyond this, for example, silicon can be extremely useful when used as the substrate for thin-film sensors, even when information processing electronics are integrated. For details see Reference 1.

Table 14.1 The effect of silicon used in sensors

Physical dimension	Effect	Application
Radiation	Photoresistive	Photoresistor
	Photointerface	Photodiode, phototransistor
	Ionisation	Nuclear radiation sensor
	Photocapacitive	Photocapacitance
Mechanical	Piezoresistive, piezojunction and piezotunnel	Piezoresistive power and pressure sensors, piezoelectric diode and transistor
Thermal	Thermal resistance	Resistance temperature sensors
	Thermojunction	Temperature sensors (diode, transistor)
	Thermoelectric	Thermopile
	Pyroelectric	Pyroelectric sensor
Magnetic signals	Magnetoresistive	Magnetoresistive sensors
	Hall	Hall generator
	Magnetic interface	Magnetic diode and transistor
Chemical signals	Charge sensitive field	ISFET

14.3 Micromechanics

The term *micromechanics*, with its obvious similarity to the term microelectronics, is used to describe a completely new discipline. Its objective is the construction of complex microsystems consisting largely of integrated sensors, a logical signal processing stage and actuators. In this connection, micromechanics refers to the fabrication of mechanical structures whose geometrical size, at least in one dimension, is so small that it no longer is sensible to use the methods of fine mechanics. Depending on the boundary conditions imposed by the desired function or by the properties of the material, this limit may be located anywhere between the millimetre and the sub-micrometre range (see Figure 14.2). In contrast to microelectronics, micromechanics is concerned with the production of three-dimensional structures.

Modern micromechanics make it possible to produce micropumps, microvalves, micro-loudspeakers and microphones and therefore it is of interest to disciplines other than sensor technology [1, 2].

14.4 Temperature sensors

The most common electronic temperature measurement devices currently available include the thermocouple, the resistance temperature detector, the thermistor, and the

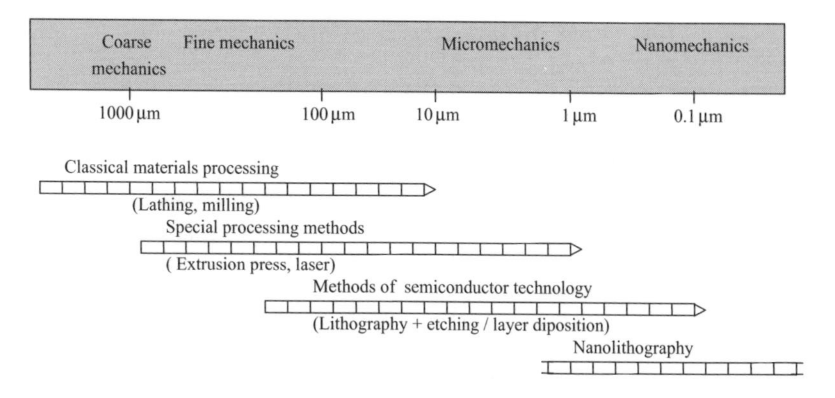

Figure 14.2 The size of micromechanics

integrated circuit temperature transducer. All have associated application benefits and limitations which are delineated in Table 14.2 [3].

14.4.1 Resistance temperature detectors

Resistance temperature detectors (RTDs) are wire windings or thin-film serpentines that exhibit changes in resistance with changes in temperature. Although metals such as copper, nickel and nickel–iron are often used, the most linear, repeatable and stable RTDs are constructed from platinum.

14.4.2 Negative temperature coefficient thermistors

Negative temperature coefficient (NTC) thermistors are composed of metal oxide ceramics, are low in cost, and are the most sensitive temperature sensors. They are also, however, the most non-linear, and have a negative temperature coefficient. Thermistors are offered in a huge variety of sizes, base resistance values and R–T curves to facilitate both packaging and output linearisation schemes.

14.4.3 Thermocouples

Thermocouples consist of two dissimilar metal wires welded together at both ends to form two junctions. Temperature differences between the junctions cause a thermoelectric potential (i.e. a voltage) between the two wires. By holding the reference junction at a known temperature and measuring this voltage, the temperature of the sensing junction can be deduced. Thermocouples have very large operating temperature ranges despite a very small size. However, they have low output voltages, susceptibility to noise pickup by the wire loop, and relatively high drift. Silicon integrated circuits are available for interface with the thermocouples. Some examples are AD 594 and AD 595 from Analog Devices [4, 5].

Table 14.2 A comparison of thermal sensors (reproduced by permission of Microswitch, Honeywell Inc.)

Characteristic	Platinum RTD		Thermistor	Thermocouple	Silicon
	Thin-film type	Wire wound type			
Active material	Platinum thin film	Platinum, wire wound	Metal oxide ceramic	Two dissimilar metals	Silicon transistor cascade
Relative sensor cost	Moderate to low	Moderate	Low to moderate	Low	Low
Relative system cost	Moderate	Moderate	Low to moderate	High	Low
Temperature range	−200 °C to 750 °C (560 °C max. typ.)	−200 °C to 850 °C (600 °C max. typ.)	−100 °C to 500 °C (125 °C max. typ.)	−270 °C to 1800 °C	−40 °C to 125 °C
Changing parameter	Resistance	Resistance	Resistance	Voltage	Voltage
Base value	100 Ω to 200 Ω	100 Ω	1 kΩ to 1 MΩ	<10 μV at 25 °C	750 mV at 25 °C
Interchangeability	±1%, ±3 °C	±0.06%, ±0.2 °C	±10%, ±2 °C typ.	±0.5%, ±2 °C	±1%, ±3 °C
Stability	Excellent	Excellent	Moderate	Poor	Moderate
Sensitivity	0.39% K^{-1}	0.39% K^{-1}	−4% K^{-1}	40 μV K^{-1}	10 mV K^{-1}
Relative sensitivity	Moderate	Moderate	Highest	Low	Moderate
Linearity	Excellent	Excellent	Logarithmic, poor	Moderate	Moderate
Slope	Positive	Positive	Negative	Positive	Positive
Noise susceptibility	Low	Low	Low	High	Low
Lead resistance errors	Low	Low	Low	High	Low
Special requirements		Lead compensation	Linearisation	Reference junction	

14.4.4 Silicon temperature sensors

Temperature sensors that utilise the temperature dependent properties of silicon are appearing in the market in a wide variety of types and the prices are reasonably low. Practical integrated circuits available in the market basically are either voltage output or current output temperature sensors.

Figure 14.3 (a) shows the use of the temperature dependence of the PN junction voltage to provide a temperature dependent voltage output, V_{be}. The voltage is related to the temperature and other parameters by the equation

$$V_{be} = \frac{2kT}{q} \ln\left(\frac{I_F}{I_S}\right), \tag{14.1}$$

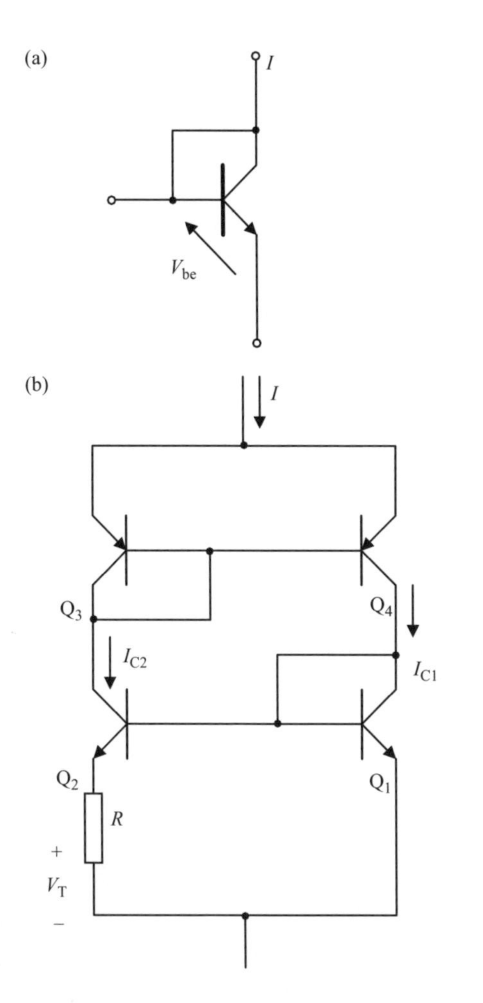

Figure 14.3 Temperature sensing using PN junction properties: (a) transistor used as temperature sensor; (b) integrated temperature sensor

where I_F is the forward current of the transistor, I_S is the saturation current of the transistor, q is the elementary charge, and k is the Boltzmann constant.

If the ratio I_F/I_S is kept constant, then the result would be a sensor exhibiting ideal linear temperature dependence of the forward voltage.

A similar relationship is found in transistors. If the collector and base are held at the same potential (Figure 14.3(a)) then the relationship of the base emitter voltage, V_{be}, to the collector current I_C is given by

$$V_{be} = \frac{kT}{q} \ln\left(\frac{I_C}{I_S}\right). \tag{14.2}$$

Here again the saturation current I_S is influenced by the temperature dependence of a number of parameters, similar to the case of diodes. Despite this, if the collector current I_C is held constant and the components are carefully selected, it is possible to obtain approximately linear behaviour for temperatures between −50 °C and 150 °C.

Motorola's MTS 10X series silicon temperature sensors are a classic example of this technique. The device family allows temperature measurement precisely in the range −40 °C to 150 °C

Modern temperature management ICs range from purely analogue voltage vs temperature devices to mixed signal VLSI chips containing logic and ADCs. Most of the ICs rely on a bandgap reference with a known temperature coefficient to provide temperature information.

14.4.4.1 Simple current output temperature transducers

Most simple current output temperature transducers use more practical forms of the circuit in Figure 14.3. The AD 590 from Analog Devices is an example of this. Referring to Figure 14.3(b), the difference in base emitter voltages of transistors Q_1 and Q_2 is given by

$$\Delta V_{be} = \frac{kT}{q} \ln\left[\frac{I_{C2}}{A_2} \bigg/ \frac{I_{C1}}{A_1}\right]. \tag{14.3}$$

The temperature dependence of V_{be} therefore is solely dependent on the ratio, r, of the two collector current densities.

$$\Delta V_{be} = \frac{kT}{q} \ln(r). \tag{14.4}$$

Provided that this ratio can be kept constant, ΔV_{be} is directly proportional to the absolute temperature. There are two ways of keeping the ratio constant. First, it is possible to operate two transistors with the same geometric dimensions on a single chip using two collector currents ($I_{C1} \neq I_{C2}$). The alternative is for a constant collector current to flow through two transistors with different emitter areas ($A_1 \neq A_2$). The second variant has been of greater practical relevance because of the simpler circuitry involved. An example of this type of integrated sensor is presented in the basic circuit diagram in Figure 14.3(b). Transistors Q_1 and Q_2 perform the detection function. The identical transistors Q_3 and Q_4 act as current mirrors. This causes a splitting of the current I into two equal collector currents I_{C1} and I_{C2}. The emitter area of Q_2 should be r times that of Q_1. Its collector current density therefore is only $1/r$ that of T_1. The difference of ΔV_{be} causes a current I_{C2} that is proportional to the temperature to flow across a resistor R. Because of the current mirroring, the value of I also must be proportional to the absolute temperature. Laser alignment of the resistance R makes it possible to adjust the constant of proportionality in eq. (14.4) to 1 $\mu A\,K^{-1}$. If the circuit is changed to allow for a voltage output signal then temperature coefficients of a few millivolts per kelvin can be achieved.

In the AD 590, this ΔV_{be}, directly proportional to absolute temperature (PTAT), is converted to a PTAT current by low temperature coefficient thin-film resistors. The total current of the device is then forced to be a multiple of this PTAT current.

Figure 14.4(a) is a schematic diagram of the AD 590. Q_8 and Q_{11} are the transistors that produce the PTAT voltage. R_5 and R_6 convert the voltage to current. Q_{10}, whose collector current tracks the collector currents in Q_9 and Q_{11}, supplies all the bias and substrate leakage current for the rest of the circuit, forcing the total current to be PTAT. R_5 and R_6 are laser trimmed on the wafer to calibrate the device at +25 °C. Figure 14.4(b) shows the typical V–I characteristic of the circuit at +25 °C and the temperature extremes.

The device features a 1 $\mu A\,K^{-1}$ linear current output over the temperature range −50 °C to 150 °C. Some applications and accuracy are discussed in References 6 and 7.

Current output temperature transducers have a number of advantages.

- They are based on a linear relationship and are highly repeatable.
- The current is independent of voltage drops, voltage noise, common mode voltage, and is practically independent of excitation voltage.
- The current can be translated to a voltage at a remote destination via an appropriate value of resistance ($V = IR$); simple offsetting circuitry may be used when necessary.
- They are easy to use; they require no linearisation circuitry, high precision voltage amplifiers, resistance measuring circuitry or cold-junction compensation.

Current output temperature sensors are widely used for cold-junction compensation of thermocouple circuitry.

When voltage drops and noise are not an important consideration, it may be more convenient to work with a voltage output temperature transducer. These provide a direct output to an analogue-to-digital converter or a comparator set point. Many practical components provide a voltage output as well as other functions.

14.4.4.2 AD 22100: ratiometric voltage output temperature sensor

The AD 22100 is a ratiometric temperature sensor IC whose output voltage is proportional to the power supply voltage. The heart of the sensor is a proprietary temperature dependent resistor, similar to an RTD, built into the IC. Figure 14.5(a) is a simplified block diagram of the AD 22100.

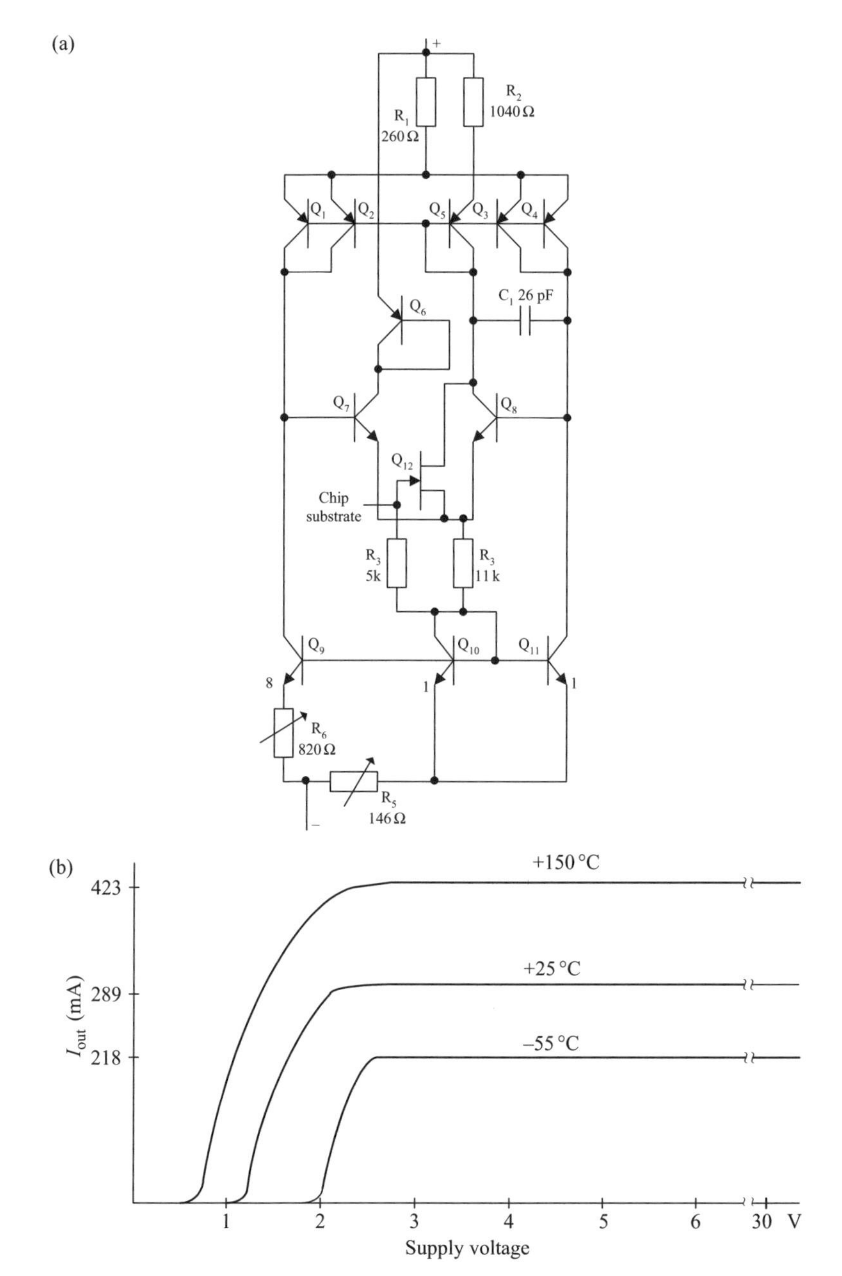

Figure 14.4 The AD 590: (a) schematic diagram; (b) V–I characteristics (reproduced by permission of Analog Devices, Inc.)

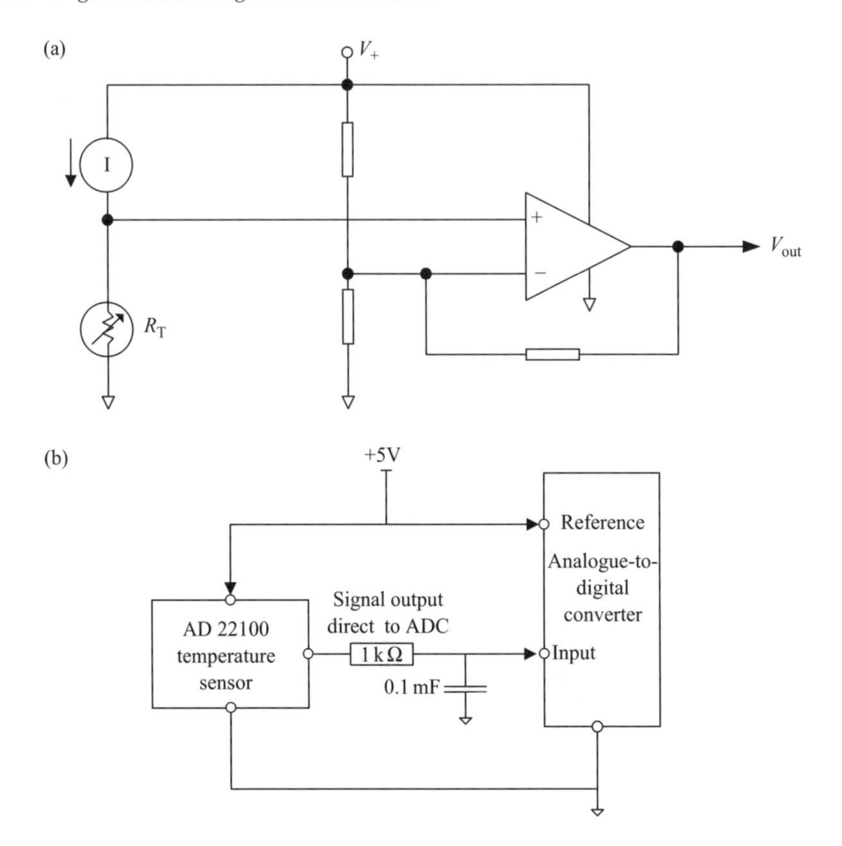

Figure 14.5 The AD 22100 voltage output temperature sensor: (a) simplified block diagram; (b) an application (reproduced by permission of Analog Devices, Inc.)

The temperature dependent resistor, R_T exhibits a change in resistance that is nearly linearly proportional to temperature. This resistor is excited with a current source that is proportional to power supply voltage (V_+). The resulting voltage across R_T therefore is both supply voltage proportional and linearly varying with the temperature (T_A). The remainder of the AD 22100 consists of an op amp signal conditioning block that takes the voltage across R_T and supplies the proper gain and offset to achieve the following output voltage function:

$$V_\mathrm{out} = \left(\frac{V_+}{5}\right) \cdot (1.375 + (22.5 \times 10^{-3} \cdot T_\mathrm{A})). \qquad (14.5)$$

Owing to its ratiometric nature, the device offers a cost effective solution when used as an interface to an analogue-to-digital converter. This is accomplished by using the ADC's +5 V power supply as a reference to both the ADC and the AD 2100 (see Figure 14.5(b)), eliminating the need for a precision reference.

The devices such as the AD 22100 provide low cost temperature measurement for microprocessor and microcontroller based systems. Many inexpensive 8-bit microprocessors now offer an onboard 8-bit ADC capability at a modest cost. Total 'cost of ownership' then becomes a function of the voltage reference and analogue signal conditioning necessary to mate the analogue sensor with the microprocessor ADC.

Such devices can provide a low cost system by eliminating the need for a precision voltage reference and any additional active components. The ratiometric nature of the device allows the microprocessor to use the same power supply as its ADC reference. Variations in the supply voltage have little effect as the sensor and the ADC use the supply as their reference. For details see Reference 8.

14.4.5 Temperature management ICs

Silicon temperature sensors easily can be combined with other circuit blocks for temperature control [9, 10]. Over temperature alarms, faulty circuitry shutdown, or initiation of corrective actions in a thermal feedback loop are ways in which temperature management ICs can prevent catastrophic failures. Devices commercially available include temperature controllers, airflow temperature sensors, serial digital output, thermostat ICs, and programmable thermostat ICs. These devices are produced on a mass scale, using the standard IC production processes and prices vary from $0.50 to $4. Most of these ICs rely on a bandgap reference with a known temperature coefficient, coupled with other analogue and digital circuitry which may include the logic and ADCs as well. Two modern trends in temperature management ICs are increasing incorporation of digital circuitry and incorporation of more management functions [9].

Temperature control ICs include a temperature sensor that generates a voltage output proportional to the absolute temperature and a control signal from one or two outputs when the device is below or above a specified temperature range. An example of these devices is the TMP 01 from Analog Devices.

The TMP 01 consists of a bandgap voltage reference combined with a pair of matched comparators. The reference provides both a constant 2.5 V output and a voltage proportional to absolute temperature (VPTAT), which has a precise temperature coefficient of $5\,\text{mV}\,\text{K}^{-1}$ and is 1.49 V (nominal) at +25 °C. The comparators compare the VPTAT with the externally set temperature trip points and generate an open-collector output signal when one of these thresholds has been exceeded. Figure 14.6(a) indicates the functional block diagram of the TMP 01.

Hysteresis also is programmed by the external resistor chain and determined by the total current drawn out of the 2.5 V reference. This current is mirrored (Figure 14.6(b)) and used to generate a hysteresis offset voltage of the appropriate polarity after a comparator has been tripped. The comparators are connected in parallel, which guarantees no hysteresis overlap and eliminates erratic transitions between adjacent trip zones.

The device utilises proprietary thin-film resistors in conjunction with production laser trimming to maintain a typical temperature accuracy of ±2 °C over the rated temperature range, with excellent linearity. The open collector outputs are capable of sinking 20 mA, enabling the TMP 01 to drive control relays directly.

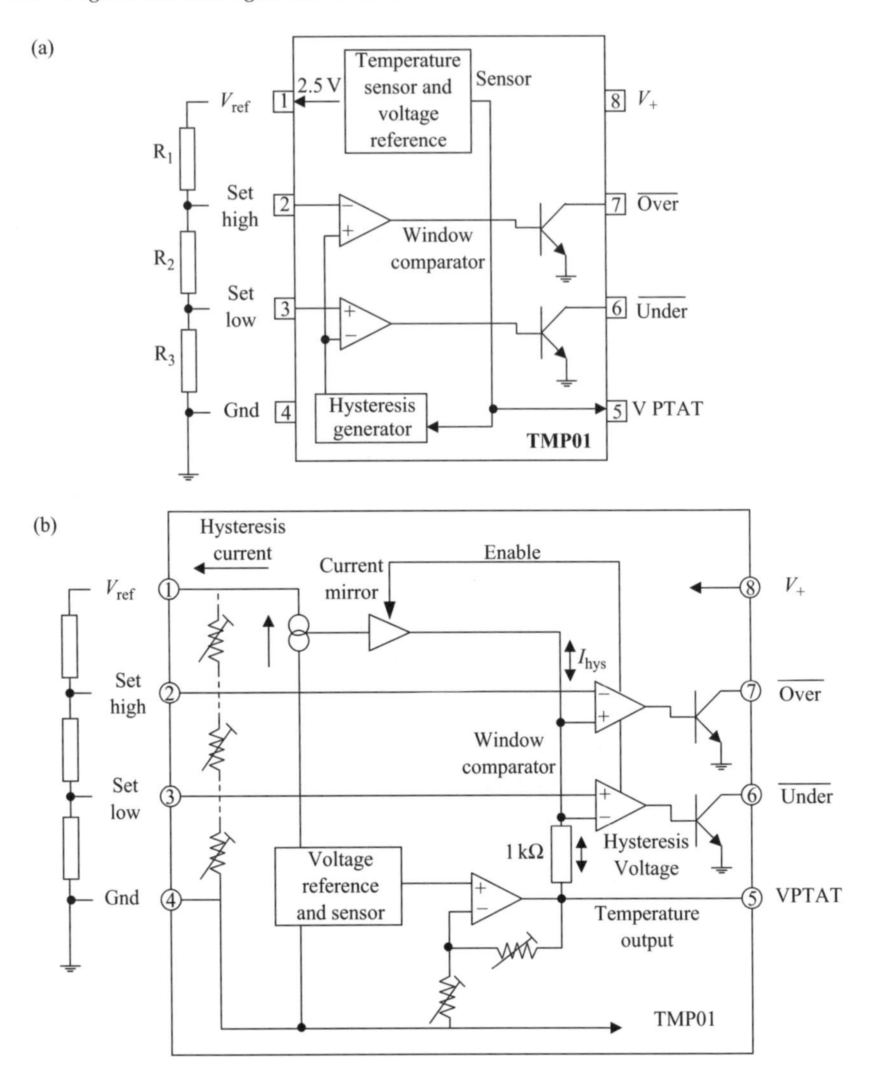

Figure 14.6 The TMP 01, a low power, programmable temperature controller: (a) functional block diagram; (b) detailed block diagram (reproduced by permission of Analog Devices, Inc.)

The TMP 01 is a very linear voltage output temperature sensor, with a window comparator that can be programmed by the user to activate one of two open collector outputs when a predetermined temperature set point voltage has been exceeded. A low drift voltage reference is available for set point programming (see Figure 14.7).

In many temperature sensing and control applications some type of switching is required. The open collector outputs (over and under) of the TMP 01 can be used for applications similar to turning on a heater or to switching off a motor. In such

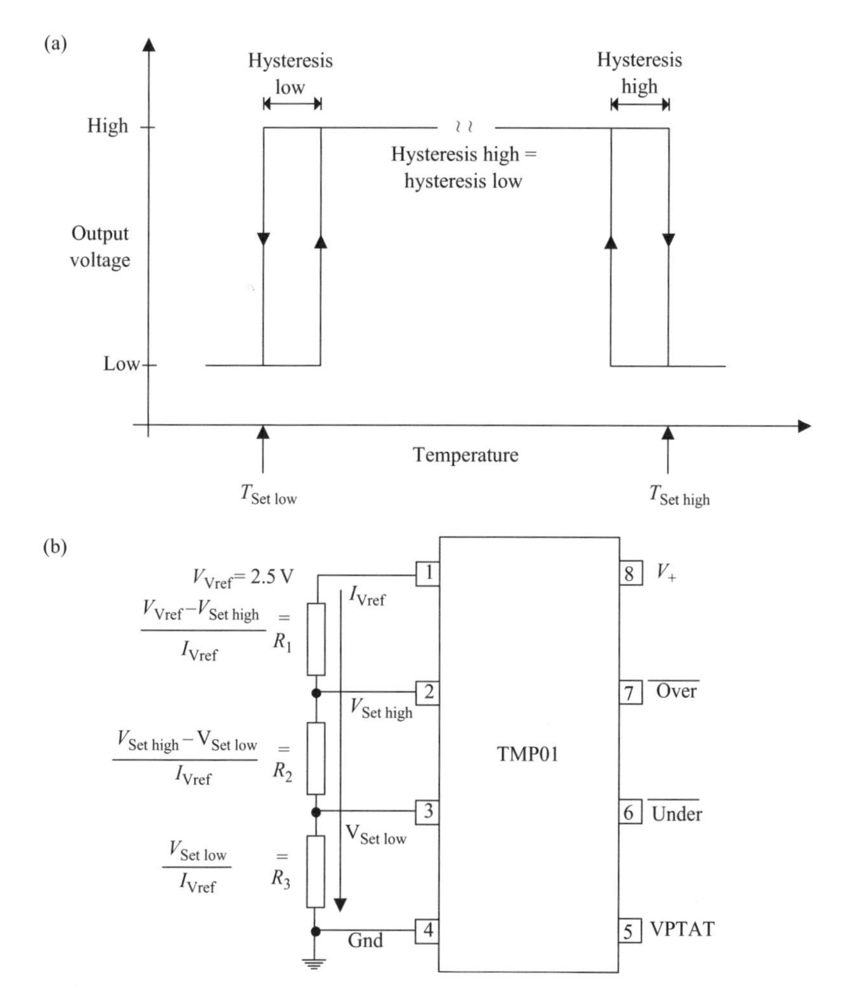

Figure 14.7 The TMP 01: (a) hysteresis profile; (b) set point programming

applications the switches need to handle large currents usually much more than 20 mA, which is the rated current of the outputs. In such cases, external switching devices such as relays, power MOSFETs, thyristors, IGBTs or Darlington can be used as shown in Figure 14.8. For further details, see Reference 8.

14.4.6 Serial digital output thermometers

Several manufacturers offer basic sensor devices coupled with analogue to digital converters. These ICs allow a series digital output from the ADC proportional to the temperature. Examples are TMP 03/04 from Analog Devices, LM 75 from National & DS 1621 from Dallas Semiconductor; see References 11–14.

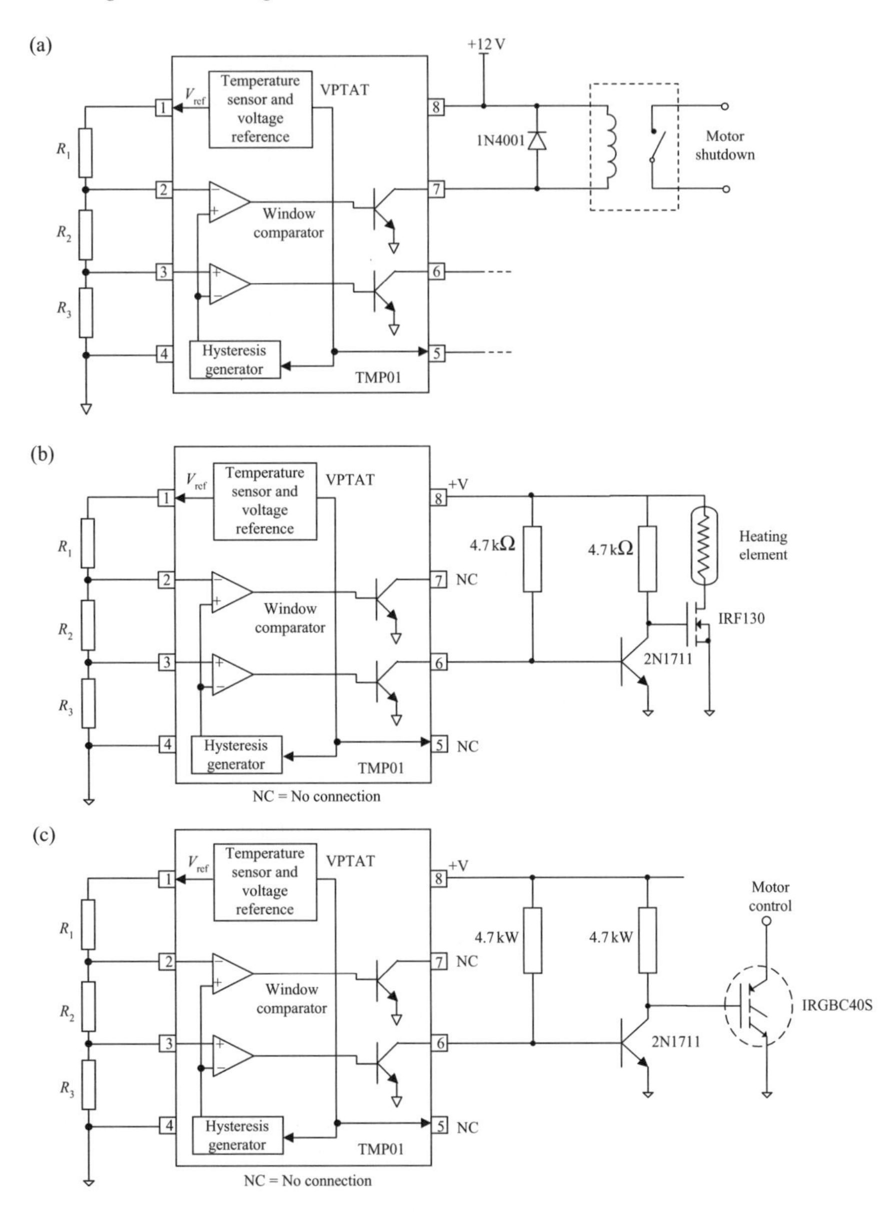

Figure 14.8 *Switching loads with the open collector outputs of the TMP 01: (a) reed relay drive; (b) driving an N-channel MOSFET; (c) driving an IGBT (reproduced by permission of Analog Devices, Inc.)*

14.4.6.1 TMP 03/04

The TMP 03/TMP 04 is a monolithic temperature detector generating a modulated serial digital output that varies in direct proportion to the temperature of the device. An onboard sensor generates a voltage precisely proportional to absolute temperature that is compared to an internal voltage reference and input into a precision digital modulator.

The sensor output is digitised by a first order Σ-Δ modulator (see Figure 14.9(a)). This type of converter utilises time domain over-sampling and a high accuracy comparator to deliver 12 bits of effective accuracy in an extremely compact circuit. Figure 14.9(a) is a basic functional block diagram and Figure 14.9(b) describes the first order modulator interacting with the VPTAT and the voltage reference source.

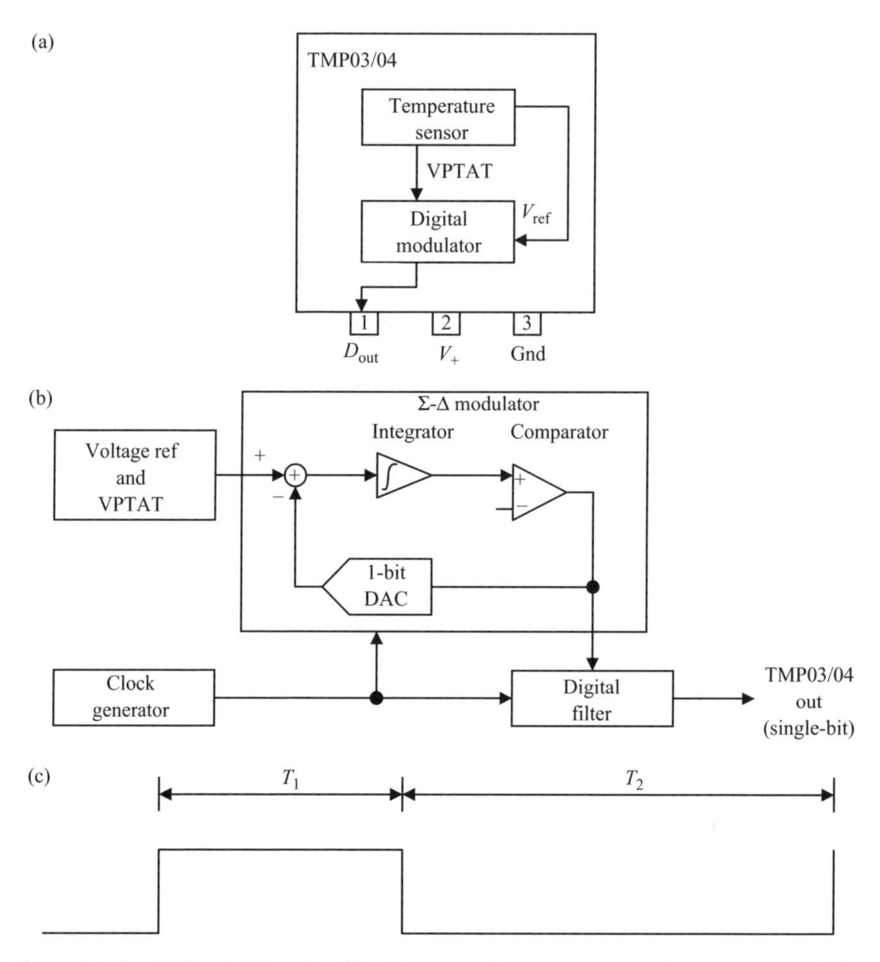

Figure 14.9 TMP 03/04 serial digital output thermometer: (a) functional block diagram; (b) block diagram showing Σ-Δ modulator; (c) output format (reproduced with permission by Analog Devices, Inc.)

The modulated output of the comparator is encoded using a circuit technique that results in a serial digital signal with a mark-space ratio format easily decoded by any microprocessor into either Celsius or Fahrenheit degrees, and readily transmitted or modulated over a single wire. Most important, the encoding method neatly avoids major error sources common to other modulation techniques, as it is clock independent.

14.4.6.1.1 Output encoding

The TMP 03/04 is designed as a low cost three terminal device with the output format shown in Figure 14.9(c). This patented design avoids an accurate external clock or high accuracy, low drift types of an internal clock system within the IC. The modulation and encoding techniques within the TMP 03/04 achieve this by using a simple, compact onboard clock and an over-sampling digitiser that are insensitive to sampling rate variations. The digitised signal is encoded into a ratiometric format in which the exact frequency of the clock is irrelevant, and the effects of clock variations are effectively cancelled on decoding by the digital filter.

The output of the TMP 03/TMP 04 is a square wave with a nominal frequency of 35 Hz (±20 per cent) at +25 °C. The output format is readily decoded by the user as in Figure 14.9(c);

$$\text{temperature } (^\circ\text{C}) = 235 - \frac{(400 \times T_1)}{T_2}. \tag{14.6}$$

The time periods T_1 (high period) and T_2 (low period) are values easily read by a microprocessor timer/counter port, with the preceding calculations performed in software. Because both periods are obtained consecutively, using the same clock, performing the division indicated in these formulas results in a ratiometric value independent of the exact frequency of, or drift in, either the originating clock of the TMP 03/TMP 04 or the user's counting clock. Figure 14.10 shows the output frequency and T_1/T_2 values vs temperature.

14.4.6.1.2 Application considerations

These types of component are quite useful in applications such as isolated sensors, environmental control systems, computer thermal monitoring, thermal protection, and industrial process control and power system monitors. The low voltage power supply (4.5–7 V) of these devices, low cost 3-pin package, low power consumption and the flexible open collector output (TMP 03) or CMOS/TTL compatible output (TMP 04) are the useful features of these devices for the wide variety of applications proposed.

Precision analogue products such as the TMP series require a well filtered power source. Because the TMP 03/04 devices operate from a single +5 V supply, it is convenient to use the logic supply. Unfortunately, the logic supply often is a switchmode design, which generates noise in the 20 kHz–1 MHz range. In addition, fast logic gates can generate glitches of hundreds to a few millivolts in amplitude due to wiring resistance and inductance.

To minimise the noise affecting the operation, the circuit arrangement in Figure 14.11(a) is proposed. Even if a separate power supply trace is not available,

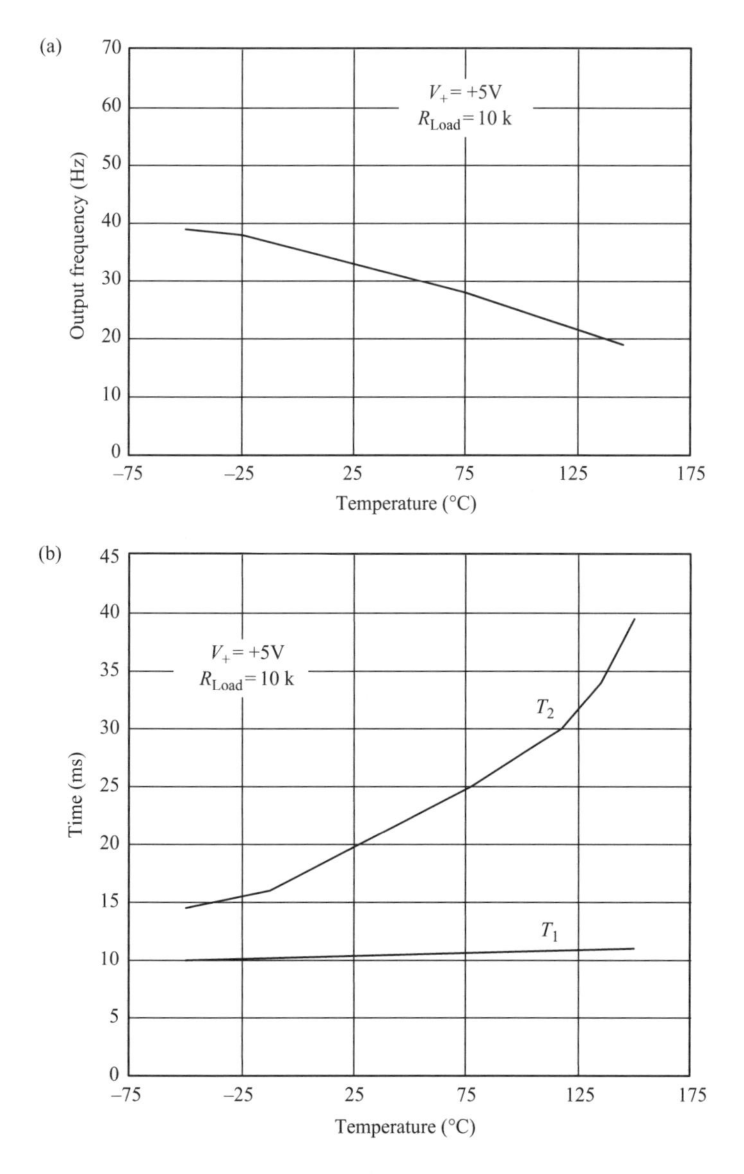

Figure 14.10 *TMP 03/04 output vs temperature: (a) output frequency vs temperature; (b) T_1 and T_2 vs temperature (reproduced by permission of Analog Devices, Inc.)*

however, generous supply by passing will reduce supply line induced errors. Local supply bypassing consisting of a 10 μF tantalum electrolytic in parallel with a 0.1 μF ceramic capacitor (Figure 14.11(b)) is recommended. As the quiescent power supply current of the device typically is 900 μA, a simple RC filter network as in Figure 14.11(c) could be used when the device drives a light load such as a CMOS gate.

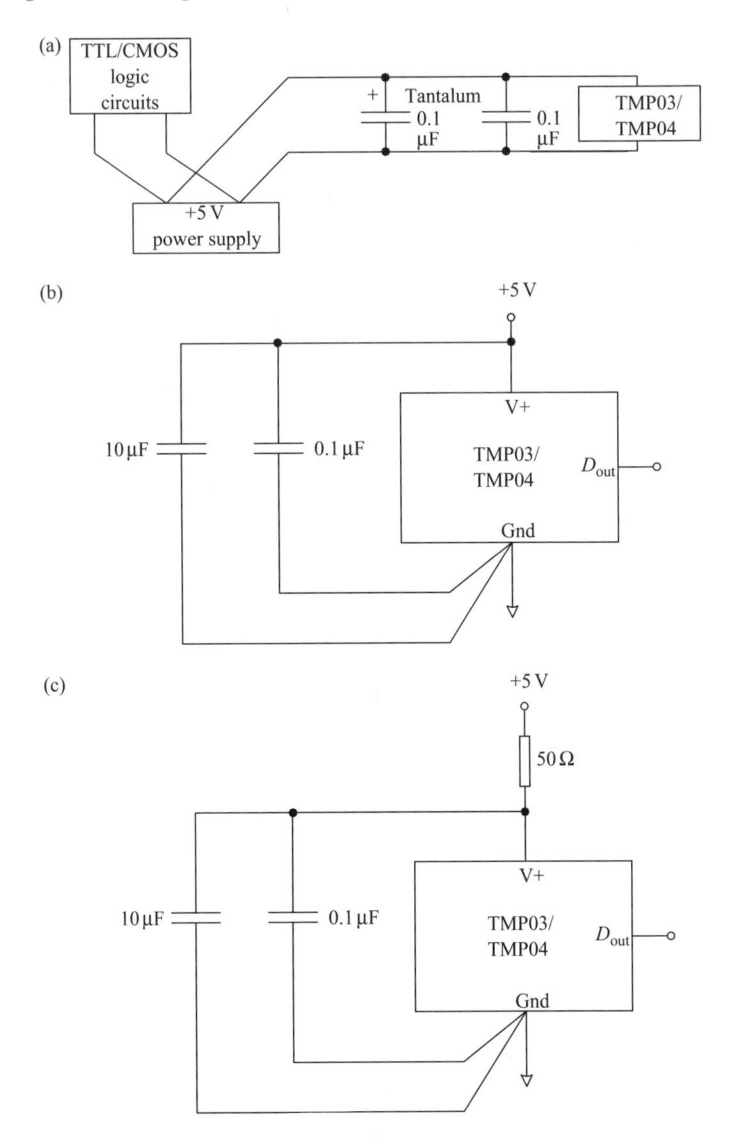

Figure 14.11 Supply bypassing techniques for TMP 03/04: (a) use of separate supply traces; (b) simple capacitor bypassing; (c) RC filter using a 50 Ω resistor and capacitors (reproduced by permission of Analog Devices, Inc.)

The TMP 03 (Figure 14.12(a)) has an open collector NPN output suitable for driving a high current load, such as an opto-isolator. Because the output source current is set by the pull-up resistor, output capacitance should be minimised in TMP 03 applications. Otherwise, unequal rise and fall times will skew the pulse width and introduce measurement errors.

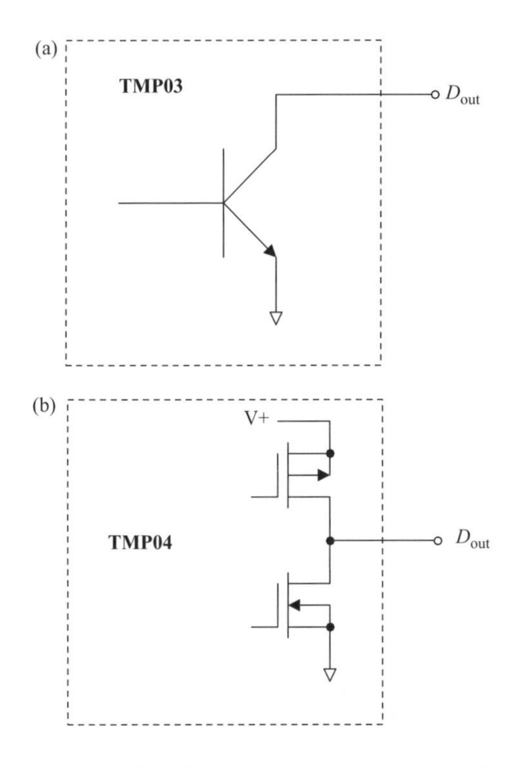

Figure 14.12 The TMP 03/04 digital output structures: (a) the TMP 03 open collector output; (b) the TMP 04 totem-pole CMOS output

The TMP 04 has a 'totem-pole' CMOS output (Figure 14.12(b)) and provides a rail-to-rail output drive for logic interfaces. The rise and fall times of the TMP 04 output are closely matched to minimise errors caused by capacitive loading. If load capacitance is large (for example when driving a long cable), an external buffer may improve accuracy. For more details on output configurations and interfaces to low voltage logic and the like, see Reference 11.

14.4.6.1.3 Microcontroller and DSP interfaces

Here is an example of an 80C51 interface. The TMP 03/TMP 04 output is easily decoded with a microprocessor. The microprocessor simply measures the T_1 and T_2 periods in software or hardware and then calculates the temperature using eq. (14.6).

Because the TMP 03/TMP 04's output is ratiometric, precise control of the counting frequency is not required. The only timing requirements are that the clock frequency is high enough to provide the required measurement resolution and that the clock source is stable. The ratiometric output of the TMP 03/TMP 04 is an advantage because the microcomputer's crystal clock frequency is often dictated by the serial baud rate or other timing considerations.

Pulse width timing usually is done with the microcomputer's on-chip timer. A typical example, using the 80C51, is shown in Figure 14.13. This circuit requires only

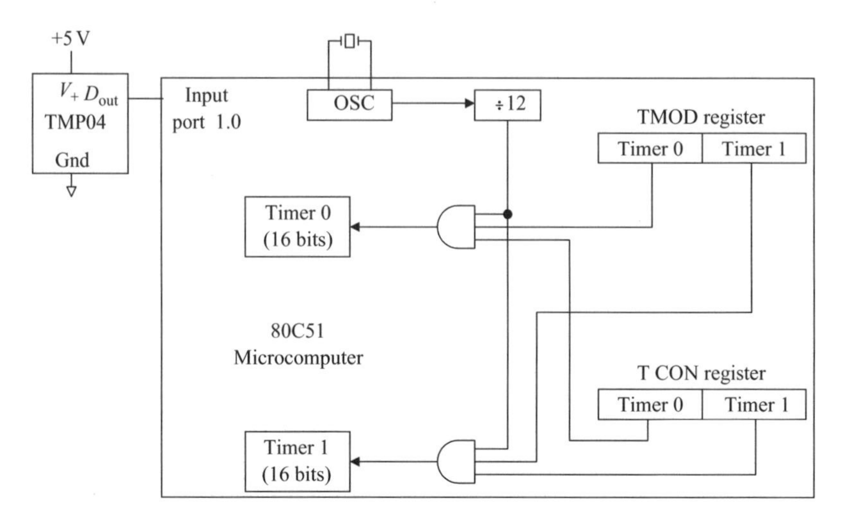

Figure 14.13 *A TMP 04 and 80C51 microcomputer interface (reproduced with permission from Analog Devices, Inc.)*

one input pin on the microcomputer, which highlights the efficiency of the TMP 04's pulse width output format. Traditional serial input protocols, with data line, clock and chip select, usually require three or more I/O pins.

The 80C51 has two 16-bit timers. The clock source for the timers is the crystal oscillator frequency divided by 12. Therefore, a crystal frequency of 12 MHz or greater will provide resolution of 1 μs or less. The 80C51 timers are controlled by two dedicated registers. The TMOD register controls the timer mode of operation, while TCON controls the start and stop times. Both the TMOD and TCON registers must be set to start the timer.

Software for the interface is shown in listing 14.1. The program monitors the TMP 04 output, and turns the counters on and off to measure the duty cycle. The time that the output is high is measured by timer 0, and the time that the output is low is measured by timer 1. When the routine finishes, the results are available in special function registers 08AH to 08DH.

Listing 14.1 *An 80C51 software routine for the TMP 04 (reproduced by permission of Analog Devices, Inc.)*

```
Test of a TMP 04 interface to the 80C51, using timer 0 and timer
1 to measure the duty cycle
This program has three steps


1.  Clear the timer registers, then wait for a low-to-high
    transition on input P1.0 (which is connected to the output
    of the TMP 04.
```

```
2.  When P1.0 goes high, timer 0 starts. The program then
    loops, testing P1.0.
3.  When P1.0 goes low, timer 0 stops & timer 1 starts. The
    program loops until P1.0 goes low, when timer 1 stops
    and the TMP 04's T1 and T2 values are stored in special
    function registers 8AH through 8DH (TL0 through TH1).

Primary controls
$ MOD51
$ TITLE(TMP 04 interface, Using T0 and T1)
$ PAGEWIDTH (80)
$ DEBUG
$ OBJECT

Variable declarations
PORT1           DATA  90H                 SFR register for port 1
TCON            DATA  88H                 timer control
TMOD            DATA  89H                 timer mode
THO             DATA  8CH                 timer 0 hi byte
TH1             DATA  8DH                 timer 1 hi byte
TL0             DATA  8AH                 timer 0 low byte
TL1             DATA  8BH                 timer 1 low byte

                ORG   100H                arbitrary start

READ_TMP 04:    MOV   A,#00               clear the
                MOV   TH0,A               counters
                MOV   TH1,A               first
                MOV   TL0,A
                MOV   TL1,A
WAIT_LO:        JB    PORT1.0, WAIT_LO    wait for TMP 04 output to go low
                MOV   A.#11H              get ready to start timer0
                MOV   TMOD,A
WAIT_HI:        JNB   PORT1.0, WAIT_HI    wait for output to go high
Timer 0 runs while TMP 04 output is high

                SETB  TCON.4              start timer 0
WAITTIMER0:     JB    PORT1.0,WAITTIMER0
                CLR   TCON.4              shut off timer 0

Timer 1 runs while TMP 04 output is low

                SETB  TCON.6              start timer 1
WAITTIMER1:     JNB   PORT1.0,WAITTIMER1
                CLR   TCON.6              stop timer 1
                MOV   A,#OH               get ready to disable timers
                MOV   TMOD,A
                RET
                END
```

When the READ_TMP 04 routine is called, the counter registers are cleared. The program sets the counters to their 16-bit mode, and then waits for the TMP 04 output to go high. When the input port returns a logic high level, timer 0 starts. The timer continues to run while the program monitors the input port. When the TMP 04 output goes low, timer 0 stops and timer 1 starts. Timer 1 runs until the TMP 04 output goes high, at which time the TMP 04 interface is complete. When the subroutine ends, the timer values are stored in their respective SFRs and the TMP 0's temperature can be calculated in software.

Because the 80C51 operates asynchronously to the TMP 04, there is a delay between the TMP 04 output transition and the start of the timer. This delay can vary between 0 μs and the execution time of the instruction that recognised the transition. The 80C51's 'jump on port-bit' instructions (JB and JNB) require 24 clock cycles for execution. With a 12 MHz clock, this produces an uncertainty of 2 μs (24 clock cycles/12 MHz) at each transition of the TMP 04 output. The worst case condition occurs when T1 is 4 μs shorter than the actual value and T2 is 4 μs longer. For a 25 °C reading ('room temperature'), the nominal error caused by the 2 μs delay is only about ±0.5 °C.

The TMP 04 also easily interacts with digital signal processors, such as the ADSP-210x series. Again, only a single I/O pin is required for the interface (Figure 14.14).

The ADSP-2101 has only one counter, so the interface software differs somewhat from the 80C51 example. The lack of two counters is no limitation, however, because the DSP architecture provides very high execution speed. The ADSP-2101 executes one instruction for each clock cycle, vs one instruction for 12 clock cycles in the 80C51, so the ADSP-2101 actually produces a more accurate conversion while using a lower oscillator frequency.

The timer of the ADSP-2101 is implemented as a down counter. When enabled by a software instruction, the counter is decremented at the clock rate divided by a programmable pre-scaler. Loading the value $n - 1$ into the pre-scaler register will divide the crystal oscillator frequency by n.

For the circuit of Figure 14.14, therefore, loading 4 into the pre-scaler will divide the 10 MHz crystal oscillator by 5 and thereby decrement the counter at a 2 MHz rate. The TMP 04 output is ratiometric, of course, so the exact clock frequency is not important.

A typical software routine for an interface between the TMP 04 and the ADSP-2101 is shown in listing 14.2. The program begins by initialising the pre-scaler and loading the counter with OFFF. The ADSP-2101 monitors the FI flag input to establish the falling edge of the TMP 04 output, and starts the counter. When the TMP 04 output goes high, the counter is stopped. The counter value is subtracted from 0FFFh to obtain the actual number of counts, and the count is saved. Then the counter is

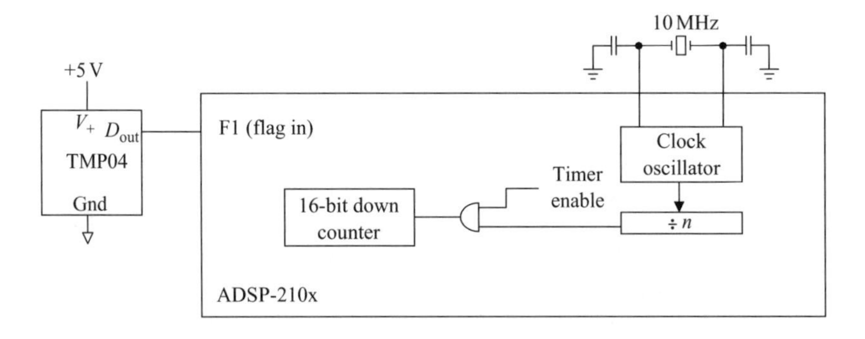

Figure 14.14 Interface between the TMP 04 to the ADSP-210x (reproduced by permission of Analog Devices, Inc).

reloaded and runs until the TMP 04 output goes low. Finally, the TMP 04 pulse widths are converted to a temperature using the scale factor of eq. (14.6).

Listing 14.2 Software routine for the TMP 04 to ADSP-210x interface (reproduced by permission of Analog Devices, Inc.)

```
{ADSP-21XX Temperature Measurement Routine      TEMPERAT.DSP
Altered registers:          ax0, ay0, af, ar,
                            si, sr0,
                            my0, mr0, mr1, mr2.
Return value:               ar -> temperature result in 14.2 format
Computation time:           2 * TMP 04 output period}
.MODULE/RAM/BOOT=0     TEMPERAT; {Beginning TEMPERAT Program}
.ENTRY TEMPMEAS;                 {Entry point of this subroutine}
.CONST PRESCALERS=4;
.CONST TIMFULSCALE=0xffff;
TEMPMEAS: si=PRESCALER;          {For timer prescaler}
          sr0=TIMFULSCALE;       {Timer counter full scale}
          dm(0x3FFB)=si;         {Timer prescaler set up to 5}
          si=TIMFULSCALE;        {CLKin=10 MHz, timer period=32.768 ms}
          dm (0x3FFC)=si;        {Timer counter register to 65535}
          dm (0x3FFD)=si;        {Timer period register to 65535}
          imask=0x01;            {Unmask interrupt timer}
TEST1:    if not fi jump TEST1;  {Check for FI=1}
TEST0:    if fi jump TEST0;      {Check for FI=0 to locate transition}
          ena timer;             {Enable timer, count at a 500 ns rate}
COUNT2:   if not fi jump COUNT2; {Check for FI=1 to stop count}
          dis timer;
          ay0=dm(0x3FFC);        {Save counter=T2 in ALU register}
          ar=sr0-ay0;
          ax0=ar;
          dm(0x3FFC)=si;         {Reload counter at full scale}
          ena timer;
COUNT1:   if fi jump COUNT1;     {Check for FI=0 to stop count}
          dis timer;
          ay0=dm(0x3FFC);        {Save counter=T1 in ALU register}
          ar=sr0-ay0;
          my0=400;
          mr=ar*my0(uu);         {mr=400*T1}
          ay0=mr0;               {af=MSW of dividend, ay0=LSW}
          ar=mr1; af=pass ar;    {ax0=16-bit divisor}
COMPUTE:  astat=0;               {To clear AQ flag}
          divq ax0; divq ax0;    {Division 400*T1/T2}
          divq ax0; divq ax0;    {with 0.3 < T1/T2 < 0.7}
          divq ax0; divq ax0;
          divq ax0; divq ax0;
          divq ax0; divq ax0;
          divq ax0; divq ax0;
          divq ax0; divq ax0;
          divq ax0; divq ax0;
          divq ax0; divq ax0;    {Result in ay0}
          ax0=0x03AC;            {ax0=235*4}
          ar=ax0-ay0;            {ar=235-400*T1/T2, result in φC}
          rts;                   {format 14.2}
.ENDMOD;                         {End of the subprogram}
```

14.4.6.1.4 Miscellaneous other applications

Sensors similar to TMP 03/04 can be used for many other useful applications. One such use is to monitor the temperature of a high power microprocessor. The TMP 04 interface depicted in Figure 14.15 could be used to measure the output pulse widths with a resolution of ±1 μs. The TMP 04 sensor T_1 and T_2 periods are measured with two cascaded 74HC520 counters. The counters, accumulating clock pulses from a 1 MHz external oscillator, have a maximum period of 65 ms.

The circuit shown in Figure 14.15 can be an ASIC application (as part of the system ASIC) so that the microprocessor would not be burdened with the overhead of timing the output pulse width. For details see Reference 11.

Another example of using such an IC to monitor a high power dissipation ULSI is shown in Figure 14.16. The device, in a surface mounted package, is mounted directly beneath the device's pin grid array (PGA) package. In a typical application the device's output could be connected to an ASIC where the pulse width could be measured (Figure 14.15 is a suitable interface).

The TMP 04 pulse output provides a significant advantage in this application because it produces a linear temperature output while needing only one I/O pin and without requiring an A/D converter.

14.4.7 Precision temperature sensors and airflow temperature sensors

14.4.7.1 Low voltage precision temperature sensors

Low voltage temperature sensors provide a voltage output directly proportional to the temperature for temperature monitoring and thermal control systems; for example, the TMP 35, TMP 36, and TMP 37 from Analog Devices (Figure 14.17) or the LM35 and LM36 from National Semiconductor. These devices require no external calibration to provide a typical accuracy of ±1 °C at 25 °C and ±20 °C over the −40 °C to 125 °C temperature range. For application details see Reference 12.

14.4.7.2 Airflow temperature sensors

Modern electronic systems and products need be incorporated with suitable airflow temperature control systems. For such systems such as low cost fan controllers and over-temperature protection, commercial silicon sensors measure the airflow temperature. These devices consist of a bandgap element (with a voltage reference source and a VPTAT) and a heating element plus the associated circuitry. An example is the TMP 12 airflow and temperature sensor from Analog Devices (Figure 14.18).

The TMP 12 incorporates a heating element, temperature sensor, and two user selectable set point comparators on a single substrate. By generating a known amount of heat, and using the set point comparators to monitor the resulting temperature rise, the TMP 12 can indirectly monitor the performance of a system's cooling fan. The TMP 12 temperature sensor section consists of a bandgap voltage reference that provides both a constant 2.5 V output and a voltage proportional to absolute temperature. The VPTAT has a precise temperature coefficient of 5 mV K^{-1} and is 1.49 V (nominal) at +25 °C. The comparators compare VPTAT with the externally

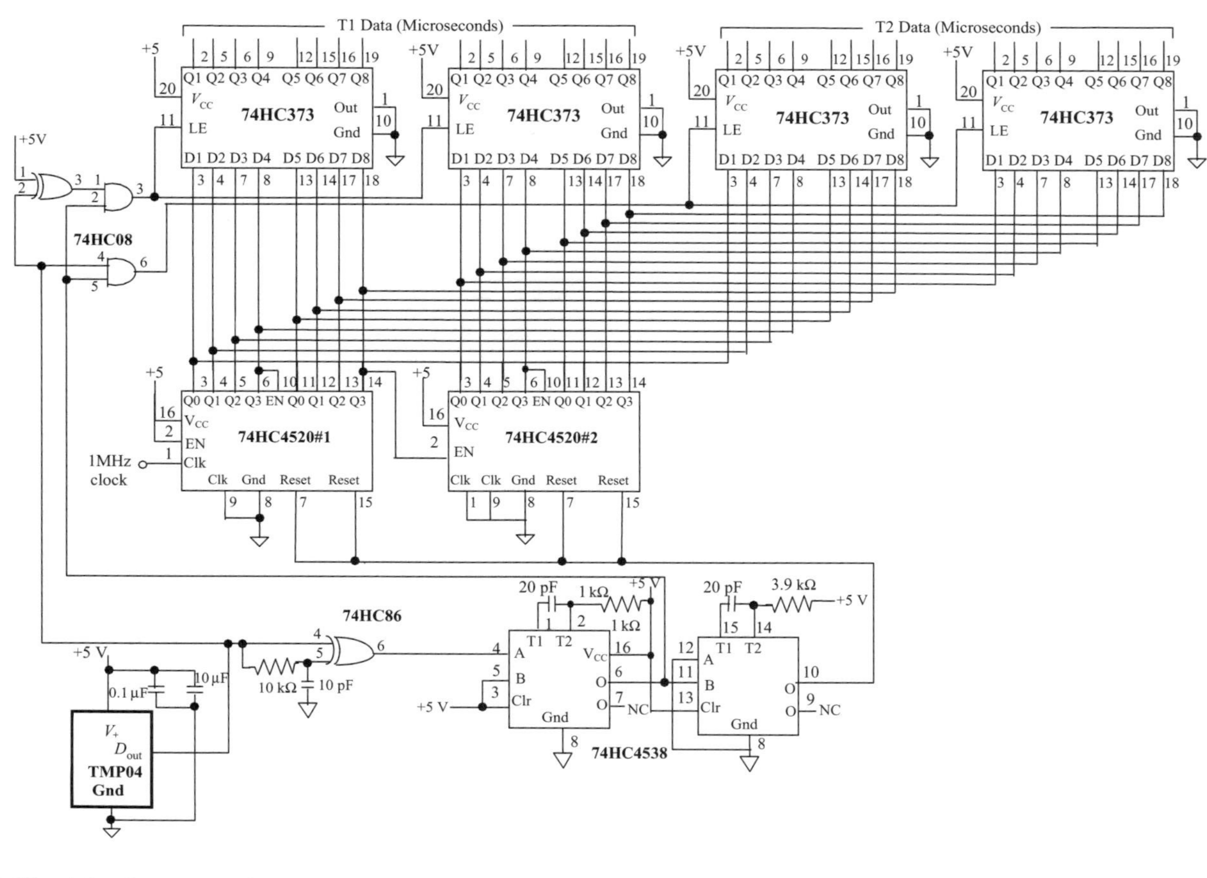

Figure 14.15 A hardware interface for TMP 04 (reproduced by permission of Analog Devices, Inc.)

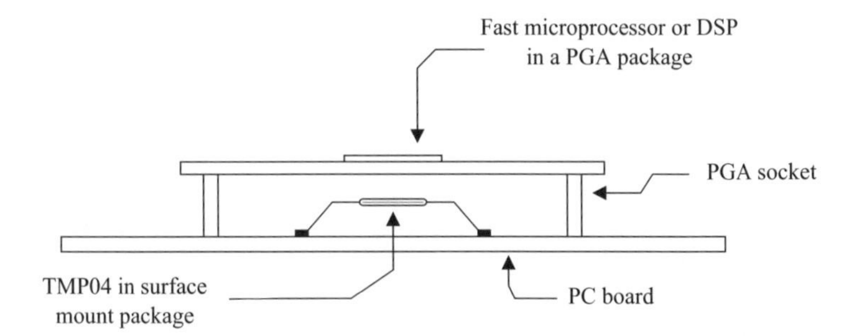

Figure 14.16 Monitoring the temperature of a ULSI using a surface mounted sensor device

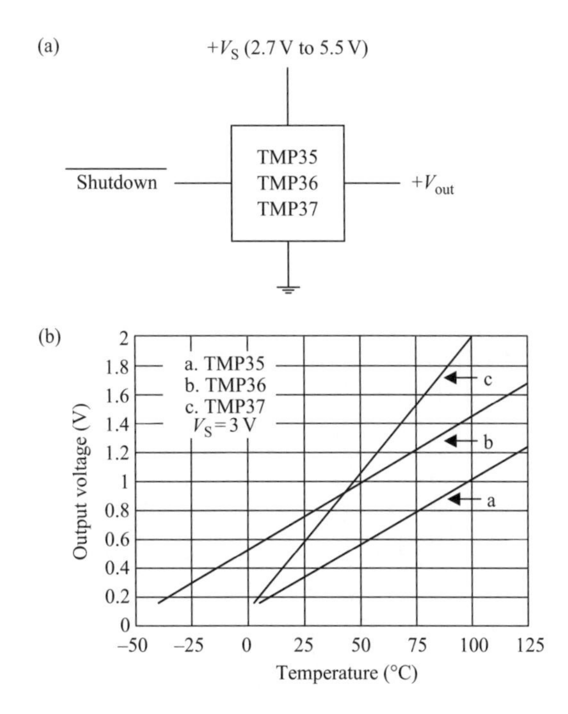

Figure 14.17 TMP 3x series: (a) functional diagram; (b) output voltage vs temperature (reproduced by permission of Analog Devices, Inc.)

set temperature trip points and generate an open collector output signal when one of the respective thresholds has been exceeded. The heat source for the TMP 12 is an on-chip 100 Ω thin-film resistor with a low temperature coefficient. When connected to a 5 V source, this resistor dissipates:

$$P_D = \frac{V^2}{R} = \frac{5^2}{100} = 0.25\,\text{W}, \tag{14.7}$$

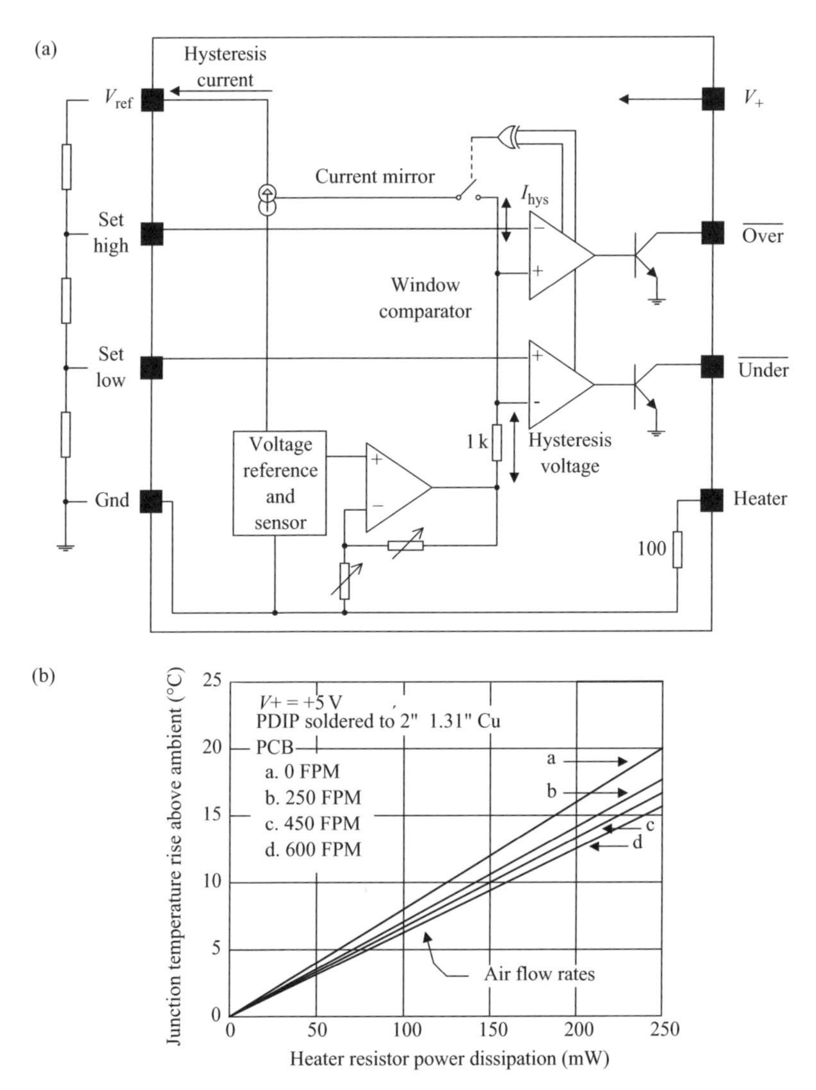

Figure 14.18 The TMP 12 airflow and temperature sensor: (a) functional block diagram; (b) temperature rise vs heater dissipation for a plastic dual inline (DIP) package (reproduced by permission of Analog Devices, Inc.)

which generates a temperature rise of about 32 K in still air for the small outline SO packaged device. With an air flow of 450 feet per minute (FPM), the temperature rise is about 22 K. By selecting a temperature set point between these two values, the TMP 12 can provide a logic level indication of problems in the cooling system.

A typical application for devices similar to TMP 12 is shown in Figure 14.19(a). The airflow sensor is placed in the same cooling airflow as a high power dissipation

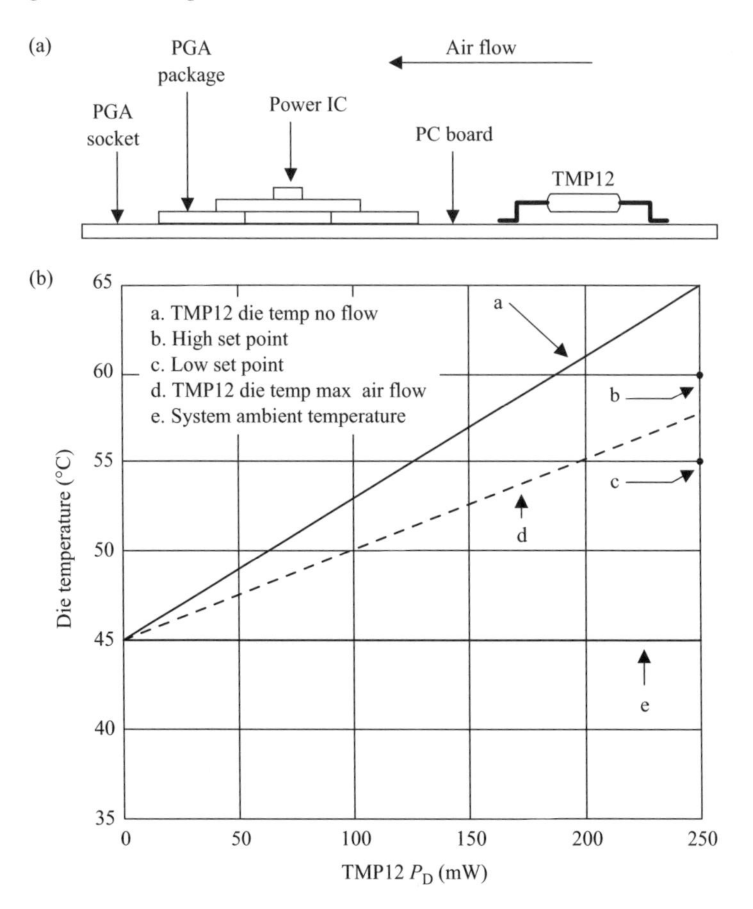

Figure 14.19 The TMP 12: (a) typical application; (b) choosing temperature set points (reproduced by permission of Analog Devices, Inc.)

IC. The sensor's internal resistor produces a temperature rise proportional to air flow, as shown in Figure 14.19(b). Any interruption in the airflow will produce an additional temperature rise when the sensor's chip temperature exceeds a user defined set point limit, the system controller can take corrective action, such as reducing clock frequency, shutting down unused peripherals or turning on an additional fan. These devices have hysteresis profiles similar to discussions in previous sections. For further details see Reference 13.

14.4.8 Sensors with built-in memories

To store set points for temperature monitoring systems, some manufacturers have developed processes to embed memory devices inside the sensor ICs. For example, the DS 1621 from Dallas Semiconductor and LM 75 from National Semiconductor.

The DS 1621 provides 9-bit (serial) temperature data and user settable thermostatic set points. As the user settings are non-volatile, the devices can be programmed before insertion into the system. For details see Reference 14.

One of the more unusual temperature management ICs is Dallas Semiconductor's DS 1820 with its digital thermometer (Figure 14.20). This device is a multidrop temperature sensor with 9-bit serial digital output. Information is sent to and from the DS 1820 over a single wire interface, so only one wire (and ground) needs to be connected from a central microprocessor to a DS 1820. Power for reading, writing, and performing temperature conversions can be derived from the data line itself with no need for an external power source.

Because each DS 1820 contains a unique silicon serial number, multiple DS 1820s can exist on the same single wire bus. This allows placing temperature sensors in many different places. Applications where this feature is useful include HVAC environmental controls, sensing temperatures inside buildings, equipment or machinery, and process monitoring and control.

The block diagram of Figure 14.20(b) shows the major components of the DS 1820. The DS 1820 has three main data components: a 64-bit ROM, a temperature

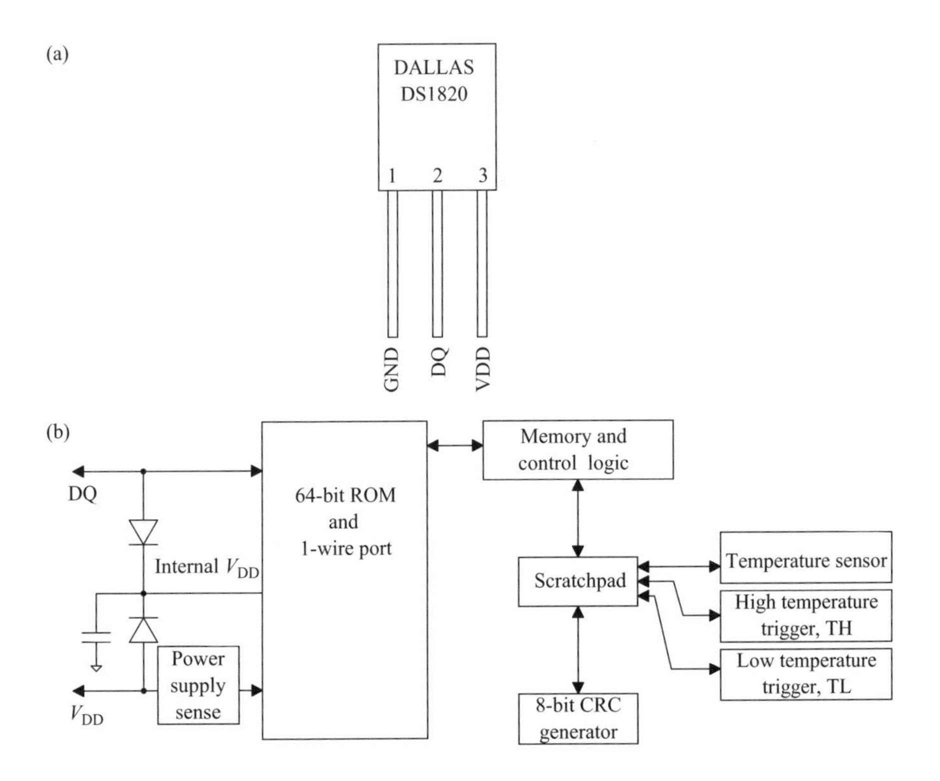

Figure 14.20 The DS 1820: (a) the device in its three terminal package; (b) block diagram package (reproduced by permission of Dallas Semiconductors)

sensor, and non-volatile temperature alarm triggers TH and TL. The device derives its power from the one wire communication line by storing energy on an internal capacitor during periods of time when the signal line is high and continues to operate off this power source during the low times of the one wire line until it returns high to replenish the parasitic (capacitor) supply. As an alternative, the DS 1820 can be powered from an external 5 V supply.

14.4.9 Thermal response time of sensors

The time required for a temperature sensor to settle to a specified accuracy is a function of the thermal mass of, and thermal conductivity between, the sensor and the object being sensed. Thermal mass often is considered equivalent to capacitance. Thermal conductivity is commonly represented by the symbol θ, and can be thought of as thermal resistance. It commonly is specified in units of degrees per watt of power transferred across the thermal joint.

The time required for the sensor IC to settle to the desired accuracy depends on the package selected, the thermal contact established in the particular application, and the equivalent power of the heat source. In most applications, the settling time is probably best determined empirically. Thermal time constants for thermal sensors can vary from a few seconds to over 100 s, depending on the package and socket used, air velocity and other factors. Practical techniques for maximum accuracy from sensors are discussed in Steele (1996).

14.5 Silicon pressure sensors

14.5.1 Background on the piezoresistive effect

The roots of silicon micromachining technology date back to Bell Laboratories. The research team developing the basics of semiconductor technology discovered a piezoresistive effect in silicon and germanium. The piezoresistive effect creates a resistance change in the semiconductor material in response to stress. This change was approximately two orders of magnitude larger than the equivalent resistance change of metals (used previously for strain gauge applications), promising an attractive option for sensors. The high sensitivity, or gauge factor, is perhaps 100 times that of wire strain gauges. Piezoresistors are implanted into a homogeneous single crystalline silicon medium. The implanted resistors thus are integrated into a silicon force sensing member. Typically, other types of strain gauges are bonded to force sensing members of dissimilar material, resulting in thermoelastic strain and complex fabrication processes. Most strain gauges are inherently unstable owing to degradation of the bond, as well as temperature sensitivity and hysteresis caused by the thermoelastic strain. Silicon is an ideal material for receiving the applied force because it is a perfect crystal and does not become permanently stretched. After being strained, it returns to the original shape. Silicon wafers are better than metal for pressure sensing diaphragms, as silicon has extremely good elasticity within its operating range. Silicon diaphragms normally fail only by rupturing.

14.5.2 Piezoresistive effect-based pressure sensor basics

The most popular silicon pressure sensors are piezoresistive bridges that produce a differential output voltage in response to pressure applied to a thin silicon diaphragm. The sensing element of a typical solid state pressure sensor consists of four nearly equal piezoresistors buried in the surface of a thin circular silicon diaphragm; see Figure 14.21.

A pressure or force causes the thin diaphragm to flex, inducing a stress or strain in the diaphragm and the buried resistors. The resistor values will change depending on the amount of strain they undergo, which depends on the amount of pressure or force applied to the diaphragm. Therefore, a change in pressure (mechanical input) is converted to a change in resistance (electrical output). The resistors can be connected in either a half bridge or a full Wheatstone bridge arrangement. For a pressure or force applied to the diaphragm using a full bridge arrangement, the resistors can be approximated theoretically as shown in Figure 14.21 (non-amplified units). Here $R + \Delta R$ and $R - \Delta R$ represent the actual resistor values at the applied pressure or force. R represents the resistor value for the undeflected diaphragm (pressure is zero) where all four resistors are nearly equal in value. And ΔR represents the change

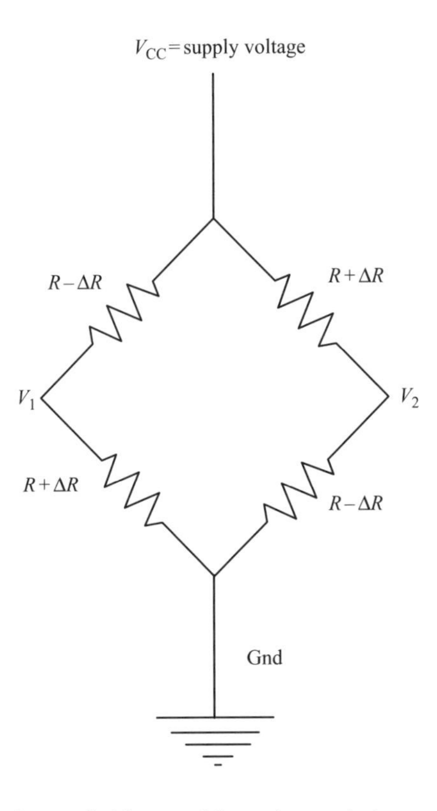

Figure 14.21 Four element bridge used in a piezoresistive pressure sensor

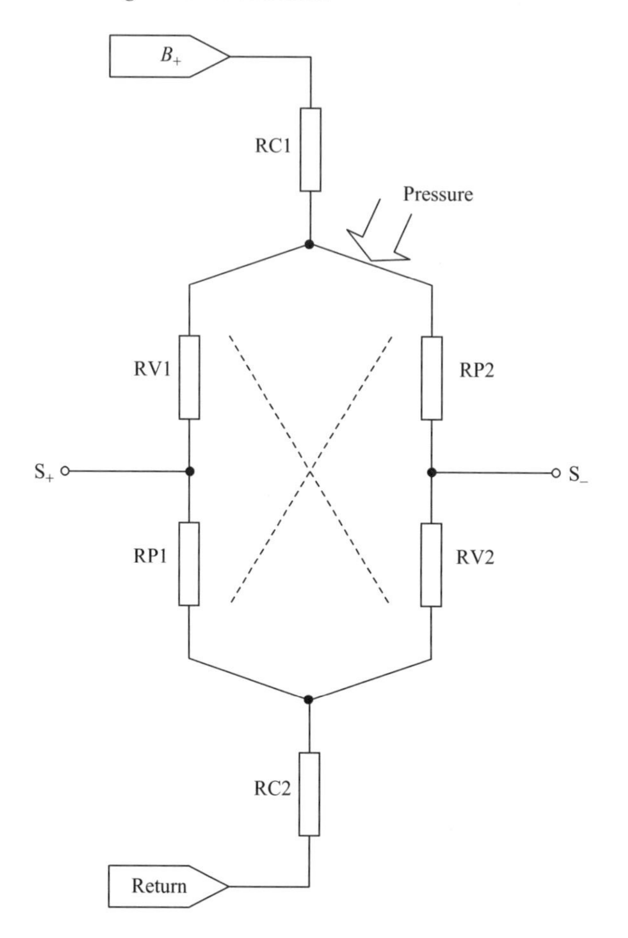

Figure 14.22 Sensor equivalent circuit (Copyright of Motorola, reproduced by permission)

in resistance due to an applied pressure or force. All four resistors will change by approximately the same value. Note that two resistors increase and two decrease depending on their orientation with respect to the crystalline direction of the silicon material. The signal voltage generated by the full bridge arrangement is proportional to the amount of supply voltage (V_{cc}) and the amount of pressure or force applied that generates the resistance change, ΔR. In a practical pressure sensor such as the Motorola MPX 2100 the Wheatstone bridge as shown in Figure 14.22 is used.

Bridge resistors RP1, RP2, RV1 and RV2 are arranged on a thin silicon diaphragm such that when pressure is applied RP1 and RP2 increase in value while RV1 and RV2 decrease a similar amount. Pressure on the diaphragm, therefore, unbalances the bridge and produces a differential output signal. A fundamental property of this structure is that the differential output voltage is directly proportional to bias voltage $B+$.

This characteristic implies that the accuracy of the pressure measurement depends directly on the tolerance of the bias supply. It also provides a convenient means for temperature compensation. The bridge resistors are silicon resistors that have positive temperature coefficients. Therefore, when they are placed in series with zero T_C temperature compensation resistors RC1 and RC2 the amount of voltage applied to the bridge increases with temperature. This increase in voltage produces an increase in electrical sensitivity which offsets and compensates for the negative temperature coefficient associated with piezoresistance.

Because RC1 and RC2 are approximately equal, the output voltage common mode is very nearly fixed at $0.5B_+$. In a typical MPX 2100 sensor, the bridge resistors are nominally 425 Ω; RC1 and RC2 are nominally 680 Ω. With these values and 10 V applied to B_+, a ΔR of 1.8 Ω at full scale pressure produces 40 mV of differential output voltage.

14.5.3 Pressure sensor types

Most pressure sensor manufacturers support three types of pressure measurement: absolute pressure, differential pressure and gauge pressure. These are illustrated in Figure 14.23.

Absolute pressure is measured with respect to a vacuum reference, an example of which is the measurement of barometer pressure. In absolute devices the P_2 port is sealed with a vacuum representing a fixed reference. The difference in pressure between the vacuum reference and the measured amount applied at the P_1 port causes the deflection of the diaphragm, producing the output voltage change (Figure 14.23(a)).

Differential pressure is the difference between two pressures. For instance, the measurement of pressure dropped across an orifice or venturi used to compute flow rate. In differential devices measurements are applied to both ports (Figure 14.23(b)).

Gauge pressure is a form of differential pressure measurement in which atmospheric pressure is used as the reference. Measurement of auto tyre pressure, where a pressure above atmosphere is needed to maintain tyre performance characteristics, is an example. In gauge devices the P_1 port is vented to atmospheric pressure and the measured is applied to the P_2 port (Figure 14.23(c)).

14.5.4 Errors and sensor performance

In practical applications, when calculating the total error of a pressure sensor, several defined errors should be used. To determine the degree of specific errors for the pressure sensor selected, it is necessary to refer to the sensor's specification sheets. In specific customer applications some of the published specifications can be reduced or eliminated. For example, if a sensor is used over half the specified temperature range, then the specific temperature error can be reduced by half. If an auto-zeroing technique is used, the null offset and null shift errors can be eliminated. The major factor affecting high performance applications is the temperature dependence of the pressure characteristics. Some of the error parameters are as follows.

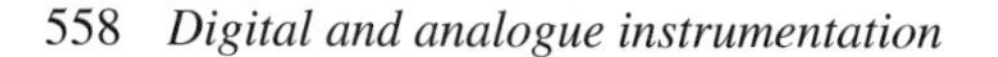

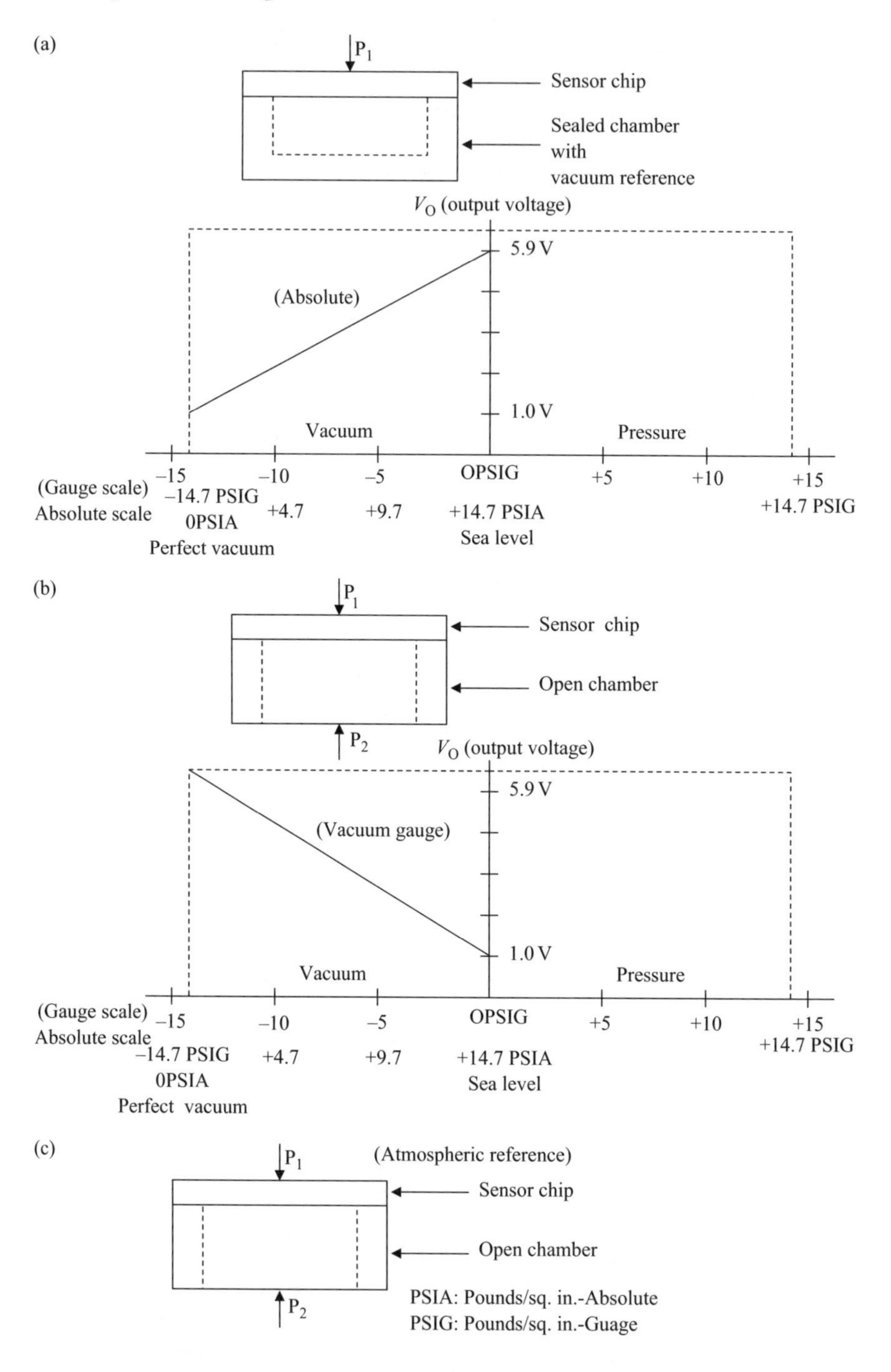

Figure 14.23 Different pressure measurements: (a) absolute; (b) differential; (c) gauge (reproduced by permission of Microswitch, Honeywell Inc., USA)

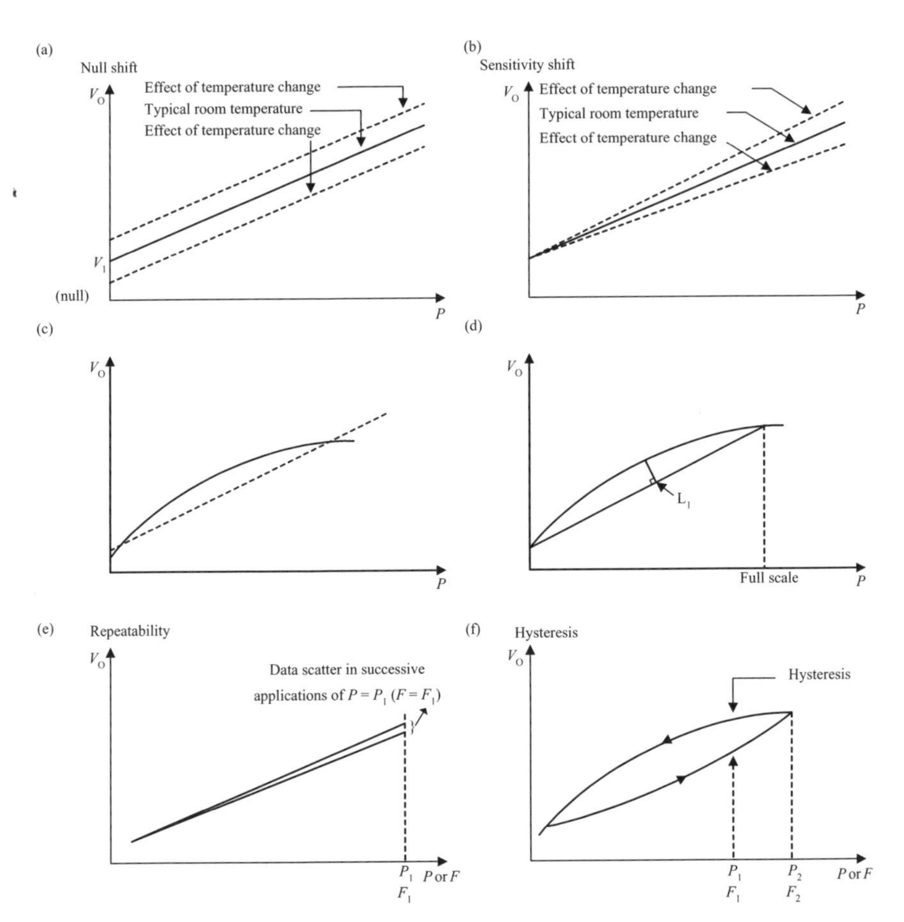

Figure 14.24 Typical error curves for pressure sensors: (a) null shift error; (b) sensitivity error; (c) best fit straight line linearity; (d) terminal base linearity; (e) repeatability; (f) hysteresis

- *Null offset*. Null offset is the electrical output present when the pressure or force on both sides of the diaphragm is equal.
- *Span*. Span is the algebraic difference between the output end points. Normally the end points are null and full scale.
- *Null temperature shift*. Null temperature shift is the change in null resulting from a change in temperature. Null shift is not a predictable error because it can shift up or down from unit to unit. A change in temperature will cause the entire output curve to shift up or down along the voltage axis (Figure 14.24(a)).
- *Sensitivity temperature shift*. Sensitivity temperature shift is the change in sensitivity due to change in temperature. A change in temperature will cause a change in the slope of the sensor output curve (Figure 14. 24(b)).

- *Linearity error.* Linearity error is the deviation of the sensor output curve from a specified straight line over a desired pressure range. One method of computing linearity error is least squares, which mathematically provides a best fit straight line to the data points (Figure 14.24(c)). Another method is terminal base linearity or end point linearity, which is determined by drawing a straight line (L1) between the end data points on the output curve. Next a perpendicular line is drawn from line L1 to a data point on the output curve. The data point is chosen to achieve the maximum length of the perpendicular line. The length of the perpendicular line represents terminal base linearity error (Figure 14.24(d)).
- *Repeatability error.* Repeatability error is the deviation in output readings for successive applications of any given input pressure or force with other conditions remaining constant (Figure 14.24(e)).
- *Hysteresis error.* Hysteresis error usually is expressed as a combination of mechanical hysteresis and temperature hysteresis. Some manufactures such as Microswitch express hysteresis as a combination of the two effects (Figure 14.24(f)). Mechanical hysteresis is the output deviation at a certain input pressure or force when that input is approached first with increasing pressure or force and then with decreasing pressure or force. Temperature hysteresis is the output deviation at a certain input, before and after a temperature cycle.
- *Ratiometricity error.* Ratiometricity implies the sensor output is proportional to the supply voltage with other conditions remaining constant. Ratiometricity error is the change in this proportion and usually is expressed as a percentage of span.

When choosing a pressure or force sensor, the total error contribution is important. Two methods take into account the individual errors and the unit-to-unit interchangeability errors: the root sum squared using maximum values, and the worst case error. The root sum squared method gives the most realistic value for accuracy. With the worst case error method, the chances of one sensor having all errors at the maximum are very remote.

14.5.5 Practical components

Pressure sensing is one of the most established and well developed areas of sensor technology. One reason for its popularity is that it can be used to measure various real world phenomena like flow, fluid level, and acoustic intensities, in addition to pressure. In the automotive industry alone, for example, pressure sensors have been identified for use in ten different applications. In guidance control and industrial control systems, pressure sensors long have been used for a number of precision pressure measurements.

Practical components available from manufacturers could be basically divided into several categories: basic uncompensated types, calibrated and temperature compensated types, and signal conditioned types.

The standard pressure ranges, from manufacturers such as Motorola, Honeywell and IC Sensors, vary between none to a few psi (pounds per square inch) up to 0–5000 psi.

Table 14.3 Uncompensated pressure sensors (Copyright of Motorola, reproduced by permission)

Device series	Pressure range kPa/psi (max)	Over-pressure (kPa)	Offset mV (typ.)	Full scale span mV (typ.)	Sensitivity (mV kPa^{-1}) (typ.)	Linearity % of FSS[1]		Temperature coefficient of span %C (typ.)	Input impedance (Ω) (typ.)
						(Min)	(Max)		
MPX 10D	1.45	100	20	35	3.5	−1	1	−0.19	475
MPX 12D	1.45	100	20	55	5.5	0	5	−0.19	475
MPX 50D	50/7.3	200	20	60	1.2	−0.1	0.1	−0.19	475
MPX 100D,A	100/14.5	200	20	60	0.6	−0.1	0.1	−0.19	475
MPX 200D,A	200/29	400	20	60	0.3	−0.25	0.25	−0.19	475
MPX 700D	700/100	2100	20	60	0.086	0.5	0.5	−0.18	475

[1] Based on end point straight line fit method. Best fit straight line linearity error is approximately half of listed value.

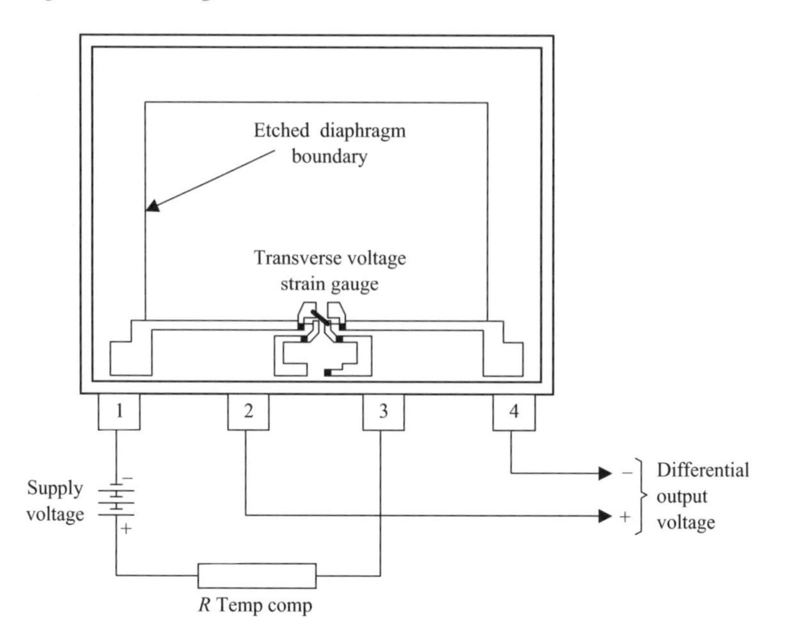

Figure 14.25 Sensor construction as applied to Motorola MPX series showing electrical connections (Copyright of Motorola, reproduced by permission)

14.5.5.1 Basic uncompensated sensors

Most of the basic uncompensated pressure sensor devices are silicon piezoresistive strain gauge designs. Some examples of these devices are listed in Table 14.3.

These uncompensated basic sensors contain a basic transducer structure as shown in Figure 14.25. Figure 14.25 illustrates the top view of the pressure sensor silicon chip, showing the strain gauge resistor diagonally placed on the edge of the diaphragm. Voltage is applied across pins 1 and 3, while the taps that sense the voltage differential transversely across the pressure sensitive resistor are connected to terminals 2 and 4. An external series resistor is used to provide temperature compensation while reducing the voltage impressed on the sensor to within its rated value.

The recommended voltage drive is 3 V d.c., and should not exceed 6 V under any operating condition. The differential voltage output of the sensor, appearing between terminals 2 and 4, will be positive when the pressure applied to the 'pressure' side of the sensor is greater than the pressure applied to the 'vacuum' side. The nominal full scale span of the transducer is 60 mV when driven by a 3 V constant voltage source.

When no pressure is applied to the sensor there will be some output voltage, called *zero pressure offset*. For the MPX 700 sensor this voltage is guaranteed to be within the range of 0–35 mV. The zero pressure offset output voltage is easily nulled out by a suitable instrumentation amplifier. The output voltage of the sensor will vary in a linear manner with applied pressure. Figure 14.26 illustrates output voltage vs pressure differential applied to the sensor, when driven by a 3 V source.

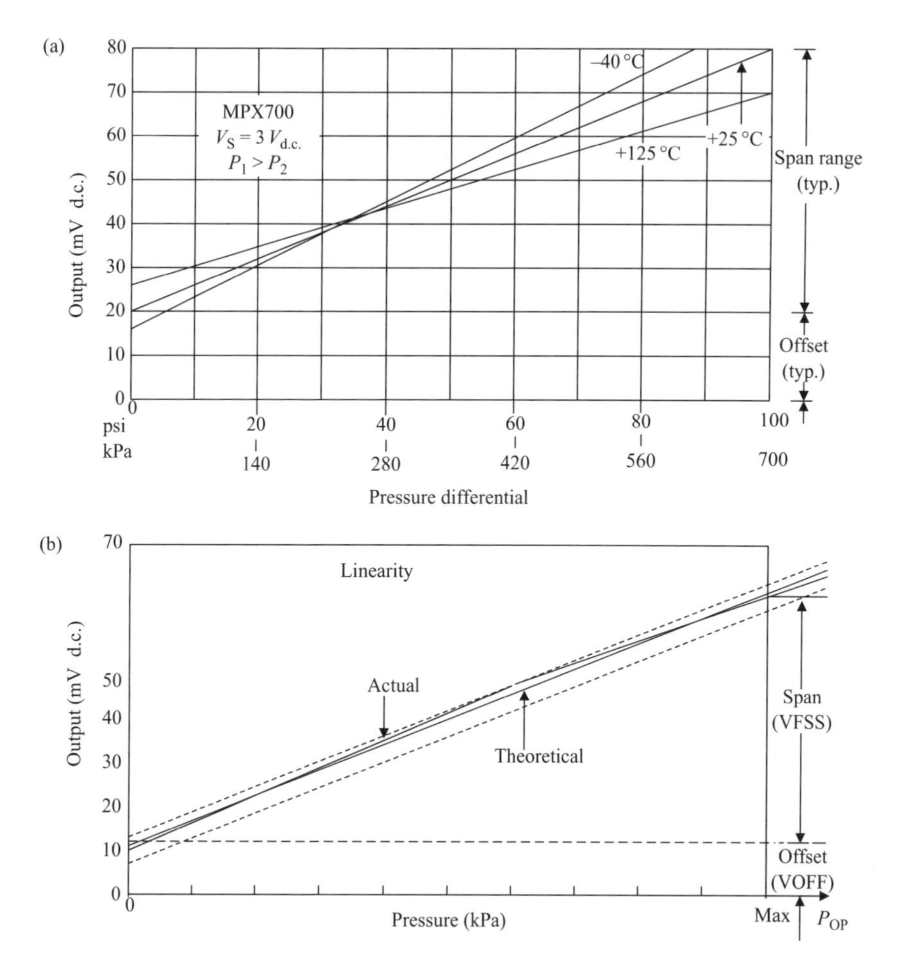

Figure 14.26 Characteristics of the MPX 700 series devices: (a) output vs pressure differential; (Copyright of Motorola, reproduced by permission) (b) linearity specification comparison

14.5.5.1.1 Temperature compensation

Because this strain gauge is an integral part of the silicon diaphragm, there are no temperature effects due to differences in the thermal expansion of the strain gauge and the diaphragm, as often are encountered in bonded strain gauge pressure sensors. However, the properties of the strain gauge itself are temperature dependent, requiring that the device be temperature compensated if it is to be used over an extensive temperature range. Temperature compensation and offset calibration can be achieved rather simply with additional resistive components. Figure 14.27 shows a practical circuit for a digital pressure gauge.

The simplest method of temperature compensation, placing a resistance (R19 and R20) in series with the sensor driving voltage, is used as shown in Figure 14.27. This

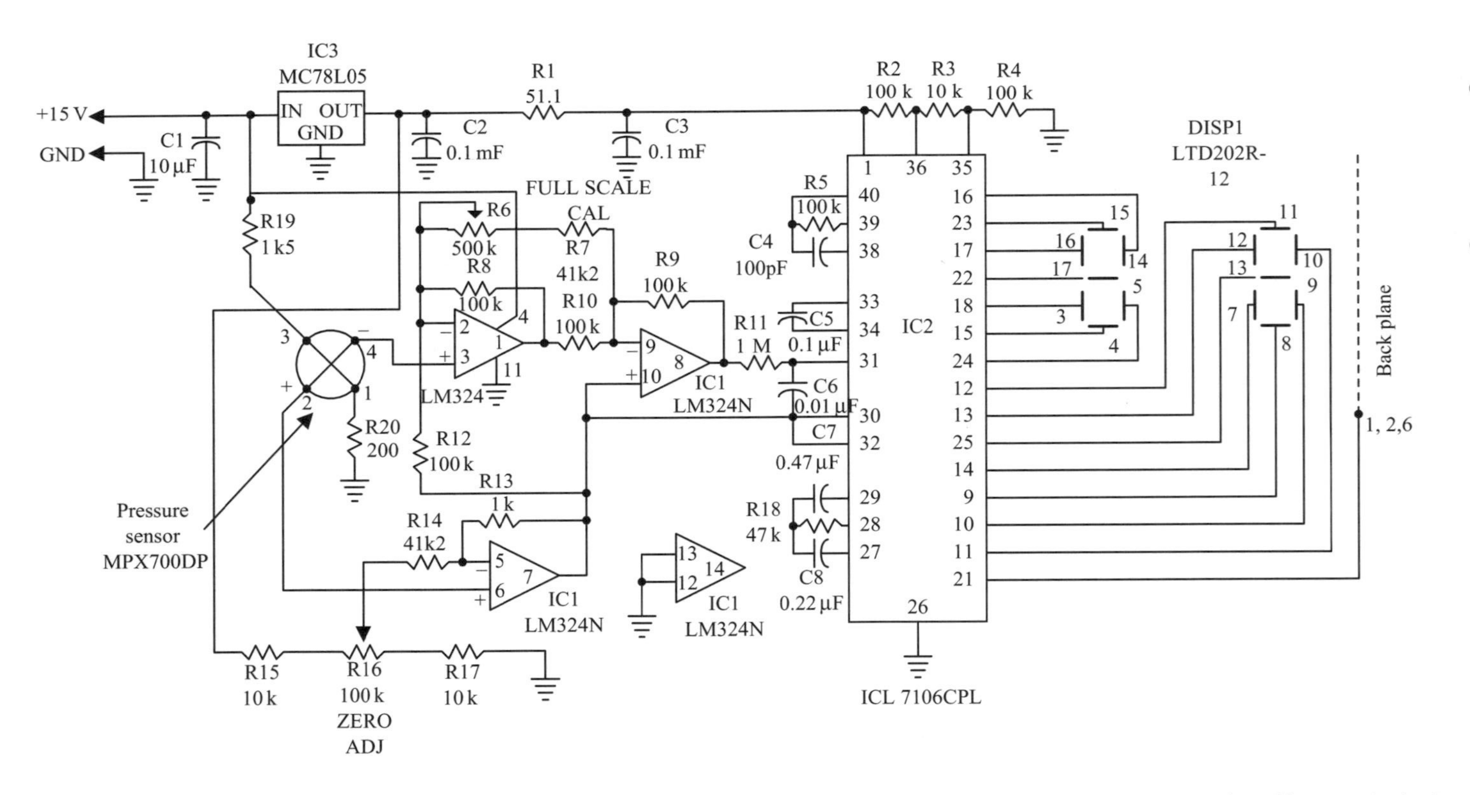

Figure 14.27 Schematic diagram of a digital pressure gauge using MPX 700 series (Copyright of Motorola, reproduced by permission)

provides good results over a temperature span of 0–80 °C, yielding a 0.5 per cent full scale span compensated device. As the desired bridge driving voltage is about 3 V, placing the temperature compensating resistor in series with the bridge circuit has the additional advantage of reducing the power supply voltage, 15 V, to the desired 3 V level. Note that the 15 V power source must be held to within a tight tolerance, because the output voltage of the transducer is ratiometric with the supply voltage. In most applications, an ordinary fixed 15 V regulator chip can be used to provide the required stable supply voltage.

The series method of compensation requires a series resistor which is equal to 3.577 times the bridge input resistance at 25 °C. The range of transducer resistance is between 400 and 550 Ω so the compensating network will be 1431–1967 Ω. If a temperature compensated span of greater than ±0.5 per cent is satisfactory or the operating temperature range of the circuit is less than 80 °C, one value of compensating resistance can be used for any sensor resistance over the range of 400–550 Ω. In the circuit of Figure 14.27 the temperature compensating network is composed of two resistors to allow the quiescent voltage of the sensor at pins 2 and 4 to be near the centre level (2.5 V) of the analogue and digital circuit that follows.

14.5.5.1.2 Signal amplification

To amplify the transducer output (60 mV at 100 psi) to a useful level that can drive subsequent circuitry, common op amps such as LM 324 could be used. The circuit in Figure 14.27 shows the application, which allows means to null out the d.c. offset output voltage of the transducer when no pressure is applied. The high input impedance of the IC_1 ensures that the circuit does not load the basic transducer. In the practical circuit of Figure 14.27, the differential output of the instrumentation amplifier is fed to the ADC (IC_2), to provide a digital readout of the pressure difference impressed upon the transducer. For further details, see Reference 16.

14.5.5.1.3 Signal conditioning for uncompensated pressure sensors

Today's unamplified solid state sensors typically have an output voltage of tens of millivolts (Motorola's basic 10 kPa pressure sensor, MPX 10 has a typical full scale output of 58 mV, when powered with a 5 V supply). Therefore, a gain stage is needed to obtain a signal large enough for additional processing. This additional processing may include digitisation by a microcontroller's analogue-to-digital converter, input to a comparator, and the like.

An instrumentation amplifier for pressure sensors should have a high input impedance, a low output impedance, differential to single ended conversion of the pressure related voltage output, and high gain capability. In addition it will be useful to have the gain adjustment without compromising common mode rejection, and both positive and negative d.c. level shifts of the zero pressure offset.

Varying the gain and offset is desirable because full scale span and zero pressure offset voltages of pressure sensors will vary somewhat from unit to unit. Therefore, a variable gain is desirable to fine tune the sensor's full scale span, and a positive or negative d.c. level shift (offset adjustment) of the pressure sensor signal is needed to translate the pressure sensor's signal conditioned output span to a specific level (e.g. with the high and low reference voltages of an ADC).

Pressure sensor interface circuits may require either a positive or a negative d.c. level shift to adjust the zero pressure offset voltage. As described previously, if the signal conditioned pressure sensor voltage is an input to an ADC, the sensor's output dynamic range must be positioned within the high and low reference voltages of the ADC; that is, the zero pressure offset voltage must be greater than (or equal to) the low reference voltage and the full scale pressure voltage must be less than (or equal to) the high reference voltage (see Figure 14.28(a)). Otherwise, voltages above

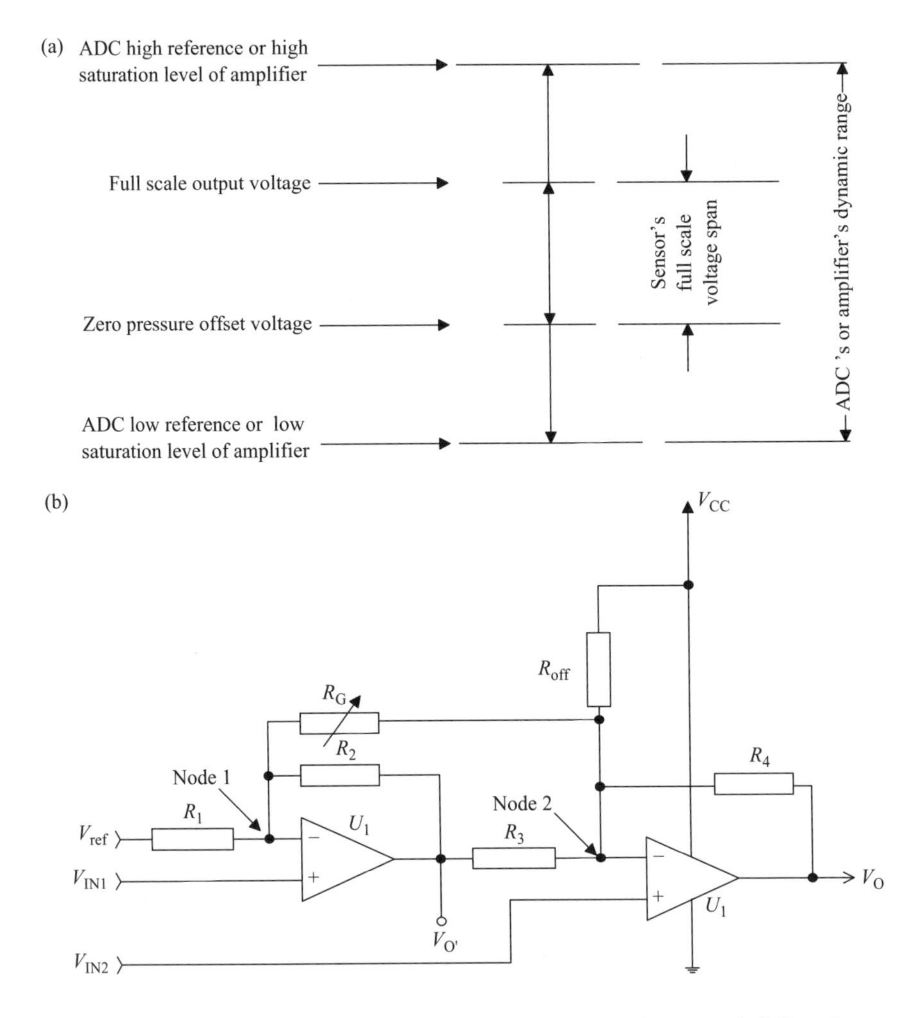

Figure 14.28 Sensor signal conditioning: (a) positioning the sensor's full scale span within the ADC's or amplifier's dynamic range; (b) a suitable two amp signal conditioning stage

the high reference will be digitally converted as 255 decimal (for 8-bit ADC), and voltages below the low reference will be converted as 0. This creates non-linearity in the analogue-to-digital conversion.

A similar requirement that warrants the use of a d.c. level shift is to prevent the pressure sensor's voltage from extending into the saturation regions of the operational amplifiers. This also would cause non-linearity in the sensor output measurements. For example, if an op amp powered with a single ended 5 V supply saturates near the low rail of the supply at 0.2 V, a positive d.c. level shift may be required to position the zero pressure offset voltage at or above 0.2 V. Likewise, if the same op amp saturates near the high rail of the supply at 4.8 V, a negative d.c. level shift may be required to position the full scale pressure voltage at or below 4.8 V.

It should be obvious that, if the gain of the amplifiers is too large, the span may be too large to be positioned within the 4.6 V window (regardless of ability to level shift d.c. offset). In such a case, the gain must be decreased to reduce the span.

Figure 14.28(b) shows a suitable two amplifier signal conditioning state with variable gain and negative d.c. level shift capability [17]. Complete analysis of the circuit is beyond the scope of the chapter. For further details, see References 19 and 20.

14.5.5.2 Calibrated and temperature compensated pressure sensors

To provide precise span, offset calibration and temperature compensation, basic sensor elements such as Motorola's X-ducer could be supplemented with special circuitry within the sensor package. An example of such a device family is the MPX 2000 series pressure transducers from Motorola. The MPX 2000 series sensors are available both as unported elements and as ported assemblies suitable for pressure, vacuum and differential pressure measurements in the range 10–200 kPa.

Figure 14.29 is a block diagram of the MPX 2000 series sensors, showing the arrangement of seven laser trimmed resistors and two thermistors used for calibration of the sensor for offset, span, symmetry and temperature compensation.

14.5.5.3 Signal conditioned pressure sensors

In this category of sensors, additional circuitry is added for signal conditioning (amplification), temperature compensation, calibration, and the like, so that the user needs fewer additional components. An example of such a sensor family from Motorola is the MPX 5000. These sensors are available in full scale pressure ranges of 50 kPa (7.3 psi) and 100 kPa (14.7 psi). With the recommended 5.0 V supply, the MPX 5000 series produces an output of 0.5 V at no pressure to 4.5 V at full scale pressure. (See Table 14.4 for the MPX 5100DP's electrical characteristics.)

These sensors integrate on-chip bipolar op amp circuitry and thin-film resistor networks to provide high level analogue output signal and temperature compensation. The small form factor and high reliability of on-chip integration make these devices suitable for automotive applications such as manifold absolute pressure sensing, etc. Figure 14.30 is a schematic of the fully integrated pressure sensor.

To explain the advantage of signal conditioning on-chip, refer to Figure 14.31. Figure 14.31(a) is a schematic of the circuitry to be coupled with an MPX 2000 series

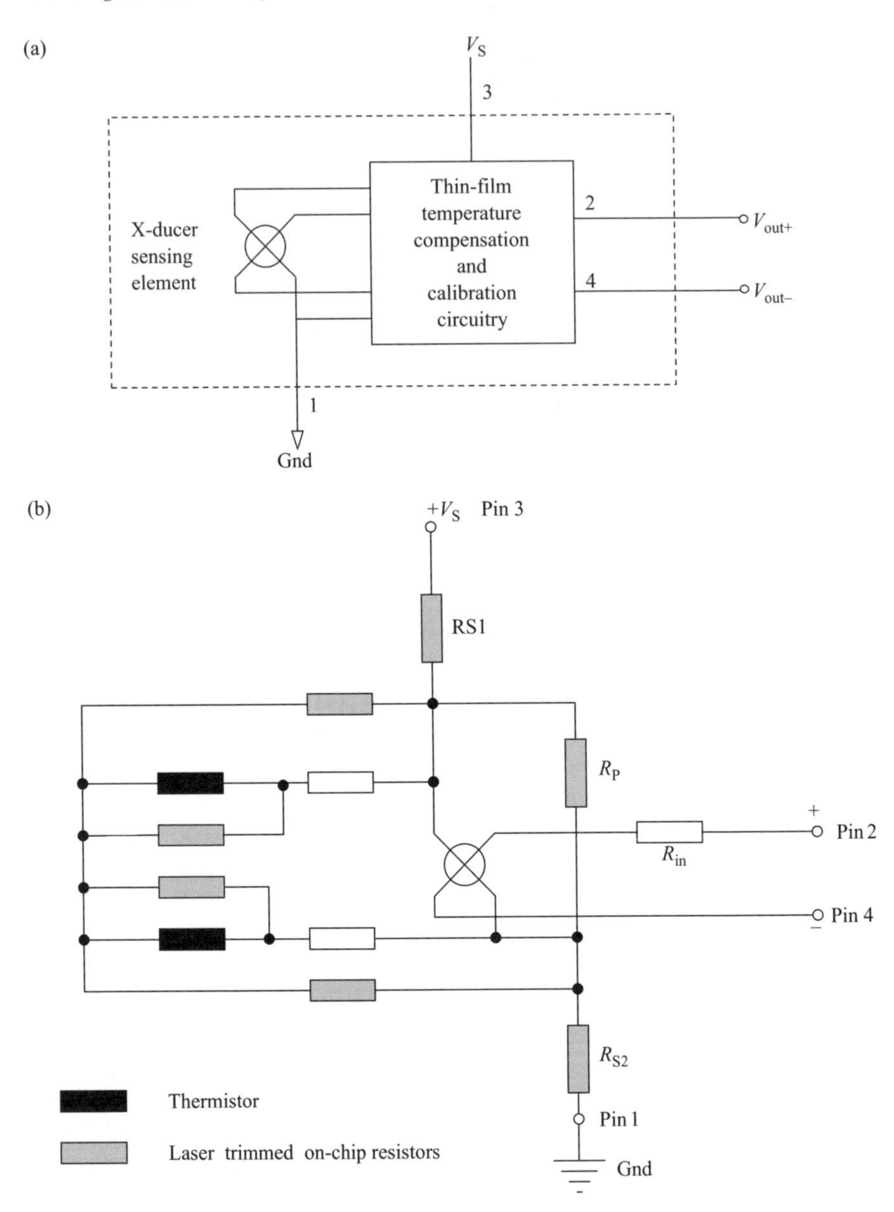

Figure 14.29 *The MPX 2000 series: (a) sensor block diagram; (b) arrangement of thermistors and laser trimmed resistor sensors (Copyright of Motorola, reproduced by permission)*

Table 14.4 MPX 5100DP electrical characteristics

Characteristics	Symbol	Min.	Typ.	Max.
Pressure range (kPa)	P_{op}	0	—	100
Supply voltage (V)	V_s	—	5.0	6.0
Full scale span (V)	V_{FSS}	3.9	4.0	4.1
Zero pressure offset (V)	V_{off}	0.4	0.5	0.6
Sensitivity (mV kPa^{-1})	S	—	40	—
Linearity (% FSS)	—	−0.5	—	0.5
Temperature effect on span (% FSS)	—	−1.0	—	1.0
Temperature effect on offset (mV)	—	−50	0.2	50

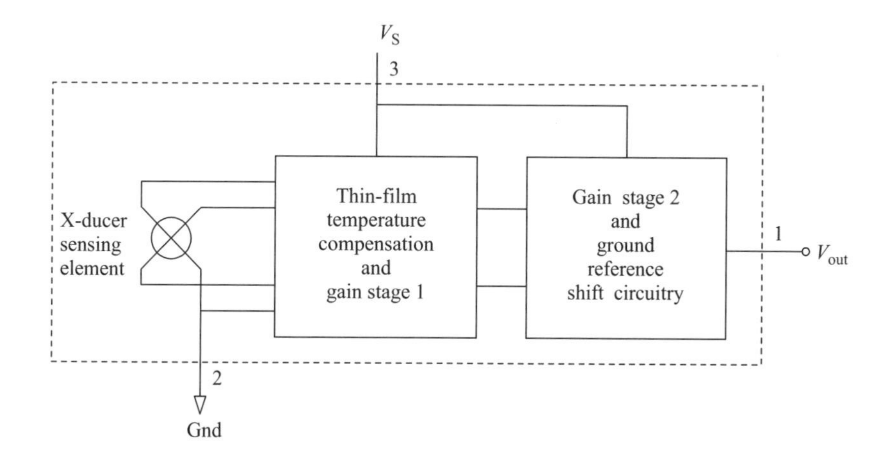

Figure 14.30 A fully integrated pressure sensor (Copyright of Motorola, reproduced by permission)

(which is compensated for temperature and calibrated for offset) to achieve ground referenced output with amplification.

Some devices similar to MPX 5100 go one step further by adding the differential-to- ground referenced conversion and the amplification circuitry on-chip. This reduces the 18-component circuit in Figure 14.31(a) to a one-signal conditioned sensor, as shown in Figure 14.31(b). Figure 14.32 is a schematic of a fully integrated pressure sensor such as MPX 5100. For details Reference [18] is suggested.

14.5.5.4 Interface between pressure sensors and microprocessors or ADCs

In many practical situations, designers face the need to provide an interface between pressure sensors and microprocessor or microcontroller based systems. In such cases,

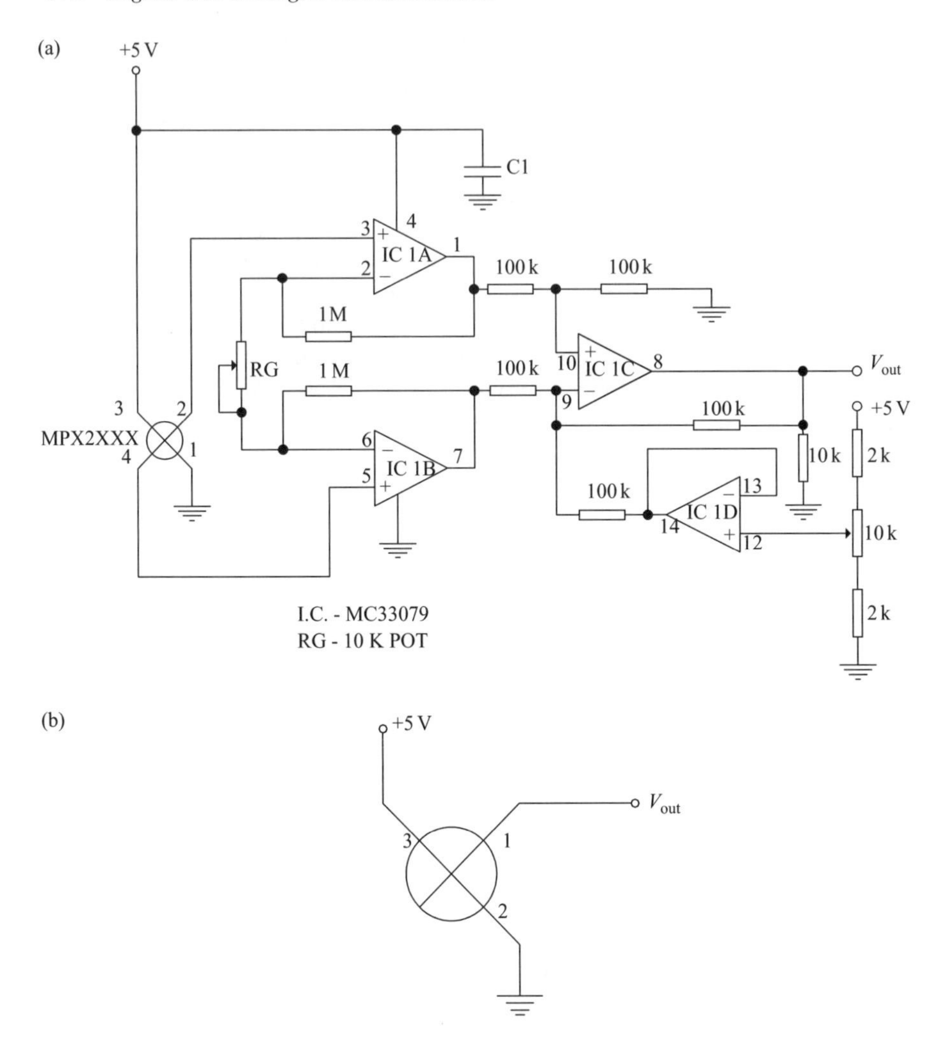

Figure 14.31 *Simplification of external circuitry by internal signal conditioning: (a) high level ground referenced output using an MPX 2000 series transducer; (b) similar output integrated device (Copyright of Motorola, reproduced by permission)*

the designer should consider the level of on-chip signal conditioning or on-chip temperature compensation/calibration available in designing the system. In sensors with on-chip calibration and temperature compensation, the basic block diagram of a system could be depicted as in Figure 14.33.

When on-chip calibration and temperature compensation are not available the gain stages shown need be designed to take care of such needs. Although processor techniques are similar to other applications such as temperature sensors, there are

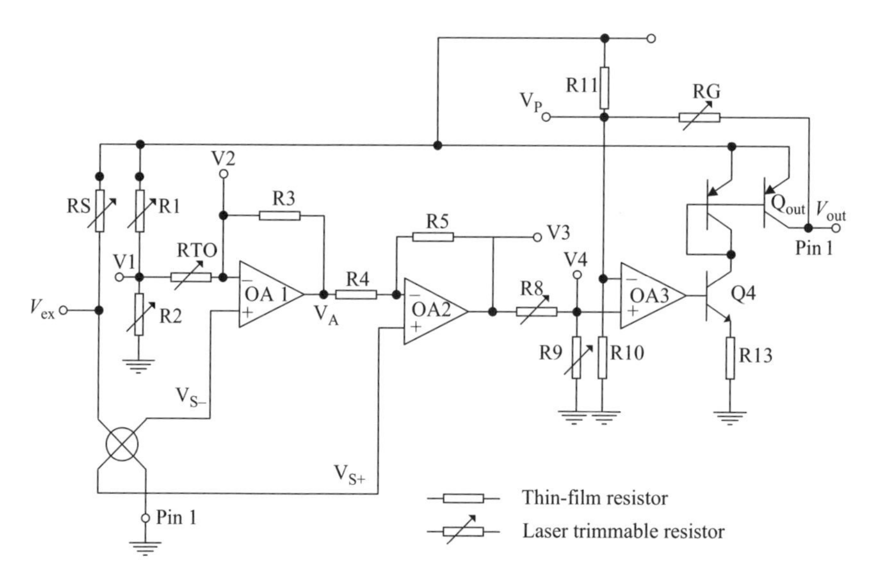

Figure 14.32 A fully integrated pressure sensor (Copyright of Motorola, reproduced by permission)

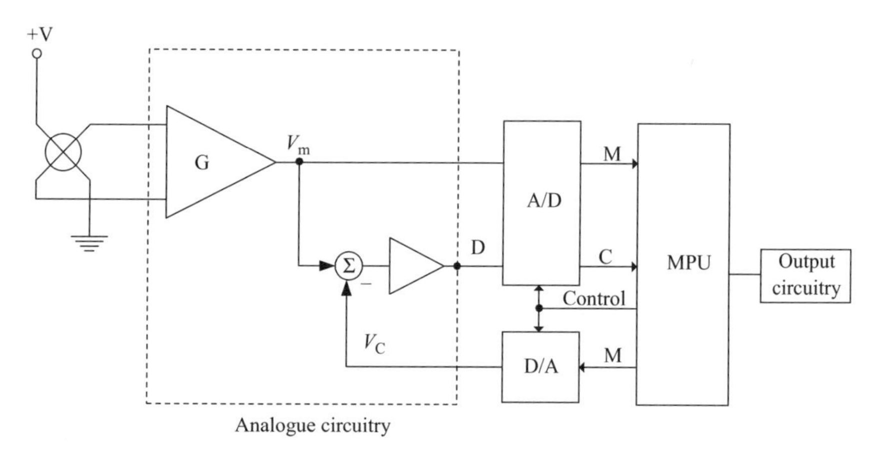

Figure 14.33 A basic block diagram for the interface between a compensated sensor and an ADC and microprocessor unit (MPU) (Copyright of Motorola, reproduced by permission)

many advanced techniques for higher resolution or compensating for the offset and temperature; for such examples see References 21–24. When configuring silicon pressure sensors with ADCs or microcontrollers with built-in ADCs, the ratiometric function of both the ADC and sensor could be used to minimise the need for additional components such as voltage reference sources. The ratiometric function of these elements makes all voltage variations from power supply rejected by the system.

The many advance techniques of using microcontroller based sensor systems are beyond the scope of this chapter. The four Motorola application notes above describe such practical and useful techniques.

General applications of pressure sensors are discussed in Reference 15 and the references cited in Further Reading at the end of this chapter.

14.6 Silicon accelerometers

With the demand from automotive and other industries in the latter 1980s, several sensor manufacturers developed micromachined silicon ICs for sensing acceleration. Today several component manufacturers such as IC Sensors, Analog Devices, and Motorola have families of silicon accelerometers. Basic sensor elements as well as signal conditioned versions are available.

14.6.1 Basic principles of sensing acceleration

The principles of acceleration sensing were simulated using a weight and spring connected to a frame to develop silicon accelerometers by using the piezoresistive properties of silicon and building capacitive structures with variation of effective capacitance between plates attached to a seismic mass of silicon. To simulate the basic mechanical analogy of accelerometers and minimise secondary effects that complicate the measuring process [25, 26] required several improvements in silicon-processing technology.

One such improvement was the advent of silicon fusion bonding [26]. Fusion bonding, which allows the bonding of two wafers while preserving the crystalline structure of silicon, permits the creation of complex three-dimensional structures without introducing mechanical discontinuities or thermal dependent stress. This structuring ability allows accelerometer manufacturers to capture the seismic mass with a sealed cavity by bonding a cap and a base plate to the frame. By controlling the space between the mass and cavity, vendors can use the air sealed inside the cavity as a viscous damping fluid for the system's motion.

Silicon fusion bonding also provides an answer to another limitation: shock resistance. Simply falling off a desk can produce a 200 g shock when the sensor hits the floor. Despite silicon's toughness and flexibility, that kind of shock could break the springs in an accelerometer unless the motion of the seismic mass is limited. Silicon fusion bonding allows the placement of bumpers and other mechanical stops to make the accelerometer much more shock resistant. Devices now routinely handle shocks as great as 2000 g.

In commercially available sensors, a single or double cantilever or a membrane supported mass is coupled with a piezoresistive or capacitive element.

14.6.1.1 Piezoresistive sensors

Figure 14.34(a) is a diagram of a single cantilevered design of an accelerometer using piezoresistive elements. Thin beams support one edge of a seismic mass, which is

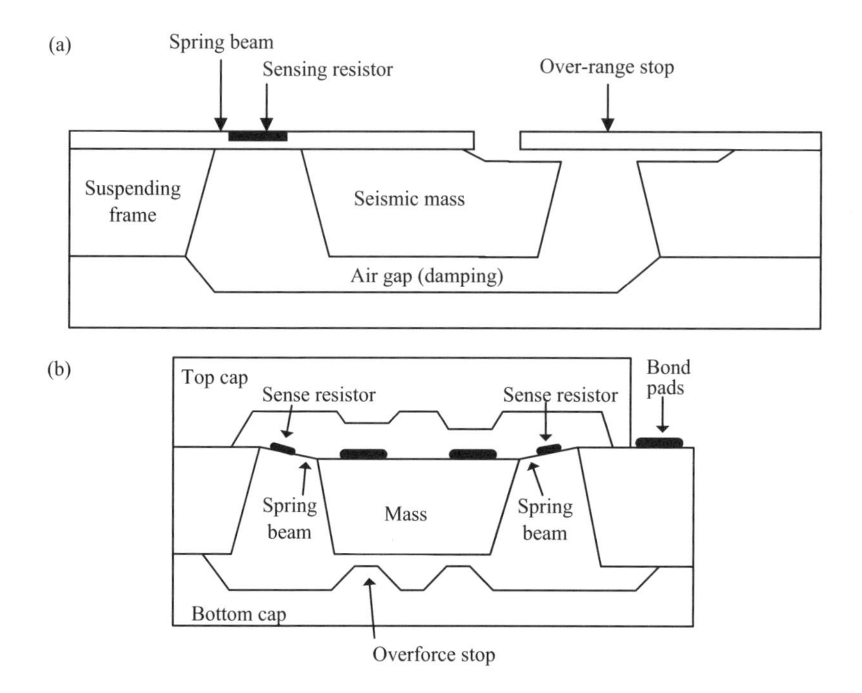

Figure 14.34 Silicon accelerometer simulating the mass and spring: (a) single cantilevered; (b) double cantilevered

free to move within a cavity created by fusion bonding two additional wafers to the one containing the mass. Piezoelectric resistors fabricated at the beams measure the displacement by changing resistance as the beams bend. The double cantilevered approach, shown in Figure 14.34(b), supports the mass from two sides.

Although single cantilevered types are simplest and sensitive, they have drawbacks such as transverse sensitivity [26]. Double cantilevered types can be designed with self-compensating effects for transverse forces. IC Sensors and Lucas Nova Sensor manufacture piezoresistive type sensors.

14.6.1.2 Capacitive sensors

Figure 14.35 shows a typical capacitive sensing device. In these devices the mass is supported on all four sides and transverse sensitivity is very much reduced. Capacitive sensors use top and bottom plates to form a capacitor divider with a seismic mass that is temperature insensitive. Sensing the change in capacitance requires relatively complex circuitry, however.

In some capacitive accelerometers such as the ADXL series from Analog Devices, the seismic mass is not a single block but a series of interdigitated fingers, as shown in Figure 14.36. This allows the sensing of acceleration in the plane of the chip, rather than in other types of sensors where sensing is normal to the surface.

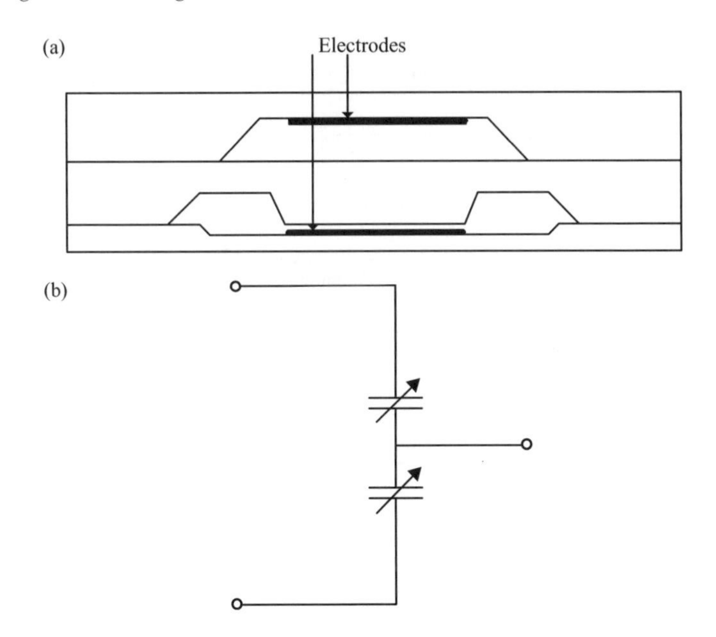

Figure 14.35 The basic arrangement of a capacitive sensor: (a) structure; (b) equivalent circuit

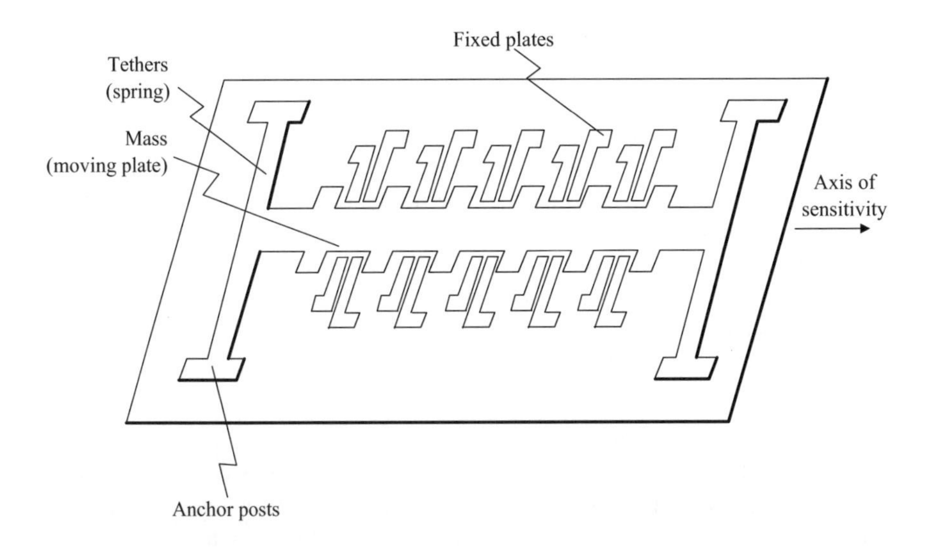

Figure 14.36 Arrangement in the ADXL 50 (reproduced by permission of Analog Devices, Inc.)

The drawback of complicated interface circuitry in a capacitive sensor is compensated for by the additional ability inherent in the capacitor structure. The presence of charge carrying plates in the sensors provides a built-in means for applying an electrostatic force on the seismic mass. This capability lets the sensor be used in a closed loop configuration.

Instead of letting the seismic mass move freely during acceleration, a closed loop system applies a restoring force to the mass, keeping it relatively motionless. Restricting the movement of the mass has two advantages. First, it improves sensor linearity by confining the motion to the linear region of the spring's restoring force. Second, it extends the range of a sensor beyond the limits imposed by its housing on the movement of the seismic mass. In such force-feedback systems the restoring force, not the actual movement, serves as the measure of acceleration.

The ability to apply a force to the proof mass has an additional advantage; it gives the sensor a self-test capability. This capability is particularly important in systems such as automotive airbags, where you cannot test the system by actually accelerating it, yet testing is necessary for safety or reliability.

14.6.2 Practical components

14.6.2.1 Piezoresistive example: model 3255 from IC Sensors

The model 3255 is a fully signal conditioned accelerometer containing two chips: the silicon sensing element and a custom integrated circuit (ASIC) for signal conditioning. The model 3255 accelerometer is available in various measurement ranges. With a supply voltage of 5 V, the output voltage is 2.5 V at no applied acceleration and the output range is 0.5–4.5 V for the full acceleration range. The output voltage is ratiometric with the supply voltage and will track the supply voltage in the range 5.0 ± 0.5 V. Only three connections need to be made to use the accelerometer: 5 V supply, ground, and signal output. Figure 14.37(a) is a photograph of the device showing the two chips and the sealed unit. Figure 14.37(b) shows the arrangement of the sensor element.

14.6.2.1.1 Sensor element

The silicon sensor element is shown in Figure 14.37(b). A seismic mass and four flexures are formed using bulk micromachining processes. Each of the four beams contains two implanted resistors that are interconnected to form a Wheatstone bridge. When the device undergoes acceleration the mass moves up or down, causing four of the resistors to increase and the other four to decrease in value. This results in an output voltage change proportional to the applied acceleration. The eight resistors are interconnected such that the effect of any off-axis acceleration is cancelled.

Silicon top and bottom caps are attached to the section containing the seismic mass and the beams. The silicon caps serve several purposes. Precision gaps are etched into the caps to provide air damping to suppress the resonant peak of the structure. Because the part is critically damped, the frequency response is flat up to several kHz with little dependence on temperature. Small elevated stops on the top and bottom caps limit the motion of the mass to a fraction of the deflection at which fracture occurs.

The caps also form a chamber around the seismic mass to provide protection during the later stages of manufacturing and its operating lifetime.

Last, the top cap allows testing the accelerometer in the absence of acceleration. When a voltage is applied to a metal electrode on the top cap, an electrostatic force moves the mass toward the top cap. This results in a change in output voltage proportional to the sensitivity and to the square of the applied voltage. It thus is possible to generate an 'acceleration' using an external voltage and check the functionality of the mechanical structure as well as the electronics.

14.6.2.1.2 Signal conditioning

The signal conditioning circuitry amplifies the output of the sensor element and corrects the sensitivity and offset changes that occur with overheating. As a result, the output signal is accurate and no trimming is required by the user. The data used to set the performance of the accelerometer are stored in fused registers within the signal conditioning IC.

The signal conditioning IC converts the differential signal from the sensor element (nominally ±5 mV) into a single ended signal in the 0.5–4.5 V range while correcting for the temperature related signal variations. Signals are processed by differential amplifiers throughout most of the circuit to minimise common mode effects and noise. Switched capacitor circuitry is used to save space and because high accuracy gain stages can be made easily. As a result the compensated accelerometers are interchangeable with a very small total error. The signal conditioning IC is made in 1.5 μm CMOS technology and is intended for 5 V operation; see Figure 14.37(c).

(a)

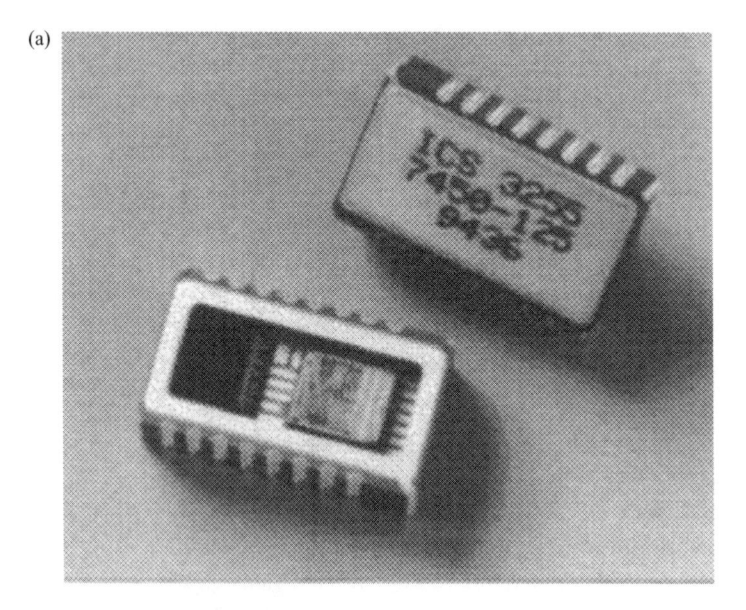

Figure 14.37 Continued overleaf

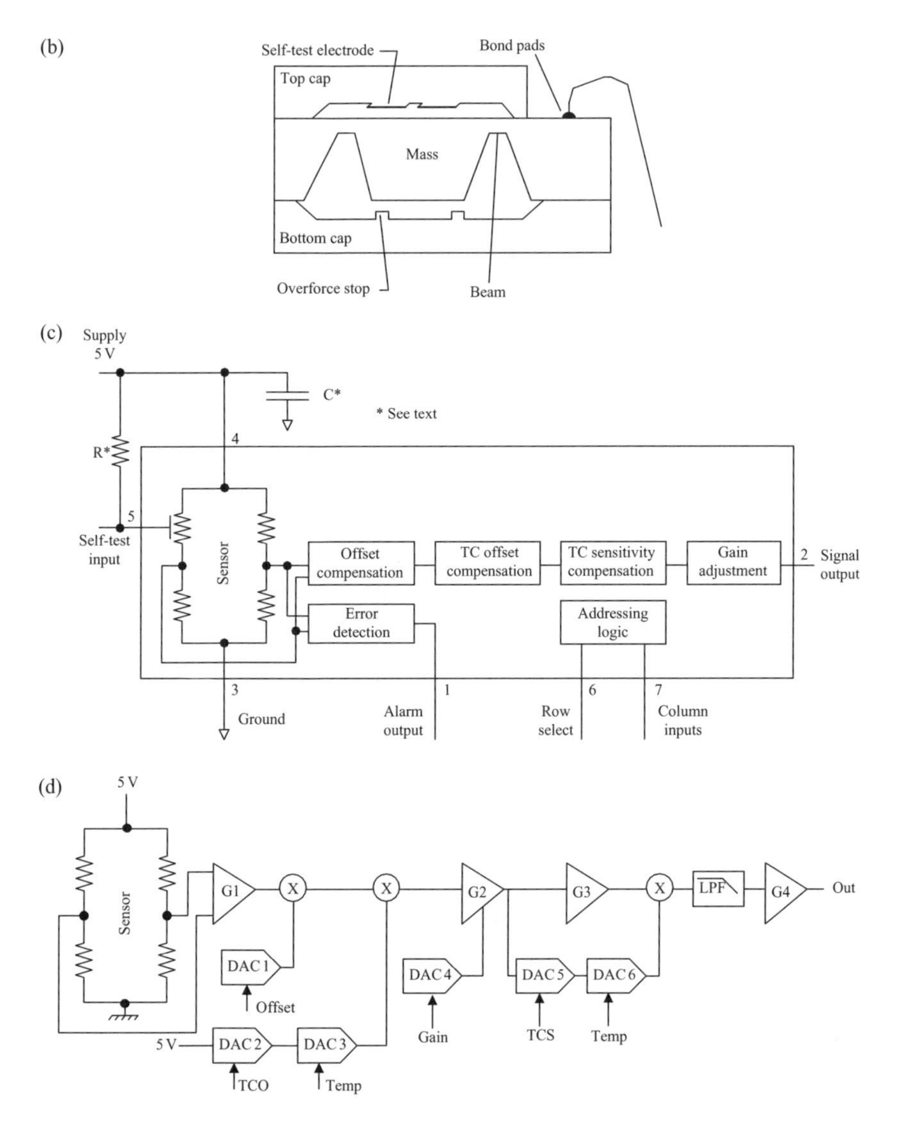

Figure 14.37 Model 3255 accelerometer: (a) photograph; (b) cross-section of accelerometer die; (c) functional block diagram; (d) simplified schematic diagram (reproduced by permission of IC Sensors)

The signal path is shown in the block diagram in Figure 14.37 (d). The output signal of the accelerometer die is processed by the following stages.

- The first stage provides a high impedance load for the sensor and amplifies the signal to maximise the dynamic range during subsequent processing.

- The offset of the sensor die is reduced to less than 0.5 per cent of full scale at room temperature by adding a voltage generated by DAC 1. This DAC is controlled by a digital word representing the programmed offset value.
- The temperature coefficient of offset (TCO) of the sensor is compensated for by adding a voltage generated by DAC 2 and DAC 3. This voltage is controlled by digital words representing the temperature and the programmed TCO value. Both the offset and TCO voltages are derived from the supply to ensure that the signal remains ratiometric with the supply voltage.
- The signal gain is set by the value in DAC 4. The gain can be varied in a 5 : 1 range to allow for different full scale specifications.
- The temperature coefficient of sensitivity (TCS) of the sensor is compensated in the next stage, built around a feed-forward loop using two DACs controlled by digital words representing the temperature and the programmed TCS value. The sensitivity decrease over temperature is compensated for by increasing the signal gain linearly with temperature.
- The output bias voltage can be set to either 0.5 or 2.5 V by connecting an input pad on the chip to ground during assembly of the part. This allows signals to be processed with either a bipolar or unipolar range.
- A two-pole passive filter removes signals generated by the internal oscillator and switched capacitor networks. Switching noise is further minimised by having separate digital and analogue internal supply lines and the differential signal processing.
- The final stage provides a low impedance output for driving resistive and capacitive loads without influencing the signal. The output will go in an impedance 'tri-state' mode if the part is not addressed.

The temperature word that controls DACs 3 and 6 is generated by an ADC that digitises the output of a temperature PTAT source driven by a bandgap reference. The temperature word therefore is linearly proportional to the temperature but does not depend on the supply voltage. For application and performance see Reference 33.

14.6.2.2 Capacitive example: the ADXL 50 from Analog Devices

The ADXL 50 is a complete acceleration measurement system on a single monolithic IC [28]. Three external capacitors and a +5 V power supply are all that are required to measure accelerations up to ±50 g. Device sensitivity is factory trimmed to 19 mV g^{-1}, resulting in a full-scale output swing of ±0.95 V for a ±50 g applied acceleration. Its 0 g output level is +1.8 V. A TTL compatible self-test function can electrostatically deflect the sensor beam at any time to verify device functionality. A functional block diagram of ADXL 50 is shown in Figure 14.38.

The ADXL 50 contains a polysilicon surface micromachined sensor and signal conditioning circuitry. The ADXL 50 is capable of measuring both positive and negative acceleration to a maximum level of ±50 g.

Figure 14.39(a) is a simplified view of the ADXL 50's acceleration sensor at rest. The actual structure of the sensor consists of 42 unit cells and a common beam. The differential capacitor sensor consists of independent fixed plates and a movable

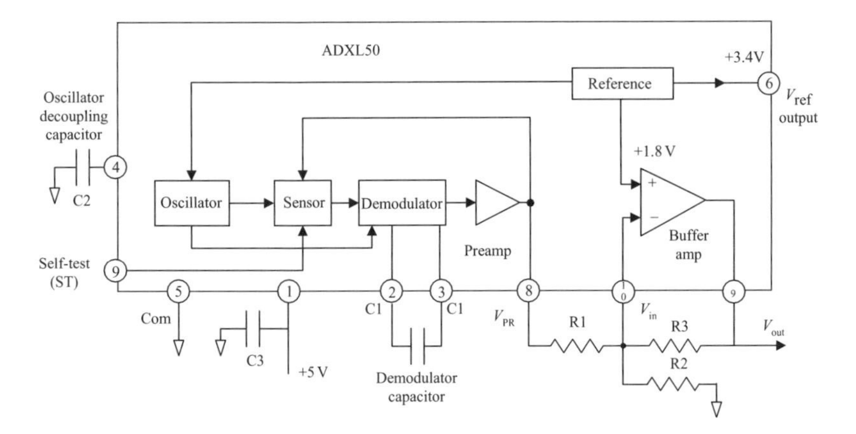

Figure 14.38 Functional block diagram of the ADXL 50 (reproduced by permission of Analog Devices, Inc.)

'floating' central plate that deflects in response to changes in relative motion. The two capacitors are series connected, forming a capacitive divider with a common movable central plate. A force balance technique counters any impending deflection due to acceleration and drives the sensor back to its 0 g position.

Figure 14.39(b) shows the sensor responding to applied acceleration. When this occurs, the common central plate or 'beam' moves closer to one of the fixed plates and further from the other. The sensor's fixed capacitor plates are driven deferentially by a 1 MHz square wave; the two square wave amplitudes are equal but 180° out of phase with one another. When at rest, the values of the two capacitors are the same and, therefore, the voltage output at their electrical centre (i.e. at the centre plate) is 0.

When the sensor begins to move, a mismatch in the value of their capacitance is created, thereby producing an output signal at the central plate. The output amplitude will increase with the amount of acceleration experienced by the sensor. Information concerning the direction of beam motion is contained in the phase of the signal with synchronous demodulation being used to extract this information. Note that the sensor needs to be positioned so that the measured acceleration is along its sensitive axis.

Figure 14.39(c) shows a block diagram of the ADXL 50. The voltage output from the central plate of the sensor is buffered and applied to a synchronous demodulator. The demodulator also is supplied with a (nominal) 1 MHz signal from the same oscillator that drives the fixed plates of the sensor. The demodulator will rectify any voltage in sync with its clock signal. If the applied voltage is in sync and in phase with the clock, a positive output will result. If the applied voltage is in sync but 180° out of phase with the clock, the demodulator's output will be negative. All other signals will be rejected. An external capacitor, C_1, sets the bandwidth of the demodulator.

The output of the synchronous demodulator drives the pre-amp – an instrumentation amplifier buffer that is referenced to +1.8 V. The output of the preamp is fed back to the sensor through a 3 MΩ isolation resistor. The correction voltage required

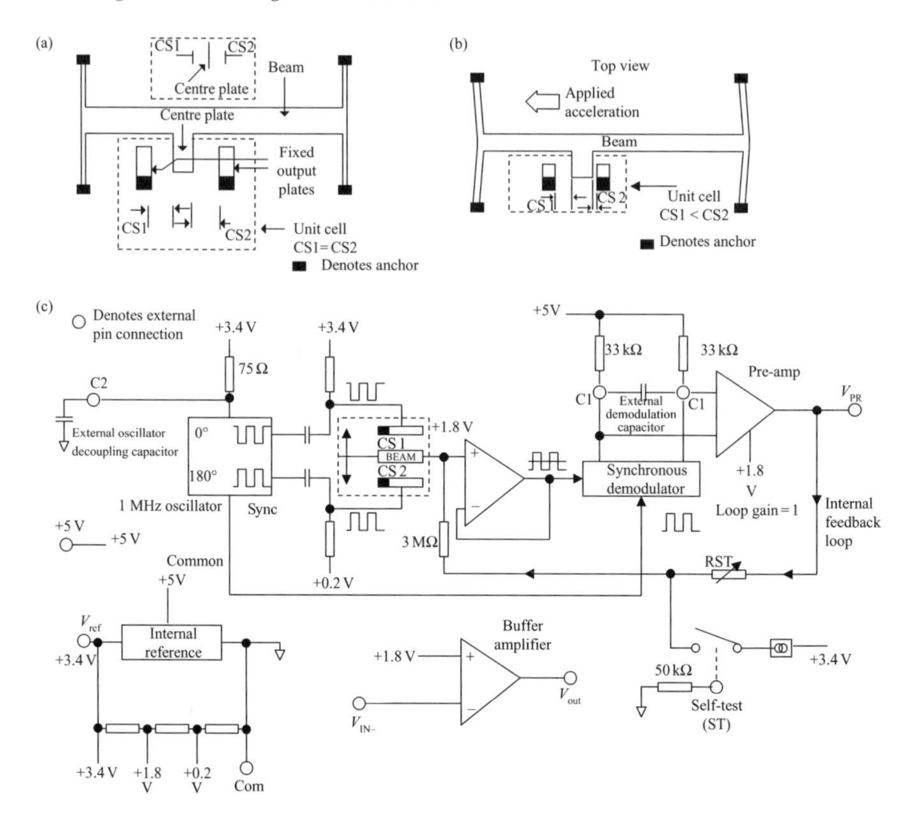

Figure 14.39 ADXL 50 operation: (a) sensor element at rest; (b) sensor momentarily responding to acceleration; (c) functional block diagram (reproduced by permission of Analog, Devices, Inc.)

to hold the sensor's centre plate in the 0 g position is a direct measure of the applied acceleration and appears at the V_{PR} pin. When the ADXL 50 is subject to acceleration, its capacitive sensor begins to move creating a momentary output signal. This is signal conditioned and amplified by the demodulator and preamp circuits. The d.c. voltage appearing at the pre-amp output then is fed back to the sensor and electrostatically forces the centre plate back to its original centre position.

At 0 g the ADXL 50 is calibrated to provide +1.8 V at the V_{PR} pin. With applied acceleration, the V_{PR} voltage changes to the voltage required to hold the sensor stationary for the duration of the acceleration and provides an output that varies directly with the applied acceleration. The loop bandwidth corresponds to the time required to apply feedback to the sensor and is set by external capacitor C_1. The loop response is fast enough to follow changes in gravitational level up to that exceeding 1 kHz. The ADXL 50's ability to maintain a flat response over this bandwidth keeps the sensor virtually motionless. This eliminates any non-linearity or aging effects due to the sensor beam's mechanical spring constant, as compared with an open loop sensor.

An uncommitted buffer amplifier provides the capability to adjust the scale factor and 0 g offset level over a wide range. An internal reference supplies the necessary regulated voltages for powering the chip and +3.4 V for external use.

Applications and further details of ADXL series devices can be found in Reference 29.

14.7 Hall effect devices

The basic Hall sensor is simply a small sheet of semiconductor material. A constant voltage source forces a constant bias current to flow in the semiconductor sheet. The output, a voltage measured across the width of the sheet, reads near 0 if a magnetic field is not present. If the biased Hall sensor is placed in a magnetic field oriented at right angles to the Hall current, the voltage output is in direct proportion to the strength of the magnetic field. This is the Hall effect, discovered by E.H. Hall in 1879 (see Figure 14.40). When a magnetic field, B, is applied to a specimen (metal or semiconductor) carrying a current I_c, in the direction perpendicular to I_c, a potential difference, V_H, proportional to the magnitude of the applied magnetic field B appears in the direction perpendicular to both I_c and B. This relationship is expressed in the form:

$$V_H = K \times I_c \times B, \tag{14.8}$$

where K represents a constant, the product sensitivity, which depends on the physical properties and dimensions of the material used for the Hall effect device.

The basic Hall sensor [30] essentially is a transducer that will respond with an output voltage if the applied magnetic field changes in any manner. Differences in the response of devices are generally related to tolerances and specifications, such as operate (turn on) and release (turn off) thresholds, as well as temperature ranges and temperature coefficients of these parameters. Also available are linear output sensors that differ in sensitivity or respond per gauss change.

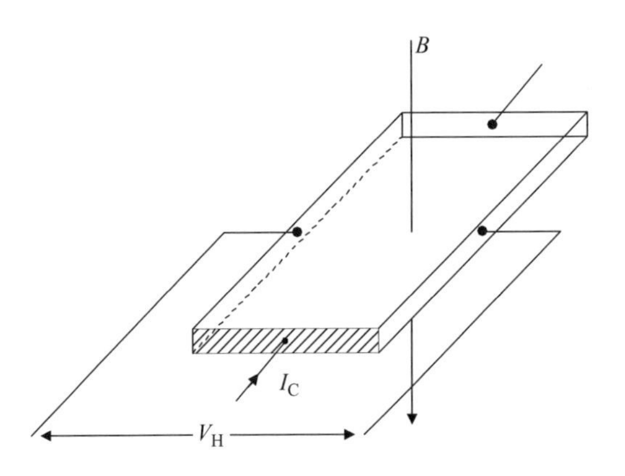

Figure 14.40 The Hall effect

14.7.1 Linear output devices

Linear output Hall effect devices are the simplest Hall sensor devices. Practical devices such as Allegro Microsystems' UGN-3605 give an output voltage response to applied magnetic field changes. Electrical connections for UGN-3605 are given in Figure 14.41(a). Applications of Hall devices are discussed in Reference 31.

The output voltage of the devices such as UGN-3605 is quite small, which can present problems, especially in an electrically noisy environment. Addition of a suitable d.c. amplifier and a voltage regulator to the circuit improves the transducer's output and allows it to operate over a wide range of supply voltages. Such combined devices are available and an example is UGN-3501 from Allegro Microsystems.

14.7.2 Digital output devices

The addition of a Schmitt trigger threshold detector with built-in hysteresis, as shown in Figure 14.42, gives the Hall effect circuit digital output capabilities. When the applied magnetic flux density exceeds a certain limit, the trigger provides a clean transition from off to on with no contact bounce. Built-in hysteresis eliminates oscillation (spurious switching of the output) by introducing a magnetic dead zone in which switch action is disabled after the threshold value is passed.

An open collector NPN output transistor added to the circuit gives the switch digital logic compatibility. The transistor is a saturated switch that shorts the output terminal to ground wherever the applied flux density is higher than the on trip point of the device. The switch is compatible with all digital families. The output transistor can sink enough current to directly drive many loads, including relays, triacs, SCRs, LEDs, and lamps.

14.8 Humidity and chemical sensors

14.8.1 Humidity sensors

Humidity, usually understood to refer to the water content of the air, could also be sensed using silicon based sensor elements. Relative humidity (RH), which is the ratio of absolute humidity to saturation humidity, has a value between 0 and 1 (or 0 per cent and 100 per cent). Several techniques are used to measure the relative humidity using capacitance, resistance, conductivity and temperature based measurements. Thermoset polymer or thermoplastic polymer based materials are used on silicon or ceramic based substrates for RH. A comparison of RH sensors is available in Reference 32.

Capacitive RH sensors dominate both atmospheric and process measurements, and are the only type of full range RH measuring devices capable of operating accurately down to 0 per cent RH. Because of their low temperature effect, they often are used over wide temperature ranges without active temperature compensation.

Thermoset polymer (as opposed to thermoplastic based polymer) capacitive sensors (see Figure 14.43) allow higher operating temperatures and provide better

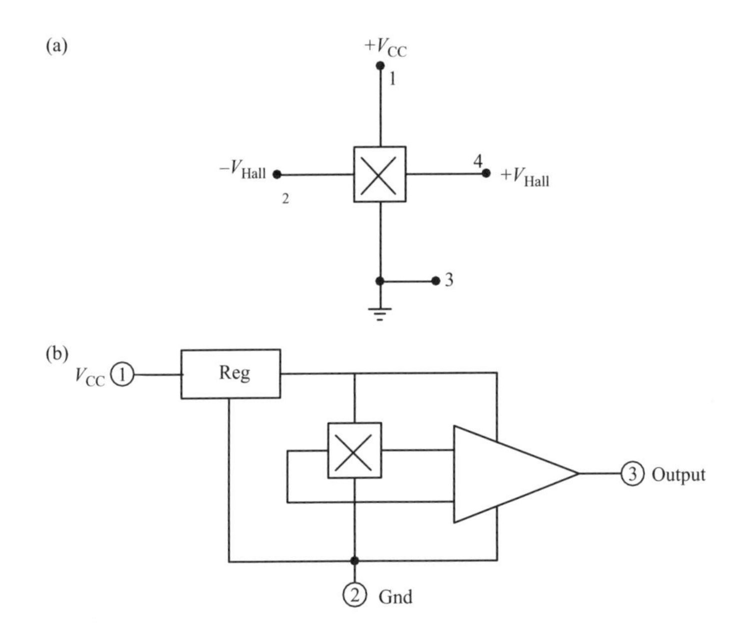

Figure 14.41 *Linear output Hall effect device: (a) device connections for UGN-3605; (b) amplified version (reproduced by permission of Allegro Microsystems)*

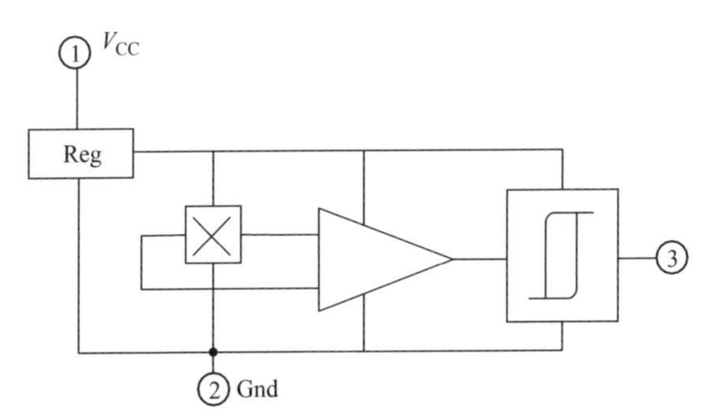

Figure 14.42 *Digital output Hall effect switch (reproduced by permission of Allegro Microsystems)*

resistivity against chemical liquids and vapours such as isopropyl, benzene, toluene, formaldehyde, oils, common cleaning agents, and ammonia vapour in concentrations common to chicken coops and pig barns. In addition, thermoset polymer RH sensors provide the longest operating life in ethylene oxide (ETO) based sterilisation processes.

An example of a thermoset polymer based capacitance RH device family is the Hycal IH36 XX series from Microswitch. These devices come with on-chip signal conditioning and provide a fairly linear ratiometric output based on the d.c. supply.

In operation, water vapour in the active capacitor's dielectric layer equilibrates with the surrounding gas. The porous platinum layer shields the dielectric response from external influences while the protective polymer overlayer provides mechanical protection for the platinum layer for contaminants such as dirt, dust and oil. A heavy contaminant layer of dirt, however, will slow down the sensor's response time, because it will take longer for water vapour to equilibrate in the sensor.

14.8.2 *Temperature and humidity effects*

The output of all absorption based humidity sensors (capacitive, bulk resistive, conductive film, etc.) are affected by both temperature and percentage RH. Because of this, temperature compensation is used in applications that call for either higher accuracy or wider operating temperature ranges. When temperature compensating a

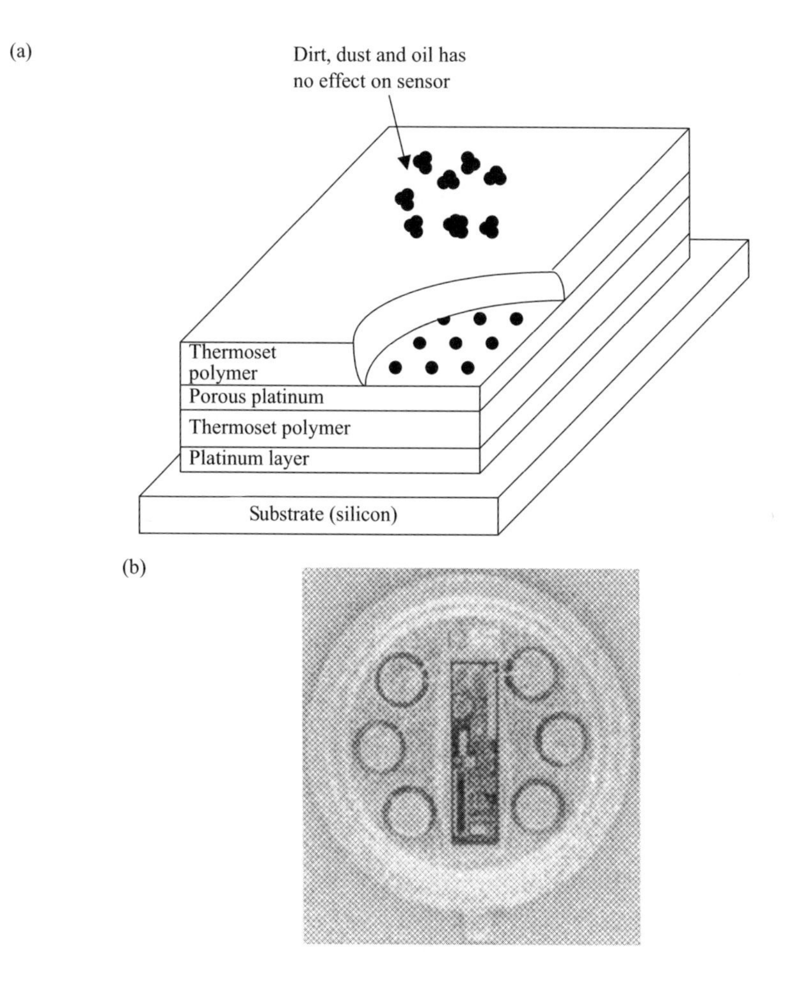

Figure 14.43 Continued overleaf

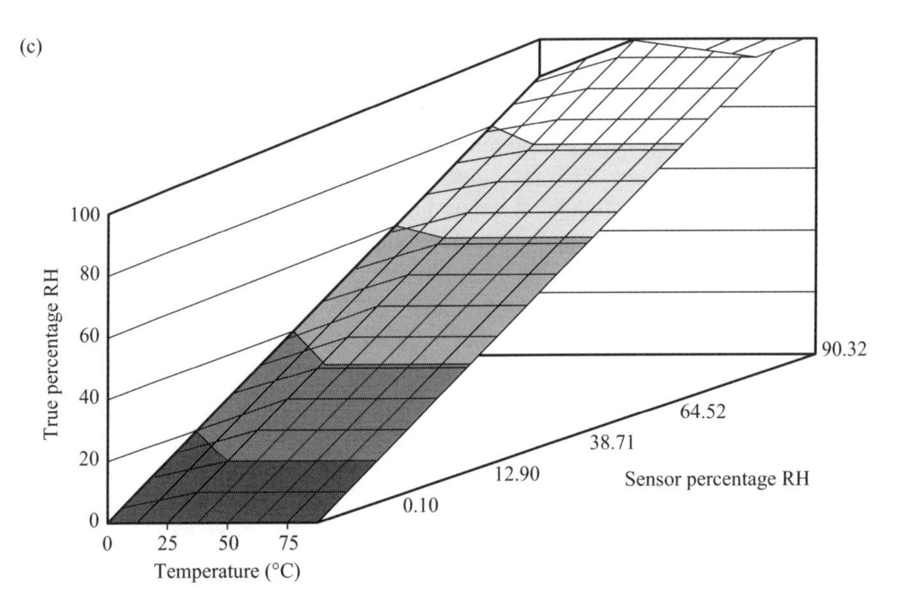

Figure 14.43 *Thermoset polymer based RH sensors: (a) basic construction; (b) relative humidity; (c) output voltage of IH-3602 vs relative humidity (reproduced by permission of Microswitch)*

humidity sensor, it is best to make the temperature measurement as close as possible to the humidity sensor's active area; that is within the same moisture micro-environment. This is especially true when combining RH and temperature as a method of measuring dew point.

HyCal's industrial grade humidity and dew point instruments incorporate a HyCal 1000 Ω platinum resistance temperature detector on the back of the ceramic sensor substrate for unmatched temperature compensation measurement integrity. No on-chip signal conditioning is provided in these high temperature sensors (Figure 14.43(b)).

14.8.3 Chemical sensors

Numerous technologies are used in the chemical sensing industry [33]. Silicon as a basic structure for chemical sensors has been investigated in numerous labs over the past 20 years. Based on metal oxide semiconductor gas sensors, some sensor manufacturers such as FiS Sensors (Japan) offer a wide range of products covering many applications, including carbon monoxide sensing, flammable gas detection, toxic gas detection, indoor air quality controls, and combustion monitoring and control.

The sensing element used in these devices is a mini-bead-type semiconductor, composed mainly of tin dioxide (SnO_2). A heater coil and an electrode wire are embedded in the element (Figure 14.44(a)). The element is installed in the metal housing, which uses double stainless steel mesh in the path of gas flow and provides an antiexplosion feature (Figure 14.44(b)). The sensor has three pins for output signal and heater power supply (Figure 14.44(c)). The SB-50 uses an active charcoal filter as shown in Figure 14.44(b). The conductivity of tin dioxide (SnO_2) based metal oxide semiconductor material changes according to gas concentration changes. This is caused by adsorption and desorption of oxygen and the reaction between surface oxygen and gases. These reactions cause a dynamic change of electric potential on the SnO_2 crystal and result in the decrease in sensor resistance under the presence of reducing gases such as CO, methane and hydrogen.

Figures 14.44(d) and (e) show the equivalent circuit and application circuit. Figure 14.44(f) shows the operating condition and output signal. The applied heater voltage regulates the sensing element temperature to obtain the specific performance

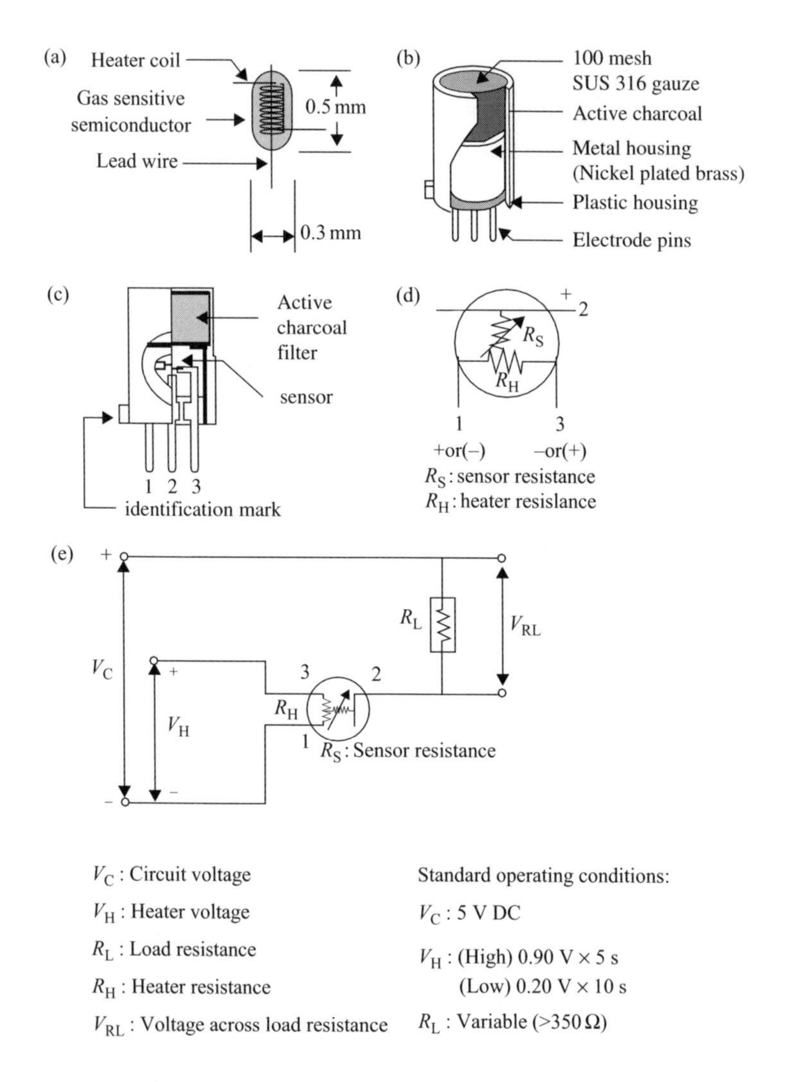

Figure 14.44 Continued overleaf

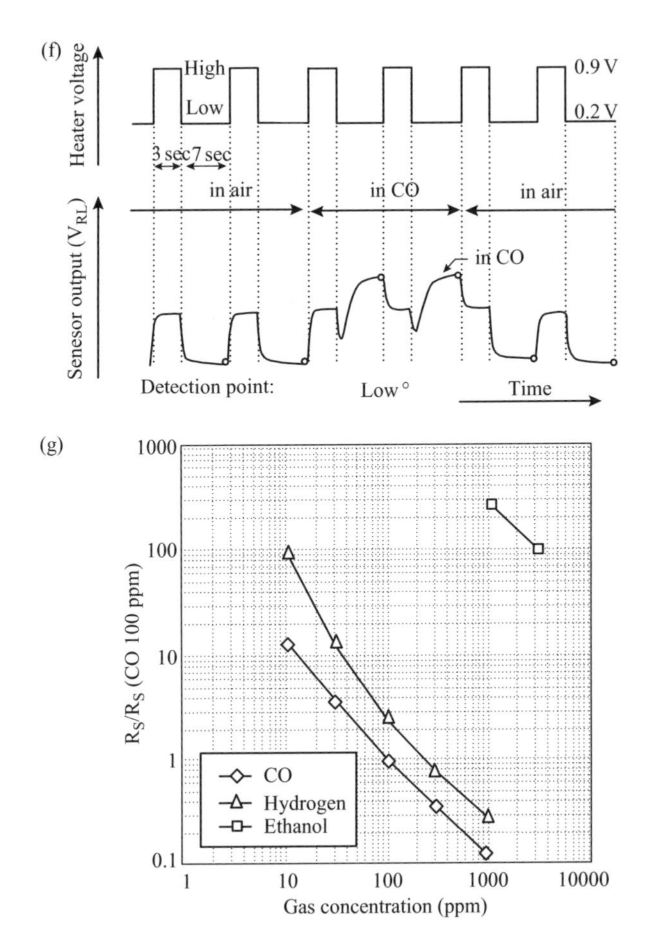

Figure 14.44 *The SB-50 gas sensor: (a) sensing element; (b) configuration; (c) pin layout; (d) equivalent circuit; (e) application; (f) operating condition & output signal; (g) sensitivity characteristics (reproduced with permission of FiS, Inc. Japan)*

of sensors. The change in the sensor resistance generally is obtained as a change in the output voltage across the fixed or variable load resistor (R_L) in series with the sensor resistance (R_S) (Figure 14.44(g)).

In general, the sensitivity characteristics of semiconductor gas sensors are shown by the relationship between the sensors resistance (R_S) and concentration of gases. The sensor resistance decreases with an increase of the gas concentration based on a logarithmic function. The standard test conditions of each are model calibrated to meet a typical target gas and concentration: for example, methane 1000 p.p.m. for flammable gas detection, hydrogen 100 p.p.m. for hydrogen detection or ethanol, and

300 p.p.m. for solvent detection. Figure 14.45 shows the typical sensitivity characteristics of the SB series. In these diagrams, the sensor resistance change is normalised by the R_S at specific conditions. For further details see Reference 34.

14.9 IEEE-P1451 Standard for smart sensors and actuators

Sensors are used in a wide range of applications from industrial automation to patient-condition monitoring in hospitals. With the advancement of silicon and micro-electromechanical systems (MEMS) technologies, more 'smarts' are integrated into sensors. The emerging of the control networks and smart devices in the marketplace may provide economical solutions for connecting transducers (hereafter specified as sensors or actuators) in distributed measurement and control applications; therefore networking small transducers is seriously considered by transducer manufacturers and users.

Control networks provide many benefits for transducers.

- Significant reduction of installation costs by eliminating many and long analogue wires.
- Acceleration of control loop design cycles, reduction of commissioning time, and reduction of downtime.
- Dynamic configuration of measurement and control loops via software.
- Addition of intelligence by leverage of the microprocessors used for digital communication.

For anyone attempting to choose a sensor interface or networking standard, the range of choices is overwhelming. Some standards are open, and some are proprietary to a company's control products. To remedy the situation, the IEEE Sensor Technology Committee TC-9 is developing the IEEE-P1451 Standard for a smart transducer interface for sensors and actuators. The sensor market comprises widely disparate sensor types. Designers consume relatively large amounts of all types of sensor. However, the lack of a universal interface standard impedes the incorporation of 'smart' features, such as an on-board electronic data sheet, on-board A/D conversion, signal conditioning, device-type identification, and communications hand shaking circuitry, into the sensors. In response to the industry's need for a communication interface for sensors, the IEEE with cooperation from the National Institute of Standards and Technology (NIST), decided to develop a hardware independent communication standard for low cost smart sensors that includes smart transducer object models for control networks [35].

The IEEE-P1451 standards effort, currently under development, will provide many benefits to the industry. P1451 – 'Draft Standard for smart transducer interface for sensors and actuators' – consists of four parts, as follows.

(i) IEEE 1451.1 – Network Capable Application Processor (NCAP) information model.
(ii) IEEE 1451.2 – Transducer to Microprocessor Protocols and Transducer Electronic Data Sheet (TEDS) formats.

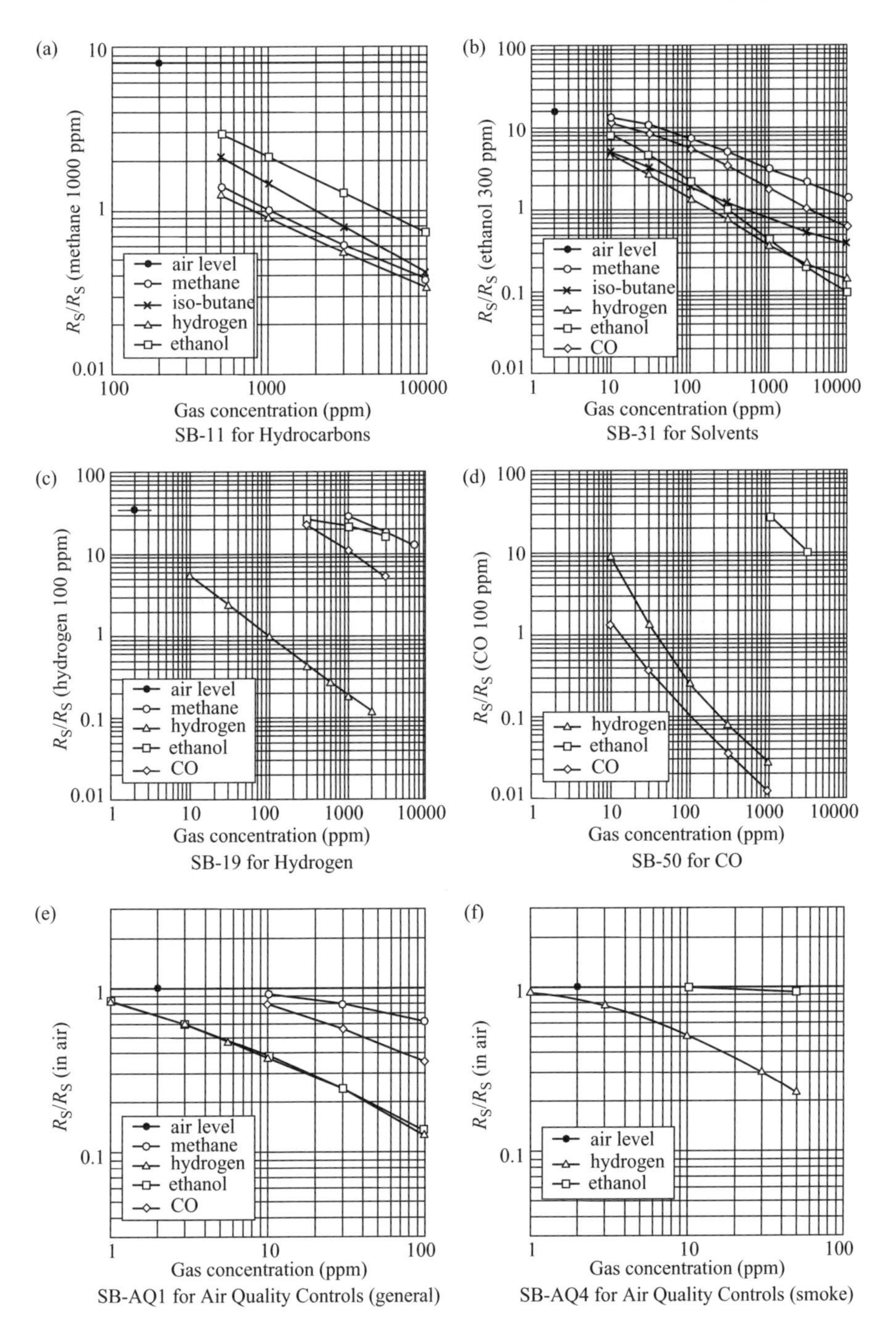

Figure 14.45 Typical sensitivity characteristics of the SB series gas sensors: (reproduced by permission of FiS, Inc.)

(iii) IEEE 1451.3 – Digital Communication and Transducer Electronic Data Sheet (TEDS) formats for distributed multi-drop systems.
(iv) IEEE 1451.4 – Mixed-mode Communication Protocols and Transducer Electronic Data Sheet (TEDS) formats.

In the process of writing the draft document, the working group has defined the Smart Transducer Interface Module (STIM), Transducer Electronic Data Sheet (TEDS), Transducer Independent Interface (TII), and a set of communication protocols between STIM and the Network Capable Application Processor (NCAP).

A system block diagram depicting the interface is shown in Figure 14.46. A STIM is specified to include up to 255 transducers, a signal converter or conditioning, a TEDS, and the necessary logic circuitry to support digital communication with NCAP. The TEDS is a small physical memory containing manufacturer's information and data for the transducer in a standardised data format. The TII, a 10-wire digital interface with provision for hot-swapping a sensor to a network, is used to access the TEDS, read sensors and set actuators.

Figure 14.47(a) depicts a STIM and the associated digital interface as described in the P1451.2-1997 hot swap. The STIM shown here is under the control of a network node microprocessor. In addition to their use in control networks, STIMs can be used with microprocessors in a variety of applications such as portable instruments and data acquisition cards as shown in Figure 14.47(b). The origin and function of each signal line of the 10-wire interface is listed in Table 14.5.

Standard 1451.2 was adopted by the IEEE as a full use standard, designated as IEEE Std. 1451.2-1997. The IEEE Std. 1451.2-1997 can be applied standalone, or it can be used with P1451.1. The two documents together will define a standard interface for networked smart sensors and actuators. Likewise, the P1451.1 information can be implemented in a sensor control or field network without 1451.2.

The IEEE Std. 1451.2-1997 standard and IEEE Std. 1451.1-1999 draft can be ordered from the IEEE customer service department by calling 1-(800)-678-4333 (IEEE) in the United States and Canada, 1-(732)-981-0600 from outside the United States and Canada, or by faxing 1-(732)-981-9667.

14.10 P1451 and practical components

Existing microcontrollers fall short of fully implementing the standard in silicon, whether because of functionality or prohibitive cost. For example, the standard transducer interface module (STIM) portion of the standard specifies the sensor interface electronics, signal conditioning, data conversion, calibration, linearisation, basic communication capability, and non-volatile 565-byte TEDS. Some microcontrollers with integrated 8- or 10-bit ADCs of comparator based slope conversion can implement most of the STIM functionality, but are limited in conversion speed and accuracy. Moreover, few available controllers have economically integrated analogue conversion together with high density EEPROM because of the additional process complexity requirements of both functions.

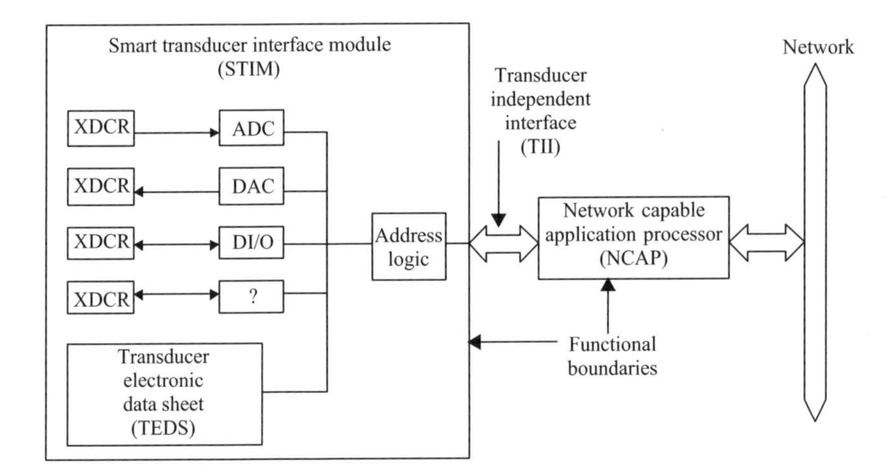

Figure 14.46 System block diagram depicting the transducer interface (courtesy of NIST)

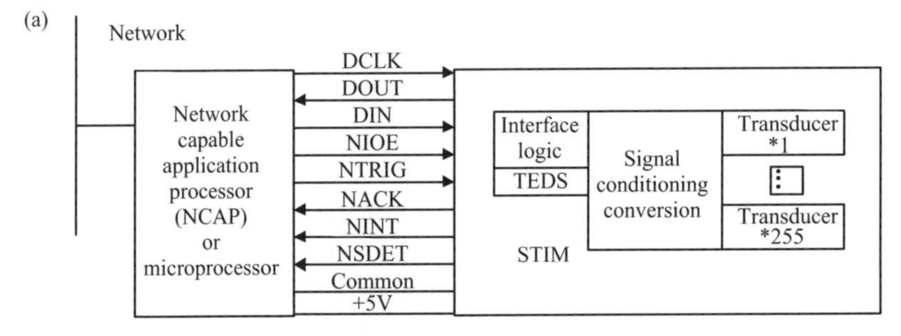

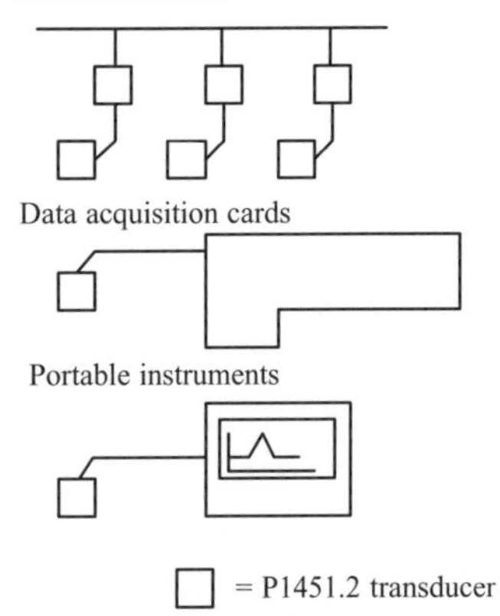

Figure 14.47 (a) Hardware partition proposed by P 1451.2 and (b) possible use for the interface (source: Reference 36)

Table 14.5 The ten lines that make up the transducer independent interface (courtesy of NIST)

Line	Driven by	Function
DIN	NCAP	Address and data transport from NCAP to STIM
DOUT	STIM	Data transport from STIM to NCAP
DCLK	NCAP	Positive-going edge latches data on both DIN and DOUT
NIOE	NCAP	Signals that the data transport is active and delimits data transport framing
NTRIG	NCAP	Performs triggering function
NACK	STIM	Serves two functions: trigger acknowledge and data transport acknowledge
NINT	STIM	Used by the STIM to request service from the NCAP
NSDET	STIM	Grounded in the STIM and used by the NCAP to detect the presence of a STIM
POWER	NCAP	Nominal 5 V power supply
COMMON	NCAP	Signal common or ground

These limitations are overcome by recently introduced components such as the AduC812 MicroConverter™ [37] from Analog Devices, which integrates key STIM elements with 12-bit, 5 μs data conversion on a single chip for high accuracy, fast conversion time applications such as battery monitoring, pressure and temperature management, gas monitoring, and leak detection. In a typical application, the AduC812 conditions and converts signals from various types of sensors, sends signals to actuators and display devices, and communicates with the host microprocessor over signal and control lines.

The AduC812 MicroConverter™ is supported by a development system that includes documentation, applications board, power supply, serial port cable, and software. Provided on a 3.5 inch floppy disk, the software consists of an assembler, simulator, debugger, serial down loader, and example code.

14.11 References

1 HAUPTMANN, P.: *Sensors – Principles and Applications*, Prentice Hall, 1991.
2 GUCKEL, H.: 'Micromechanisms fabrication: challenge in micromechanisms and microelectricity', *ISSCC Proceedings*, 1992, pp. 14–17.
3 MICROSWITCH: 'Temperature and moisture sensors', May 1997, pp A1–A10.
4 LE FORT, B. and RIES, B.: 'Taking the uncertainty out of thermocouple temperature measurement (with AD 594/AD 595)', Application Note AN-274, Analog Devices.

5 MARCIN, J.: 'Thermocouple signal conditioning using the AD 594/AD 595', Application Note AN-369, Analog Devices.
6 KLONOWSKI, P.: 'Use of the AD 590 temperature transducer in remote sensing applications', Application Note AN-273, Analog Devices Inc.
7 ANALOG DEVICES: 'Accessories of AD 590', Application Note AN-272.
8 ANALOG DEVICES: 'Design – in reference manual', 1994, Analog Devices Inc.
9 TRAVIS, B.: 'Temperature-management ICs combat system meltdown', *EDN*, August 15, 1996, pp. 38–48.
10 FREEMAN, W.: 'Solid state temperature sensors protect & control', *PCIM*, November 1993, pp. 39–45.
11 ANALOG DEVICES DATA SHEET: Serial Output Thermometers – TMP 03/04, Rev. 0, 1995.
12 ANALOG DEVICES – DATA SHEET: 'Low voltage temperature sensors – TMP 35/TMP 36/TMP 37'. Rev. 0, 1996.
13 ANALOG DEVICES – DATA SHEET: 'Airflow & temperature sensor – TMP 12'. Rev. 0, 1995.
14 DALLAS SEMICONDUCTOR: 'System extension data book', 1994–1995.
15 BRY ZEK, J.: 'Characterization of MEMS industry in Silicon Valley and its impact on sensor technology'. Proceedings of Sensors Expo, October 1996, pp. 1–13.
16 CARISTI, A. J.: 'A digital pressure gauge using the Motorola MPX 700 series differential pressure sensors', Motorola Application Note AN-1105.
17 JACOBSEN, E. and BAUM, J.: 'The A-B-C's of signal conditioning amplifier design for sensor applications', Motorola Application Note AN 1525.
18 WILLIAMS, D.: 'A simple pressure regulator using semiconductor pressure transducer', Motorola Application Note AN1307, Rev1, 1997.
19 JACOBSON, E. and BAUM, J.: 'Optimize sensor systems using fixed components', *EDN*, July 6, 1995, pp. 85–95.
20 JACOBSON, E.: 'Designing amplifiers for sensor applications: A cookbook approach', *EDN*, January 4, 1996, pp. 119–128.
21 SCHULTZ, W.: 'Interfacing semiconductor pressure sensors to microcomputers'. Motorola Application Note – AN 1318.
22 BURRI, M.: 'Calibration free pressure sensor system', Motorola Application Note – AN1097.
23 LUCUS, B.: 'An evaluation system for direct interface of MPX 5100 pressure sensor with a microprocessor', Motorola Application Note AN 1305.
24 WINKLER, C. and BAUM, J.: 'Barometric pressure measurement using semiconductor pressure sensors', Motorola Application Note AN 1326.
25 IC SENSORS, INC.: 'Understanding accelerometer technology', Application Note (UATRO – 9111).
26 QUINNELL, R.A.: 'Silicon accelerometers tackle cost-sensitive applications', *EDN*, September 3, 1992, pp. 69–76.
27 IC SENSORS: 'Model 3255 accelerometer', Application Note TN – 010, April 1995.
28 ANALOG DEVICES, Inc.: 'Data sheet – ADXL 50', Rev. B, 1996.

29 ANALOG DEVICES, Inc.: 'Accelerometer application guide' (G 2112A).
30 WOOD, T.: 'The Hall effect sensor', *Sensors*, March 1986.
31 SWAGER, A. W.: 'Hall effect sensors – improved ICs find broad application', *EDN*, May 11, 1989, pp. 75–92.
32 MICROSWITCH: 'Moisture tutorial – relative humidity sensors'. Temperature and Moisture Sensors Catalog, May 1997, pp. B1–B7.
33 WALTERS, D.: 'Chemical sensing : an emergent MEMS technology', *Proceedings of Sensors Expo*, USA, October 1996, pp. 173–186.
34 FiS Inc.: 'SB products review: sensors & systems technology', 1996.
35 TRAVIS, B.: 'Smart-sensor standard will ease networking woes', *EDN*, June 22, 1995, pp. 49–52.
36 WOODS, S. P.: 'IEEE – P 1451.2 smart interface module', *Proceedings of Sensors Expo*, USA, October 1996, pp. 25–46.
37 ANALOG DEVICES, Inc.: 'The AduC812 as an IEEE 1451.2 STIM: Microconvertor™ Technical note uc0003 (version 1.0, Sept, 1999).

Further reading

AJLUNI, C.: 'Pressure sensors strive to stay on top', *Electronic Design*, October 3, 1994, pp. 67–74.

DEMINGTON, C.: 'Compensating for non linearity in the MPX 10 series pressure transducer', Motorola Application Note AN 935.

IC Sensors: 'Temperature compensation – IC pressure sensors', Application Note TN-002, March 1985.

IC Sensors: 'Signal conditioning for IC pressure sensors', Application Note TN-001, April 1988.

MICROSWITCH: 'Pressure, force & airflow sensors', Catalogue – 15, August 1996, pp. 74–91.

MOTOROLA, INC.: 'Sensor device data book', DL200D/Rev 4, Q3/98.

MOTOROLA, INC.: 'Analog to digital converter resolution extension using a Motorola pressure sensor', Application Note AN 1100.

STEELE, J.: 'Get maximum accuracy from temperature sensors', *Electronic Design*, August 19, 1996, pp. 99–110.

Chapter 15

Calibration of instruments

15.1 Introduction

Almost all measuring instruments provide the user with a quantitative measurement. The user always expects a known level of confidence in that measured value. The ultimate aim of a measurement is to have accuracy, reliability and confidence in the exercise of measurement and its quantitative output.

Calibration is the process that ensures accuracy in a measurement, and makes the measurement and its process traceable to standards discussed in chapter 1. With the growing adoption of quality standards such as ISO9000, many institutes are paying more attention to calibration of instruments, because all measurements have a direct bearing on product quality or service quality.

Since the end of World War II, technical advances have permited scientists to reduce uncertainty in measurements dramatically. In the 1940s, working measurements were made with an analogue iron-vane or D'Arsonval meter. With care, uncertainties were in the range of 0.5 per cent of full scale, or 5000 p.p.m. As shown in Figure 15.1, that 5000 p.p.m. in the 1940s became 100 p.p.m. by 1960, 10 p.p.m. by 1970 and 2 p.p.m. today [1].

This chapter provides an introduction to calibration of common instruments and the calibrators.

15.2 Metrology and calibration

Simply stated, metrology is the science of measurement. Everything that has to do with measurement, be it designing, conducting, or analysing the results of a test, exists within the realm of metrology. These things cover the range from the abstract, comparing statistical methods, for example, to the practical, such as deciding which scale of a ruler to read.

Calibration is the comparing of a measurement device (an unknown) against an equal or better standard. A standard in a measurement is considered the reference; it is the one in the comparison taken to be the more correct of the two. One calibrates

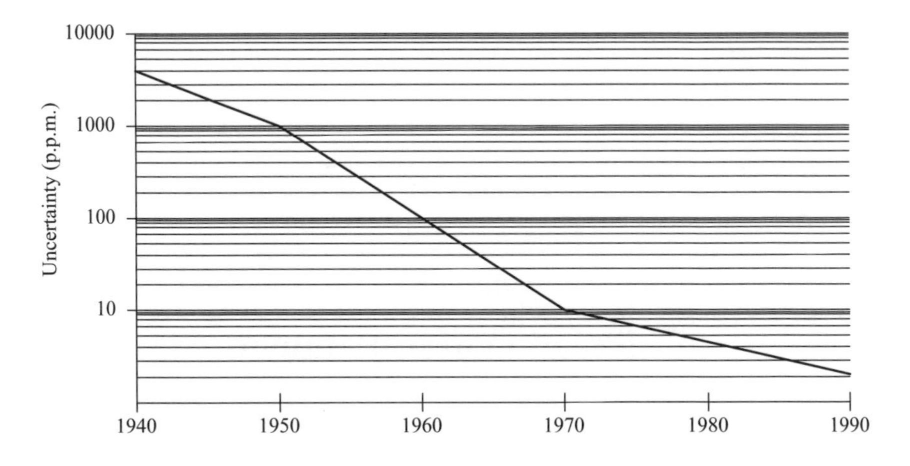

Figure 15.1 *The trend towards lower uncertainties in working measurements (source: Reference 1)*

to find out how far the unknown is from the standard. A comprehensive discussion on the philosophy of calibration and metrology in a practical sense, as applied to electrical measurements, is in Reference 2.

15.3 Traceability

Traceability refers to procedures and records that are used and kept to demonstrate that calibrations made in a local laboratory accurately represent the quantities of interest. The International Standards Organisation (ISO) defines traceability as the 'property of the result of a measurement or the value of a standard whereby it can be related to stated references, usually national or international standards, through an unbroken chain of comparisons all having stated uncertainties'. The scientific aspects of traceability involve principles of metrology used locally and independently by scientists, engineers, and technicians. The legal aspects of traceability involve a governmentally established and maintained infrastructure within which the measurements are made. The infrastructure that is in general use today has three major constituents.

- World-wide legal adoption of the International System of Units (SI) as the basic system of units of weights and measure.
- The establishment of national laboratories such as National Institute of Standards and Technology (NIST), chartered to maintain representations of the SI units (standards) and to disseminate their values to calibration laboratories.
- Definition, implementation, and use of methods and procedures that allow individual calibration laboratories to compare their local standards with those of the national laboratories.

15.3.1 Establishing and maintaining traceability

Traceability includes keeping in-house standards that are compared with one of the previously described standards. More often than not, an unbroken chain of calibrations ends with comparison standards maintained by national authorities such as NIST in the USA. Figure 15.2 is a diagram for direct voltage that traces the measurement of a d.c. voltage source such as Fluke 8842A back to an intrinsic standard at a national laboratory. Similar diagrams can be made for other units such as resistance, alternating voltage, and so forth.

Within an organisation, measurement traceability is maintained by a hierarchy of comparisons up to national standards. A local primary standard is required for this process. Accurate documentation of the process is essential.

A simplified picture in Figure 15.3 shows the typical links in the traceability chain for d.c. voltage from a DMM to a national laboratory such as NIST in the USA. For details, Reference 3 is suggested.

15.3.2 Test uncertainty ratio

Test uncertainty ratio (TUR) is defined as the specified uncertainty of the test instrument divided by the uncertainty of the calibrating instrument. When the accuracy of the instrument under test reaches the accuracy of the calibrator TUR becomes 1 : 1. Generally TURs in metrology exercises should be kept below 4 : 1. For example, the calibration of a DMM by a multifunction calibrator (MFC) is discussed below.

The immediate purpose of calibration is to gain confidence that a UUT is able to make measurements within its specifications. For example, suppose one wants to check a DMM's performance as specified on the 10 V range: $\pm$20 p.p.m. of reading, +1.6 p.p.m. of range. The calibration of this DMM can be checked by applying a known 10 V from an MFC to its input. If exactly 10 V is applied to this DMM, any reading within $\pm$216 μV of 10 V will be within specifications.

Calibration laboratories generally do not have access to exact value of stimulus to test DMMs or other equipment. Instead, they must use commercially available calibrators such as MFCs, which also have uncertainty specifications. For example, data for the Fluke 5700A MFC specifies that the outputs on its 11 V range are within $\pm$ (5 p.p.m. of output $\pm$4 μV) of the true value. When this specification is converted to units, the uncertainty in the MFC's 10 V output is $\pm$54 μV or $\pm$5.4 p.p.m. This is its 99 per cent confidence level (2.6-σ value). This corresponds to $1\sigma = 5.4/2.6 = 2.08$ p.p.m. For details, chapter 20 of Reference 2 is suggested.

15.4 Calibration of practical instruments in the field

Given the above definitions and traceability of a measurement, practical calibration work in the field can be done by using more accurate and traceable instruments with a higher order accuracy than the unit under calibration. As it is beyond the scope of a brief chapter of this kind to discuss calibration of all electrical parameters and instruments, we will now discuss the calibration of common instruments such as multimeters and oscilloscopes from a field viewpoint.

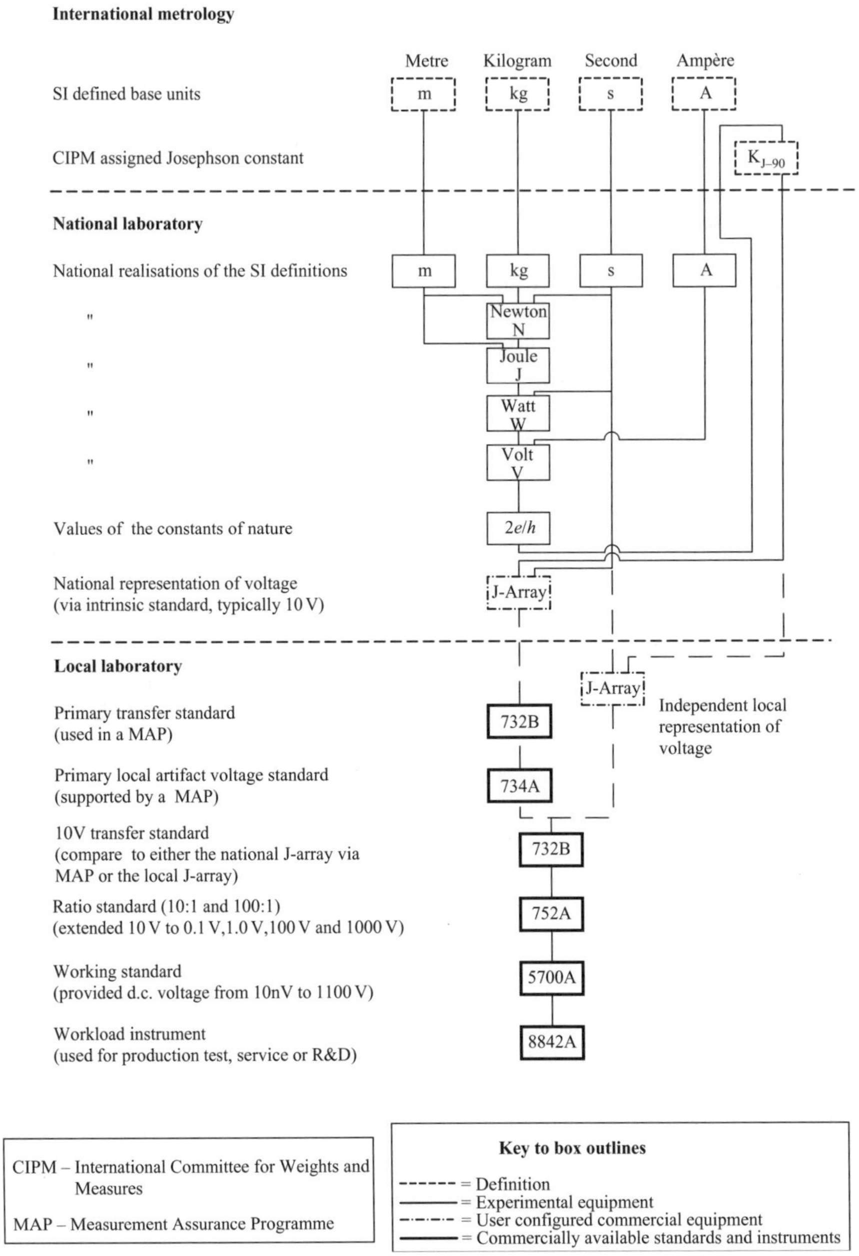

International metrology

Metre Kilogram Second Ampère

SI defined base units — m, kg, s, A

CIPM assigned Josephson constant — K_{J-90}

National laboratory

National realisations of the SI definitions — m, kg, s, A

" — Newton N

" — Joule J

" — Watt W

" — Volt V

Values of the constants of nature — $2e/h$

National representation of voltage (via intrinsic standard, typically 10 V) — J-Array

Local laboratory

J-Array — Independent local representation of voltage

Primary transfer standard (used in a MAP) — 732B

Primary local artifact voltage standard (supported by a MAP) — 734A

10V transfer standard (compare to either the national J-array via MAP or the local J-array) — 732B

Ratio standard (10:1 and 100:1) (extended 10 V to 0.1 V, 1.0 V, 100 V and 1000 V) — 752A

Working standard (provided d.c. voltage from 10nV to 1100 V) — 5700A

Workload instrument (used for production test, service or R&D) — 8842A

CIPM – International Committee for Weights and Measures

MAP – Measurement Assurance Programme

Key to box outlines

------ = Definition
——— = Experimental equipment
–·–·– = User configured commercial equipment
▬▬▬ = Commercially available standards and instruments

Figure 15.2 Traceability diagram for direct voltage (courtesy of Fluke Corporation, USA)

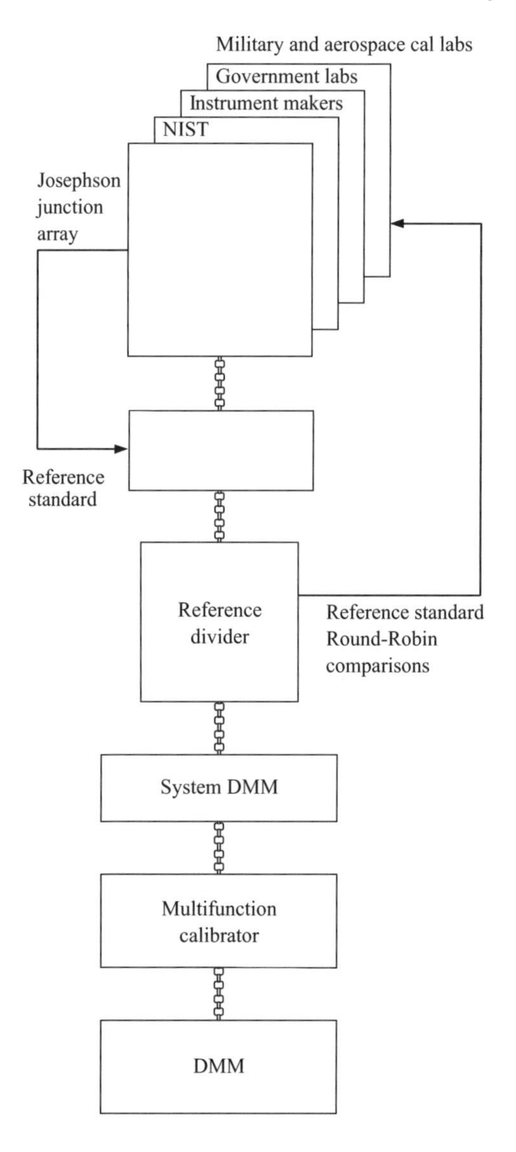

Figure 15.3 A pictorial chain of instruments in a DMM calibration chain/d.c. voltage calibrations (courtesy of Test & Measurement World magazine, December 1999)

15.4.1 Multimeters and their calibration

To maintain the confidence of field measurements, multimeters need be calibrated on a regular basis. Effective field calibrations require a certain amount of knowledge of the DMM technology.

15.4.2 Types of DMM

DMMs typically display electrical quantities on their digital displays. However, the method of sensing electrical quantities may alter their response from the calibrated value. This is particularly true when complex a.c. current and voltage waveforms are measured. Therefore it is important for the test engineer to understand the DMM's characteristics in order to conduct a competent calibration.

Of the wide variety of DMMs on the market, three of the most popular types will be discussed in the following paragraphs. These are:

- laboratory DMMs,
- bench/systems DMMs, and
- handheld DMMs.

A fourth type of DMM is emerging, the DMM on a card. Because of their relative newness, they are not discussed here but their calibration is similar to the calibration of bench/systems DMMs.

15.4.2.1 Laboratory DMMs

Laboratory DMMs are usually five-function DMMs that measure d.c. and a.c. voltage, resistance, and d.c. and a.c. current. They typically offer the highest level of accuracy and resolution and may approach the accuracy of the multifunction calibrators (MFC) used to calibrate them. They display up to $8\frac{1}{2}$ digits in their readouts and are often automatically calibrated by the closed loop method using their IEEE-488 bus compatible systems interface.

These DMMs use similar techniques to measure d.c. voltage, d.c. current and resistance. However, they do not necessarily use the same principles of operation to measure a.c. voltage and current. Therefore, it is necessary that very pure sinewave quantities are generated by the calibrator, because the a.c. reading displayed assumes and displays a value based on an ideal sinewave.

The modern laboratory DMMs make use of state-of-the-art components such as microprocessor and computer memory chips, which allow them to perform complex mathematical computations. These include storing all corrections on all functions and ranges of a DMM, thus eliminating the need for opening the case of the DMM physically to make corrective adjustments.

15.4.2.2 Bench/systems DMMs

Bench/systems DMMs usually measure the same five functions as laboratory DMMs but with less accuracy and resolution, typically $4\frac{1}{2}$ or $5\frac{1}{2}$ digits. They may or may not have an IEEE-488 bus or RS-232 interface. If no interface is installed, they are manually calibrated.

In some cases, a manual calibration only involves selecting the range and function and applying the nominal stimulus to which the DMM adjusts itself. Otherwise, the DMM's response is calibrated via adjustment of potentiometers, rheostats, and variable capacitors. These DMMs also use sophisticated components and state-of-the-art circuits and may be as complex as the top end laboratory DMMs.

In many calibration laboratories, the bulk of the workload is composed of bench/systems DMMs, thus lending them to automated, closed loop calibration, such as with the Fluke 5700A controlled by a PC running software such as MET/CAL discussed later. It is not unusual for a user to purchase an IEEE-488 or RS-232 interface option with his DMM to implement automated calibration even through the DMM is used in a manually operated application.

15.4.2.3 Handheld DMMs

Handheld DMMs are the most commonly used version of the digital multimeter. They are truly multifunctional units that include the five electrical functions plus additional functions such as frequency, continuity, diode test, peak voltage measure and hold, temperature, capacitance, and waveform measurement.

These units typically have a $3\frac{1}{2}$ or $4\frac{1}{2}$ digit display and some also include an analogue readout and audible tones for continuity measurements. They tend to be used in a broad variety of applications from very sophisticated electronic circuit testing to automotive maintenance and hobbyist use, etc.

These DMMs tend to be small, rugged, and battery operated. Most of these typically do not have a computer interface to automate fully their calibration. However, with the growing emphasis on quality, and compliance with ISO 9000, it is likely that future handhelds will have some form of interface for calibration to optimise cost of ownership and to comply with quality standards.

Calibration of handhelds is not a problem from the standpoint of accuracy. However, owing to the increasing number of functions being placed in handhelds, future MFCs will no doubt include more functions to calibrate them.

15.4.3 Anatomy of a DMM and ADC range

In calibration activities it is important to understand the overall arrangement of blocks, and of the ADC in particular.

A DMM is really two integrated devices: the measurement section and the control section (see Figure 15.4). The control section is typically made up of a microprocessor and its support circuitry. The measurement section is made up of the analog conditioning circuits and the ADC. The control section requires no calibration adjustment, but is often checked internally during DMM power-up and self-tests. The measurement section does require calibration adjustment and verification.

15.4.3.1 Analogue-to-digital converters (ADC)

Every input to a DMM is eventually processed by the ADC, the heart of any DMM. The purpose of the ADC, is to digitally represent an analogue signal. The ADCs used today typically operate over a range of ±200 mV for the less accurate meters: more accurate DMMs have ADCs operating over a ±2, or ±20 V range. There are two major types of ADCs in widespread use today, namely integrating, and successive approximation types.

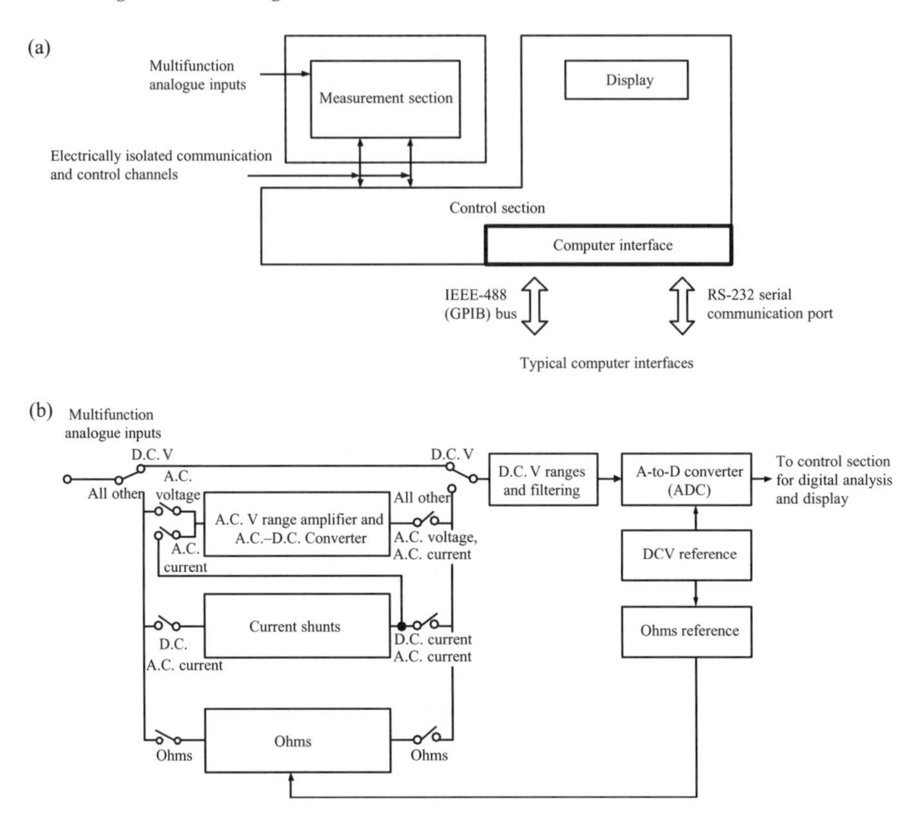

Figure 15.4 DMM block diagram and a typical measurement section: (a) block diagram; (b) typical measurement section (courtesy of Fluke Corporation, USA)

Integrating ADCs can operate on dual-slope, multi-slope, or charge-balance schemes. Because of their simple design, and also their ability to deliver high precision, integrating converters appear in every type of DMM from the lowest cost handheld DMMs, to systems DMMs with precision of $8\frac{1}{2}$ digits. Integrating ADCs are quite effective in rejecting noise, because the analogue integrator acts as a low pass filter.

15.4.3.1.1 Successive approximation ADCs and the R^2 converter

Another common type of ADC is the successive approximation type. The Fluke recirculating remainder (R^2) scheme, a variation on successive approximation, provides the high speed and high precision in many of today's popular DMMs, for example in the Fluke 8506A and 8840A. Combined with external analogue and digital filtering, the R^2 converter can have noise rejection characteristics comparable to those of the integrating converter. For details of the R^2 converter, chapter 17 of Reference 2 is suggested.

15.4.3.2 Signal conditioning circuits and converter stages

A DMM's d.c. voltage section is used to measure a wide range of input voltages. This section's d.c. output is scaled to be compatible with the ADC input. Additional signal conditioning circuits are used for the other measurement functions. These circuits consist of amplifiers, attenuators, and filters for each DMM range. Each range scales the input signal to the proper level for the ADC. The calibration of each range corrects for zero and gain variations of the amplifier and attenuators. Most DMM filters do not require calibration because filter topologies usually have little effect on uncertainty.

The converter sections such as (i) a.c.-to-d.c. or r.m.s.-to-d.c., (ii) resistance converter, (iii) current converters, and (iv) miscellaneous other converters also have an effect on calibration. The user and the test engineer should be able to get a better overall achievement in a calibration exercise if these stages are understood practically.

15.4.4 General DMM calibration requirements

Regardless of the type of DMM, calibration adjustments are performed to reduce instability in the offset, gain and linearity of the transfer functions of the signal processing circuits. Each of the functional blocks in a DMM's measurement section is subject to these sources of variation in performance. A DMM is designed to operate under a relationship such as

$$Y = mx + b. \tag{15.1}$$

where Y is the instrument reading and X is the signal input.

The instrument is designed in such a way that m has an exact nominal value such as 10, and b should be zero. Figure 15.5 depicts the variation in gain, offset and linearity for a typical DMM. Traditionally, these deviations are referred to as gain, offset and linearity errors.

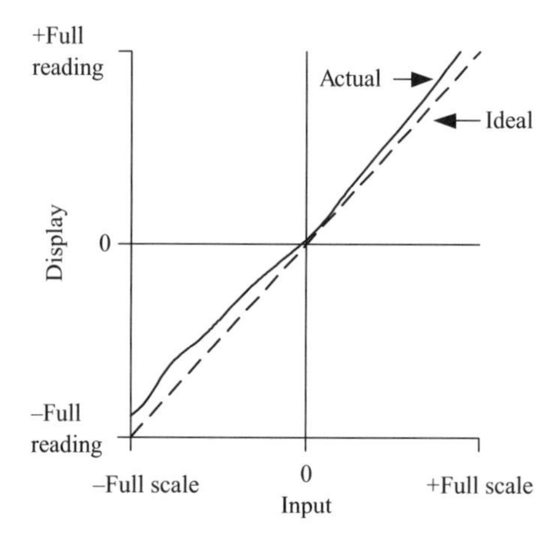

Figure 15.5 Offset, gain and linearity variations

Zero uncertainty is caused by voltage offsets in the ADC amplifiers and comparators. Gain uncertainty is tied to inaccuracies of the d.c. voltage reference and any gain determining networks of the ADC. Linearity uncertainty is the result of secondary error sources, such as mismatches in resistive ladders of the DAC in R^2 converters, or dielectric absorption errors by storage capacitors, resulting in a drooping response at full scale.

In addition to these sources of uncertainty, some DMMs have spurious voltage spikes, d.c. bias currents, and pump-out currents present at their input terminals. These can introduce errors in the output of an instrument such as an MFC calibrating the DMM.

15.4.4.1 Internal references

Every DMM has a d.c. voltage reference. References range from 1 to 15 V. The d.c. voltage reference is used as the reference for the ADC and is the limiting factor for the DMM's best accuracy for all voltage and current measurements. DMMs may or may not require separate calibration of the d.c. voltage reference. Often, the DMM design is such that an actual measurement (using another DMM) of the d.c. voltage reference is not required. Instead, a known input voltage is applied to the unit under test (UUT), and an adjustment is made until an accurate representation of the known input voltage appears on the UUT display.

Also, in any DMM, there are one or more resistor references that determine the accuracy of the resistance converter. Most DMMs do not require a separate measurement of the reference resistors during calibration. Instead, a standard resistor of known value is applied to the DMM, and the DMM is adjusted for the correct readout.

15.4.4.2 ADC adjustment considerations

Because all functions use the ADC, it must be calibrated as one of the first steps. Some DMM calibration schemes call for separate ADC and reference calibrations; other schemes merely call for d.c. voltage range calibrations and do not burden the test engineer with the specific details of the ADC calibration. DMMs whose procedures do not call for separate ADC adjustment do have internal corrections. The calibration procedure simply calls for an adjustment in the straight-in d.c. voltage range first. This range, going directly into the ADC, has unity gain and thus has no significant error in scale factor. The calibration procedure will often call for a zero offset adjustment of the d.c. range, and then the ADC is adjusted. For details of ADC adjustment, Reference 2 is suggested.

15.4.4.3 D.C. voltage range calibration

DMMs usually have full scale range of 200, 300, or 1000 V d.c., typically divided by factors of 10 to determine the lower ranges. Normal calibration practice limits the testing of ranges to just less than full scale. This is because most DMMs are autoranging, and will go to the next higher range if the input is a certain percentage of full scale. So the calibration voltages for DMMs with full scale ranges of 200 V d.c. usually follow a 190 mV, 1.9 V, 19 V, 190 V and 1000 V d.c. sequence.

Modern MFCs take advantage of this structure. For example, the lower four of the ranges in the preceding list are very quickly calibrated using a calibrator such as the Fluke 5700A by setting up its output on the lowest range and then using its 'Multiply Output By 10' key to step through the next three ranges. The zero on each range may also need to be adjusted or recorded. This is usually done by applying an external short to the DMM input and taking the appropriate action. After the zero is adjusted or recorded, the full scale input voltage is applied and the range gain is adjusted.

For example, assume that a DMM's ADC operates within the range of ±20 V. Its 200 mV d.c. range scales the input signal ×100 using an amplifier stage to keep the full scale ADC input to within the proper range. The amplifier used in the 200 mV d.c. range may have a voltage offset error, and the gain provided by the feedback may not be exactly 100, causing a scale factor error.

The calibration procedure is to apply a short circuit to the input to the amplifier and adjust the DMM reading to exactly 0 V d.c. This would be followed by an input near full scale, in this case, 190 mV d.c. The DMM is then adjusted to display the exact input voltage. In effect, the user is making the gain exactly 100. A DMM calibration may also require an input of opposite polarity, −190 mV. This is to correct either for secondary linearity errors of the range amplifier, or for linearity errors within the ADC.

15.4.4.4 A.C.–D.C. converter calibration

A.C.–D.C. converter calibration is similar for all types of converters. Amplitude adjustments are often made at two or more frequencies on each range. Zero adjustments are not usually made.

15.4.4.4.1 Average responding a.c.–d.c. converters

Calibration adjustment of average responding converters corrects for deviations in the equation $y = mx + b$. However, the b term is solved not for inputs at 0 V, but typically for inputs at 1/10th or 1/100th of full scale. This is because precision rectifiers tend not to work well at zero volts input. Near 0 V, they rectify noise that would otherwise ride on the input waveform and average to zero. By moving the b term calibration off zero, this problem is avoided. For more information, chapter 33 of Reference 2 is suggested.

15.4.4.4.2 True r.m.s. a.c.–d.c. converters

True r.m.s. converters need similar calibration of ranges at full scale and at near zero input levels. But true r.m.s. circuits have an even greater problem near 0 V than average responding converters. For example, analogue log/antilog schemes (chapter 17 of Reference 2) employ a precision rectifier, resulting in noise rectification problems like those associated with an average responding converter. Also, since the displayed output voltage is:

$$V_{out} = \sqrt{V_{noise}^2 + V_{sig}^2}, \qquad (15.2)$$

where:

$$V_{\text{noise}} = \text{r.m.s. value of the noise voltage,}$$
$$V_{\text{sig}} = \text{r.m.s. value of the signal being measured.}$$

Very small input signals cause the noise term to dominate: consequently, true r.m.s. calibrations are accomplished with inputs well above 0 V.

There are additional adjustments to correct for errors in linearity of r.m.s. converters. If the DMM has a d.c.-coupled a.c. voltage function, there may be zero input adjustments to correct for offset errors of the scaling amplifiers.

The a.c. voltage converter can have a separate calibration for the true r.m.s. converter, but often its calibration is combined into the main a.c. voltage function calibration. For example, if the true r.m.s. module can handle a 2 V full scale waveform, the 2 V a.c. range should be calibrated first; then adjust the true r.m.s. module. The other ranges are then adjusted to correct for their gain errors. In some cases, the true r.m.s. module needs more than two different inputs to correct for linearity errors. For example, an a.c. voltage calibration adjustment may call for inputs of 19 mV, 190 mV and 1.9 V on the 2 V range.

Scaling sections, averaging converters, and true r.m.s. converters have resistive and capacitive errors. A major consideration when calibrating a.c. voltage is frequency response. The general requirement is to use a constant amplitude calibrator that covers the entire frequency range of the DMM. The conversion takes some time, and slewing through the pass band is time consuming. Because of this, spot frequency checks are usually made instead. Low frequency calibration is generally performed at 400 Hz to 1 kHz. The high frequency adjustments are done afterwards. The frequencies used for high frequency adjustments are largely determined by the capacitive vs resistive characteristics of the scaling circuits. For details, chapter 17 of Reference 2 is suggested.

15.4.4.5 Resistance converter calibration

Calibration of resistance converters generally consists of zero and gain adjustments. For example, each range of the DMM's resistance function is first corrected for proper zero reading. Then a near full scale input is applied to each range, and adjustments are made if necessary.

When calibrating DMMs with 4-wire resistance capability, resistance sources should have remote sensing capability, as with the Fluke 5450A Resistance Calibrator and the Fluke 5700A MFC.

In the 2-wire setup, the sense path and ohms current path are on the same set of terminals. Resistance errors due to the connecting leads can be significant. In general, proper connection of 2-wire resistance calibration for a meter with even 100 mΩ resolution can be a problem. Fluke Corporation has addressed this problem in the 5700A, which has its 2-wire ohms compensation feature. The 2-wire ohms compensation feature of the 5700A virtually eliminates the effects of lead resistance in a 2-wire hookup.

Separate 2- and 4-wire resistance calibrations could be specified in calibration procedures. Because the 4-wire sense path is the same as the d.c. voltage signal conditioning path, all that is required is to correct the gain of the current source for resistance converters with current sources. However, second order effects cause discrepancies in apparent gain when the d.c. voltage amplifiers are configured to sense either a d.c. voltage input directly or the voltage across the unknown resistor. Because of this, many of the higher precision DMMs require calibration for both 2- and 4-wire resistances.

15.4.4.6 Calibration of current converters

Both a.c. and d.c. current converters are calibrated by applying a known current and adjusting for the correct reading. Direct current converters correct for zero and gain; alternating current converters correct for down scale and full scale response. Typically, there are no high frequency adjustments for alternating current, primarily because current shunts are not inductive enough to cause significant errors relative to the DMM's specifications.

15.4.4.7 Miscellaneous other calibrations

There are other characteristics of a DMM that require calibration. The DMM must not only measure d.c. voltage to a high degree of precision; it must also have a minimal loading effect on the source voltage. One aspect of circuit loading is the bias current error of the DMM's d.c. voltage amplifiers; for example, how much current does it draw from the circuit under test? Many high end DMMs, with bias currents in the 10 pA region, require a bias current adjustment of the d.c. voltage scaling section. These adjustments are made by comparing the DMM reading with a low impedance zero input against a high impedance zero input. The adjustment itself adds or subtracts an actual bias current into the main d.c. voltage amplifiers.

15.4.5 Simplified DMM calibrations

15.4.5.1 Closed case calibration

The calibration of microprocessor controlled IEEE-488 bus or RS-232 communicating DMMs has been simplified by automation. The DMM itself may provide a great deal of the automated capabilities. However, external software, running in a PC, provides even advanced calibration automation. Over the past two decades many instruments and modules with software for closed case (or closed box) calibration has entered the market. Few early examples are HP 3455A, Fluke 8500A and Datron 1061. In a simplified process, Fluke model 8840A allows the prompting at the user through exact calibration steps via a firmware procedure built into the instrument design. In this case average calibration time is approximately 12 min in an automated system.

Closed case calibration, along with the advent of IEEE-488 bus controllers and calibration instruments controlled by the IEEE-488 bus, revolutionised the traditional DMM calibration by fully automating it. This automated process is often referred to as closed loop calibration. Close loop calibration is one of the most significant developments in the past two decades. The loop is the connection between the measurement of an instrument's performance and the actual adjustment of its operating characteristics. For example, in a manual calibration the operator applies short circuits and known voltages respectively for adjusting zero and full scale readings. The loop begins with the operator's interpretation of values and his/her physical actions in training potentiometers inside the instrument and the necessary re-iterations.

Closed loop calibration eliminates the need for any kind of arm in the loop. New digital techniques use a microprocessor to store zero and gain corrections into non-volatile memory. These are then applied in real time to modify the displayed reading. Instead of solving the $y = mx + b$ equation by trimming resistor values or d.c. voltage references, the equation is solved and corrected in software. Electrical parameters of the d.c. voltage scaling circuits, for instance, are no longer changed; the DMM displays the correct reading y by using the internal microprocessor to apply calibration constants m and b to the digitised value of input x.

Closed loop calibration is also possible for those circuits that do require a change in electrical performance. For example, high frequency a.c. voltage adjustments can consist of DAC reprogramming of varactor diodes. Or a DAC can be reprogrammed to physically readjust the d.c. voltage sense amplifier bias current. The programming information is then stored in non-volatile memory. For more details, chapter 19 of Reference 2 is suggested.

For more comprehensive details of DMM calibration and associated metrology, References 2 and 4 are suggested.

15.5 Calibration of oscilloscopes

Based on the discussion in chapters 5 and 6 related to oscilloscope designs and techniques, the following sections provide the essentials in calibrating an oscilloscope using working standards and MFCs, etc. Oscilloscope calibration, unlike DMM calibration, can be a lengthy process if each range and function is checked. Although the accuracy requirements are much lower than those for DMMs, the test engineer must make specification judgements based on visual displays on the display. This is particularly true of analogue oscilloscopes where the fidelity of the display is so important. The calibration of DSOs is as exacting, but they lend themselves better to automated calibration because of their interface, typically IEEE-488, and their relatively high degree of accuracy.

Traditional oscilloscope calibration consisted of gathering the appropriate primary equipment, usually a time marker, constant amplitude, and pulse generator to make up a workstation. These instruments used to appear as modules in a single instrument such as the Tektronix 4051, which were more convenient to work with. Whichever instruments are selected, it is important that their specifications and characteristics adequately support the oscilloscope (workload) to be calibrated. Wherever possible, a test uncertainty ratio (TUR) of 4 : 1 should be maintained between the calibrator and the scope being calibrated.

15.5.1 Calibration requirements

Despite the burgeoning increase in oscilloscope functionality, the essential features of faithful and accurate representation of waveform parameters remain few, as given below:

- vertical deflection coefficients,
- horizontal time coefficients,
- frequency response, and
- trigger response.

Techniques and procedures for calibration must measure these features, while coping with the functional conditions which surround them. Good metrological practice must be used to ensure that an oscilloscope's performance at the time of use is comparable with that observed and measured during calibration. This will provide confidence in certificates of traceability and documentation which result from calibration. Compared with a multimeter, an oscilloscope is used to observe waveforms with confidence with regard to its shape and any fine details that may be not straightforward in observations.

Manual calibration methods are well established, and for analogue oscilloscopes there is possibly no cost effective alternative, although techniques are being developed that employ oscilloscope calibrators together with memorised calibration procedures directed at individual oscilloscope models. For example, these procedures, now incorporated into the Wavetek Model 9100 Option 250 using procedure mode, use a form of 'prompted' manual calibration.

For DSOs, which are based on programmable digital techniques, and may already be programmed to respond to remote signals (say via the IEEE-488 interface), automated calibration can be achieved, with great benefits to repeatability, productivity, documentation generation, and statistical control.

15.5.2 Display geometry

Before it is possible to calibrate the main parameters it is necessary to ensure that the essential geometry of the oscilloscope is set up correctly. This may, in fact, be regarded as part of the calibration process, as the parameter measurements are dependent on visual observations. In real time (analogue) oscilloscopes, the graticule is a separate entity from the screen images. This means that if the graticule is to be used as a measurement tool, alignment to it must be included in the calibration process.

The innovation of the electronic graticule in DSOs has largely removed the need to establish geometrical links between screen data and the graticule.

15.5.3 Parameters to be calibrated

Expanding the four essential calibrations described in section 15.5.1, the following is a list of items to be checked during the calibration process:

- accuracy of vertical deflection,
- range of variable vertical controls,

- vertical channel switching,
- accuracy of horizontal deflection,
- accuracy of any internal calibrator,
- pulse edge response,
- vertical channel bandwidth,
- Z axis bandwidth,
- X axis bandwidth,
- horizontal timing,
- time base delay accuracy,
- time magnification,
- delay time jitter,
- standard trigger functions, and
- X–Y phase relationship.

Because the oscilloscope provides the user with a visual indication of the wave shape on a CRT or an LCD display, setting up of the display geometry is important. Examples of geometry features for a CRT version are:

- CRT alignment,
- earth's field screening or compensation,
- range of focus and intensity controls,
- barrel distortion,
- pincushion distortion, and
- range of X and Y axis positioning controls.

In general before calibration process is carried out these items must be examined, and adjusted if necessary, to ease the measurement process.

15.5.4 Amplitude calibration and vertical defection system

The Y axis of an oscilloscope is almost exclusively used for displaying the amplitude of incoming signals. As with the descriptions given in chapters 5 and 6, signals are processed through channel amplitude and associated switching circuitry providing the alternate and chop modes. A few basic setup features to be checked before calibrating the vertical system are:

- zero alignment to graticule (offset),
- vertical amplifier balance,
- vertical channel switching, and
- operation of alternate/chopped presentation.

There are five main parameters to be checked in calibrating each vertical amplifier system. Those are: (i) offset, (ii) gain, (iii) linearity, (iv) bandwidth, and (v) pulse response. These parameters are crucial to achieve accurate representation of signal. For effective comparisons between signals applied through different channels, their channel parameters must be equalised. Measurement of a channel amplifier's gain is usually performed by injection of a standard signal and measurement against the display graticule. Because the amplifier coupling may be switched between a.c./d.c.

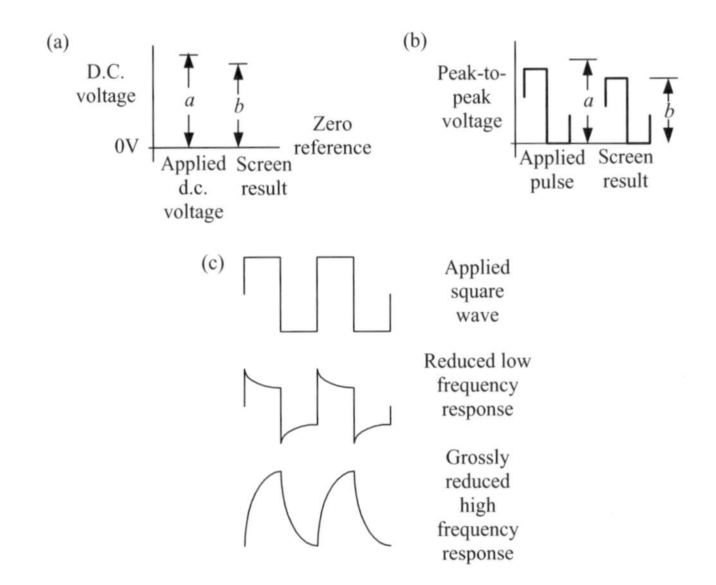

Figure 15.6 *Gain and distortion adjustments: (a) d.c. voltage gain; (b) low frequency square wave gain; (c) low frequency square wave distortion*

and often 50 Ω/1 MΩ, it will be necessary to inject signals that test the operation of each of these forms of coupling.

15.5.4.1 Amplitude adjustments

Two standard signals for measuring an amplifier's gain are usually employed.

(i) With d.c. coupling, either a d.c. signal as in Figure 15.6(a) or a square wave (as in Figure 15.6(b)) is injected, and the channel's response is measured against graticule divisions or cursor readings. Figure 15.6(a) includes offset and Figure 15.6(b) can be manipulated to remove the offset. As in Figure 15.6(a), d.c. gain can be measured by comparing values (a and b) on screens, where

$$b \div a = \text{d.c. gain.} \tag{15.3}$$

Commercial calibrators such as the Wavetek model 9100-Option 250 provide d.c. voltage and 1 kHz square wave outputs for testing the gain and offset of d.c.-coupled amplifiers.

(ii) With a.c. coupling, a square wave signal is injected at 1 kHz, and again the response of the channel is measured against graticule divisions or cursor readings (Figure 15.6(b)). The peak-to-peak value shown on the screen (b) is compared with the known value (a), where

$$b \div a = \text{gain at 1 kHz.} \tag{15.4}$$

Using a low frequency pulse waveform approximate check of low frequency and high frequency response can be made as in Figure 15.6(c). This is only a very rough test of gross distortion. A result that may appear to be square must still be checked for pulse response and band width using the calibrator's 1 kHz. A channel amplifier's linearity can be tested by injecting either a d.c. or a square wave signal, varying the amplitude and checking the changes against the graticule or cursor readings.

15.5.4.2 Pulse response

Viewing the rise time of pulse fast edges is one of two complimentary methods of measuring the response of the vertical channel to pulsed inputs (the amplifier's bandwidth should also be measured).

Response to fast edges depends on the input impedance of the oscilloscope. Two standard input impedance values are generally in use: 50 Ω and 1 MΩ (paralleled typically with a 15 pF). A value of 1 MΩ is the industry standard input generally used with passive probes. Where the 50 Ω input is provided it gives optimal matching to HF signals.

To measure the rise time (see Figure 15.7(a)), the pulse signal is injected into the channel to be tested; the trigger and time base are adjusted to present a measurable screen image; and the rise/fall time is measured against the graticule or cursor readings. The observed rise/fall time is a combination of the rise/fall times of the applied signal and the channel under test. They are combined as the root of the sum of squares, and to calculate the time for the UUT channel, a formula must be used:

$$\text{UUT rise/fall time} = \sqrt{(\text{observed time})^2 - (\text{applied signal time})^2}.$$

In some oscilloscopes the vertical graticule is specially marked with 0, 10, 90 and 100 per cent to ease the process.

15.5.4.2.1 Measurement

In commercial calibrators such as the Wavetek model 9100, two different sorts of pulse are used.

- Low edge function: a low voltage amplitude pulse matched into 50 Ω with a rise/fall time $\leq$1 ns. When using the formula to calculate the UUT rise/fall time, the applied signal rise time must be that certified at the most recent calibration of the calibrator, closest to the amplitude of the applied pulse.
- High edge function: a high voltage amplitude pulse matched into 1 MΩ with a rise time $\leq$100 ns. This function is used mainly to calibrate the response of the oscilloscope's channel attenuators.

15.5.4.2.2 Leading edge aberration

In Figure 15.7(b), some leading edge aberrations (overshoot and undershoot) are shown at the top end of the edge, before the voltage settles at its final value (which is

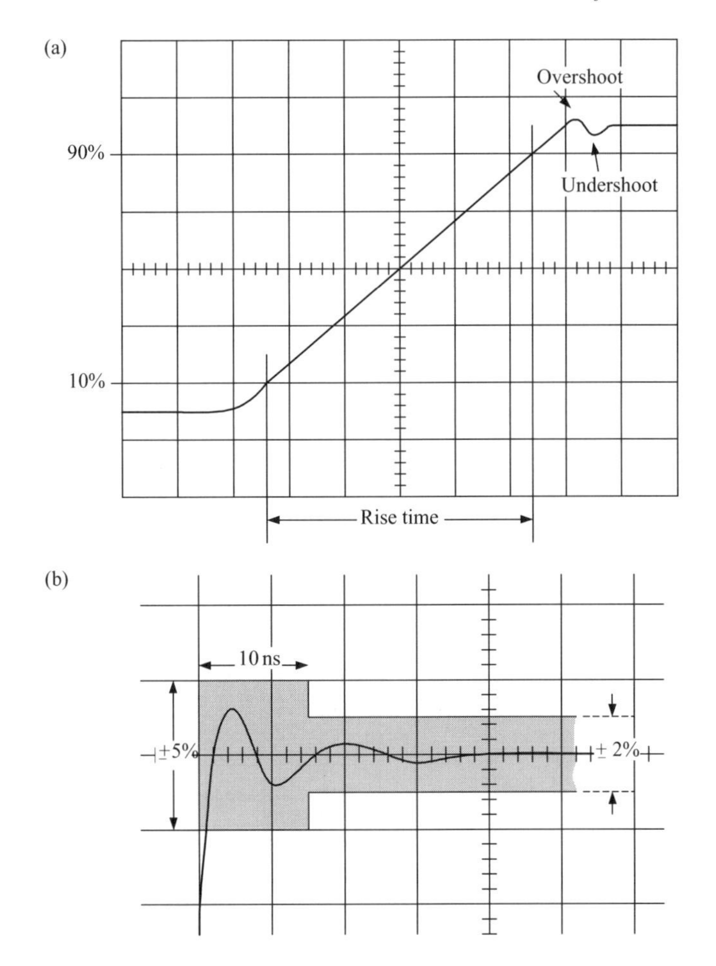

Figure 15.7 Pulse response measurements and adjustments: (a) measurement of rise time; (b) leading edge aberration

the value defined as 100 per cent amplitude level). Where oscilloscope specifications include aberrations, the specification limits can be expressed as shown in the shaded area of the magnified Figure 15.7(b) with typical limits shown as ±5 per cent and ±2 per cent.

15.5.4.3 Channel bandwidth

An oscilloscope vertical channel should provide a flat response within its specified bandwidth. Measurements and adjustments are made using a 'levelled sinewave source'. This is done at an input impedance of 50 Ω, to maintain the integrity of the 50 Ω source and transmission system. For high input impedance oscilloscopes, an in-line 50 Ω terminator is used to match the line at the oscilloscope input.

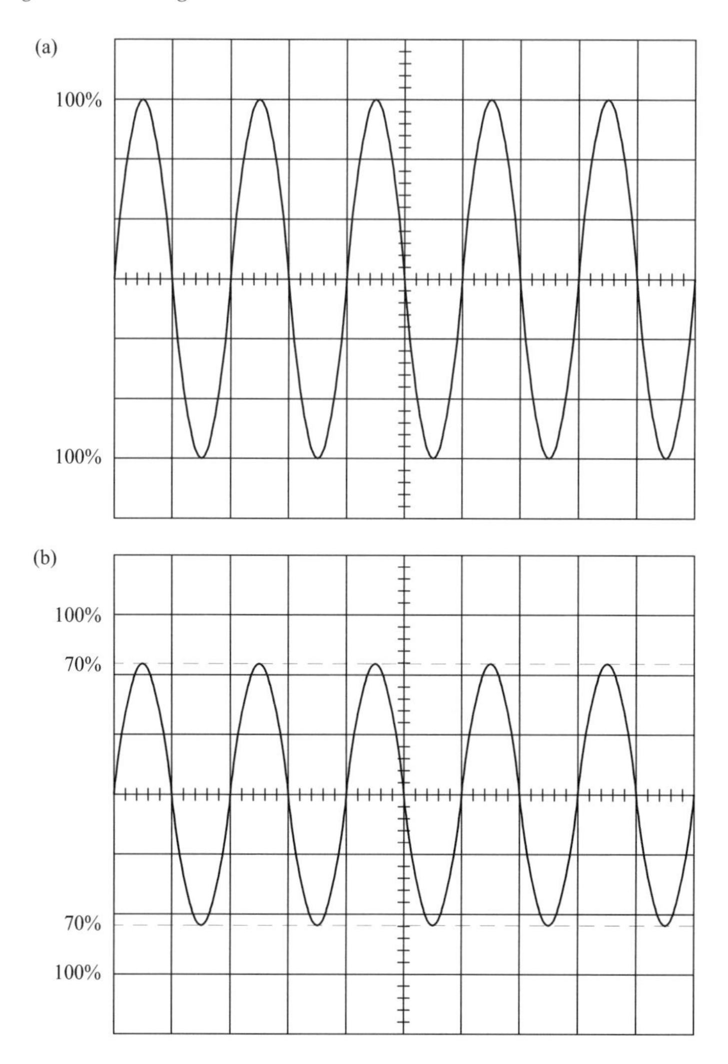

Figure 15.8 *Channel bandwidth measurement: (a) setting the amplitude at the reference frequency; (b) measurement of amplitude at the 3 dB point*

First the displayed amplitude of the input sinusiodal wave is measured at a reference frequency (usually 50 kHz), then the frequency is increased, at the same amplitude, to the specified 3 dB frequency of the channel (Figure 15.8(a)). The displayed amplitude is measured again as in Figure 15.8(b).

The bandwidth is correct if the observed 3 dB point amplitude is equal to or greater than 70 per cent of the value at the reference frequency. If it is needed to establish the actual 3 dB point, the frequency should be increased until the peak-to-peak amplitude is approximately 70 per cent of the value at the reference frequency.

15.5.5 Time base and horizontal system

In modern oscilloscopes there is a minimum of two vertical channels, and there will often be two time bases: main and delayed. These two time bases may be achieved in DSOs by two independent sampling rates, or via a positioned 'zoom' window. When determining the accuracy of horizontal deflection, where applicable, the geometry of the display must have first been set up. It is assumed that this will be included as part of the initial geometry setup. Once this has been done, the following adjustment or checks can be attempted:

- X-axis bandwidth,
- horizontal timing,
- time base delay accuracy,
- time magnification,
- delay time jitter,
- trigger functions, and
- X–Y phase relationship.

15.5.5.1 *X*-axis bandwidth

For analogue oscilloscopes, the horizontal amplifier's bandwidth will be checked using a 'levelled sinewave', similar to the checks of vertical channels, but with the time base turned off. This consists first of measuring the displayed length of the horizontal trace (Figure 15.9), for a sinusoidal wave provided as the X input at reference frequency (usually 50 kHz). The frequency is then changed, at the same amplitude, to the specified 3 dB point of the horizontal amplifier and the displayed trace length is measured again (Figure 15.9(b)). The bandwidth is correct if the observed 3 dB point trace length is equal to or greater than 70 per cent of the length at the reference frequency.

DSOs generally employ a vertical channel amplifier as the horizontal amplifier. Therefore it is adequate to measure the vertical channel bandwidth.

15.5.5.2 Horizontal timing accuracy

15.5.5.2.1 Test setup and timing calibration accuracy

In this test the time base is switched to the sweep speed (or time/div setting) to be checked, and the output from a timing marker generator is input via the required vertical channel. From the Wavetek model 9100 these are square waves up to 112.5 MHz, changing to sinewaves between 112.5 and 250 MHz.

A timing accuracy of 25 p.p.m. will be sufficient to calibrate most real time oscilloscopes and many DSOs, although a timing accuracy better than 2 p.p.m. is required for higher performance DSOs. The basic accuracy of a commercial calibrator model such as Wavetek model 9100 is 25 p.p.m. With Option 100 fitted, this improves to 0.25 p.p.m.

15.5.5.2.2 Use of square waves versus comb waveforms

In the past, timing markers have taken the form of a 'comb' waveform, consisting of a series of differentiated edges in one direction, with the return edges suppressed.

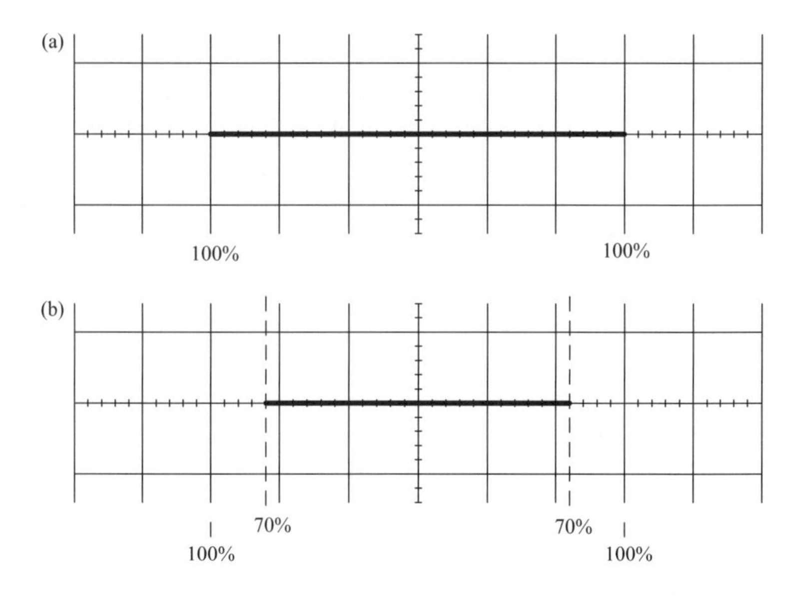

Figure 15.9 X-axis bandwidth calibration of analogue/real time scopes: (a) setting trace length at reference frequency; (b) measuring trace length at the 3 dB point frequency

This leads to difficulties in DSOs owing to sampling, in which the comb peak can fall between samples, leading to amplitude variations and difficulty in judging the precise edge position. The use of timing markers in the form of square or sinewaves significantly reduces the inaccuracies due to this one-dot jitter.

15.5.5.2.3 Measurement

The marker timing is set to provide one cycle per division if the horizontal timing is correct. By observation, the marker generator's deviation control is adjusted to align the markers on the screen behind their corresponding vertical graticule lines, and the applied deviation is noted. The applied deviation should not exceed the oscilloscope's timing specification. The operation is repeated for all the sweeps and time base settings designated for calibration by the oscilloscope manufacturer; see Figure 15.10.

15.5.5.3 Time base delay accuracy

For this test it is assumed that the delayed time base is indicated as an intensification of the main time base, and can be switched to show the delayed time base alone. For all oscilloscopes, it is necessary to ensure that the retrigger mode is switched off.

The output from a timing marker generator (e.g. Wavetek model 9100, Option 250) is input via the required vertical channel, and the oscilloscope is adjusted to display one cycle per division as illustrated in Figure 15.11(a). The mode switch is set to intensify the delayed portion of the main time base over a selected marker edge as shown (this may require some adjustment of the oscilloscope's delay control). The

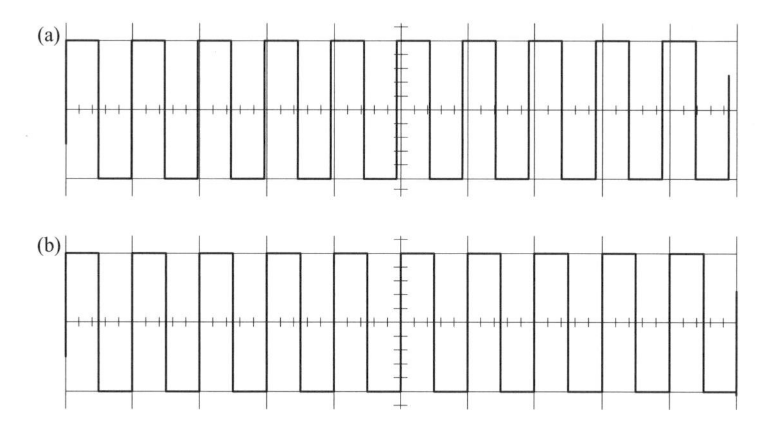

Figure 15.10 *Adjusting the marker generator's deviation for correct alignment: (a) initial state (before deviation adjustment); (b) aligned state (after deviation adjustment)*

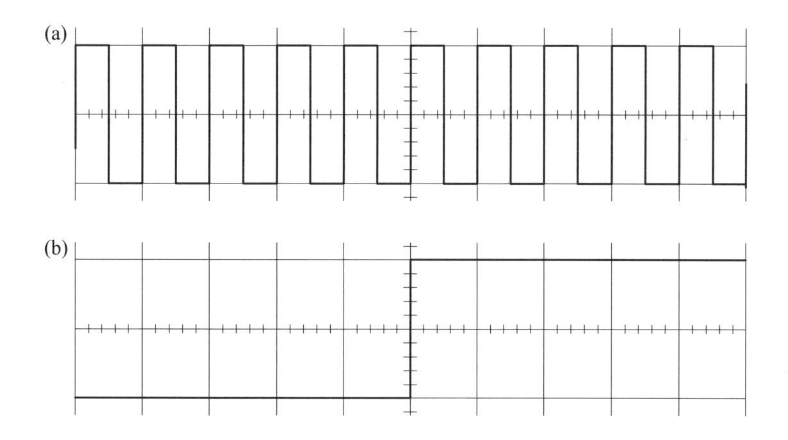

Figure 15.11 *Adjusting the delayed time base to the first datum marker: (a) delayed time base intensified on the main time base; (b) adjusting the delayed time base to the first datum marker*

oscilloscope delay mode switch is set to display the delayed sweep alone, and the delay control is adjusted to align the time marker edge to a chosen vertical datum line (e.g. centre graticule line as shown at Figure 15.11(b)). The setting of the oscilloscope's delay is noted.

Then the oscilloscope mode switch is set to intensify the delayed portion of the main time base over a different selected marker edge (Figure 15.12(a)). The oscilloscope delay mode switch is again set to display the delayed sweep alone, and the delay control is adjusted to align the time marker edge to the same vertical datum line (Figure 15.12(b)). The setting of the oscilloscope's delay is again noted.

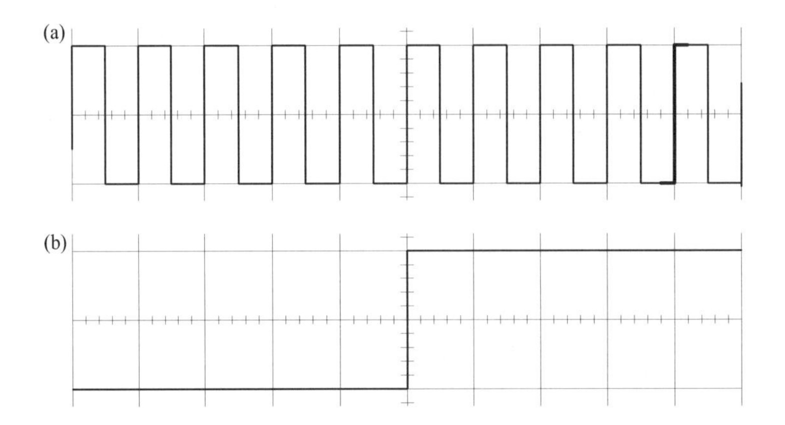

Figure 15.12 *Adjusting the delayed time base to the second datum marker: (a) delayed time base intensified on the main time base; (b) adjusting the delayed time base to the second datum marker*

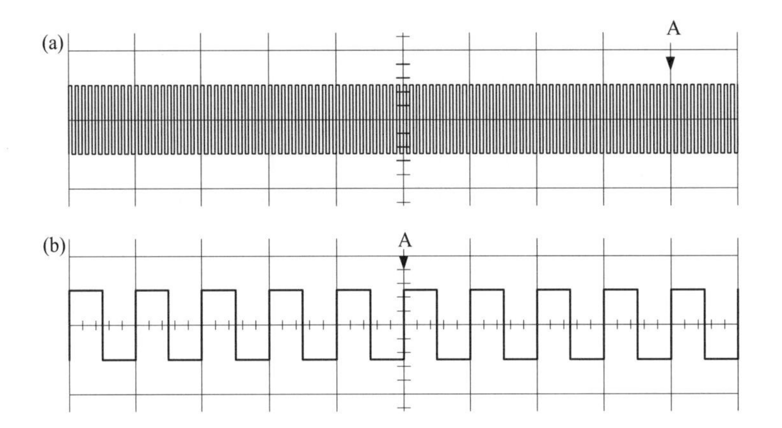

Figure 15.13 *Checking the effect of ×10 magnification: (a) markers set at ×1 magnification; (b) markers set at 10/div at ×10 magnification*

Finally, the two settings of the oscilloscope delay are compared, to check that their difference is the same as the time between the two selected markers, within the specified limits for the oscilloscope.

15.5.5.4 Horizontal ×10 magnification accuracy

The output from a timing marker generator (e.g. 9100, Option 250) is input via the required vertical channel, and the oscilloscope is switched to display 10 cycles per division as illustrated in Figure 15.13(a). The timing marker generator frequency/period is adjusted to give exactly 10 cycles per division.

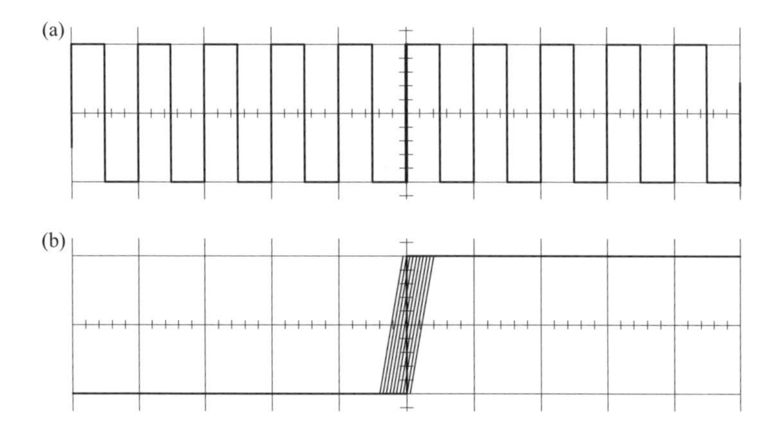

Figure 15.14 Measurement of delay time jitter: (a) delayed time base intensified on main time base; (b) edge showing jitter on delayed time base

The errors are likely to be greatest on the right of the trace (the longest time after the trigger), so the oscilloscope's horizontal position control is adjusted to place the marker edge at 'A' at the centre of the screen.

The oscilloscope is set to display the $\times 10$ sweep, and the horizontal position control is adjusted to align the marker edge 'A' exactly to the centre graticule line. The marker generator frequency/period deviation control is adjusted to align the marker edges exactly to the graticule lines as shown at Figure 15.13(b). The marker generator frequency/period deviation setting is noted. This setting should be within the specified limits for the oscilloscope.

Similarly, for a DSO, the range of available 'zoom' or '$\times$-magnification' factors are calibrated as designated by the manufacturer.

15.5.5.5 Delay time jitter

The delay jitter on an oscilloscope is often measured under time magnifications of the order of 20 000 : 1.This means that the delayed time base must run 20 000 times faster than the main time base (for example, a main time base running at 20 ms div^{-1}, the delayed time base must run at 1 μs div^{-1}). For this test the intensification of the main time base is adjusted onto the edge at the centre graticule line (with such a difference between the speeds of the main and delayed time bases, a very small part of the main time base is intensified, and adjustment may be difficult).

The 20 ms period output from a timing marker generator is input to the required vertical channel, and the oscilloscope is adjusted to display one cycle per division (20 ms div^{-1}) as illustrated in Figure 15.14(a).

The delayed time base is set to run at 1 μs div^{-1}, and the mode switch is set to intensify the delayed portion of the main time base over the centre marker edge as shown using the oscilloscope's delay time control. The oscilloscope delay mode switch is set to display the delayed sweep alone, and the delay control is adjusted to align the time marker edge to a chosen vertical datum line (e.g. centre graticule line as shown in Figure 15.14(b)).

The width of the vertical edge (which displays the jitter) of the displayed portion of the waveform, measured along a horizontal axis, should not exceed the oscilloscope's specified jitter limits (i.e. in this example, for 20 000 : 1 specification, the oscilloscope's contribution to the width should be less than one division).

15.5.6 Trigger system calibration

For most oscilloscopes, a wide variety of trigger modes exist, being sourced either via a nominated Y-input channel, or from a separate external trigger input. The functionality of the trigger modes allows for a.c. or d.c. coupling, repetitive – or single – sweep, and trigger level control operations.

These tests check the operation of:

- internal trigger sensitivity in both polarities, from each of the available Y-input channels,
- operation of the trigger level control for a sinewave external trigger input,
- effect of vertical position on trigger sensitivity,
- minimum trigger levels for normal 'trigger view' modes,
- bandwidth of trigger circuits , and effect of HF rejection filters,
- LF and d.c. performance of the trigger circuits, and
- single sweep performance and response to position controls.

Tests performed on one channel should be repeated on all other input channels, which may practically vary from two to eight in most commercial scope models.

15.5.6.1 Internal triggers – trigger level operations

15.5.6.1.1 Initial setup and the test procedure

A standard 4 V p-p (50 Ω) reference sinusoidal signal is input via a.c. coupling into the vertical input channels in turn. Using internal triggers and d.c. trigger coupling other than 'AUTO', the positive and negative slopes are selected in turn. The sweep speed setting is 10 μs div^{-1}; the Y-channel sensitivity is 0.5 V div^{-1} so that the input signal occupies eight divisions. Table 15.1 provides details of checking process for internal triggers.

In the following trigger operations, during tests on a DSO, the trace will not disappear as a result of the interruption of the trigger (or reduction of its amplitude below the threshold). Instead, the trace will remain but not be refreshed, and this is the condition to be detected.

15.5.6.2 External triggers

15.5.6.2.1 Initial setup

These tests start with the 200 mV signal applied to the external trigger input to the oscilloscope. (This is similar to the 'Display triggers' feature in Table 15.1.) Table 15.2 indicates the steps and the methodology.

Table 15.1 Internal triggers and trigger level operation check

Process/step	Methodology
Trigger level adjustment	Over almost all of its range of adjustment, the trigger level control must be shown to produce a stable trace, moving the starting point over a range of levels up and down the selected slope of the displayed sinewave.
Trigger sensitivity	With the input signal reduced to 10 per cent of its amplitude, adjustment of the trigger level control must be shown to reacquire stable triggering. With trigger coupling switched to a.c., and using vertical positioning to place the trace at extreme upper and lower limits of the CRT screen in turn, stable triggering must be maintained.
'Display triggers' (or 'trigger view') feature	If the oscilloscope has a 'display triggers' or 'trigger view' feature, this is selected to display the trigger region of the waveform. Using a 200 mV sinusoidal signal input to the channel, the trigger region is checked for correct amplitude on the display.

Table 15.2 External trigger test procedure

Process/step	Methodology
Presence of a trace	Adjustment of the oscilloscope's trigger level control should be able to produce a trace. The external trigger input is disconnected and reconnected again, while checking that the trace disappears and is then reinstated.
Trigger sensitivity	The input signal reduced to the minimum amplitude specified by the manufacturer, adjustment of the trigger level control must be shown to regain stable triggering.
Trigger bandwidth	With the input signal set to the minimum amplitude and maximum frequency specified by the manufacturer, adjustment of the trigger level control must be shown to sequre stable trigging. The external trigger input is disconnected and reconnected again, while checking that the trace disappears and is then reinstated.
ACHF rejection trigger mode	With the input signal set as for the trigger bandwidth check, the ACHF reject feature is activated then deactivated again, while checking that the trace disappears and is then reinstated.

15.5.6.3 Internal triggers – d.c. coupled operation

15.5.6.3.1 Initial setup and test procedure

With the external trigger input disconnected, and the Y-channel input externally grounded, the oscilloscope Y-channel is set to 'd.c. coupling' and trigger mode for

Table 15.3 Test procedure for the internal triggers – d.c. coupled operation

Process/step	Methodology
D.C. triggering	By adjusting the vertical positioning control to pass through a point in its range corresponding to the trigger level setting and selected slope direction, a single trace should appear then disappear.
ACLF rejection trigger mode	With the input signal set as for the trigger bandwidth check, the ACLF reject feature (if available) is activated and the previous step is repeated. The single trace action should not occur.

'internal triggers' from the Y-channel. There should be no trace on the CRT. Table 15.3 provides details.

15.5.6.4 External triggers – single sweep operation

15.5.6.4.1 Initial setup and test procedure

With the external trigger input connected, the oscilloscope is set to 'single sweep', and trigger mode for 'internal triggers' from the Y-channel. There should be no trace on the CRT. This procedure applies only to those oscilloscopes with single sweep capability. Pressing the 'reset' or 'rearm' switch should produce a single trace. This action should not produce a trace when the external trigger input is disconnected.

15.5.6.5 Low frequency triggers

15.5.6.5.1 Initial setup and test procedure

A 30 mV, 30 Hz sinewave signal is input simultaneously to channel 1, channel 2, external trigger sweep A (main time base) and external trigger sweep B (delayed time base). The oscilloscope is set for: trigger mode to 'internal triggers', channels 1 and 2 sensitivity to 10 mV div^{-1}, and sweep speed to 5 ms div^{-1}. Both main and delayed time bases should be displayed when selected, for both channels. Table 15.4 provides details.

15.5.6.6 Z-axis adjustments

15.5.6.6.1 Z-axis input

If provided, the Z-axis input is usually positioned on the rear panel, but sometimes can be found near the CRT controls on the front panel. DSOs generally do not have a Z-input.

15.5.6.6.2 Z-axis bandwidth

Initial setup and test procedure A 3.5 V p-p, 50 kHz sinewave is applied to both channel 1 and external trigger inputs. The sweep speed, trigger slope and trigger level controls are set to provide a stable display of 1 cycle per division. Table 15.5 provides details.

Table 15.4 Low frequency triggers test procedure

Process/step	Methodology
Channel 2 grounded	With channel 2 input grounded, and channel 1 set for 0.1 V div^{-1} with the trigger selector set to channel 1, stable displays should appear as expected.
Channel 1 grounded	With channel 1 input grounded, and channel 2 set for 0.1 V div^{-1} with the trigger selector set to channel 2, stable displays should appear as expected.
ACHF reject	With channel 2 input grounded, and channel 1 set for 50 mV div^{-1} with the trigger selector set to channel 1, the ACHF reject feature is activated for both sweeps A and B. Adjusting the trigger level control should acquire a stable display.
Positive and negative slope operation	With channel 2 input grounded, and channel 1 set for 10 mV div^{-1} with the trigger selector set to channel 1, adjusting the trigger level control should acquire a stable display for both positive and negative slope selections.
ACLF reject	With channels 1 and 2 set for 10 mV div^{-1} with the trigger selector set to either channel, the ACLF reject feature is activated for both sweeps A and B. Adjusting the trigger level control should not be able to secure a stable display for either positive or negative slope selection.

Table 15.5 Test procedure for the Z-axis input

Process/step	Methodology
Signal transfer to Z input	The signal input to Channel 1 is disconnected and transferred to the Z-axis input. The trace should collapse to a series of bright and dim sections. Using the oscilloscope brightness control, the trace is dimmed so that the brightened portions just disappear.
Bandwidth check	The frequency of the input sinewave is increased to the exact specified Z-axis bandwidth point. The amplitude of the sinewave is increased to 5 V p-p. Adjustment of the sweep speed and trigger level controls should acquire a dotted, or intermittently brightened, trace.

15.5.7 X–Y phasing

15.5.7.1 *X* input

Depending on the type of oscilloscope, the X input will be applied either via the external trigger connector, or via channel 1, with suitable switching. In either case, the same signal of 50 mV, 50 kHz will be applied to both X and Y inputs.

15.5.7.2 Phasing test

15.5.7.2.1 Initial setup

The oscilloscope controls are set as follows:

Vertical mode: X–Y
Sensitivity: 5 mV div^{-1}, both channels
Channel 1 or X: a.c. coupled
Channel 2 or Y: grounded
Vertical mode: X–Y
Vertical position: central
Horizontal position: central

During X–Y phasing tests on a DSO, maximum sampling rate would be used. Even so, the visible extent of any captured Lissajou is limited to interrupted segments by the store length, until the test frequency is high enough for an entire cycle to be captured.

15.5.7.2.2 Trace acquisition

The display intensity is adjusted until a horizontal trace is just visible (should be 10 divisions long). After the X and Y position controls have been used to centre the trace, the intensity and focus controls are adjusted for best display.

15.5.7.2.3 Phasing check

The common input signal is reduced until the trace is 8 divisions long. Channel 2 (or Y) input mode is switched to d.c., and X and Y position controls are used to centre the (now sloping) trace. If the X and Y channels do not introduce any phase error, then the centre of the trace will pass through the origin. Phase error between X and Y channels will cause the sloping trace to split into an ellipse, which for small phase errors will be apparent only close to the origin. The trace separation at the origin, along the centre horizontal graticule line, should not be greater than 0.4 division for a commonly specified phase-shift of three degrees.

15.6 Calibrating high speed DSOs

The past decade has been the development maturity era for the high performance, high sampling rate DSOs with sophisticated DSP techniques, real time capability, and interfacing and waveform mathematics capability. Sample rates achieved between 500 Msamples s^{-1} to 4 Gsamples s^{-1}, etc., now demand more advanced calibrators and time base performance checks reaching picosecond ranges.

Dedicated oscilloscope calibrators are special signal generators that source levelled sinewave, fast rise time pulse and time mark signals. Until recently, the most commonly used instruments were individual, manually operated plug-in generators manufactured by manufacturers such as Tektronix. The growing interest in oscilloscope calibration has spawned a new generation of integrated oscilloscope calibrators offered as an option in general purpose calibrators. The Fluke 5500A is an example of this type of instrument.

Most basic oscilloscope calibrators available today can accommodate the largest cross-section of the oscilloscope calibration workload – instruments up to 250 MHz bandwidth. With the switch over to more powerful DSOs and the availability of low cost 500 MHz instruments, additional equipment and techniques are required to extend the capabilities of the oscilloscope calibration equipment already in use. It is worth citing the availability of the IEEE Std. 1057-1994 titled 'IEEE Standard for digitizing waveform recorders' [6].

15.6.1 Warranted vs typical specifications

As with analogue oscilloscopes, high speed digital oscilloscopes have warranted specifications for d.c. vertical gain, bandwidth and time base uncertainty. Conspicuously absent are specifications for rise time and aberrations. Unlike their analogue predecessors, these parameters for digital oscilloscopes are stated as 'typical' or 'calculated'.

For example, a 1 GHz oscilloscope will provide a typical rise time of 350–450 ps (based on a bandwidth-rise time product of 0.35 to 0.45). This variability is significant.

In rise time measurements, the test engineer should keep in mind that the pulse we observe on screen is a combination of the rise time of the source and the rise time of the oscilloscope (eq. (5.7)). For example, when a 350 ps pulse is applied to an oscilloscope with a rise time of 350 ps, the resulting waveform should have a rise time of <495 ps.

For these reasons, expensive calibrators use special attachments such as tunnel diode pulsers and special pulse heads, etc. A discussion on these is beyond the scope of the chapter and for details References 7–9 are suggested.

15.6.2 Dynamic testing

Many manufacturers of digital oscilloscopes use the term called 'effective bits' (discussed in chapter 3) to describe the dynamic performance – non-linearity and distortion – of the analogue-to-digital converter. Although this is seldom a warranted specification, if it is specified at all, an effective bits test can tell you a lot about the dynamic performance of the vertical section of a digital oscilloscope. A sinewave curve fitting method is used to test effective bits. A sinewave is digitised and then compared with a perfect sine wave. Using an algorithm supplied by the manufacturer or other industry source, the effective bits resolution is calculated. While a typical high speed digital oscilloscope may use an 8-bit ADC, the actual resolution may be 4.5–6 bits depending on the dynamic response of the ADC. To perform this test, you need a signal generator that can output a sinewave between 200 and 300 mV with better than −48 dB of distortion, assuming an 8-bit ADC. Many oscilloscope calibrators have sufficient performance to do this test. If an RF signal generator is used, external narrow bandpass filters may be required to remove harmonics and noise. Tests are normally done at one vertical gain setting, typically 0.05 V per division.

15.7 Multifunction calibrators

Between 1970 and 1980 almost all medium and high price oscilloscopes carried the circuitry to display almost all basic parameters we measure using a multimeter. During the 1980s and 1990s, the portable products such as DMMs developed towards acquiring oscilloscope functions. With the standards such as ISO 9000 being widely accepted across the worldwide industry, fast and traceable calibration became a mandatory requirement while minimisation of cost and turnaround time also became important.

In the typical calibration laboratory, calibrators perform a number of different functions, ranging from precise calibration of DMMs to being used as a low noise power supply. For some applications, such as meter calibration, the ability to supply a large burden current is not important but a wide range of output voltage is a primary requirement. For other applications, such as the calibration of a.c.–d.c. thermal voltage converters (TVC), appreciable current may be necessary at many voltages. DMM and meter calibrators provide the functions needed to calibrate DMMs and other meters. As DMMs have grown in functionality, metrology laboratories found that an ensemble of single function calibrators (SFCs) was required to calibrate the DMMs. This led to the introduction of newer models such as multifunction calibrators (MFCs) that provide all or nearly all of the functions needed to calibrate most DMMs and oscilloscopes.

A typical multifunction calibrator provides direct and alternating voltage and current stimulus. It also provides resistance stimulus, and may provide wideband RF stimulus too. SFCs can be used to complement the MFC to provide additional functions, such as capacitance, inductance, and frequency stimulus as well as thermocouple and resistance temperature detection (RTD) emulation.

15.8 Multiproduct calibrators

A traditional meter calibrator will support many of the voltage, current and resistance requirements. But two of them phaselocked together, plus a phase meter, are required for power calibration. A thermocouple simulator is needed for temperature. Capacitance is traditionally supported through the use of standard capacitors or RCL meters. For multiple waveforms and oscilloscope calibration, several specialised function generators or signal generators are needed.

Each calibrator itself requires calibration support. Some can be automated, but most can only be used in manual mode. For a new laboratory setup purchase of all the required equipment can be quite expensive. Clearly, to meet the workload requirements of a rather diverse number of calibration labs, a new kind of calibrator is required. This new class of instrument will need to:

- integrate an unprecedented number of calibration functions to maximise its workload coverage;
- provide tools for dealing with documentation and control of procedures, results and traceability to nationally recognised standards; and

- be cost effective in terms of purchase, support and operator training.

To cater for such diverse calibration needs a new class of calibrator called multiproduct calibrators (MPC) have entered the market. An example of such a calibrator is the Fluke 5500A. The 5500A is slightly over one cubic foot, weighs less than 20 kg, and integrates a very high level of functionality in one package, as follows:

- d.c. volts, 100 nV to 1000 V with 40 p.p.m. uncertainty,
- a.c. volts, 1 mV to 1000 V, 10 Hz to 500 kHz,
- d.c. and a.c. current to 11 A,
- variable resistance, 0–330 Ω,
- variable capacitance, 330 pF to 1 mF,
- temperature simulation and measurement, nine thermocouple types,
- RTD simulation, three types,
- power into phantom loads, up to 11 kW,
- multiple waveforms and extended bandwidth from 0.01 Hz to 2 MHz,
- complete oscilloscope calibration, including a 250 MHz levelled sinewave generator.

The design of the 5500A is similar to the multifunction calibrators of the past, but with some important additions. Key elements in the 5500A are precision digital-to-analogue converters (DACs), arbitrary waveform generators, voltage and current amplifiers, and circuits used for resistance and capacitance generation.

15.9 Automated calibration and calibration software

The successful automation of the calibration process requires an understanding of the actions in the process and their relationship to each other. Some of these actions are more suitable for automation than others. This may lead the laboratory manager to a piecemeal approach that can evolve into islands of automation incompatible with each other.

The calibration process, automated or not, involves many tasks on the part of the calibration laboratory manager, test engineers, and staff. These tasks can be divided into three categories namely:

- setup activities,
- day-to-day operations, and
- adjustments in workload.

Figure 15.15 illustrates the typical cycle of the overall calibration process. There are various activities in each category, as are itemised in the following lists. These lists can serve as guides for calibration laboratory managers who are considering automated calibration in their facilities.

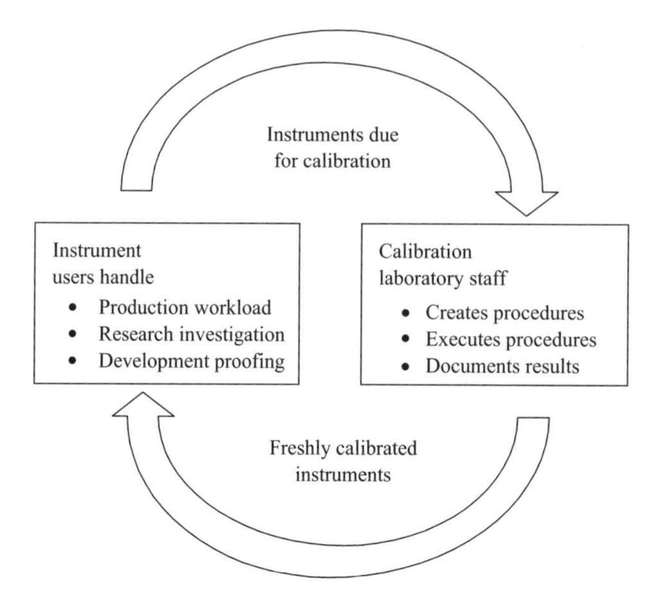

Figure 15.15 Typical instrument calibration cycle (courtesy of Fluke Corporation, USA)

15.9.1 Setup activities

This category of activities establishes the general calibration protocols and procedures, such as:

- identifying the instruments to be calibrated,
- establishing calibration intervals,
- setting up an acceptable quality level (AQL) for the instruments,
- documenting the instrument's location and use,
- establishing the kind of calibration procedure to be used to calibrate the various units under test (UUTs),
- determining whether to calibrate instruments on-site or in the calibration laboratory,
- determining the skill levels required to manage the workload and do the calibrations, and
- determining the environment and supporting equipment required.

15.9.2 Day-to-day operations

Day-to-day operations involve the regular routine of the calibration laboratory. Activities that can be put into the day-to-day category include:

- bringing instruments to the calibration laboratory or calibrating them on-site,
- locating the appropriate test procedure,
- inspecting and repairing test instruments,

- making as-found or as-left calibration verification (in some environments),
- making calibration adjustments,
- maintaining measurement standards,
- issuing reports, certificates, stickers, and other pertinent calibration documents to meet the needs of instrument owners and auditors, and
- returning completed instruments to service.

15.9.3 Adjustments in workload

From time to time, the calibration laboratory reviews and adjusts its activity. A short list of some of these actions includes:

- identifying consistently marginal devices and removing them from the workload,
- optimising calibration intervals,
- refining calibration procedures, and
- updating instrument location and user files.

15.9.4 Benefits of automation

Automation of the calibration process has distinct advantages. These include consistency of measurements, increased productivity, automatic documentation, and cost reduction. Automation can assist in complying with regulations such as MIL-STD-45662A; it can also contribute significantly to quality programs such as ISO 9000. Specific advantages exist for each member of the calibration organisation: managers, metrologists, technicians, and customers. For details, Reference 2 is suggested.

15.9.5 Computers and software

Automated calibration relies heavily on computers to assist calibration laboratory personnel to do their jobs. PCs are the modern system of choice for automated calibration applications. They are widely available and offer an abundance of commercially available software that can be used in conjunction with automated calibration software to carry out a calibration laboratory's mission. At this time, there are two broad categories of commercially available metrology software: calibration software, and workload management software.

Calibration software is used to conduct the actual calibrations. This helps to maintain an unbroken chain of measurements traceable to national standards. Workload management software provides the documents and records that serve as the evidence of an operating quality system or of contractual compliance for traceable measurements. This helps to prove that the chain of measurements is indeed unbroken to national standards.

When an appropriate IEEE-488 bus control card is installed, the PC can be used with commercial calibration software such as Fluke MET/CAL to control calibrators and UUTs for computer aided and closed loop calibrations. MET/CAL software can also be used for operator prompted calibrations in conjunction with manually operated calibration equipment. For details of MET/CAL, Reference 10 is suggested, in addition to Reference 2.

15.10 Calibration intervals

Calibration increases your confidence that an instrument is operating within a tolerance. When the time since calibration is close to zero, you assume that an instrument is operating within tolerance; your confidence in the instrument's performance is high. As time passes, you lose confidence and your uncertainty increases. An increasing uncertainty does not mean that the instrument's performance must degrade over time, only that the probability that the measurements are out-of-tolerance has increased. The calibration laboratory may find that the instrument is well within tolerance, but your confidence was nonetheless lost.

A calibration interval is a tradeoff between a given in-tolerance probability and cost. Calibrate too infrequently (calibration interval too long), and you increase the probability that an instrument will operate out of tolerance. (Statistically speaking, that is 'failure'.) Calibrate too frequently (calibration interval too short), and you will pay unnecessary lab charges and lose productivity while the instrument is in the calibration laboratory.

Figure 15.16 shows how confidence reduces (uncertainty increases) with time. At time 0, most of the area under the curve is between the tolerance limits. That area represents confidence that the instrument's true reading is within the tolerance limits. Your confidence is high and uncertainty (the area outside the limits) is low. At time 1, you have lost some of that confidence (uncertainty has increased); the area under the curve that is outside the tolerance limits has grown. At time 2, you may have lost enough confidence to warrant a calibration.

15.10.1 Manufacturers' recommendations

The easiest calibration interval to implement is the individual manufacturer's recommended interval. The manufacturer bases its recommended calibration interval on experience, experimental conditions in a test lab, and engineering judgment. Unfortunately, sticking with the recommended calibration interval may mean that your calibration interval is too short or too long. Many instrument's users modify both instrument specifications and calibration procedures to fit their needs. From having calibrated many instruments of the same model, your calibration laboratory may know that your instrument is likely to be well within tolerance or be out of tolerance a year from now. Therefore, the laboratory may shorten or lengthen the calibration interval. In effect, the calibration laboratory is using history from a pool of similar instruments to predict that your instrument will remain in tolerance throughout the calibration interval.

The National Conference of Standards Labs (NCSL, Boulder, CO) Recommended Practice 1 (RP-1) describes several alternative methods for calculating and adjusting intervals. For details, References 11 and 12 are suggested; see also Reference 13.

Let us take an example such as the DMM you use every day. The meter's calibration gives you confidence that its measurements fall within the manufacturer's specifications. That confidence comes from your belief that the standards used to calibrate your meter are more accurate (have a smaller uncertainty) than your meter.

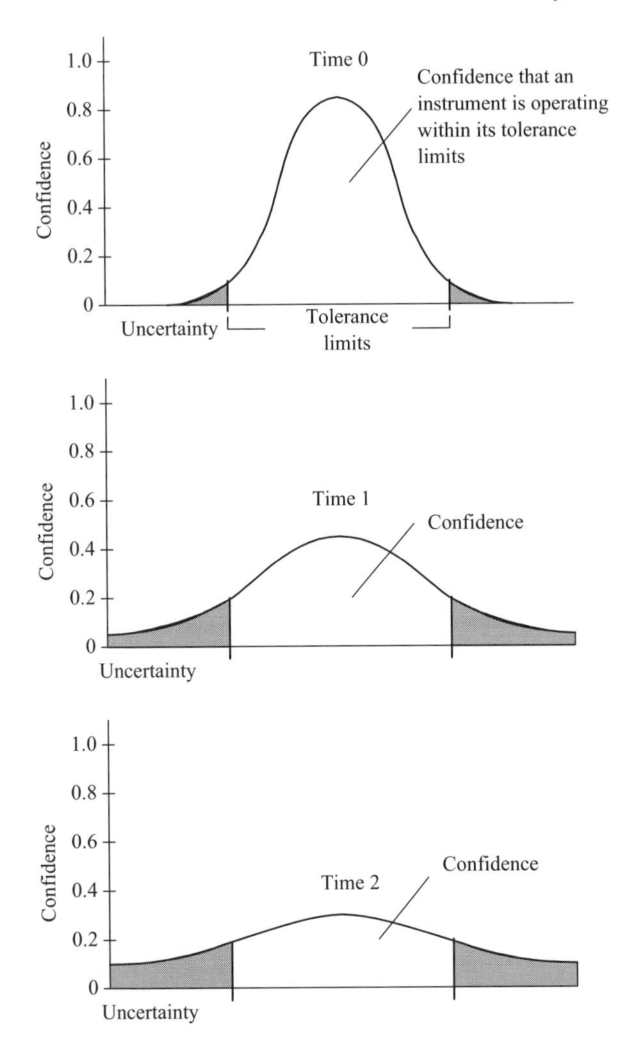

Figure 15.16 Confidence intervals

As in the example in Figure 15.5, each equipment must have its documented proof of traceability to National Institute of Standards & Technology (NIST). For details, Reference 3 is suggested.

15.11 Conclusion

This chapter is inserted as a special one because calibration and metrology have reached an important dimension recently with management standards such as the ISO 9000 series, etc. The chapter is prepared not as a comprehensive document describing all possible parameters and common instruments, but as a guideline for a test engineer to be familiar with modern concepts and practices in metrology, calibration and traceability, etc. The list of references is comprehensive enough if one needs a broad idea of calibration and metrology practices as applicable in the field.

15.12 References

1. FLUKE, INC.: 'How many calibrators do you need to meet ISO 9000?' Application note 5/99, 1999.
2. FLUKE, INC.: *Calibration: Philosophy in Practice*, Fluke Corporation, 1994.
3. ROWE, M.: 'Follow the chain to NIST traceable calibrations', *Test & Measurement World*, December 5 1999, pp. 19–20.
4. CRISP, P.B.: 'Getting the best out of long scale DMMs in metrology applications', *Proceedings of NCSL*, 1997.
5. WAVETEK LTD.: 'Customer guidance note. – Calibration of Oscilloscopes', DS 186, July 1997.
6. IEEE: 'IEEE Standard for Digitizing Waveform Records', IEEE Std. 1057-1994.
7. MYERS, B.: 'A practical look at calibrating high speed DSOs', *Evaluation Engineering*, May 1997.
8. ASCHCROFT, M. and ROBERTS, P.: 'Developments in fully automated oscilloscope calibration to 3.2 GHz', Wavetek Inc.
9. RODDIS, R.: 'Implementing Automated Oscilloscope Calibration Systems', *Proceedings of NCSL*, July 1997.
10. FLUKE CORPORATION: 'MET/CAL plus version 6.0' Fluke Application note – 1267 836D-ENG – N-Rev A.
11. ROWE, M.: 'Cal intervals weigh tradeoffs', *Test & Measurement World*, April 1993, pp. 36–43.
12. Recommended Practice 1 (RP-1), Establishment and Adjustment of Calibration Intervals, National Conference of Standards Laboratories, Boulder, CO, January 1996.
13. NCSL Recommended Practice 12 (RP-12) – Determining & Reporting Measurement Uncertainties. National Conference of Standards Laboratories, Boulder, CO, www.ncsl-hq.org, April 1995.

Index